DEVELOPMENTAL BIOLOGY

DEVELOPMENTAL BIOLOGY SECOND EDITION

Leon W. Browder
UNIVERSITY OF CALGARY, CANADA

SAUNDERS COLLEGE PUBLISHING

Philadelphia New York Chicago
San Francisco Montreal Toronto
London Sydney Tokyo Mexico City
Rio de Janeiro Madrid

Address orders to:
383 Madison Avenue
New York, NY 10017

Address editorial correspondence to:
West Washington Square
Philadelphia, PA 19105

Text Typeface: 10/12 Trump
Compositor: Hampton Graphics
Acquisitions Editor: Michael Brown
Project Editor: Diane Ramanauskas
Copyeditor: Elaine L. Honig
Managing Editor & Art Director: Richard L. Moore
Art/Design Assistant: Virginia A. Bollard
Text Design: Emily Harste
Cover Design: Lawrence R. Didona
Production Manager: Tim Frelick
Assistant Production Manager: Maureen Iannuzzi

Cover credit: Embryos: early blastula. (Photograph courtesy of Photo Researchers, Inc.)
Title page credit: Development II by M.C. Escher © Beeldrecht, Amsterdam/VAGA, New York.
Collection Haags Gemeentemuseum the Hague.

Library of Congress Cataloging in Publication Data

Browder, Leon W.
Developmental biology.

Includes bibliographies and index.

1. Developmental biology. I. Title. [DNLM:
1. Biology. 2. Embryology. 3. Genetics. 4. Growth.
5. Morphogenesis. QH 491 B877d]
QH491.B77 1984 574.3 83-19580

ISBN 0-03-063506-3

DEVELOPMENTAL BIOLOGY   ISBN 0-03-063506-3

56   032   9876543

CBS COLLEGE PUBLISHING
Saunders College Publishing
Holt, Rinehart and Winston
The Dryden Press

To Sandy once again for the same reasons and more.

Preface

This book is an attempt to capture the spirit of a dynamic, expanding, and exciting discipline. The subject is elusive, however, because the limits of developmental biology are variable. I visualize developmental biology as the study of progressive changes that occur within cells, tissues, and organisms themselves during their life span. Development may be studied at a variety of levels: molecular, biochemical, genetic, morphological, or physiological—terms that describe biological points of view that also imply methodology. Much current biological research focuses on the cellular and subcellular levels, mainly because the primary controls over development are exerted there. This research has yielded an incredible amount of new information about development in the four years since the first edition of this book appeared, and the pace of discovery shows no signs of abating. Much of the recent progress has resulted from the application of new techniques, such as those utilizing recombinant DNA, to problems of development. Recombinant DNA studies have opened many new avenues of investigation that allow us to examine the composition of the genome and its utilization in regulating development with much more precision than was possible when the first edition of this book was written. Another area of active investigation has concerned the roles of the cytoskeleton and the cell surface in mediating cellular morphogenesis. Accordingly, I have added a chapter entitled "The Cellular Basis of Morphogenesis" (Chap. 5), which deals with the function of these entities in cell motility, cellular shape changes, and cell-cell interactions. These themes also recur in later chapters and reflect an increased emphasis on morphogenesis at the cellular level throughout the text.

Contemporary research has provided us with improved insight into the developmental process and a more sophisticated understanding of the developmental phenomena that are described in the classical literature. I have discussed many of the key experiments upon which our current level of understanding of development is based. This approach gives the student an appreciation for the role of experimentation in science and a fuller understanding of the experimental results. Familiarity with these experiments should also facilitate the understanding of additional new research results introduced by instructors, help students to read research papers in the literature, and, hopefully, stimulate them to pose questions and design experiments about the many unanswered aspects of development.

This book contains several boxed essays with details on experimental procedures and other topics related to the textual material. The use of boxes allows these topics to be presented without interrupting the flow of the text. The titles and locations of these essays are indicated in the Contents.

The Index also serves as a glossary. All boldfaced page numbers refer to definitions of terms.

In such a multifaceted subject as developmental biology, the choice of material to be included in a text is discretionary and will certainly not satisfy everyone. I have attempted to select material that illustrates the major principles of development. The instructor will undoubtedly choose to use additional examples and introduce other topics. Ideally, the framework of the text will provide the flexibility that facilitates substitution and amplification by instructors.

Since principles of plant and animal development have much in common, an understanding of these principles is enhanced when they are treated together rather than separately. Accordingly, plant development is included in the text. However, I assume that most readers will have stronger zoological than botanical interests, and the ratio of animal to plant material reflects this assumption.

This text has resulted from the author's experience in teaching developmental biology to upper division undergraduate students who normally have a background in cell biology and genetics. I present an extensive review of the principles of molecular biology near the beginning of this course, which prepares most students to understand the molecular approach to development. This review is included in Chapters 3 and 4 of this text. Most students in my course are familiar with animal development at the descriptive level. For those readers who lack this background, I have included a brief overview of animal development in Chapter 1, which should provide a sufficient foundation for the early chapters. More detailed analyses of morphogenesis are incorporated into subsequent chapters.

I believe that visual material is an important aid to the learning process and have, therefore, included numerous illustrations in the text. I am grateful to the many investigators and publishers who have granted permission to reproduce these figures. Complete citations for the sources of previously published figures can be found in the reference list at the end of each chapter.

I have been encouraged and assisted in this project by my colleagues and associates at the University of Calgary. My Department Head, Dr. Dennis Parkinson, has nurtured two editions of this book. Every textbook author should be so fortunate as to have such an understanding Head. I would like to express my appreciation to my secretary, Ms. Maragaret Hunik, for her assistance. The graphic art for both editions has been prepared in the university's Commedia Department. Mrs. Marilyn Croot, who prepared the art for the first edition, contributed some art for the second edition. Most of the art was prepared by Mr. Bill Matheson, who has an exceptional ability to present data and scientific concepts in an understandable and aesthetically pleasing way.

My laboratory has managed to function in spite of (or because of) my preoccupation with this book. Mrs. Jillian Wilkes, Dr. Malgorzata Kloc, and Mr. Jeff Bury have been very patient with me. I am grateful for their understanding and assistance. Dr. Kloc also assisted in proof-reading, which was very useful.

My wife, Sandy, has been a partner in this enterprise, often working evenings, weekends, and holidays. I have relied on her assistance, her judgement, and her encouragement to complete this task. My daughters, Teri and Diana, have also provided assistance, frequently helping out when deadlines were to be met. My father-in-law, Mr. Lowell O'Connor, also assisted in proofreading. No author has had a more supportive family.

Leon W. Browder

Acknowledgments

The editorial departments at Saunders have done a great deal to ease the burden of preparing the second edition. I am pleased to have worked with such competent professionals and fine people. I would like to express my gratitude to Mr. Michael Brown, Biology Editor, Ms. Margaret Mary Kerrigan, his assistant, and Ms. Diane Ramanauskas, Project Editor.

Many investigators have kept me informed about recent advances in their fields. This information has been invaluable in ensuring that the text is as up to date as possible. I have also had the benefit of advice on the manuscript given by several respected developmental biologists. I am pleased to acknowledge their contribution. The burden for errors of omission, fact, or style, however, rests with me. I am especially grateful to my colleague at the University of Calgary, Dr. Michael J. Cavey. He is the perfect reviewer—totally honest, meticulous, and knowledgeable. I am very fortunate to have had his help. Thanks also to Dr. Carol Erickson, University of California, Davis; Dr. Gordon P. Moore, University of Michigan; and Dr. Peter Meyerhof, State University of New York at Buffalo. Their extensive comments on the manuscript were very useful.

The following reviewers have given valuable advice on topics relating to their areas of expertise:

Dr. Everett Anderson, Harvard Medical School
Dr. J. Derek Bewley, University of Calgary
Dr. Bruce Brandhorst, McGill University
Dr. Howard Ceri, University of Calgary
Dr. Marie A. DiBerardino, Medical College of Pennsylvania
Dr. Jacob D. Duerksen, University of Calgary
Dr. Martin Hammond, University of Washington
Dr. Merrill Hille, University of Washington

Dr. William A. Jensen, University of California, Berkeley
Dr. Elias Lazarides, California Institute of Technology
Dr. Charles Metz, University of Miami
Dr. Gilbert A. Schultz, University of Calgary
Dr. Steven Subtelny, Rice University
Dr. Eva Turley, University of Calgary
Dr. J. Richard Whittaker, Boston University
Dr. Gary E. Wise, Texas College of Osteopathic Medicine
Dr. Edward Yeung, University of Calgary
Dr. Sara Zalik, University of Alberta

Contents

PART ONE
INTRODUCTION

1 The Origins of Developmental Biology

Developmental biology is a multidisciplinary science concerned with analyzing the progressive acquisition of specialized structure and function by organisms and their various components. The multidisciplinary approach to the study of development first emerged before the turn of the century as an integration of embryology with cytology and later with the new science of genetics.

Literally, **embryology** means "the study of embryos," but the word cannot be rigidly defined, because it implies a point of view and is frequently used to connote the study (descriptive or experimental) of changes in the form or shape (**morphogenesis**) of animals during their embryonic phase. The cytologists of the late 1800s saw the development of the embryo as a manifestation of the changes occurring in the individual cells that compose the embryo. They believed that the cell is the key to all ultimate biological problems and that the fundamental principles underlying development would emerge by studying the properties and behavior of cells.

Every multicellular organism begins life as a single cell, the fertilized egg, which is essentially similar to all other cells but differs by its potential to divide and produce all the cells of the body. The cells diverge from one another structurally and functionally and become organized into an adult organism. Accordingly, these basic processes of development must first be understood at the cellular level before an overall understanding of development is possible.

Since the fertilized egg (or **zygote**) is derived from fusion of the male and female gametes, the cytologists assumed that the factors that control development could be identified by tracing the elements of the gametes and following their behavior in the cells of the developing organism.

3

The foremost proponent of the cytological approach was E.B. Wilson, and the most eloquent statements of this philosophy were contained in the three editions (1896, 1900, and 1925) of his classic, *The Cell in Development and Inheritance* (retitled *The Cell in Development and Heredity* in 1925). He wrote in the first edition:

> Every discussion of inheritance and development must take as its point of departure the fact that the germ is a single cell similar in its essential nature to any one of the tissue-cells of which the body is composed. That a cell can carry with it the . . . heritage of the species, that it can in the course of a few days or weeks give rise to a mollusk or a man, is the greatest marvel of biological science. In attempting to analyze the problems that it involves, we must from the onset hold fast to the fact . . . that the wonderful formative energy of the germ is not impressed upon it from without, but is inherent in the egg as a heritage from the parental life of which it was originally a part. The development of the embryo is nothing new. It involves no breach of continuity, and is but a continuation of the vital processes going on in the parental body. What gives development its marvelous character is the rapidity with which it proceeds and the diversity of the results attained in a span so brief.

Thus, Wilson understood that the characteristics of the organism emerge during development by utilization of inherited information, and the only way to understand development fully is to comprehend the nature of that information and the ways in which it is utilized. These concepts have had a rocky history, both before and after Wilson's time.

1–1.   GERM CELLS: BRIDGING THE GENERATION GAP

The idea that the form of an embryo gradually emerges during development originated with Aristotle. However, nearly 2000 years later, in the seventeenth and eighteenth centuries, many embryologists rejected this concept and proposed that the egg contains a miniature, fully formed embryo. Development to them was analogous to the unfolding of a flower bud. Bonnet (1745) formalized this concept in the theory of *emboîtement*, or encasement. He stated that since the egg contains the complete embryo, it must also contain similar preformed eggs for all future generations, like an infinite series of boxes encased one inside the other. Others of his contemporaries believed that a preformed embryo exists inside the sperm, and many microscopists of the time claimed to have seen a tiny creature (homunculus) curled up in the sperm head (Fig. 1.1).

These theories of **preformation** were not universally held, however. Caspar Wolff (1759) championed the theory of an alternative mechanism of development—**epigenesis.** Epigenesis means that the adult gradually develops from a rather formless egg as originally proposed by Aristotle. Wolff made careful observations on the development of the chick and showed that the early embryo is entirely different from the adult and that development is progressive, with new parts being continually formed. Wolff's conclusions were rejected by most of his contemporaries, and the concept of epigenesis was not accepted by a majority of biologists until the early part of the nineteenth century.

Even after epigenesis was determined to be the mechanism for the formation of the external structure of the organism during development, the nature and the significance of the germ cells remained unclear for nearly a century. The fact that the egg itself is a cell was recognized by Schwann in 1839. Likewise, the cellular nature of sperm was determined in 1865 by Schweigger-Seidel and St. George. Another decade elapsed before Oscar Hertwig (1876) established that fertilization results from the union of the egg and sperm. Thus, each sex contributes a single cell from its own body.

These cells, which carry the complete set of instructions for production of another generation similar to the preceding one, fuse to form a single cell, the zygote, which undergoes division or **cleavage,** to produce

FIGURE 1.1 Homunculus in human sperm. (After Hartsoeker. Redrawn from J.A. Moore. 1972. *Heredity and Development,* 2nd ed. Copyright © 1972 by Oxford University Press, Inc., New York, p. 257. Reprinted by permission.)

the cells that form the embryo. Within the fertilized egg, two nuclei, one derived from the egg and the other derived from the sperm, combine to form a single zygote nucleus, which gives rise by division to all nuclei of the body. In contrast to the equal nuclear contribution to the zygote made by both gametes, the cytoplasmic contributions are decidedly uneven; the egg contributes virtually all of the cytoplasm. These observations led Hertwig to the conclusion that the germ cell nuclei, and not the cytoplasm, are the vehicles of inheritance.

Hertwig's discoveries marked the beginning of a new era of investigation into the role of the nucleus in fertilization and development. Much of the attention of cytologists during this era was directed to the primary nuclear constituents, the **chromosomes.** The first detailed description of the behavior of the chromosomes in fertilized eggs was by van Beneden (1883), who used the nematode, *Ascaris megalocephala.* This species was a splendid choice for chromosome study, since it has only four chromosomes, which are large and stain intensely. Van Beneden observed that the egg nucleus and the sperm nucleus each provides two of the four chromosomes that align on the metaphase plate at the first cleavage division. Each chromosome splits lengthwise, and the daughter chromosomes are transported to the opposite poles of the spindle where they are incorporated into the nuclei of the two-celled stage. Therefore, each of these nuclei receives an equal number of maternal and paternal chromosomes.

Soon after publication of van Beneden's work on *Ascaris*, reports based upon studies using several species of animals and plants established the rule that each gamete nucleus contributes one half the number of chromosomes that is characteristic of somatic cells. The behavior of chromosomes at fertilization led Hertwig and three other German scientists—Strasburger, Kolliker, and Weismann—to conclude independently that chromosomes are the means for transmission of inherited information. As Wilson stated in the 1896 edition of *The Cell*, these observations on chromosomes during fertilization could not logically lead to any other conclusion:

> These remarkable facts demonstrate the two germ-nuclei to be in a morphological sense precisely equivalent, and they not only lend very strong support to Hertwig's identification of the nucleus as the bearer of hereditary qualities, but indicate further that these qualities must be carried by the chromosomes; for their precise equivalence in number, shape, and size is the physical correlative of the fact that the two sexes play, on the whole, equal parts in hereditary transmission. And we are finally led to the view that chromatin is the physical basis of inheritance. . . .

1–2. THE ROUX–WEISMANN THEORY

The genetic material was beginning to take on a definite physical form in the theories of biologists. However, the absence of hard information about the nature of genetic material and its function in development led to speculation that was based on logic that Wilson characterized as bordering on the metaphysical. Wilson reserved most of his contempt for a theory of development that originated with Wilhelm Roux in 1883. Roux believed that the hereditary material represents different characteristics of the organism. He assumed that the fertilized egg receives all of these substances, which, as cell division ensues, become linearly aligned on the chromosomes. The substances are then distributed unequally to daughter cells, Roux proposed. This "qualitative division" fixes the fate of the cells and their descendants, since a portion of determinants is lost to a cell at each division.

August Weismann later elaborated on Roux's ideas. Weismann developed a highly complex scheme for the hereditary material, or **germ plasm.** The primary hereditary units were biophores, which aggregated to form determinants, the determinants to form ids, the ids to form the larger idants, which were equivalent to chromosomes. He believed that there are two kinds of division: qualitative and quantitative. Weismann proposed that the id gradually disintegrates during development, splitting into smaller and smaller groups of determinants, which are isolated into different daughter cells during cell division. Finally, only one kind of determinant remains in a particular cell or group of cells. The determinant then breaks up into its constituent biophores and imparts specific characteristics on the cell. To account for the formation of germ cells that must contain the entire set of heritable information, Weismann proposed that the germ cell line is set aside early in development by quantitative, rather than qualitative division. By equal distribution of nuclear constituents, the germ plasm remains intact.

This separate developmental pathway for the germ cells was an important concept. Weismann astutely concluded that although the body is derived from the germ plasm, the germ plasm itself is passed on without modification (the concept of mutation was unknown at that time) from one generation to another. Hence, he concluded that inheritance of acquired characteristics, as proposed by Lamarck, was impossible. The usefulness of this concept far outlived that of the Roux–Weismann theory of qualitative division, which was soon disproven.

For his part, Roux (1888) appeared to have confirmed his theories by an experiment he conducted on frog eggs. He used a hot needle to destroy one of the two cells that result from the first division after fertilization. In some cases (Fig. 1.2A) the uninjured half continued developing to form a half-embryo that lacked structures corresponding to the side that was damaged. Roux concluded that this result supported his theories, since the undamaged half appeared to lack the information that was necessary to produce the other half of the embryo.

Hans Driesch (1892) approached the problem differently with sea urchin embryos, dissociating them by mechanical shaking at the two-cell stage. These half-embryos developed into normally formed dwarf larvae. Driesch subsequently modified his technique, separating the cells

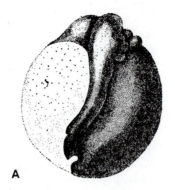

A

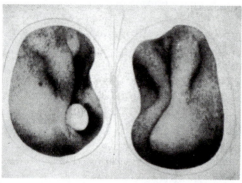

B

FIGURE 1.2 Developmental potential of half embryos of the frog. *A,* Half embryo produced by Roux following destruction of one blastomere with a hot needle. (From T.H. Morgan. 1927. *Experimental Embryology.* Columbia University Press, New York, p. 382.) *B,* Two whole embryos produced by Schmidt following ligation at the two-cell stage. (From G.A. Schmidt. 1933. Schnurungs- und Durchschneidungsversuche am Amphibienkeim. Wilh. Roux' Archiv. Entwicklungsmechanik der Organismen, *129:* 11. Reprinted with permission of Springer-Verlag, Heidelberg. *A* and *B* reproduced from B.I. Balinsky. 1975.)

in calcium-free seawater, and found that isolated cells at the four-cell stage also develop normally. Thus, Driesch concluded that each cell retains all the developmental potential of the zygote.

The conflict between these two opposing views of development has been settled in favor of Driesch's interpretation by numerous cell separation experiments on several animal species. These have included experiments on the frog embryo, the same kind of embryo that Roux used. When the cells of the two-cell frog embryo are separated with a ligature, both halves proceed to develop normally (Fig. 1.2*B*).

The experiment conducted by Roux illustrates the importance of proper experimental design. Roux had introduced an artifact into his experiment by allowing the damaged half of the embryo to remain attached to the uninjured half, interfering with its development. This fact was demonstrated by Hertwig, who repeated Roux's experiment, but instead of leaving the damaged cell attached to the normal half, he removed it. Development of the remainder of the embryo resulted in a complete, half-sized embryo. The interference caused by the death of one cell in Roux's original experiments was only temporary, for Roux himself admitted that the missing half of the damaged embryo was restored during later development, indicating that the frog embryo has the ability to regulate its development by formation of missing parts. Roux should have realized that this observation contradicted his hypothesis. He instead attempted to rationalize the result. Perhaps this was his greatest failure. Wilson dismissed Roux's rationalization as "an artificial explanation."

Roux's work did have one major lasting effect on developmental biology, however: the *experimental* approach to development. For the first time, an embryologist had manipulated embryos and observed the effects of these manipulations on them. For this reason, many embryologists consider Roux to be the "Father of Experimental Embryology."

1–3. THE MENDELIAN ERA

It was now clear that in cell division the inherited information is equally distributed to all cells of the embryo. But the central question still remained: How does the inherited information participate in development? Frustration began to replace the optimism generated by the discoveries of the significance of the nucleus and chromosomes in inheritance. The year 1900 was the turning point. During that year, the paper that Mendel had presented to an unreceptive audience in 1865 was discovered. But now the scientific world was more receptive, for scientists needed new concepts. Mendel's major new insight was that characteristics of organisms are determined by factors that retain their identity through generations of breeding. Each individual inherits a single factor for a trait from each parent. If the parents have "antagonistic" characters, the offspring (hybrid) displays the unaltered trait of one parent, while the trait of the other parent is missing. However, if the hybrid generation is bred with itself, the previously missing secondary character could reappear unchanged in the next generation.

Mendelism did not immediately take the scientific world by storm, since existing ideas about heredity were often held tenaciously. William Bateson, a Cambridge biologist, was one of the most active proselytizers of the new faith, arguing convincingly and passionately in favor of Mendelism. Bateson coined the term "genetics" and worked hard to establish this new biological discipline. Gradually, nebulous terms such as "biophores" and "idants" were to be replaced with genes and linkage groups. Mendelism made possible a new precision in thought, and soon the modern concepts of genetics were formulated.

One other experimental observation was also important in setting the stage for this new era. This was a paper written by Theodor Boveri and published in 1902. Although published two years after the rediscovery of Mendelian inheritance, Boveri was apparently unaware of Mendelism. Boveri's work established more precisely the role of chromosomes in development.

His experiments utilized sea urchin embryos with variable numbers of chromosomes (a condition that today is called **aneuploidy**). This abnormal situation can be produced when sea urchin eggs are fertilized by two sperm. The frequency of the **dispermy** is increased by adding an excess number of sperm when fertilizing sea urchin eggs *in vitro*. The two sperm nuclei and the egg nucleus form an abnormal division figure, usually with four centrioles, so that the first mitotic division produces four cells that rarely receive the normal diploid number of chromosomes (Fig. 1.3). Since the distribution of chromosomes to these four cells is random, cells with variable numbers of chromosomes are produced from a single zygote.

Boveri separated the four cells from one another by placing the dispermic egg in calcium-free seawater and observed the development of each. Subsequent cleavage divisions were normal, thus fixing the abnormal chromosome number in the embryos developing from the separated cells. Never did these embryos develop normally. But *most importantly*, each typically developed in a different way, arresting at a different stage. Boveri concluded that the different developmental patterns were due to the different chromosomal combinations produced in the abnormal first division. Thus, normal development is dependent upon the *normal combination of chromosomes*. Clearly, each chromosome must have qualitatively unique effects on development.

That conclusion seems self-evident to us today—each chromosome contains a linear sequence of genes, and the gene sequences on the chromosomes are unique. But only a minority of biologists of the early 1900s (Wilson among them) accepted the conclusion that the hereditary factors are on the chromosomes. In 1914, Wilson presented a lecture in London

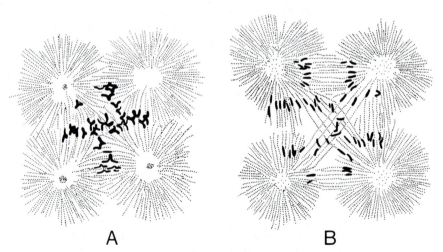

A   B

FIGURE 1.3 Mitosis in the first cleavage of dispermic sea urchin eggs. *A*, metaphase; *B*, anaphase. Each sperm contributed two centrioles, each of which attracts chromosomes at random. (*A* redrawn from E.B. Wilson, 1925; *B* after F. Baltzer, redrawn from E.B. Wilson, 1925. Reprinted with permission of Macmillan Publishing Company from *The Cell in Development and Heredity*, third edition by E.B. Wilson. Copyright 1925 by Macmillan Publishing Company, renewed 1953 by Anna M.K. Wilson.)

in which he gently chided those who did not accept the chromosome theory:

> I am well aware that some eminent students of genetics are still reluctant to accept this theory, at least in its more detailed applications. I am not disposed to reproach them for such scepticism. The cytologist suffers under the disadvantage of working in so unfamiliar a field that some of his conclusions, even among those most certainly established and most readily verifiable, are apt to give a certain expression of unreality, even to his fellow naturalists. It is undeniable, too, that in this subject, for better or for worse, hypothesis and speculation have continually run far in advance of observation and experiment. It is quite possible that some of my hearers may consider some of the views I have touched upon as a fresh illustration of that fact. If so, I beg them to bear in mind that no conclusion which I have considered has been reached as a merely logical or imaginative construction. I have endeavoured to limit myself to matters of observed fact, and to conclusions that are either demonstrated by facts or directly and naturally suggested by them. To those who have had opportunity to come into intimate touch with both cytological and genetic research the conclusion has become irresistible that the chromosomes are the bearers of the "factors" or "gens" with the investigation of which genetics is now so largely occupied.

The evidence for the chromosome theory was based on both cytological and genetic data. Although Wilson enthusiastically promoted Boveri's conclusions, they were dependent upon indirect evidence; more convincing proof was needed. Walter Sutton, a graduate student working with Wilson, was the first to make a clear correlation between the behavior of Mendel's genetic factors and the behavior of chromosomes. Sutton pointed out that the parallel behavior of hereditary factors and chromosomes must mean that the factors are localized on the chromosomes. Thus, in a haploid gamete, genes and chromosomes are singly represented. Following fertilization the cells of the diploid zygote contain a set of chromosomes that are derived from each parent; the chromosomes are present in homologous pairs, and the members of each chromosome pair possess the alleles, which specify the maternal and paternal traits, respectively. During the formation of gametes in the mature diploid organism, the homologous chromosomes pair, and as a result of meiosis, every gamete receives one chromosome of each homologous pair (Mendel's law of segregation). Furthermore, since the behavior of each homologous pair of chromosomes during meiosis is independent of every other pair, the chromosomes are distributed randomly to the gametes, resulting in several combinations of maternal and paternal chromosomes (Mendel's law of independent assortment). Sutton also noted that if two genes were located on the same chromosome, they would be inherited together and would not behave according to Mendelian principles. He had, therefore, foreshadowed the concept of linkage.

Regular text continued on page 16

MEIOSIS

After van Beneden's discovery that the nuclei of egg and sperm have half the number of chromosomes of somatic cell nuclei, attention soon was directed to the mechanism for reduction of chromosome number during production of the germ cells. Once again, *Ascaris* proved to be excellent material for these investigations, conducted during the 1880s by van Beneden, Boveri, and Hertwig. They discovered that two unusual cell divisions occur during gamete formation, reducing the diploid chromosome number to the haploid condition. The two divisions are known as **meiotic divisions** and the process itself as **meiosis.** As we shall discuss in more detail in section 1–8, the germ cell precursors become established in the gonad early in development and increase in number by normal mitotic division, which maintains the diploid chromosome number of four. These stem cells are called oogonia in the ovary and spermatogonia in the testis. Meiosis in the female *Ascaris* is illustrated in the following figure, which is based upon drawings made by Boveri (after Moore, 1972). Only the upper portion of the egg is shown.

At the beginning of meiosis the four long chromosomes shorten to form tiny spheres. Homologous chromosomes then form pairs in a process called **synapsis.** Following synapsis, each chromosome of the two chromosome pairs splits. Each pair contains four chromatids, and each group of four is called a **tetrad.** In *a*, the

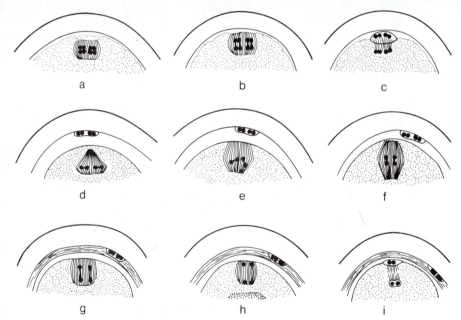

Redrawn from J.A. Moore. 1972. *Heredity and Development,* 2nd ed. Copyright © 1972 by Oxford University Press, Inc., New York, p. 38. Reprinted by permission.

tetrads are aligned on the metaphase plate. During the first meiotic division (b,c), the tetrads separate to form **dyads.** Although the nuclear division is equal, the cytoplasmic division is not. During the division process the egg nucleus is displaced to the periphery of the egg, and when the division occurs, nearly all of the cytoplasm remains with the egg. The other cell receives only a small amount of cytoplasm; it is called the first polar body. During the second meiotic division the chromosomes do not duplicate themselves. As shown in d through h, the dyads separate, and at the completion of the second meiotic division (i) the egg nucleus receives two chromosomes, while two chromosomes enter the second polar body. Meanwhile, the first polar body may also undergo a second meiotic division. Thus, the potential result of meiosis is one haploid egg and three haploid polar bodies. The unequal cytoplasmic divisions of meiosis in the female ensure that the egg will retain the vast majority of the cytoplasm and yolk built up during oogenesis (see section 1–8).

Shortly after female meiosis was described, the analagous process was also discovered in the male, again utilizing *Ascaris.* This process is illustrated in the following figure, based on drawings by Brauer, which were first published in 1893 (after Moore, 1972).

A diploid spermatogonium is shown in a. The four long chromosomes are readily visible. The chromosomes pair near the beginning of meiosis, and as meiosis continues, the chromosomes shorten to form small spheres. A nucleus during synapsis is shown in b. As in the female, the synapsed chromosomes divide to reveal their tetrad configuration (c). During the first meiotic division (d,e) the nucleus is centrally located so that two equal-sized daughter cells are formed (f,g), each receiving two dyads. The second meiotic division (h,j and i,k) results in separation of the dyads and distribution of two chromosomes to each of the four cells formed (l,m and n,o). Once again the cytoplasm is equally divided. Thus, four equal cells are produced, each of which develops into a functional sperm (see section 1–8). The utilization of all four haploid cells in the male (in contrast to the female where only one of four is functional) is significant, since the testis is required to produce millions of sperm simultaneously. The loss of three fourths of the cells during meiosis would make this task monumental.

The essential modification of meiosis that results in production of haploid cells is that there is only one duplication of chromosomes for the two divisions. In contrast, each mitotic division is accompanied by chromosomal division, maintaining a constant chromosome number at each division.

The basic elements of meiosis were soon confirmed for a large number of animals and plants. Detailed studies of meiosis revealed that prophase of the first meiotic division is specialized to facilitate exchange of material between homologous chromosomes. This exchange occurs by **crossing-over,** which results in recombination of genes in a linkage group. Because of its complexity, prophase I is subdivided into a number of arbitrary stages. The figure on page 15 illustrates the chromosomal events of meiosis and compares meiosis with mitosis. The first stage of

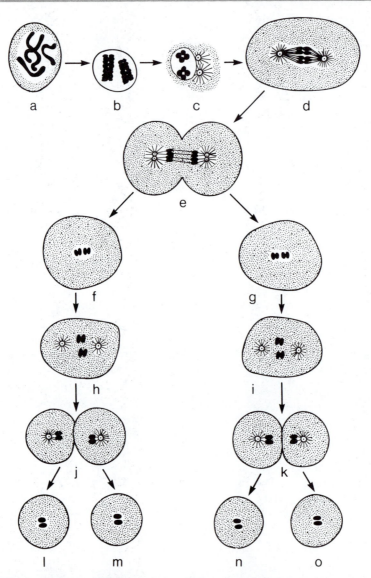

Redrawn from J.A. Moore. 1972. *Heredity and Development,* 2nd ed. Copyright © 1972 by Oxford University Press, Inc., New York, p. 40. Reprinted by permission.

prophase I is **leptonema.*** During leptonema the chromosomes appear as long threadlike structures. Although they appear to be singular, they actually consist of two chromatids. Homologous chromosomes pair during **zygonema** to form **biva-lents.** The homologues are represented by lines of different densities; thus, each

---

*Stages of prophase I end in *"-nema"* when used as a noun and in *"-tene"* when used as an adjective (e.g., leptonema, leptotene).

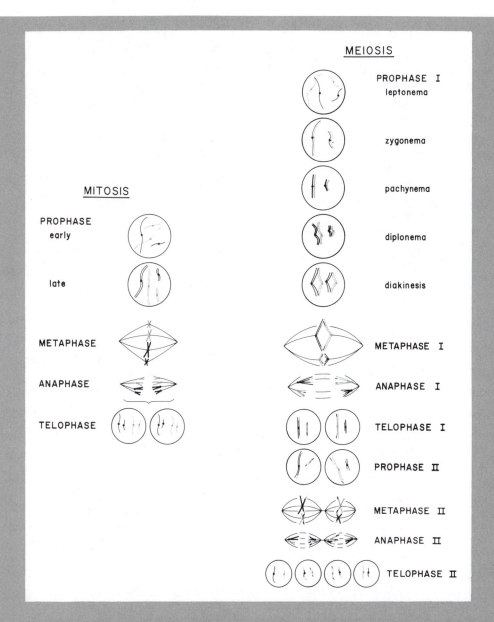

MEIOSIS

PROPHASE I
leptonema

zygonema

pachynema

diplonema

diakinesis

METAPHASE I

ANAPHASE I

TELOPHASE I

PROPHASE II

METAPHASE II

ANAPHASE II

TELOPHASE II

MITOSIS

PROPHASE
early

late

METAPHASE

ANAPHASE

TELOPHASE

homologue can be followed through meiosis. The pairing process results in close apposition between the homologues, which is essential for crossing-over.

The next stage, **pachynema,** is characterized by a shortening of the chromosomes. The chromatids become individualized during **diplonema,** as each bivalent is seen to be composed of four chromatids and is now called a tetrad. This stage is often of extremely long duration, particularly in the female. Diplotene chromosomes in some species become highly modified, with numerous loops extending laterally from the chromosome axis. These **lampbrush chromosomes** are very active synthetically and may play an important role in differentiation of the gamete. We shall discuss lampbrush chromosomes in more detail in Chapter 7. During diplonema the homologues separate but remain joined at chiasmata, which are the points near the sites of physical exchange between chromatids during crossing-over. The chiasmata move to the ends of the homologues during **diakinesis** in a process called terminalization. During this stage the chromosomes shorten, the nuclear envelope breaks down, and the chromosomes begin moving toward the metaphase plate. The first division is characterized by separation of homologues from one another. A short interphase separates the two meiotic divisions. The second division is much more rapid and less complicated than the first. As in mitosis the chromosomes do not pair but become aligned in tandem on the metaphase plate, and the centromere of each chromosome divides, allowing separation of sister chromatids.

Regular text continued from page 11

The final proof of the chromosome theory of inheritance resulted from experiments conducted in the laboratory of Thomas Hunt Morgan, utilizing the fruit fly, *Drosophila melanogaster*. Mostly as a result of work by Morgan and his associates A. H. Sturtevant, C. B. Bridges, and H. J. Muller, the basic concepts of transmission genetics were rapidly discovered. These concepts have been summarized by Moore (1972). They are:

1. Inheritance is the transmission of genes from parents to offspring.
2. Genes are located on chromosomes.
3. Each gene occupies a specific site (locus) on a chromosome.
4. Each chromosome has many genes that are arranged in linear order.
5. Somatic cells of diploid organisms contain two of each kind of chromosome (homologues), and thus each gene locus will be represented twice.
6. During each mitotic cycle, each gene is replicated.
7. Genes can exist in several alternative states (alleles). The change from one state to another is a mutation.
8. Genes can be transferred from one homologous chromosome to the other by crossing-over during meiosis.
9. Every gamete receives one chromosome of each homologous pair; the distribution of chromosomes to the gametes is random.
10. The distribution of chromosomes of one homologous pair to the gametes has no effect on the distribution of chromosomes of other pairs.

11. At fertilization, gametes unite randomly; the zygote receives one chromosome of each homologous pair from its father and one from its mother.
12. When the cells of an organism contain two different alleles of the same gene (heterozygous condition), one allele (the dominant) usually has a greater phenotypic effect than the other (the recessive).

The choice of *Drosophila* by Morgan was indeed a fortunate one. These remarkable creatures are tailor-made for genetics research: They are easily bred and reproduce rapidly (a single pair can produce several hundred progeny in a couple of weeks), and large numbers of them can be maintained in small vials or bottles containing simple inexpensive "fly food." Although *Drosophila* are small, their external characteristics are simple to score with no more than a hand lens, which Morgan used in his early work.

Geneticists and cytologists were soon to discover other qualities that made *Drosophila* a nearly ideal organism for genetic research. For example, the haploid chromosome number is four, meaning only four linkage groups. Furthermore, certain larval cells have giant chromosomes called **polytene chromosomes.** These structures were described in the larval salivary glands by T. S. Painter in 1934. The salivary gland chromosomes are about 100 times longer and thousands of times wider than chromosomes of normal body cells. Actually, Painter had "rediscovered" this; it was known to the cytologists in 1881, and a drawing of polytene chromosomes can be found in the 1896 edition of *The Cell.* Although one would expect salivary gland nuclei to contain eight chromosomes, Painter found only four. This is due to pairing of homologous chromosomes. The most important aspect of polytene chromosomes is the presence of cross-bands on them, as shown in Painter's first drawing of salivary gland chromosomes in Figure 1.4. The cross-banding extends across both paired chromosomes; the pairing of the homologues is so exact that each band appears as a single structure. An explanation of these chromosomes in molecular terms can be found in Chapter 4. But for the early *Drosophila* geneticists, the significance of polytene chromosomes was that chromosome markers were now available.

The bands are in various shapes and sizes, and interband distances are variable. However, in different cells of the same larva, and even in different larvae, the band pattern itself is constant. The pattern apparently represents the essential organization of the chromosome, and as was soon discovered, it marks the organization of genes within the chromosomes. Once again, cytology and genetics joined forces, and a new branch of genetics called **cytogenetics** emerged. For the first time, the concepts of transmission genetics, such as gene deletions and duplications, translocations, and inversions, could be correlated with chromosome markers. These correlations have resulted in very elaborate chromosome maps showing genes in a linear order and in definite sequence

Chromocenter

FIGURE 1.4 Painter's first drawing of the salivary gland chromosomes of *Drosophila melanogaster.* The chromosomes radiate out from the chromocenter. The X is attached to the chromocenter by one end so it appears as a single long structure. Both the II and III chromosomes are attached by their middle portions. Consequently, both of these chromosomes have two arms extending from the chromocenter. The tiny IV chromosome is attached by its end to the chromocenter. The approximate location of several X chromosome genes (*B, f, sd,* etc.) is shown. (From T.S. Painter. 1934. A new method for the study of chromosome aberrations and the plotting of chromosome maps in *Drosophila melanogaster.* Genetics, *19:* 179.)

(Fig. 1.5). Genes have been localized to a portion of the chromosome within a single band, and genes have been detected only in regions containing bands. These observations led to the hypothesis that bands represent single gene loci (Judd et al., 1972). Although this hypothesis has been shown to be an oversimplification (Young and Judd, 1978), it is still thought to be generally valid.

1–4.   EMBRYOLOGY: LOSING THE FAITH

One of the most important concepts to emerge from Wilson's writing concerns the nature of the relationship among genes, cells, and the developing embryo. Gene function determines cellular characteristics in a process called **cell differentiation.** The characteristics of the whole organism are, in turn, the net result of the characteristics of its individual cells. Thus, genes function at the cellular level to cause development of the organism, or as stated by Wilson, development is "the appearance of hereditary traits in regular order of space and time."

What is the specific role of the genetic material in this process? How is temporal and spatial coordination of cell differentiation regulated by the nucleus? Wilson pointed out that the nucleus itself does not cause

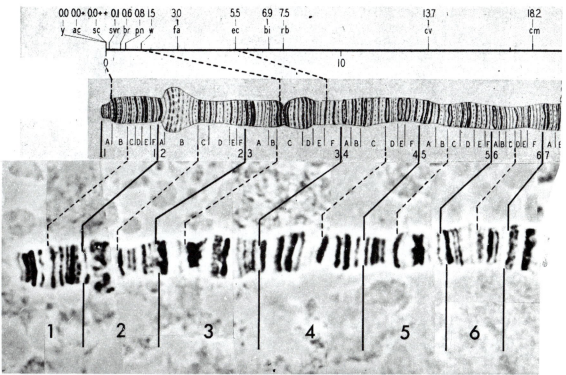

FIGURE 1.5 Photographic map of a portion of the *Drosophila* X chromosome, correlated with Bridges' (1935) drawings. The cytological positions of several of the common sex-linked genes are indicated. (From G. Lefevre, Jr. 1976. Reproduced with permission from *The Genetics and Biology of Drosophila*, Vol. 1a, M. Ashburner and E. Novitski, Eds. Copyright: Academic Press Inc. (London) Ltd.; Bridges' chromosome map reproduced by permission of the American Genetic Association.)

development. Although the nucleus alone suffices to transmit inherited information, it must work in concert with the cytoplasm. Wilson proposed that the specific role of the nucleus is the regulation of "constructive metabolism" (i.e., synthetic metabolism) of the cell, which is fundamentally a problem of biochemistry and was well beyond the level of understanding of scientists of that time. Wilson and his contemporaries had expected that genetics and cytology would together lead the way to a solution of the "how" and "why" questions of development. However, Wilson admitted that they had underestimated the magnitude of the problem of development. Just as the questions of heredity required the Mendelian principles and the chromosome theory, explanation of the mechanisms of development would require the understanding of new, and as yet undiscovered, principles. Studies on the role of genes in development could not progress significantly until the nature of the gene was discovered and advances were made in understanding how genes function to influence cell biochemistry. An era had ended.

The embryologists' disillusionment with genetics was complete. Many embryologists considered genetics to be peripheral to the fundamental mechanism of development. They believed that the basic structure of an embryo is produced by *embryological* mechanisms and that the only function of genes is to add the nonessential finishing touches, such as eye and hair color, number of bristles on a leg segment, and color of flower petals. Emphasis during the 1920s, 1930s, and 1940s was on the descriptive and comparative aspects of embryonic development. Experimentation was concentrated on the interactions and rearrangements of the cells and tissues that form embryos. Chemical analyses of embryos were largely devoted to descriptions of the chemical components of embryos, comparisons of chemical components among embryos of different taxonomic groups, and attempts to establish the chemical basis for embryonic induction (see Chap. 13). The importance of this era of classical embryology should not be underestimated. It has provided us with a vast amount of information about embryonic development, without which cellular differentiation is an isolated phenomenon that loses much of its significance. The student of development should remain aware that cell differentiation is a process that occurs during a phenomenon of much greater consequence—the formation of a living, functioning adult organism.

1–5. GENETICS: KEEPING THE FAITH

Since very few embryologists concerned themselves with gene function in development, it became incumbent upon geneticists to keep alive the concept that development occurs under genetic control. That this was truly a "rear guard action" can be deduced from this quote by Richard Goldschmidt (1958): "In spite of such isolationism or possessiveness [by embryologists], the geneticists will continue to worry about the problem of genetic action and take the risk of climbing over the fence erected by some jealous embryologists, who, while claiming the kingdom for themselves, do not set out to till its soil."

It is not surprising that geneticists became strong advocates of the idea that genes play an important role in basic developmental processes, because the new mutations they were discovering provided strong evidence for this fundamental principle. In *Drosophila*, for example, there are mutants that affect virtually every aspect of development from the most general down to the smallest detail. In fact, mutants can even alter the basic body plan that characterizes *Drosophila* as a fly belonging to the insect class Diptera. The body of an adult dipteran consists of a head, thorax, and abdomen. The thorax is made up of three segments: the **prothorax,** the **mesothorax,** and the **metathorax.** Each of these segments bears a pair of legs; in addition, the mesothorax bears the wings, and the metathorax has a pair of rudimentary appendages called **halteres.** The halteres are balancing organs that maintain the fly in an upright position while flying. The combination of two wings and two halteres is unique to the dipterans; most insects have four wings—one pair on the mesothorax and a second pair on the metathorax. A mutation called *bithorax (bx)* converts *Drosophila* from a typical dipteran to a four-winged creature by producing a pair of wings on the metathorax in place of the halteres. Combination of *bithorax* with certain other mutant genes (e.g., *postbithorax* and *ultrabithorax*) can change the metathorax so that it resembles a second mesothorax (Fig. 1.6). These genes can apparently bring about a major switch when the body plan is being laid out, causing the group of cells that normally form a metathorax to form mesothoracic structures instead.

The completion of any organ from the initial structure formed during organization of the basic body plan to the definitive organ is under genetic control at virtually every step. Literally hundreds of genes have been discovered that affect the detailed development of *Drosophila* body parts. The detection of a mutant gene affecting a developmental process implies that the normal allele of the gene is involved in control over that developmental step. The nature of the defect often indicates which de-

FIGURE 1.6 Photographs of wild-type (*A*) and bithorax (*B*) *Drosophila.* The fly in *B* has the genotype bx^3pbx/Ubx^{105}. Code: *bx, bithorax; pbx, postbithorax; Ubx, Ultrabithorax.* (From E.B. Lewis. 1963. Genes and developmental pathways. Am. Zool., *3:* 39.)

A

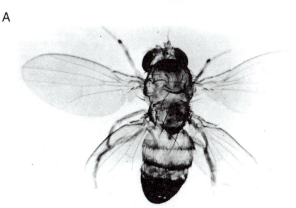

B

velopmental process is affected. The sequence of gene-controlled steps in development of a body part can be reconstructed by careful analysis of a series of mutants that affects its development. Analysis of development in this manner is called **genetic dissection.**

1–6. PHYSIOLOGICAL GENETICS

Many geneticists who were interested in development during the 1930s and 1940s were asking how genes produce their effects. The most prominent among these scientists was Richard Goldschmidt. Although most of Goldschmidt's specific ideas about the role of genes in development have been disproved, he served an important role by drawing attention once again to this basic problem. A new field emerged—**physiological genetics.**

As we have already noted, biologists had long suspected that genes are involved in regulating cellular metabolism. Therefore, the discovery of the specific function of genes could be found only by utilizing a biochemical approach. New insight into the nature of cellular metabolism began emerging from biochemical studies at approximately the same time that the new science of genetics was making its rapid discoveries about the mechanism of inheritance. These critical biochemical discoveries involved the role of enzymes in metabolism and the concept of intermediary metabolism. Biochemical reactions in cells were found to require specific enzymes that temporarily combine with the reactants and accelerate their conversion. The synthesis or degradation of complex molecules was found to occur by elaborate metabolic pathways in which a series of consecutive reactions (each controlled by a specific enzyme) are linked in sequence. The study of these complex pathways (intermediary metabolism) required the development of techniques to study the rapid intermediate reactions. Poisons were found that would destroy or inhibit specific enzymes. Consider the following sequence of reactions:

$$A \xrightarrow{\text{Enzyme 1}} B \xrightarrow{\text{Enzyme 2}} C$$

If the action of enzyme 1 is prevented, B and C will not be formed, but A may accumulate so that it can be detected. Similarly, if the action of enzyme 2 is prevented, C will not be formed and B will accumulate. In this way the intermediates formed during metabolism could be detected and the complex pathways clarified.

How could genes be involved in regulating these pathways? Could the regulation be mediated by the enzymes? This possibility was first suggested by an English physician, Sir Archibald Garrod. He was interested in inherited metabolic diseases of humans, which he called "inborn errors of metabolism." In 1909, Garrod published a book in which he discussed the rare disease alkaptonuria. The disease is detected early in life, since the urine of alkaptonuric babies stains their diapers black. The staining results from the presence in the urine of homogentisic acid (or alkapton), which blackens upon exposure to the air. Homogentisic acid is normally converted to acetoacetic acid. Garrod proposed that this conversion is accomplished by an enzyme, and this enzyme is missing in the affected individual. The effect is similar to that of a metabolic poison—the metabolic conversion is blocked, resulting in the buildup of homogentisic acid, which is excreted. Garrod also observed that parents of affected children were often first cousins, and he speculated that the disease is inherited. He consulted the geneticist Bateson, who concluded that alkaptonuria is inherited as a simple Mendelian recessive gene. Thus, Garrod correlated the metabolic defect with the mutant gene. He did not categorically state that the mutant gene resulted in the absence of the enzyme, but he certainly implied it. Unfortunately, Garrod's work lay dormant in the medical literature and was ignored by a generation of geneticists.

Serious experimental analysis of the role of genes in metabolism began in the 1930s. The initial efforts utilized mutant genes that had an obvious effect on metabolism: eye color mutants of *Drosophila*. The modification of color by mutation suggested that the genes were involved in some way in the synthesis of the pigments. The task was to identify the pigments, clarify the synthetic pathways, pinpoint the metabolic block in the mutant, and thus establish the relationship between the block and the mutant gene. Several mutants were known to affect the production of brown eye pigment. George Beadle and Boris Ephrussi reasoned that some of these mutants must be involved in regulating the various steps in the synthesis of brown pigment from its precursor. Through a series of ingenious experiments, Beadle and Ephrussi identified the genes affecting the synthetic pathway and established the sequence in which the mutant genes blocked the synthesis of the brown pigment. As we have seen, the blockage of a synthetic step implies that the enzyme for that conversion is missing. If the mutation removes the enzyme, it stands to reason that the function of the normal allele of that gene is to produce the enzyme. Therefore, they suggested the hypothesis: **one gene—one enzyme,** that is, the primary function of a gene is to produce a specific enzyme. They could not, however, test their hypothesis on *Drosophila*, since they were unable to unravel the pathway for brown pigment synthesis. It was impossible to say with certainty that a gene specifies the enzyme that catalyzes a reaction, because they did not know what the reaction was, let alone which enzyme was involved.

A new approach to gene function was needed. Beadle, working with Edward Tatum, decided to reverse the usual procedure. Instead of beginning with known mutants and attempting to identify the biochemical reactions they affect, they would start with known reactions, induce mutations, and find the genes that control the reactions. The organism they selected for this work was the red bread mold, *Neurospora crassa*. Just as *Drosophila* is an ideal organism for transmission genetics, *Neurospora* is tailor-made for biochemical genetics. Two aspects of *Neuro-*

spora particularly contribute to this desirability: its life cycle and its ease of maintenance in the laboratory.

Neurospora is haploid throughout most of its life. Hence, every mutant gene will be expressed; none will be suppressed by a dominant allele. The haploid organism exists as a colony that profusely reproduces asexually by fragmentation or by forming asexual spores. In addition to the asexual mode of reproduction, *Neurospora* also reproduces sexually. There are two *Neurospora* mating types—*A* and *a*. If *A* and *a* colonies are grown together, parts of the colonies will fuse with one another, and *A* nuclei will fuse with *a* nuclei to form diploid zygote nuclei. The diploid stage is abbreviated, and each zygote nucleus immediately undergoes meiosis to produce four haploid nuclei. These nuclei then undergo mitosis, and a cell wall forms around each nucleus to form eight haploid **ascospores.** Each ascospore will then form a separate colony.

Beadle and Tatum soon found that the nutritional requirements of *Neurospora* are modest; the mold can be maintained *in vitro* on a **minimal medium** of water, inorganic salts, sucrose, and the vitamin biotin. The mold is capable of synthesizing all of its chemical constituents from this minimal medium (i.e., it is **autotrophic**). Since the pathways for the synthesis of basic metabolic compounds such as amino acids were well known, Beadle and Tatum decided to search for mutations in genes that regulate these metabolic reactions. Since such basic compounds are essential to life, mutants could survive only if they had an external source of the missing compound. Thus, these mutants would be **conditional lethals;** that is, they would be lethal unless compensation was made for the defect.

The experimental procedure Beadle and Tatum adopted has become the standard way to select for conditional lethals. First, the *Neurospora* were irradiated with x-ray to induce mutation. Next, the irradiated colony was mated with a colony of opposite mating type. Ascospores from this cross were individually placed in separate vials containing a **complete medium,** which includes all 20 essential amino acids plus certain other biologically important organic compounds. The ascospores formed colonies in these vials. To determine whether a colony carried a metabolic defect, bits of it were removed to a vial containing the minimal medium. If the colony could survive on the complete medium but not on the minimal medium, it meant that it was unable to produce some component of the complete medium, presumably owing to a mutation induced by the radiation. These nutritional mutants are called **auxotrophs.** To determine whether the deficiency was in amino acid metabolism, a bit of the colony was placed on minimal medium supplemented with a mixture of essential amino acids. If the colony grew, it meant that it was unable to synthesize one of the 20 amino acids.

The pathway for the synthesis of the amino acid arginine had been discovered shortly before the *Neurospora* work was begun. Arginine is formed from citrulline, citrulline from ornithine, and ornithine from an unknown precursor. Each step requires a specific enzyme. The reaction is written as follows:

$$\text{precursor} \xrightarrow{\text{E}_1} \text{ornithine} \xrightarrow{\text{E}_2} \text{citrulline} \xrightarrow{\text{E}_3} \text{arginine}$$

Amino acid auxotrophs were tested systematically to determine whether any of them required arginine for growth. Several arginine-requiring mutants were found, and all of them fell into one of three categories (Srb and Horowitz, 1944): One class could grow only if arginine was added to minimal medium. Another would grow with either arginine or citrulline. The third would grow with arginine, citrulline, or ornithine. The three classes of mutant genes affected different steps in the arginine synthetic pathway. The mutants that would survive only with the addition of arginine were unable to convert citrulline to arginine. The mutants that would survive with citrulline were able to make arginine from citrulline but could not convert ornithine to citrulline. Likewise, the mutants that would survive with ornithine lacked the ability to convert a precursor into ornithine. Thus, the wild-type genes specify the enzymes (E_1–E_3) for the reactions, and the mutant genes lack this ability. The one gene–one enzyme hypothesis had been verified.

Subsequently, the rule was generalized to "one gene–one protein," since all proteins, not only enzymes, are gene-dependent. However, proteins are frequently composed of unlike subunits (polypeptides), which are specified by different genes. For example, the hemoglobin molecule is composed of four subunits representing two different polypeptide chains, each specified by a different gene. Hence, the rule can be restated as "one gene–one polypeptide."

1–7. THE MODERN ERA

Now that the function of genes in the cell had been established, one of the key stumbling blocks to a clear understanding of the role of genes in development was removed. It now remained to determine the composition of genes and the way that the genetic information is utilized. During the late 1940s, 1950s, and early 1960s, several important breakthroughs emerged that laid the groundwork for a renewed onslaught of investigations into the role of genes in development and the mechanisms involved in cell differentiation. These contributions resulted mainly from work in biochemistry and two new disciplines—cell biology and molecular biology. Biochemists gathered valuable data on the relationships between genes and proteins and the role of enzymes in cellular metabolism. Cell biologists described the structure and function of cellular components, aided by a powerful new tool, the electron microscope. But most importantly, molecular biologists determined the nature and structure of the genetic material, the genetic code was broken, and the protein synthetic machinery was unraveled.

Nearly all of the basic principles of molecular biology were obtained from work on bacteria and viruses. But these principles also apply to multicellular plants and animals, with certain variations. The rapid developments in molecular biology are too numerous to discuss here in detail, but the contemporary student of development should be intimately familiar with the discoveries that form the basic foundations of molecular biology. Some of the more important ones are the following:

1. Genetic information is coded in deoxyribonucleic acid (DNA) as two antiparallel polynucleotide strands in a double-helical structure (the Watson-Crick double helix).
2. Genetic information is stored in a linear sequence of purine and pyrimidine bases. The genetic alphabet consists of four letters (bases): **A**denine, **T**hymine, **G**uanine, **C**ytosine.
3. During replication of DNA, each strand serves as a template for the formation of the complementary strand. Each base of the template strand specifies a nucleotide bearing the complementary base: A is complementary to T, G is complementary to C, and vice versa.
4. Genetic information in the chromosomes is expressed by transcription of the sequence of bases in DNA into a complementary sequence of bases in ribonucleic acid (RNA). During transcription, each base of the DNA template strand specifies a ribonucleotide bearing the complementary base: A is complementary to **U**racil, T is complementary to A, G is complementary to C, and C is complementary to G.
5. The genetic information for synthesis of a specific protein is a **structural gene.** Structural genes are transcribed into messenger RNA, which is transported into the cytoplasm, where it is translated into protein.
6. Proteins are composed of a linear sequence of amino acids. The placement of an amino acid in protein is designated by a triplet of bases in mRNA called a **codon.**
7. The genetic code is universal (mitochondria are exceptions), non-overlapping (except in some viruses), and degenerate (i.e., redundant), and it contains codons that punctuate protein synthesis.
8. Protein synthesis occurs on ribosomes, which are composed of protein and ribosomal RNA. The genetic code is read by transfer RNA molecules. These molecules recognize a specific codon and bear the corresponding amino acid, which is added to the sequence of amino acids by formation of a peptide bond. Like messenger RNA, ribosomal and transfer RNA are transcribed from genes. Thus, not all genes code for polypeptides.
9. Recent investigations have revealed the presence of nucleotide sequences whose function is to regulate the transcription of structural genes. These regulatory sequences may not produce transcripts themselves.

By the late 1950s and early 1960s, enough of the critical pieces were in place. Molecular biologists began asking how gene function can be controlled to produce the wide variety of cell types in adult organisms. Biochemists began analyzing the changes in cellular biochemistry that occur during development. Cell biologists began monitoring the structural and functional changes that accompany cell differentiation. Genes that modify development were recognized as valuable tools in understanding normal developmental events. Developmental biology emerged as a vital, exciting, broadly based science. Interdisciplinary barriers fell as investigators came to realize that plant and animal development have much in common and that simple organisms such as algae and slime molds are excellent model systems for studying cell differentiation. Horizons were broadened with the realization that developmental events occur during all phases in the life span of an organism, not only during embryogenesis.

In such a broadly based science the student is expected to understand concepts that are diverse in nature but have a common goal. That goal is to interpret the processes that produce a tree, a frog, or a human from a fertilized egg. Although the analysis is multileveled, each approach analyzes the same phenomenon from its own particular perspective. The formation of an eye is viewed differently by the biochemist, the electron microscopist, and the molecular biologist. Yet the eye develops nevertheless, and it is up to the developmental biologist to integrate the information provided by these diverse analyses and describe how and why the eye develops.

Developmental biology has made remarkable strides in its short life. It has been a robust science, benefiting immensely from periodic transfusions of new technological advances originating in its contributory disciplines. Among these technological advances are:

1. *In vitro* analyses of development.
2. Improvements in ultrastructural analysis.
3. Radiotracer technology.
4. Refinements in separation sciences, allowing for resolution and identification of small amounts of cell-specific molecules.
5. Ingenious procedures to trace the fates of specific cells during development.
6. Nucleic acid hybridization technology.
7. Isolation of messenger RNA molecules and their use as templates for synthesis of DNA.
8. Recombinant DNA technology, allowing for isolation and amplification of DNA nucleotide sequences of developmental significance.
9. Nucleic acid sequencing techniques, which allow the evaluation of possible control sequences in DNA.
10. Application of immunochemistry for the identification and quantification of cell-specific molecules.
11. Cell-cell hybridization techniques, allowing for introduction of chromosomes into foreign nuclei.
12. Microinjection, enabling analyses of the function of exogenous genes.

13. Acquisition of a new understanding of the cell surface, facilitating the study of the surface in cellular interactions.
14. Sophisticated monitoring devices to trace the movements of minute amounts of ions during development.

The rapid development of technology in recent years and the judicious application of this technology to the study of development have helped to make developmental biology one of the most exciting and dynamic sciences in recent years. In their ready acceptance of the ideas and technology of allied sciences, however, contemporary students of development owe a debt of gratitude to E. B. Wilson, who championed the idea that development is best understood from a multidisciplinary approach.

1–8. OVERVIEW OF ANIMAL DEVELOPMENT

Development is reflected by continual change in functional and structural properties, not only in the organism itself but also in its individual cells. To study development, we devise means of monitoring these changes, either quantitatively or qualitatively. When a number of changes are occurring simultaneously to form a pattern of events that is unique to that period of development, we call it a **developmental stage.** Although stages are not real entities, they are useful aids for discussing development because they allow us to refer to embryos at particular times without describing all of the separate events or morphological features that characterize the embryo at those times. Thus, it is worthwhile to describe briefly the stages of animal development at this point. We shall discuss later in this book the detailed structural changes that occur in developing animal embryos and the mechanisms that produce those changes.

1. **Gametogenesis** is arbitrarily designated the first stage of animal development. The question of "Which came first, the chicken or the egg?" is one that has intrigued man for centuries. In embryology, the gametes (eggs and sperm) are usually discussed first, since they provide both the blueprint and the raw material from which the embryo is formed. Gamete formation in the two sexes is tailored to the roles of their gametes in reproduction. The male gametes are usually small and mobile. They are dispensed from the male reproductive organ— often into a hostile environment—and they must locate the female gamete, make contact, and fuse with it. The female gamete is usually less mobile than the sperm and larger, often by several orders of magnitude. The female gamete must be "competent" to be fertilized, which means that it must develop a number of specialized properties to enable it to interact with the sperm. Both classes of gametes make an equal contribution to the nucleus of the zygote, each providing a haploid genome. However, the male gamete makes a minimal con-

tribution to the cytoplasm; the female gamete provides the zygote with virtually all of the cytoplasm, which contains the constituents from which the embryo is fashioned.

Reproduction is a prime concern for any species, since it ensures the species' survival. The **germ cell line** is, therefore, a precious commodity, and its formation is an important developmental event, which is often one of the first orders of business for the embryo after fertilization. The germ cell line may derive its specificity from a specialized cytoplasmic constituent, the germ plasm, which may preexist in the egg prior to fertilization and become segregated into the germ cell line during cleavage (see Chap. 11). Determination of the germ cell line results in two distinct categories of cells in the embryo (Fig. 1.7). The nongerm cells are called the **somatic cells.** This distinction is retained throughout the life of the organism.

The initial cells in the germ line are called **primordial germ cells.** The primordial germ cells of both sexes are indistinguishable from one another. The acquisition by germ cells of sex-specific characteristics occurs at a later stage of development and is culminated by the formation of mature sex cells with distinctly different shapes and organelles. The primordial germ cells may arise at some distance from the presumptive gonads to which they migrate, become established, and increase in number by mitosis. The establishment of germ cells in the gonads often involves a close association between the germ cells and the somatic cells of the gonad. These somatic cells may serve to support and protect the germ cells and to provide them with nutritive material.

In the female, the somatic cells surrounding the germ cell are called **follicle cells.** In the male, various terms have been used for them. Among the most familiar examples are the **Sertoli cells** in mammalian testes. During the proliferative phase the germ cells are called **gonia** (spermatogonia in the testis and oogonia in the ovary) and act as a stem cell population for the production of cells that will differentiate into functional gametes. The gonial cell divisions may be incomplete so that the daughter cells remain in communication with one another via intercellular bridges. Successive incomplete divisions produce very large clones of interconnected cells. This intercellular communication may serve to synchronize the development of the conjoined cells. The formation of sperm from spermatogonia is **spermatogenesis,** and the formation of ova (or eggs) from oogonia is **oogenesis.** These processes involve the reduction in chromosome number by meiosis and acquisition of the structural and functional characteristics of the distinct sex cells.

In the male, meiosis precedes sex cell differentiation. In the female, however, differentiation may occur early in meiosis, which is completed after ovulation, and in some cases after the sperm has entered the egg at fertilization. It is important to keep in mind the

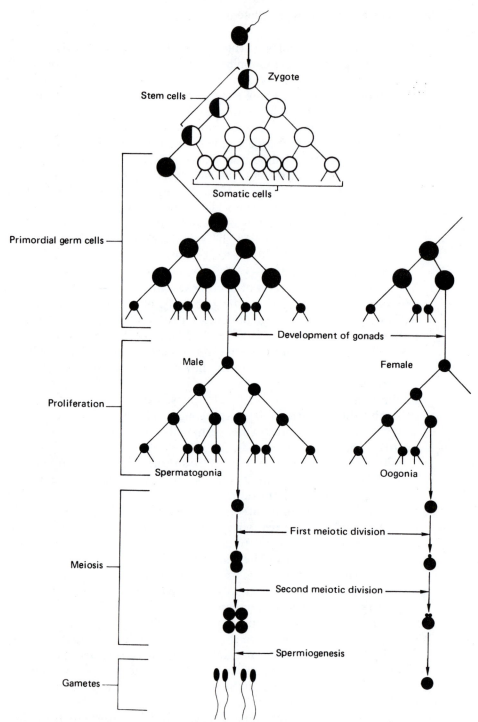

FIGURE 1.7 Origin and fate of the male gametes (left) and the female gametes (right). The number of mitotic divisions is larger than depicted. (After E.B. Wilson, 1925. From F.J. Longo and E. Anderson. 1974. Gametogenesis. *In* J. Lash and J.R. Whittaker (eds.), *Concepts of Development,* 1st ed. Sinauer Associates, Inc., Sunderland, Mass., p. 4.)

difference in the number of sex cells that result from meiosis in the male and female. In the male, a single spermatogonium enters the first meiotic division as a **primary spermatocyte.** This division produces two **secondary spermatocytes,** each of which divides to form two haploid **spermatids.** Consequently, *four* haploid cells result from each diploid spermatogonium. Each spermatid differentiates into a **spermatozoon** by the elaboration of structural and functional specializations that enable the sperm to fertilize the egg. By contrast, in the female, each of the meiotic divisions is uneven, producing only one full-sized cell. During the first meiotic division the **primary oocyte** divides to produce one small polar body and one **secondary oocyte.** The latter enters the second meiotic division to produce the second polar body and the haploid **ovum,** which is the only functional sex cell to result from meiotic reduction of an oogonium.

2. **Fertilization** is the union of male and female gametes, which activates, or initiates, embryonic development and restores the diploid condition.

3. **Embryogenesis** is the phase that encompasses most of the developmental events in animals. Most animal species pass through comparable embryonic stages, although the details vary considerably from one group to another. Since amphibian embryos have been used extensively for research in experimental embryology, the stages of embryogenesis are illustrated here with amphibian embryos:

    a. **Cleavage.** In order for the single-celled zygote to produce a multicellular organism, a number of mitotic divisions must occur in rapid succession. During cleavage the size and shape of the embryo is retained, while the cleavage cells, or **blastomeres,** become smaller at each division. At the completion of each division, the blastomeres are separated from one another by the formation of **cleavage furrows.**

    In the frog embryos shown in Figure 1.8, the first cleavage furrow is seen to begin at one pole of the egg and spread to the opposite pole. The egg consists of a dark half and a light half. The dark half of the egg is the **animal hemisphere,** whereas the light half is the **vegetal hemisphere.** The dark color of the animal hemisphere is due to a layer of pigment granules below the surface of the egg. The nucleus and most of the egg cytoplasm are located in this half of the egg. The unpigmented vegetal hemisphere contains virtually all of the yolk and very little cytoplasm. The first cleavage begins at the **animal pole** of the egg, spreading around both sides to meet at the **vegetal pole.** The second cleavage furrow also begins at the animal pole, at right angles to the first furrow. The third cleavage is in the **equatorial plane** of the egg, at right angles to both of the first two cleavages. This cleavage is displaced slightly toward the animal pole, causing a size disparity in the cells; animal hemisphere cells are smaller than those in the vegetal hemisphere. Note

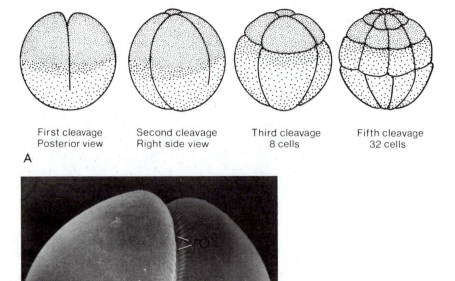

First cleavage	Second cleavage	Third cleavage	Fifth cleavage
Posterior view	Right side view	8 cells	32 cells

A

FIGURE 1.8 Cleavage in the frog embryo. *A,* Pattern of first five divisions. (After R. Rugh. 1951. *The Frog: Its Reproduction and Development.* McGraw-Hill Book Co., New York, p. 19.) *B,* First cleavage furrow visualized with scanning electron microscopy (SEM). Stress lines or folds (FO) radiate from the cleavage groove (CG). SEM, × 400. (From H.W. Beams and R.G. Kessel. 1976. Cytokinesis: A comparative study of cytoplasmic division in animal cells. *Am. Sci., 64:* 284.)

that the size disparity between animal and vegetal hemisphere cells is retained throughout cleavage.

Small fluid-filled spaces appear between the blastomeres during early cleavage stages. As cleavage proceeds, these spaces coalesce to form a large central cavity, the **blastocoele,** surrounded by a layer of cells, the **blastoderm.** The embryo at this time is called a **blastula.** The frog blastula is represented in Figure 1.9, both in a surface view and sectioned through the animal–vegetal axis to show internal organization.

b. **Gastrulation.** The cells of the blastoderm undergo extensive rearrangement during this stage to produce three layers of cells known as **germ layers,** from which the various organs of an animal's body

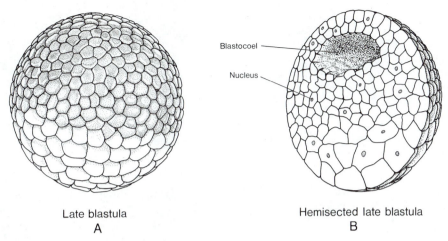

Late blastula
A

Hemisected late blastula
B

FIGURE 1.9 The frog blastula. *A,* Surface view of late blastula. *B,* Hemisected blastula of comparable stage. (Redrawn with permission of Macmillan Publishing Company, from *Fundamentals of Comparative Embryology of the Vertebrates,* Revised Edition, by A.F. Huettner. Copyright © 1949 by Macmillan Publishing Company; renewed 1977 by Mary R. Huettner, Richard A. Huettner, and Robert J. Huettner.)

are derived. The three layers are the outer **ectoderm,** which gives rise to the epidermis and the nervous system; the intermediate **mesoderm,** which produces the circulatory system, muscle, skeletal system, and connective tissue; and the inner **endoderm,** which produces the gut and its associated organs. The cell movements of gastrulation involve the inward displacement of endoderm and mesoderm cells and the complete envelopment of the internal cells by the ectoderm. As we shall see in Chapter 13, the cell rearrangement mechanisms employed by animal embryos vary considerably from group to group. The organization of the embryo during gastrulation often leads to the formation of a new embryonic cavity, the **archenteron,** which eventually gives rise to the cavity of the alimentary tract. The external opening of the archenteron is the **blastopore.** The rearrangement of frog embryo cells during gastrulation is represented in Figure 1.10. A detailed discussion of frog gastrulation may be found in Chapter 13.

c. **Establishment of the basic body plan.** Following gastrulation, the basic body plan of the embryo is laid out along the axis of body symmetry. In a bilaterally symmetrical animal such as an amphibian, the embryo organizes along the anterior–posterior axis, with distinct developmental regions forming along the axis. Externally, the conspicuous changes in the embryo during this time are caused by the formation of the future nervous system, which extends along the entire dorsal midline of the elongating embryonic axis. The frog embryo during this phase of development (Fig. 1.11) is called a **neurula,** since the formation of the nervous system

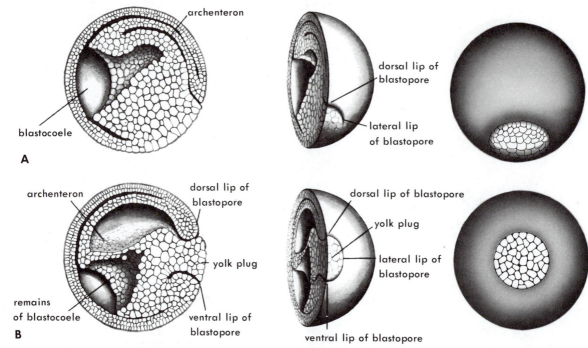

FIGURE 1.10 Gastrulation in the frog. Drawings on the left represent embryos cut in the median plane. Drawings in the center represent the same embryos viewed at an angle from the dorsal side (*A*) or from the posterior end (*B*). (From B.I. Balinsky. 1981. *An Introduction to Embryology*, 5th ed. Saunders College Publishing, Philadelphia, p. 167.) Comparable stages of intact embryos are shown on the right. Stages of gastrulation: *A*, middle gastrula stage; *B*, late gastrula stage.

dictates the shape of the embryo. Internally, the archenteron organizes a tubular gut with an anterior and a posterior opening. The germ layers may dissociate and form clusters of cells that are the precursors, or **rudiments,** from which definitive organs and tissues will be formed. These clusters often consist of cells derived from more than one germ layer, such as endoderm plus mesoderm for internal structures and ectoderm plus mesoderm for peripheral structures.

d. **Organogenesis.** The embryonic rudiments that form after gastrulation acquire the functional and structural characteristics of organs and distinct body parts during the subsequent phase of organogenesis.

4. **Postembryonic development** normally begins when the embryo hatches from its protective coats to become a free-living **larva** (indirect development). Development of adult features is completed during or near the termination of the larval period. Larvae also serve one or both of two additional functions. They are often the chief means for geographic distribution of the species. This is particularly true in species whose adults are more or less **sessile,** or stationary. Most larvae, there-

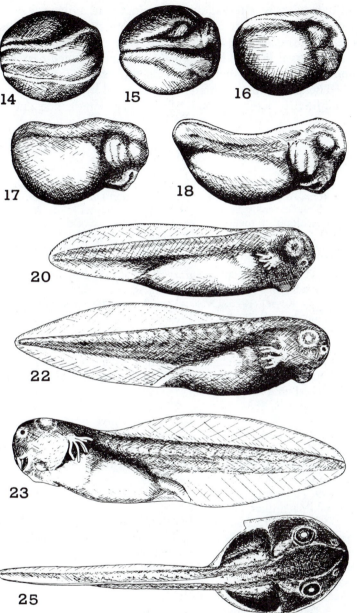

FIGURE 1.11 Late stages in the development of the leopard frog, *Rana pipiens,* beginning with the neurula stage. Hatching occurs at stage 20, while stage 25 is a feeding tadpole (After R. Rugh, 1962. From B.I. Balinsky. 1981. *An Introduction to Embryology,* 5th ed. Saunders College Publishing, Philadelphia, p. 298.)

fore, have well-developed locomotory organs. The larval period may also serve a nutritional function. Larvae may be voracious feeders, usually subsisting on a different diet from that of the adult. Food reserves accumulate and provide raw material and energy for the construction of the adult. Transformation of a larva into an adult is called **metamorphosis.** The amphibian larva (Fig. 1.11) is called a **tadpole,** which is a free-swimming aquatic organism.

In animals without a larva (direct development) a juvenile stage may follow embryogenesis. Thus, a miniature organism emerges that closely resembles the adult in appearance (e.g., nematode worms). In other cases the larval life has been replaced by an extended period of development within an egg shell (e.g., birds, reptiles) or a uterus (e.g., mammals). The adult phase of life is characterized by growth, sexual maturity, and finally, deterioration by a process of senescence, ending in death.

REFERENCES

The history of developmental biology is thoroughly documented in a very readable book by John A. Moore, *Heredity and Development.* The author has used Dr. Moore's book extensively in developing this chapter. The reader is referred to it for more details concerning the fascinating story of this science.

Balinsky, B.I. 1975. *An Introduction to Embryology,* 4th ed. W.B. Saunders, Philadelphia.

Balinsky, B.I. 1981. *An Introduction to Embryology,* 5th ed. Saunders College Publishing, Philadelphia.

Beams, H.W., and R.G. Kessel. 1976. Cytokinesis: A comparative study of cytoplasmic division in animal cells. Am. Sci., *64:* 279–290.

Bonnet, C. 1745. *Traité d'Insectologie.* Paris.

Bridges, C.B. 1935. Salivary chromosome maps. J. Heredity, *26:* 60–64.

Driesch, H. 1892. Entwicklungsmechanisme Studien. I. Der Werth der beiden ersten Furchungszellen in der Echinodermentwicklung. Experimentelle Erzeugen von Theil–und Doppelbildung. Zeitschrift für wissenschaftliche Zoologie, *53:* 160–178; 183–184.

Garrod, A.E. 1909. *Inborn Errors of Metabolism.* Oxford University Press, Oxford.

Goldschmidt, R.B. 1958. *Theoretical Genetics.* University of California Press, Berkeley.

Hertwig, O. 1876. Beiträge zur Kenntnis der Bildung, Befruchtung und Teilung des tierischen Eies. Morph. Jahrb., *1:* 347–434.

Huettner, A.F. 1949. *Fundamentals of Comparative Embryology of the Vertebrates,* rev. ed. Macmillan, New York.

Judd, B.H., M.W. Shen, and T.C. Kaufman. 1972. The anatomy and function of the X chromosome of *Drosophila melanogaster.* Genetics, *71:* 139–156.

Lefevre, G., Jr. 1976. A photographic representation and interpretation of the polytene chromosomes of *Drosophila melanogaster* salivary glands. *In* M. Ashburner and E. Novitski (eds.), *The Genetics and Biology of Drosophila,* Vol. 1a. Academic Press, London, pp. 31–66.

Lewis, E.B. 1963. Genes and developmental pathways. Am. Zool., *3:* 33–56.

Longo, F.J., and E. Anderson. 1974. Gametogenesis. *In* J. Lash and J.R. Whittaker (eds.), *Concepts of Development.* Sinauer Associates, Sunderland, Mass., pp. 3–47.

Moore, J.A. 1972. *Heredity and Development,* 2nd ed. Oxford University Press, New York.

Morgan, T.H. 1927. *Experimental Embryology.* Columbia University Press, New York.

Painter, T.S. 1934. A new method for the study of chromosome aberrations and the plotting of chromosome maps in *Drosophila melanogaster.* Genetics, *19:* 175–188.

Roux, W. 1888. Beiträge zur Entwicklungsmechanik des Embryo. Ueber die künstliche Hervorbringung halber Embryonen durch Zerstörung einer der beiden ersten Furchungskugeln, sowie über die Nachentwicklung (Postgeneration) der fehlenden Köperhälfte. Virchows Arch. Path. Anat. Physiol. *114*: 113–153; 289–291.

Rugh, R. 1951. *The Frog. Its Reproduction and Development.* McGraw-Hill Book Company, New York.

Rugh, R. 1962. *Experimental Embryology. Techniques and Procedures,* 3rd ed. Burgess Publishing, Minneapolis, Minn.

Schmidt, G.A. 1933. Schnürungs-und Durchschneidungsversuche am Amphibienkeim. Wilh. Roux' Archiv f. Entwicklungsmechanik, *129*: 1–44.

Srb, A.M., and N.H. Horowitz. 1944. The ornithine cycle in *Neurospora* and its genetic control. J. Biol. Chem., *154*: 129–139.

van Beneden, E. 1883. Recherches sur la maturation de l'oeuf, la fécondation et la division cellulaire. Arch. de Biol., *4*: 265–640.

Wilson, E.B. 1896. *The Cell in Development and Inheritance.* Reprinted by Johnson Reprint Corp., New York. (1966).

Wilson, E.B. 1900. *The Cell in Development and Inheritance,* 2nd ed. Macmillan, New York.

Wilson, E.B. 1914. The bearing of cytological research on heredity. Proc. R. Soc. Lond. (Biol.), *88*: 333–352.

Wilson, E.B. 1925. *The Cell in Development and Heredity,* 3rd ed. Macmillan, New York.

Wolff, C.F. 1759. *Theoria generationis.* Halle.

Young, M.W., and B.H. Judd. 1978. Nonessential sequences, genes, and the polytene chromosome bands of *Drosophila melanogaster.* Genetics, *88*: 723–742.

PART TWO
THE CELL
IN DEVELOPMENT

2 Gene Function in Cell Determination and Differentiation

During the development of a multicellular organism, numerous functional cell types are produced from a single cell—the fertilized egg, or zygote. The magnitude of this feat and the challenge of explaining development become apparent if we examine the starting material that the zygote utilizes to form a complex organism: a single diploid nucleus, which is embedded in the egg cytoplasm and contains the genes that designate the properties of each and every cell in the body.

How does this single genome produce the myriad cell types that constitute the completed organism? The Roux–Weismann theory proposed that the nucleus contains determinants that are sorted out during mitosis and, by this means, gradually segregated into different cells; the presence of different determinants in different cells was the apparent answer to the question of how cell determination occurs. The cell separation experiments by Driesch were the undoing of Roux and Weismann; clearly, there is no loss of genetic information by early cleavage stage nuclei.

However, the Driesch experiments did not completely quell the beliefs of many developmentalists that modifications in the genetic material itself could occur during development. The task of the nucleus would be much simpler if unnecessary genetic information could be dispensed with or modified in some way so that it becomes nonfunctional. In other words, the nucleus itself would become highly adapted to perform a singular role in development, for example, the production of a blood cell or a bone cell or a skin cell. Such adaptation would commit a nucleus permanently to a restricted developmental role. In the jargon of embryology, the nucleus would lose **potency.** The zygote nucleus is said to be **totipotent**—it has complete potential to produce the entire range of cell types that make up the organism. We shall now examine

41

the results of experiments that were designed to test the potency of nuclei of differentiated cells.

2-1. GENOME EQUIVALENCE

One way to determine whether the genomes of differentiated cells retain genetic information that is not utilized in those cells is to challenge their nuclei experimentally to reinitiate development and promote differentiation of alternate cell types. The cell separation experiments discussed in Chapter 1 indicate clearly that early cleavage stage nuclei are totipotent. But what happens to nuclei as development continues past early cleavage? Are they irreversibly changed during later development? Cell separation experiments become impractical in later stages, since the volume of cytoplasm is reduced during each cleavage division, and the reduction can affect the ability of nuclei to function properly. Plants are a definite exception to animals in this regard. As we shall see later in this chapter, isolated plant cells—even from adults—may be induced to form an entire plant.

An alternative (and highly ingenious) procedure for analyzing nuclear potential in later cleavage stage newt embryos was developed by Hans Spemann (Fig. 2.1). He partially constricted a *Triturus* egg with a hair loop just before cleavage, confining the nucleus to one half of the egg. A narrow cytoplasmic bridge joined the nucleated and enucleated halves. At the first cleavage (Fig. 2.1*A*) the two cells that formed were decidedly unequal. One contained approximately one fourth of the cytoplasm, whereas the other was in the shape of an asymmetrical dumbbell with one fourth of the cytoplasm on the nucleated side and one half of the cytoplasm on the enucleated side. The two lobes of this cell remained in communication via the cytoplasmic bridge, but the nucleus stayed in the smaller lobe, unable to cross the cytoplasmic bridge because it remained too large to enter. At each subsequent division the small lobe of the dumbbell divided, always forming one small complete cell and one cell that remained continuous with the cytoplasmic bridge. During cleavage, nuclei became progressively reduced in size; by the 16-cell stage, the nuclei were so small that during the next division, one of the daughter nuclei was able to slip through the cytoplasmic bridge and into the large lobe of the dumbbell (Fig. 2.1*B*). Since a cell membrane forms between daughter nuclei after mitosis, a membrane appeared in the bridge, forming a complete cell in the previously enucleated half (Fig. 2.1*C*).

After renucleation the undivided half of the embryo began to cleave, and Spemann drew the hairloop tight, completely separating the two halves. Both halves continued to develop normally, the renucleated side somewhat more slowly (Fig. 2.1*D* and *E*). The nucleus that entered the enucleated half of the embryo was equivalent to a nucleus at the 32-cell

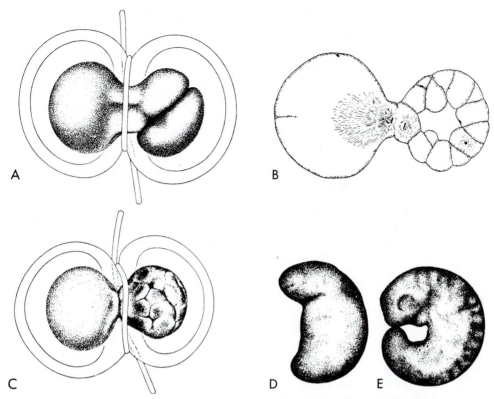

FIGURE 2.1 Spemann's delayed nucleation experiment. *A*, Ligature confines nucleus to right half, which cleaves. *B* and *C*, Nucleus adjacent to intracellular bridge enters enucleate half and cleavage furrow appears in bridge. *D* and *E*, Embryos following completion of constriction. The embryo developing from the side with delayed nucleation (*D*) developed more slowly. (*A*, *C*, *D*, and *E* from H. Spemann. 1928. Die Entwicklung seitlicher und dorso-ventraler Keimhälften bei verzögerter Kernversorgung. Zeitschrift für wissenschatfliche Zoologie, *132:* 105–134; *B* from G. Fankhauser. 1930. Zytologische Untersuchungen an geschnürten Triton-Eiern. 1. Die verzögerte Kernversorgung einer Hälfte nach hantelförmiger Einschnürung des Eis. Wilh. Roux' Arch. für Entwicklungsmechanik der Organismen, *122:* 135. All reprinted with permission of Springer-Verlag, Heidelberg.)

stage. Upon separation of the renucleated half of the embryo, the 32-cell-stage nucleus was shown to contain the complete set of genetic information for the formation of an entire embryo. Clearly, there was no loss of developmental potential during the cell divisions that led up to the 32-cell stage.

Nuclear Transplantation

In Spemann's constriction experiments a nucleus was transmitted through a bridge of cytoplasm to a region of the embryo that was previously devoid of a nucleus. Several years later, Spemann (1938) suggested a more sophisticated method for placing a nucleus in enucleated cytoplasm and testing its potency:

> Probably the same effect could be attained if one could isolate the nuclei of the morula and introduce one of them into an egg or an egg fragment without an egg nucleus. The first half of this experiment, to provide an isolated nucleus, might be attained by grinding the cells between two slides, whereas for the second, the introduction of an isolated nucleus into the protoplasm of an egg devoid of a nucleus, I see no way for the moment. If it were found, the experiment would have to be extended, so that older nuclei of various cells could be used. This experiment might possibly show that even nuclei of differentiated cells can initiate normal development in the egg cytoplasm.

Spemann was suggesting the possibility of **nuclear transplantation** from a somatic cell to an enucleated egg but could see no way to perform the experiment. However, 14 years later, Briggs and King (1952) reported successful transplantation of nuclei of the leopard frog, *Rana pipiens*. Their procedure (Figs. 2.2 and 2.3) involves two main steps: (1) preparation of the recipient egg and (2) isolation of a donor cell and transfer of its nucleus. The recipient eggs are obtained from a female by induced ovulation. Since normal fertilization has two components—activation of development and formation of a diploid nucleus—successful nuclear transplantation must also mimic these events. Activation of the egg is normally caused by the interaction of the sperm with the egg surface. Activation for nuclear transplantation is achieved by pricking the surface of the egg with a clean glass needle. As a result, the egg nucleus moves to the animal pole in preparation for the formation of the second polar body. The location of the nucleus is indicated by a depression in the egg surface at the animal pole, facilitating its removal with a second glass needle (Porter, 1939). Donor cells are obtained by dissociating the cells of an embryo in a special solution. Normal cell adhesion requires the presence of calcium and magnesium ions. The dissociation solution contains no Ca^{++} or Mg^{++} and may contain ethylenediamine tetra-acetic acid (EDTA), which binds the Ca^{++} and Mg^{++} located on the surface of the embryo cells, causing them to separate from one another. An individual cell is drawn into a micropipette that has slightly smaller diameter than the cell itself, causing the cell membrane to break. The nucleus, together with its surrounding cytoplasm, is then injected into the activated enucleated egg with the same pipette.

When nuclei from blastulae were transplanted to enucleated eggs, approximately 55% promoted normal cleavage and blastula formation, and 80% of the resulting blastulae continued to develop to the tadpole stage. Of these, 75% reached metamorphosis, at which stage the experiment was terminated (Briggs and King, 1960). McKinnell (1962) obtained normal postmetamorphic frogs from blastula nuclear transplants, providing additional evidence that there is no reduction of developmental potential in blastula nuclei. In a series of investigations Briggs and King

Nuclear Transplantation in *Rana*

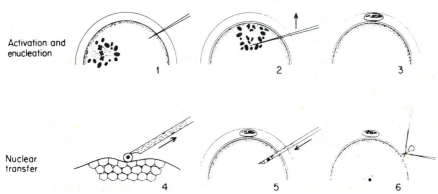

Activation and enucleation

Nuclear transfer

FIGURE 2.2 Nuclear transplantation technique of Briggs and King. 1, Eggs are activated with a clean glass needle. 2, Activated eggs rotate and are enucleated with a second glass needle. 3, Exovate containing the nucleus is trapped by the vitelline envelope surrounding the egg. 4, Donor cells are dissociated. An individual cell is drawn into a micropipette of slightly smaller diameter than the cell, bursting it and releasing the nucleus. 5, Donor nucleus is inserted into enucleated egg. 6, Leakage is minimized by severing the connection that forms between the egg surface and the vitelline envelope. (From T.J. King, 1966. Reproduced with permission from: Nuclear transplantation in amphibia. *In* D.J. Prescott (ed.), *Methods in Cell Physiology, 2:* 3. Copyright: Academic Press Inc., New York.)

used donor nuclei from endodermal cells of progressively later stages of development in order to examine the possible loss of developmental potential in nuclei (Briggs and King, 1957, 1960; King and Briggs, 1956). They indeed found that the percentage of transplants showing normal development declined as the age of the donor nuclei increased. Early gastrula nuclei were nearly as effective in promoting normal develop-

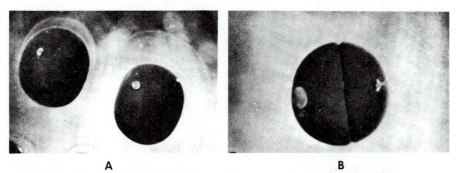

A　　　　　　　　　　　　　**B**

FIGURE 2.3 Nuclear transplantation in *Rana pipiens. A,* Enucleated eggs approximately 30 minutes after activation and enucleation. Exovate containing nucleus is visible above the eggs. *B,* Nuclear transplant recipient at the four-cell stage. (From T.J. King, 1966. Reproduced with permission from: Nuclear transplantation in amphibia. *In* D.J. Prescott (ed.), *Methods in Cell Physiology, 2:* 5, 10. Copyright: Academic Press Inc., New York.)

ment as the cleavage nuclei, but nuclei from late gastrula stages showed a definite restriction in developmental potential, as indicated by a dramatic reduction in the number of nuclear transplant recipients that developed normally. Transplants with nuclei from postgastrula stages exhibited a progressive loss of the ability to develop past the gastrula stage. In those transplants receiving late gastrula and postgastrula endodermal cell nuclei that did develop further, the most pronounced deficiencies were found in the size and extent of differentiation of ectodermal and mesodermal derivatives, whereas endodermal derivatives were hardly affected. This condition is termed the "endoderm syndrome." Tailbud stage endoderm nuclei were incapable of supporting normal development. These data suggest a progressive restriction in the ability of endodermal nuclei to promote normal development and further indicate that the nuclei have become irreversibly specialized as "endodermal" nuclei.

The early Briggs and King experiments with postgastrula nuclei were conducted with nuclei from only one type of cell of one species, endoderm of *Rana pipiens*. Perhaps some peculiarity of these endoderm nuclei causes this restriction, which may not be typical of other nuclei. This may be partially true, since occasionally nuclei from other postgastrula cells have been found to promote normal development. For example, DiBerardino and King (1967) obtained a low percentage of normal larvae from neural plate cell nuclei derived from *Rana pipiens* neurulae. It is necessary to point out, however, that neural nuclear transplants are similar to endodermal nuclear transplants in that the percentage of normal development decreases dramatically when the nuclei are derived from progressively older donor embryos. Furthermore, the developmental defects found in some abnormal neural nuclear transplants reflect the origin of the nuclei, since the extent of ectodermal development exceeds the development of endodermal and mesodermal derivatives. This "ectoderm syndrome" is the exact opposite of the pattern seen in transplants with nuclei of endodermal origin.

The results of nuclear transplantation experiments with embryonic nuclei of a number of amphibian species have confirmed that as cells become more highly differentiated, their nuclei progressively become more restricted in their ability to promote development of recipient eggs. Figure 2.4 compares the success of nuclear transplantation that was obtained with *Rana pipiens* and *Xenopus laevis*, the South African clawed frog, as age of the donor embryo increases. The tendency for loss of developmental potential with age is similar in both species. Nuclear transplantation experiments with nonamphibian species are less extensive (for review, see DiBerardino [1980]). Recently, success has been reported in examining the developmental potential of mammalian embryonic nuclei. Illmensee and Hoppe (1981) have indicated success in transplanting nuclei of mouse embryonic cells into mouse eggs and ob-

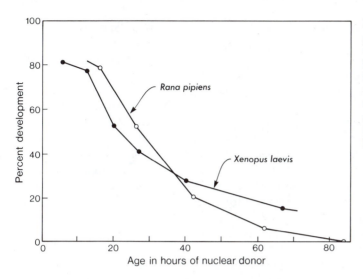

FIGURE 2.4 A comparison of the capacities of *Rana* and *Xenopus* nuclei to promote normal development as the ages of donor nuclei increase. (After R.G. McKinnell. 1972. Nuclear transfer in *Xenopus* and *Rana* compared. *In* R. Harris, P. Allin, and D. Viza (eds.), *Cell Differentiation.* Munksgaard International Publishers, Copenhagen, p. 61.)

tained live-born nuclear transplant recipients. Their procedure is outlined in Figure 2.5.

Early embryos are removed from a donor parent. One of two mouse strains is used as a donor. Both strains contain genes for identifiable coat colors. The LT/Sv strain has a grey coat, and the CBA/H-T6 strain has an agouti coat. In addition, the LT/Sv strain carries a gene for a variant of the enzyme glucose phosphate isomerase, which can be used to identify biochemically cells containing LT/Sv nuclei. The other strain, CBA/

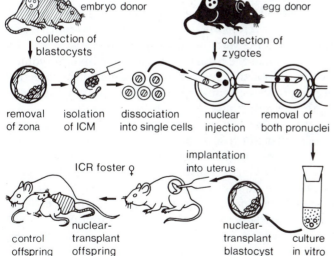

FIGURE 2.5 Procedure for nuclear transplantation in the mouse. (From K. Illmensee and P.C. Hoppe. 1981. Nuclear transplantation in Mus musculus: Developmental potential of nuclei from preimplantation embryos. Cell, *23:* 10. Copyright © Massachusetts Institute of Technology; published by the MIT Press.)

H-T6, has an unusual chromosome constitution (**karyotype**) that enables these nuclei to be identified by microscopic examination.

The 4-day-old embryos used in these experiments are at the blastocyst stage (see Chap. 10). The blastocyst is surrounded by an acellular layer, the zona pellucida, and the blastocyst itself contains two recognizable regions—an outer layer of trophectoderm cells and an inner cluster called the inner cell mass (ICM). Trophectoderm cells contribute to the formation of the placenta, whereas the inner cell mass gives rise to the embryo proper. Trophectoderm and ICM are separated, and the cells whose nuclei are to be transplanted are dissociated into single cells. As with amphibian nuclear transplantation, the donor cell is broken by the transplantation micropipette. The liberated nucleus is then injected into a fertilized egg. The male and female pronuclei, which have not yet formed the zygote nucleus, are removed with the same pipette. The eggs are obtained from another mouse strain, C57BL/6. The coat color (black) and either the glucose phosphate isomerase or the karyotype can be compared to those of the donor nuclei as proof of successful removal of host nuclei and replacement by the transplanted nucleus. The nuclear transplant recipients are cultured until the blastocyst stage, at which time they are implanted into the uterus of a foster mother. The foster mother is of yet another genotype (strain ICR, white coat) so that the nuclear transplant offspring can be identified. Control (i.e., uninjected) blastocysts of strains producing a white coat color are transferred along with the transplant recipients.

The results of these experiments showed a marked contrast between the capacities of the trophectoderm and ICM cell nuclei to promote development. No nuclear transplant recipients with trophectoderm nuclei developed, whereas a substantial number of recipients with ICM nuclei formed blastocysts. Of these, 16 were transferred to foster mothers along with 44 control blastocysts. Three nuclear transplant recipients (18.8%) and 32 (72.7%) control mice were born alive (Fig. 2.6). The identities of the nuclear transplant recipients were confirmed by their coat color. In addition, the two grey mice had the form of glucose phosphate isomerase typical of the LT/Sv strain, and the single agouti mouse had the CBA/H-T6 karyotype. Unfortunately, the agouti mouse died before reaching sexual maturity, but both grey mice have successfully mated.

Nuclei of blastocyst ICM cells are apparently developmentally totipotent. However, we must not lose sight of two important aspects of these experiments. First, the trophectoderm cell nuclei lack potency. Hence, these nuclei have undergone restrictions during development. Second, the blastocyst stage is very early (roughly comparable to the amphibian blastula stage). Preliminary results with nuclei from cells of later stage (day 7) embryos are very intriguing. The day 7 embryo consists of several separable cell types: an outer layer of distal endoderm, an inner layer of proximal endoderm, and—still more internally—embryonic and extraembryonic ectoderm and the ectoplacental cone (a derivative of tro-

FIGURE 2.6 Live-born mice from ICM nuclear transplantation experiments. Each photograph shows an ICR (white) foster mother with her nuclear transplant and control offspring at four weeks of age. The control offspring are white, and the nuclear transplant offspring are grey (*A* and *B*) or agouti (*C*). (From K. Illmensee and P.C. Hoppe. 1981. Nuclear transplantation in Mus musculus: Developmental potential of nuclei from preimplantation embryos. Cell, *23:* 14. Copyright © Massachusetts Institute of Technology; published by the MIT Press.)

phectoderm). Nuclei from distal endoderm, extraembryonic ectoderm, and the ectoplacental cone were unable to promote normal development of recipient eggs. On the other hand, nuclei from cells of embryonic ectoderm and proximal endoderm were capable of promoting normal development. Some of these nuclear transplant recipients have developed into live-born mice (Illmensee et al., 1981). The donor embryos in these experiments were in pregastrula stages. We await with anticipation confirmation of these preliminary experiments. It will also be interesting to learn the results of nuclear transplantation studies with still later stages to know whether nuclei in the derivatives of the embryonic ectoderm and proximal endoderm show restrictions or retain their potency.

In all species in which nuclear transplantation experiments have been done with nuclei of cells of advanced developmental stages, there is a tendency for increased developmental restrictions as donor age increases. In spite of this tendency, experiments with amphibians have shown that some nuclei of cells from postembryonic stages can promote considerable development of enucleated eggs. One series of experiments utilized nuclei that were obtained from cells of a common renal tumor, the Lucké adenocarcinoma. A small percentage of these nuclear transplants developed into tadpoles that, although not perfectly normal, did contain a large variety of functional cell types—an indication that the tumor nucleus has a large repertoire of genetic information that is potentially utilizable to direct the formation of numerous divergent cell types (King and DiBerardino, 1965; DiBerardino and King, 1965). Similar results have been reported by McKinnell et al. (1969), using nuclei from Lucké tumor cells derived from juvenile frogs (Fig. 2.7). The latter experiments utilized a nuclear marker to ensure that the development that was obtained following nuclear transplantation was directed by the donor nucleus. The tumor cells were derived from frogs with an extra chromosome set. These **triploid** nuclei are readily distinguished from normal diploid nuclei, since they have a larger nuclear diameter. In addition, they contain 39, rather than 26, chromosomes. Confirmation of the triploidy of nuclear transplant individuals is an important control for these experiments, since it provides further support for the conclusion that development is directed by the tumor nucleus.

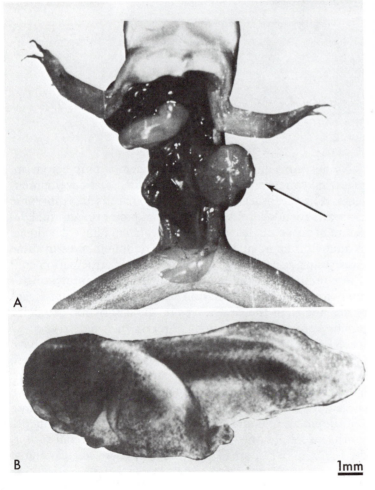

FIGURE 2.7 Transplantation of *Rana pipiens* renal tumor nuclei. *A*, Left renal tumor (arrow) of recently metamorphosed triploid frog. *B*, Tumor nuclear transplant tadpole with well-formed head, body, and tail. (From R.G. McKinnell, B.A. Deggins, and D.D. Labat. 1969. Transplantation of pluripotential nuclei from triploid frog tumors. Science, *165:* 394–396. Copyright 1969 by the American Association for the Advancement of Science.)

A

B   1mm

Nuclei of the germ cell line have also been transplanted to enucleated eggs. In normal development, germ cells form mature gametes that participate in zygote formation and are therefore *genetically totipotent*. This does not mean, however, that germ cell nuclei are *developmentally totipotent* at stages during their development. The germ cell nuclear transplantation experiments were designed to examine this question. When primordial germ cell nuclei of young tadpoles were used as donors, 40% of the complete blastulae developed into normal tadpoles (Smith, 1965). However, when spermatogonial nuclei of juvenile and adult *frogs* were transplanted, most of the complete blastulae arrested before finishing gastrulation (DiBerardino and Hoffner, 1971). Of 13 complete blastulae, three developed past the gastrula stage, and one formed an abnormal larva that commenced feeding but did not survive.

Although the germ cell line retains the complete genome, germ cell nuclei apparently undergo a developmental restriction after the primordial germ cell stage, limiting the utilization of the genome for development. These experiments are important for interpreting the results of somatic cell nuclear transplantation. Clearly, genetic totipotence is not sufficient for developmental totipotence. The nucleus apparently can undergo changes that affect the utilization of the genetic material without altering its essential nature.

The developmental potency of nuclei from postembryonic cells of *Xenopus laevis* has also been examined by extensive nuclear transplantation studies. In the initial experiments of this type, Gurdon (1962) transplanted nuclei from *Xenopus* tadpole intestinal epithelium cells to enucleated *Xenopus* eggs. The techniques used for transplantation of *Xenopus* nuclei (Fig. 2.8) are somewhat different from those used for *Rana*. Instead of mechanical enucleation, the nucleus is destroyed by ultraviolet irradiation. The irradiation treatment also activates the egg to begin developing following implantation of the donor nucleus. The donor nucleus is derived from a cell obtained by chemical dissociation of tadpole intestinal epithelium. As with *Rana*, the cell containing the donor nucleus bursts as it is sucked into the transplant pipette. The liberated nucleus is then injected into the enucleated host egg. A genetic marker found in certain mutant *Xenopus* allows the investigator to determine whether the nuclei of the nuclear transplant individuals are indeed derived from the donor nucleus (the reliability of this marker has been challenged by Du Pasquier and Wabl, 1977). Donor nuclei are obtained from a strain of *Xenopus* that has only one, rather than the normal two, nucleoli per nucleus. The host eggs are from a wild-type strain with two nucleoli. Microscopic examination of cells of the nuclear transplant individuals is conducted routinely to confirm the origin of the nucleus. (For a detailed discussion of this mutation, see Chap. 12.) A very low percentage (approximately 1.5%) of intestinal cell nuclear transplants actually developed into normal feeding tadpoles. Some of these nuclear transplant tadpoles were successfully reared in the laboratory into adult frogs.

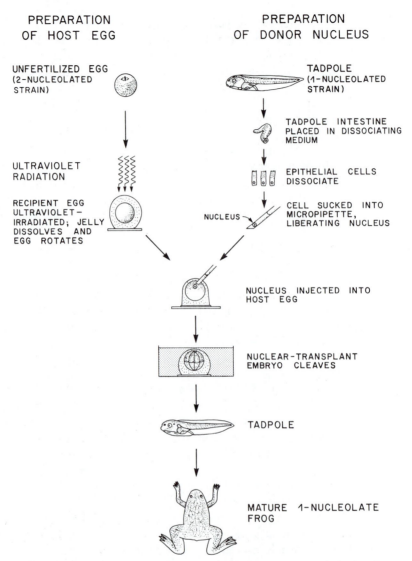

PREPARATION
OF HOST EGG

PREPARATION
OF DONOR NUCLEUS

UNFERTILIZED EGG
(2-NUCLEOLATED
STRAIN)

TADPOLE
(1-NUCLEOLATED
STRAIN)

TADPOLE INTESTINE
PLACED IN DISSOCIATING
MEDIUM

ULTRAVIOLET
RADIATION

EPITHELIAL CELLS
DISSOCIATE

RECIPIENT EGG
ULTRAVIOLET-
IRRADIATED; JELLY
DISSOLVES AND
EGG ROTATES

NUCLEUS

CELL SUCKED INTO
MICROPIPETTE,
LIBERATING NUCLEUS

NUCLEUS INJECTED INTO
HOST EGG

NUCLEAR-TRANSPLANT
EMBRYO CLEAVES

TADPOLE

MATURE 1-NUCLEOLATE
FROG

FIGURE 2.8 Nuclear transplantation in *Xenopus laevis*. Donor tadpole is from a mutant strain with one nucleolus per nucleus. Host egg nucleus contains two nucleoli. Presence of nuclei with one nucleolus in nuclear transplant individual proves that development is due to donor nucleus. (Adapted from J.B. Gurdon. 1966. The cytoplasmic control of gene activity. Endeavour, *25:* 96, and from Transplanted nuclei and cell differentiation, by J.B. Gurdon. Copyright © 1968 by Scientific American, Inc. All rights reserved.)

Gurdon (1962) has argued that technical difficulties with *Xenopus* nuclear transplantation may restrict the ability of some nuclear transplant recipients to develop normally. For example, recipient embryos frequently undergo partial cleavage. Normal nuclei may be restricted to the cleaving portion of the embryo, but their potential to promote dif-

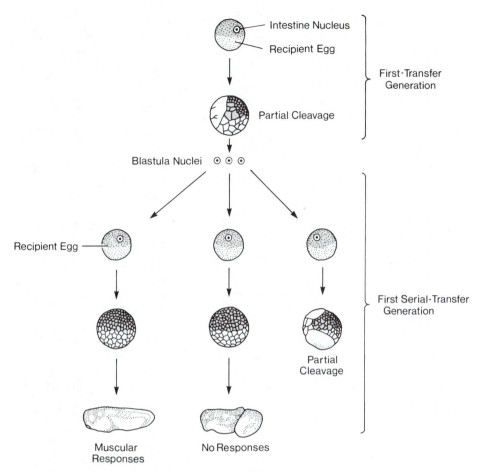

FIGURE 2.9 Serial nuclear transplantation. Abnormal first-transfer embryos were dissociated, and nuclei from individual cells were transplanted to enucleated eggs. Many serial-
transfer embryos developed more normally than the first-transfer donor. The presence of
muscular responses indicates development of neuromuscular function. (Adapted from
Transplanted nuclei and cell differentiation, by J.B. Gurdon. Copyright © 1968 by Scientific
American, Inc. All rights reserved.)

ferentiation would never be realized because gastrulation would be abnormal. Therefore, the percentage of normal development may not reflect the developmental capabilities of the transplanted nuclei accurately.
For this reason, **serial nuclear transplantation** (Fig. 2.9) has been conducted
to maximize the opportunities for transplanted nuclei to promote normal
development (Gurdon, 1962). This technique was developed earlier by
King and Briggs (1956) with *Rana pipiens.*

Abnormal first-transfer *Xenopus* embryos were selected at random
and dissociated. Several nuclei from each embryo were then injected
singly into enucleated eggs. All serial-transfer embryos produced from
a single first-transfer embryo are a **clone** of genetically identical individ-

uals, since they are all derived from nuclei that descended from a single intestinal nucleus. In every case, some of the serial-transfer embryos developed more normally than their first-transfer embryo nuclear donor, indicating that normal nuclei can be present in embryos that exhibit abnormal development. Some of the serial-transfer embryos developed into feeding tadpoles. When the results of the first-transfer generation were combined with those of the first serial-transfer generation, the percentage of intestinal cell nuclei that was capable of promoting development to the tadpole stage was raised to 7%. The most normal serial-transfer embryo was then used to provide nuclei for a second serial-transfer generation. This generation did not show any further improvement in development over that of the first serial-transfer generation. Thus, one serial-transfer generation is sufficient to demonstrate the maximal developmental capacity of the original intestinal cell nucleus. If one of the serial-transfer embryos is used subsequently as a source of nuclei for transplantation, and this process is repeated in each generation, the nuclear clone could be perpetuated indefinitely, and the developmental pattern would remain fixed throughout the lifetime of the clone.

Nuclei from another larval cell type of *Xenopus* also have been shown to be totipotent. Kobel et al. (1973) obtained a mature adult frog following serial transplantation of nuclei from epidermal cells of hatching tadpoles. Interestingly, only the nuclei from nonciliated epidermal cells promoted development; nuclei from ciliated epidermal cells, which appear to be more highly differentiated, never promoted development beyond the blastula stage.

The results of nuclear transplant experimentation have led to two generalizations:

1. As cells differentiate, their nuclei become progressively restricted in their ability to promote the development of enucleated eggs.
2. The nuclei of some differentiated cells can promote at least partial development of enucleated eggs.

These statements reflect two different points of view concerning the significance of the results of nuclear transplantation: The first emphasizes the general developmental restriction, whereas the second emphasizes the exceptions to the rule. To evaluate the relative merits of the two points of view, we should attempt to understand why developmental restriction occurs. Is it due to a normal developmental *loss* of genetic information, as proposed by Roux and Weismann? This possibility is unlikely, particularly in light of the germ cell nuclear transplants previously discussed. In those experiments, nuclei that are known to retain the complete genome nevertheless exhibit developmental restrictions during certain phases of their life cycle. Is developmental restriction due to a *modification* in the genetic material during differentiation that prevents its utilization for alternative modes of development? Finally, is there an *incompatibility* between nuclei of most differentiated cells and the cytoplasm of recipient eggs? There are no completely satisfactory answers to these questions. However, chromosomal analyses of nuclear transplants do provide some insight into the difficulty of obtaining normal development from nuclei of differentiated cells.

Numerous investigators have reported that abnormal nuclear transplants often have an aberrant karyotype; that is, the chromosome number may be modified, or the chromosomes themselves may appear abnormal when studied with the microscope. Furthermore, the severity of chromosomal abnormalities appears to be directly correlated with the extent of the developmental restriction shown by nuclear transplants (DiBerardino, 1979). Thus, those with the most severe chromosomal abnormalities arrest earliest, whereas those with less severe aberrations can develop to larval stages. Careful studies of nuclei transplanted from late gastrula endoderm cells demonstrate that chromosomes of these nuclei frequently are unable to replicate properly when transferred to the cytoplasm of eggs (DiBerardino and Hoffner, 1970). The most obvious cause of this restriction appears to be the failure of some of the chromatin to attain the proper state for replication. As we shall see in Chapter 3, chromatin exists in different degrees of condensation. Decondensation of chromatin is necessary before normal replication can occur. Variable amounts of condensed chromatin may remain in transplanted nuclei, preventing those regions from replicating. This causes both a loss of that portion of the genome and also the consequent chromosomal abnormalities that frequently are found in transplants of late gastrula and postgastrula stage nuclei.

During normal development the mitotic rate of embryo cells is reduced considerably following the blastula stage. The continued reduction in the normal mitotic rate during development roughly parallels the decrease in the ability of nuclei from progressively later stages to promote normal development after transplantation. Furthermore, many differentiated cells never undergo mitosis. Thus, when a nucleus from such a cell is placed in the egg cytoplasm, it is in an extremely difficult situation. It is adapted to directing the specialization of one particular cell type that does not divide, but it is suddenly called upon to replicate its DNA and divide at the rate demanded during cleavage. It is not surprising that chromosomal aberrations would arise under such circumstances. Following this line of reasoning, the extent of abnormal development would depend upon the amount of damage to the genome *after transplantation.*

The hypothesis that mitotic adaptability of donor nuclei affects their capacity to promote development is supported by experiments with adult *Xenopus* erythroblast and erythrocyte nuclei. Erythroblasts proliferate extensively by mitosis and are the precursors of erythrocytes, which are no longer proliferative. Thus, the effects of mitotic capacity of nuclei on their developmental potential can be tested by comparing the results of transplantation of nuclei of these two cell types. In fact, erythrocyte nuclei are incapable of promoting development beyond the early gastrula stage, whereas some erythroblast nuclei promote the development of eggs to abnormal early tadpoles (Brun, 1978). Although these nuclei differ in ways other than their mitotic activity, the correlation between mitotic activity and developmental potential is striking.

The endoderm and ectoderm syndromes previously described appear to be exceptions to the correlation between abnormal development of nuclear transplant recipients and aberrant chromosome constitution. Briggs et al. (1961), Subtelny (1965), and DiBerardino and King (1967) observed that nuclear transplant recipients displaying either the endoderm or ectoderm syndrome apparently had normal karyotypes, whereas only those with generalized developmental abnormalities could be correlated with chromosome abnormalities. These results suggest that specific nuclear differentiation occurs during embryonic development. This nuclear differentiation is stable, as demonstrated by the fact that clones can be established by serial transfer of nuclear descendants from single endoderm nuclei in which the majority of the members of the clone have the endoderm syndrome and a normal or a nearly normal karyotype. One possible explanation for these observations should be considered. Post-transplantation genetic damage that cannot be seen by microscopic examination of the chromosomes could be occurring at the gene level rather than at the gross chromosomal level, possibly as a result of the inability of specific genes to replicate properly following transplantation. This inability to replicate might be a consequence of the mechanisms that regulate gene expression (see section 2–2). Thus, different regions of the genome may be susceptible to damage in different nuclear types.

Reversal of Nuclear Differentiation

In spite of three decades of intense investigation, nuclear transplantation has yet to provide an unequivocal answer to the question of nuclear totipotency. The experiments of Briggs and King demonstrate that nuclei do become progressively restricted in promoting development of enucleated eggs. Is it possible to reverse these changes experimentally so that differentiated cell nuclei can express their complete genetic repertoire? A number of experimental approaches have been used in attempts to reprogram nuclei of differentiated cells to function as substitute zygotic nuclei. One approach is to allow nuclei to adapt to a more rapid division rate *before* transplantation. This procedure attempts to maximize the opportunities for nuclei to make the transition to the rapid mitotic rate of cleavage and thus minimize or eliminate posttransplantation genetic damage. In these experiments, cells obtained from adult frogs are grown in tissue culture where they undergo mitosis prior to nuclear transfer.

This approach has been used with cells of the adult *Xenopus* kidney, heart, lung, and skin. Their nuclei have been shown to promote development of abnormal larvae, provided the cells are grown in tissue culture before nuclear transplantation (Laskey and Gurdon, 1970). Most first-transfer eggs either fail to cleave or cleave abnormally, with the production of abnormal blastulae. However, if the abnormal blastulae are dissociated and used as a source of donor nuclei for serial nuclear transfer, the extent of development is increased tremendously. An extensive series of experiments have been conducted with cultured adult *Xenopus* skin cells. The differentiated state of these cells was confirmed by the detection of keratin, the protein that is specific for differentiated epidermal cells. Abnormal swimming tadpoles have been obtained in serial transfers of nuclei of these cells. Histological examination of the tadpoles has revealed that these nuclei can promote the development of a number of diverse tissues and organs, including the heart, striated muscle, brain, nerve cord, notochord, pronephros, intestine, and eye (Gurdon et al., 1975). Similar studies have shown that adult *Xenopus* lymphocyte nuclei can promote the development of abnormal swimming tadpoles (Du Pasquier and Wabl, 1977).

These various experiments illustrate that mitotic preadaptation of donor nuclei can increase their developmental potential, but the recipients are still not *completely* normal. Thus, mitotic incompatibility may not be the only problem to overcome.

Another approach has been to modify nuclei chemically to enhance their developmental potential. These experiments are based upon a natural phenomenon. As we shall discuss in detail in Chapter 6, the DNA of sperm nuclei is associated with the protein protamine. After fertilization, the egg cytoplasm promotes the removal of protamine from the sperm nucleus and its replacement by histones, which associate with DNA in all somatic cell types (see Chap. 3). Thus, if DNA in nuclei of cells in advanced stages of development were complexed with protamine, perhaps the egg cytoplasm could remove the protamine, which would prepare the nuclei to function as substitute zygote nuclei. In such an attempt to "remodel" chromatin, Briggs (1979) exposed *Rana pipiens* endoderm nuclei to protamine before and during transplantation and obtained a two- to threefold increase in development of recipients to larval stages. Hennen (1970) used another DNA binding molecule, spermine, in conjunction with lowered temperature when transplanting *Rana pipiens* endoderm nuclei. These conditions caused a vast increase in the ability of these nuclei to promote development to the larval stage. Whether the spermine produces its effects by actually binding to DNA has not been demonstrated experimentally. The effect of low temperature is presumably to lengthen the first cleavage cycle, which would allow the donor nuclei more time to complete DNA synthesis before the first cleavage division. These experiments demonstrate that changing the conditions of nuclear transplantation can enhance the developmental potential of nuclei. Thus, the restricted developmental potential of differentiated cell nuclei in nuclear transplantation may reflect the limited ability of egg cytoplasm to prepare these nuclei to function as unrestricted zygotic nuclei. For example, the egg cytoplasm may be more efficient at remodeling chromatin if the chromatin itself has been modified so that it resembles sperm chromatin.

Recently, experiments have been initiated to attempt nuclear reprogramming by oocyte, rather than egg, cytoplasm. The ability of oocyte cytoplasm to influence injected nuclei is firmly established (see p. 66). In initial experiments to test the developmental potential of nuclei transplanted to oocytes, Hoffner and DiBerardino (1980) injected tailbud stage endoderm nuclei into oocytes that were subsequently activated and had their own nuclei removed. A significant percentage (21%) of the injected oocytes formed gastrulae, and some of the gastrulae developed even further. Improvements in this technique may produce even better results and may provide the means to test the developmental potentials of adult cell nuclei that have not been amenable to transplantation to eggs.

Recent attempts to reprogram nuclei have produced promising results. However, nuclear totipotency must remain an uncertainty until some means is found to produce a high frequency of perfectly normal nuclear transplants from nuclei of adult differentiated cells. But, *at the very least*, nuclear transplantation experiments demonstrate that the nuclei of some differentiated cells can promote the differentiation of a variety of diverse cell types and are therefore **pluripotent.**

Totipotency in Plants

A far more satisfactory demonstration of nuclear potency comes from experiments with plant cells. The system for demonstrating totipotency was developed by F. C. Steward and evolved out of efforts during the 1940s to stimulate growth (i.e., cell division) in quiescent plant tissue. The tissue selected for these experiments was carrot root phloem. Explants of this tissue contain cells that would not normally divide again. The explants were cultured in a medium that contained coconut milk, which is the liquid endosperm of the coconut seed and normally functions to nourish and stimulate the growth of the developing coconut plant. A special culture flask was designed to allow large-scale culture of up to 100 carrot explants. The culture flasks were rotating constantly to expose the explants alternately to air and liquid (Fig. 2.10A).

Under these conditions the explants grew as disorganized tissue masses known as **calluses.** During rotation of the flasks, free cells were rubbed off the calluses and became suspended in the culture medium (Fig. 2.10B). Steward and his associates also observed small multicellular structures called **embryoids,** which resembled the normal stages in carrot embryonic development from the zygote (Fig. 2.10C to H). They interpreted the embryoids to be the products of division and differentiation

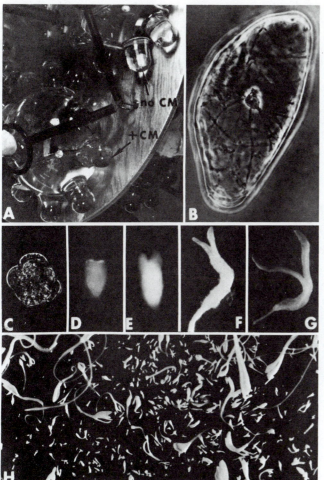

FIGURE 2.10 Growth of carrot explants and free cells and their embryogenesis. *A,* Culture flasks containing basal medium, with and without coconut milk (CM), showing the effect on the growth of the original explants (2.5 mg). In the presence of coconut milk these explants grew to approximately 250 mg and were green, in contrast to those in the basal medium without coconut milk, which remained small and orange in color. (Photography by M.O. Mapes.) *B,* A carrot cell (300 × 125 µm) freely suspended in the coconut milk medium as it might have originated in the flask at *A.* (Photograph by M.O. Mapes.) *C* to *G,* Stages in embryogenesis that developed in free cell suspensions; the magnification decreases from the microscopic globular form at *C* to the cotyledonary stage at *G.* *H,* A random sample of a crop of carrot embryos, slightly magnified, as developed from free cells by a sequence of treatments. The field shows all stages of development from globular embryos to plantlets, and, as cells are sloughed off, embryos are repeatedly formed. (From F.C. Steward. 1970. From cultured cells to whole plants: The induction and control of their growth and morphogenesis. Proc. R. Soc. Lond. [Biol.] Ser. B, *175:* 7.)

of the single cells that were sloughed from the calluses. The embryoids continued developing to form small plantlets.

Presumably, then, single cells derived from carrot root phloem are totipotent, since they can be caused to divide and then recapitulate normal development. However, some controversy has arisen over the interpretation of these experiments, since the complete progression from single cell to embryoid to plantlet was not actually observed. Formal proof that single plant cells can give rise to complete mature plants requires that a single cell be isolated from a callus and produce a complete plant. This procedure circumvents the possibility that the embryoids are actually formed from cell aggregates.

Regular text continued on page 63

PLANT PROTOPLASTS

An extension of plant somatic cell culture is culture of **plant protoplasts.** Protoplasts are single cells whose cell walls have been removed enzymatically. Leaf tissue is frequently used as a source of cells for making protoplasts. When protoplasts are transferred to enzyme-free medium, the cell walls regenerate. Under appropriate conditions these cells can, in turn, be stimulated to divide and be transformed into

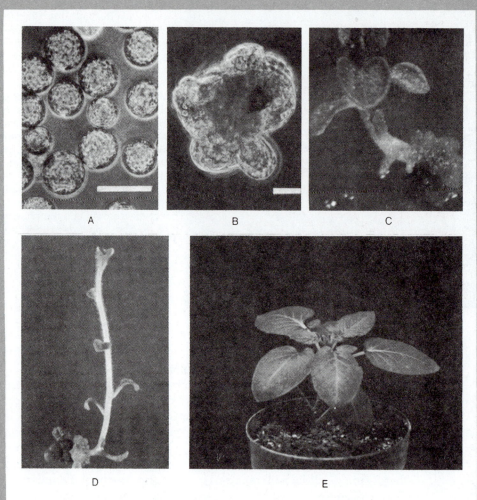

A, Protoplasts isolated from potato plant leaflets. Bar equals 30 μm. *B,* Small callus derived from a single protoplast. Bar equals 150 μm. *C,* Shoot, 7 mm tall, regenerated from callus. *D,* Elongating shoot, 24 mm tall. *E,* Fully developed potato plant, 10 cm tall, regenerated from a single protoplast. (From J.F. Shepard and R.E. Totten. 1977. Mesophyll cell protoplasts of potato. Isolation, proliferation, and plant regeneration. Plant Physiol., *60:* 315.)

mature plants. Single protoplasts isolated from a population of cells have been shown to redifferentiate whole plants, confirming the totipotency of plant somatic cells. The sequence of development from a single protoplast derived from a potato leaf to a fully developed potato plant is shown on the preceding page (from Shepard and Totten, 1977).

The absence of cell walls allows protoplasts to fuse with one another. The efficiency of fusion is vastly improved by the use of polyethylene glycol (Kao and Michayluk, 1974). Occasionally, the nuclei of fused protoplasts will also fuse, producing a **somatic hybrid.** Cell wall regeneration and culture of the hybrids can lead to the formation of hybrid plants. The hybrid nature of such plants can be demonstrated by detection of genetically determined characteristics of both parental plants in the hybrids (Dudits et al., 1980; Douglas et al., 1981). The figure below illustrates a hybrid cell that was obtained after fusion of soybean and pea protoplasts.

One example of the formation of hybrid plants by the protoplast method is the fusion of two species of tobacco *(Nicotiana langsdorffii* and *Nicotiana glauca)*. The techniques used in this experiment (Carlson et al., 1972) are illustrated on page 62.

Protoplast techniques provide an alternative to traditional hybridization by breeding methods. The potential implications for agriculture are amazing. Production of hybrids by normal breeding methods is slow. Furthermore, hybridization of distantly related species by breeding techniques is frequently impossible because of biological barriers associated with pollination and embryo development. Therefore, production of hybrids that combine agriculturally desirable characteristics of two species might be impossible by traditional methods. Somatic hybridization might allow these barriers to be by-passed.

The absence of cell walls makes protoplasts permeable to DNA and even whole chromosomes (Hughes et al., 1979; Szabados et al., 1981). This property of pro-

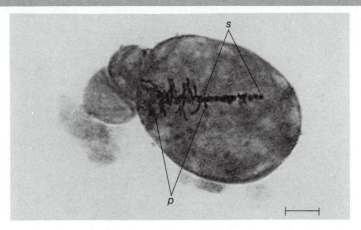

Metaphase of soybean-pea hybrid. Note pea chromosomes (*p*) on the left and soybean chromosomes (*s*) on the right. Bar represents 10 μm. (From F. Constabel et al. 1975. Nuclear fusion in intergeneric heterokaryons. A note. Can. J. Bot., *53:* 2094.)

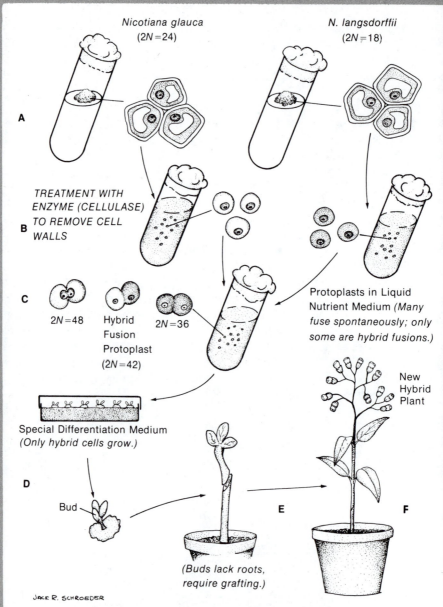

Nicotiana glauca (2N = 24)

N. langsdorffii (2N = 18)

A

TREATMENT WITH ENZYME (CELLULASE) TO REMOVE CELL WALLS

B

C

2N = 48   Hybrid Fusion Protoplast (2N = 42)   2N = 36

Protoplasts in Liquid Nutrient Medium (*Many fuse spontaneously; only some are hybrid fusions.*)

Special Differentiation Medium (*Only hybrid cells grow.*)

D

Bud

(*Buds lack roots, require grafting.*)

E

New Hybrid Plant

F

JACK R. SCHROEDER

A method for obtaining hybrid plants from protoplasts. *A*, Culture of cells from two species. *B*, Removal of cell walls by enzyme treatment. *C*, Combining of the two kinds of protoplasts to obtain fusion. *D*, Isolation of fusion products and induction of cell division followed by bud formation. *E*, Grafting of bud onto stock of one of the parental plants. *F*, Maturation of hybrid. (After P.S. Carlson, H.H. Smith, and R.D. Dearing, 1972. From K. Norstog and R.W. Long. 1976. *Plant Biology.* W.B. Saunders, Philadelphia, p. 548.)

toplasts opens up the possibility of **genetic transformation** of protoplasts by introducing foreign genes into them. Regeneration of plants from transformed protoplasts could lead to production of plants with altered genetic compositions.

Genetic engineering on plants could have enormous economic benefits. One possibility is equipping various food-producing, nonleguminous plants with nitrogen-fixing capabilities. Imagine the improved efficiency of crop production if wheat could fix nitrogen from the air! Such exotic results may never prove feasible, but it is likely that certain genetic improvements in plants will be introduced by protoplast fusion and protoplast transformation.

Regular text continued from page 59

An experiment of this kind has been conducted by Vasil and Hildebrandt (1965). These investigators removed explants of stem pith from tobacco plants *(Nicotiana)* and placed the explants in tissue culture, where they formed calluses. By shaking the calluses, they dislodged large numbers of single cells and groups of cells. Single cells were removed from these suspensions and isolated to microcultures (Fig. 2.11*A*). The isolated single cells underwent cell division and formed disorganized cell masses rather than embryoids (Fig. 2.11*B* and *C*). Cell masses from the single cells were then placed on semisolid culture medium containing plant growth factors. Plantlets with roots and aerial shoots with leaves formed (Fig. 2.11*D*). These plantlets were then transferred to soil, where mature flowering plants developed (Fig. 2.11*E* and *F*).

The formation of a whole plant from a single adult somatic cell confirms that the nuclei of these cells contain all the genetic information that was originally present in the zygote. Furthermore, plant cells contain the cytoplasmic machinery that is necessary to reprogram the genome to produce a mature plant, if the proper growth factors and nutritional requirements are provided.

2–2. DIFFERENTIAL GENE EXPRESSION

Although nuclear totipotency still remains the subject of much experimental investigation, it is clear that loss of genetic information does not necessarily accompany cell differentiation. One of the most convincing examples of the retention of the genome is the polytene chromosome map of *Drosophila* (see Chap. 1). The *entire genome* is represented in certain differentiated larval cells; the band pattern (which represents the genes) is identical in different cells of the same larva and even in different larvae. Molecular analyses of the genome also support the concept of

FIGURE 2.11 Development of mature plants from single cultured cells. *A,* Single cell after one day in microculture. *B* and *C,* Stages in the formation of a mass of cells from the single cell in *A. D* (bottom), Callus formation. *D* (top), Shoot formation from callus. *E,* Plantlet after transfer to soil. *F,* Mature flowering plant. (From V. Vasil and A.C. Hildebrandt. 1965. Differentiation of tobacco plants from single isolated cells in microcultures. Science, *150:* 889–892. Copyright 1965 by the American Association for the Advancement of Science.)

genomic integrity. Genetic information is encoded in DNA in the sequence of its nucleotides. As we shall discuss in Chapter 4, the nucleotide sequence for specific genes does not vary from cell to cell in an adult organism. There is no loss or gain of genetic information during cell differentiation. It is necessary, however, to note that there are some exceptions to this rule. These are discussed in the box on page 136.

It is worthwhile at this point to consider some of the ramifications of the developmental integrity of the genome. The primary function of the genome is to direct the synthesis of proteins. Theoretically, every cell has the capacity to produce the same proteins. However, the capacity to synthesize certain proteins becomes restricted to specialized cells during development. For example, red blood cells are specialized to synthesize hemoglobin, a protein that is not found in nerve cells.

Restrictions on protein synthesis must mean that certain mechanisms are operating to select portions of the genome to be utilized in different cells. Therefore, the variability in protein composition in differentiated cells must result from **differential gene expression.** The mechanisms that control development must exert their effects by controlling differential gene expression. When the developmental pathway of a cell is specified, these control mechanisms must somehow select which genes are to be expressed and which genes are to remain latent. This is the process of **cell determination.** The processes involved in utilizing those restricted portions of the genome to produce a specialized cell are cell differentiation.

Since individual cells of a developing organism are assumed to have identical genomes, determination of the pathway of differentiation to be followed in any particular cell must be imposed upon the genome from outside the nucleus. This, of course, implicates the cytoplasm in regulation of gene expression, since the nucleus is, in a sense, the captive of the cytoplasm, which completely surrounds it. E. B. Wilson referred to the cell as a "reaction system" in which the nucleus and cytoplasm each play a role in determination of cellular characteristics. Nucleus and cytoplasm are in dynamic interaction, resulting in an altered physiological state (differentiation) in which new nucleocytoplasmic interactions come into operation.

Cytoplasmic effects on nuclei are demonstrated by experiments in which nuclei are exposed to foreign cytoplasm. Two techniques have been particularly useful for combining nuclei and cytoplasms from different cells. These are nuclear transplantation and **somatic cell hybridization.** Oocytes are particularly good targets for the nuclear transplantation experiments, since they have a large amount of cytoplasm and are nondividing. Injected oocytes can be cultured *in vitro* for long periods of time, and the functions of the transplanted nuclei can then be assessed. Somatic cell hybridization is a technique for fusion of cells. If cells of different types are fused, the effects of the cytoplasm on the nuclei can be determined.

An example of nuclear transplantation to oocytes is the implantation of *Xenopus* blastula nuclei in the cytoplasm of a *Xenopus* oocyte. As we shall see in Chapter 7, oocytes are extremely active in RNA synthesis and inactive in DNA synthesis. The blastula nucleus, on the other hand, is adapted to a high rate of mitosis and therefore synthesizes DNA but little RNA. Blastula nuclei transplanted to oocyte cytoplasm undergo changes in both morphology and nucleic acid synthesis that cause them to resemble oocyte nuclei. Among the morphological changes observed, there are an enlargement of nuclear volume, dispersion of chromatin, and the appearance of nucleoli. Molecular analyses demonstrate that RNA synthesis is activated in the transplanted blastula nuclei and that DNA synthesis is inactivated (Gurdon, 1968a).

The ability of the oocyte cytoplasm to regulate gene expression has been demonstrated in an elegant experiment by De Robertis and Gurdon (1977). Nuclei from cultured *Xenopus* kidney cells were injected into oocytes of the newt *Pleurodeles waltlii*. The proteins synthesized by the recipient oocytes were compared to those synthesized by the kidney cells and control oocytes. It was found that the recipient oocytes synthesize *Xenopus* proteins, indicating that the *Xenopus* nucleus is active in the foreign cytoplasm. However, most surprisingly, the *Xenopus* proteins include some normally synthesized by *Xenopus* oocytes, but not by cultured kidney cells. Apparently, genes that remain unexpressed in the cultured cells are expressed when the nuclei are exposed to the oocyte cytoplasm. Clearly, the oocyte cytoplasm has the ability to regulate the qualitative expression of genes.

Similar experiments have demonstrated that nuclei from adult differentiated cells can be modified by exposure to oocyte cytoplasm. Etkin (1976) injected nuclei from adult liver cells of the urodele amphibian *Ambystoma texanum* into oocytes of a related species, *Ambystoma mexicanum*. Liver cells produce the enzyme alcohol dehydrogenase (ADH), whereas oocytes do not produce ADH. Both cells produce another enzyme, lactate dehydrogenase (LDH). In the case of both enzymes, the *A. texanum* enzyme is distinguishable from the *A. mexicanum* enzyme by the technique of electrophoresis (see p. 92 for a discussion of this technique). When injected oocytes are assayed for these enzymes, it is found that they have synthesized *A. texanum* (donor) LDH but not ADH. Thus, liver cell nuclei in the oocyte cytoplasm produce proteins that are typical of the oocyte, not of the liver. This result provides further evidence for the influence of oocyte cytoplasm on nuclei.

Cytoplasmic modulation of nuclear function is also illustrated by somatic cell hybridization. The principles of animal cell fusion are similar to those of plant protoplast fusion, although the techniques are quite different. Animal cell fusion may be facilitated by use of a virus (Sendai) that has been killed with ultraviolet light. The virus, although noninfective, binds to cell membranes and causes cells to adhere to one another. At the point of cell-to-cell contact, the membranes fuse, and the cytoplasms coalesce to form a single cell with multiple nuclei. When the cells are from different species, the hybrid cell that is formed is called a **heterokaryon.**

A particularly interesting heterokaryotic combination is between a human tissue culture cell, called a HeLa cell, and a hen erythrocyte. These cells were selected for fusion because of their contrasting patterns of DNA and RNA synthesis. HeLa cells are highly active in both DNA and RNA synthesis, whereas hen erythrocytes are inactive in DNA synthesis and synthesize very little RNA (the absolute levels of RNA synthesis in hen erythrocytes are controversial; see MacLean [1976]). Hen erythrocytes are particularly useful for cell fusion experiments, since the Sendai virus causes the cells to rupture, releasing the cytoplasm. When these acytoplasmic, nucleated cells fuse with a HeLa cell, all of the cytoplasm of the heterokaryon is derived from the HeLa cell (Fig. 2.12).

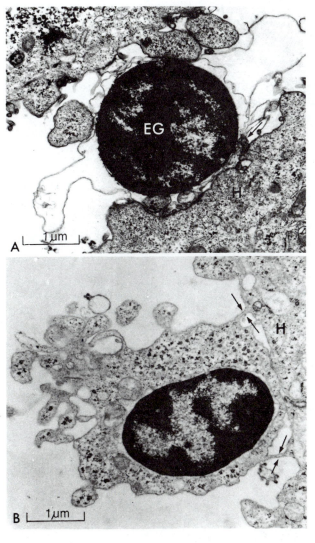

A

B

FIGURE 2.12 Fusion of an erythrocyte ghost (EG) and a HeLa cell (H). *A,* The arrow shows a virus particle wedged between the two cell membranes. *B,* The HeLa cytoplasm has flowed into the erythrocyte ghost. Note the highly condensed chromatin in the erythrocyte nucleus. (From E.E. Schneeberger and H. Harris. 1966. An ultrastructural study of interspecific cell fusion induced by inactivated Sendai virus. J. Cell Sci., *1:* 401–405.)

Therefore, the effects on the erythrocyte nuclei are caused solely by the HeLa cytoplasm. After fusion the erythrocyte nuclei undergo morphological changes that resemble those seen in blastula nuclei transplanted to oocytes—they increase in volume and their chromatin disperses (Fig. 2.13A-C). These enlarged nuclei become highly active in both RNA and DNA synthesis, apparently in response to signals from the HeLa cytoplasm that regulate the level of RNA and DNA synthesis (Harris, 1965, 1968).

RNA and DNA synthesis are analyzed by use of radioactive tracers. The most specific metabolic precursor of RNA is uridine; it is not incorporated into any other macromolecule in the cell. Hence, if a source of radioactive uridine (^{3}H-uridine) is supplied to the oocyte, it will be

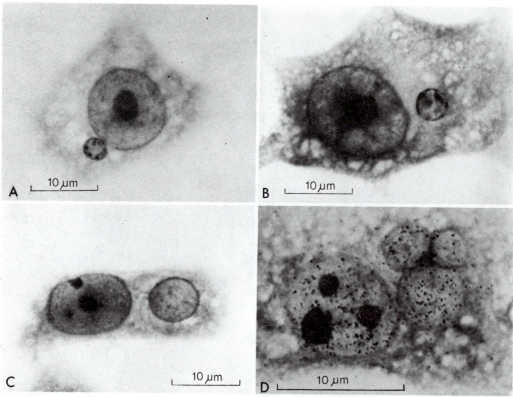

FIGURE 2.13 Heterokaryons with HeLa and erythrocyte nuclei. *A* to *C*, Morphological changes in erythrocyte nuclei. *A*, A dikaryon immediately after fusion. *B*, Erythrocyte nucleus has begun to enlarge. *C*, Further stage of enlargement. Note dispersion of erythrocyte chromatin. *D*, Autoradiograph of heterokaryon after incorporation of ^{3}H-uridine. The silver grains represent synthesis of RNA. The cell contains one HeLa nucleus and three erythrocyte nuclei in various stages of enlargement. Note that the labeling of erythrocyte nuclei increases as they enlarge. (*A* to *C* from H. Harris. 1967. The reactivation of the red cell nucleus. J. Cell Sci., *2*: 23–32. *D* from H. Harris. 1968. *Nucleus and Cytoplasm.* Clarendon Press, Oxford.)

incorporated into newly synthesized RNA. By detecting ^{3}H-uridine in nuclei, investigators can demonstrate RNA synthesis. DNA synthesis is detected by incorporation of the labeled precursor, ^{3}H-thymidine. Incorporation of labeled precursors is detected by **autoradiography.** This technique uses a photographic emulsion placed over a cytological preparation to detect radioactivity. Particles emitted from the radioisotope activate the emulsion so that silver grains are produced when the emulsion is developed by normal darkroom procedures. The silver grains are seen as tiny black dots when observed with the microscope and are located immediately over the source of radiation. RNA synthesis in the erythrocyte nuclei is shown in Figure 2.13*D,* which is an autoradiograph of a heterokaryon after incorporation of ^{3}H-uridine.

Cytoplasmic regulation of gene expression provides an answer to the question of how numerous cell types are differentiated from the same basic genetic information. However, it also points to an apparent enigma. Multicellular organisms begin life as a single cell, the zygote. If the cytoplasm controls gene expression, how can the derivatives of this single cell ever generate the cytoplasmic diversity that will in turn produce distinct patterns of gene expression? There are apparently two general solutions to this problem. The first is the **intrinsic heterogeneity** in the egg cytoplasm itself. If the egg cytoplasm consists of regions that differ from one another, cleavage would cause fixation of these differences, producing cells with distinct cytoplasmic constituents. The second solution involves **extrinsic influences** on the embryo. In this developmental strategy the different regions of the embryo receive positional information that establishes their fates. These two modes of determination are discussed in detail in Chapter 11.

Extrinsic factors are also important in regulating cell differentiation during later stages of development in both plants and animals. Formation of organs or tissues is often dependent upon developmental signals produced elsewhere in the developing organism. They could be in the form of interactions between adjacent cells, such as in embryonic induction (see Chap. 13), or they could result from long-distance interactions, such as those mediated by animal hormones or plant growth factors. However, those interactions that occur during later development are all dependent upon the initial events that establish cytoplasmic heterogeneity during early development. Since heterogeneity often results from inherent regional differences within the egg, its establishment is in turn dependent upon processes that occur during oogenesis.

The great challenge of development is to understand how it is controlled. Since the control mechanisms exert their effects by regulating gene expression, it is desirable to explain interactions of the genome with the cytoplasm, hormones, growth factors, and other regulatory factors in molecular terms. Such an endeavor requires that we have an understanding of the fundamental nature of the genetic material and its utilization by the cell. Thus, before beginning a detailed analysis of development, we shall review molecular biology in the context of differentiation. Chapter 3 deals with the organization and cellular processing of genetic information, whereas the basic mechanisms involved in control of gene expression are discussed in Chapter 4.

REFERENCES

Briggs, R. 1979. Genetics of cell type determination. *In* J.F. Danielli and M.A. DiBerardino (eds.), *Nuclear Transplantation.* Int. Rev. Cytol. Suppl. No 9. pp. 129–160.

Briggs, R., and T.J. King. 1952. Transplantation of living nuclei from blastula cells into enucleated frogs' eggs. Proc. Natl. Acad. Sci. U.S.A., *38*: 455–463.

Briggs, R., and T.J. King. 1957. Changes in the nuclei of differentiating endoderm cells as revealed by nuclear transplantation. J. Morphol., *100*: 269–312.

Briggs, R., and T.J. King. 1960. Nuclear transplantation studies on the early gastrula *(Rana pipiens).* I. Nuclei of presumptive endoderm. Dev. Biol., *2*: 252–270.

Briggs, R., T.J. King, and M.A. DiBerardino. 1961. Development of nuclear-transplant embryos. *In* S. Ranzi (ed.), *Symposium on the Germ Cells and Earliest Stages of Development.* Fond. A. Baselli, Milan, pp. 441–477.

Brun, R.B. 1978. Developmental capacities of *Xenopus* eggs, provided with erythrocyte or erythroblast nuclei from adults. Dev. Biol., *65*: 271–284.

Carlson, P.S., H.H. Smith and R.D. Dearing. 1972. Parasexual interspecific plant hybridization. Proc. Natl. Acad. Sci. U.S.A., *69*: 2292–2294.

Constabel, F. et al. 1975. Nuclear fusion in intergeneric heterokaryons. A note. Can. J. Bot., *53*: 2092–2095.

De Robertis, E.M., and J.B. Gurdon. 1977. Gene activation in somatic nuclei after injection into amphibian oocytes. Proc. Natl. Acad. Sci. U.S.A., *74*: 2470–2474.

DiBerardino, M.A. 1979. Nuclear and chromosomal behavior in amphibian nuclear transplants. *In* J.F. Danielli and M.A. DiBerardino (eds.), *Nuclear Transplantation.* Int. Rev. Cytol. Suppl., No. 9, pp. 129–160.

DiBerardino, M.A. 1980. Genetic stability and modulation of metazoan nuclei transplanted into eggs and oocytes. Differentiation, *17*: 17–30.

DiBerardino, M.A., and N. Hoffner. 1970. Origin of chromosomal abnormalities in nuclear transplants—A reevaluation of nuclear differentiation and nuclear equivalence in amphibians. Dev. Biol., *23*: 185–209.

DiBerardino, M.A., and N. Hoffner. 1971. Development and chromosomal constitution of nuclear-transplants derived from male germ cells. J. Exp. Zool. *176*: 61–72.

DiBerardino, M.A., and T.J. King. 1965. Transplantation of nuclei from the frog renal adenocarcinoma. II. Chromosomal and histogenic analysis of tumor nuclear-transplant embryos. Dev. Biol., *11*: 217–242.

DiBerardino, M.A., and T.J. King. 1967. Development and cellular differentiation of neural nuclear transplants of known karyotype. Dev. Biol., *15*: 102–128.

Douglas, G.C. et al. 1981. Somatic hybridization between *Nicotiana rustica* and *N. tabacum.* III. Biochemical, morphological, and cytological analysis of somatic hybrids. Can. J. Bot., *59*: 228–237.

Dudits, D. et al. 1980. Intergeneric gene transfer mediated by plant protoplast fusion. Molec. Gen. Genet., *179*: 283–288.

Du Pasquier, L., and M.R. Wabl. 1977. Transplantation of nuclei from lymphocytes of adult frogs into enucleated eggs. Special focus on technical parameters. Differentiation, *8*: 9–19.

Etkin, L.D. 1976. Regulation of lactate dehydrogenase (LDH) and alcohol dehydrogenase (ADH) synthesis in liver nuclei following their transfer into oocytes. Dev. Biol., *52*: 201–209.

Fankhauser, G. 1930. Zytologische Untersuchungen an geschnürten Triton-Eiern. 1. Die verzögerte Kernversorgung einer Hälfte nach hantelförmiger Einschnürung des Eis. Wilh. Roux' Arch. für Entwicklungsmechanik der Organismen, *122*: 116–139.

Gurdon, J.B. 1962. The developmental capacity of nuclei taken from intestinal epithelium cells of feeding tadpoles. J. Embryol. Exp. Morphol., *10*: 622–641.

Gurdon, J.B. 1966. The cytoplasmic control of gene activity. Endeavour, *25*: 95–99.

Gurdon, J.B. 1968a. Changes in somatic cell nuclei inserted into growing and maturing amphibian oocytes. J. Embryol. Exp. Morphol., *20*: 401–414.

Gurdon, J.B. 1968b. Transplanted nuclei and cell differentiation. Sci. Am., *219*(6): 24–35.

Gurdon, J.B., R.A. Laskey, and O.R. Reeves. 1975. The developmental capacity of nuclei transplanted from keratinized skin cells of adult frogs. J. Embryol. Exp. Morphol., *34*: 93–112.

Harris, H. 1965. Behaviour of differentiated nuclei in heterokaryons of animal cells from different species. Nature (Lond.), *206*: 583–588.

Harris, H. 1967. The reactivation of the red cell nucleus. J. Cell Sci., *2*: 23–32.

Harris, H. 1968. *Nucleus and Cytoplasm*. Clarendon Press (Oxford University Press), New York.

Hennen, S. 1970. Influence of spermine and reduced temperature on the ability of transplanted nuclei to promote normal development in eggs of *Rana pipiens*. Proc. Natl. Acad. Sci. U.S.A., *66*: 630–637.

Hoffner, N.J., and M.A. DiBerardino. 1980. Developmental potential of somatic nuclei transplanted into meiotic oocytes of *Rana pipiens*. Science, *209*: 517–519.

Hughes, B.G., F.G. White, and M.A. Smith. 1979. Fate of bacterial plasmid DNA during uptake by barley and tobacco protoplasts: Protection by poly-L-ornithine. Plant Sci. Lett., *14*: 303–310.

Illmensee, K. et al. 1981. Nuclear and gene transplantation in the mouse. *In* D.D. Brown and C.F. Fox (eds.), *Developmental Biology Using Purified Genes.* ICN-UCLA Symposia on Molecular and Cellular Biology, Vol. 23. Academic Press, New York, pp. 607–619.

Illmensee, K., and P.C. Hoppe. 1981. Nuclear transplantation in Mus musculus: Developmental potential of nuclei from preimplantation embryos. Cell, *23*: 9–18.

Kao, K.N., and M.R. Michayluk. 1974. A method for high-frequency intergeneric fusion of plant protoplasts. Planta (Berl.), *115*: 355–367.

King, T.J. 1966. Nuclear transplantation in amphibia. *In* D.J. Prescott (ed.), *Methods in Cell Physiology*, Vol. 2. Academic Press, New York, pp. 1–36.

King, T.J., and R. Briggs. 1956. Serial transplantation of embryonic nuclei. Cold Spring Harbor Symp. Quant. Biol., *21*: 271–290.

King, T.J., and M.A. DiBerardino. 1965. Transplantation of nuclei from the frog renal adenocarcinoma. I. Development of tumor nuclear-transplant embryos. Ann. N.Y. Acad. Sci., *126*: 115–126.

Kobel, H.R., R.B. Brun, and M. Fischberg. 1973. Nuclear transplantation with melanophores, ciliated epidermal cells, and the established cell-line A-8 in *Xenopus laevis*. J. Embryol. Exp. Morphol., *29*: 539–547.

Laskey, R.A., and J.B. Gurdon. 1970. Genetic content of adult somatic cells tested by nuclear transplantation from cultured cells. Nature (Lond.), *228*: 1332–1334.

McKinnell, R.G. 1962. Intraspecific nuclear transplantation in frogs. J. Heredity *53*: 199–207.

McKinnell, R.G. 1972. Nuclear transfer in *Xenopus* and *Rana* compared. *In* R. Harris, P. Allin, and D. Viza (eds.), *Cell Differentiation*. Munksgaard International Publishers, Copenhagen, pp. 61–64.

McKinnell, R.G., B.A. Deggins, and D.D. Labat. 1969. Transplantation of pluripotential nuclei from triploid frog tumors. Science, *165*: 394–396.

MacLean, N. 1976. *Control of Gene Expression*. Academic Press, New York.

Norstog, K., and R.W. Long. 1976. *Plant Biology*. W.B. Saunders, Philadelphia.

Porter, K.R. 1939. Androgenetic development of the egg of *Rana pipiens*. Biol. Bull., *77*: 233–257.

Schneeberger, E.E., and H. Harris. 1966. An ultrastructural study of interspecific cell fusion induced by inactivated Sendai virus. J. Cell Sci., *1*: 401–405.

Shepard, J.F., and R.E. Totten. 1977. Mesophyll cell protoplasts of potato. Isolation, proliferation, and plant regeneration. Plant Physiol., *60*: 313–316.

Smith, L.D. 1965. Transplantation of the nuclei of primordial germ cells into enucleated eggs of *Rana pipiens*. Proc. Natl. Acad. Sci. U.S.A., *54*: 101–107.

Spemann, H. 1928. Die Entwicklung seitlicher und dorso-ventraler Keimhälften bei verzögerter Kernversorgung. Zeitschrift für wissenschafliche Zoologie *132*: 105–134.

Spemann, H. 1938. *Embryonic Development and Induction*. Yale University Press, New Haven, Conn. Reprinted by Hafner Press (Macmillan, Inc.), New York. (1962)

Steward, F.C. 1970. From cultured cells to whole plants: The induction and control of their growth and morphogenesis. Proc. R. Soc. Lond. (Biol.), *175*: 1–30.

Subtelny, S. 1965. On the nature of the restricted differentiation-promoting ability of transplanted *Rana pipiens* nuclei from differentiating endoderm cells. J. Exp. Zool., *159*: 59–92.

Szabados, L., Gy. Hadlaczky, and D. Dudits. 1981. Uptake of isolated plant chromosomes by plant protoplasts. Planta (Berl.), *151*: 141–145.

Vasil, V., and A.C. Hildebrandt. 1965. Differentiation of tobacco plants from single isolated cells in microcultures. Science, *150*: 889–892.

3

The Nature of Genetic Information and Its Utilization During Cell Differentiation

SECTION ONE
THE NATURE OF GENETIC INFORMATION

The nucleus of a fertilized egg contains a vast amount of genetic information—the genome—that specifies the nature of each of the phenotypic traits, which together constitute the functional organism. Utilization of this information must be controlled so that the proper genes function at the right time and in the right cells to produce the various cells that constitute the different parts of the complete organism. The genome contains a number of different categories of information; these include ribosomal RNA genes, transfer RNA genes, and the structural genes that act as templates for messenger RNA. Additional evidence has accumulated in recent years, indicating that the genome also contains regions that do not fit the classical definition of genes. Some of these regions do not code for RNA, whereas others code for RNA that apparently does not function in protein synthesis.

The utilization of the genome for protein synthesis is a multi-step process that begins with selection of the appropriate regions of the genome, their transcription into RNA, the processing of RNA, and the translation of mRNA into protein. Investigation of the mechanisms involved in the control of information flow from the genome constitutes one of the major thrusts of contemporary developmental biology, and this chapter lays the foundation for understanding these analyses by describing the physical basis for the genome and the processes of utilization of genomic information.

3–1. THE STRUCTURE OF CHROMATIN

The DNA of multicellular organisms is not simply a free double helix within the nucleus. Instead, it is combined with proteins and RNA to form a complex called **chromatin,** which in turn is organized into many individual chromosomes containing compacted DNA. The condensation of DNA makes the long strands manageable within the cell nuclei. Human chromosomes, for example, contain enough DNA to form structures that would vary from 1.4 to 7.3 cm in length if the DNA were not condensed.

The degree of compaction of DNA is dependent upon the functional state of the cell. During mitosis the chromosomes are highly compact and form convenient packages that facilitate transmission of the genome to daughter cells. The most compact state of DNA occurs during metaphase (Fig. 3.1), when the lengths of the chromosomes range from 2 to 10 μm (Lewin, 1980). When the chromatin is in this form, it is inactive in synthetic activity. Between divisions (i.e., during interphase) chromosomes are partially unwound, forming a loose fibrous chromatin net-

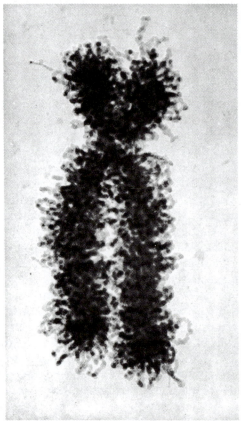

FIGURE 3.1 Human metaphase chromosome as seen with the electron microscope. Note individual chromatin fibers. ×29,600. (From E.J. DuPraw. 1970. *DNA and Chromosomes.* Holt, Rinehart & Winston, New York, p. 144.)

work. It is in this configuration that the bulk of transcription and replication of the genome occurs, and thus, this configuration is the most significant to development. At the completion of interphase, chromatin condenses to form recognizable chromosomes. Following mitosis, chromosomes once again unravel, and fibrous chromatin is formed.

Each round of this continuous process is designated the **cell cycle,** which is the interval between the completion of one cell division and the completion of the next division. The cell cycle is represented diagrammatically in Figure 3.2. In most cells, interphase occupies most of the cell cycle. In actively dividing cells, interphase is further subdivided into three phases—G_1, **S,** and G_2. The S phase is the period of DNA synthesis and is bracketed between the two G (for gap) phases. Differentiated cells may withdraw from the cycle and not undergo mitosis. Since their function is to produce specialized proteins, their nuclei usually have diffuse chromatin, which facilitates RNA synthesis.

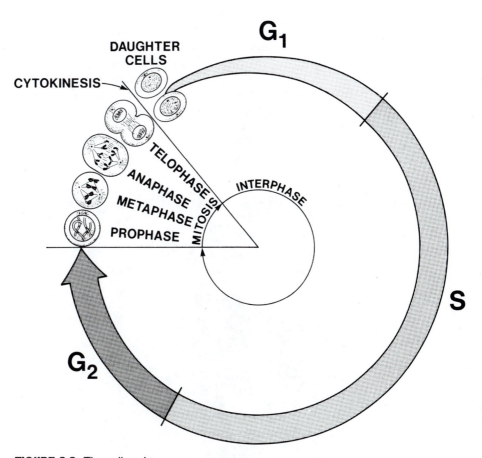

FIGURE 3.2 The cell cycle.

Not all chromatin behaves according to the idealized pattern we have outlined. A portion of it remains condensed during interphase. It is called **heterochromatin,** in contrast to **euchromatin,** which refers to the part that undergoes the normal decondensation during interphase. These two states of chromatin have a different appearance in the interphase nucleus. Most of the central region of the nucleus is filled with diffuse euchromatin, whereas compact heterochromatin is prominent near the nuclear envelope and in association with the nucleolus (Fig. 3.3).

Cytologists have long proposed that heterochromatin is inactive in RNA synthesis during interphase. This concept of heterochromatin has been confirmed by autoradiography. ^{3}H-uridine is primarily incorporated into RNA in euchromatin, whereas the genetically inactive heterochromatin shows little evidence of RNA synthesis (Littau et al., 1964). Since heterochromatin remains condensed when most DNA synthesis occurs, it is incapable of replicating at that time. Instead, it briefly decondenses during the latter part of the S phase and undergoes **delayed replication.**

Cytologists generally recognize two kinds of heterochromatin— **constitutive** and **facultative.** Constitutive heterochromatin is formed by an obligatory condensation of regions of the genome that have specialized nucleotide sequences. Most constitutive heterochromatin is located adjacent to the centromeres, at the ends of chromosomes, and next to the nucleolus. This portion of the genome is *permanently heterochromatic.* Facultative heterochromatin, on the other hand, is formed *during development* by condensation of euchromatin. This is one means of inactivating unnecessary portions of the genome. Once this state is imposed upon a portion of the genome, it remains heterochromatic throughout the life of the organism.

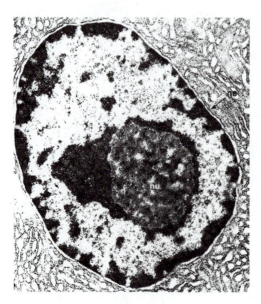

FIGURE 3.3 Electron micrograph of the nucleus from a mouse pancreatic cell. The euchromatin is present as diffuse fibers occupying the interior of the nucleus. The condensed heterochromatin is associated with the nucleolus (nu) and the inner surface of the nuclear envelope (ne). ×7800. (Micrograph by J. Andre. From E.D.P. De Robertis and E.M.F. De Robertis, Jr. 1980. *Cell and Molecular Biology,* 7th ed. Saunders College Publishing, Philadelphia, p. 23.)

TABLE 3–1
Relative Amounts of the Major Amino Acids in Calf Thymus Histones

Histone	Major amino acids		
H1	28.7% lysine	+ 25.1% alanine	+ 10.1% proline
H2A	12.5% lysine	+ 13.2% alanine	+ 10.1% leucine
H2B	16.7% lysine	+ 10.2% alanine	+ 10.9% serine
H3	13.6% arginine	+ 13.5% alanine	+ 10.2% glutamic acid
H4	9.8% lysine	+ 13.9% arginine	+ 15.9% glycine

After Hnilica, L.S. 1972. Reprinted with permission from *The Structure and Biological Function of Histones.* Copyright The Chemical Rubber Co., CRC Press, Inc.

The physical properties of chromatin are determined by both the nature of the DNA and the kinds of proteins with which the DNA is combined. Proteins are integral chromatin constituents that affect not only the gross appearance of chromatin, such as the extent of condensation, but also subtle characteristics that influence the function of particular nucleotide sequences. There are two general classes of protein constituents in chromatin: **histones** and **non-histone chromosomal proteins (NHC proteins).** The histones are positively charged molecules with a high proportion of basic amino acids, such as lysine and arginine. The percent composition of the major amino acids of the five histones of calf thymus tissue is shown in Table 3–1.

The five histone classes differ primarily in the relative amounts of lysine and arginine; histone H1 is lysine-rich histone, H2A and H2B are moderately lysine-rich histones, and H3 and H4 are arginine-rich histones. These five histones are nearly universal among eukaryotes. The association of the histones with DNA appears to account to a large extent for the basic structure of chromatin. This structure is revealed in electron micrographs, which show chromatin as a series of repeating subunits resembling a "string of beads" (Fig. 3.4). The "beads" are spherical structures called **nucleosomes.** They are approximately 10 nm in diameter, and the "string" is a thin 1.5- to 2.5-nm filament of DNA. When nucleosomes are treated with the proteolytic enzyme trypsin, their structure unfolds, indicating that they are held together by proteins (Sahasrabuddhe and Van Holde, 1974). When the proteins of the beaded subunits are analyzed, it is found that they are composed of an octamer of histones with two molecules each of histones H2A, H2B, H3, and H4 (Thomas and Kornberg, 1975).

One of the most valuable techniques for clarifying the relationship of DNA to the histones is treatment of chromatin with certain nuclease enzymes, such as micrococcal nuclease. With brief digestion, DNA fragments are produced that are approximately 200 base pairs in length. With further digestion, the enzyme cleaves the DNA fragments from their ends and progressively reduces their length. A transitory impediment to digestion is reached at 166 base pairs. This particle is subject to further

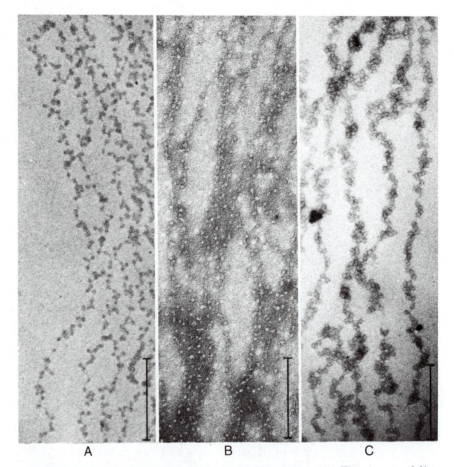

FIGURE 3.4 Chromatin fibers spilling out of ruptured nuclei. The degree of fiber swelling and the proximity of individual nucleosomes to each other varies within different regions of a single nucleus. Scale bars, 0.2 μm. *A*, Rat thymus chromatin, positively stained. *B*, Rat thymus chromatin, negatively stained. *C*, Chicken erythrocyte chromatin, negatively stained as in *B*. Connecting strands are most easily seen in *B*. (From A.L. Olins and D.E. Olins. 1974. Spheroid chromatin units [ν bodies]. Science, *183:* 330–331. Copyright 1974 by the American Association for the Advancement of Science.)

degradation until a length of 146 base pairs is reached. The products of these digestion stages differ not only in DNA length but in their histone composition. The 200 base pair and 166 base pair fragments retain the octamer of histones as described previously, plus one molecule of the fifth histone, H1. However, digestion beyond the 166 base pair length releases histone H1 from the DNA (Noll and Kornberg, 1977). The release of histone H1 upon digestion of the 166 base pair intermediate particle suggests that histone H1 is associated with the DNA that is released during digestion. We shall return subsequently to the relationship between histone H1 and DNA.

The 200 base pair-long fragment is a **complete nucleosome.** The DNA that is lost during digestion is called the **linker,** which is the strand that joins adjacent "beads"; it can vary in length among different cell types and between species. The set of eight histone molecules and its associated 146 base pairs of DNA compose the **core particle** (Kornberg, 1977; Lutter, 1979; Prunell et al., 1979). The length of DNA associated with core particles does not vary significantly among species. Experiments designed to determine the relationship between the DNA and the histone octamers have demonstrated that the continuous thread of DNA is periodically coiled into a helix that wraps around the *outside* of the octamers (Baldwin et al., 1975). Since the DNA molecule itself is a helix, this coiled configuration of DNA is called a **superhelix.** The DNA makes 1¾ turns around the protein particle. The histone octamer itself is a wedge-shaped plate; the helical path of the DNA follows surface contours on the particle. Thus, the octamer determines the architecture of the nucleosome (Klug et al., 1980). The nucleosomes form a repeat pattern in chromatin, with linkers alternating with cores. This alternating pattern gives chromatin the string-and-bead appearance.

The intermediate-sized 166 base pair fragments provide clues to the role of histone H1 in chromatin. These fragments contain two full superhelical turns of DNA, slightly longer than the 1¾ turns in the core particle itself. Thus, the two-turn superhelix is completed in the region flanking the core particle. The DNA that completes the two turns constitutes two 10 base pair–long strands that project from each side of the core particle. Histone H1 is associated with both these strands, binding them at a common point (Simpson, 1978; Thoma et al., 1979; Thoma and Koller, 1981). A model of the 166 base pair fragment is shown in Figure 3.5*A*. Examination of the model reveals that the presence of histone H1 compacts the DNA. Figure 3.5*B* shows that DNA would be extended in the absence of histone H1. According to this model, the role of histone H1 is to compact DNA. Evidence in favor of this function for histone H1 is provided by electron microscopic examination of chromatin with and without histone H1. In the presence of histone H1, chromatin is in a compact configuration in which neighboring nucleosomes are close together (Fig. 3.6*A*). In the absence of histone H1, chromatin assumes the extended "beads on a string" appearance with long internucleosome stretches of DNA (Fig. 3.6*B*).

The 10-nm chromatin fibers we have discussed so far are not seen in electron micrographs of sectioned nuclei. Instead, the micrographs reveal 20- to 30-nm chromatin fibers (Ris and Kubai, 1970). Thus, the 20- to 30-nm fibers—and not the 10-nm fibers—appear to represent the native configuration of chromatin in nuclei. The larger diameter fibers are now known to be formed by the folding of the 10-nm configuration. This folding can be simulated *in vitro* by subjecting 10-nm fibers to high ionic strength solution. It is dependent upon the presence of histone H1 and is thought to occur by the polymerization of histone H1 molecules that are associated with nucleosomes, which cross-link the nucleosomes

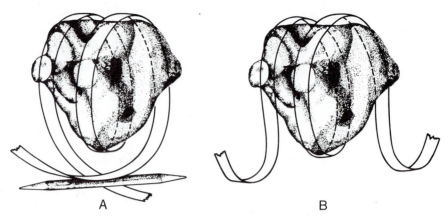

A   B

FIGURE 3.5 The relationship of histone H1 to the nucleosome. *A,* In the presence of histone H1, DNA completes a two-turn superhelix around the wedge-shaped histone octamer. Histone H1 appears to link the strands projecting from each end of the core particle, thus sealing the two-turn superhelix. *B,* In the absence of histone H1, the superhelix completes only 1¾ turns, and DNA is extended from the core particle. The histone octamers represented here are based upon the model of Klug et al. (1980). The actual structure of histone H1 is not known. However, it has been drawn here as an elongate structure with two free ends. These ends can interact with other histone H1 molecules to cross-link nucleosomes and, thus, compact the chromatin. The nucleosome core particle model is reprinted by permission from *Nature,* Vol. 287, No. 5782, pp. 509–516. Copyright © 1980 Macmillan Journals Limited.

and thus form the higher-order structure (Thoma et al., 1979; Thomas and Khabaza, 1980).

The extent of nucleosome compaction undoubtedly varies during the cell cycle and is highest during metaphase and lowest during interphase. Furthermore, interphase chromatin probably has variable regions

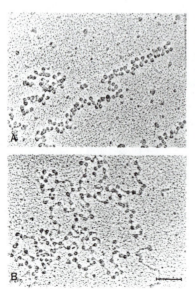

FIGURE 3.6 Effects of histone H1 on the structure of chromatin. *A,* Rat chromatin with histone H1 present has nucleosomes that lie close to each other. *B,* In the absence of histone H1, the distance between nucleosomes is increased. Scale bar equals 0.1 μm. (From F. Thoma, Th. Koller, and A. Klug. 1979. Reproduced from *The Journal of Cell Biology,* 1979, vol. 83, pp. 403–427 by copyright permission of The Rockefeller University Press.)

of nucleosome compaction, with heterochromatin being more tightly compacted than euchromatin. The regulation of nucleosome compaction, therefore, must play an important role in modulating chromatin function. Perhaps histone modification or the nonhistone chromosomal protein content of chromatin is involved in this regulation.

The functional implications of the nucleosome configuration are not well understood. However, since the nucleosome is the basic structural unit of chromatin, the processes of replication and transcription must be affected by nucleosomes, and any models of chromatin function must be compatible with this structural organization. In fact, the nucleosome structure of transcribed chromatin is modified as compared to transcriptionally inactive chromatin. This topic will be discussed in detail in Chapter 4.

The other major class of proteins in chromatin is the nonhistone chromosomal (NHC) proteins, which fall into three categories: (1) **enzymes** such as RNA polymerase, DNA polymerase, and nucleases; (2) **structural proteins** that may modify the packaging of chromatin; and (3) **regulatory proteins** that affect the expression of specific genes. The NHC proteins are a more heterogeneous group of molecules than the histones, and thus, their characterization is more difficult. Some of the NHC proteins appear to vary from species to species and even from tissue to tissue. These variable proteins could be involved in regulating the expression of genes. This possible function of nonhistone proteins is discussed in detail in Chapter 4. NHC proteins also differ from histones in their rapid rate of turnover. Whereas histones remain associated with DNA for long periods of time with a low turnover rate, the NHC proteins are in a state of flux. Modulations in the relative amounts of specific NHC proteins will in turn modify the composition of chromatin.

The RNA associated with chromatin falls into two categories. One is **nascent RNA.** This minor chromatin constituent consists primarily of transcripts prior to their release from DNA templates. The second category includes so-called **small nuclear RNAs (snRNAs),** some of which associate with chromatin during prophase and anaphase of mitosis (Goldstein et al., 1977).

3–2. ORGANIZATION OF DNA IN CHROMATIN

A great deal of effort has been expended by molecular biologists in recent years in formulating detailed characterizations of the eukaryotic genome. This rather tedious chore is considered by most molecular biologists to be a necessary first step in understanding the function of the eukaryotic genome. This expectation results in large part from the findings that regulatory control of bacterial and bacteriophage genomes is facilitated by the structure of the genome itself. For example, genes for proteins that function together may be organized in a cluster. Regulatory control of transcription in some gene clusters is mediated by specialized nucleotide sequences adjacent to them. Although certain differences in the basic structure and function of prokaryotic and eukaryotic genomes indicate that the regulatory process is different in eukaryotes, the search for regulatory control elements within the genome remains a prime objective of contemporary molecular biology.

When the genomes of eukaryotic organisms are compared to those of prokaryotes, one of the most obvious differences is the incredible size discrepancy (Lewin, 1980). For example, the human haploid genome consists of approximately 3.3×10^9 base pairs, compared to 4.2×10^6 base pairs in the bacterium *Escherichia coli*. This difference of nearly 1000 times is generally considered to be much in excess of the actual difference in informational content that is necessary to produce even a complex organism such as man. A conservative estimate of the number of genes in the human genome is 50,000. However, there is enough DNA in the genome to form almost two million genes. Apparently, eukaryotes contain a vast amount of DNA that does not fit the classic concept of a gene. This conclusion is also borne out by comparisons of the amount of DNA among eukaryotes.

It is generally believed that evolution is accompanied by an increase in genetic complexity; that is, more complexity is necessary to produce a man than an amoeba. If the minimum haploid amounts of DNA (usually referred to as **C-values**) found in each taxonomic group are compared, there is a good correlation between evolutionary complexity and genome size (Fig. 3.7). However, *within any group* there is a tremendous amount of variation in C-values. As shown in Figure 3.8, most classes show a variation in genome size of approximately tenfold. The Amphibia have

Minimum DNA content reported for any species in category, mammalian DNA content (%)

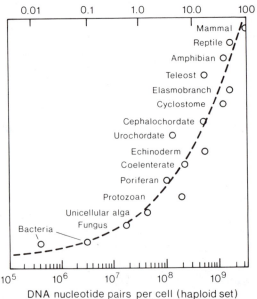

FIGURE 3.7 Minimum content of haploid genome in each class of species. DNA content increases with complexity, but other members of each class may have appreciably greater DNA contents, as shown in Figure 3.8. (After R.J. Britten and E.H. Davidson. 1969. Gene regulation for higher cells: A theory. Science, *165:* 349–357. Copyright 1969 by the American Association for the Advancement of Science.)

particularly variable amounts of DNA, with a nearly hundredfold difference. Some amphibia have as much as 25 times more DNA than man. There is no reason to assume that 25 times more genes are necessary to produce an amphibian than a man. The extra DNA is presumably not

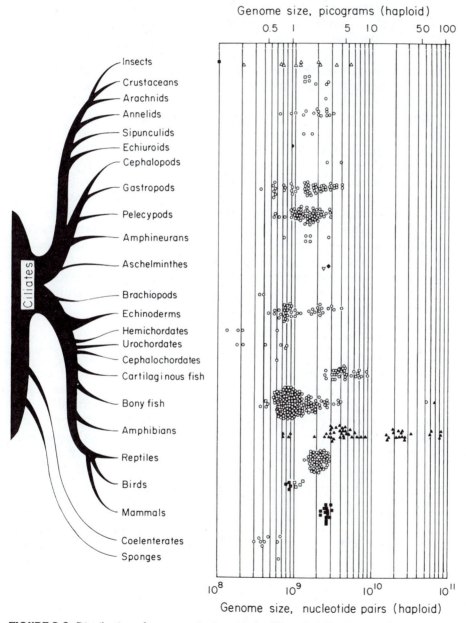

FIGURE 3.8 Distribution of genome size in animals. (From R.J. Britten and E.H. Davidson. 1971. Repetitive and non-repetitive DNA sequences and a speculation on the origins of evolutionary novelty. Q. Rev. Biol., *46:* 116.)

functioning as genes in the classical sense. The eukaryotic genome is obviously highly complex, with several kinds of DNA.

The basic organization of eukaryotic genomes has been determined largely by the technique of DNA **reassociation, reannealing,** or **renaturation.** This technique takes advantage of the complementary relationships between nucleotide pairs on the two strands of the DNA double helix. When the strands are separated (**denatured**) by experimental means, they will recognize one another and reassociate under appropriate experimental conditions. Denaturation of DNA is accomplished by heating a solution of short segments of purified DNA in a salt solution. The short segments are produced by shearing the DNA with ultrasound or high pressure. Denaturation generally occurs at 80 to 90° C as a result of the disruption of the hydrogen bonds that join the two strands. The midpoint of this transition is T_m, or the **melting temperature** of DNA. When a solution of denatured DNA is subsequently cooled, renaturation or reannealing occurs at about 20 to 25 C° below the T_m.

Annealing between two strands of DNA depends upon the complementarity between bases of the two strands. Although denatured DNA of a single species will renature readily, denatured DNA of two species will renature only if they have nucleotide sequences in common, and the extent of renaturation depends upon the degree of similarity between nucleotide sequences. Renaturation results from a random collision between two complementary strands. Thus, the **concentration of the DNA** and the **duration of the reaction** must be sufficient to allow the matching to occur. The renaturation reaction is controlled by the product of these two factors, and this is expressed as **Cot.**

Cot = moles of nucleotides × seconds/liter

Cot values vary over several orders of magnitude, and therefore, Cot is expressed by a logarithmic scale. An ideal Cot curve is shown in Figure 3.9, in which the percent of reassociated DNA is plotted against Cot. The reassociation properties of a given type of DNA are described by the Cot at 50% reassociation. This is called **Cot½.** Since reassociation is the result of random collisions between single strands of DNA, the reaction rate is inversely proportional to the *number* of different sequences pres-

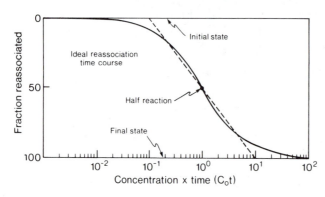

FIGURE 3.9 Ideal Cot curve. The percent reassociation is plotted against the product of total concentration and time on a logarithmic scale. (After R.J. Britten and D.E. Kohne. 1968. Repeated sequences in DNA. Science, *161*: 529–540. Copyright 1968 by the American Association for the Advancement of Science.)

ent. Thus, the larger the size of the genome, the more difficult it is for complementary strands to find one another, since each sequence is diluted by all the others. Renaturation will therefore occur more readily with a small genome than with a large genome. For example, the *E. coli* genome contains 4.2×10^6 nucleotide pairs, and the MS-2 virus genome contains 4×10^3 nucleotide pairs. Therefore, the concentration of any particular sequence is reduced a thousandfold in *E. coli* as compared to MS-2, requiring a Cot½ approximately 10^3 greater (Fig. 3.10).

Cot½ also indicates the combined length of *different* sequences present in a genome. This measure is called the **complexity** of the genome, which is usually expressed in nucleotide pairs. For example, if a genome consists of sequences A, B, and C, the complexity is the sum of the different sequences present; that is, A + B + C. If each of the preceding sequences were 10^2 base pairs in length, the complexity would be 3×10^2. For *E. coli* every sequence is represented once in the genome. Therefore, its complexity is identical with its genome size; that is, 4.2 $\times 10^6$ nucleotide pairs.

When reassociation of two genomes is compared, each should have a Cot½ proportional to its complexity. *E. coli* is usually the standard used for such comparisons. Thus, the complexity of MS-2 (4×10^3 nucleotide pairs) is considerably less than that of *E. coli*. However, such straightforward comparisons are not possible with advanced eukaryotic organisms. We shall use the mammalian genome as an example. Since, as we discussed previously, the mammalian genome is considerably larger than that of *E. coli*, it should be more complex and have a Cot½ commensurate with its complexity. For example, the calf haploid genome contains 3.2×10^9 nucleotide pairs. Since the Cot½ of *E. coli* DNA is approximately 10, the Cot½ of calf DNA should be approximately 10,000. However, eukaryotic DNA reanneals in a more complicated way than does prokaryotic DNA.

As shown in Figure 3.11, a portion of calf thymus DNA is very rapidly reassociating, with a Cot½ of 0.03 and low complexity. This

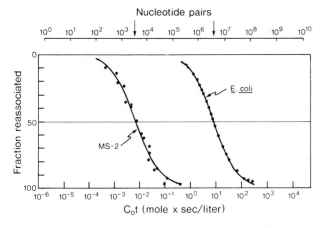

FIGURE 3.10 Reassociation of double stranded nucleic acids from MS-2 virus and *E. coli*. (After R.J. Britten and D.E. Kohne. 1968. Repeated sequences in DNA. Science, *161*: 529–540. Copyright 1968 by the American Association for the Advancement of Science.)

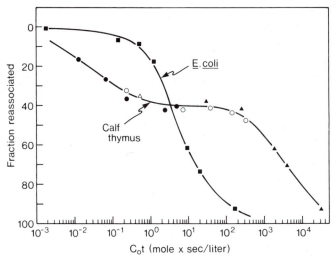

FIGURE 3.11 Reassociation of calf thymus DNA compared to *E. coli* DNA. (After R.J. Britten and D.E. Kohne. 1968. Repeated sequences in DNA. Science, *161:* 529–540. Copyright 1968 by the American Association for the Advancement of Science.)

rapid renaturation indicates that a portion of the DNA is present in numerous copies. Repetition of nucleotide sequences increases their concentration and consequently causes more rapid reannealing. There is a clear separation between this DNA and a slow annealing, high complexity fraction. The latter fraction has a $Cot\frac{1}{2}$ of 3×10^3, which is close to the predicted value for calf DNA present in single copies. Since the reassociation of the rapidly renaturing DNA is 100,000 times as rapid as the single copy DNA, the nucleotide sequences in the repetitive DNA are repeated on the average of 100,000 times. There is an extremely broad spectrum of sequence repetition in eukaryotic DNA. This is illustrated very clearly by the Cot curve for mouse DNA (Fig. 3.12). Approximately 70% of the DNA has a $Cot\frac{1}{2}$ value corresponding to high complexity, unique sequence DNA. An intermediate component, which corresponds to about 15% of the DNA, apparently contains sequences repeated 1,000 to 100,000 times, whereas a rapidly annealing component that constitutes 10% of the genome seems to consist of sequences repeated 1,000,000 times.

The presence of both single copy DNA and sequences with variable levels of repetition is a universal property of eukaryotic genomes—including both plants and animals. There is considerable variation among organisms as to the relative amounts of the different categories of DNA; that is, single copy, moderately repetitive, and highly repetitive. However, there are some generalizations that can be made about the functions that these portions of the genome perform in cell metabolism.

The high complexity—or single copy—DNA contains most of the structural genes; that is, the genes that code for messenger RNA. Although it is unclear whether all unique sequences are structural genes, it can be said with a great deal of certainty that most structural genes are present as single copies. This conclusion has important repercussions for regulation of gene expression, since it means that if a cell becomes

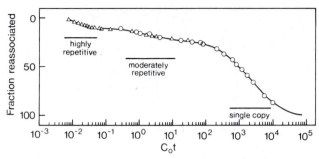

FIGURE 3.12 Reassociation of mouse DNA. (After B.L. McConaughy and B.J. McCarthy. 1970. Related base sequences in the DNA of simple and complex organisms. VI. The extent of base sequence divergence among the DNAs of various rodents. Biochem. Genet., 4: 425–446. Reprinted with permission of Plenum Publishing Corporation.)

specialized for the production of a protein, it can rely upon only a single template (two in a diploid cell) for production of the messenger RNA for that protein.

Moderately repetitive DNA is composed of two categories of nucleotide sequences: (1) identifiable genes that occupy a limited number of chromosomal sites (histone genes, ribosomal RNA genes, and transfer RNA genes) and (2) sequences that do not fit the classic concept of genes and are scattered throughout the genome. Highly repetitive DNA consists of short, repetitive nucleotide sequences that do not code for protein. This category of DNA is usually called **satellite DNA.** We shall discuss these various functional categories subsequently.

Structural Genes

One of the spectacular achievements of molecular biology is the development of techniques for characterization of structural genes, most of which are present only once in the haploid genome. A unique sequence is like the proverbial needle in a haystack. It may be as short as 1000 nucleotides long embedded in a genome containing more than 10^9 nucleotides. The information in the molecular biology literature on structural genes has mushroomed since the early 1970s with the development of techniques for preparing **molecular probes** to locate and quantify genes and for isolating and analyzing the genes themselves (see the following box). Molecular probes are molecules that can seek out and identify a specific nucleotide sequence from DNA. One of the most common probes is **messenger RNA (mRNA).** Since mRNA is produced as a complementary copy of DNA, it can be annealed with single-stranded DNA to form an RNA-DNA hybrid. The formation of hybrids is dependent upon a random association of RNA-DNA fragments. Thus, like DNA-DNA reassociation, hybridization can be characterized by the time required to form an RNA-DNA duplex and by the concentration of the reactants to determine whether the mRNA hybridizes with unique or repetitive DNA. A substantial fraction of total mRNA has been found to hybridize almost exclusively with nonrepetitive DNA. Hence, most of the mRNA molecules consist of transcripts from this type of DNA (see, e.g., Goldberg et al., 1973).

Regular text continued on page 102.

TECHNIQUES IN MOLECULAR BIOLOGY

Knowledge of the techniques used in molecular biology research is a necessity for understanding the principles of this exciting field of study. These principles have been garnered from experiments that utilize elegant, but simple, technology. We shall briefly review some of the most important techniques and instrumentation used in molecular biology research.

ULTRACENTRIFUGATION.   Ultracentrifuges are among the most valuable instruments used by molecular biologists. Essentially, very high centrifugal force is generated to move particles in a centrifuge tube, and the distribution of particles in the tube is measured either while the particles are being displaced or after particle distribution has reached equilibrium. These data yield information about the physical properties of molecules, such as shape, molecular weight, and density. Furthermore, since particles of different sizes sediment at different rates under centrifugal force, ultracentrifugation can be used to isolate particles from a mixture. There are two types of ultracentrifuge in general use—analytical and preparative.

The Analytical Ultracentrifuge.   These elaborate instruments are equipped with an optical system to determine particle concentration while centrifugation is in progress. The very high speeds achieved by ultracentrifuge rotors (up to 70,000 rpm) are made possible by the creation of a vacuum in the rotor chamber, which reduces friction on the rotor. The chamber is also refrigerated to protect samples during centrifuge runs, which may take several hours. A schematic diagram of an ultracentrifuge is shown in the following drawing.

One type of analysis that can be conducted with this instrument is **velocity centrifugation.** This technique is used to determine the **sedimentation coefficient** of a particle. This value is expressed in **Svedberg (S)** units and is related to the molecular weight and shape of the particle. Macromolecules—particularly RNA molecules—are frequently described by their S value: 18S, 28S RNA. S value is determined by centrifuging highly purified molecules at a constant speed in a liquid that has a uniform density lower than that of the test molecules. The molecules sediment through the liquid, and the rate of sedimentation is observed through the optical system by detecting changes in their concentration along the length of the tube. The observed sedimentation velocity is then used to calculate the sedimentation coefficient.

Analytical ultracentrifuges are also used frequently for analysis of the **buoyant density** of molecules. This involves centrifuging the molecules in a solution with a density gradient (i.e., the density increases from the top to the bottom of the centrifuge tube). Centrifugal force causes the molecules to sediment in the tube. This continues until the molecules reach the level where the buoyant density of the solution counteracts the centrifugal force, causing the molecules to remain suspended. The molecules are said to have reached **sedimentation equilibrium.**

This kind of analysis is especially valuable for the study of DNA. The buoyant density of DNA is dependent on the relative content of guanine and cytosine and on

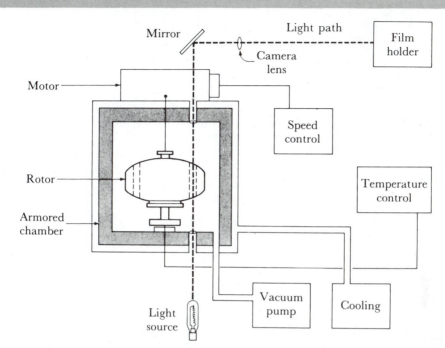

A Beckman analytic ultracentrifuge. The rotor is in an evacuated and cooled chamber and is suspended on a wire coming from the drive shaft of the motor. The tip of the rotor contains a thermistor for measuring temperature. Electrical contact of the thermistor to the control circuit is by means of a pool of mercury, which the rotor tip touches. The rotor chamber contains an upper and a lower lens. The lower lens collimates the light so that the sample cell is illuminated by parallel light. The upper lens and the camera lens focus the light on the film. (From *Physical Biochemistry*, Second Edition, by David Freifelder. W.H. Freeman and Company. Copyright © 1982.)

whether it is single stranded or double stranded. Buoyant density may be determined by centrifuging DNA in a solution of cesium chloride. When the solution is centrifuged, the cesium ions are so heavy that they are displaced by the centrifugal force, and a cesium chloride gradient is formed. The different concentrations of cesium along the length of the tube produce a density gradient. Consequently, the DNA will concentrate as a band in the gradient at a position where its density matches that of the cesium salt. The location of the DNA in the gradient is determined by illuminating the centrifuge tube with ultraviolet (UV) light. Nucleic acids absorb UV light—the greater the concentration is, the higher will be the absorbance. Therefore, the level of absorbance gives a quantitative measure of the amount of DNA. A photograph is made of the centrifuge tube with UV illumination to provide a permanent record of the relative position of DNA and its absorbance.

A representative cesium chloride gradient analysis of a DNA mixture is shown in the following diagram. The cesium chloride density in the tube is represented by stippling. The DNA bands are seen when the tube is illuminated with UV light at a wavelength of 260 nm. The intensity of absorbance is shown on the graph, as are the corresponding buoyant densities of the three peaks.

The Preparative Ultracentrifuge. Measurements are made of particle distribution *during* analytical ultracentrifugation, but preparative ultracentrifuges are used to determine the distribution of the separated components in the centrifuge tube *after* centrifugation. The preparative instrument is much simpler than an analytical ultracentrifuge. It is constructed very much like an ordinary laboratory centrifuge, except that it is equipped with a vacuum pump and refrigeration.

The preparative ultracentrifuge is often used to isolate cell constituents for further analysis. The isolation procedure takes advantage of the principle that different levels of centrifugal force are required to displace various cellular components to the bottom of a centrifuge tube. Only the ultracentrifuge can produce the very high centrifugal forces that are needed to isolate very small cell organelles. These **cell fractionation** procedures begin by breaking the cells with either osmotic shock, ultrasound, high pressure, grinding, or detergents. This complex mixture can then be separated into its individual components by **differential centrifugation.**

Large components such as nuclei and unbroken cells are removed by centrifuging the mixture at relatively low speeds (less than $1000 \times$ gravity). The sediment at the bottom of the centrifuge tube is called the **pellet.** It may be discarded at this

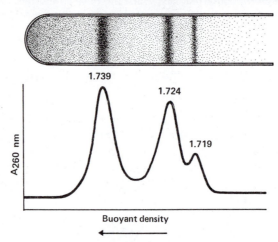

Diagram of a continuous gradient of cesium chloride (CsCl) showing the position of three bands with the corresponding buoyant density expressed in grams per cm^2. Since these bands correspond to nucleic acid molecules the concentration of the molecules is measured by absorbancy at 260 nm ($A_{260\ nm}$). (From E.D.P. De Robertis and E.M.F. De Robertis, Jr. 1980. *Cell and Molecular Biology*, 7th ed. Saunders College Publishing, Philadelphia, p. 54.)

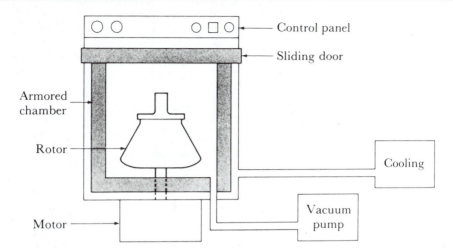

A Beckman preparative ultracentrifuge. (From *Physical Biochemistry*, Second Edition, by David Freifelder. W.H. Freeman and Company. Copyright © 1982.)

stage, or it may be purified further for the analysis of isolated nuclei. The liquid containing the material that did not sediment is the **supernatant.** By centrifuging the supernatant at even higher speeds, a new pellet and supernatant are obtained. This procedure can be repeated several times, progressively pelleting smaller and less dense components at each step. Ribosomes are among the smallest cell organelles isolated by differential centrifugation. They remain in the supernatant during centrifugation at low and moderate speeds and pellet only with very high centrifugal forces (100,000 × gravity). The resultant supernatant contains the soluble portion of the protoplasm and very small particles that do not sediment at that speed. It is called the **postribosomal supernatant.**

   The term *preparative ultracentrifuge* is a slight misnomer, since this instrument can also be used for experimental analysis of organelles or macromolecules. This is usually accomplished by centrifuging the sample in a preformed density gradient. The most commonly used material for constructing a gradient is sucrose, with the sugar less concentrated at the top of the tube than it is at the bottom. The sample is layered at the top of the gradient, the tube is centrifuged, and the components of a complex sample are sedimented in layers of the gradient. They will remain in the layers after spinning, and the components of each layer can then be analyzed separately by punching a hole in the bottom of the tube and collecting the material drop by drop. The drops are combined into fractions containing a set volume of sucrose solution.

   Sucrose gradients provide an excellent means for analyzing synthesis of the different classes of RNA. Cells are allowed to incorporate ^{3}H-uridine into RNA, and the RNA is extracted, placed on a sucrose gradient, and centrifuged. When the material is collected from the tube after centrifugation, the amount of RNA in each

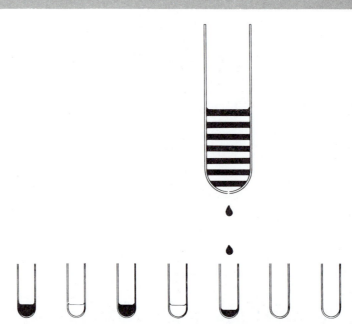

Fractionating the contents of a centrifuge tube by drop collection. The bottom of the tube is pierced with a needle. As long as the system is stabilized against convection (e.g., by a concentration gradient), the drops represent successive layers of liquid. These layers are shown schematically as alternating black with white. (From *Physical Biochemistry,* Second Edition, by David Freifelder. W.H. Freeman and Company. Copyright © 1982.)

drop is determined by measuring UV absorbance. The absorbance is then plotted on a graph according to fraction number (the bottom of the tube is fraction 1). The RNA is distributed in definite peaks (e.g., 4S, 18S, 28S). The radioactivity in each peak can then be determined by analysis with a **scintillation counter.** The radioactivity (in counts per minute or cpm) is plotted on the same graph as absorbance to obtain a radioactivity profile. A comparison of these two curves indicates whether synthesis of particular classes of RNA has occurred during the labeling period (p. 93).

ELECTROPHORESIS. Electrophoresis is another extremely valuable technique used to separate complex mixtures of macromolecules. Most macromolecules are electrically charged and will, therefore, migrate in an electrical field; this migration is electrophoresis. The molecules are placed in an electrical field where they will migrate to either the positive electrode (**anode**) or the negative electrode (**cathode**), that is, negatively charged molecules to the anode and positively charged molecules to the cathode. The greater the charge is, the more rapid the migration will be. Mobility is also affected by other physical properties of the molecules—large molecules move more slowly than small ones, and certain shapes migrate more readily than others. In a complex mixture of macromolecules, the combination of these different variables causes molecules to migrate at different rates and results in a sorting out of different components. Electrophoresis is quite versatile and can be

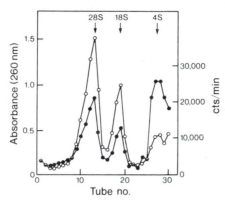

Sedimentation pattern of RNA from *Xenopus* embryo. The line with open circles represents $A_{260\ nm}$; the line with solid circles represents radioactivity (Redrawn with permission from D.D. Brown and E. Littna. 1964. RNA synthesis during the development of *Xenopus laevis,* the South African clawed toad. Journal of Molecular Biology *8:* 669–687. Copyright: Academic Press Inc. [London] Ltd.)

adapted for a number of different separation requirements, including resolution of complex mixtures of proteins or nucleic acids.

The sample containing a solution of macromolecules is normally placed on a semisolid support medium, an electrical field is applied, and the molecules migrate through the medium. This is called **zone electrophoresis,** since the molecules separate and migrate as discrete **zones** or **bands.** The support medium may act as a molecular sieve that assists in separating molecules by differentially impeding their movement. Proteins and nucleic acids are usually separated on a gel composed of either **starch, polyacrylamide,** or **agarose.** The gels are made with buffers that are selected to produce a pH that facilitates molecular separation. Following separation the molecules are localized by staining the gel with a specific reagent, by scanning the gel with ultraviolet light (to localize nucleic acids), or, if the sample is radioactive, by determining the radioactivity distribution in the gel. In actual practice, many variations of the electrophoretic technique are used for separation of different types of macromolecules. A description of specific applications of electrophoresis follows.
Polyacrylamide Gel Electrophoresis of Nucleic Acids. Separation of nucleic acids is mainly achieved by molecular sieving, since the charge-to-mass ratio is nearly the same for all nucleic acids. This means that small molecules will migrate faster than large ones. Either RNA molecules or DNA fragments can be separated in this way. The following figure compares the separation of RNA by sucrose gradients with polyacrylamide electrophoretic separation.

Polyacrylamide gel electrophoresis has gained new prominence in recent years because it is a key element of the new techniques for rapid sequencing of nucleic acids. We shall discuss this application of electrophoresis on page 100.
Starch Gel Electrophoresis of Proteins. The use of starch gels for protein separation preceded that of polyacrylamide. Following solidification of the gel, slots are cut to

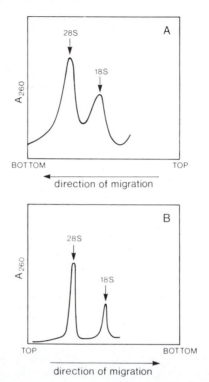

Separation of ribosomal RNA by (*A*) sucrose gradients and by (*B*) polyacrylamide gels. Note that in *A* the 28S species is displaced toward the bottom of the tube, while in *B* the 28S RNA remains near the top.

hold the protein sample. After sample application the voltage is applied, and the proteins migrate as bands through the gel. After electrophoresis the gel is stained to locate the proteins. Figure 4.10 shows an example of starch gel separation of glucose-6-phosphate dehydrogenase enzymes. Localization of the enzyme bands is accomplished by coupling an enzyme reaction to a reaction that deposits a stain in the gel where the enzyme is located.

Polyacrylamide Disc Gel Electrophoresis of Proteins. This is a high-resolution separation technique. "Disc" refers to the fact that different regions of the gel are made with buffers that vary in pH and ionic strength. The pH *discontinuity* increases the sharpness of the protein bands, thus improving resolution. The gel shown here is a disc gel separation of lactate dehydrogenase (LDH) enzymes from a six-day rabbit embryo.

Isoelectric Focusing of Proteins. This technique is based upon the fact that a protein will not migrate in an electrical field at the pH at which it is electrically neutral. This pH is called the **isoelectric point (pI)**. Since the isoelectric points of proteins differ from one another, proteins can be separated by electrophoresing a mixture of proteins in a stable pH gradient, with the lowest pH at the anode and the highest pH at

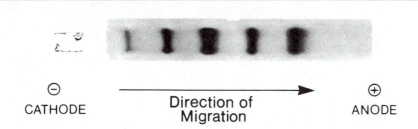

CATHODE   Direction of Migration   ANODE

Disc-gel separation of lactate dehydrogenase (LDH) enzymes from a six-day rabbit embryo.

the cathode. If the proteins are applied to the gel at the position in the gradient where the pH is 7, those with a pI below 7 bear a net negative charge and migrate toward the anode. Conversely, those with a pI above 7 bear a net positive charge and migrate toward the cathode. The proteins continue migrating until they reach a pH in the gradient that corresponds to their pI. They then lose electrophoretic mobility and become focused in a narrow zone on the gel.

SDS-Gel Electrophoresis. This procedure is used to determine the molecular weights of proteins. Electrophoresis is done in polyacrylamide gels containing the detergent sodium dodecyl sulfate (SDS). Before electrophoresis, the protein mixture is treated with SDS and mercaptoethanol (or dithiothreitol). Either of the latter reagents reduces disulfide bonds, causing polypeptide subunits to dissociate and erasing secondary structure. The SDS binds to the polypeptides, forming complexes that migrate as though they have uniform shapes. The charge on the complexes is determined solely by the SDS. Since all polypeptides treated in this way have similar

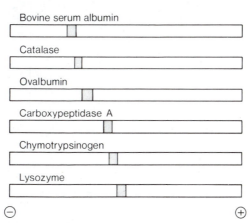

Bovine serum albumin

Catalase

Ovalbumin

Carboxypeptidase A

Chymotrypsinogen

Lysozyme

Drawings of SDS-polyacrylamide gels stained with Coomassie blue to show the relative migrations of polypeptide chains of a number of proteins. (After A.L. Shapiro, E. Viñuela, and J.V. Maizel, Jr. 1967. Molecular weight estimation of polypeptide chains by electrophoresis in SDS-polyacrylamide gels. Biochem. Biophys. Res. Comm., *28:* 817.)

charge-to-mass ratios, and shape differences are eliminated, the only factor that influences migration is molecular weight—due to molecular sieving. A plot of the distance migrated versus log molecular weight gives a straight line. The molecular weight of any protein can be calculated by subjecting it and two proteins of known molecular weight to electrophoresis, plotting molecular weight versus distance migrated for the knowns, and extrapolating the molecular weight for the unknown.

Two-Dimensional Gel Electrophoresis of Proteins. When a mixture of proteins is separated in one-dimensional gel electrophoresis, each band on the gel may correspond to a mixture of proteins that migrate the same distance under a single set of conditions. The resolution of electrophoresis can be increased tremendously if the separated proteins are electrophoresed subsequently under a different set of conditions at a direction that is perpendicular to the first dimension. The most commonly used procedure for two-dimensional electrophoresis of proteins is one developed by O'Farrell (1975).

In the O'Farrell procedure, proteins are separated initially by isoelectric focusing. The linear gel is then placed across the top of a slab gel where it is subjected to SDS electrophoresis. In the second dimension, the components in each zone on the isoelectric focusing gel migrate into the slab gel and separate according to molecular weight. Two-dimensional electrophoresis is a powerful technique for resolving individual polypeptides in a large, heterogeneous protein mixture. An example of the kind of results obtained with this technique is shown in Figure 12.10.

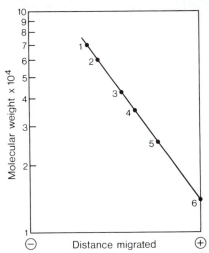

Typical semilogarithmic plot of molecular weight vs. distance migrated in an SDS-polyacrylamide gel. The proteins are: (1) bovine serum albumin; (2) catalase; (3) ovalbumin; (4) carboxypeptidase A; (5) chymotrypsinogen; (6) lysozyme. (After *Physical Biochemistry*, Second Edition, by David Freifelder. W.H. Freeman and Company. Copyright © 1982.)

THE NEW GENE TECHNOLOGY.   Recent advances have revolutionized the study of DNA structure. The implications of these advances for developmental biology are becoming very significant as eukaryotic genomes yield to systematic analysis, much as bacterial and bacteriophage genomes were analyzed a few years ago.

1. **Preparation of molecular probes.** Messenger RNA can be obtained from cells by several methods. One common procedure is to isolate the messenger in the process of translation. As we shall discuss later in this chapter, mRNA is concurrently translated by several ribosomes. Thus, the translational complex consists of one mRNA molecule and a number of ribosomes. This complex is called a polysome. Treatment of polysomes with EDTA or puromycin will cause the complex to dissociate, releasing the messenger, which can be separated from the ribosomes. Confirmation of the identity of isolated messengers is normally done by *in vitro* protein synthesis. This technique involves addition of the messenger to a preparation of cytoplasmic components that are necessary for messenger translation. These preparations are called "cell-free protein-synthesizing systems." Common sources of such preparations are reticulocytes or wheat germ. Radioactive amino acids are usually added to cell-free systems so that newly synthesized proteins can be detected by incorporation of radioactive label.

   Another useful probe is prepared from the messenger. This is **cDNA (complementary DNA),** which is a single strand of DNA made by copying mRNA with the enzyme **reverse transcriptase.** This enzyme, which is isolated from RNA tumor viruses, uses RNA to make a complementary DNA copy.

2. **Preparing synthetic genes.** The single-stranded cDNA can be used as a template for the production of a complementary DNA strand, with the assistance of the enzyme DNA polymerase I. After being copied, both strands remain associated and form a double helix, which is a completely synthetic gene. A number of structural genes have been synthesized in this way, including the gene for rabbit globin (Liu et al., 1977) and the gene for chicken ovalbumin (O'Malley et al., 1976). Recently, the chemical synthesis of completely synthetic genes has been achieved. The codon sequence is constructed, based on knowledge of the amino acid sequence of the protein. This approach avoids the necessity to isolate messenger RNA first and then copy it with reverse transcriptase. In addition to the scientific applications of synthetic genes, the medical and industrial uses of these genes could be quite significant. For example, human insulin genes have now been chemically synthesized (Crea et al., 1978; Sung et al., 1979), a feat that may make possible the commercial manufacture of human insulin (see subsequent discussion). Test tube structural genes, which were only a dream a few years ago, are now a reality.

3. **Recombinant DNA.** No single development in molecular biology has caused more excitement or anxiety than the discovery of procedures to clone eukaryotic DNA by inserting it into bacterial plasmids. This technique has made it possible to produce virtually unlimited amounts of specific DNA sequences. **Plasmids** are

small, circular DNA molecules that replicate independently in bacteria, such as the common laboratory organism *E. coli.* The plasmid is a vehicle used to carry the eukaryotic DNA, which is inserted into the plasmid by **enzymatic recombination.** Essentially, this technique involves breaking the circular plasmid DNA and inserting the desired DNA. The latter becomes a "passenger" carried by the plasmid vehicle and is replicated together with the plasmid. Insertion of the passenger DNA was made possible by the discovery of a class of enzymes called **restriction endonucleases.** These are enzymes isolated from microorganisms that cut DNA internally at sequence-specific sites. These "restriction sites" usually consist of four or six base pairs. Restriction sites invariably have **palindromic symmetry;** that is, the base sequence on one strand is the inverse of the other (see the following table).

The names of restriction enzymes are based upon a shorthand notation that identifies the microorganism from which they are derived. For example, *Eco* RI is derived from *E. coli.* Some commonly used restriction enzymes and their restriction sites are shown in the table (Roberts, 1982).

The enzymes listed here produce staggered cuts in the double-stranded DNA, which leave single-stranded ends on the fragment. These ends are "sticky" in the sense that they could anneal with one another. Thus, a single fragment produced in this way could become circular by the annealing of its two complementary sticky ends. However, if two different molecules are treated with the same restriction enzyme, the sticky ends of each can anneal, producing a single, circular recombinant DNA molecule.

The utilization of this technique to produce recombinant plasmids is illustrated in the following drawing. Both the plasmid and the eukaryotic DNA are treated with the same restriction enzyme, producing single-stranded ends in each. The plasmid is now a linear molecule. The two kinds of DNA are combined, and the sticky ends anneal, causing formation of a circular, recombinant molecule.

Enzyme	Sequence[1]
Eco RI	↓ 5' GAATTC 3' 3' CTTAAG 5' ↑
Hind III	↓ 5' AAGCTT 3' 3' TTCGAA 5' ↑
Hpa II	↓ 5' CCGG 3' 3' GGCC 5' ↑

[1]Arrows indicate cleavage sites of the enzyme.

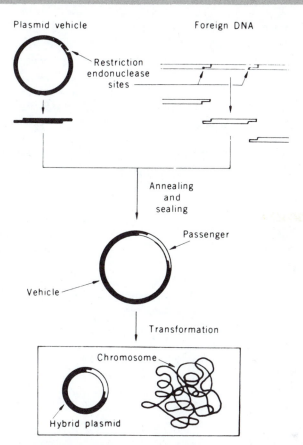

Scheme of events in recombinant DNA experiments. (From J. Abelson. 1977. Recombinant DNA: Examples of present-day research. Science *196:* 159–160. Copyright 1977 by the American Association for the Advancement of Science.)

Another enzyme, DNA ligase, is used to seal gaps remaining in the two strands of the hybrid molecule.

After formation of the hybrid plasmid, it is introduced into a host *E. coli* cell, where it can replicate. This process is called **transformation.** The DNA is added to a culture of bacteria, and it is taken up by a small proportion of the cells. The plasmids are constructed in a way that gives the transformed cells a selective advantage. Thus, the small minority of transformed cells can be selected from the nontransformed cells. One common way to do this is to place a gene for resistance to an antibiotic in the plasmid. Thus, only the transformed bacteria will be able to multiply and form colonies in the presence of the antibiotic. Every cell in a colony contains the same passenger DNA. Replication of this DNA by plasmids results in "cloning" of the DNA. The amplified foreign DNA can be recovered from the plasmids by treating them with a restriction enzyme.

Recombinant plasmids are excellent vehicles for amplification of specific DNA sequences. But one of the most exciting possible uses for transformed bacteria involves their ability to transcribe inserted eukaryotic structural genes and synthesize protein from the transcripts. The possibility of the synthesis of human proteins such as insulin from bacterial "factories" is a potential boon to humankind. Evidence now becoming available indicates that eukaryotic genes can be expressed in *E. coli* containing recombinant plasmids. For example, a protein with the properties of chicken ovalbumin is produced in *E. coli,* transformed by plasmids containing the ovalbumin gene (Mercereau-Puijalon et al., 1978; Fraser and Bruce, 1978). Human insulin also has been produced in *E. coli* containing plasmids constructed with chemically synthesized insulin genes (Goeddel et al., 1979).

4. **Southern blotting.** This technique, which is named for the individual who developed it (Southern, 1975), is one of the most powerful tools used in molecular biology research. It is a method for eluting electrophoretically separated fragments of DNA from polyacrylamide or agarose gels and identifying specific fragments by hybridization to a radioactive probe. The eluted DNA fragments are bound to nitrocellulose paper, which retains the exact pattern of separated fragments and is a suitable medium for the hybridization reactions to take place.

Before transferring the DNA to nitrocellulose, the DNA is denatured (i.e., treated with sodium hydroxide to make it single-stranded) while in the gel. The gel is then placed on filter paper that is soaking in buffer. A sheet of nitrocellulose is placed over the gel, and dry blotting paper is laid over the nitrocellulose. Capillary action draws the buffer through the gel, transferring an exact replica of the DNA pattern in the gel into the nitrocellulose.

A radioactive single-stranded probe (RNA or DNA) for a specific gene is then hybridized to complementary single-stranded DNA fragments on the nitrocellulose paper. Unhybridized probe is washed off, and autoradiography is conducted by placing photographic film next to the nitrocellulose in the dark. As with cytological autoradiography (see p. 69), silver grains are produced when the film emulsion is activated by particles emitted from the radioisotope. Thus, a replica of the band or bands containing the specific DNA is formed on the autoradiograph.

Southern blotting is used when one needs to detect specific DNA fragments from a large, heterogeneous mixture of fragments, as, for example, following restriction enzyme treatment of DNA. The mixture of fragments is resolved into a smear of overlapping bands of fragments of different size by electrophoresis. The band corresponding to a specific gene then can be detected from the smear of overlapping bands by transfer to nitrocellulose, followed by hybridization with a radioactive probe for that gene. An example of such an experiment can be found in Figure 3.13.

5. **DNA sequencing.** We shall use the procedure of Maxam and Gilbert (1977) as our example of this technique.

DNA fragments are produced by treating DNA with restriction enzymes. By the appropriate combination of enzymes, DNA fragments of a length convenient for sequencing (about 100 bases long) are produced. Because of the specificity of the enzymes, the fragments are homogeneous. Next the DNA fragments are labeled at their 5' end with ^{32}P and denatured, and the single strands are separated from one another and sequenced separately.

Single-stranded fragments are divided into four subsamples, which are subjected to different chemical reactions. These reactions cleave the fragments at specific bases. The reactions are controlled so that a single cleavage occurs per molecule, cutting it into one labeled and one unlabeled fragment. For example, the reaction that cleaves at cytosine will cleave some fragments at the first C, some at the second C, and some at the third C, producing a set of labeled fragments of different length, all beginning at the labeled 5' end but terminating at various positions, depending upon the sites of the cytosine bases.

The four sets of fragments are subjected to electrophoresis in parallel, and the fragments separate according to their length. The longest fragments will band at the top of the gel, and fragments with fewer nucleotides will band at progressively lower positions. The bands containing labeled fragments are then located by autoradiography. A dark band in a column identifies a fragment cleaved at a specific base. The gel is read from bottom to top. An example of sequencing is shown in the gel in the figure. The sequence of this fragment is GCGCTCACTGCCCGCTTTCC.

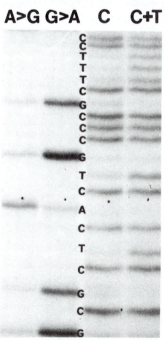

High resolution polyacrylamide gel autoradiograph from a DNA sequencing experiment. (From A.M. Maxam and W. Gilbert. 1977. A new method for sequencing DNA. Proc. Natl. Acad. Sci. U.S.A., 74: 563.)

Regular text continued from page 87

Similar experiments have been done with purified specific messengers as well. Specific messengers are usually obtained from cells that are committed to the predominant synthesis of a single protein. For example, red blood cells synthesize predominantly hemoglobin protein. Consequently, the messenger RNA that is isolated from these cells is relatively pure globin mRNA. A number of specific messengers have been purified from specialized cells. Nearly all of them reassociate with unique sequences (Davidson and Britten, 1973). There is one clear-cut exception, however: the genes that code for histones. Histone messenger hybridizes to DNA at a rate that is considerably faster than expected for single copy sequences. Careful studies of RNA-DNA hybridization have been done with sea urchin histone mRNA and DNA and indicate that the histone genes are repeated several hundred times (Kedes, 1979).

Another powerful molecular probe is synthetic cDNA, which is synthesized *in vitro* from an mRNA template utilizing the enzyme reverse transcriptase (see box, p. 88). The cDNA is identical to that portion of the gene that is transcribed into the mRNA, and therefore, its rate of hybridization with cellular DNA can also be used to determine gene frequency. Experiments of this kind have confirmed that most structural genes are present in the genome as unique sequence DNA.

Since most structural genes are present as single copies, their analysis is difficult: They are present in an incredibly complex mixture of DNA sequences. Recombinant DNA methods have simplified this problem. A specific gene can be isolated and inserted in a plasmid vector, where it is amplified numerous times. Once a sufficient amount of material is available, the amplified gene can be subjected to enzymatic analysis and its nucleotide sequence determined. This approach has proved to be most revealing when the amplified gene is a natural (instead of synthetic) gene excised from the genome, since natural genes retain sequences that may be functionally significant but are not reflected in the nucleotide sequences of their messengers.

Analysis of natural genes has led to the surprising discovery that the protein coding regions of structural genes can be interrupted by noncoding regions. The noncoding regions are called **intervening sequences,** or **introns.** The coding regions are called **exons.** Introns and exons are both transcribed, but the introns are excised from the RNA during maturation of the transcript to messenger RNA (section 3–4). Such messengers are, therefore, formed by splicing together noncontiguous nucleotides.

Strong circumstantial evidence for the presence of internal noncoding regions in genes has been provided by digestion of DNA with restriction endonuclease enzymes (see box, p. 88). These enzymes cleave DNA at sites containing specific nucleotide sequences. If the double-stranded cDNA for the chicken ovalbumin gene is treated with the enzyme *Hind* III, the enzyme has no effect on the gene, as determined by the Southern blotting procedure (this technique is also discussed in the box on p. 88). This means that there are no *Hin*d III restriction sites in the cDNA nucleotide sequence. This led to the expectation that one should be able to obtain fragments of DNA containing complete ovalbumin genes by treating the chicken genome with *Hin*d III. However, when this experiment was done, three fragments of DNA were found to hybridize to an ovalbumin mRNA probe (Fig. 3.13). This result indicates that the genomic ovalbumin gene is fragmented by *Hin*d III, which recognizes internal restriction sites and cleaves the gene. Clearly, the natural gene has

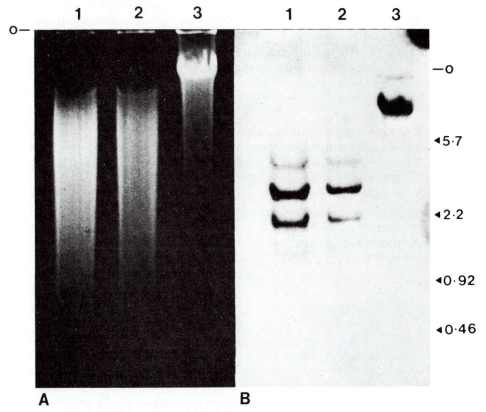

FIGURE 3.13 Digestion of chicken DNA with *Hin*d III. *A*, Agarose gel of DNA stained with ethidium bromide. DNA shows bright against a dark background. Tracks 1 and 2: *Hin*d III-treated DNA. The DNA shows up as a smear of numerous overlapping bands of various sizes. Track 3: unrestricted DNA shows up as high molecular weight fragments. *B*, Autoradiograph of nitrocellulose filter to which DNA from the gel in *A* was transferred and hybridized with ^{32}P-labeled ovalbumin mRNA. Dark bands indicate bands containing ovalbumin coding sequences. Tracks 1 and 2: *Hin*d III-treated DNA. Three discrete fragments of DNA contain ovalbumin coding sequences. Track 3: unrestricted DNA in which the ovalbumin gene is present in a single, high molecular weight fragment. Molecular sizes of DNA fragments in kilobase pairs are indicated by the numbers on the right. (From M.T. Doel, et al. 1977. The presence of ovalbumin mRNA coding sequences in multiple restriction fragments of chicken DNA. Nucleic Acids Res., *4:* 3708.)

internal nucleotide regions that are missing from the cDNA. Since ovalbumin cDNA is a faithful copy of the ovalbumin mRNA, we can conclude that internal regions of the gene are not represented in the messenger. Extensive analysis of the ovalbumin gene with several restriction
enzymes has led to the discovery of seven introns in this gene. This
means that the coding portion of the gene is in eight separate pieces
(Dugaiczyk et al., 1979).

Further evidence for the presence of introns in the ovalbumin gene
was provided by electron microscopic examination of the structures
formed by hybridizing ovalbumin mRNA with cloned ovalbumin genes.
In this technique, the DNA duplex is partially denatured to allow a complementary RNA strand to hybridize to one DNA strand and displace a
single-stranded region of DNA over a comparable distance. As shown in
Figure 3.14, hybridization of the messenger with the gene causes regions
of the gene corresponding to the introns to loop out because they are
lacking in the messenger. All seven introns are clearly visible in this
micrograph.

Introns are now known to be present in a wide variety of eukaryotic
genes. Notable exceptions to this generalization are the histone genes.
These genes, which have been highly conserved in evolution and are
present in large numbers in every genome studied, universally lack introns (Hentschel and Birnstiel, 1981).

A more complete understanding of the organization of structural
genes is provided by DNA-sequencing techniques (see box, p. 88). These
techniques have been used to reveal details about both the internal organization of genes and the regions flanking the genes. Thus, introns
have been precisely localized in genes, and their nucleotide sequences

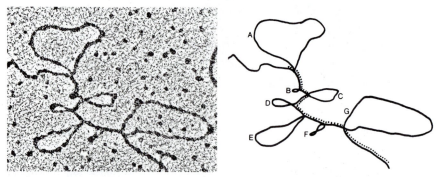

FIGURE 3.14 Visualization of the seven introns of the chicken ovalbumin gene. Left,
Electron micrograph of the structure formed by incubation of partially denatured, cloned
ovalbumin gene with ovalbumin mRNA. Seven loops are formed since the mRNA lacks
complementary nucleotide sequences in those regions. Right, Interpretive drawing. The
dotted line indicates the mRNA hybridized to the DNA (solid line). Loops are lettered
A–G. (From A. Dugaiczyk et al. 1979. The ovalbumin gene: Cloning and molecular
organization of the entire natural gene. Proc. Natl. Acad. Sci. U.S.A., 76: 2256.)

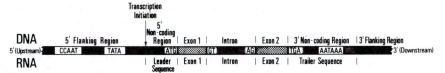

FIGURE 3.15 Eukaryotic structural gene organization.

have been determined. Furthermore, comparisons of the noncoding regions within and adjacent to different genes have led to the discovery of so-called **consensus sequences,** which are sequences that are most frequently found in these regions (for review, see Breathnach and Chambon [1981]). The discovery of related sequences in a wide variety of genes has led to the suspicion that they play important functional roles in gene expression. This suspicion is based upon previous discoveries that RNA codons provide signals for the initiation and termination steps of translation (see section 3–5) and that transcription initiation in prokaryotes is signaled by specific nucleotide sequences.

The organization of a hypothetical eukaryotic structural gene is shown in Figure 3.15. The orientation of genes in such drawings is based on the direction of transcription. Transcription proceeds from left to right. The nucleotide sequence of DNA is always given in terms of the sequence of the RNA it produces and therefore is the sequence of the strand that is complementary to the coding strand itself. This convention is followed since the RNA sequence can be obtained directly from the DNA sequence by substituting uracil (U) for thymine (T). As we shall discuss in section 3–3, the synthesis of RNA begins at the 5' end of the RNA molecule. Thus, the nucleotide sequence of DNA is represented from 5' to 3'. The 5' flanking region, which is involved in transcription initiation, precedes the gene, and the 3' flanking region follows the gene.

The consensus sequence that is farthest upstream from the point of transcription initiation is CCAAT. Closer to the gene is the sequence TATA. The latter is usually called the Goldberg-Hogness box, after the investigators who first recognized it. The site of transcription initiation precedes the translation initiation codon ATG (AUG in the RNA). The portion of the transcript upstream from the first exon is called the **leader sequence.** The gene represented in Figure 3.15 has two exons and one intron. Each structural gene intron begins with the dinucleotide GT and ends with AG. The final exon is followed by a translation termination codon, TGA or TAA (UGA or UAA in RNA). Transcription continues past the translation termination codon until the point of transcription termination is reached. The segment of RNA produced from the region downstream from the last exon is called the **trailer sequence.** It may contain a hexanucleotide sequence AAUAAA (AATAAA in DNA). Downstream from the transcribed portion of the gene is the 3' flanking region.

The consensus sequences we have described are found in many structural genes, but not in all of them. Variations are found in some cases. For example, the Goldberg-Hogness box in some β-globin genes is CATA, rather than TATA (Efstratiadis et al., 1980). Some genes lack certain consensus sequences. For example, histone genes lack the AATAAA hexanucleotide in the 3' noncoding region (Kedes, 1979). We shall discuss the functions of the various consensus sequences in subsequent sections of this chapter.

Even though individual structural genes are present as single copies, many of them are accompanied by closely related sequences that are often clustered in the same chromosomal region. These closely related genes constitute **multigene families.** Families of genes probably arose during evolution by repeated duplication of ancestral genes followed by divergence of nucleotide sequences (Smithies, et al., 1981). These descendants may code for related polypeptides (e.g., β and δ globin—see Chap. 14). Other descendants seem to be evolutionary relics with no apparent function (**pseudogenes**). Pseudogenes have acquired changes in nucleotide sequence that are so deleterious as to render them nonfunctional. The presence of pseudogenes helps to explain a portion of the apparent excess of DNA in eukaryotic genomes.

Ribosomal RNA Genes

There are four kinds of ribosomal RNA in eukaryotic cells—5S, 5.8S, 18S, and 28S rRNA. The names given to rRNA indicate the relative sizes of the molecules (see box, p. 88). The three larger molecules are transcribed in tandem from closely linked genes, with the gene for 5.8S rRNA situated between the 18S and 28S genes (Spiers and Birnstiel, 1974); the 5S RNA is transcribed from a separate, unrelated region of the genome and will be discussed later. Each complex of 18S–5.8S–28S rRNA genes is present in several hundred copies in most eukaryotes. Therefore, these genes are categorized as moderately repetitive genes. They are located at the nucleolar organizer (NO) locus of eukaryotes, which is the site of formation of the nucleolus. Many organisms have one NO region per haploid chromosome set and therefore two NO regions in diploid cells. This situation is seen, for example, in *Drosophila melanogaster* and in *Xenopus laevis*. There are certain exceptions, however. For example, the human karyotype has NO regions on several chromosomes. Genetic manipulation of both *Xenopus laevis* and *Drosophila melanogaster* has been used to produce organisms with variable numbers of nucleoli. The normal diploid cell in both these organisms has two nucleoli, but variants of both species can have either no nucleoli or unusual numbers of nucleoli: one or two in *Xenopus* and from one to four in *Drosophila*.

Hybridization of rRNA with DNA from organisms with variable numbers of nucleoli has established that the level of hybridization is proportional to the number of NO regions, indicating that the genes for rRNA are located in those regions (Ritossa and Spiegelman, 1965; Wallace and Birnstiel, 1966). Another means of establishing the identity of the NO as the site of the rRNA genes is RNA-DNA *in situ* hybridization, whereby cytological preparations of chromosomes that have been treated with alkali to denature the DNA are incubated with radioactive rRNA. The labeled RNA will hybridize with the DNA containing complementary nucleotide sequences. Autoradiography is then used to localize the site of radioactivity, which corresponds to the site of the ribosomal genes. Following *in situ* hybridization, autoradiographs show that the rRNA genes are indeed located at the nucleolar organizer (Fig. 3.16).

As previously mentioned, ribosomal DNA is organized into gene complexes, each of which contains one 18S gene, one 5.8S gene, and one 28S gene in the order listed from the 5' to the 3' end (as before, the polarity of the DNA reflects the polarity of the RNA it produces; that is, the 5' end of RNA is said to be produced from the 5' end of the gene). The multiple ribosomal gene complexes are separated from one another by **spacer DNA,** which is not transcribed. This kind of gene arrangement is called **tandem repetition.** As we shall discuss in detail in section 3–3, each gene complex is transcribed as a unit, resulting in a large precursor molecule containing the 18S, 5.8S, and 28S elements. The large molecule is later processed to yield the three smaller molecules.

The repetition of rRNA genes provides cells with multiple coding units for the efficient production of this class of RNA. During oogenesis,

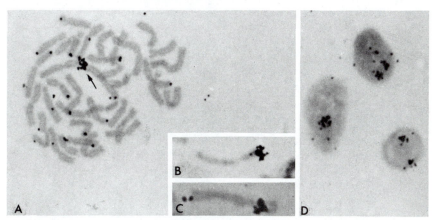

FIGURE 3.16 Autoradiographs of cytological preparations from *Xenopus borealis* and *X. laevis* hybridized *in situ* with ³H-rRNA. *A,* Metaphase plate from *X. borealis* showing hybridization of the RNA to the nucleolar organizer region. × 1230. *B,* Nucleolar organizer chromosome from *X. borealis* showing localization of sequences coding for 18S, 5.8S, and 28S RNA. × 2900. *C,* Nucleolar organizer chromosome from *X. laevis.* × 2900. *D,* Interphase nuclei from *X. borealis* showing hybridization over nucleoli. × 1070. (From M.L. Pardue. 1973. Localization of repeated DNA sequences in *Xenopus* chromosomes. Cold Spring Harbor Symposium Quant. Biol., *38:* 476.)

even this level of repetition may be insufficient to produce the required amount of rRNA, and numerous extra rRNA genes are produced by a selective replication of the nucleolar organizer (see Chap. 7). This process is called **gene amplification** and has been observed in some fish, insects, and amphibians. Apparently, amplification involves replication of the nucleolar organizer. These copies are released from the chromosomes, and the copies themselves are used as templates for the production of still more replicates. Each free NO region then forms a small nucleolus that is free within the oocyte nucleus and located just below the nuclear envelope. The amphibian oocyte nucleus may have more than 1000 of these nucleoli (Fig. 3.17*A*). The small nucleoli that are produced consist of a fibrous core surrounded by a granular cortex (Fig. 3.17*B*). The core contains the ribosomal DNA and is, therefore, the site of synthesis of ribosomal RNA.

After synthesis the rRNA detaches from the DNA and is transferred to the cortex, where it undergoes processing prior to its packaging into ribosomes. The structure of the DNA within the core is difficult to discern because the core is so compact; its structure is best studied by breaking open the nucleolus to allow the DNA to unravel. This can be accomplished by placing the nucleoli in distilled water. The nucleolar DNA is then found to be a single closed circular molecule.

Besides providing the oocyte with additional rRNA coding units for rapid synthesis of rRNA, amplification provides investigators with a con-

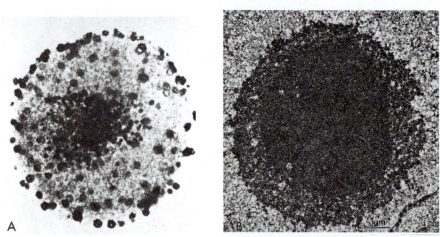

FIGURE 3.17 *A*, Isolated nucleus of *Xenopus* oocyte. Darkly staining structures are nucleoli. Diameter of nucleus is about 400 μm. (From D.D. Brown and I.B. Dawid. 1968. Specific gene amplification in oocytes. Science, *160:* 272–280. Copyright 1968 by the American Association for the Advancement of Science.) *B*, Electron micrograph of nucleolus from amphibian oocyte nucleus, showing granular cortex and fibrous core. (From O.L. Miller, Jr., and A.H. Bakken. 1972. Morphological studies of transcription. *In* E. Diczfalusy (ed.), *Karolinska Symposia on Research Methods in Reproductive Endocrinology, 5th Symposium. Gene Transcription in Reproductive Tissue.* Periodica, Copenhagen, p. 159.)

venient source of nucleoli for experimental analysis. Consequently, much of the information we have concerning organization of rRNA genes and the mechanisms of their transcription (section 3–3) has been obtained with oocyte nucleoli.

5S and Transfer RNA Genes

These two categories of genes produce small-sized RNA. However, in addition to size, there is another reason for considering them together: These genes are organized differently from the other genes of eukaryotes, and they are transcribed by an RNA polymerase that is selective for them.

The DNA that encodes 5S and transfer RNA is moderately repetitive. Furthermore, the tRNA genes are actually a heterogeneous group of genes that encode the several kinds of transfer RNA. In *Xenopus*, each of the 43 different tRNA genes is repeated an average of 200 times—making a total of about 8600 genes (Clarkson et al., 1973). There may also be heterogeneity among the 5S RNA genes. In *Xenopus laevis*, for example, there are three kinds of genes that code for 5S RNA: somatic, oocyte, and trace oocyte. The genes that code for each kind of 5S RNA are present as multigene clusters, with units consisting of a gene plus spacer DNA tandemly repeated hundreds to thousands of times. A repeat unit may also contain a pseudogene (Jacq et al., 1977). The oocyte 5S RNA genes are present in the highest number in the *X. laevis* genome. They are expressed only during oogenesis and are inactive in somatic cells (see Chap. 7). The oocyte 5S RNA genes are present as approximately 24,000 tandemly repeated 120 base pair coding units per haploid genome (Brown and Sugimoto, 1973). Clusters of oocyte 5S RNA genes have been localized to the telomeres of the long arms of most, if not all, of the *Xenopus* chromosomes by *in situ* hybridization (Pardue et al., 1973).

As already mentioned, the individual gene units that encode the various kinds of tRNA and 5S RNA have a novel organization that is related to their mode of transcription. We shall return to this intriguing topic in section 3–3.

Interspersed Repetitive DNA

In addition to the well-characterized repetitive sequences previously discussed, the eukaryotic genome also contains repetitive sequences that are scattered throughout the genome. These sequences are interspersed between the unique sequences (at least some of which are structural genes). It is now thought that most, if not all, unique DNA sequences are flanked on either side by moderately repetitive DNA. The initial evidence for this genome organization was obtained from renaturation studies on relatively long fragments (20,000 nucleotides) of *Xenopus* DNA. Since unique sequences account for the majority of DNA, one would expect most of these fragments to consist of unique sequences and therefore reanneal under high Cot conditions. However, they actually reassociate under low Cot conditions that allow moderately repetitive DNA to react. This result indicates that the fragments must contain moderately repetitive sequences in addition to unique sequence DNA and led to the proposal that these moderately repetitive sequences are interspersed between unique sequences. This proposal has since been supported by a large body of evidence from studies with cloned genomic DNA. Cloned fragments containing structural genes have frequently been shown to contain repetitive DNA in the regions flanking the genes. For example, a fragment containing the rabbit β-globin gene possesses a repetitive segment downstream from the gene, which is also present at several scattered sites throughout the genome. The rabbit genome contains about 5000 copies of this particular sequence (Hoeijmakers-van Dommelen et al., 1980)

Recently, repetitive sequences themselves have been cloned and used as hybridization probes. In addition, the nucleotide order of some of these sequences has been determined. One result of studies with repetitive sequence probes has been the discovery of families of sequences that hybridize to the same probe. Members of a **repetitive sequence family** have high sequence homology with one another and may share consensus sequences at specific sites. For example, a single 300 nucleotide-long, repetitive sequence family predominates in the human genome. This family of approximately 300,000 members has been called the ***Alu* family,** because each member possesses a restriction site for the enzyme *Alu* I (Houck et al., 1979; Rubin et al., 1980). Other mammals have recently been shown to have *Alu* family equivalents (Jelinek et al., 1980).

The function of interspersed repetitive DNA is as yet unknown. An important question is whether these sequences are transcribed and, if so, what the function of the transcripts is. Cloned sea urchin repetitive sequences have been used to probe the RNA for such transcripts. The results of these experiments indicate that the sequences are transcribed. However, the transcripts are restricted to the nucleus. Furthermore, there are tissue-specific differences in the relative frequencies of different repeat family transcripts (Scheller et al., 1978). These two characteristics suggest that the repetitive sequence transcripts may serve some regulatory function in the nuclei, such as the regulation of structural gene expression or processing of transcripts.

In a similar experiment a cloned probe for human *Alu* family sequences has been demonstrated to hybridize with a high proportion of human messenger RNA (Calabretta et al., 1981). Thus, these sequences are transcribed and remain an integral part of the transcripts that can function as messenger RNA. The roles the *Alu* sequences might play in messenger RNA are unknown.

An interesting and possibly highly significant class of repetitive sequences has been identified whose members are capable of being inserted at several locations in the genome. These sequences, which are found in a wide variety of organisms, are called **transposable elements** or **mobile dispersed elements.** Their transposable nature is indicated by the fact that they occupy quite different positions in the genome among closely related strains of the same species (Strobel et al., 1979). Thus, they are capable of excision from sites they occupy and of reinsertion at alternate sites. Although the presence of a transposable element may affect the expression of genes in its vicinity (McClintock, 1980; Bingham and Judd, 1981), the functional roles of these elements are poorly understood.

Satellite DNA

Rapidly reannealing DNA consists of short, tandemly repeated nucleotide sequences that do not code for protein. This category of DNA is sometimes called satellite DNA. This term refers to the fact that the highly repetitive DNA has a buoyant density that is different from that of bulk DNA when subjected to analytical density gradient centrifugation (see box, p. 88). Thus, during centrifugation the bulk DNA forms a major band, whereas the smaller amount of repetitive DNA forms a small satellite (Fig. 3.18). The difference in buoyant density is due to differences in base composition between the two classes of DNA. Since the rapidly reassociating DNA consists of short sequences that are repeated approximately one million times, it is enriched for certain bases; therefore, its base composition differs greatly from the remainder of the DNA, resulting in different buoyant density.

The correspondence between satellite DNA and rapidly reassociating DNA is demonstrated by density gradient analysis of purified rapidly

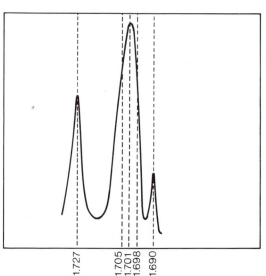

FIGURE 3.18 Analytical ultracentrifugation of mouse DNA in cesium chloride. The peak at 1.727 is a density marker. Main band mouse DNA reaches equilibrium at 1.701 and a satellite peak forms at a density of 1.690. (After B.L. McConaughy and B.J. McCarthy. 1970. Related base sequences in the DNA of simple and complex organisms. VI. The extent of base sequence divergence among the DNAs of various rodents. Biochem. Genet., *4:* 425–446. Reprinted with permission of Plenum Publishing Corporation.)

1.727   1.705 1.701 1.698   1.690

reassociating DNA. If denatured DNA is renatured under very low Cot conditions, the double-stranded rapidly reassociating DNA can be separated from the remaining DNA (unique and moderately repetitive sequences) by passing the solution over a column of hydroxyapatite (calcium phosphate). The hydroxyapatite selectively binds double-stranded DNA, but single-stranded fragments pass through the column. The bound DNA can then be removed from the hydroxyapatite by raising the salt concentration of the buffer. Under high salt conditions the DNA no longer binds to the hydroxyapatite and is flushed (**eluted**) from the column. The highly repetitive DNA prepared by this technique of **affinity chromatography** has a buoyant density that is identical to that of satellite DNA (Waring and Britten, 1966).

Satellite DNA is localized to specific chromosomal regions where it is present as large blocks of tandemly repeated nucleotide sequences. These locations have been established by *in situ* hybridization. A cytological preparation of chromosomes on a microscope slide is treated with alkali to denature the DNA, and the preparation is incubated with radioactively labeled RNA that is synthesized *in vitro* from satellite DNA. Silver grains in developed autoradiographs indicate the presence of radioactive RNA that has hybridized with the chromosomal DNA (Fig. 3.19). The silver grains are localized predominantly over the constitutive

FIGURE 3.19 Autoradiograph of a mouse chromosome preparation after *in situ* hybridization with radioactive RNA copied *in vitro* from mouse satellite DNA. The RNA has bound to the centromeric heterochromatin. ×1460. (From M.L. Pardue and J.G. Gall. 1972. Chromosome structure studied by nucleic acid hybridisation in cytological preparations. *Chromosomes Today, 3:* 47–52.)

heterochromatin surrounding the centromeres of metaphase chromosomes. Hence, the major DNA component of constitutive heterochromatin is the rapidly reassociating satellite DNA. A variety of studies using *in situ* hybridization as well as other techniques have provided evidence that satellite sequences may also be present in nonheterochromatic regions of the genome (Steffenson et al., 1981; Sealy et al., 1981; Stambrook, 1981).

It is generally assumed that satellite DNA is inactive in RNA synthesis. However, this DNA is transcribed in amphibian oocyte lampbrush chromosomes (see Chap. 7). Transcription in oocytes is an apparent exception to the rule of genetic inactivity for this DNA. In other cell types, satellite DNA may serve a structural, rather than informational, role. Because of its location near the centromere, it is possible that this DNA serves a mechanical function during mitosis or meiosis.

The Genome: Summing Up

The eukaryotic genome is obviously a heterogeneous assortment of nucleotide sequences, with genes coding for a variety of types of RNA and some sequences that are not transcribed. Some of the RNA that is transcribed may remain in the nucleus and not participate directly in protein synthesis. Various levels of repetition are found for nucleotide sequences, from those that are present singly to those repeated hundreds of thousands of times. Thus, the apparent excess DNA in eukaryotes over and above the amount needed to code for proteins has an explanation, but its significance to the phenotypes of organisms that contain it is only partially understood and remains a challenge to those concerned with the role of the genome in development.

SECTION TWO
THE UTILIZATION OF GENETIC INFORMATION DURING CELL DIFFERENTIATION

3–3. RNA SYNTHESIS

The initial step in cellular utilization of the information stored in the genome is transcription of DNA to form RNA. The enzymes responsible for this activity are a group collectively called DNA-dependent RNA polymerases. Three distinct polymerase enzymes are responsible for transcription in eukaryotes; each functions on different classes of DNA. RNA polymerase I is found preferentially in the nucleolus and is utilized in transcription of rDNA. The two other polymerases, II and III, are described as being "nucleoplasmic"; that is, they are found in the nonnucleolar regions of the nucleus, presumably associated with chromatin. Polymerase II transcribes structural genes, whereas type III transcribes the genes for transfer RNA and 5S RNA.

The differential inhibition of polymerase function is often an important experimental manipulation in the study of cell differentiation. It is useful, for example, in determining whether *de novo* transcription is essential for the appearance of a particular differentiated trait or for the synthesis of a particular protein. The most commonly used inhibitor is **actinomycin D**, which at low concentrations inhibits rRNA synthesis and at higher concentrations inhibits synthesis of all classes of RNA. Another inhibitor, α-**amanitin**, is more specific; low concentrations selectively prevent polymerase II function, higher concentrations inhibit polymerase III, whereas polymerase I is unaffected by these concentrations of this inhibitor.

Polymerase molecules transcribe DNA in a linear fashion by sequential addition of ribonucleotides to a growing RNA chain. Each nucleotide in RNA has the same orientation, giving the chain a definite polarity, which is opposite that of the DNA template. The end terminated by the 5'-carbon atom and its associated $(PO_4)_3$ group is called the 5' end, whereas the end containing the 3'-carbon atom and its OH group is the 3' end.

During transcription a polymerase molecule attaches to the beginning of the region of DNA to be transcribed (a **transcriptional unit**). The site at which the polymerase initiates transcription is called the **promoter**. The promoter also influences the efficiency of initiation. Attachment of the polymerase to the DNA causes the double helix in the attachment region to unwind. The polymerase then moves along the DNA, causing local unwinding and reformation of the helix as it proceeds. Only one DNA strand is transcribed by the polymerase, which always moves in the 3'→5' direction of this strand. This produces an RNA molecule with opposite polarity—it begins at the 5' end and terminates at the 3' end. The growing RNA strand remains attached to the polymerase during the entire transcriptional process, and when this complex reaches the 5' end of the transcriptional unit, both the polymerase and the completed transcript are released from the DNA. As we discussed in section 3–2, the polarity of a gene is given in terms of the noncoding (or antisense) strand. Thus, the beginning of a gene is said to be the 5' end, whereas the terminus is the 3' end. Unless stated otherwise, we shall follow this convention when discussing gene organization.

The structure and function of promoters is best understood in prokaryotes. The prokaryotic RNA polymerase recognizes and binds to two regions of DNA, located approximately 10 nucleotides and 35 nucleotides, respectively, upstream from the mRNA start site. The former, which is called **Pribnow's box**, has the sequence TATAATG. The segment at 35 nucleotides upstream has the consensus sequence TGTTGA-CAATTT, although there is considerable variation of the latter sequence among prokaryotes (Rosenberg and Court, 1979). These promoter regions are remarkably similar to consensus sequences found in the 5' flanking regions of eukaryotic polymerase II genes (see Fig. 3.15). These are the TATA, or Goldberg-Hogness, box and the CCAAT box, which is located farther upstream.

Analysis of the roles of the eukaryotic consensus sequences in transcription of polymerase II genes involves observing the effects on transcription of modification or deletion of these sequences. Genes with aberrant 5' flanking regions are cloned, and transcription of the cloned sequences is assessed by either *in vitro* transcription with polymerase II or the technique of **surrogate genetics**. The latter procedure involves introducing the cloned genes into the nucleus of another cell and allowing the endogenous RNA polymerase II of the host nucleus to transcribe them. Exogenous genetic material may either be injected into the nucleus of a living cell (usually a *Xenopus* oocyte) or introduced into tissue culture cells by transformation (Birnstiel and Chipchase, 1977). Experimental analysis of the TATA box indicates that it directs the polymerase to initiate transcription about 30 nucleotides downstream from it. Thus, it specifies the RNA start site. Regions located farther upstream—among them the CCAAT box—exert an influence on the efficiency of transcription. Therefore, the eukaryotic polymerase II promoter is a multisite entity in the 5' flanking region (Wasylyk et al., 1980; Grosschedl and Birnstiel, 1980; Dierks et al., 1981; Sassone-Corsi et al., 1981; Grosveld et al., 1982).

The promoter of ribosomal RNA gene transcription by RNA polymerase I is also located in the 5' flanking region preceding each 18S–5.8S–28S rRNA gene complex (Grummt, 1981). However, genes transcribed by RNA polymerase III (5S RNA and transfer RNA genes) are unique in that the sequences directing specific initiation of transcription are *intragenic* and are flanked by coding sequences (Bogenhagen et al., 1980; Sakonju et al., 1980; Sharp et al., 1981; Hofstetter et al., 1981; Galli et al., 1981). Thus, the polymerase III promoter specifies that transcription begin at a site far upstream from it.

One of the most powerful means of studying transcription is actually to observe chromatin in the act of RNA synthesis with either the light or electron microscope. The initial observations of transcription were light-microscopic studies of so-called giant chromosomes. One kind is the lampbrush chromosome, which appears during the diplotene stage of meiosis in a wide variety of plant and animal species. We shall discuss lampbrush chromosomes in detail in Chapter 7. The other kind of giant chromosome is found in certain enlarged cells of the larvae of *Drosophila, Chironomus* and some other dipteran flies. These polytene chromosomes have provided considerable information concerning regulation of transcription and will be covered in Chapter 4.

A technique developed by Oscar Miller allows observation of transcription with the electron microscope. Essentially, this procedure in-

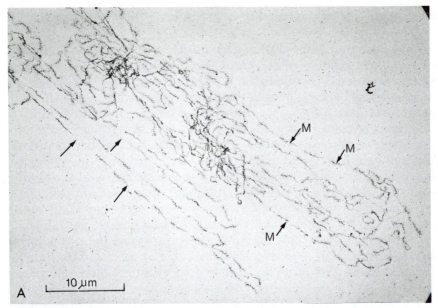

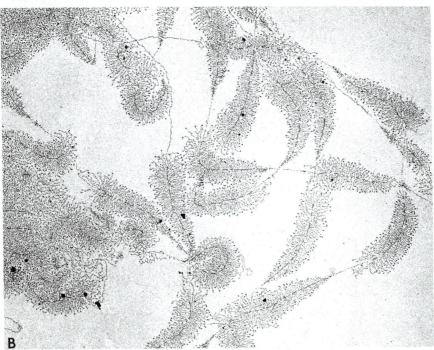

FIGURE 3.20 Visualization of nucleolar transcription. *A,* Portion of a nucleolar core isolated from a *Triturus* oocyte, showing matrix units (M) separated by matrix free spacers (arrows). (From O.L. Miller, Jr., and B.R. Beatty. 1969a. Extra-chromosomal nucleolar genes in amphibian oocytes. Genetics (Suppl.), *61:* 136.) *B,* Higher magnification (×10,900) electron micrograph of oocyte nucleolar transcription. Each matrix unit consists of a gradient of fibrils extending from the chromatin. Note that each matrix unit has the same polarity. (From O.L. Miller, Jr., and B.R. Beatty. 1969b. Visualization of nucleolar genes. Science, *164:* 955–957. Copyright 1969 by the American Association for the Advancement of Science.)

volves a **gentle dispersion** of nuclear contents to separate constituents without breakage of chromatin or loss of the nascent RNA. This step is followed by **fixation**, the **spreading** of the chromatin on electron microscope grids, and **staining**, which allows identification of the chromatin and RNA. Ribosomal RNA synthesis is perhaps the most visually pleasing transcriptional process that has been observed with the electron microscope. The initial observations of ribosomal DNA transcription were made on chromatin prepared from amphibian oocyte nucleoli. Numerous observations have subsequently been made on nucleolar transcription in a wide variety of species, and the pattern is virtually identical wherever it is observed.

Under low magnification, nucleolar chromatin has the appearance of a series of "fuzzy" regions alternating with smooth chromatin (Fig. 3.20A). The fuzzy regions are called **matrix units**. At higher magnification, the matrix units are seen to consist of a sequence of progressively longer fibrils attached to chromatin by a small granule (Fig. 3.20B). The fibrils are nascent RNA transcripts, and the granules at the base of the fibrils are thought to be RNA polymerase I molecules (Labhart and Koller, 1982). Transcription apparently begins at the narrow end of the matrix unit with the binding of a polymerase molecule to the chromatin. The polymerase then moves along the DNA, progressively adding ribonucleotides to the growing RNA chain until it reaches the end of the matrix unit and detaches from the chromatin. As soon as one polymerase has moved a sufficient distance along the DNA, another polymerase attaches, and so on. This "assembly line" process allows for the matrix DNA to be transcribed simultaneously by several polymerase molecules.

As we discussed earlier in this chapter, the 18S, 5.8S, and 28S ribosomal RNA genes are organized as gene complexes, repeated numerous times, and the gene complexes are separated from one another by nontranscribed spacer DNA. The DNA within each matrix unit is one gene complex, and the smooth chromatin represents the nontranscribed spacers. Transcription is initiated at the 5' end of the gene complex (narrow end of the matrix unit), as polymerase I binds to the 5' end of the complex—as specified by the polymerase I promoter located there—and progressively transcribes the 18S, 5.8S, and 28S genes, producing a single, large precursor molecule that contains all three kinds of ribosomal RNA. The functional 18S, 5.8S, and 28S molecules are released from the precursor by processing (see section 3–4). Not only is each gene complex transcribed simultaneously by several polymerase molecules, but all genes complexes are transcribed at the same time. Ribosomal transcription is obviously a highly efficient process.

Transcription of nonribosomal interphase chromatin of somatic cells has been observed with the electron microscope only in recent years. The few nonribosomal transcriptional units that have been seen are similar to ribosomal transcriptional units in that they also have a series of progressively longer fibrils associated with the chromatin (Fig. 3.21). However, in contrast to ribosomal transcription, the transcrip-

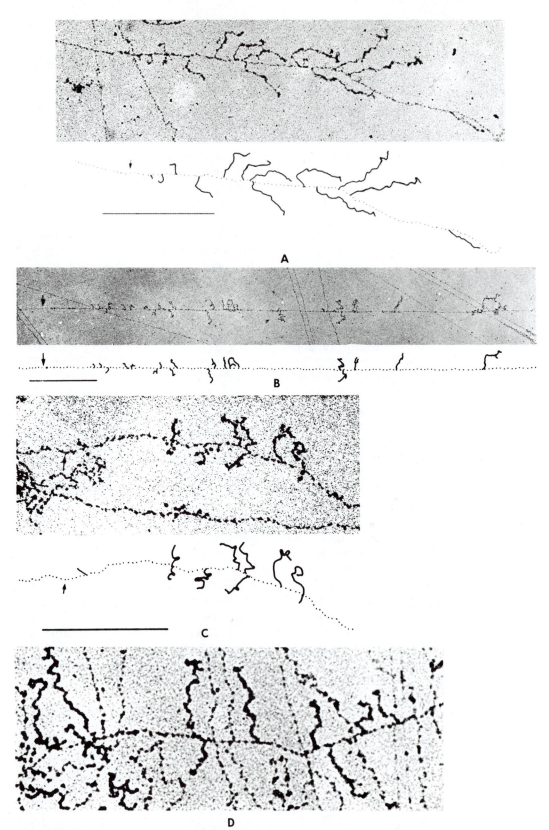

FIGURE 3.21 *(See legend on opposite page)*

◀ FIGURE 3.21 Electron micrographs of nonribosomal transcription. *A* to *C*, Complete transcription units and interpretive drawings. Arrows indicate inferred sites of transcription initiation for each unit. *A* and *B*, Chromatin from mealy bug (*Oncopeltus fasciatus*) embryos. *C*, Chromatin from *Drosophila* embryo. Bars represent 1 μm. *D*, High magnification (×67,000) electron micrograph of nonribosomal transcription unit of *O. fasciatus*. Note beaded chromatin. (*A* and *D* from V.E. Foe, L.E. Wilkinson, and C.D. Laird. 1976. Comparative organization of active transcription units in *Oncopeltus fasciatus*. Cell, *9:* 136, 142. Copyright © Massachusetts Institute of Technology; published by the MIT Press. *B* from C.D. Laird, et al. 1976. Analysis of chromatin-associated fiber arrays. Chromosoma (Berl.), *58:* 179. *C* from C.D. Laird and W.Y. Chooi. 1976. Morphology of transcriptional units in *Drosophila melanogaster*. Chromosoma (Berl.), *58:* 199. *B* and *C* reprinted with permission of Springer-Verlag, Heidelberg.)

tional units are found not as tandem repeats but as isolated regions of transcriptional activity. Apparently, they are unique nucleotide sequences and may represent structural genes. The nascent RNA fibrils in nonribosomal transcriptional units are farther apart than those in ribosomal units, indicating that fewer RNA molecules are synthesized simultaneously on these genes. Careful examination of these transcriptional units at high magnification (Fig. 3.21*D*) reveals that the transcribed chromatin is beaded.

The presence of nucleosomes in transcriptionally active chromatin makes it necessary to account for this configuration in any explanation of the transcriptional process. How does the polymerase molecule traverse the nucleosome? Do nucleosomes in active chromatin differ from those in inactive chromatin? Although we know little about the movement of the polymerase, there is strong evidence that the chromatin of transcribed DNA has a somewhat different nucleosomal configuration than that of nontranscribed DNA. This topic is discussed in detail in Chapter 4.

From the moment of its synthesis, RNA is combined with protein to form a complex called **ribonucleoprotein (RNP)**. In fact, the growing fibrils seen with the Miller electron microscopy technique are visible because they are stained for protein, not RNA. Following transcription, completed transcripts are released from chromatin and transported to the cytoplasm as RNP complexes. The type of complex formed depends upon the nature of the RNA. The ribosome is a complex of the four kinds of rRNA—5S, 5.8S, 18S, and 28S—with specialized ribosomal proteins. Structural gene transcripts are combined with proteins to form messenger RNP particles. Considerable evidence suggests that messenger RNA remains combined with protein throughout its functional life in the cell.

3–4. PROCESSING OF RNA

Transcription is only the first step in the sequence of events that must occur so that a gene may exert its influence on cellular function. The raw transcript is a diamond in the rough; it must undergo a number of modifications in the nucleus and then be transported to the cytoplasm, where it becomes functional. These modifications include addition, deletion, and alteration of nucleotides. The primary transcript may be longer

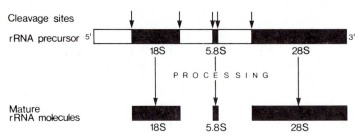

Cleavage sites

rRNA precursor 5'

18S   5.8S   28S

PROCESSING

Mature
rRNA molecules

18S   5.8S   28S

FIGURE 3.22 Processing of ribosomal RNA precursor molecule. The precursor has the following regions from the 5' to 3' end: spacer–18S–spacer–5.8S–spacer –28S. The spacers are cleaved away, releasing the individual rRNA molecules.

than its cytoplasmic derivative. Enzymatic cleavage of the precursor is necessary to trim away excess nucleotides. For example, the ribosomal RNA precursor is a complex molecule containing 18S, 5.8S, and 28S segments, separated from one another by spacer regions (Fig. 3.22). Nuclease enzymes in the nucleolus trim away the spacer regions, which releases the 18S, 5.8S, and 28S molecules. A number of nucleotides of ribosomal RNA are also altered by the addition of methyl groups to the ribose moieties. The primary ribosomal transcript is, therefore, subject to extensive **post-transcriptional modification.**

Nonribosomal transcripts also undergo posttranscriptional processing. Like the rRNA precursor, nonribosomal transcripts in the nucleus are very large molecules. In mammalian cells, for example, nuclear transcripts vary in length from about 5000 nucleotides to greater than 50,000, whereas messenger RNA molecules in the cytoplasm range in size from 500 to 5000 nucleotides (Jelinek et al., 1974). Nuclear RNA is usually called **heterogeneous nuclear RNA,** abbreviated as **HnRNA.** It is a mixture of several molecular size classes. Most of the HnRNA is very short lived and is degraded in the nucleus, never reaching the cytoplasm (Brandhorst and Humphreys, 1972). At least a portion of the RNA that is restricted to the nucleus includes transcripts of the interspersed moderately repetitive DNA (see p. 109).

There is a large body of evidence that at least some of the HnRNA molecules are large precursors from which mRNA molecules are derived by processing. HnRNA and mRNA have many properties in common, which suggests a possible relationship. For example, both are called DNA-like RNA because their base compositions are similar to that of the average DNA composition, an indication that both kinds of RNA are transcribed from generalized DNA. Ribosomal RNA, on the other hand, is transcribed from a limited portion of the genome with a high ratio of guanine and cytosine, which is reflected in the base composition of the RNA.

Establishing a precursor-product relationship between HnRNA and mRNA has been more difficult than determining the relationship between nucleolar RNA and ribosomal RNA. Strong evidence for this relationship is the discovery of messenger RNA sequences in HnRNA molecules with cDNA probes. For example, globin cDNA has been used to detect the high molecular weight nuclear precursors to mammalian globin mRNA and to trace the processing of the globin gene transcript. The nuclear transcript is a large 15S molecule, whereas its cytoplasmic messenger derivative sediments at 10S (Ross, 1976; Curtis and Weissmann, 1976; Bastos and Aviv, 1977). The latter is derived from the precursor by removal of excess nucleotides, modification of existing nucleotides, and posttranscriptional addition of certain nucleotides.

The processing of HnRNA from its synthesis to the formation of mRNA is now becoming clear. As we discussed previously, the beginning of the transcript (the 5′ end) is marked by a triphosphate group, which is part of the initial nucleoside triphosphate at the transcription initiation site (Fig. 3.23A). After initiation has occurred the 5′ end is

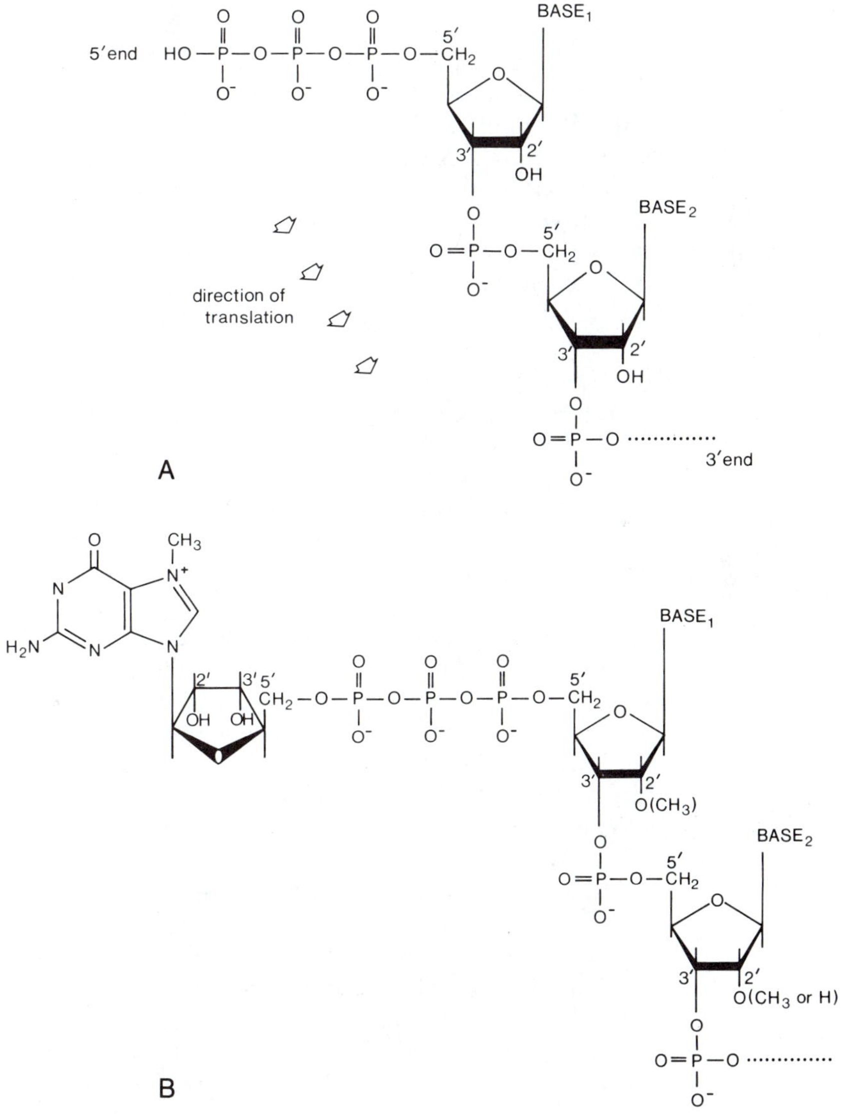

FIGURE 3.23 The 5′ terminus of mRNA. *A,* Unmodified. *B,* Capped configuration.

modified by the addition of a guanosine nucleoside that is attached in opposite polarity to the rest of the RNA molecule. In addition to having reversed polarity, this guanosine is also unusual in that it contains a methyl group bound to the guanine portion of the molecule. This methylated guanosine is called the **5′ cap.** Transcripts may also be methylated in the nucleotide adjacent to the capping nucleotide (i.e., in the initial nucleotide of the transcript) or in the first two nucleotides. A capped transcript is shown in Figure 3.23B. Compare it to the unmodified terminus shown in Figure 3.23A. The capping of the initial nucleotide has inspired a new term for molecular biology jargon: The site of transcription initiation by RNA polymerase II is called the **cap site.** As we have learned, this site is specified by the TATA box, some 30 nucleotides upstream.

The function of the 5′ cap appears to be related directly to the role of the 5′ end of the molecule in translation. Since translation is initiated at the 5′ end of the messenger, the integrity of this portion of the molecule is crucial. A growing body of evidence suggests that 5′ caps may be essential for the initiation process (see Shatkin [1976] for review). An alternative function for the cap also has been proposed. Gedamu and Dixon (1978) suggest that the cap may protect messenger RNA from digestion by ribonuclease enzymes that degrade RNA from the 5′ end. If degradation is rapid, initiation could not occur, and protein synthesis would be blocked.

The opposite, or 3′ hydroxyl, end of a significant fraction of HnRNA and mRNA molecules has a series of about 150 to 250 adenylate residues. These regions are called **poly (A) tracts.** The poly (A) tract is added to the trailer sequence of the primary transcript, downstream from the site of the translation termination codon. Before addition of the poly (A), the trailer sequence is cleaved internally, and the poly (A) tract is added to the newly created 3′ end of the transcript (Hofer and Darnell, 1981; Weintraub et al., 1981). A consensus hexanucleotide sequence AAUAAA is located upstream from the polyadenylation sites of transcripts (Proudfoot and Brownlee, 1976). This sequence is thought to be the signal for cleavage/polyadenylation. This interpretation is supported by the observations that (1) mutational transposition of the AAUAAA sequence results in a corresponding transposition of the polyadenylation site, which is 11 to 19 nucleotides downstream from the sequence; (2) mutational deletion of the AAUAAA sequence abolishes cleavage/polyadenylation (Fitzgerald and Shenk, 1981); and (3) histone messengers, which lack poly (A) tracts, also lack the AAUAAA sequence (Kedes, 1979).

The functional significance of poly (A) is not well understood. Its widespread occurrence in organisms of all phylogenetic levels in both the plant and animal kingdoms suggests that its presence on messenger molecules is important to their proper functioning. Most speculation concerning the role of poly (A) has centered on the possibility that it (1) functions in transport of mRNA from the nucleus to the cytoplasm, (2) helps to protect the messenger from degradation and hence extends its functional lifetime, (3) increases the efficiency of translation, or (4) functions in processing the HnRNA to mRNA. Defining a functional role for poly (A) has been complicated by the indisputable fact that some messengers lack poly (A) but are functionally normal. They are called poly (A)⁻ messengers, as distinguished from poly (A)⁺ messengers. One well-defined poly (A)⁻ messenger is histone mRNA (see preceding discussion).

Although a functional role for poly (A) in the cell has not been established, it has proved to be of inestimable value for investigators who need to prepare mRNA for experimental work. Because of the natural complementarity of adenine with uracil and thymine, the poly (A) associated with messengers will bind selectively either poly (U) or oligo (dT). In practice, affinity chromatography is used to isolate polyadenylated mRNA. A column of either poly (U) bound to Sepharose or oligo (dT) bound to cellulose is prepared. A solution of RNA is then passed through the column. Polyadenylated messengers bind, whereas other RNAs pass through it. The messengers can then be eluted from the column by passing a reagent through it to break the bonds between poly (A) and poly (U) or oligo (dT). The messengers are collected in the column eluate.

One reason for the excess size of initial transcripts as compared to their mRNA derivatives is the presence of introns in the former, which are lacking in mature messengers. The mouse β-globin gene transcript is an example. The nuclear transcript is about three times the size of the mature messenger, which is sufficient to encompass both the β-globin coding sequence and the two introns. This suggests that the precursor is a full-length gene transcript, whereas the messenger is obtained by removal of the introns from the precursor. Confirmation of this hypothesis has been obtained by ultrastructural comparisons of the structures formed by hybridizing the mouse β-globin gene with either the 15S precursor or 10S mRNA. This is the same technique used for the experiment described in Figure 3.14. As shown in Figure 3.24A, the *entire* 15S β-globin RNA hybridizes with the β-globin gene. However, when the messenger RNA is hybridized with the gene (Fig. 3.24B), regions of the gene corresponding to the introns loop out because they are lacking in the messenger.

The removal of introns involves their excision, followed by splicing of the coding sequences. The excision and splicing steps must be performed with great precision to ensure the sequence integrity of the resulting messenger. The process is assumed to involve the bringing together of the 3' end of one coding sequence and the 5' end of the adjacent coding sequence. The intron would form a loop structure that could be excised enzymatically so that the adjoining coding regions could be ligated. As we discussed in section 3–2, each intron begins with GU and ends with AG. These nucleotides are thought to provide the signal for precise excision/splicing.

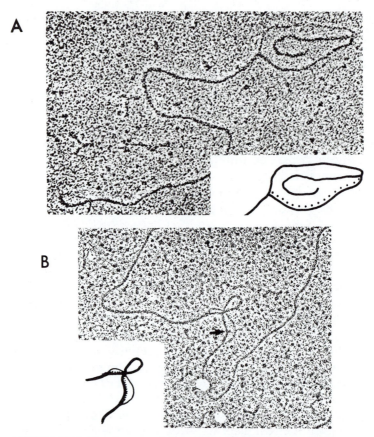

A

B

FIGURE 3.24 The introns of the mouse β-globin gene. *A,* Electron micrograph of structure formed by incubation of partially denatured globin DNA with 15S transcript. Inset: Interpretive drawing. × 120,000. (From S.M. Tilghman et al. 1978b. The intervening sequence of mouse β-globin gene is transcribed within the 15S β-globin mRNA precursor. Proc. Natl. Acad. Sci. U.S.A., *75:* 1311.) *B,* Structure formed between globin DNA and globin mRNA. The large intron loops out because there is no complementary region in the mRNA. The arrow indicates a small loop corresponding to the small intron. Inset: Interpretive drawing. The location of the small intron is not indicated on this drawing. Heavy lines represent double-stranded DNA; narrow lines represent single-stranded DNA; combined dotted lines–heavy lines indicate the DNA-RNA hybrids. × 78,400. (From S.M. Tilghman et al. 1978a. Intervening sequence of DNA identified in the structural portion of a mouse β-globin gene. Proc. Natl. Acad. Sci. U.S.A., *75:* 727.)

Excision/splicing may involve the participation of one of the small nuclear RNAs (Yang et al., 1981). This RNA, which is called U1-RNA has near its 5′ end the base sequence ACCU. As shown in Figure 3.25, this portion of U1-RNA could juxtapose by base pairing the splice junctions of HnRNA to facilitate cleavage and splicing.

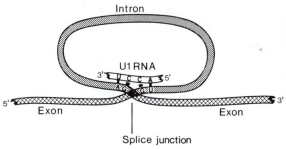

Intron

U1 RNA

3' U C C A 5'

5' Exon   Exon 3'

Splice junction

FIGURE 3.25 Proposed mechanism for removal of introns from transcripts. The ACCU sequence of U1-RNA aligns with the GU and AG dinucleotides at the 5' and 3' ends, respectively, of the intron, thus bringing exons into juxtaposition for ligating. (Based upon the mechanism proposed by M.R. Lerner et al. 1980. Are snRNPs involved in splicing? Nature, *283*: 220–224.)

3–5. PROTEIN SYNTHESIS

Utilization of the messenger RNA template for the synthesis of protein is called **translation.** This process has three major steps, which are summarized in Figure 3.26: (1) **initiation,** (2) **elongation,** and (3) **termination.** During initiation an initiation tRNA (a specific methionyl tRNA called Met-tRNA$_i$), guanosine triphosphate (GTP), and an initiation factor (eIF-2) first combine with one another and then are transferred to a small ribosome subunit. The next stage in initiation is the association of a messenger RNA with this complex in a reaction that requires ATP and initiation factors. Finally, the large ribosome subunit is added, completing the **initiation complex.** This reaction requires GTP and another initiation factor (Grunberg-Manago and Gros, 1977).

The Met-tRNA$_i$ recognizes an AUG codon near the 5' end of the messenger RNA. The initiation codon marks the start of the translatable portion of the mRNA molecule and specifies placement of a methionine amino acid at the beginning of the peptide. Each subsequent amino acid is specified as the codons are read in the 5'→3' direction. Addition of these amino acids is elongation. The transfer RNA bearing the next amino acid enters the ribosome, and a peptide bond is formed between this amino acid and the preceding one by the enzyme **peptidyl transferase.** The growing peptide chain is then bound to the newly arrived tRNA, and the tRNA that was bound to the preceding amino acid is ejected from the ribosome. The peptide-tRNA complex is then translocated to the site formerly occupied by the ejected tRNA by movement of the mRNA. Translocation brings a new codon into the ribosome so that the next amino acid–tRNA complex can enter. The elongation process continues until a termination codon enters the ribosome. Termination is designated by either UAG, UAA, or UGA. The termination codon represents the 3' limit to the translatable portion of the mRNA. The entrance of the termination codon into the ribosome causes the completed polypeptide chain and the ribosome to be released from the messenger.

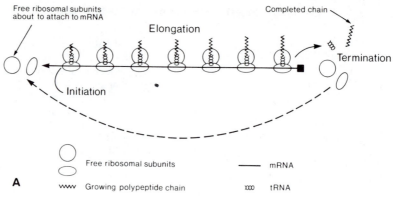

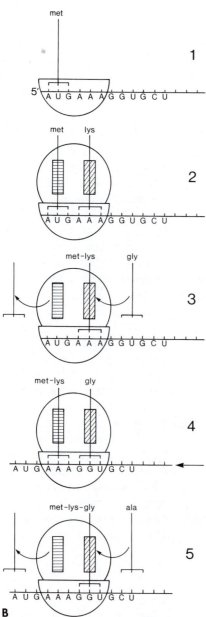

FIGURE 3.26 *A*, Schematic picture of a polyribosome during protein synthesis. The mRNA molecule is moving from right to left. (After J.D. Watson. 1976. *Molecular Biology of the Gene,* 3rd ed. Benjamin/Cummings Publishing Co., Menlo Park, Calif., p. 340.) *B*, Diagram representing the early stages of translation of messenger RNA (5' to 3'). The peptidyl, or donor, and aminoacyl, or acceptor, sites on the large sub-unit are indicated by horizontal and oblique stripes, respectively. 1: Initiation complex in which Met-tRNA$_i$ binds to the first codon in mRNA (AUG). 2: The complete ribosome has been formed, and the second aminoacyl-tRNA (lys-tRNA) binds the second codon (AAA). 3: The tRNA is eliminated from the peptidyl site, and the first peptide bond is formed. 4: Translocation of the mRNA and of the peptidyl-tRNA has occurred, and a new aminoacyl-tRNA (gly-tRNA) binds to the third codon (GGU). 5: The molecular events of 3 are now repeated. (After S. Ochoa, 1968. Translation of the genetic message. Die Naturwissenschaften, *55:* 511. Reprinted with permission of Springer-Verlag, Heidelberg. Modified from E.D.P. De Robertis and E.M.F. De Robertis, Jr. 1980. *Cell and Molecular Biology,* 7th ed. Saunders College Publishing, Philadelphia, p. 530.)

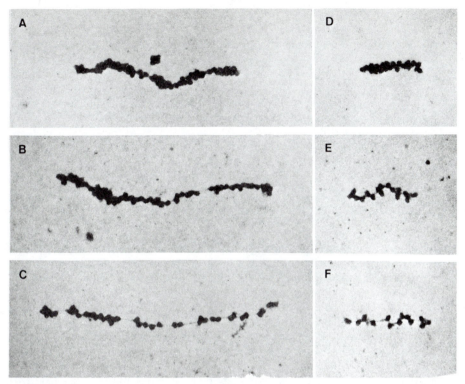

FIGURE 3.27 Electron micrographs of polysomes from *Chironomus tentans* salivary glands. Polysomes were isolated and sedimented in sucrose gradients to obtain a fraction of heavy polysomes (*A, B, C*) and light polysomes (*D, E, F*), which reflect the translation of two size classes of mRNA. ×85,400. (From B. Daneholt, K. Anderson, and M. Fagerlind. 1977. Reproduced from *The Journal of Cell Biology*, 1977, vol. 73, pp. 149–160 by copyright permission of The Rockefeller University Press.)

Protein synthesis is generally accomplished by several ribosomes attached to a single mRNA. As translocation moves the initiation codon out of a ribosome, a new initiation complex can be formed. This process can be repeated several times; the number of ribosomes that can simultaneously read a messenger is roughly proportional to the length of the messenger. The complex of mRNA and associated ribosomes is a **polyribosome** or **polysome** (Fig. 3.27).

As with transcription, inhibition of translation is an important experimental manipulation. Two commonly used inhibitors of eukaryotic translation are **puromycin** and **cycloheximide.** Puromycin is also frequently used in the preparation of messenger RNA from polysomes. This molecule enters the ribosome, causing premature termination of translation and release of ribosomes from the messenger. Thus, treatment of purified polysomes with puromycin allows for separation of the mRNA molecules from the ribosomes.

REFERENCES

Abelson, J. 1977. Recombinant DNA: Examples of present-day research. Science, *196*: 159–160.

Baldwin, J.P. et al. 1975. The subunit structure of the eukaryotic chromosome. Nature (Lond.), *253*: 245–249.

Bastos, R.N., and H. Aviv. 1977. Globin mRNA precursor molecules. Biosynthesis and processing in erythroid cells. Cell, 7: 641–650.

Bingham, P.M., and B.H. Judd. 1981. A copy of the *copia* transposable element is very tightly linked to the w^a allele at the *white* locus of D. melanogaster. Cell, *25*: 705–711.

Birnstiel, M.L., and M. Chipchase. 1977. Current work on the histone operon. Trends Biochem. Sci., *2*: 149–152.

Bogenhagen, D.F., S. Sakonju, and D.D. Brown. 1980. A control region in the center of the 5S RNA gene directs specific initiation of transcription: II. The 3′ border of the region. Cell, *19*: 27–35.

Brandhorst, B.P., and T. Humphreys. 1972. Stabilities of nuclear and messenger RNA molecules in sea urchin embryos. J. Cell Biol., *53*: 474–482.

Breathnach, R., and P. Chambon. 1981. Organization and expression of eucaryotic split genes coding for proteins. Ann. Rev. Biochem., *50*: 349–383.

Britten, R.J., and E.H. Davidson. 1969. Gene regulation for higher cells: A theory. Science, *165*: 349–357.

Britten, R.J., and E.H. Davidson. 1971. Repetitive and non-repetitive DNA sequences and a speculation on the origins of evolutionary novelty. Q. Rev. Biol., *46*: 111–138.

Britten, R.J., and D.E. Kohne. 1968. Repeated sequences in DNA. Science, *161*: 529–540.

Brown, D.D., and I.B. Dawid. 1968. Specific gene amplification in oocytes. Science, *160*: 272–280.

Brown, D.D., and E. Littna. 1964. RNA synthesis during the development of *Xenopus laevis*, the South African clawed toad. J. Mol. Biol., 8: 669–687.

Brown, D.D., and K. Sugimoto. 1973. 5S DNAs of *Xenopus laevis* and *Xenopus mulleri*: Evolution of a gene family. J. Mol. Biol., *48*: 397–415.

Calabretta, B. et al. 1981. mRNA in human cells contains sequences complementary to the Alu family of repeated DNA. Proc. Natl. Acad. Sci. U.S.A., 78: 6003–6007.

Clarkson, S.G., M.L. Birnstiel, and V. Serra. 1973. Reiterated transfer RNA genes of *Xenopus laevis*. J. Mol. Biol., 79: 391–410.

Crea, R. et al. 1978. Chemical synthesis of genes for human insulin. Proc. Natl. Acad. Sci. U.S.A., 75: 5765–5769.

Curtis, P.J., and C. Weissmann. 1976. Purification of globin mRNA from dimethylsulfoxide-induced Friend cells and detection of a putative globin mRNA precursor. J. Mol. Biol., *106*: 1061–1075.

Daneholt, B., K. Anderson, and M. Fagerlind. 1977. Large-size polysomes in *Chironomus tentans* salivary glands and their relation to Balbiani ring 75S RNA. J. Cell Biol., *73*: 149–160.

Davidson, E.H., and R.J. Britten. 1973. Organization, transcription, and regulation in the animal genome. Q. Rev. Biol., *48*: 565–613.

De Robertis, E.D.P., and E.M.F. De Robertis, Jr. 1980. *Cell and Molecular Biology.* 7th ed. Saunders College, Philadelphia.

Dierks, P. et al. 1981. DNA sequences preceding the rabbit β-globin gene are required for formation in mouse L cells of β-globin RNA with the correct 5′ terminus. Proc. Natl. Acad. Sci. U.S.A., 78: 1411–1415.

Doel, M.T. et al. 1977. The presence of ovalbumin mRNA coding sequences in multiple restriction fragments of chicken DNA. Nucleic Acids Res., 4: 3701–3713.

Dugaiczyk, A. et al. 1979. The ovalbumin gene: Cloning and molecular organization of the entire natural gene. Proc. Natl. Acad. Sci. U.S.A., 76: 2253–2257.

DuPraw, E.J. 1970. *DNA and Chromosomes.* Holt, Rinehart & Winston. New York.

Efstratiadis, A. et al. 1980. The structure and evolution of the human β-globin gene family. Cell, *21*: 653–668.

Fitzgerald, M., and T. Shenk. 1981. The sequence 5′-AAUAAA-3′ forms part of the recognition site for polyadenylation of late SV40 mRNAs. Cell, *24*: 251–260.

Foe, V.E., L.E. Wilkinson, and C.D. Laird. 1976. Comparative organization of active transcription units in Oncopeltus fasciatus. Cell, *9*: 131–146.

Fraser, T.H., and B.J. Bruce. 1978. Chicken ovalbumin is synthesized and secreted by *Escherichia coli.* Proc. Natl. Acad. Sci. U.S.A., *75*: 5936–5940.

Freifelder, D. 1982. *Physical Biochemistry.* 2nd ed. W. H. Freeman, San Francisco.

Galli, G., H. Hofstetter, and M.L. Birnstiel. 1981. Two conserved sequence blocks within eukaryotic tRNA genes are major promoter elements. Nature (Lond.), *294*: 626–631.

Gedamu, L., and G.H. Dixon. 1978. Effect of enzymatic "decapping" on protamine mRNA translation in wheat germ S-30. Biochem. Biophys. Res. Commun., *85*: 114–124.

Goeddel, D.V. et al. 1979. Expression in *Escherichia coli* of chemically synthesized genes for human insulin. Proc. Natl. Acad. Sci. U.S.A., *76*: 106–110.

Goldberg, R.D. et al. 1973. Nonrepetitive DNA sequence representation in sea urchin embryo messenger RNA. Proc. Natl. Acad. Sci. U.S.A., *70*: 3516–3520.

Goldstein, L., G.E. Wise, and C. Ko. 1977. Small nuclear RNA localization during mitosis: An electron microscope study. J. Cell Biol., *73*: 322–331.

Grosschedl, R., and M.L. Birnstiel. 1980. Identification of regulatory sequences in the prelude sequences of an H2A histone gene by the study of specific deletion mutants *in vivo.* Proc. Natl. Acad. Sci. U.S.A., *77*: 1432–1436.

Grosveld, G.C. et al. 1982. DNA sequences necessary for transcription of the rabbit β-globin gene *in vitro.* Nature (Lond.), *295*: 120–126.

Grummt, I. 1981. Mapping of a mouse ribosomal DNA promoter by *in vitro* transcription. Nucleic Acids Res., *9*: 6093–6102.

Grunberg-Manago, M. and F. Gros. 1977. Initiation mechanisms of protein synthesis. Prog. Nucleic Acid Res. Mol. Biol., *20*: 209–284.

Hentschel, C.C., and M.L. Birnstiel. 1981. The organization and expression of histone gene families. Cell, *25*: 301–313.

Hnilica, L.S. 1972. *The Structure and Biological Function of Histones.* CRC Press, Cleveland, Ohio.

Hoeijmakers-van Dommelin, H.A.M. et al. 1980. Localization of repetitive and unique DNA sequences neighbouring the rabbit β-globin gene. J. Mol. Biol., *140*: 531–547.

Hofer, E., and J.E. Darnell, Jr. 1981. The primary transcription unit of the mouse β-major globin gene. Cell, *23*: 585–593.

Hofstetter, H., A. Kressmann, and M.L. Birnstiel. 1981. A split promoter for a eukaryotic tRNA gene. Cell, *24*: 573–585.

Houck, C.M., F.P. Rinehart, and C.W. Schmid. 1979. A ubiquitous family of repeated DNA sequences in the human genome. J. Mol. Biol., *132*: 289–306.

Jacq, C., J.R. Miller, and G.G. Brownlee. 1977. A pseudogene structure in 5S DNA of Xenopus laevis. Cell, *12*: 109–120.

Jelinek, W., et al. 1974. Secondary structure in heterogeneous nuclear RNA: Involvement of regions from repeated DNA sites. J. Mol. Biol., *82*: 361–370.

Jelinek, W.R., et al. 1980. Ubiquitous, interspersed repeated sequences in mammalian genomes. Proc. Natl. Acad. Sci. U.S.A., *77*: 1398–1402.

Kedes, L.H. 1979. Histone genes and histone messengers. Ann. Rev. Biochem., *48*: 837–870.

Klug, A. et al. 1980. A low resolution structure for the histone core of the nucleosome. Nature (Lond.), *287*: 509–516.

Kornberg, R.D. 1977. Structure of chromatin. Ann. Rev. Biochem., *46*: 931–954.

Labhart, P., and T. Koller. 1982. Structure of the active nucleolar chromatin of Xenopus laevis oocytes. Cell, *28*: 279–292.

Laird, C.D., and W.Y. Chooi. 1976. Morphology of transcriptional units in *Drosophila melanogaster*. Chromosoma, *58*: 193–218.

Laird, C.D. et al. 1976. Analysis of chromatin-associated fiber arrays. Chromosoma, *58*: 169–192.

Lerner, M.R. et al. 1980. Are snRNPs involved in splicing? Nature (Lond.), *283*: 220–224.

Lewin, B. 1980. *Gene Expression*, Vol. 2. *Eucaryotic Chromosomes*. 2nd ed. John Wiley & Sons, New York.

Littau, V.C. et al. 1964. Active and inactive regions of nuclear chromatin as revealed by electron microscope autoradiography. Proc. Natl. Acad. Sci. U.S.A., *52*: 93–100.

Liu, A.Y. et al. 1977. Nucleotide sequences from a rabbit alpha globin gene inserted into a chimeric plasmid. Science, *196*: 192–195.

Lutter, L.C. 1979. Precise location of DNase I cutting sites in the nucleosome core determined by high resolution gel electrophoresis. Nucleic Acids Res., *6*: 41–56.

Maxam, A.M., and W. Gilbert. 1977. A new method for sequencing DNA. Proc. Natl. Acad. Sci. U.S.A., *74*: 560–564.

McClintock, B. 1980. Modified gene expressions induced by transposable elements. *In* W.A. Scott et al. (eds.), *Mobilization and Reassembly of Genetic Information*. 17th Miami Winter Symposium. Academic Press, New York, pp. 11–19.

McConaughy, B.L., and B.J. McCarthy. 1970. Related base sequences in the DNA of simple and complex organisms. VI. The extent of base sequence divergence among the DNAs of various rodents. Biochem. Genet., *4*: 425–446.

Mercereau-Puijalon, O. et al. 1978. Synthesis of an ovalbumin-like protein by *Escherichia coli* K12 harbouring a recombinant plasmid. Nature (Lond.), *275*: 505–510.

Miller, O.L., Jr., and A.H. Bakken. 1972. Morphological studies of transcription. *In* E. Diczfalusy (ed.), *Karolinska Symposia on Research Methods in Reproductive Endocrinology. 5th Symposium, Gene Transcription in Reproductive Tissue*. Periodica, Copenhagen, pp. 155–173.

Miller, O.L., Jr., and B.R. Beatty. 1969a. Extra-chromosomal nucleolar genes in amphibian oocytes. Genetics (Suppl.), *61*: 133–143.

Miller, O.L., Jr., and B.R. Beatty. 1969b. Visualization of nucleolar genes. Science, *164*: 955–957.

Noll, M., and R.D. Kornberg. 1977. Action of micrococcal nuclease on chromatin and the location of histone H1. J. Mol. Biol., *109*: 393–404.

Ochoa, S. 1968. Translation of the genetic message. Die Naturwissenchaften, *55*: 505–514.

O'Farrell, P.H. 1975. High resolution two-dimensional electrophoresis of proteins. J. Biol. Chem., *250*: 4007–4021.

Olins, A.L., and D.E. Olins. 1974. Spheroid chromatin units (*v* bodies). Science, *183*: 330–331.

O'Malley, B.W. et al. 1976. The synthesis, isolation, amplification and transcription of the ovalbumin gene. *In* D.P. Nierlich, W.J. Rutter, and C.F. Fox (eds.), *Molecular Mechanisms in the Control of Gene Expression*. Academic Press, New York, pp. 309–329.

Pardue, M.L. 1973. Localization of repeated DNA sequences in *Xenopus* chromosomes. Cold Spring Harbor Symp. Quant. Biol., *38*: 475–482.

Pardue, M.L., D.D. Brown, and M.L. Birnstiel. 1973. Location of the genes for 5S ribosomal RNA in *Xenopus laevis*. Chromosoma, *42*: 191–203.

Pardue, M.L., and J.G. Gall. 1972. Chromosome structure studied by nucleic acid hybridisation in cytological preparations. Chromosomes Today, *3*: 47–52.

Proudfoot, N.J., and G.G. Brownlee. 1976. 3' Non-coding region sequences in eukaryotic messenger RNA. Nature (Lond.), *263*: 211–214.

Prunell, A. et al. 1979. Periodicity of deoxyribonuclease I digestion of chromatin. Science, *204*: 855–858.

Ris, H., and D.F. Kubai. 1970. Chromosome structure. Ann. Rev. Genet., *4*: 263–294.

Ritossa, F.M., and S. Spiegelman. 1965. Localization of DNA complementary to ribosomal RNA in the nucleolar organizer region of *Drosophila melanogaster*. Proc. Natl. Acad. Sci. U.S.A., *53*: 737–745.

Roberts, R.J. 1982. Restriction and modification enzymes and their recognition sequences. Nucleic Acids Res., *10*: r117–r144.

Rosenberg, M., and D. Court. 1979. Regulatory sequences involved in the promotion and termination of RNA transcription. Ann. Rev. Genet., *13*: 319–353.

Ross, J. 1976. A precursor of globin messenger RNA. J. Mol. Biol., *106*: 403–420.

Rubin, C.M. et al. 1980. Partial nucleotide sequence of the 300-nucleotide interspersed repeated human DNA sequences. Nature (Lond.), *284*: 372–374.

Sahasrabuddhe, C.G., and K.E. Van Holde. 1974. The effect of trypsin on nuclease-resistant chromatin fragments. J. Biol. Chem., *249*: 152–156.

Sakonju, S., D.F. Bogenhagen, and D.D. Brown. 1980. A control region in the center of the 5S RNA gene directs specific initiation of transcription: I. The 5' border of the region. Cell, *19*: 13–25.

Sassone-Corsi, P. et al. 1981. Promotion of specific *in vitro* transcription by excised "TATA" box sequences inserted in a foreign nucleotide environment. Nucleic Acids Res., *9*: 3941–3958.

Scheller, R.H. et al. 1978. Specific representation of cloned repetitive DNA sequences in sea urchin RNAs. Cell, *15*: 189–203.

Sealy, L. et al. 1981. Characterization of a highly repetitive sequence DNA family in rat. J. Mol. Biol., *145*: 291–318.

Shapiro, A.L., E. Viñuela, and J.V. Maizel, Jr. 1967. Molecular weight estimation of polypeptide chains by electrophoresis in SDS-polyacrylamide gels. Biochem. Biophys. Res. Commun., *28*: 815–820.

Sharp, S. et al. 1981. Internal control regions for transcription of eukaryotic tRNA genes. Proc. Natl. Acad. Sci. U.S.A., *78*: 6657–6661.

Shatkin, A.J. 1976. Capping of eukaryotic mRNAs. Cell, *9*: 645–653.

Simpson, R.T. 1978. Structure of the chromatosome, a chromatin particle containing 160 base pairs of DNA and all the histones. Biochemistry, *17*: 5524–5531.

Smithies, O. et al. 1981. Co-evolution and control of globin genes. *In* S. Subtelny and U.K. Abbott (eds.), *Levels of Genetic Control in Development*. Thirty-ninth Symposium of the Society for Developmental Biology. Alan R. Liss, New York, pp. 185–200.

Southern, E.M. 1975. Detection of specific sequences among DNA fragments separated by gel electrophoresis. J. Mol. Biol., *98*: 503–517.

Spiers, J., and M.L. Birnstiel. 1974. Arrangement of the 5.8S RNA cistrons in the genome of *Xenopus laevis*. J. Mol. Biol., *87*: 237–256.

Stambrook, P.J. 1981. Interspersion of mouse satellite deoxyribonucleic acid sequences. Biochemistry, *20*: 4393–4398.

Steffenson, D.M., R. Appels, and W.J. Peacock. 1981. The distribution of two highly repeated DNA sequences within *Drosophila melanogaster* chromosomes. Chromosoma, *82*: 525–541.

Strobel, E., P. Dunsmuir, and G.M. Rubin. 1979. Polymorphisms in the chromosomal locations of elements of the *412, copia* and *297* dispersed repeated gene families in Drosophila. Cell, *17*: 429–439.

Sung, W.L. et al. 1979. Synthesis of the human insulin gene. Part II. Further improvements in the modified phosphotriester method and the synthesis of seventeen deoxyribooligonucleotide fragments constituting human insulin chain B and mini-C DNA. Nucleic Acids Res., *7*: 2199–2212.

Thoma, F., and Th. Koller. 1981. Unravelled nucleosomes, nucleosome beads and higher order structures of chromatin: Influence of non-histone components and histone H1. J. Mol. Biol., *149*: 709–733.

Thoma, F., Th. Koller, and A. Klug. 1979. Involvement of histone H1 in the organization of the nucleosome and of the salt-dependent superstructure of chromatin. J. Cell Biol., *83*: 403–427.

Thomas, J.O., and A.J.A. Khabaza. 1980. Cross-linking of histone H1 in chromatin. Eur. J. Biochem., *112*: 501–511.

Thomas, J.O., and R.D. Kornberg. 1975. An octamer of histones in chromatin and free in solution. Proc. Natl. Acad. Sci. U.S.A., *72*: 2626–2630.

Tilghman, S.M. et al. 1978a. Intervening sequence of DNA identified in the structural portion of a mouse β-globin gene. Proc. Natl. Acad. Sci. U.S.A., *75*: 725–729.

Tilghman, S.M. et al. 1978b. The intervening sequence of mouse β-globin gene is transcribed within the 15S β-globin mRNA precursor. Proc. Natl. Acad. Sci. U.S.A., *75*: 1309–1313.

Wallace, H., and M.L. Birnstiel. 1966. Ribosomal cistrons and the nucleolar organizer. Biochim. Biophys. Acta, *114*: 296–310.

Waring, M., and R.J. Britten. 1966. Nucleotide sequence repetition: A rapidly reassociating fraction of mouse DNA. Science, *154*: 791–794.

Wasylyk, B. et al. 1980. Specific *in vitro* transcription of conalbumin gene is drastically decreased by single-point mutation in T-A-T-A box homology sequence. Proc. Natl. Acad. Sci. U.S.A., 77: 7024–7028.

Watson, J.D. 1976. *Molecular Biology of the Gene*. 3rd ed. Benjamin/Cummings, Menlo Park, CA.

Weintraub, H., A. Larsen, and M. Groudine. 1981. α-Globin-gene switching during the development of chicken embryos: Expression and chromosome structure. Cell, *24*: 333–344.

Yang, V.W. 1981. A small nuclear ribonucleoprotein is required for splicing of adenoviral early RNA sequences. Proc. Natl. Acad. Sci. U.S.A., 78: 1371–1375.

4 Control of Gene Expression During Development

The evidence presented in Chapter 2 establishes that cells of an embryo, which are differentiating along distinct developmental pathways, may possess equivalent genetic information. This principle leads to the inevitable conclusion that the appearance of morphological and functional differences during development is due to differential utilization of genetic information. However, as with any generalization concerning the living world, there are exceptions to the rule of genome constancy during development. We shall discuss these exceptions in the box on page 136. Having outlined in the previous chapter the modes of information utilization, we shall now describe how the utilization of different portions of the genome can be modulated to produce divergent cell types.

As we pointed out earlier, the uniqueness of any cell is due to the functional proteins it contains. Obviously, every cell requires certain essential enzymes and structural proteins to carry on the "housekeeping" functions of normal metabolism. But the *raison d'être* of any type of cell in a multicellular organism is to carry out some special function that is not shared by any other cell type, and every specialized function requires a special set of enzymes and structural proteins. Cell specialization therefore requires a developmental commitment to restrict the protein synthetic machinery to the production of a special protein "set." Consequently, cell differentiation is marked by dynamic changes in the messenger RNA population in the cytoplasm.

The ultimate change in protein synthetic patterns is specialization for the synthesis of one major functional protein. This kind of restriction occurs when the messenger population becomes dominated by a single messenger species. For example, differentiation of an erythroid cell is accompanied by the preferential accumulation of globin messenger. A corollary to the presence of specific messengers in specialized cells is that these messengers are lacking in other specialized cell types; for example, globin messenger is unique to erythroid cells, which lack the messengers that characterize other cell types. The appearance and accumulation of specialized messengers provide a useful means of monitoring cell differentiation.

The translation of unique messengers by specialized cells is not the cause of differentiation, but its result. There is no single mechanism that causes these restrictions—regulation of gene expression could be exerted at each step in cellular information processing. In this chapter we shall examine these regulatory mechanisms. The most decisive step in information processing is transcription. Obviously, if a gene were not transcribed, there would be no need for posttranscriptional regulation. What mechanisms determine which genes are to be transcribed into RNA? In fact, is there any discrimination at all at this level, or are all genes transcribed, only to be thwarted at a later step? Since cytoplasmic RNA is derived from a more complex population of nuclear RNA, some discrimination process must determine which RNA can enter the cytoplasm. How are specific RNA sequences selected from the precursor population and thus saved from degradation by the nuclease enzymes? Finally, does a cell translate all of its messenger RNA, or are there mechanisms that channel specific ones onto ribosomes?

The answers to these questions are not all known, but enough is known to make an exciting story. Although much of this information has been gathered by molecular biologists, the results of their investigations are vital to the developmental biologist's understanding of cell differentiation. It is important to keep in mind that what is significant to developmentalists is the process of *change*. Development is a dynamic process, and comparisons of static differences are relevant to development only insofar as they relate to how different cell types are produced and how progressive stages of differentiation are achieved.

4–1. EVIDENCE FOR DIFFERENTIAL GENE TRANSCRIPTION

The majority of the nucleotide sequences in the genome are inactive in transcription in most cells. Selection of the "active" versus the "inactive" genes is considered by many investigators to be a critical step in controlling cell differentiation. However, experimental demonstration of differential transcription is difficult. The evidence for differential transcription has been derived from cytological studies and, more recently, from the fast-moving field of molecular biology.

Regular text continued on page 137

EXCEPTIONS TO THE GENE CONSTANCY RULE

The basic premise upon which this chapter is based is that the integrity of the genome is retained during development and that the appearance of specialized traits in particular cells is due to differential gene expression. This assumption is generally true, but there are some important exceptions. These include **gene deletion, gene amplification,** and **gene rearrangement.**

GENE DELETION.   The entire somatic cell line of some animals may lose a portion of its genome during early cleavage. This **chromosome diminution** is known to occur during development of some nematodes, crustaceans, and insects. The germ cell line in these embryos retains the complete genome. Some determinative event in very early development discriminates between the presumptive somatic cells and germ cells, enabling the latter cells to retain the complete genome while causing selective loss of a portion of the genome in somatic cells. An example of such a mechanism occurs in the nematode *Ascaris megalocephala* and will be discussed in Chapter 11.

Chromosome diminution is a singular and dramatic developmental event that lacks the subtlety of differential gene expression. There is no evidence that further selective loss of genes occurs in cell determination in species exhibiting this phenomenon. They presumably must resort to more conventional means of information selection for subsequent cell determination.

GENE AMPLIFICATION.   Multiple copies of certain genes may be selectively replicated in some cells to provide additional templates for the exaggerated production of transcripts from them. This is one mechanism that is employed by cells that require a large amount of a given gene product in a very short time. An example of gene amplification is the differential replication of 18S, 5.8S, and 28S ribosomal RNA genes in amphibian oocytes, which will be discussed in Chapter 7.

Recently, an example of structural gene amplification has been discovered: Genes for chorion proteins in *Drosophila melanogaster* are amplified in ovarian follicle cells (Spradling and Mahowald, 1980). As with ribosomal RNA gene amplification, the multiple copies of the chorion protein genes satisfy the need to produce a great deal of product in a very short time. The polytene chromosomes of *Drosophila* and other dipterans are an example of total genome amplification. There is recent evidence, however, that not all genes are replicated to the same extent (Laird et al., 1980). Perhaps this is a mechanism for modulation of production of numerous gene products in these cells.

GENE REARRANGEMENT.   In 1965, Dryer and Bennett proposed that the means for the incredible diversity of immunoglobulins produced by the vertebrate immune system is generated during development by rearrangement of separate genetic elements into functional immunoglobulin genes. This prediction has been confirmed by contemporary molecular biology research (Hozumi and Tonegawa, 1976; Seidman and Leder, 1978). Clearly, immunoglobulin-producing cells undergo irreversible changes in their genomes during their development—a definite exception to the gene constancy rule. However, this remains a singular exception and does not nullify the rule. We shall discuss rearrangement of immunoglobulin genes in detail in Chapter 14.

Much speculation has been generated about a possible role for transposable elements in modifying gene expression during development. As we discussed in Chapter 3, the insertion of one of these elements can affect the expression of adjacent genes. However, rearrangement of transposable elements has only been detected between successive generations of organisms rather than during the lifetimes of single organisms. Thus, we must assume that these elements are fixed in position during development.

Regular text continued from page 135

Polytene Chromosomes

One of the most dramatic demonstrations of differential transcription is the morphological change occurring in certain loci of dipteran polytene chromosomes when they become functionally active. Polytene chromosomes are found in certain enlarged cells of the larvae of some dipterans, including *Drosophila* and *Chironomus*. Dipterans are somewhat unusual, since homologous chromosomes pair and remain perfectly aligned in somatic cells during interphase. In larval salivary glands, midgut, and malpighian tubules of *Drosophila* and *Chironomus*, repeated DNA replication of both homologues follows somatic pairing. The numerous replicates (chromatids) do not separate but remain attached side by side, forming one extremely large structure—a polytene chromosome. Polytene chromosomes can attain a cross-sectional width as much as 10,000 times that of a normal interphase chromosome. The increased size and the parallel alignment of the chromatids reveal structural details that cannot be seen in normal interphase chromatin.

Figure 4.1 shows the polytene chromosomes of the *Drosophila* larval salivary glands. The four chromosomes radiate out from a structure called the **chromocenter,** which results from an aggregation of the centromeres of all the chromosomes. The chromosomes have a striated appearance of bands alternating with interband regions. The bands are visible because the concentration of DNA is greater than in the interbands, due to local folding of each chromatid. Each of these folded domains is called a **chromomere.** The parallel alignment of chromomeres on the individual strands produces the dense bands.

Beermann (1952) observed that some bands may exhibit a swollen or **puffed** appearance under certain conditions. A puff results from an

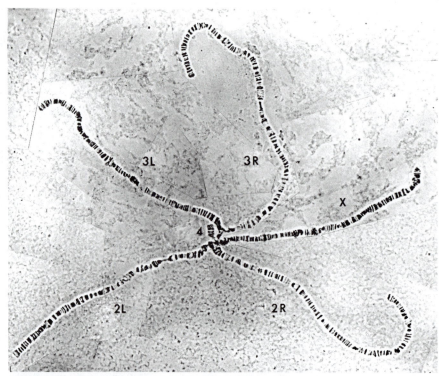

FIGURE 4.1 Composite photomicrograph of *Drosophila* salivary gland polytene chromosomes. The four chromosomes radiate from the chromocenter, which is formed by an aggregation of centromeres. The centromere of the X chromosome is at one end of the chromosome, so it appears as a single element. The centromeres of the second and third chromosomes are centrally located, so these chromosomes each have two arms—designated L (left) and R (right). The small fourth chromosome is located adjacent to the centromere. (From G. Lefevre, Jr. 1976, with permission from *The Genetics and Biology of Drosophila*, Vol. la. M. Ashburner and E. Novitski, Eds. Copyright: Academic Press Inc. [London] Ltd.)

unfolding of the chromomeres that constitute a band (Fig. 4.2). Puffing may be slight, with the only apparent change being a small enlargement and diffusion of the band, or it may be extreme, with the uncoiling of the chromomeres to form large loops. The larger puffs found in *Chironomus* are called **Balbiani rings.** Beermann proposed that the puffs represent sites of intense transcriptional activity. This hypothesis is supported by several lines of evidence. For example, examination of *Chironomus* salivary gland chromosomes, using the Miller technique for observing transcription by electron microscopy (see Chap. 3), reveals regions that are assumed to be Balbiani rings that show chromatin loops bearing highly active transcription units (Fig. 4.3). These loops are presumably formed by the unfolding of individual chromomeres. Also, puffs readily incorporate radioactive RNA precursors, as demonstrated by au-

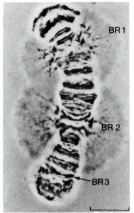

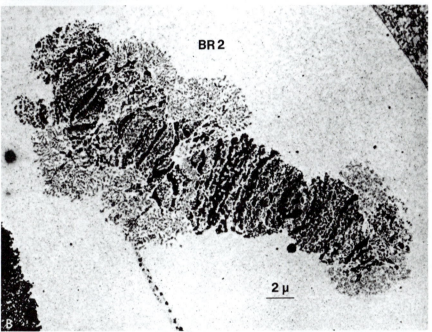

FIGURE 4.2 Polytene chromosome puffing. *A,* Phase contrast micrograph of chromosome 4 from the salivary gland of *Chironomus tentans.* The three Balbiani rings are indicated on the micrograph. Scale bar equals 7 μm. (From H. Sass. 1981. Effects of DMSO on the structure and function of polytene chromosomes of *Chironomus.* Chromosoma (Berl.), *83:* 623. Reprinted with permission of Springer-Verlag, Heidelberg.) *B,* Electron micrograph of chromosome 4 from a salivary gland cell of *Chironomus tentans.* The individual longitudinal strands, the chromatids, of the polytene chromosome are not resolved in this micrograph, but the transverse bands, each formed from homologous chromomeres, are easily observed. Moreover, the three giant puffs, the Balbiani rings (BR 2 being the intermediate one), are readily recognized. In the upper right corner, part of the cytoplasm is displayed. Nucleolar material comes into sight in the lower left corner. (From B. Daneholt. 1975. Transcription in polytene chromosomes. Cell, *4:* 1. Copyright © Massachusetts Institute of Technology; published by the MIT Press.)

Illustration continued on the following page

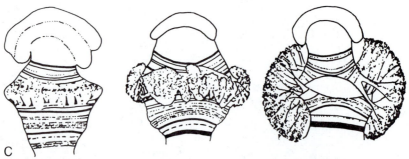

C

FIGURE 4.2 *(Continued)* C, Diagrammatic representation of one of the three large puffs (Balbiani rings) of chromosome 4 characteristic of all salivary gland nuclei of *Chironomus tentans* and *Chironomus pallidivittatus.* Three different stages of puffing are shown. Magnification approximately ×890. (From W. Beermann. 1963. Cytological aspects of information transfer in cellular differentiation. Am. Zool., *3:* 24.)

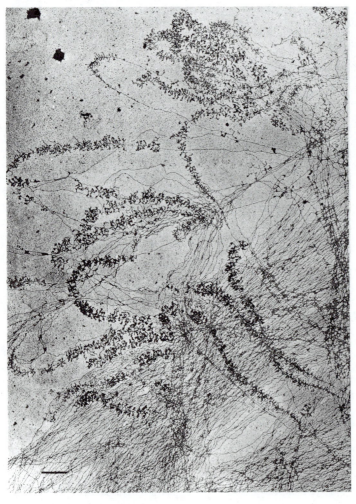

FIGURE 4.3 Electron micrograph of active transcription units from chromosome 4 of *Chironomus tentans.* Scale bar equals 1 μm. (From M.M. Lamb and B. Daneholt. 1979. Characterization of active transcription units in Balbiani rings of Chironomus tentans. Cell, *17:* 838. Copyright © Massachusetts Institute of Technology, published by the MIT Press.)

toradiography (Fig. 4.4). Furthermore, one can correlate the amount of incorporation of label with the extent of puffing; a puff site that has undergone experimentally induced puff regression demonstrates a commensurate reduction in ^{3}H-uridine incorporation. Regression of puffing can be induced with a variety of agents. This is illustrated by incubation of *Chironomus* salivary glands with the RNA polymerase II and III inhibitor, α-amanitin, which causes regression of Balbiani rings. As shown in Figure 4.5, incorporation of ^{3}H-uridine in the regions of puff regression (as with all nonnucleolar chromosome regions) has been eliminated. These results suggest that these regions are transcriptionally active when they are decondensed (i.e., puffed) and suppressed when they are condensed.

Confirmation of this interpretation is provided by experiments in which puffing is induced, and this morphological change is correlated with the synthesis of specific transcripts. An experimental procedure that facilitates this kind of analysis is treatment of *Drosophila* with high temperatures (Ritossa, 1962). Briefly shifting the organisms from normal temperatures (approximately 25°C) to 37°C elicits the formation of a discrete set of puffs (Fig. 4.6) and the synthesis of a specialized set of mRNAs and proteins. These proteins are called **heat shock proteins.** At the same time, the expression of other gene loci is inhibited; all preexisting puffs regress, most nonheat shock RNAs are no longer produced, and the translation of preexisting messengers is halted. The same effects on RNA and protein synthesis are also observed in other *Drosophila* tissues and cells as well as in tissue culture cells.

The role of heat shock puffs in the synthesis of heat shock RNA in salivary glands is demonstrated by *in situ* hybridization of a cloned heat shock gene to nascent RNA on the puffs. This technique, which was developed by Pukkila (1975), differs from that used for *in situ* hybridization of probes to chromosomal DNA (see Chap. 3) in that the chromosomes are not denatured. Thus, the chromosomal DNA remains double

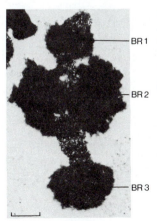

BR 1

BR 2

BR 3

FIGURE 4.4 Autoradiograph of chromosome 4 of *Chironomus tentans* illustrating ^{3}H-uridine incorporation by Balbiani rings. Scale bar equals 7μm. (From H. Sass. 1981. Effects of DMSO on the structure and function of polytene chromosomes of *Chironomus.* Chromosoma (Berl.), *83:* 634. Reprinted with permission of Springer-Verlag, Heidelberg.)

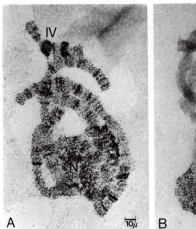

FIGURE 4.5 Effects of α-amanitin on ³H-uridine incorporation by *Chironomus pallidivittatus* salivary gland chromosomes. *A*, Control. *B*, Chromosomes from α-amanitin treated salivary gland. Note that the nucleolus (N) remains enlarged and incorporates label, while the Balbiani rings on chromosome 4 (IV) have regressed and have ceased incorporation. Scale bar equals 10 μm. (From W. Beermann. 1971. Effect of α-amanitine on puffing and intranuclear RNA synthesis in *Chironomus* salivary glands. Chromosoma (Berl.), *34:* 160. Reprinted with permission of Springer-Verlag, Heidelberg.)

stranded and inaccessible to the DNA probe. The hybridization detected by autoradiography is therefore due to annealing between the probe DNA and the nascent RNA. As shown in Figure 4.7, two of the heat shock-induced puffs contain RNA that hybridizes with the cloned probe. No hybridization is detected without heat shock. Consequently, heat shock induces transcription at these loci.

Comparisons of puffing patterns of polytene chromosomes during normal development illustrate that transcription is subject to both spatial and temporal regulation. Examination of puffs in different cell types reveals that each type has a unique puffing pattern. For example, Balbiani ring 2 is formed in salivary gland chromosomes of *Chironomus tentans* but not in other tissues having polytene chromosomes, such as the malpighian tubules (Daneholt et al., 1978). Temporal regulation is illustrated by examination of puffs of *Drosophila* in the same tissue at different

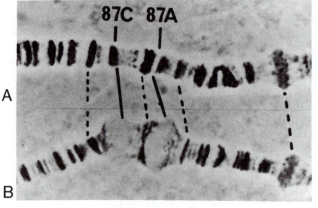

FIGURE 4.6 Induction of puffs on a *Drosophila melanogaster* salivary gland chromosome by heat shock (40 minutes at 37°C). *A*, Control. *B*, Heat-shocked. Note puffs formed by bands at 87C and 87A. (From M. Ashburner and J.J. Bonner. 1979. The induction of gene activity in Drosophila by heat shock. Cell, *17:* 242. Copyright © Massachusetts Institute of Technology; published by the MIT Press.)

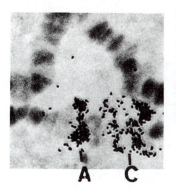

FIGURE 4.7 Hybridization *in situ* of a cloned heat shock gene to RNA on a salivary gland chromosome from a *Drosophila melanogaster* larva kept at 36°C for 15 minutes. This autoradiograph shows hybridization at puffs formed by bands at 87A and 87C (labeled A and C, respectively). ×1175. (From K.J. Livak et al. 1978. Sequence organization and transcription at two heat shock loci in *Drosophila*. Proc. Natl. Acad. Sci. (U.S.A., *75:* 5617.)

stages of development (Fig. 4.8); the pattern of transcription changes as development proceeds. Developmental regulation of puffing is due to the molting hormone **ecdysone,** which causes puffing of some gene loci and regression of puffs at other sites. We shall discuss this function of ecdysone in section 4–2.

Facultative Heterochromatin

Visual evidence for differential transcription is also provided by facultative heterochromatin, in which one chromosome of a pair becomes heterochromatic during development, and hence, only one gene of each allele pair is active in transcription. In addition to its genetic inactivity, facultative heterochromatin has other properties in common with constitutive heterochromatin; that is, both types of heterochromatin remain condensed during interphase and are late replicating. Their fundamental difference is that constitutive heterochromatin is composed of highly repetitive DNA, whereas facultative heterochromatin is not. Instead, the latter type behaves as heterochromatin because of a still undefined developmental event that renders it heterochromatic.

A classical example of facultative heterochromatin involves one of the X chromosomes of female mammals. Although females contain two X chromosomes (and hence two copies of each X-linked gene) and males carry only one, both sexes produce similar amounts of proteins that are coded by genes located on the X chromosome. This is achieved by a mechanism called **dosage compensation.** Mary Lyon (1961) proposed that dosage compensation occurs by random heterochromatization of either the paternal or maternal X chromosome during development (the **Lyon hypothesis**). After the initial inactivation occurs in an embryonic cell, the same X chromosome remains inactive in that cell and all its mitotic derivatives during the lifetime of the organism. The heterochromatic X can be seen as a darkly staining structure called the **Barr body** lying near the nuclear envelope in interphase nuclei (Fig. 4.9).

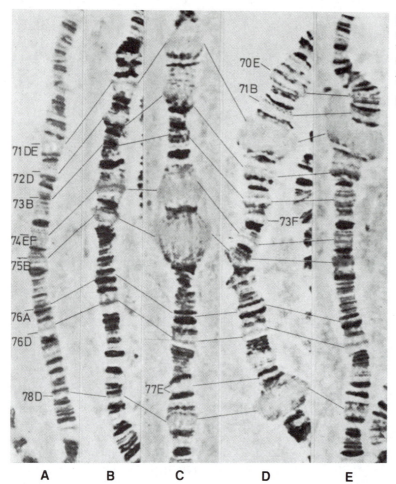

FIGURE 4.8 Puffing sequence of a portion of salivary gland chromosome 3 of *Drosophila melanogaster*. *A* and *B*, 110 hr. larva. *C*, 115 hr. larva, *D*, 0 hr. prepupa. *E*, 4 hr. pre-pupa.

70E

71B

71DE

72D

73B

74EF

75B

73F

76A

76D

78D

77E

A   B   C   D   E

Illustration continued on the opposite page

Since heterochromatization is random, females that are heterozygous for genes on the X chromosome have a **mosaic** phenotype; that is, some cells will have the paternal chromosome inactivated, whereas others will lose the function of the maternal chromosome. For example, the gene for glucose-6-phosphate dehydrogenase (G-6-PD) is located on the X chromosome in humans. Two different alleles of the G-6-PD gene have been found in the American black population. These alleles produce enzymes with different electrophoretic mobilities. The alleles are designated A and B; A produces an enzyme that migrates rapidly during electrophoresis, whereas B produces a slowly migrating enzyme.

Individual cells of a female heterozygous for the two electrophoretic forms produce either the A form or the B form of the enzyme but not both. She is, therefore, a phenotypic mosaic, with small patches of cells that produce either form of the enzyme scattered throughout her body.

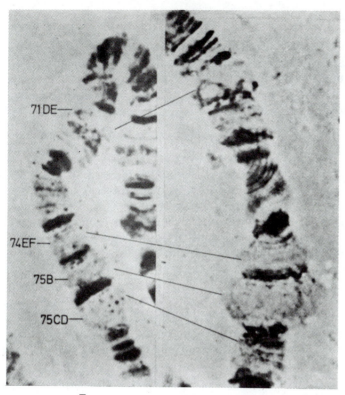

71DE—

74EF—

75B—

75CD—

F G

FIGURE 4.8 *(Continued) F,* 8 hr. pre-pupa. *G,* 12 hr. pre-pupa. (From M. Ashburner. 1972b. Puffing patterns in *Drosophila melanogaster* and related species. *In* W. Beermann (ed.), *Developmental Studies on Giant Chromosomes.* Springer-Verlag, Berlin, pp 119–120.)

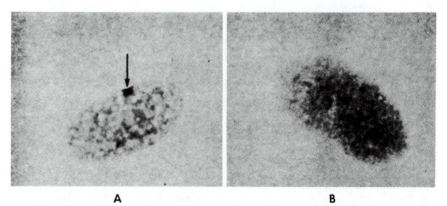

A B

FIGURE 4.9 Barr body in human tissue culture cell nuclei. This darkly staining structure, which is the heterochromatic X chromosome, is present in female cell nuclei (*A*) but absent in male cell nuclei (*B*). Approximately ×2300. (From S.W. Brown. 1966. Heterochromatin. *Science 151:* 417–425. Copyright 1966 by the American Association for the Advancement of Science.)

This can be demonstrated by sampling skin cells, which can be conveniently removed and placed in culture. After the cells are established and have multiplied *in vitro*, single cells are selected to form new cultures, which are clones of genetically identical cells. Cells from the clones are then homogenized, and the G-6-PD enzymes are resolved by electrophoresis. As shown in Figure 4.10, each clone has only the A or the B form of the enzyme, never both.

Molecular Evidence

Differential transcription is dramatically demonstrated by polytene chromosome puffing patterns and by formation of facultative heterochromatin. Most chromatin, however, is not so obliging as to provide such obvious clues of transcriptional activity. Therefore, we require molecular techniques to reveal modulation of transcription so that we may examine transcriptional regulation in detail. Differential transcription is indicated by the presence of different transcripts in different cell types. We know, for example, that specialized cells preferentially accumulate certain transcripts. Molecular probes, such as cloned cDNA, have the specificity and sensitivity to detect low levels of transcripts, and thus can distinguish between their presence and absence. Hybridization analyses between probes and RNA extracted from cells reveal that (1) the transcripts for cell-specific proteins increase in amount during cell dif-

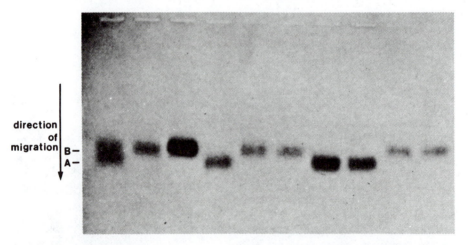

FIGURE 4.10 Electrophoretic pattern of G-6-PD from homogenates of cultured cells. Samples were run singly, starting from the origin at the top of the figure. From left to right are: the AB phenotype of the mixed cell culture from Mrs. De. prior to cloning; and the single bands of nine clones derived from the original cell lines. Variation in intensity of staining is due to inequality of enzyme concentration applied to the starch gel. (After R.G. Davidson, H.M. Nitowsky, and B. Childs. 1963. Demonstration of two populations of cells in the human female heterozygous for glucose-6-phosphate dehydrogenase variants. Proc. Natl. Acad. Sci. U.S.A., *50:* 484.)

ferentiation and (2) cells specialized for other functions do not have significant numbers of these same transcripts. For example, globin transcripts are present in chick embryo erythropoietic tissue but are not found in significant amounts in either the nuclear or cytoplasmic RNA of nonerythropoietic tissue (Groudine et al., 1974). The results of a large number of experiments with probes for transcripts encoding cell-specific proteins strongly suggest that genes for such proteins are regulated at the transcriptional level.

4–2. TRANSCRIPTIONAL CONTROL MECHANISMS

The preceding evidence indicates that gene expression *can* be regulated at the transcriptional level. We shall now examine *how* transcriptional regulation could occur. This discussion will center on chromatin itself and on extrinsic factors that affect cell differentiation.

Properties of Transcribed Chromatin

The discovery of the basic structure of chromatin has made it possible to attempt to correlate chromatin structure with its function in transcription. Specifically, does the configuration of chromatin differ between transcribed and nontranscribed chromatin? If so, what is the nature of the difference, and how is it regulated? The logistical problems of transcription suggest that modifications to chromatin would facilitate RNA synthesis. For example, the RNA polymerase molecule is itself about the same size as a nucleosome. The DNA is wound around the outside of the nucleosome core particle. The movement of the large polymerase molecule along the DNA would probably be facilitated if some modification of this configuration were effected. Furthermore, the two strands of the DNA helix are separated in the region that is transcribed by the polymerase. Strand separation may be facilitated by a modified nucleosomal configuration. Alternatively, strand separation itself would most likely alter nucleosomes.

The most useful tools for distinguishing between transcribed and nontranscribed chromatin have been nuclease enzymes such as micrococcal nuclease and DNase I. The use of these enzymes is illustrated by studies on chick globin and ovalbumin genes. These genes are relatively resistant to digestion by micrococcal nuclease in all tissues, demonstrating that they are packaged into nucleosomelike particles. However, brief treatment of erythrocyte nuclei with DNase I (Fig. 4.11) degrades much of the DNA that hybridizes with globin cDNA (i.e., the globin genes). As a control, ovalbumin cDNA was used to probe the presence of ovalbumin genes in nuclease-treated erythrocyte and fibroblast nuclei. These genes are inactive in both cell types, and as shown in the graph, they are not digested by the enzyme. The globin gene is also inactive in fibroblasts. In a result that is consistent with the ovalbumin data, DNase I

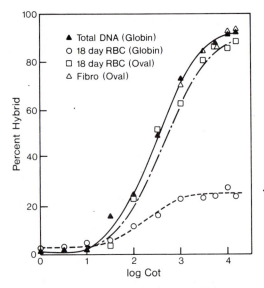

FIGURE 4.11 Preferential digestion of active genes by pancreatic deoxyribonuclease I. Red blood cells (RBC) were obtained from 18-day chick embryos, and fibroblasts (fibro) were dissected from 11-day chick embryos and grown in culture. Nuclei were isolated from these cells and digested with DNase I until 10 to 20% of the DNA was soluble in trichloroacetic acid. The remaining DNA was extracted for hybridization studies. Total DNA was prepared directly from 18-day red cell nuclei. DNA samples were hybridized with either ovalbumin cDNA or globin cDNA. The percentage of hybridized cDNA is plotted as a function of the Cot. At the completion of the reaction between total DNA and globin cDNA, the curve plateaus since all of the cDNA has reacted with complementary sequences. However, after treatment of red cell nuclei with the nuclease enzyme, the hybridization reaction plateaus at a lower level. The lower plateau indicates that fewer globin sequences are available to react with the cDNA. The difference between plateau levels is a measure of the globin sequences digested by the enzyme. By comparison, the plateau levels with ovalbumin cDNA are not reduced in either red cell or fibroblast DNA from nuclease-treated nuclei. Therefore, the ovalbumin sequences are not digested by the enzyme. (Redrawn from H. Weintraub and M. Groudine. 1976. Chromosomal subunits in active genes have an altered chromatin. Science *193:* 848–856. Copyright 1976 by the American Association for the Advancement of Science.)

was found to have no effect on globin genes of fibroblast nuclei. Differential sensitivity to DNase I reflects not only the transcriptional activity of genes but also latent transcriptional competence; that is, globin genes of mature erythrocytes that are no longer active in transcription retain this property (Weintraub and Groudine, 1976).

Garel and Axel (1976) investigated the sensitivity of ovalbumin genes to DNase I in hen oviduct nuclei and liver nuclei; the ovalbumin genes are transcribed in the former tissue but not in the latter. In agreement with the previously described studies, they found that brief treatment with DNase I preferentially digests the ovalbumin genes in nuclei in which they are transcriptionally active (i.e., in oviduct nuclei, but not in liver nuclei). The preferential sensitivity of chromatin composed of transcribed genes suggests that the nucleosomes of this chromatin are in a modified configuration that renders the DNA more susceptible to nuclease digestion. Perhaps this configuration facilitates transcription by making the DNA accessible to RNA polymerase, but this remains to be verified.

The demonstration of a modified configuration for transcriptionally active chromatin has stimulated a search for the structural basis of this difference. It is hoped that this search will yield important clues regarding the regulation of gene expression. Attention has been focused on the nonhistone chromosomal proteins, specifically a fraction known as the **high mobility group (HMG).** These small proteins have high mobility on polyacrylamide gels—hence their name. They have a high percentage of both acidic and basic amino acids, making them highly charged molecules. They are easily extracted from chromatin with 0.35 M NaCl. Two HMG proteins have received considerable attention. These are HMG 14 and HMG 17, which are found in a variety of cell types in a large number of organisms (Goodwin et al., 1978).

Experimental evidence correlates the presence of HMGs 14 and 17 with the differential sensitivity of transcriptionally active genes to DNase I. When embryonic chick erythrocyte nuclei are treated with DNase I, HMGs 14 and 17 are solubilized (Weisbrod and Weintraub, 1979). Thus, they are released from erythrocyte chromatin by the same treatment that solubilizes globin genes. To test the relationship between HMG proteins and DNase I sensitivity, Weisbrod and Weintraub pretreated chick erythrocyte chromatin with 0.35 M NaCl—the procedure that removes HMGs 14 and 17 from chromatin. As shown in Figure 4.12, chromatin treated in this way displays no preferential DNase I sensitivity

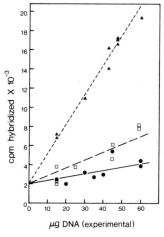

μg DNA (experimental)

FIGURE 4.12 Reconstitution of DNase I sensitivity of chick embryo erythrocyte chromatin with HMG proteins. The chromatin was treated with 0.35M NaCl to remove HMGs 14 and 17 and then was separated into separate aliquots. One portion was treated with DNase I (line with solid triangles), and the other was reconstituted with HMGs 14 and 17 before DNase I treatment (line with open squares). As controls, nuclei were treated with DNase I to 10% trichloroacetic acid solubility (line with closed circles). Each sample was then hybridized with excess globin cDNA (total globin cDNA was 100,000 cpm) to measure relative numbers of globin genes in the chromatin. There is a clear correlation between the presence of HMGs 14 and 17 and DNase I sensitivity. (After S. Weisbrod and H. Weintraub. 1979. Isolation of a subclass of nuclear proteins responsible for conferring a DNase I-sensitive structure on globin chromatin. Proc. Natl. Acad. Sci. U.S.A., 76: 632.)

of the globin gene as assayed by hybridization with globin cDNA. However, when purified HMGs 14 and 17 are recombined with HMG-depleted chromatin, DNase I sensitivity is restored. This result indicates that the globin genes in HMG-depleted chromatin retain some property that allows them to bind HMG proteins selectively, which renders them DNase I sensitive. Furthermore, HMGs 14 and 17 from brain nuclei will restore DNase I sensitivity to HMG-depleted erythrocyte chromatin, but erythrocyte HMGs will not render the globin genes of brain chromatin sensitive to DNase I. Thus, the HMG proteins are not responsible for the selectivity; this property is retained by some other component of the chromatin. In a more recent study, Weisbrod et al. (1980) demonstrated that actively transcribed genes *in general* are rendered sensitive to DNase I by HMG proteins—an indication that this property is not unique to globin genes.

The experiments we have just discussed suggest that differential nuclease sensitivity is dependent upon the association of HMGs 14 and 17 with active genes. However, this association itself must be a consequence of some more basic property of transcribed genes that causes HMGs 14 and 17 to interact with them. Two possibilities to be considered are modifications to histones and to DNA itself. Perhaps significantly, the known modifications to histones occur in regions of the molecules that interact with DNA. Thus, changes in histones could reduce the compaction of DNA within nucleosomes and facilitate transcription. One histone modification that is known to be correlated with transcriptional capacity of chromatin is acetylation of the lysine residues of the core histones (Allfrey, 1977). Chromatin containing a high level of histone acetylation is very sensitive to DNase I, which we have learned is a property of transcriptionally active chromatin (Simpson, 1978; Vidali et al., 1978; Davie and Candido, 1980). It has been proposed that HMGs 14 and 17, which are also associated with transcriptionally active chromatin, may help to maintain the high level of acetylation in this chromatin (Reeves and Candido, 1980).

Recently, a great deal of interest has been generated concerning the possibility that modifications to DNA bases may be correlated with transcriptional activity of genes. The most common modification is post-replication methylation of cytosine to produce 5-methylcytosine (m^5C). Methylation occurs preferentially in the dinucleotide sequence m^5CG and is detected by a comparison of the cleavage patterns of the restriction enzymes *Hpa* II and *Msp* I. Both these enzymes recognize sites containing CCGG sequences. However, *Hpa* II cannot cleave Cm^5CGG, whereas *Msp* I cleaves both CCGG and Cm^5CGG. Thus, *Msp* I is used to identify CCGG sites in genes, and methylation of these sites is detected by comparison with the digestion pattern obtained with *Hpa* II.

In a typical experiment, total DNA is restricted with these enzymes, the DNA fragments are resolved by agarose gel electrophoresis and transferred by Southern blotting to nitrocellulose. The DNA fragments containing genes of interest are then visualized by hybridization to a labeled probe. The results of such an experiment are shown in Figure 4.13. The figure compares the restriction patterns obtained for the α-fetoprotein (α-FP) gene from yolk sac endoderm and mesoderm (see Chap. 10 for a discussion of the mammalian yolk sac). The α-FP gene is expressed in endoderm but not in the mesoderm. *Msp* I cleaves the α-FP genes from both tissues into five fragments (labeled a–e). The same five fragments are generated with *Hpa* II treatment of endoderm DNA. However, treatment of mesoderm DNA with *Hpa* II releases only one fragment. These contrasting results demonstrate that the CCGG sites are methylated in mesoderm but hypomethylated (i.e., the level of methylation is reduced) in endoderm—the tissue in which the gene is expressed.

Results with a number of genes indicate that in cells expressing a specific gene, CCGG sites are hypomethylated, whereas the highest levels of methylation are found in tissues in which the gene is not expressed (Razin and Riggs, 1980). These observations support the hypothesis that demethylation may be necessary for gene expression during development.

The demethylation hypothesis is also supported by the results of experiments with the nucleoside analogue 5-azacytidine. This drug, which causes demethylation of cytosine residues in DNA, has been shown to promote the expression of previously unexpressed genes (Groudine et al., 1981; Compere and Palmiter, 1981) and *in vitro* differentiation of cultured cells (Jones and Taylor, 1980; Taylor and Jones, 1982). An-

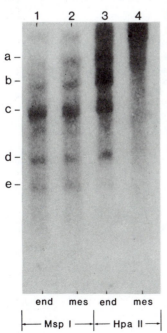

FIGURE 4.13 Methylation of the α-fetoprotein gene in the mouse yolk sac. DNA was restricted with either *Hpa* II or *Msp* I and analyzed by Southern blotting using α-FP cDNA as a probe. Tracks 1 and 2 show *Msp* I-restricted DNA from yolk sac endoderm and mesoderm respectively, while tracks 3 and 4 show *Hpa* II-restricted DNA from these same tissues. (From G.K. Andrews, M. Dziadek, and T. Tamaoki. 1982. Expression and methylation of the mouse α-fetoprotein gene in embryonic, adult, and neoplastic tissues. J. Biol. Chem., *257*: 5151.)

other dramatic example of the effects of 5-azacytidine is the activation of genes on heterochromatic mammalian X chromosomes in somatic cell hybrids with this drug (Mohandas et al., 1981).

One of the most intriguing consequences of the demethylation hypothesis for regulation of gene expression is that it satisfies the important requirement for stable transmission of the pattern of gene expression of a cell to its progeny (Wigler et al., 1981; Stein et al., 1982). Methylation of the appropriate cytosine residues occurs enzymatically after replication of DNA, conserving the methylation pattern of the parental strands in each daughter strand (Riggs, 1975; Holliday and Pugh, 1975; Bird, 1978). This is represented in Figure 4.14. This figure illustrates another important property of methylated CG sequences; that is, the methylation pattern is symmetrical on the two strands of the DNA double helix. The symmetrical methylation makes possible the clonal inheritance of the exact methylation pattern at cell division, presumably by means of a methylase enzyme that recognizes the half-methylated sites that exist after replication and restores the fully methylated symmetrical pattern.

We shall now summarize the demethylation hypothesis. Genes may be retained in the unexpressed state by maintenance of the methylation of certain cytosine residues within the genes or in their vicinity.

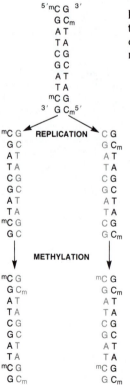

FIGURE 4.14 Restoration of the parental methylation pattern following DNA replication. Note that of the three CG dinucleotides only those methylated before replication show methylation in their complementary strands after replication.

When cells with methylated cytosines divide, the methylation pattern is inherited by their progeny. The cell determination events that lead to differentiation of specific cell types would cause demethylation of these critical cytosine residues in particular genes, allowing these genes to be expressed. When determined cells divide, they retain the pattern of methylation–demethylation that maintains the differentiation state in their progeny. Perhaps as a consequence of demethylation, HMGs 14 and 17 associate with the chromatin containing these genes, which in turn displays enhanced DNase I sensitivity. Conversely, demethylation might be a consequence of the presence of HMGs 14 and 17.

In addition to the generalized sensitivity of transcribed genes to DNase I as previously described, it has been shown that such genes are flanked at their 5' ends (and sometimes at their 3' ends) by so-called **hypersensitive sites,** which are preferentially cleaved by light digestion with the nuclease (Wu, 1980; Stalder et al., 1980; Keene et al., 1981). For example, the appearance of hypersensitive sites of globin genes has been correlated with the expression of these genes during cell differentiation. The reasons for hypersensitivity are not known. One possibility is that a special nucleotide sequence that recognizes and interacts with regulatory proteins is preferentially susceptible when the proteins are present. Alternatively, the nucleotide sequence may have a peculiar base order that is nuclease sensitive when the chromatin is in the active configuration (Wu, 1980).

Specific Gene Regulation

The preceding section of this chapter outlines some of the generalized properties of transcriptionally active chromatin that may facilitate transcription and maintenance of the transcriptionally active state of genes. However, we have not addressed the problem of the regulation of *differential* gene expression. How are specific genes selected to be transcribed from the vast library of available sequences? It is this selectivity that allows for the differentiation of cells with specialized properties. It is generally assumed that specific nonhistone chromosomal proteins interact with gene regulatory sequences to facilitate gene expression. However, proof of this hypothesis is a very difficult proposition. One problem is that many, if not most, of the NHC proteins show little tissue specificity (e.g., HMGs 14 and 17). Moreover, since most structural genes are present as single copies, putative regulatory proteins may be present in amounts too small to be detected by current analytical techniques. The 5S RNA genes of *Xenopus laevis* are highly reiterated (see Chap. 3) and for this reason (among others) have provided a favorable system to search for a specific regulatory protein.

Transcription of 5S RNA genes is regulated by a protein called TF IIIA. The function of TF IIIA is demonstrated by *in vitro* transcription analyses. Cloned 5S RNA genes are transcribed *in vitro* only in the presence of TF IIIA (Honda and Roeder, 1980; Pelham et al., 1981). When TF IIIA is combined with the 5S RNA genes it binds to the gene promoters to facilitate transcription (Engelke et al., 1980). The specificity of TF IIIA is demonstrated by the fact that it has no effect on transcription of tRNA genes, which—like the 5S RNA genes—are transcribed by RNA polymerase III. As we shall discuss in more detail in Chapter 7, the presence of TF IIIA is thought to be necessary for the transcription of 5S RNA genes during oogenesis. TF IIIA is reduced to very low levels in eggs after ovulation (Honda and Roeder, 1980; Pelham et al., 1981); this deficiency is apparently responsible for the absence of 5S RNA synthesis following ovulation and for preventing the activation of the zygote's oocyte-type 5S RNA genes during early embryonic development (Honda and Roeder, 1980; Bogenhagen et al., 1982; Gottesfeld and Bloomer, 1982). Thus, regulation of the cellular level of TF IIIA may indirectly regulate expression of the oocyte-type 5S RNA genes. Additional work with 5S RNA genes and the discovery of other gene-specific regulatory proteins and their effects on chromatin will surely yield additional valuable clues about regulation of transcription.

The distinct properties of transcriptionally active chromatin we have described must be taken into consideration in any experiments attempting to modify previously established patterns of gene expression, such as in nuclear transplantation and somatic cell hybridization studies (see Chap. 2 and the next section of this chapter). The failure of a nucleus to show an alternate pattern of gene expression does not necessarily mean that the DNA nucleotide sequence has been changed. It may simply mean that the pattern of expression imprinted on the chromatin (by the presence of NHC proteins, acetylation of histones, or methylation of cytosine) is stable and cannot be reversed under the particular experimental conditions. A more complete understanding of the regulation of gene expression will be forthcoming when we discover more not only about the establishment of patterns of gene expression but about reversal of that process as well.

Coordination of Transcription

Although immediate control over transcription may be mediated by chromosomal proteins, it is still necessary to account for the chronological and spatial patterns of gene expression that occur during orderly development. The transcription of a gene or set of genes may be mediated by the presence of a factor produced by yet another gene in the same cell. For example, the transcription of 5S RNA genes is dependent upon the presence of TF IIIA, which is itself a product of gene expression. As we shall discuss in Chapter 7, the transcription of the oocyte-type 5S RNA genes during oogenesis and after fertilization appears to be regulated by the cellular level of TF IIIA.

The functional properties of cells in embryos may be influenced by specialized cytoplasm that is heterogeneously distributed in eggs (**nucleocytoplasmic interactions**) and by interactions between adjacent cells (**cell-cell interactions**). These effects could be exerted at any level, and demonstrating regulation at the level of transcription is not an easy task. One experimental system that is amenable to such an analysis is somatic cell hybridization. In one such experiment, Deisseroth and Hendrick (1979) have shown that human α-globin genes from nonerythroid cells are transcribed when fused with mouse erythroid cells. Thus, some cytoplasmic factor or process in the mouse erythroid cells must have the capacity to activate quiescent α-globin genes.

Another significant influence over cell function during development is the endocrine system. Hormones have a dramatic effect on organisms and the state of differentiation of their cells. Experimental analysis of hormonal effects is facilitated by their ease of administration, both *in vivo* and *in vitro*. Consequently, transcriptional regulation by hormones has been studied extensively. We shall analyze the results of some of these experiments subsequently.

REGULATION OF PUFFING IN POLYTENE CHROMOSOMES BY ECDYSONE

Direct hormonal control of transcription is illustrated by the modulation of the puffing activity of polytene chromosomes in dipteran larvae. Detailed examination of the polytene chromosomes has revealed the existence of specific patterns of puffing activity that vary with cell type and stage of development. The larval phase of the dipteran life cycle is characterized by a series of stages, each terminated by a **molt.** The final molt (sometimes called metamorphosis) results in formation of the immobile pupa. A complex pattern of regression of some puffs and appearance of new puffs occurs just preceding larval molts or metamorphosis.

These abrupt changes in puffing pattern correspond to the time of release of the hormone ecdysone, which triggers molting. This correlation led Becker (1959) to propose that ecdysone is directly responsible for regulation of puffing activity. Since ecdysone is released from the prothoracic gland, which is located in the first thoracic segment, Becker (1962) was able to ligature a *Drosophila* larva prior to hormone release so that the anterior portion of the elongate salivary gland would be exposed to the hormone, but the posterior portion would be isolated from the source of the hormone. When Becker examined the polytene chromosomes, he found that the chromosomes in the anterior part of the gland showed a normal stage-specific puffing pattern after ecdysone release, whereas those in the posterior portion retained the pattern existing before ligation. In a further series of experiments Becker transplanted salivary glands between animals of different stages and found that puffs characteristic of the host animals were induced. These experiments confirmed Becker's proposal that stage-specific puffs are regulated by ecdysone.

The role of ecdysone has been demonstrated more directly by experiments in which the hormone was injected into larvae, resulting in premature appearance of stage-specific puffs (Clever and Karlson, 1960; Clever, 1961). More recently, isolated *Drosophila melanogaster* salivary glands have been exposed to ecdysone *in vitro*. The chromosomes of hormone-treated glands apparently progress through a normal puffing pattern. Figure 4.15 shows a chromosomal region from salivary glands treated with ecdysone *in vitro*. The region corresponds to one whose normal *in vivo* puffing pattern was shown in Figure 4.8. A comparison of these two figures shows that hormone treatment results in a pattern that is virtually identical to that found *in vivo*. Note in particular the sequence of puffing-regression-puffing shown by bands 74EF and 75B.

Investigators have recently determined that certain ecdysone-responsive genes code for so-called glue polypeptides that are secreted from the salivary glands to attach the pupal case to the substrate (Korge, 1975, 1977; Akam et al., 1978; Velissariou and Ashburner, 1980, 1981). Furthermore, some of these genes have now been cloned (Muskavitch and Hogness, 1980; Meyerowitz and Hogness, 1982), providing probes for transcripts of these genes and a means for detailed analyses of the genes. Therefore, this should prove to be an excellent experimental system for analyzing regulation of transcription at the molecular level. Recent evidence suggests that ecdysone binds specifically to ecdysone-responsive chromosomal regions (Gronemeyer and Pongs, 1980). However, the subsequent events in transcriptional regulation are unknown. The cloning of ecdysone-responsive genes may be the technological breakthrough that will facilitate this analysis.

REGULATION OF TRANSCRIPTION BY VERTEBRATE STEROID HORMONES

Another important system for analyzing hormonal regulation of transcription is the steroid hormone response of vertebrates. Steroids evoke changes in patterns of protein synthesis. These changes are apparently mediated, at least in part, by modifications in transcriptional patterns. This conclusion was originally based upon the fact that steroid responses are inhibited by α-amanitin and actinomycin D (O'Malley and Means, 1974; Edelman, 1975) and is strongly supported by more direct evidence, which we shall discuss later.

One of the most popular experimental systems for studying the effects of hormones on transcription is the chick oviduct response to estrogen and progesterone. Estrogen is required for the differentiation of the tubular gland cells of the oviduct, the cells that produce the eggwhite proteins—ovalbumin, conalbumin, lysozyme, and ovomucoid.

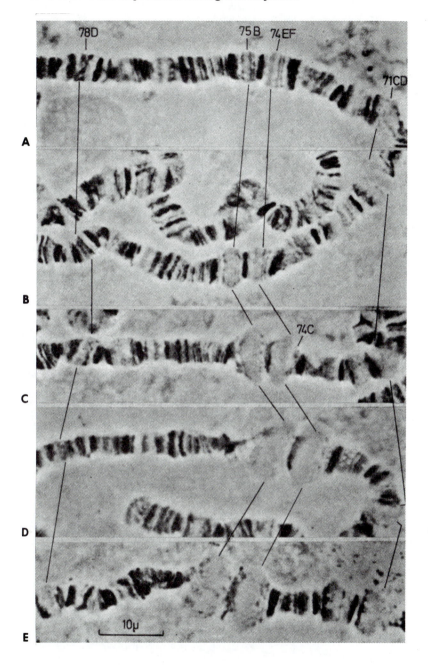

FIGURE 4.15 Ecdysone-induced puffing cycle in cultured *Drosophila* salivary glands. This chromosomal region corresponds to that shown in Figure 4.8. *A,* Unincubated control; *B,* 25 minutes; *C,* 1 hour; *D,* 2 hours; *E,* 4 hours.

Illustration continued on the following page

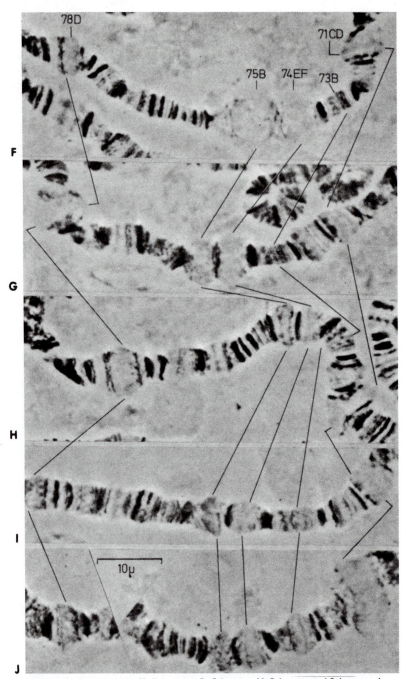

FIGURE 4.15 *(Continued)* *F,* 5 hours; *G,* 6 hours; *H,* 8 hours; *I,* 10 hours; *J,* 12 hours. (From M. Ashburner. 1972a. Patterns of puffing activity in the salivary gland chromosomes of *Drosophila.* VI. Induction by ecdysone in salivary glands of *D. melanogaster* cultured *in vitro.* Chromosoma (Berl.), *38:* 272–273. Reprinted with permission of Springer-Verlag, Heidelberg.)

The major egg-white protein is ovalbumin, which makes up as much as 50% of the protein that is synthesized by the gland cells. Estrogen is not present in measurable amounts in immature chicks. Hence, the effects of the hormone on gland cell differentiation and ovalbumin synthesis can be studied by hormonal injection. The oviducts of immature chicks can be induced to develop by daily administration of estrogen, which causes proliferation and differentiation of the tubular gland cells.

Ovalbumin synthesis (Fig. 4.16) is first detectable in the oviduct about 18 hours after hormone administration and reaches a plateau after about ten days. If hormone treatment is interrupted, ovalbumin synthesis declines to undetectable levels, although some of the tubular gland cells remain. Resumption of estrogen administration or administration of progesterone (**secondary stimulation**) reinitiates ovalbumin synthesis in the preexisting gland cells after about four hours. Secondary stimulation provides some experimental advantages over primary stimulation for studying hormonal effects: The oviduct is much larger than it is during primary stimulation, and the initial events in ovalbumin synthesis are unaffected by cell division, unlike in primary stimulation (Palmiter, 1975).

The increased ovalbumin synthesis promoted by steroids is the direct result of an accumulation of ovalbumin messenger RNA in the cells of hormone-treated tissues. This is shown by hybridization analysis with ovalbumin cDNA. One of the advantages of cDNA studies is the ability to quantify the presence of specific messenger RNA molecules. Prior to estrogen stimulation, there are approximately 30 molecules of ovalbumin mRNA per oviduct epithelial cell (Moen and Palmiter, 1980). However,

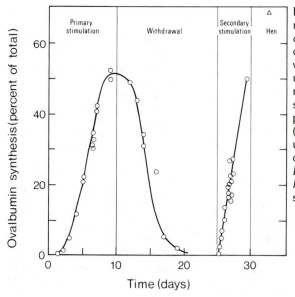

FIGURE 4.16 Effect of estrogen on the relative rate of synthesis of ovalbumin in chick oviduct during primary stimulation, withdrawal, and secondary stimulation. Immature 4-day old chicks were injected with estrogen daily (primary stimulation), and after 10 days without estrogen administration (withdrawal), administration was resumed (secondary stimulation). Results are presented as per cent of total protein synthesis. (After R.T. Schimke et al. 1977. Hormone regulation of egg-white protein synthesis in chick oviduct. *In* J. Dumont and J. Nunez (eds.), *First European Symposium on Hormones and Cell Regulation.* Elsevier Biomedical Press B.V., Amsterdam, p. 210.)

daily injection of estrogen results in the accumulation of 50,000 ovalbumin mRNA molecules per cell after 18 days (O'Malley et al., 1975). Discontinuation of hormone treatment results in a dramatic reduction of ovalbumin mRNA to approximately 60 molecules per tubular gland cell (McKnight et al., 1975). Readministration of either estrogen or progesterone causes renewed accumulation of ovalbumin mRNA. The presence of ovalbumin messenger can be demonstrated by *in situ* hybridization with radioactive ovalbumin cDNA. As shown in Figure 4.17*A*, tubular gland cells of chronically stimulated oviducts are heavily labeled, reflecting hybridization of the cDNA with ovalbumin transcripts. Cells of oviducts prepared from chicks that were withdrawn from estrogen for three days show no hybridization (Fig. 4.17*B*). When withdrawn chicks were restimulated with progesterone for ten hours, labeled cells were detected once again (Fig. 4.17*C*).

Do the differences in ovalbumin messenger accumulation reflect changes in the rates of ovalbumin mRNA synthesis in these different physiological states, or could the differences have other explanations such as different turnover rates for the transcripts, with transcripts in cells from hormone-withdrawn chicks being degraded rapidly following their synthesis? The synthetic rates have been determined by hybridization of radioactively labeled messenger RNA to cloned ovalbumin cDNA. Chronic injection of estrogen leads to high levels of ovalbumin messenger synthesis that remain constant as long as hormone administration is continued. The synthetic rate drops rapidly during estrogen withdrawal to become undetectable by 60 hours. Readministration of hormone causes a rapid return to the maximal level of synthesis (Swaneck et al.,

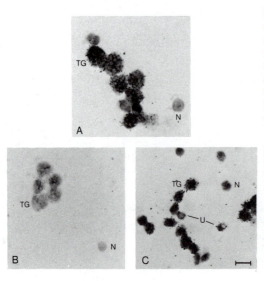

FIGURE 4.17 Accumulation of ovalbumin transcripts revealed by *in situ* hybridization with ovalbumin cDNA. *A*, Tubular gland cells (TG) from estrogen-stimulated chicks are heavily labeled. Nontubular gland cells (N) remain unlabeled. *B*, Cells from hormone-withdrawn chicks show little label. *C*, Heavy label is present over gland cells from chicks restimulated with progesterone. Two unclassified cells (U) are also seen here. Scale bar equals 2 μm. (From J.H. Shepherd et al., 1980. Reproduced from *The Journal of Cell Biology*, 1980, vol. 87, pp. 142–151 by copyright permission of The Rockefeller University Press.)

1979a,b). Thus, the steroids exert their effects on the level of ovalbumin gene transcription.

These demonstrated effects of steroids on transcription provide investigators with the opportunity to study specificity of gene regulation; that is, why do hormones affect only certain cells in the body and then exert their effects on specific genes in those cells? Vertebrates have evolved an elaborate system to transport hormones from the circulation to cell nuclei, where the steroid response is generated. This system also serves to ensure that the hormones will influence only specific organs—a necessity, since all cells of the body are exposed to hormones in the circulation. Organs that respond to a particular steroid hormone are called **target organs.** This specificity is due to a difference in the cytoplasms of target and nontarget cells. Target cells are "competent" to respond to steroids because of the presence of specific cytoplasmic receptors that combine with the hormones to form hormone-receptor complexes, which are the functional entities within the target cells that are responsible for the effects of the hormones.

The receptor concept was initially based upon observations that injected radioactive steroids are retained only by hormone target tissues. After injection of radioactive steroid, both cytoplasmic and nuclear binding sites are found in target cells. However, in time, the labeled steroid accumulates in the nucleus. Therefore, steroids reach their site of action by a "two-step" mechanism. *First,* the hormone diffuses into cells from the circulation. In nontarget cells, the hormone is not retained; in target cells, the hormone binds with a cytoplasmic receptor to form a hormone-receptor (H-R) complex, concentrating the hormone in these cells. The cytoplasmic receptors are proteins that have high affinity for specific hormones. A number of receptors have been identified, including those for estrogens, progesterone, androgens, and corticosteroids; the receptors for these hormones are, of course, present only in their respective target tissues (O'Malley et al., 1971). The *second step* is translocation of the H-R complex to the nucleus, where it binds with a nuclear acceptor that has a high affinity for the complex. This results in its accumulation within the nucleus and consequent depletion from the cytoplasm. Nuclear accumulation is also a specific property of target cells, since nontarget nuclei will not accumulate hormone-receptor complexes (O'Malley et al., 1971).

A determination of the specific nuclear site of the steroid acceptors is important in establishing the role of steroids in regulation of cell function. This site has been located by reextraction of radioactive hormone from target cell nuclei. The vast majority of the hormone is discovered bound to chromatin as part of a hormone-receptor complex (Edelman, 1975). These results suggest that the nuclear function of H-R complexes is to influence directly the pattern of transcription on chromatin. Therefore, discovering the identity of the chromatin acceptor and the general mode of interaction of the H-R complexes with chromatin has become a primary objective of research on the steroid response. An important experimental advantage in these investigations is that *isolated chromatin* will bind hormone-receptor complexes. Furthermore, isolated chromatin retains its cell specificity; that is, only target cell chromatin will bind these complexes (Spelsberg et al., 1971).

The interaction of H-R complexes with target chromatin allows for the possibility of directly analyzing transcriptional regulation of hormone-responsive genes. Such an analysis for the ovalbumin gene has been greatly facilitated by the availability of cloned fragments of the gene. A recent investigation with such fragments has demonstrated a 19 base-pair sequence that is present several times in the ovalbumin gene and in its 5′ flanking region that binds specifically with the progesterone-receptor complex. Significantly, the sequence is also found upstream from other progesterone-responsive egg white protein genes (Mulvihill et al., 1982). Thus, the selectivity of action of the hormone on certain genes within the genome appears to result from an interaction between the hormone-receptor complex and sequences flanking those genes that respond. It will be very instructive to learn how this interaction facilitates their transcription.

4–3. POSTTRANSCRIPTIONAL CONTROLS

Cellular utilization of structural gene transcripts is not necessarily automatic. The cell is not merely an assembly line that takes any transcript that is synthesized and transports it to the cytoplasm to be used in protein synthesis. This conclusion is obvious when we compare nuclear RNA with messenger RNA. Hybridization analyses have demonstrated that considerably more structural gene transcripts are contained in the nucleus than are associated with polysomes.

The number of different gene transcripts is calculated from the **RNA sequence complexity.** As with DNA complexity (see Chap. 3) RNA complexity is the total length of different sequences, as measured in nucleotides (Davidson, 1976). One method of measuring the complexity of structural gene transcripts is determination of the extent of hybridization of RNA to total single copy DNA (Galau et al., 1974), the class of DNA that contains most structural genes (see Chap. 3). For example, tobacco leaf cell nuclei have a complexity of 1.19×10^8 nucleotides. These transcripts result from transcription of approximately 19% of the single copy sequences in the genome. However, polysomal RNA (i.e., mRNA) has a complexity of 3.33×10^7 nucleotides, which represents only about 5% of the total single copy sequence complexity of the genome (Goldberg et al., 1978). This fourfold difference between nuclear and polysomal RNA complexity indicates that the majority of nuclear RNA is not represented in polysomal RNA. Similar measurements on sea urchin gastrula RNA have revealed that nuclear RNA has a complexity that exceeds that of polysomal RNA by tenfold (Galau et al., 1974; Hough et al., 1975). The differences in complexity between nuclear and polysomal RNA suggest that posttranscriptional events may regulate the appearance of some structural gene transcripts on the polysomes.

The necessity of posttranscriptional regulation is also apparent from comparisons of RNA complexity *between* cells of the same organism. These measurements have revealed that the complexity differences between the *nuclear RNAs* of any two cell types are considerably less than the differences between the *polysomal RNAs* of these cells (Galau et al., 1976; Chikaraishi et al., 1978). Additional hybridization studies, using a sea urchin single copy DNA tracer, have revealed that some transcripts that are on polysomes of a given cell type and absent from the mRNA of other cell types are nevertheless present in nuclear RNA of both categories of cells (Wold et al., 1978). These results have been interpreted to mean that transcripts may be subject to selective utilization as messengers. This interpretation is supported by recent experiments with specific cloned sequences (Knöchel and John, 1982).

The results of these various hybridization studies indicate that cells must rely very heavily upon posttranscriptional mechanisms to ensure that specific proteins are produced at the right time, in the right cells, and in the correct amounts. In Chapter 3 we summarized the processes involved in utilizing transcripts for the production of functional proteins. These include the series of modifications that RNA must undergo after transcription, the transport of messengers to the cytoplasm, and their utilization by the translational machinery. These processes are subject to regulation that can delay, prevent, or affect the rate of messenger translation. Each of these modes of posttranscriptional regulation is known to be utilized during plant and animal development. Some clearcut examples of posttranscriptional regulation have been elucidated in the literature and will be discussed in this book. Some of these are reviewed on the following pages.

DELAYED TRANSLATION.   Storage of RNA for translation at a later stage of development is one mechanism of posttranscriptional regulation. Very early animal embryos provide an excellent example of delayed mRNA translation. As we shall discuss in detail in Chapter 12, the fertilized egg is involved in very intense protein synthesis. Nuclear function, however, is not necessary for this protein synthesis, since translation continues unabated when the nucleus is removed or transcription is inhibited. These eggs contain the complete protein synthetic machinery in the cytoplasm—including mRNA, tRNA, and ribosomes—for immediate utilization following fertilization. The stored messengers are called "masked oogenic messengers"; they are synthesized during oogenesis and stored as mRNP particles until their utilization after the development of the egg begins.

Delayed utilization of transcripts is also typical of organisms that exhibit an interruption in development during early developmental stages. In these organisms, transcripts are stored until development resumes. This mechanism is typical of the early development of seed plants and certain animals. Following plant embryogenesis the seed coat surrounding the embryo hardens, and the embryo dries up, halting development. Typically, the mature seed is shed from the plant, and development does not resume until conditions that are favorable for plant growth and development have been reestablished. After this hiatus, development is reinitiated by **germination.** Protein is synthesized during germination, but synthesis of the RNA templates and the ribosomes for this protein synthesis occurred during embryogenesis, and they were stored in the seed for later utilization. A similar kind of situation is found in pollen. When pollen lands on the style of the flower, a pollen tube may develop. Production of the tube requires protein synthesis, which utilizes stored mRNA and ribosomes.

Dormancy can also occur during postfertilization animal development. For example, the development of the brine shrimp embryo *(Artemia)* can be interrupted at the gastrula stage by rather long dormant periods, during which the embryo becomes dessicated and metabolic functions cease. Dormancy is ended by hydration of these encysted gastrulae, and development resumes, utilizing stored protein synthetic machinery including mRNA.

DIFFERENTIAL LONGEVITY OF MESSENGERS. In contrast to prokaryotic messengers, which are typically degraded very rapidly after transcription, eukaryotic messengers may enjoy much longer life spans. This is particularly true for messengers of differentiated cells. Cell differentiation is often accompanied by stabilization in the life span of cell-specific messengers.

Evidence for this differentiation mechanism was first obtained with actinomycin D experiments. Treatment of undifferentiated cells with actinomycin D generally results in rapid inhibition of protein synthesis. However, differentiated cells are much less sensitive to actinomycin D, and protein synthesis in these cells may continue for some time in the absence of RNA synthesis. An additional important observation made during these investigations is that the messengers for cell-specific proteins are particularly stable. When differentiated cells are treated with actinomycin D, the synthesis of cell-specific proteins may continue unabated for some time, whereas the synthesis of other proteins declines rapidly. It appears that one mechanism for narrowing the range of protein synthetic capacities during cell differentiation is differential messenger decay rates. Cell-specific proteins are made on long-lived mRNA templates, and therefore, each template is available for numerous rounds of protein synthesis. However, messengers for nonspecific proteins may be translated only a few times before their decay.

These actinomycin experiments have some major drawbacks, however. Inhibition of RNA synthesis may be incomplete, allowing continued production of mRNA. Furthermore, actinomycin has been found to influence translation as well as transcription (Singer and Penman, 1972).

Messenger stability has been demonstrated by other techniques that do not rely upon inhibitors. One clear-cut example is the globin mRNA of mammalian reticulocytes. The reticulocyte has no nucleus but is active in protein synthesis, utilizing RNA that was synthesized prior to nuclear loss. The pattern of protein synthesis in these enucleated cells might be expected to change as a result of the loss of short-lived messengers during cell maturation. This process would produce an enrichment for long-lived messengers, causing protein synthesis to become more homogeneous in time. In fact, the protein synthetic pattern of reticulocytes does exhibit this phenomenon. As the cells mature, the synthesis of nonglobin proteins declines relative to that of globin (see Chap. 14 for details of this experiment). The reticulocyte is atypical in that it lacks a nucleus. However, many differentiated cells undergo considerable reduction in RNA synthetic capacity as they mature. Hence, differential messenger decay rates may play a significant role in specialization of protein synthesis in a great variety of cell types. As we discussed in Chapter 3, one factor affecting transcript stability might be the 5' cap, since uncapped transcripts may be more susceptible to degradation by certain nuclease enzymes. Thus, regulation of capping might affect protein synthesis indirectly by altering transcript stability.

REGULATION OF TRANSLATION. Utilization of messengers by the translational machinery depends upon successful completion of the initiation step. Regulation of initiation could therefore have a profound effect upon protein synthesis. We shall examine three examples of regulation at this level.

1. The globin messengers of the rabbit reticulocyte provide clear-cut evidence that regulation of initiation can differentially affect the production of specific proteins. Two kinds of globin polypeptide chains are being made in these cells: α and β. The synthesis of these polypeptides is directed by two distinct mRNAs. The rates of initiation of translation of these messengers are different (Lodish, 1976): The α-globin mRNA initiates protein synthesis only 60% as often as β-globin mRNA. Thus, ribosomes are loaded onto the β-globin messenger more rapidly than onto the α-globin messenger. This results in more ribosomes being present per unit length on the β-globin messenger than on the α-globin messenger. Thus, even though both messengers are approximately the same length, the β-globin polysomes average five ribosomes, whereas α-globin is made on polysomes with an average of three ribosomes.

   These results clearly demonstrate that initiation can occur at different rates on two kinds of messengers that are simultaneously present in the same cell. Initiation rates could be affected by any mechanism that alters the formation of the initiation complex. For example, messengers might have specialized nucleotide sequences upstream from the AUG codon, which could affect initiation. Another possibility is the existence of messenger-specific initiation factors that could alter the rate of initiation complex formation. There is at present no clear-cut evidence for α- or β-globin mRNA-specific translation factors. Nucleotide sequencing techniques may reveal whether the messenger sequences play a role in initiation.

2. Although messenger-specific translation factors have not been demonstrated for the α-or β-globin messengers, such factors do appear to be involved in modulating the rates of translation in some cells. Evidence for such factors comes from assessing the effects of initiation factors from different cells on cell-free translation of messengers. One example involves experiments utilizing initiation factors of two types of embryonic chick muscle. Both red muscle and white muscle synthesize myosin, but only red muscle synthesizes myoglobin. *In vitro* translation of myosin and myoglobin mRNA reveals differences in translation factors between these two cell types. Myosin mRNA can be translated *in vitro* only when initiation factors from muscle cells are added to the cell-free translation system. Both red muscle and white muscle have these initiation factors, which is consistent with the ability of both cell types to produce myosin *in vivo*. On the other hand, when myoglobin mRNA is added to a cell-free system, red muscle initiation factors are effective in promoting translation of the messenger, whereas white muscle factors are not (Thompson et al., 1973). These results suggest that the factor that is necessary for translation of myoglobin mRNA is present only in red muscle. The results do not imply that every messenger has a specific initiation factor that is necessary for its translation. They do suggest, however, that certain differentiated cells have initiation factors that recognize cell-specific messengers and make the utilization of those messengers more efficient in the large-scale production of cell-specific proteins.

3. Mammalian reticulocyte lysates (see Chap. 3) have long been a favored system for *in vitro* studies of translation. Experiments with this system have demonstrated yet another mechanism of regulation of initiation. Protein synthesis in these lysates declines very rapidly unless hemin is added to the incubation mixture (Zucker and Schulman, 1968). This decline is due to a block in the formation of initiation complexes.

   This effect is produced by a translational inhibitor that forms from a latent proinhibitor in the absence of hemin. The presence of hemin greatly retards the conversion of proinhibitor to inhibitor, apparently by interfering with the normal cellular mechanisms that promote this conversion (for reviews, see Ochoa [1977]; Revel and Groner [1978]). It remains to be seen whether this mechanism can discriminate for or against specific messengers.

REFERENCES

Akam, M.E. et al. 1978. Drosophila: The genetics of two major larval proteins. Cell, *13*: 215–225.

Allfrey, V.G. 1977. Post-synthetic modifications of histone structure: A mechanism for the control of chromosome structure by the modulation of histone-DNA interaction. *In* H.J. Li and R.A. Eckhardt (eds.), *Chromatin and Chromosome Structure*. Academic Press, New York, pp. 167–191.

Andrews, G.K., M. Dziadek, and T. Tamaoki. 1982. Expression and methylation of the mouse α-fetoprotein gene in embryonic, adult and neoplastic tissues. J. Biol. Chem., *257*: 5148–5153.

Ashburner, M. 1972a. Patterns of puffing activity in the salivary gland chromosomes of Drosophila. VI. Induction by ecdysone in salivary glands of *D. melanogaster* cultured *in vitro*. Chromosoma, *38*: 255–281.

Ashburner, M. 1972b. Puffing patterns in *Drosophila melanogaster* and related species. *In* W. Beermann (ed.), *Developmental Studies on Giant Chromosomes*. Springer-Verlag, Berlin, pp. 101–151.

Ashburner, M., and J.J. Bonner. 1979. The induction of gene activity in Drosophila by heat shock. Cell, *17*: 241–254.

Becker, H.J. 1959. Die Puffs der Speicheldrüsenchromosomen von *Drosophila melanogaster*. I. Beobachtungen zum Verhalten des Puffmusters im Normalstamm und bei zwei Mutanten *giant* und *lethal-giant-larvae*. Chromosoma, *10*: 654–678.

Becker, H.J. 1962. Die Puffs der Speicheldrüsenchromosomen von *Drosophila melanogaster*. II. Die Auslösung der Puffbildung, ihre Spezifität und ihre Beziehung zur Funktion der Ringdrüse. Chromosoma, *13*: 341–384.

Beermann, W. 1952. Chromomerenkonstanz und spezifische modifikationen der Chromosomenstruktur in der Entwicklung und Organdifferenzierung von *Chironomus tentans*. Chromosoma, *5*: 139–198.

Beermann, W. 1963. Cytological aspects of information transfer in cellular differentiation. Am. Zool., *3*: 23–32.

Beermann, W. 1971. Effect of α-amanitine on puffing and intranuclear RNA synthesis in *Chironomus* salivary glands. Chromosoma, *34*: 152–167.

Bird, A.P. 1978. Use of restriction enzymes to study eukaryotic DNA methylation: II. The symmetry of methylated sites supports semi-conservative copying of the methylation pattern. J. Mol. Biol., *118*: 49–60.

Bogenhagen, D.F., W.M. Wormington, and D.D. Brown. 1982. Stable transcription complexes of Xenopus 5S RNA genes: A means to maintain the differentiated state. Cell, *28*: 413–421.

Brown, S.W. 1966. Heterochromatin. Science, *151*: 417–425.

Chikaraishi, D.M., S.S. Deeb, and N. Sueoka. 1978. Sequence complexity of nuclear RNAs in adult rat tissue. Cell, *13*: 111–120.

Clever, U. 1961. Genaktivitäten in den Riesenchromosomen von *Chironomus tentans* und ihre Beziehungen zur Entwicklung. I. Genaktivierungen durch Ecdyson. Chromosoma, *12*: 607–675.

Clever, U., and P. Karlson. 1960. Induktion von Puff-veranderüngen in den Speicheldrüsenchromosomen von *Chironomus tentans* durch Ecdyson. Exp. Cell. Res., *20*: 623–626.

Compere, S.J., and R.D. Palmiter. 1981. DNA methylation controls the inducibility of the mouse metallothionein-I gene in lymphoid cells. Cell, *25*: 233–240.

Daneholt, B. 1975. Transcription in polytene chromosomes. Cell, *4*: 1–9.

Daneholt, B. et al. 1978. The 75S RNA transcription unit in Balbiani ring 2 and its relation to chromosome number. Philos. Trans. R. Soc. Lond. (Biol.), *283*: 383–389.

Davidson, E.H. 1976. *Gene Activity in Early Development,* 2nd ed. Academic Press, New York.

Davidson, R.G., H.M. Nitowsky, and B. Childs. 1963. Demonstration of two populations of cells in the human female heterozygous for glucose-6-phosphate dehydrogenase variants. Proc. Natl. Acad. Sci. U.S.A., *50:* 481–485.

Davie, J.R., and E.P.M. Candido. 1980. DNase I sensitive chromatin is enriched in the acetylated species of histone H4. FEBS Lett., *110:* 164–168.

Deisseroth, A., and D. Hendrick. 1979. Activation of phenotypic expression of human globin genes from nonerythroid cells by chromosome-dependent transfer to tetraploid mouse erythroleukemia cells. Proc. Natl. Acad. Sci. U.S.A., 76: 2185–2189.

Dryer, W.J., and J.C. Bennett. 1965. The molecular basis of antibody formation. A paradox. Proc. Natl. Acad. Sci. U.S.A., *54:* 864–869.

Edelman, I.S. 1975. Mechanism of action of steroid hormones. J. Steroid Biochem., 6: 147–159.

Engelke, D.R. et al. 1980. Specific interaction of a purified transcription factor with an internal control region of 5S RNA genes. Cell, *19:* 717–728.

Galau, G.A., R.J. Britten, and E.H. Davidson, 1974. A measurement of the sequence complexity of polysomal messenger RNA in sea urchin embryos. Cell, *2:* 9–20.

Galau, G.A. et al. 1976. Structural gene sets active in embryos and adult tissues of the sea urchin. Cell, 7: 487–505.

Garel, A. and R. Axel. 1976. Selective digestion of transcriptionally active ovalbumin genes from oviduct nuclei. Proc. Natl. Acad. Sci. U.S.A., *73:* 3966–3970.

Goldberg, R.B. et al. 1978. Sequence complexity of nuclear and polysomal RNA in leaves of the tobacco plant. Cell, *14:* 123–131.

Goodwin, G.H., J.M. Walker, and E.W. Johns. 1978. The high mobility group (HMG) nonhistone chromosomal proteins. *In* H. Busch (ed.), *The Cell Nucleus,* Vol. VI. Academic Press, New York, pp. 181–219.

Gottesfeld, J., and L.S. Bloomer. 1982. Assembly of transcriptionally active 5S RNA gene chromatin in vitro. Cell, *28:* 781–791.

Gronemeyer, H., and O. Pongs. 1980. Localization of ecdysterone on polytene chromosomes of *Drosophila melanogaster.* Proc. Natl. Acad. Sci. U.S.A., 77: 2108–2112.

Groudine, M., R. Eisenman, and H. Weintraub. 1981. Chromatin structure of endogenous retroviral genes and activation by an inhibitor of DNA methylation. Nature (Lond.), *292:* 311–317.

Groudine, M. et al. 1974. Lineage-dependent transcription of globin genes. Cell, 3: 243–247.

Holliday, R., and J.E. Pugh. 1975. DNA modification mechanisms and gene activity during development. Science, *187:* 226–232.

Honda, B.M., and R.G. Roeder. 1980. Association of 5S gene transcription factor with 5S RNA and altered levels of the factor during cell differentiation. Cell, *22:* 119–126.

Hough, B.R. et al. 1975. Sequence complexity of heterogeneous nuclear RNA in sea urchin embryos. Cell, *5:* 291–299.

Hozumi, N., and S. Tonegawa. 1976. Evidence for somatic rearrangement of immunoglobulin genes coding for variable and constant regions. Proc. Natl. Acad. Sci. U.S.A., *73:* 3628–3632.

Jones, P.A., and S.M. Taylor. 1980. Cellular differentiation, cytidine analogs and DNA methylation. Cell, *20*: 85–93.

Keene, M.A. et al. 1981. DNase I hypersensitive sites in *Drosophila* chromatin occur at the 5′ ends of regions of transcription. Proc. Natl. Acad. Sci. U.S.A., *78*: 143–146.

Knöchel, W., and M.E. John. 1982. Cloning of *Xenopus laevis* nuclear poly (A)-rich RNA sequences. Evidence for post-transcriptional control. Eur. J. Biochem., *122*: 11–16.

Korge, G. 1975. Chromosome puff activity and protein synthesis in larval salivary glands of *Drosophila melanogaster*. Proc. Natl. Acad. Sci. U.S.A., *72*: 4550–4554.

Korge, G. 1977. Direct correlation between a chromosome puff and the synthesis of a larval saliva protein in *Drosophila melanogaster*. Chromosoma, *62*: 155–174.

Laird, C. et al. 1980. Proposed structural principles of polytene chromosomes. Chromosomes Today, *7*: 74–83.

Lamb, M.M., and B. Daneholt. 1979. Characterization of active transcription units in Balbiani rings of Chironomus tentans. Cell, *17*: 835–848.

Lefevre, G., Jr. 1976. A photographic representation and interpretation of the polytene chromosomes of *Drosophila melanogaster* salivary gland. *In* M. Ashburner and E. Novitski (eds.), *The Genetics and Biology of Drosophila*, Vol. 1a. Academic Press, London, pp. 31–66.

Livak, K.J. et al. 1978. Sequence organization and transcription at two heat shock loci in *Drosophila*. Proc. Natl. Acad. Sci. U.S.A., *75*: 5613–5617.

Lodish, H.F. 1976. Translational control of protein synthesis. Ann. Rev. Biochem., *45*: 39–72.

Lyon, M. 1961. Gene action in the X-chromosome of the mouse (*Mus musculus* L.). Nature (Lond.), *190*: 372–373.

McKnight, G.S., P. Pennequin, and R.T. Schimke. 1975. Induction of ovalbumin mRNA sequences by estrogen and progesterone in chick oviduct as measured by hybridization to complementary DNA. J. Biol. Chem., *250*: 8105–8110.

Meyerowitz, E.M., and D.S. Hogness. 1982. Molecular organization of a Drosophila puff site that responds to ecdysone. Cell, *28*: 165–176.

Moen, R.C., and R.D. Palmiter. 1980. Changes in hormone responsiveness of chick oviduct during primary stimulation with estrogen. Dev. Biol., *78*: 450–463.

Mohandas, T., R.S. Sparkes, and L.J. Shapiro. 1981. Reactivation of an inactive human X chromosome: Evidence for X inactivation by DNA methylation. Science, *211*: 393–396.

Mulvihill, E.R., J.-P. LePennac, and P. Chambon. 1982. Chicken oviduct progesterone receptor: Location of specific regions of high-affinity binding in cloned DNA fragments of hormone-responsive genes. Cell, *28*: 621–632.

Muskavitch, M.A.T., and D.S. Hogness. 1980. Molecular analysis of a gene in a developmentally regulated puff of *Drosophila melanogaster*. Proc. Natl. Acad. Sci. U.S.A., *77*: 7362–7366.

Ochoa, S. 1977. Regulation of translation. *In* A.B. Legocki (ed.), *Translation of Natural and Synthetic Polynucleotides*. Poznań Agricultural Univ. Poznań, Poland, pp. 9–23.

O'Malley, B.W., and A.R. Means. 1974. Female steroid hormones and target cell nuclei. Science, *183*: 610–620.

O'Malley, B.W., D.O. Toft, and M.R. Sherman. 1971. Progesterone-binding components of chick oviduct. J. Biol. Chem., *246*: 1117–1122.

O'Malley, B.W. et al. 1975. Steroid hormone regulation of specific messenger RNA and protein synthesis in eucaryotic cells. J. Cell. Physiol., *85*: 343–356.

Palmiter, R.D. 1975. Quantitation of parameters that determine the rate of ovalbumin synthesis. Cell, *4*: 189–197.

Pelham, H.R.B., M. Wormington, and D.D. Brown. 1981. Related 5S RNA transcription factors in *Xenopus* oocytes and somatic cells. Proc. Natl. Acad. Sci. U.S.A., *78*: 1760–1764.

Pukkila, P.J. 1975. Identification of the lampbrush chromosome loops which transcribe 5S ribosomal RNA in *Notophthalmus (Triturus) viridescens.* Chromosoma, *53*: 71–89.

Razin, A., and A.D. Riggs. 1980. DNA methylation and gene function. Science, *210*: 604–610.

Reeves, R., and E.P.M. Candido. 1980. Partial inhibition of histone deacetylase in active chromatin by HMG 14 and HMG 17. Nucleic Acids Res., *8*: 1947–1963.

Revel, M., and Y. Groner. 1978. Post-transcriptional and translational controls of gene expression in eukaryotes. Ann. Rev. Biochem., *47*: 1079–1126.

Riggs, A.D. 1975. X-inactivation, differentiation, and DNA methylation. Cytogenet. Cell Genet., *14*: 9–25.

Ritossa, F.M. 1962. A new puffing pattern induced by heat shock and DNP in Drosophila. Experientia, *18*: 571–573.

Sass, H. 1981. Effects of DMSO on the structure and function of polytene chromosomes of *Chironomus.* Chromosoma, *83*: 619–643.

Schimke, R.T. et al. 1977. Hormone regulation of egg-white protein synthesis in chick oviduct. *In* J. Dumont and J. Nunez (eds.), *First European Symposium on Hormones and Cell Regulation.* Elsevier/North Holland Biomedical Press, Amsterdam, pp. 209–221.

Seidman, J.G., and P. Leder. 1978. The arrangement and rearrangement of antibody genes. Nature (Lond.), *276*: 790–795.

Shepherd, J.H. et al. 1980. Commitment of chick oviduct tubular gland cells to produce ovalbumin mRNA during hormonal withdrawal and restimulation. J. Cell Biol., *87*: 142–151.

Simpson, R.T. 1978. Structure of chromatin containing extensively acetylated H3 and H4. Cell, *13*: 691–699.

Singer, R.H., and S. Penman. 1972. Stability of HeLa cell mRNA in actinomycin. Nature (Lond.), *240*: 100–102.

Spelsberg, T.C., L.S. Hnilica, and A.T. Ansevin. 1971. Proteins of chromatin in template restriction. III. The macromolecules in specific restriction of the chromatin DNA. Biochim. Biophys. Acta, *228*: 550–562.

Spradling, A.C., and A.P. Mahowald. 1980. Amplification of genes for chorion proteins during oogenesis in *Drosophila melanogaster.* Proc. Natl. Acad. Sci. U.S.A., *77*: 1096–1100.

Stalder, J. et al. 1980. Tissue-specific DNA cleavages in the globin chromatin domain introduced by DNase I. Cell, *20*: 451–460.

Stein, R. et al. 1982. Clonal inheritance of the pattern of DNA methylation in mouse cells. Proc. Natl. Acad. Sci. U.S.A., *79*: 61–65.

Swaneck, G.E. et al. 1979a. Absence of an obligatory lag period in the induction of ovalbumin mRNA by estrogen. Biochem. Biophys. Res. Commun. *88:* 1412–1418.

Swaneck, G.E. et al. 1979b. Effect of estrogen on gene expression in chicken oviduct: Evidence for transcriptional control of ovalbumin gene. Proc. Natl. Acad. Sci. U.S.A., *76:* 1049–1053.

Taylor, S.M., and P.A. Jones. 1982. Changes in phenotypic expression in embryonic and adult cells treated with 5-azacytidine. J. Cell. Physiol., *111:* 187–194.

Thompson, W.C., E.A. Buzash, and S.M. Heywood. 1973. Translation of myoglobin messenger ribonucleic acid. Biochemistry, *12:* 4559–4565.

Velissariou, V., and M. Ashburner. 1980. The secretory proteins of the larval salivary gland of *Drosophila melanogaster.* Cytogenetic correlation of a protein and a puff. Chromosoma, *77:* 13–27.

Velissariou, V., and M. Ashburner. 1981. Cytogenetic and genetic mapping of a salivary gland secretion protein in *Drosophila melanogaster.* Chromosoma, *84:* 173–185.

Vidali, G. et al. 1978. Butyrate suppression of histone deacetylation leads to accumulation of multiacetylated forms of histones H3 and H4 and increased DNase I sensitivity of the associated DNA sequences. Proc. Natl. Acad. Sci. U.S.A., *75:* 2239–2243.

Weintraub, H., and M. Groudine. 1976. Chromosomal subunits in active genes have an altered chromatin. Science, *193:* 848–856.

Weisbrod, S., M. Groudine, and H. Weintraub. 1980. Interaction of HMG 14 and 17 with actively transcribed genes. Cell, *19:* 289–301.

Weisbrod, S., and H. Weintraub. 1979. Isolation of a subclass of nuclear proteins responsible for conferring a DNase I-sensitive structure on globin chromatin. Proc. Natl. Acad. Sci. U.S.A., *76:* 630–634.

Wigler, M., D. Levy, and M. Perucho. 1981. The somatic replication of DNA methylation. Cell, *24:* 33–40.

Wold, B.J. et al. 1978. Sea urchin embryo mRNA sequences expressed in the nuclear RNA of adult tissues. Cell, *14:* 941–950.

Wu, C. 1980. The 5' ends of *Drosophila* heat shock genes in chromatin are hypersensitive to DNase I. Nature (Lond.), *286:* 854–860.

Zucker, W.V., and H.M. Schulman. 1968. Stimulation of globin-chain initiation by hemin in the reticulocyte cell-free system. Proc. Natl. Acad. Sci. U.S.A., *59:* 582–589.

5 The Cellular Basis of Morphogenesis

The cell is the basic unit of morphogenesis. Therefore, an understanding of how cells change shape is important for understanding the processes of morphogenesis. However, morphogenesis is a highly ordered process, and cells change shape in concert with their neighbors, rather than at random. Thus, the consequences of cellular shape changes are highly dependent upon intercellular relationships.

During animal embryogenesis, cells are either tightly bound to their neighbors in a sheet of cells (**epithelial cells**) or exist as individual cells that are free to move about and acquire new locations in the embryo. Virtually all morphogenic processes can be explained by examining the reorganization and rearrangements of epithelia and free cells. The behavior of epithelia depends in turn upon the cells of which they are composed; that is, changes in the form of epithelia occur because of shape changes in their component cells, which cause deformation of the sheet. Conversely, if cells are present as free cells, shape changes are responsible for their individual movement. At the completion of their migratory phase, free cells can form associations with their new neighbors to which they may become tightly bound. Consequently, in addition to changes in shape, the nature of the interaction between cells is an important factor in morphogenesis.

A major distinction between the behaviors of animal and higher plant cells during morphogenesis is the immobility of plant cells due to the presence of cellulose cell walls. The rigidity with which plant cells are maintained in position precludes the dramatic rearrangements of cells that characterize animal development. Hence, the shapes of the bodies of higher plants must be formed by mechanisms that regulate the shapes of immobile cells.

The ultimate regulation of cellular structure and function in both animal and plant development emanates from the expression of the genome, which directs the synthesis of molecules that regulate cellular shape and affect cellular interactions. Contemporary research in morphogenesis is aimed at identifying those molecules and examining their roles in cellular dynamics.

5–1. THE CYTOSKELETON

The production of cell shape is mediated by a complex array of fibers in the cytoplasm that is variously referred to as the **cytoskeleton** or the **cytoplasmic fiber system.** The term cytoskeleton defines the role of the fibers in maintaining cell shape but it has the unfortunate connotation of rigidity and permanence. On the contrary, the cytoskeleton constitutes an exceedingly dynamic system that is capable of considerable change and provides the means whereby the cells of the embryo can change their shapes and thereby produce the form of the adult organism.

In a recent review of the cytoskeleton, Cohen (1979) quoted from a 1733 essay by Alexander Pope to emphasize that this complex system of fibers has an underlying structural basis that functions in cellular morphogenesis. Pope wrote: "A mightly maze! but not without a plan." In this section we shall attempt to untangle the maze and demonstrate how it functions in cellular morphogenesis. The cytoskeleton is composed of **microtubules, microfilaments,** and **intermediate filaments** (Fig. 5.1). Cellular shape change is brought about primarily by the microtubules and microfilaments.

Microtubules

Microtubules are present in all cells at some point in their life cycles. They are important functional elements in the mitotic spindle, are involved in translocation of cellular organelles from one location to another in the cytoplasm, and are responsible for the movement of cilia and flagella. But they play a very specific role in generation and maintenance of cell asymmetry. In asymmetrical cells they are often aligned in parallel bundles in the cellular long axis. In later chapters we shall discuss several examples of cellular asymmetry that is produced by microtubules, including amphibian neural ectoderm cells and the filopodia of sea urchin primary mesenchyme cells (Chap. 13).

Microtubules are hollow cylindrical rods, 25 nm in diameter, formed of 13 rows of **protofilaments** that run parallel to the microtubule long axis. The protofilaments are composed of two similar kinds of protein subunits, called **α-** and **β-tubulins.** *In vitro* studies of microtubules have been very instructive in revealing how the tubulins are assembled into microtubules. One molecule of α-tubulin combines with one molecule of β-tubulin to form a **tubulin dimer.** The dimers, in turn, poly-

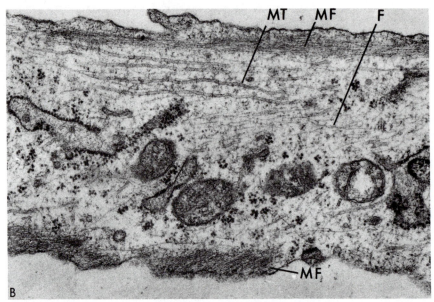

FIGURE 5.1 Electron micrograph of a portion of a motile tissue culture cell showing microtubules (MT), bundles of microfilaments (MF), and intermediate filaments (F). ×40,900. (From R.D. Goldman, 1971. Reproduced from *The Journal of Cell Biology*, 1971, vol. 51, pp. 752–762 by copyright permission of The Rockefeller University Press.)

merize end to end to form the protofilaments. Concomitant side-by-side assembly of the 13 protofilaments produces a microtubule. Each microtubule is polarized such that all the dimers are oriented in the same direction with respect to their α- and β-tubulin subunits. This mode of assembly is called head-to-tail assembly.

In vitro microtubule assembly will occur rapidly at 37°C if sufficient subunits are present, but depolarization occurs at low temperatures. Other factors involved in microtubule assembly are the nucleotide GTP and the divalent cation Mg^{++}. Addition and loss of dimers can only occur at the ends of microtubules. Margolis and Wilson (1978) demonstrated that assembly and disassembly occur at both ends of microtubules under equilibrium conditions *in vitro*. However, these opposing events occur at different rates at the two ends; that is, one end favors subunit addition (the **assembly end**), and the other end favors subunit loss (the **disassembly end**) (Fig. 5.2). A consequence of this mode of polymerization is **tubulin flux;** that is, subunits within the microtubule are continually displaced toward the disassembly end where they are eventually lost. This phenomenon is called **treadmilling** (Wegner, 1976).

The results of *in vitro* studies have important implications for *in vivo* behavior of microtubules and their regulation. Clearly, elongation of microtubules can occur only by addition of dimers to the ends of microtubules and only if the equilibrium favors assembly over disassem-

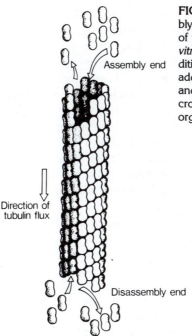

FIGURE 5.2 Microtubule assembly and disassembly. Tubulin dimers are added and lost at both ends of the microtubule under equilibrium conditions *in vitro*. At the assembly end, the rate of subunit addition exceeds the rate of subunit loss. Subunits added at this end treadmill through the microtubule and are lost at the opposite end. Note that a microtubule is composed of 13 rows of protofilaments organized into a hollow cylinder.

Assembly end

Direction of
tubulin flux

Disassembly end

bly. Conversely, microtubules will shorten if disassembly is favored. Although polymerization of dimers will occur spontaneously *in vitro* by addition of the correct components at a favorable temperature, microtubule assembly *in vivo* is a highly regulated process, with both the length and orientation of microtubules under close control. A major difference between *in vivo* and *in vitro* microtubule assembly is that there are no free microtubules in cells. All cellular microtubules have one end embedded in a structure called the **microtubule organizing center (MTOC),** which corresponds to the centrosome and contains centrioles and surrounding pericentriolar material. The MTOC serves as the point of initiation of growth for microtubules and may specify the number, distribution, and length of microtubules that emerge (Brinkley et al., 1981). Once microtubules are initiated, their continued growth occurs by tubulin dimer addition to their distal tips (Borisy, 1978; Bergen et al., 1980; Heidemann et al., 1980). A critical level of available tubulin subunits is a prerequisite for growth. Other factors that may be involved in control of microtubule growth are uncertain. One possibility is that **microtubule-associated proteins (MAPs)**, which have been shown to promote microtubule assembly *in vitro*, are involved in growth control. These proteins associate with the surface of the microtubule after dimers are incorporated into the growing polymer. The MAPs stabilize the polymer and thus shift the equilibrium of the polymerization reaction toward assembly and away from disassembly (Sloboda, 1980). However, it remains to be seen whether MAPs play this same role or, indeed, any functional role in regulating microtubule growth in the intact cell.

The occurrence of treadmilling by *in vivo* microtubules is as yet uncertain. One reason for the uncertainty is that it has yet to be determined whether cellular microtubules are free to exchange subunits at both ends as occurs *in vitro*; the anchoring of microtubules at one end in the MTOC may preclude this. If treadmilling does occur, it is easy to visualize how it could be used by cells to transport materials from place to place in the cell. Cellular components that are attached to treadmilling subunits would be displaced toward the disassembly end of a microtubule. Such a mechanism might be involved, for example, in chromosome movement on the spindle during mitosis (Margolis and Wilson, 1981).

The functional properties of microtubules in morphogenesis are due primarily to the cell's ability to elongate the microtubules rapidly, to maintain them once they are formed, and to shorten them when necessary. Drugs that affect microtubules are important tools for determining whether microtubules are involved in a particular morphogenic process. **Colchicine** is the most commonly used drug. It has been used for a number of years as a reversible inhibitor of mitosis. We now know that its effect on mitosis is due to its binding to tubulin subunits, which upon incorporation into a microtubule prevent additional polymerization (Margolis and Wilson, 1977). A relatively new synthetic antimicrotubule drug called **nocodazole** has recently been added to the developmental biologist's arsenal. This drug also binds to tubulin subunits and inhibits polymerization of microtubules (Hoebeke et al., 1976). When either colchicine or nocodazole is applied to developing systems, it serves as an excellent probe for determining whether microtubules are involved in a morphogenic process. An alternative to the use of drugs is low temperature, which causes reversible depolymerization of microtubules.

Because microtubules play a key role in shaping cells, their orientation in the cytoplasm must be precisely regulated. The microtubule organizing centers may affect the orientation of microtubules; if so, it still would be necessary to account for the orientation of the centers themselves in explaining how orientation is regulated. Furthermore, the formation of organized structures during development requires that the shapes of individual cells in a multicellular array be coordinated with one another. Hence, it would also be necessary to account for the coordinated orientation of organizing centers of contiguous cells. Clearly, we are only just beginning to understand the regulation of microtubules.

Microfilaments

Microfilaments are also universal animal cell constituents, as shown by their roles in the separation of daughter cells at cytokinesis (see Chap. 10) and in cytoplasmic streaming in plants. In morphogenesis they are utilized in cells to cause localized contractions, such as the shortening of pseudopodia or the narrowing of cell diameters. Microfilaments can be resolved with the electron microscope as 6-nm diameter threads. Contractile microfilaments are usually seen just below the cell membrane. They make functional contact with the membrane, which would account for their ability to cause cellular deformations when they contract. Microfilaments are present as close parallel arrays. These arrays can assume a number of configurations: bundles (the larger bundles as shown in Fig. 5.3B were called **stress fibers** by Buckley and Porter [1967]), sheaths (Fig. 5.3A), or meshes, networks, or lattices. None of these configurations are considered permanent; the microfilament system is capable of rapid rearrangement as well as local assembly and disassembly.

The configuration of microfilaments has been correlated with cellular activity of tissue culture cells. Cells that are nonmotile and stretched on the substrate have microfilaments that are organized into stress fibers. The microfilaments in this configuration may be primarily structural, assisting in maintaining cell shape. In motile cells, a few microfilament bundles are seen. However, most microfilaments are in a diffuse network. These microfilaments are considered to be primarily contractile. Two lines of evidence indicate that microfilaments are composed of the contractile protein actin, which is also one of the contractile proteins in muscle: (1) Microfilaments interact with heavy meromyosin (HMM) to form complexes that are visible with the electron microscope (Fig. 5.4). HMM is a fragment of myosin that is produced by proteolytic cleavage of the larger molecule. (2) Antibodies to actin will bind to microfilaments (see box, p. 179).

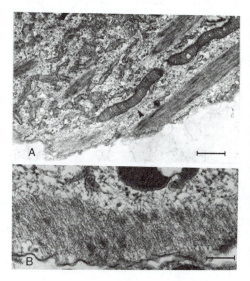

FIGURE 5.3 Electron micrographs showing microfilaments of human tissue culture cells. A, Portions of three microfilament bundles, which make contact with the plasma membrane. The membrane appears dense at the site of insertion of the lowermost bundle. Microtubules (arrowheads) run parallel to the bundles. Scale bar equals 1.0 μm. B, Obliquely oriented parallel arrays of microfilaments of the sheath system. Scale bar equals 0.25 μm. (From G.C. Godman and A.F. Miranda. 1978. Cellular contractility and the visible effects of cytochalasin. In S.W. Tanenbaum (ed.), *Cytochalasins—Biochemical and Cell Biological Aspects.* Elsevier Biomedical Press B.V., Amsterdam, p. 292.)

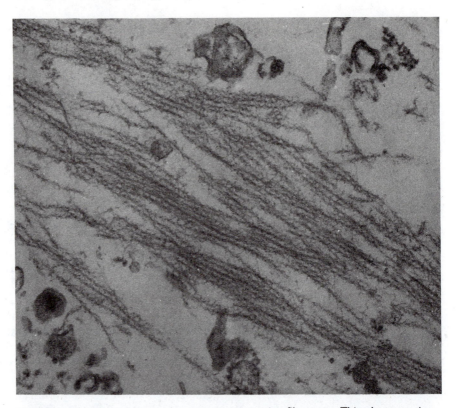

FIGURE 5.4 Demonstration of actin-containing microfilaments. This electron micrograph shows a portion of a chick embryo fibroblast treated with heavy meromyosin (HMM). HMM is produced by proteolytic treatment of the muscle contractile protein myosin. The HMM fragment is known to interact with muscle actin and is a specific probe for the presence of actin. The microfilaments are "decorated" with HMM arrowheads, an indication of their actin composition. ×76,000. (Courtesy of Dr. H. Holtzer.)

Like microtubules, microfilaments are polymers, in this case formed from actin subunits. The polymers are frequently called **F-actin** (for filamentous actin), whereas the unpolymerized subunits are **G-actin** (for globular actin). This sort of organization imparts dynamic properties to the actin network, which allow it to adapt rapidly to the changing requirements of cells during morphogenesis. As with tubulin polymerization, *in vitro* polymerization of actin subunits occurs in a head-to-tail fashion onto both ends of microfilaments. Polymerization is favored by the presence of Mg^{++}. The G-actin molecules add more readily to one end (the assembly end) of the microfilament than to the other. At equilibrium, subunit loss at the disassembly end is balanced by subunit addition at the other end, and filament length remains constant. Actin subunits would, therefore, treadmill through the filaments (Wegner, 1976). Hydrolysis of ATP is necessary to maintain actin monomer treadmilling. Growth of microfilaments occurs when the rate of subunit addition exceeds the rate of subunit loss.

Regular text continued on page 182

VISUALIZATION OF THE CYTOSKELETON BY IMMUNOFLUORESCENCE MICROSCOPY

The electron microscope is a powerful tool for studying the intricate details of cellular components. However, it is not always the instrument of choice for studying the overall pattern of distribution of a particular component so that one may correlate its distribution with its function. This is particularly true for components that may be responsible for generating *changes* in cellular configuration, since the components themselves are in a dynamic state while generating these changes. The cytoskeletal components are prime examples. Correlations of cytoskeletal configurations with changing cell morphology could assist in understanding the cytoplasmic basis for cellular dynamics. A technique that makes such an analysis possible is **immunofluorescence microscopy.**

Immunofluorescence microscopy is a technique for detecting a protein in a cell by visualizing with the fluorescence microscope antibodies to that protein that have been coupled to a fluorescent dye. The fluorescence microscope uses ultraviolet light, which causes bound fluorescent antibody to appear bright against a dark background. In *indirect* immunofluorescence, two kinds of antibodies are employed. One is an antibody to the cell protein, whereas the one that is actually visualized is a fluorescent-labeled antibody to the antibody bound to the cell protein. In a common research strategy the antibody to the cell protein is elicited in rabbits, whereas the fluorescent antibody is an antirabbit antibody that is prepared in goats and coupled to the dye. A major advantage to the indirect method is the amplified fluorescence due to the binding of several fluorescent antibodies to a single cell protein antibody.

Immunofluorescence reveals the presence and location of all major cytoskeleton constituents. Antitubulin antibodies are used to reveal microtubules (see figure, top of p. 180).

Microfilaments are apparent when cells are treated with antiactin. The appearance of the microfilament network differs in motile cells and immotile cells. In motile cells (see left figure, bottom of p. 180) most of the microfilaments are in a diffuse network. The sparse microfilament bundles point toward the center of the cell. In immotile cells (right) microfilaments are organized into stress fibers that may traverse the entire cell.

The association of myosin and tropomyosin with microfilaments is confirmed by immunofluorescence. This technique also helps to ascertain the functional relationships between actin and these proteins. Myosin has been localized to stress fibers in immotile cells (see left figure, top of p. 181) and spread throughout the cytoplasm in motile cells. Tropomyosin is localized to stress fibers in immotile cells (right).

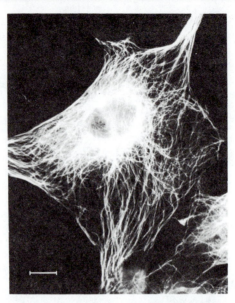

Indirect immunofluorescence of microtubules using tubulin antibody. Scale bar equals 10 μm. (From K. Weber, P.C. Rathke, and M. Osborn. 1978. Cytoplasmic microtubular images in glutaraldehyde-fixed tissue culture cells by electron microscopy and by immunofluorescence microscopy. Proc. Natl. Acad. Sci. U.S.A., *75:* 1822.)

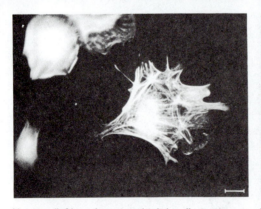 

Moving (left) and resting (right) cells in tissue culture. Indirect immunofluorescence using actin antibody reveals distribution of microfilaments. Scale bar equals 10 μm. (From E. Lazarides and J.P. Revel. 1979. The molecular basis of cell movement. Sci. Am., *240*(5): 101.)

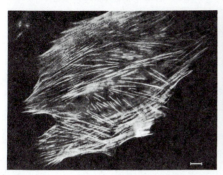
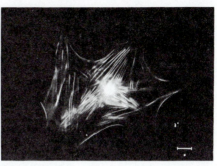

Distribution of myosin (left) and tropomyosin (right). Myosin in immotile cell is demonstrated by immunofluroescence with anti-myosin. Scale bar equals 10 μm. (From K. Fujiwara and T.D. Pollard. 1976. Reproduced from *The Journal of Cell Biology,* 1976, vol. 71, pp. 848–875 by copyright permission of The Rockefeller University Press.) Tropomyosin distribution is revealed by indirect immunofluorescence with antitropomyosin. Scale bar equals 30 μm. (From E. Lazarides. 1975. Reproduced from *The Journal of Cell Biology,* 1975, vol. 65, pp. 549–561 by copyright permission of The Rockefeller University Press.)

Intermediate filaments that are composed of the protein vimentin are visualized as a wavy network in the cytoplasm of cultured vertebrate cells. This network, shown in the figure that follows, is visualized with indirect immunofluorescence.

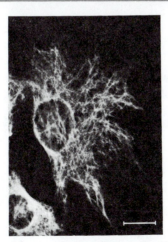

Indirect immunofluorescence of mouse tissue culture cell with antivimentin. "Vimentin" is taken from the Latin word *vimentus* (meaning "wavy"), because of the wavy pattern of these filaments. Scale bar equals 20 μm. (Courtesy of Dr. W.W. Franke.)

Regular text continued from page 179

Microfilaments *in vivo* emerge from sites on membranes. Elongation apparently occurs by addition of subunits at these membrane-associated nucleating sites and not at the free, cytoplasmic tips (Tilney et al., 1981). Thus, the direction of microfilament growth is opposite that of microtubules, in which subunits are added at the free distal tips rather than at the ends that are embedded in the MTOCs. As previously discussed, the association of microfilaments with the cell membrane accounts for their ability to deform the membrane when they contract. The membrane association is thought to be mediated by a protein called **vinculin,** which has been localized to sites of microfilament-membrane association (Geiger et al., 1980).

The generation of contractile forces by microfilaments is based upon an actin-myosin sliding-filament mechanism that is similar to that in smooth muscle. Low levels of myosin are found to be associated with microfilaments in nonmuscle cells (see box, p. 179). Contraction is regulated by Ca^{++} interaction with the calcium-binding protein, **calmodulin.** Calcium causes conformational changes in calmodulin, which can alter its biological function. A rise in the concentration of free Ca^{++} results in formation of the Ca^{++}-calmodulin complex. Calcium-bound calmodulin activates a cyclic AMP-dependent kinase enzyme, which phosphorylates the myosin light chain. This, in turn, leads to activation of myosin ATPase by actin, and contractility ensues (for review see Adelstein [1982]).

Two other proteins that are components of the contractile mechanism of muscle may also be found associated with microfilament bundles. These are **tropomyosin** and **α-actinin.** They may provide interfilament crossbridges within the bundles. Alpha-actinin is particularly intriguing because its ability to cross-link filaments is regulated by calcium: High Ca^{++} levels cause the cross-links to break, whereas low Ca^{++} levels allow them to reform rapidly. The calcium sensitivity of α-actinin may help facilitate rapid change in the cellular microfilament network: Cross-linking of microfilaments would occur under low calcium conditions, and a calcium flux would release the filaments to allow their rapid reorganization (Burridge and Feramisco, 1981). Tropomyosin is found primarily in association with microfilaments in bundles of immotile cells but not in association with microfilaments of highly motile cells. This correlation suggests that tropomyosin has a stabilizing influence on structural microfilaments, rather than an involvement in contraction (Lazarides and Revel, 1979).

Numerous other proteins have also been reported to associate with actin microfilaments in various cell types. These actin-binding proteins fall into three functional categories: (1) proteins that cross-link microfilaments to form bundles; (2) proteins that modulate actin polymerization; and (3) proteins that affect microfilament contractility (Korn, 1982; Weeds, 1982).

The microfilament network can be altered experimentally by exposing cells to **cytochalasin B,** a drug that blocks actin polymerization. Thus, it can be used to determine whether microfilaments are involved in producing particular shape changes. Unfortunately, the effects of cytochalasin on the cell are not so specific as those of colchicine; it causes side effects, including alterations of membrane transport functions (Tanenbaum, 1978). Consequently, when evaluating the effects of cytochalasin, the investigator must consider the possibility of nonspecific side effects.

Intermediate Filaments

The third category of cytoskeletal components is the intermediate filament. Intermediate filaments have a mean diameter of 10 nm. Unlike microtubules and microfilaments, intermediate filaments are not composed of the same molecular component in all cell types. Instead, the intermediate filament composition varies according to cell type and is dependent upon cell differentiation. The subunit structure varies among the five major classes of intermediate filaments. The five classes and the cell types in which they are found are: (1) keratin filaments (epithelial cells); (2) neurofilaments (neurons); (3) glial filaments (glial cells); (4) vimentin filaments (most differentiating cells, tissue culture cells, and in certain fully differentiated cells); and (5) desmin filaments (muscle and certain nonmuscle cells) (Lazarides, 1982). Filaments of different types can co-exist in the same cell. For example, both keratin and vimentin filaments have been found in cultured epithelial cells (Henderson and Weber, 1981).

It has been proposed recently that intermediate filaments interact with cellular organelles so as to help organize and maintain their three-dimensional arrangement in the cytoplasm (Lazarides, 1980). Vimentin filaments are closely associated with the nucleus and during mitosis form a cage that surrounds the mitotic spindle. During mitosis the nuclear envelope, cytoplasmic microtubules, and microfilament bundles break down. Perhaps the vimentin filaments, which are the only cytoskeletal elements retained throughout mitosis, provide the framework that helps to reconstruct cytoplasmic geometry after division (Zieve et al., 1980).

5–2. THE CELL SURFACE

The cell surface is the region of a cell that interfaces with other cells. Consequently, the surface must mediate and integrate cellular activities that involve cell-cell associations. Since cellular interactions are dynamic, it follows that cell surface properties are also dynamic. The surface also must be able to accommodate changes in its configuration under various circumstances. For example, it must be able to expand to allow cell growth. It also must be capable of fusion with other membranes (e.g., fertilization or fusion of the plasmalemma with intracellular vesicles) and can accommodate removal of a portion of itself (e.g., formation of vesicles during endocytosis). These dynamic properties of the cell surface are due to its molecular composition and organization.

The current working model for the cell surface (Fig. 5.5) is based upon the **fluid mosaic membrane model** of Singer and Nicolson (1972). According to this model, the cell membrane is composed of a bilayer of phospholipids into which numerous proteins are inserted. The lipids are arranged as a double row of molecules whose hydrophilic heads are on the bilayer's outer face and whose hydrophobic tails are pointed toward the center of the bilayer. The membrane proteins are of two types, based upon the extent of their interaction with the membrane lipids. The **integral (intrinsic) membrane proteins** penetrate both the hydrophilic and hydrophobic zones of the membrane and are tightly bound to the lipid phase, located primarily in the hydrophilic zones. Some of these proteins protrude from both the inner and outer faces of the membrane, whereas others are intercalated into either the inner or outer face. The proteins that protrude from the outer face are usually covalently bound to carbohydrates and are, therefore, **glycoproteins.** As we shall see, many of the morphogenic properties of the cell surface involve these carbohydrates, which mediate interactions with other cells, such as cell-cell adhesion. The proteins that protrude from the membrane into the cytoplasm can interact with cytoplasmic constituents, such as elements of the cytoskeleton.

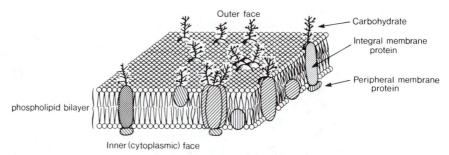

FIGURE 5.5 Fluid mosaic membrane model of Singer and Nicolson. (After G.L. Nicolson. 1979. Topographic display of cell surface components and their role in transmembrane signaling. Curr. Topics Dev. Biol., *13:* 308.)

The other category of membrane proteins (most of which are also glycoproteins) constitutes the **peripheral** (extrinsic) **membrane proteins.** These proteins are bound to either the inner or outer face of the membrane via weak bonds with either the lipids or the integral membrane proteins. On the outer surface they could mediate cellular interactions. On the inner surface some of them help to mediate the attachment of the cytoskeleton to the integral membrane proteins, whereas others serve to stabilize membrane structure by restricting mobility of the lipids and intrinsic membrane proteins (see subsequent discussion).

The spatial arrangement of membrane components is dynamic; that is, they are laterally mobile in the membrane. Rearrangements of membrane components will allow for spatial alterations in membrane properties and can indirectly affect cytoplasmic structure via the linkages between membrane proteins and the cytoskeleton. Since the lateral mobility of membrane components can have dramatic effects on membrane properties, order and stability must be maintained in the membrane except when change is required. Mobility is restricted by peripheral membrane proteins or by binding of membrane proteins to the cytoskeleton or to components of the extracellular matrix (see following discussion). Likewise, rearrangements of membrane components may be caused by a change in their interaction with these intracellular and extracellular elements.

The **extracellular matrix (ECM)** is the material that surrounds connective tissue (i.e., free) cells and underlies epithelia. Molecules of ECM interact with the cell membrane in such an intimate way that the membrane and the molecules form a functional and structural continuum (Hay, 1981). Hence, it is impossible to consider the role of the cell surface in development without considering the ECM along with the plasma membrane.

The ECM is synthesized by cells, secreted by them, and affects cell function by its interactions with them. The ECM is composed primarily of **collagens, glycosaminoglycans (GAG)**, and noncollagenous glycoproteins. Collagens form a fibrous network in the ECM, and the other components compose the ground substance that is intercalated among the collagen fibrils. GAG consist of long, unbranching chains of repeating disaccharides, one of which is a hexosamine, which may be sulfated, and the second is hexuronate or D-galactose. The glycosidic bond between these units may be in the β- or α-configuration. With the exception of hyaluronic acid, GAG (e.g., chondroitin, chondroitin sulfate, dermatan sulfate, keratan sulfate, heparan sulfate, and heparin) are linked to proteins to form **proteoglycans.** A proteoglycan molecule contains a long core protein from which numerous GAG chains extend. Variations in the number, type, and distribution of GAG on the core protein as well as variations in the core protein itself can generate a great many different proteoglycans (Hascall and Hascall, 1981).

Hyaluronic acid serves to link together individual proteoglycans to form an extensive network that associates with collagen in the ECM. The physical properties of the collagen in the matrix can be modified by certain proteoglycans. Hence, the structure of the matrix may exhibit local specialization and undergo changes during development that are controlled by the proteoglycan composition. These variations in the matrix may affect such cellular functions as their migratory properties, since cells migrate through the ECM, which they use as a scaffold (see section 5–3). GAG also have been shown to interact directly with cell surface molecules. Such interactions may influence cell differentiation by altering various parameters of cell function (Toole, 1981).

A variety of glycoproteins are constituents of the ECM. One of the best studied of these is **fibronectin,** which is a high molecular weight molecule that is widely distributed in connective tissue, at cell surfaces, and in **basal laminae** (the collagenous zones that underly epithelia and surround muscle cells). Fibronectin has binding sites for cell surface receptors as well as collagen and GAG (Yamada et al., 1980a) and may be involved, therefore, in mediating cellular interactions with these components of the extracellular matrix. Another glycoprotein, **laminin,** which is more limited in distribution, is found primarily in basal laminae and is involved in attachment of epithelia to collagen (Terranova et al., 1980).

5–3. THE CELL SURFACE AND CYTOSKELETON IN MORPHOGENESIS OF ANIMAL EMBRYOS

Morphogenesis involves a number of distinct cellular functions; the embryo forms as a consequence of the integration and control of these processes in the separate cells of which it is composed. Research in morphogenesis is aimed at identifying these cellular functions, understanding their molecular basis, and clarifying the mechanisms that control and integrate cellular morphogenic functions to produce the organism. The formation of a distinct, recognizable organ during development depends upon the mobilization of the cells that will form it. In most cases those cells arise at some distance from where the organ will be formed. They must dissociate from their original position, migrate through the embryonic connective tissue, recognize their correct new position, and become fixed in these new positions. After fixation, the individual cells of the aggregate change shapes and undergo differentiation to become functional elements of the forming organ. In this section we shall discuss the roles of the cell surface and the cytoskeleton in the changing relationships among cells during development. The role of the cytoskeleton in shaping immotile cells in the formation of tissues and organs will be discussed in Chapter 13.

Cell Motility

Cell motility plays a key role in cellular reorganizations that are necessary in morphogenesis of the animal embryo. We shall discuss several examples of migratory cells in later chapters. These will include: (1) sea urchin mesenchyme; (2) inward moving cells of vertebrate gastrulae; (3) vertebrate neural crest cells; (4) primordial germ cells of vertebrate embryos, which originate in endoderm and migrate to presumptive gonads; and (5) vertebrate mesenchyme cells, which migrate within the organizing embryo and condense around epithelia. In most of these cases, analysis of the mechanisms of migration is difficult because the cells are inaccessible to observation. For this reason, much of our knowledge of cell motility is derived from studies of vertebrate fibroblast cells in tissue culture. Although movement of fibroblasts *in vitro* is somewhat artificial, it provides a useful model for analyzing *in vivo* cell movement.

Fibroblasts move by spreading along a substrate, with which they form temporary adhesions at several points. The forward movement of the cell occurs by the protrusion of the leading edge of the cell to form a thin, fanlike structure called the **leading lamella.** After protruding forward, the leading lamella attaches to the substrate near its leading edge at points of adhesion, called **focal contacts.** As the leading edge thrusts forward, the cell elongates in that direction because the trailing edge retains its attachment for a time. The consequent shape assumed by the cell is shown in Figure 5.6. The breaking of the adhesion at the trailing edge occurs as the cell body is retracted up to the point of the next adhesion, and the cell is displaced forward. Thus, continuation of movement depends upon a repeated process of forward thrust of the leading edge, adhesion of the leading edge to the substrate, contraction of the cell, and loss of adhesion at the trailing edge. The adhesion points themselves do not move; as old adhesions fade, they are replaced by new focal contacts that are formed near the leading edge of the lamella in front of the existing ones (Trinkaus, 1976).

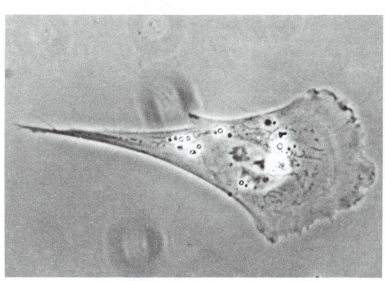

FIGURE 5.6 Phase contrast micrograph of a living chick embryo heart fibroblast migrating on a glass surface. (From M. Abercrombie, G.A. Dunn, and J.P. Heath. 1977. The shape and movement of fibroblasts in culture. *In* J.W. Lash and M.M. Burger (eds.): *Cell and Tissue Interactions.* Raven Press, New York, pp. 57–70. Courtesy of Raven Press.)

The motile force for migrating fibroblasts is provided by micro-filaments. Microfilaments have two distinct kinds of distributions in these cells. The leading lamellae are flattened projections of the cell cortex containing a lattice-work of microfilaments between the upper and lower cell membranes. In time-lapse cinemicrography the lamellae are seen to be highly active structures, going through cycles of random protrusion and withdrawal. Forward movement occurs because protrusion is more frequent and lasts longer than the withdrawal phases (Trinkaus, 1976). The anterior margin of the lamella also undulates by an upfolding that lifts a portion of the margin above the substrate. This undulating activity is called "ruffling" (Fig. 5.7). Since the activities of the leading edge are frozen by cytochalasin B (Wessells et al., 1971), it is likely that the activities of the lamella are produced by the microfilament network that is located there.

A quite different organization of microfilaments is found in the bundles, or cables, of microfilaments that are anchored on the cyto-plasmic faces of the focal contacts (Fig. 5.8), presumably by vinculin. These bundles are oriented parallel to the direction of movement and project backward to merge into fibrillar material that surrounds the nucleus. They are thought to be responsible for drawing the posterior end of the fibroblast forward by their contraction. As the microfilament bundles contract, the front edge of the cell extends forward and lays down new focal contacts, which acquire new microfilament bundles. Thus, with the contractile forces of a succession of microfilament bundles from progressively more anterior positions, the cell is pulled forward (Aber-crombie, 1980). The contractile function of the microfilament bundles is supported by observations that cytochalasin B inhibits fibroblast lo-comotion (Spooner et al., 1971).

Microtubules are thought to play a structural role in migrating cells. Microtubules are oriented parallel to the long axis of migrating fibro-blasts (see Fig. 5.1). They are considered to be responsible for the elon-gation of the cells, an interpretation that is supported by the rounding up of fibroblasts in the presence of colchicine (Trinkaus, 1976).

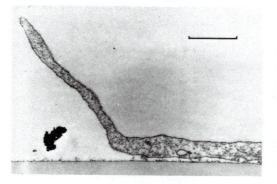

FIGURE 5.7 Electron micro-graph of section through the leading edge of a chick heart fi-broblast showing a "ruffle." Bar equals 1.0 μm. (From M. Aber-crombie. 1980. The Croonian Lecture, 1978. The crawling movement of metazoan cells. Proc. R. Soc. Lond. [Biol.] Ser. B, *207*: 129–147.)

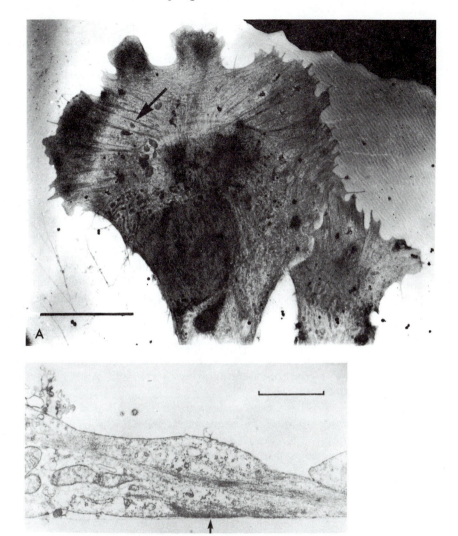

FIGURE 5.8 Microfilaments in motile fibroblasts. *A*, High-voltage electron micrograph of a whole chick heart fibroblast with numerous microfilament bundles (arrow) in the lamella. Bar equals 10 μm. (From M. Abercrombie, G.A. Dunn, and J.P. Heath. 1977. The shape and movement of fibroblasts in culture. *In* J.W. Lash and M.M. Burger (eds.): *Cell and Tissue Interactions.* Raven Press, New York, pp. 57–70. Courtesy of Raven Press.) *B*, Electron micrograph of section through a chick embryo fibroblast showing a focal contact (arrow). Note the electron-dense plaque and associated bundles of microfilaments. Bar equals 1.0 μm. (From M. Abercrombie. 1980. The Croonian Lecture, 1978. The crawling movement of metazoan cells. Proc. R. Soc. Lond. [Biol.] Ser. B, *207:* 129–147.)

The protrusive activity of the leading edge of a migrating fibroblast is inhibited by adhesion to another fibroblast. This phenomenon is called **contact inhibition of migration** (Abercrombie, 1970). Once a leading lamella is suppressed by making contact with another cell, protrusive activity is initiated elsewhere, and a new front end is formed as the cell changes its direction of migration. If cell density is sufficiently high, cells contact one another on all sides and are unable to migrate further. The result is the formation of a confluent monolayer of cells that are inhibited from movement by their mutual contacts. Cells in a confluent monolayer have a reduced rate of cell division. This is called **contact inhibition of mitosis.**

The movement of tissue culture cells is not necessarily the same as that of embryonic cells migrating within the embryo. Individually migrating cells *in vivo* are not flattened with thin leading lamellae. Instead, they appear to move by means of narrow pseudopodia and blunt protrusions. These differences probably arise because cells *in vitro* are migrating along a flat inanimate substrate, whereas cells *in vivo* migrate through the fibrous extracellular matrix. It was long ago pointed out by Harrison (1914) that cells assume different shapes *in vitro* according to the mechanical conditions to which they are exposed. Weiss (1959) stated that the shape of a migrating cell is "the result of a distinctive behavioral reaction of a living cell to its environment." Hence, the different shapes of cells migrating *in vivo* and *in vitro* may be of secondary importance.

Cell migration in embryogenesis is highly precise. Groups of precursor cells migrate in streams from their origin to their target sites. For example, the primordial germ cells migrate from the endoderm to the genital ridges and populate them to become gametes (see Chap. 11). Also, neural crest cells migrate from their middorsal position to form visceral cartilage, cartilage of the embryonic skull, adrenal medulla, spinal ganglia, and other structures (see Chap. 13). The mechanisms that direct migratory cell populations toward their target sites are very important for the proper spatial organization of the embryo.

Weiss (1959) demonstrated that cells tend to follow fibrous pathways during their migration. This phenomenon is called **contact guidance.** Dunn and Ebendal (1978) observed that a three-dimensional meshwork of fibers is more effective in guiding cells than is a flattened mat of fibers. Since embryonic cells migrate through a three-dimensional meshwork of fibers in the extracellular matrix, it has been proposed that this meshwork guides the cells toward their intended sites. Perhaps local variations in the matrix, such as glycosaminoglycan heterogeneity can affect the shape of the matrix, and hence, the path of migration. Other factors that could affect the path of migration are: (1) the nature of the adhesions that the cell forms with the substrate; (2) physical barriers that block the migratory path; and (3) molecular substances that attract cells (**chemotaxis**). The latter mechanism is utilized to orient migrating slime mold cells (see box, p. 198).

Since cellular interactions with the substrate are so important in cell motility, we shall now examine that relationship in more detail. Cell membrane glycoproteins function as receptors that bind to molecules of the extracellular matrix. One of the most important components of the matrix is fibronectin, which has been demonstrated to promote cell motility *in vitro* (Ali and Hynes, 1978), possibly because fibronectin is involved in formation of the focal contacts with the substrate and intracellular organization of the bundles of contractile elements that emanate from these sites (Virtanen et al., 1982). Its role in cell adhesion may be explained by the fact that it can bind both to cell membrane receptors and to components of the extracellular matrix such as collagen and glycosaminoglycans (Yamada et al., 1980b). Thus, it could serve to link a cell to the extracellular matrix. Fibronectin forms extracellular fibrils that are organized parallel to the long axes of leading lamellae (Hynes et al., 1978). Therefore, it is possible that fibronectin fibrils influence the direction of migration. This might occur by affecting the organization of the microfilament bundles originating at the focal contacts. The microfilament bundles may give directionality to cell movement by serving as guides along which the nucleus and the bulk of the cytoplasm move during forward displacement (Albrecht-Buehler, 1977; Pouysségur and Pastan, 1979).

Although fibronectin promotes motility of fibroblasts *in vitro*, it is uncertain whether it performs this function in the embryo. Fibronectin has been detected in the extracellular matrix along pathways of cell migration in various embryos (Hynes, 1981). Thus, it is present at the right time and place to promote *in vivo* cell migration. Further investigation is necessary to establish a functional role for fibronectin in this process.

Intercellular Adhesion

One of the most important factors governing the behavior of an individual cell of the embryo and the role of that cell in morphogenesis is the nature of its interactions with its neighboring cells. The animal embryo takes shape as its cells sort out and associate into specific multicellular assemblages that are the primordia for the organs and tissues of the adult organism. During this period, cells change their relative positions as a result of modifications in the adhesive interactions between cells. For example, adhesion between cells maintains the integrity of epithelial layers, and conversely, reduced adhesion allows cells to dissociate from an epithelium and form free cells. The latter may migrate from their site of origin to a distant location where they reassociate to form once again a cohesive unit. Condensation of free cells frequently occurs around an epithelial vesicle, bringing cells of diverse origin together into organ primordia. The assembly of such organ primordia involves cell recognition and cell adhesion. The dynamics of intercellular contacts are, therefore, highly significant for organizing cells into a definitive embryo.

The specialized adhesive properties of cells are demonstrable by experiments in which tissues are dissociated and allowed to reassemble into aggregates. H.V. Wilson (1907) was the first to perform this type of experiment. Wilson dissociated adult sponges into a suspension of cell fragments, individual cells, and small clusters. The cells sink to the bottom of the culture dish, where they actively migrate until they adhere to one another to form clusters that increase in size by addition of individual cells and by fusion with adjacent clusters. The aggregates eventually produce small, individual sponges. Specificity of sponge cell aggregation was demonstrated by mixing cells obtained from dissociated sponges of different species that have cells that are distinguishable by their pigmentation. These mixed cell suspensions at first associate at random, but the cells later sort out as separate aggregates, each containing cells of only one species. The cells sort out into species-specific aggregates because cells form more stable contacts with cells of the same species than they do with cells of another species. Thus, sponge cells have a selective affinity for cells of their own species.

Similar experiments have demonstrated that cells of developing vertebrate embryos will segregate in a mixed aggregate according to type and reestablish their former associations. These studies were initiated by Johannes Holtfreter and his associates with amphibian embryos. In 1955, Townes and Holtfreter presented the first critical evidence that embryonic cells with distinct developmental fates have different adhesive properties. They took advantage of the size and pigmentation differences of ectoderm, mesoderm, and endoderm cells to follow the cells' behavior in aggregates. The cells reorganize within aggregates to assume positions that correspond to the normal germ layer arrangement in the intact embryo. Thus, in a mixed aggregate of mesoderm and presumptive epidermis the mesoderm sinks to the inside to form mesenchyme, whereas the epidermal cells accumulate at the periphery and spread over the surface to form an epithelium (Fig. 5.9A). When cells of all three germ layers are combined, the endodermal cells form a compact ball, whereas cells of presumptive epidermis form a surface epithelium, and mesoderm produces an intermediate mass of **mesenchyme**, which is a term applied to free embryonic cells and usually refers to cells of mesodermal origin (Fig. 5.9B). Even cells of the same germ layer with different developmental fates will segregate from one another. Thus, presumptive neural ectoderm and epidermis in a mixed aggregate will sort out, with the epidermis forming a surface epithelium and the neural ectoderm sinking to the interior to form structures resembling brain vesicles (Fig. 5.9C).

The cell-sorting experiments demonstrate nicely that selective cell adhesion is an important force in embryonic organization. Two important aspects of cellular segregation are evident from these experiments. One is the preferential association of like cells. This undoubtedly plays a role in maintaining the integrity of embryonic cell layers. Likewise,

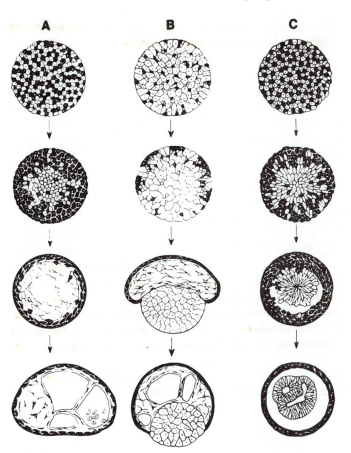

A **B** **C**

FIGURE 5.9 Rearrangement of disaggregated and reaggregated embryonic cells. *A*, Combined epidermal and mesodermal cells. *B*, Combined epidermal, mesodermal, and endodermal cells. *C*, Combined epidermal and neural plate cells. (From P.L. Townes and J. Holtfreter. 1955. Directed movements and selective adhesion of embryonic amphibian cells. J. Exp. Zool., *128*: 53–120. Reproduced from B.I. Balinsky, 1975.)

changes in adhesiveness will allow cells from a single layer to form separate cell groupings. The other aspect is the spatial arrangement of cell groups with respect to one another. Thus, during gastrulation, endoderm sinks to the inside of the embryo, while ectoderm spreads over the periphery.

The relative positioning of tissues within an aggregate has also been studied in cell-sorting experiments by using cells from the organ rudiments of chick embryos, which have been dissociated by trypsinization (Steinberg, 1963). Dissociated cells of different origins combine to form a common aggregate and then sort out according to tissue type, usually with one tissue surrounding the other. The advantage of using organ primordia is that the tissues formed after cell-sorting will differentiate to reveal their developmental fate. According to Steinberg, the relative positions of cells in a mixed aggregate result from random motility of the cells and quantitative differences in the adhesiveness between them. Thus, cells with the stronger mutual attraction will aggregate in the center, whereas cells with weaker attraction will remain at the surface. This is the **differential cellular adhesiveness** hypothesis (Steinberg, 1970).

Consider the results of the following tissue combinations (Fig. 5.10). In a mixture of limb-bud cartilage and heart, the heart cells surround the cartilage. When heart and liver cells are combined, the liver cells surround the heart cells. Finally, in a mixture of liver cells and cartilage cells the liver surrounds the cartilage. Steinberg has proposed that the relationship between liver and cartilage is predictable from the previous two combinations. Heart is less adhesive than cartilage, and liver is less adhesive than heart; therefore, liver should envelop cartilage. Thus, a hierarchy can be arranged whereby those at the top of the scale will segregate internal to those immediately below, whereas those at the bottom of the scale segregate external to all others.

Differential adhesiveness provides a probable explanation for the ordering of the germ layers during amphibian gastrulation (see Chap. 13). However, Steinberg (1970) has emphasized that differential adhesiveness does not completely explain cellular arrangement during the organization of cells into tissues and organs and its maintenance in the adult. Furthermore, the hypothesis does not deal with the chemical or physical mechanisms of cell-cell adhesion. It merely explains the consequences of differences in the strengths of adhesion between various kinds of cells (Steinberg, 1970).

Moscona (1960) has proposed that **specific cellular adhesiveness** is an important factor that determines cellular associations. According to this

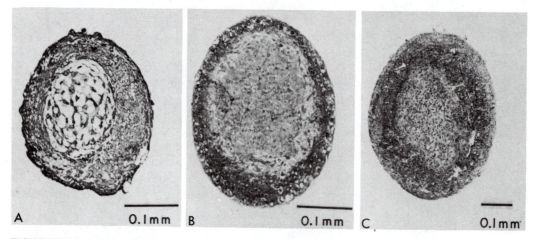

FIGURE 5.10 Sections through aggregates formed by disaggregated and reaggregated chick embryo cells. *A,* Combined limb-bud chondrogenic cells and heart ventricle cells. The reconstructed heart tissue surrounds the cartilage. *B,* Combined heart ventricle cells and liver cells. The reconstructed liver tissue surrounds the heart tissue. *C,* Combined limb-bud chondrogenic cells and liver cells. The reconstructed liver tissue surrounds the cartilage. (From M.S. Steinberg. 1963. Reconstruction of tissues by dissociated cells. Science *141:* 401–408. Copyright 1963 by the American Association for the Advancement of Science.)

hypothesis, cell adhesion is a property of specific cell surface macro-molecules that allow cells to recognize like cells and bond with them. Thus, cells belonging to different tissues will segregate from one another in a mixed aggregate because they recognize and preferentially associate with their own kind. This mechanism could also function to cause cells of different origins to come together to form organ primordia. Moscona has proposed that cell-surface constituents function as cell-cell recognition sites and cell-cell ligands that bridge between specific cells, and thus hold them together (for review, see Moscona and Hausman [1977]). Specificity of cell recognition would be determined by the molecular characteristics of the receptors and ligands and also by their topographical arrangements on the cell surface.

Cell Surface Molecules in Adhesion

The proposal that there are specific adhesion factors on cell surfaces has prompted extensive studies aimed at discovering whether such factors regulating cellular associations can be demonstrated. Cell surface aggregation factors were first detected by analyzing the supernatant fluid in which cell dissociation has occurred. The rationale for these experiments was the assumption that dissociation releases the cells' ligands into the medium. Consequently, under conditions that are favorable for reassociation, the addition of the supernatant medium would cause cells to aggregate.

Sponge cell aggregation was examined in just this way (Humphreys, 1963). Dissociation of sponge cells occurs in the absence of the divalent cations Ca^{++} and Mg^{++}. Thus, sponges placed in Ca^{++}- and Mg^{++}-free seawater will dissociate. The presence of an aggregation factor in the dissociation medium is demonstrated by adding the medium back to dissociated cells, along with the divalent cations. The factor also accounts for the species-specificity of reaggregation, since it will cause adhesion only of cells of the same species. An additional point about these experiments should be made. That is, the cells are maintained at low temperatures at all times to inhibit metabolic activity. If the temperature is raised, the cells will aggregate without added supernatant. This implies that at the higher temperature the cells are synthesizing a new aggregation factor to replace that lost during dissociation.

Aggregation in sponges is a two-step process. First, the extracellular aggregation factor binds to specific cell surface receptors. Second, the aggregation factors bound to adjacent cells interact with one another to link the cells together. The second step requires calcium ions, which explains why aggregation can be controlled by the amount of Ca^{++} (Jumblatt et al., 1980). The properties of the aggregation factor are those of an intercellular ligand.

Similar experiments have been conducted to examine adhesion between vertebrate embryonic cells (Moscona and Hausman, 1977). Chick neural retina cells can be dissociated with trypsin; these cells can then be maintained in culture, where they will reaggregate to reconstruct retinal tissue. When mixed with other chick embryo cells, the cells sort out to form tissue-specific aggregates. As with sponge cell aggregation, chick retina cells will not reaggregate when they are maintained at low temperatures. Inhibitors of macromolecular synthesis will also prevent reaggregation. Thus, metabolic activity is required, presumably to resynthesize the ligand.

When retina cells are maintained in culture as a monolayer, the ligand produced by the cells apparently accumulates in the culture medium, since the medium can be used to promote retina-specific cell aggregation. This retina-specific factor has been purified and characterized as a glycoprotein. It is called retina cognin to indicate its presumed role in recognition and association of embryonic retina cells. Retina cognin has been isolated directly from the membranes of embryonic neural retinal tissue (Hausman and Moscona, 1976), indicating that it is indeed a component of the cell surface. Another important property of the retina-specific factor is that it is produced only by retina cells from pre-13-day embryos. This suggests that it is developmentally regulated and produced by retina cells when it is needed for a specific developmental event (Moscona and Hausman, 1977). Antiserum to retina cognin has also been used as a probe to detect cognin and to investigate its functional role. The surfaces of chick embryo retina cells treated with rabbit anticognin followed by fluorescent goat antirabbit serum (see box, p. 179, for discussion of this technique) fluoresce in ultraviolet light. Thus, the anticognin recognizes the cognin on the cell surfaces (Fig. 5.11). Furthermore, treatment of retina cells with anticognin abolishes the ability of the cells to aggregate (Hausman and Moscona, 1979). Clearly, this surface-associated molecule is involved in retina cell aggregation.

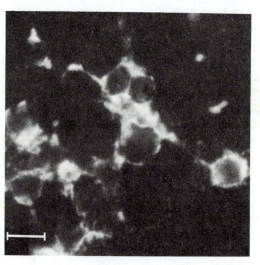

FIGURE 5.11 Indirect immunofluorescence of neural retina cells from chick embryos with anti-cognin. Scale bar equals 10 μm. (From R.E. Hausman and A.A. Moscona. 1979. Immunologic detection of retina cognin on the surface of embryonic cells. Exp. Cell Res., *119*: 197.)

Chick neural retina cell adhesion has been the subject of numerous investigations. These studies have revealed a number of cell surface molecules involved in adhesion of these cells. One example is cell adhesion molecule (CAM), which has been studied intensively with anti-CAM antibodies (Rutishauser et al., 1978a, b). Another series of investigations have revealed the existence of two retina cell adhesion mechanisms that can coexist in the same cell (Magnani et al. 1981; Thomas and Steinberg, 1981; Thomas et al., 1981). One of these mechanisms requires ionic calcium for adhesion, whereas the other is calcium independent. Studies of the possible involvement of cognin, CAM, and other adhesion molecules in these two mechanisms should help to clarify the total number of adhesion mechanisms for neural retina cells as well as how the adhesion molecules function to bind cells together. Antibodies against cell-surface molecules will be important tools in these investigations.

Research on cell adhesion has mushroomed in recent years and has led to numerous demonstrations of selective cell affinity and to the identification of additional developmentally regulated cell surface molecules that are involved in cell-cell recognition. One commonly encountered cell recognition and adhesion mechanism involves the interaction of cell surface carbohydrate-binding proteins (lectins) with complementary carbohydrates on adjacent cell surfaces. Lectins are proteins and glycoproteins that have multiple carbohydrate-binding sites and can therefore link cells together by binding to the carbohydrate moieties of cell surface glycoproteins. Lectins preferentially bind to certain carbohydrates; this specificity can be demonstrated experimentally by using specific sugars, which if they bind the lectin will block lectin-mediated cell-cell association. For example, galactose will block adhesion mediated by a galactose-specific lectin.

The carbohydrate-binding property is also used to purify lectins by passing a mixture of cell surface molecules through an affinity chromatography column (see Chap. 3 for a discussion of affinity chromatography) containing the appropriate carbohydrate that is bound to support media. The lectin will bind to the carbohydrate and may be eluted from the column by using the appropriate reagents. Lectins are frequently assayed by their ability to agglutinate trypsin-treated erythrocytes, which contain surface carbohydrates that bind the lectins. The carbohydrate specificity of the lectin is examined by determining which sugar will block the agglutination.

Lectin-mediated aggregation has been demonstrated in a number of developing systems. One example is slime mold aggregation (see box on p. 198). Lectins have also been demonstrated in chick embryonic tissues (Kobiler et al., 1978; Cook et al., 1979; Barondes, 1980). Chick embryo lectins are developmentally regulated; that is, they are prevalent at certain stages in the development of particular tissues. Their probable developmental significance is illustrated by the observation that interference with lectin activity inhibits the cohesion of chick embryo cells

(Milos and Zalik, 1982), induces changes in cell morphology, and promotes cell migratory activity *in vitro* (Milos and Zalik, 1981). Lectins are also found in amphibian embryos (Roberson and Armstrong, 1980; Harris and Zalik, 1982). Lectins are not restricted to embryos; in some cases lectin activity associated with particular embryonic cells can be demonstrated for different cells in the adult (Ceri et al., 1981). These lectins apparently play certain roles in development and are utilized for other purposes in the adult.

Lectins have also been used as tools to study cell surface glycoproteins by their selective binding to the carbohydrate moieties of glycoproteins. An example of a commonly used lectin is **concanavalin A,** which is specific for glucose and mannose residues. Lectin-binding properties of cells change during development, which implies that cell surface glycoproteins have changed. Modulations of both the glycoprotein and lectin compositions of cell surfaces likely play an important role in mediating the changing associations between cells that occur during development.

We have discussed here only a few of many examples of specific cell adhesion mechanisms that have been described in the literature. It will be interesting to learn from further investigation whether these mechanisms are unique and unrelated to one another or if they are permutations of a small number of universal mechanisms of cell-cell association.

Regular text continued on page 202

INTERCELLULAR ADHESION IN SLIME MOLDS

The slime mold *Dictyostelium discoideum* exists as a solitary, amoeboid vegetative cell in forest detritus, feeding primarily on bacteria and dividing by binary fission. However, when the food supply is limited, a series of events occur that simulate morphogenic processes in multicellular organisms. The cells migrate in large streaming patterns in which cells adhere to one another. The streams of cells form aggregates containing up to 10^5 cells. A surface sheath is deposited around the aggregate, preventing it from incorporating additional cells. The aggregate rises above the surface as a fingerlike structure, which falls over on its side and migrates over the surface as a sheath-enclosed **slug.** After variable lengths of time, slugs cease movement and undergo dramatic morphological changes, whereby the anterior tip rises from the surface in preparation for the terminal differentiations that form the fruiting body. Spores released from the fruiting body germinate to emerge as amoebae, completing the life cycle.

Aggregation of *Dictyostelium* has proved to be an excellent experimental system for the study of cellular interactions in morphogenesis. The attraction of thou-

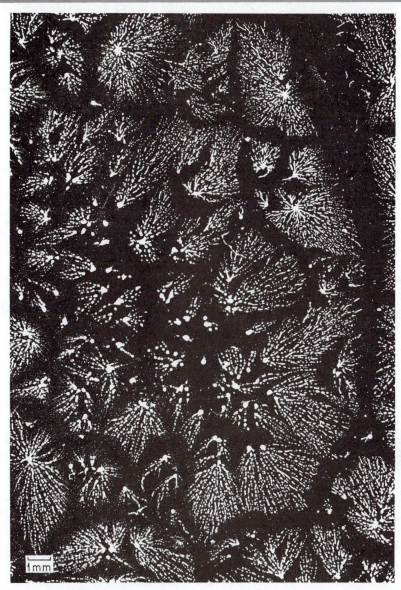

Dictyostelium aggregation patterns. Each individual amoeba appears as a small white dot; aggregation centers appear as clear areas. (From W.F. Loomis. 1975. *Dictyostelium discoideum. A Developmental System.* Academic Press, New York, p. 2.)

sands of cells to a common point involves chemotaxis to a compound excreted by the cells. The chemical attractant has been identified as cyclic AMP (Bonner et al., 1969). The cyclic AMP affects the direction of amoeboid movement, thus causing the cells to migrate toward regions of high concentration of the substance, which exist in the central cluster. The formation of aggregates is initiated by cells that spontaneously emit pulses of cyclic AMP, which diffuses through the medium. As a pulse of cyclic AMP encounters other cells, they are stimulated to move in the direction of the source of the pulse. Shortly after receiving a pulse of cyclic AMP, a responding cell emits its own pulse, thus amplifying the cyclic AMP level of the central cluster and attracting still more cells (Loomis, 1975). These cells move toward the center in well-defined waves.

End-to-end adhesion seen during the aggregation phase is a capability that the cells acquire as they differentiate from the vegetative stage. Thus, cells can be compared before and after the development of aggregation competence to determine what regulates the ability of cells to adhere to one another. The adhesive properties of *Dictyostelium* cells are studied by rotation of cell suspensions in a salt solution containing EDTA (see Chap. 2), which prevents nonspecific aggregation and allows the investigator to study selectively the specific aggregation ability that is acquired in development. Under these conditions, vegetative cells remain separate, but aggregation-competent cells form clusters.

Aggregation of these cells is made possible by changes in the composition of the membrane. The change in membrane composition is indicated by the use of antibodies that recognize membrane components. Native antibodies have two antigen recognition sites. Treatment with proteolytic enzyme produces univalent antibody fragments with a single antigen recognition site. These fragments are called **Fab fragments.** Fab fragments of antibodies prepared against membrane components of aggregation-competent cells will recognize the cell adhesion molecules on cell surfaces and render them non-functional in adhesion. However, fragments of antibodies prepared against vegetative cells will not block aggregation of aggrega-

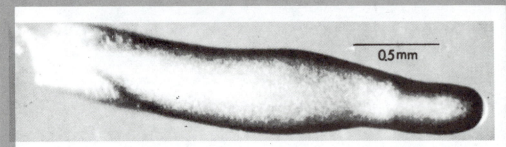

A migrating slug seen from above. The tapered anterior tip is on the right (From W.F. Loomis. 1975. *Dictyostelium discoideum. A Developmental System.* Academic Press, New York, p. 3.)

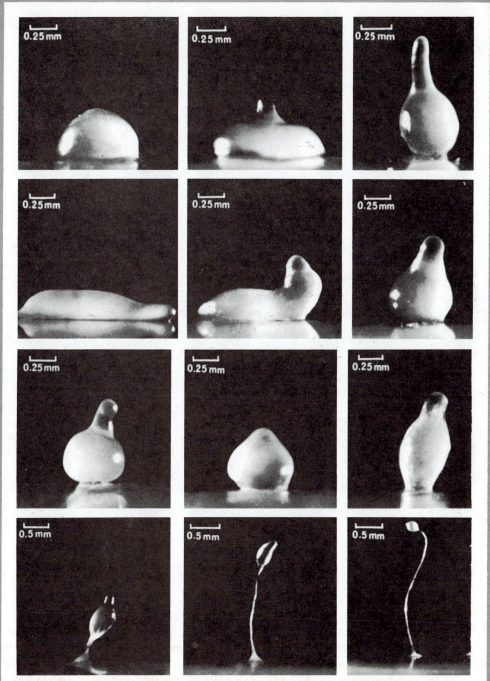

Formation of the fruiting body from the slug. (From W.F. Loomis. 1975. *Dictyostelium discoideum. A Developmental System.* Academic Press, New York, pp. 8, 9.)

tion-competent cells (Beug et al., 1973a, b). Hence, new antigens must appear on the membrane with the development of aggregation competence.

The membrane change is primarily due to the insertion in the cell surface of a specific glycoprotein. This glycoprotein occupies the sites where adhesion occurs; they have been called **contact sites A** (Müller and Gerisch, 1978). Linkage between cells may be mediated by an interaction between contact sites A of adjacent cells. Alternatively, contact sites A might have a more indirect role in adhesion. For example, they might activate other sites on the cell surface that are responsible for linking cells together (Gerisch, 1980).

There is strong evidence that aggregation also involves lectins (Barondes, 1980). These cell surface-associated carbohydrate-binding proteins have been called **discoidins.** Discoidins are absent from vegetative cells but are synthesized by aggregation-competent cells (Barondes, 1981). Discoidin is shown to be a cell agglutinin by its ability to agglutinate erythrocytes—a function that is inhibited by a sugar. The erythrocyte agglutination function implies that discoidin is involved in aggregation of slime mold cells. This implication is supported by genetic studies. Mutants that either lack the major discoidin (Siu et al., 1976) or produce a nonfunctional discoidin (Ray et al., 1979) fail to aggregate.

Purified discoidin binds to the surfaces of aggregation-competent cells of *D. discoideum* but will not bind to surfaces of vegetative cells. This observation implies that a carbohydrate-containing discoidin receptor appears on the surfaces of cells as they develop aggregation competence. It is tempting to speculate that discoidin binds to the carbohydrate portions of contact site A glycoproteins on adjacent cells, thus linking them together. However, there is no evidence that the contact sites are discoidin receptors (Gerisch, 1980). Another discoidin-binding protein that is developmentally regulated has recently been discovered (Breuer and Siu, 1981). Additional research is required to ascertain the roles of these various molecules in mediating slime mold aggregation.

Regular text continued from page 198

Intercellular Junctions

As cells that have been migrating in an embryo adhere to one another and become fixed into a cohesive unit, they form intercellular contacts that are important to the function of the unit itself. Gap junctions between cells are channels of communication through which ions and small molecules may pass. Thus, there is a potential for exchange of molecules that carry developmental information through sheets of cells that are interconnected. Whether gap junctions actually play such a role in development remains to be seen. Tight junctions are found between cells of an epithelium. These junctions seal the spaces between cells and prevent passage of material across the epithelial sheet. **Desmosomes** are points of adhesive contact between epithelial cells. Inside the cell they serve as points of attachment for a class of intermediate filaments called **tonofilaments,** which (at least in certain cultured cells) are composed of keratin (Franke et al., 1978).

Changes in Surface Properties of Malignant Transformed Cells

The formation of a malignant tumor is a case of abnormal development; tumor cells grow without restraint and may dissociate from the tumor and spread to various parts of the body (**metastasize**) to establish secondary tumors. Considerable evidence exists that the properties that characterize normal cells become modified in characteristic ways in malignant cells. Much of this evidence has been obtained from *in vitro* studies. Cells growing *in vitro* that are derived from tumors or that are capable of forming tumors in animals are called **transformed cells.** The study of transformed cells has led to an improved understanding of the properties of tumor cells.

As previously discussed, normal cells in tissue culture move along the substrate until they encounter another cell; contact results in inhibition of movement. If normal cells are surrounded on all sides by other cells, movement ceases, and a monolayer is produced in which the cells have a reduced rate of cell division. On the other hand, when migrating transformed cells contact one another, migration may not be inhibited; instead, transformed cells may crawl under one another (underlap) and continue to multiply to form multiple layers of cells on the substrate (Abercrombie, 1979). This failure of transformed cells to respond to normal controls over migration and growth may explain why cancerous cells can invade other tissues and multiply at an abnormally high rate.

The migratory behavior of normal tissue culture cells is highly dependent upon the adhesive contacts they make with their substrate. Transformed cells are much less adhesive than normal tissue culture cells. As a consequence, transformed cells do not spread out on the substrate when they migrate but remain more compact and rounded with numerous blebs and microvilli adorning their surfaces. This difference is clearly evident when comparing the morphologies of normal cells with cells of the same type that have been transformed with viruses (Fig. 5.12).

If transformed cells are grown on a more adhesive substrate, they revert to a more normal morphology (Erickson and Trinkaus, 1976). This observation is consistent with the hypothesis that transformed cells are less adhesive than normal cells, and that this difference in adhesiveness is responsible for the morphological and behavioral differences between normal and transformed migratory cells.

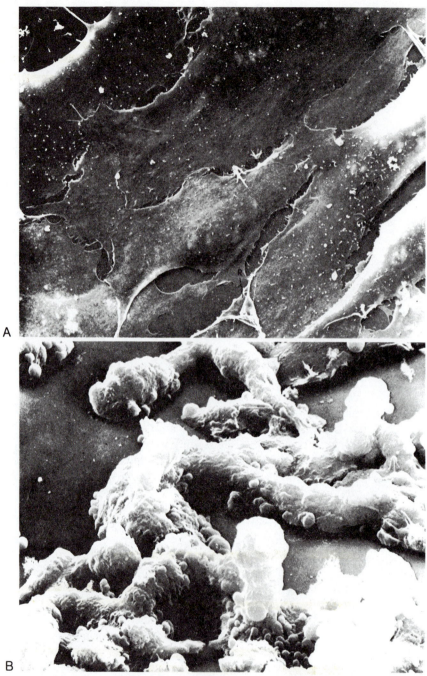

A

B

FIGURE 5.12 Scanning electron micrographs of cultured cells. *A,* Well-spread normal hamster embryo cells. A few microvilli *(mv)* and ruffles *(R)* are observed. ×1620. *B,* The same cell type after transformation by human adenovirus. Observe that the cells have blebs on the surface and tend to make several layers. ×3145. (Courtesy of R.D. Goldman. From E.D.P. De Robertis and E.M.F. De Robertis, Jr. 1980. *Cell and Molecular Biology,* 7th ed. Saunders College Publishing, Philadelphia, p. 170.)

As we have learned, adhesive properties of cells are dependent upon cell surface characteristics. Therefore, we would expect significant differences in cell surface composition of transformed cells as compared to normal cells. Such differences have been well documented (Nicolson et al., 1977). They include changes in glycoprotein composition and distribution, an increased mobility of glycoproteins in the phospholipid bilayer, the appearance of new surface antigens, and the loss of certain cell membrane proteins. One of the most dramatic changes is the reduction in the amount of fibronectin on the surfaces of some transformed cells (Hynes, 1973; Yamada and Olden, 1978; Pearlstein et al., 1980). Addition of fibronectin to cultures of transformed cells causes these cells to display more normal adhesiveness, contact inhibition of movement, and many aspects of normal morphology (Fig. 5.13). This result confirms that the absence of fibronectin in transformed cells plays an important role in the abnormal cell behavior. A more complete understanding of the roles of cell surface molecules such as fibronectin in regulation of cell behavior should assist in our understanding of both normal development and tumorigenesis.

5–4. MORPHOGENESIS OF PLANT CELLS: ROLE OF MICROTUBULES

The presence of rigid walls around plant cells places constraints on the cells that prevent plants from constructing new organs by forming new cell associations or changing the shapes of preexisting cells. The formation of plant organs such as roots, stems, and leaves is directly dependent upon mitosis. Preexisting cells in the growing tips of stems and roots divide by formation of new cell walls between the daughter cells at telophase, and the daughter cells and their enveloping cell walls expand. The mode of expansion and the modeling of the walls of cells produced by mitosis determine the final shapes of the cells, and the combined shapes of cells produced at the growing tips of plants determine the shapes of the structures that are formed. Thus, the morphogenesis of plants depends upon control of the plane of cell division, the directions of expansion of cells, and the shapes of the cell walls of daughter cells after division. These processes are all influenced by microtubules.

A   B

FIGURE 5.13 Scanning electron micrographs showing effects of fibronectin on tumor cell morphology. A, Mouse SV1 tumor cells in tissue culture. B, SV1 cells treated with fibronectin for 48 hours. Scale bar equals 10 μm. (From K.M. Yamada et al. 1978. Transformation-sensitive cell surface protein: Isolation, characterization, and role in cellular morphology and adhesion. Ann. N.Y. Acad. Sci., 312: 264. Reproduced with permission of the New York Academy of Sciences.)

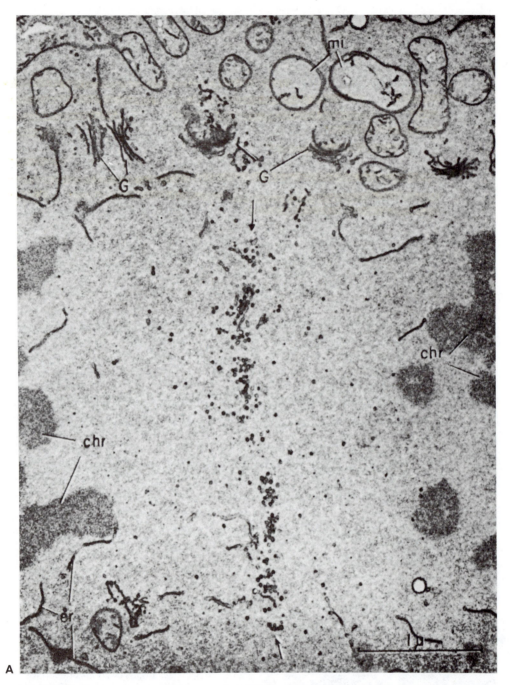

FIGURE 5.14 Cell division in plant cells. *A*, Electron micrograph of root cells of *Zea mays* at telophase. This region corresponds to the cell plate. Note at the top the marginal mitochondria (mi) and the Golgi complex (G). Between the arrows the vesicles are aligned to form the first evidence of a cell plate. chr: telophase chromosomes in the two daughter cells; er: endoplasmic reticulum. ×45,000. (Micrograph by W. Gordon Whaley and H.H. Mollenhauer. From E.D.P. De Robertis and E.M.F. De Robertis, Jr. 1980. *Cell and Molecular Biology,* 7th ed. Saunders College Publishing, Philadelphia, p. 303.)

Division of plant cells is illustrated in Figure 5.14. This process begins with the formation of a **phragmoplast,** a region of vesicles that associate with the microtubules that persist after chromosomal separation. These vesicles, apparently derived from the Golgi apparatus, fuse with one another to form a membranous **cell plate,** within which the primary cell wall forms. The apposing membranes of the cell plate become the plasmalemmae of the daughter cells. The cell wall is formed from microfibrils of cellulose that are laid down on the apposing surfaces of the daughter cells. Permanent gaps remain in the cell plate to become the **plasmodesmata,** through which adjacent cells retain functional continuity.

Microtubules play key roles in the plant cell cycle (for review, see Gunning and Hardham [1979]). Prior to mitosis a circumferential band

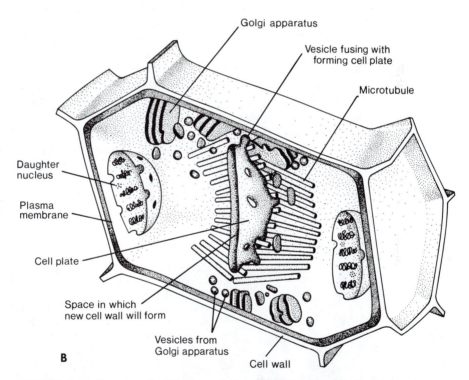

FIGURE 5.14 *(Continued)* B, Schematic representation of a cell of a higher plant as seen at telophase in mitosis. In the phragmoplast region, consisting of membranes and microtubules, a cell plate forms and grows until it separates the cytoplasm into two daughter cells. The cell plate develops as a membrane-delimited structure enclosing a space in which a new cell wall will form. The Golgi apparatus contributes many vesicles to the phragmoplast membrane. The vesicle membranes apparently are incorporated into the membrane of the cell plate, and the vesicle contents enter the forming cell wall. (After M.C. Ledbetter and K.R. Porter. 1970. *Introduction to the Fine Structure of Plant Cells.* Springer-Verlag, Berlin, p. 44. Modified from the figure in *Cells and Organelles,* Second edition by A.B. Novikoff and E. Holtzman. Copyright © 1976 by Holt, Rinehart and Winston. Reprinted by permission of Holt, Rinehart and Winston, CBS College Publishing.)

of microtubules forms in the periphery, or cortex, of the cell (Fig. 5.15). This band, which is called the **preprophase band,** girdles the cell in a plane perpendicular to its long axis. Nearly all microtubules disappear from the remainder of the cell at this time. The site of the next division plane is predicted by this band, since the new cell plate will later bisect the zone that the microtubule band occupies. When the cell enters mitosis all peripheral microtubules are lost, and microtubules are now present in the spindle. Next, microtubules are found in the phragmoplast, in which the cell plate later develops (at the previously determined site). These microtubules may be responsible for guiding the Golgi vesicles to the division plane. Immediately after cytokinesis very few microtubules are present in the daughter cells. After a short hiatus they appear in the cortex aligned parallel to the plasmalemma. These short, overlapping, parallel microtubules underlie most of the cell surface. They are oriented perpendicular to the direction of cell elongation and form extensive microtubule networks called **interphase cortical arrays** (Fig. 5.16).

The interphase microtubules play a key role in the shaping of cells after division. Cell walls are forced to expand by the turgor pressure that builds inside the daughter cells. When the cell wall is being stretched, additional cellulose fibers are deposited. Their orientation is determined by the interphase cortical microtubules that underlie the cell membrane; that is, the cellulose fibrils are laid down parallel to the microtubules (Ledbetter and Porter, 1963). The orientations of these cellulose fibrils, in turn, determine the direction of any future cell wall expansion. Cellulose fibers are hard to stretch but can be more readily separated from one another. Thus, cells tend to expand at right angles to the orientation of cellulose microfibrils. Since their orientation was previously determined by the orientation of microtubules, the forces that orient the mi-

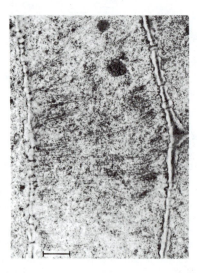

FIGURE 5.15 High-voltage electron micrograph of microtubules in the preprophase band. The band is perpendicular to the longitudinal axis of the cell. Scale bar equals 0.5 μm. (From B.E.S. Gunning and A.R. Hardham. 1979. Microtubules and morphogenesis in plants. Endeavour, N.S., *3*(3): 115.)

FIGURE 5.16 High-voltage electron micrograph of plant cell showing interphase microtubules as viewed from the center looking outward to the cell cortex. This grazing section of the cell surface shows cell wall fibrils and (where the underlying cytoplasm has been sectioned) interphase cortical microtubules. Note that the microtubules are parallel to the cell wall fibrils. This cell is oriented so that the longitudinal axis is up and down the page. Scale bar equals 0.1 μm. (From B.E.S. Gunning and A.R. Hardham. 1979. Microtubules and morphogenesis in plants. Endeavour, N.S., *3*(3): 113.)

crotubules play a key role in plant cell morphogenesis. We shall return to this topic in Chapter 13.

The utilization of common cell constituents such as microtubules for specific purposes in modeling cellular shape indicates that morphogenesis occurs by controlling the function of universal cell machinery rather than by forming a specialized mechanism. We are reminded of E.B. Wilson's statement, quoted previously in Chapter 1, that development

> . . . is nothing new. It involves no breach of continuity, and is but a continuation of the vital processes going on in the parental body. What gives development its marvelous character is the rapidity with which it proceeds and the diversity of the results attained in a span so brief. (Wilson, 1896.)

REFERENCES

Abercrombie, M. 1970. Contact inhibition in tissue culture. In Vitro, 6: 128–142.

Abercrombie, M. 1979. Contact inhibition and malignancy. Nature (Lond.), *281*: 259–262.

Abercrombie, M. 1980. The Croonian lecture, 1978. The crawling movement of metazoan cells. Proc. R. Soc. Lond. B., *207*: 129–147.

Abercrombie, M., G.A. Dunn, and J.P. Heath. 1977. The shape and movement of fibroblasts in culture. *In* J.W. Lash and M.M. Burger (eds.), *Cell and Tissue Interactions*. Raven Press, New York, pp. 57–70.

Adelstein, R.S. 1982. Calmodulin and the regulation of the actin-myosin interaction in smooth muscle and nonmuscle cells. Cell, *30*: 349–350.

Albrecht-Buehler, G. 1977. Phagokinetic tracks of 3T3 cells: Parallels between the orientation of track segments and of cellular structures which contain actin or tubulin. Cell, *12*: 333–339.

Ali, I.U., and R.O. Hynes. 1978. Effects of LETS glycoprotein on cell motility. Cell, *14*: 439–449.

Balinsky, B.I. 1975. *An Introduction to Embryology,* 4th ed. W.B. Saunders, Philadelphia.

Barondes, S.H. 1980. Endogenous cell-surface lectins: Evidence that they are cell adhesion molecules. *In* S. Subtelny and N.K. Wessells (eds.), *The Cell Surface: Mediator of Developmental Processes.* 38th Symposium of the Society for Developmental Biology. Academic Press, New York, pp. 349–363.

Barondes, S.H. 1981. Lectins: Their multiple endogenous cellular functions. Ann. Rev. Biochem., *50*: 207–231.

Bergen, L.G., R. Kuriyama, and G.G. Borisy. 1980. Polarity of microtubules nucleated by centrosomes and chromosomes of Chinese hamster ovary cells in vitro. J. Cell Biol., *84*: 151–159.

Beug, H., F.E. Katz, and G. Gerisch. 1973a. Dynamics of antigenic membrane sites relating to cell aggregation in *Dictyostelium discoideum.* J. Cell Biol., *56*: 647–658.

Beug, H. et al. 1973b. Quantitation of membrane sites in aggregating *Dictyostelium* cells by use of tritiated univalent antibody. Proc. Natl. Acad. Sci. U.S.A., *70*: 3150–3154.

Bonner, J.T. et al. 1969. Acrasin, acrasinae, and the sensitivity to acrasin in *Dictyostelium discoideum.* Dev. Biol., *20*: 72–87.

Borisy, G.G. 1978. Polarity of microtubules of the mitotic spindle. J. Mol. Biol., *124*: 565–570.

Breuer, W., and C.-H. Siu. 1981. Identification of endogenous binding proteins for the lectin discoidin-I in *Dictyostelium discoideum.* Proc. Natl. Acad. Sci. U.S.A., *78*: 2115–2119.

Brinkley, B.R. et al. 1981. Tubulin assembly sites and the organization of cytoplasmic microtubules in cultured mammalian cells. J. Cell Biol., *90*: 554–562.

Buckley, I.K., and K.R. Porter. 1967. Cytoplasmic fibrils in living cultured cells. A light and electron microscope study. Protoplasma, *64*: 349–380.

Burridge, K., and J.R. Feramisco. 1981. Non-muscle α-actinins are calcium-sensitive actin-binding proteins. Nature (Lond.), *294*: 565–567.

Ceri, H., D. Kobiler, and S.H. Barondes. 1981. Heparin-inhibitable lectin. Purification from chicken liver and embryonic chicken muscle. J. Biol. Chem., *256*: 390–394.

Cohen, C. 1979. Cell architecture and morphogenesis. I. The cytoskeletal proteins. Trends Biochem. Sci., *4*(5): 73–77.

Cook, G.M.W. et al. 1979. A lectin which binds specifically to β-D-galactose groups is present at the earliest stages of chick embryo development. J. Cell Sci., *38*: 293–304.

De Robertis, E.D.P., and E.M.F. De Robertis, Jr. 1980. *Cell and Molecular Biology,* 7th ed. Saunders College, Philadelphia.

Dunn, G.A., and T. Ebendal. 1978. Contact guidance on oriented collagen gels. Exp. Cell Res., *111*: 475–479.

Erickson, C.A., and J.P. Trinkaus. 1976. Microvilli and blebs as sources of reserve surface membrane during cell spreading. Exp. Cell Res., *99*:375–384.

Franke, W.W. et al. 1978. Antibody to prekeratin. Decoration of tonofilament-like arrays in various cells of epithelial character. Exp. Cell Res., *116*: 429–445.

Fujiwara, K., and T.D. Pollard. 1976. Fluorescent antibody localization of myosin in the cytoplasm, cleavage furrow, and mitotic spindle of human cells. J. Cell Biol., *71*: 848–875.

Geiger, B. et al. 1980. Vinculin, an intracellular protein localized at specialized sites where microfilament bundles terminate at cell membranes. Proc. Natl. Acad. Sci. U.S.A., *77*: 4127–4130.

Gerisch, G. 1980. Univalent antibody fragments as tools for the analysis of cell interactions in *Dictyostelium. In* M. Friedlander (ed.), *Immunological Approaches to Embryonic Development and Differentiation. Part II. Curr. Topics Dev. Biol., 14*: 243–270.

Godman, G.C., and A.F. Miranda. 1978. Cellular contractility and the visible effects of cytochalasin. *In* S.W. Tanenbaum (ed.), *Cytochalasins—Biochemical and Cell Biological Aspects.* Elsevier/North-Holland, Amsterdam., pp. 277–429.

Goldman, R.D. 1971. The role of three cytoplasmic fibers in BHK-21 cell motility. I. Microtubules and the effects of colchicine. J. Cell Biol., *51*: 752–762.

Gunning, B.E.S., and A.R. Hardham. 1979. Microtubules and morphogenesis in plants. Endeavour, N.S., *3*(3): 112–117.

Harris, H.L., and S.E. Zalik. 1982. The presence of an endogenous lectin in early embryos of *Xenopus laevis.* Wilh. Roux's Arch., *191*: 208–210.

Harrison, R.G. 1914. The reaction of embryonic cells to solid structure. J. Exp. Zool., *17*:521–544.

Hascall, V.C., and G.K. Hascall. 1981. Proteoglycans. *In* E.D. Hay (ed.), *Cell Biology of Extracellular Matrix.* Plenum Press, New York, pp. 39–63.

Hausman, R.E., and A.A. Moscona. 1976. Isolation of retina-specific cell-aggregating factor from membranes of embryonic neural retina tissue. Proc. Natl. Acad. Sci. U.S.A., *73*: 3594–3598.

Hausman, R.E., and A.A. Moscona. 1979. Immunologic detection of retina cognin on the surface of embryonic cells. Exp. Cell Res., *119*: 191–204.

Hay, E.D. 1981. Introductory remarks. *In* E.D. Hay (ed.): *Cell Biology of Extracellular Matrix.* Plenum Press, New York, pp. 1–4.

Heidemann, S.R., G.W. Zieve, and J.R. McIntosh. 1980. Evidence for microtubule subunit addition to the distal end of mitotic structures in vitro. J. Cell Biol., *87*: 152–159.

Henderson, D., and K. Weber. 1981. Immuno-electron microscopical identification of the two types of intermediate filaments in established epithelial cells. Exp. Cell Res., *132*: 297–311.

Hoebeke, J., G. Van Nijen, and M. De Brabander. 1976. Interaction of oncodazole (R 17934), a new antitumoral drug, with rat brain tubulin. Biochem. Biophys. Res. Commun., *69*: 319–324.

Humphreys, T. 1963. Chemical dissociation and *in vitro* reconstruction of sponge cell adhesions: Isolation and functional demonstration of the components involved. Dev. Biol., *8*: 27–47.

Hynes, R.O. 1973. Alteration of cell-surface proteins by viral transformation and by proteolysis. Proc. Natl. Acad. Sci. U.S.A., *70*: 3170–3174.

Hynes, R.O. 1981. Fibronectin and its relation to cellular structure and behavior. *In* E.D. Hay (ed.), *Cell Biology of Extracellular Matrix.* Plenum Press, New York, pp. 295–333.

Hynes, R.O. et al. 1978. A large glycoprotein lost from the surfaces of transformed cells. Ann. N.Y. Acad. Sci., *312*: 317–342.

Jumblatt, J. E., V. Schlup, and M.M. Burger. 1980. Cell-cell recognition: Specific binding of *Microciona* sponge aggregation factor to homotypic cells and the role of calcium ions. Biochemistry, *19*: 1038–1042.

Kobiler, D., E.C. Beyer, and S.H. Barondes. 1978. Developmentally regulated lectins from chick muscle, brain, and liver have similar chemical and immunological properties. Dev. Biol., *64*: 265–272.

Korn, E.D. 1982. Actin polymerization and its regulation by proteins from non-muscle cells. Physiol. Rev., *62*: 672–737.

Lazarides, E. 1975. Tropomyosin antibody: The specific localization of tropomyosin in nonmuscle cells. J. Cell Biol., *65*: 549–561.

Lazarides, E. 1980. Intermediate filaments as mechanical integrators of cellular space. Nature (Lond.), *283*: 249–256.

Lazarides, E. 1982. Intermediate filaments: A chemically heterogeneous, developmentally regulated class of proteins. Ann. Rev. Biochem., *51*: 219–250.

Lazarides, E., and J.P. Revel. 1979. The molecular basis of cell movement. Sci. Am., *240* (5): 100–113.

Ledbetter, M.C., and K.R. Porter. 1963. A 'microtubule' in plant cell fine structure. J. Cell Biol., *19*: 239–250.

Ledbetter, M.C., and K.R. Porter. 1970. *Introduction to the Fine Structure of Plant Cells.* Springer-Verlag, Berlin.

Loomis, W.F. 1975. *Dictyostelium discoideum. A Developmental System.* Academic Press, New York.

Magnani, J.L., W.A. Thomas, and M.S. Steinberg. 1981. Two distinct adhesion mechanisms in embryonic neural retina cells. I. A kinetic analysis. Dev. Biol., *81*: 96–105.

Margolis, R.L., and L. Wilson. 1977. Addition of colchicine-tubulin complex to microtubule ends: The mechanism of substoichiometric colchicine poisoning. Proc. Natl. Acad. Sci. U.S.A., *74*: 3466–3470.

Margolis, R.L., and L. Wilson. 1978. Opposite end assembly and disassembly of microtubules at steady state in vitro. Cell, *13*: 1–8.

Margolis, R.L., and L. Wilson. 1981. Microtubule treadmills—Possible molecular machinery. Nature (Lond.), *293*: 705–711.

Milos, N., and S.E. Zalik. 1981. Effect of the β-D-galactoside-binding lectin on cell to substratum and cell to cell adhesion of cells from the extraembryonic endoderm of the early chick blastoderm. Wilh. Roux's Arch., *190*: 259–266.

Milos, N., and S.E. Zalik. 1982. Mechanisms of adhesion among cells of the early chick blastoderm. Role of the β-D-galactoside-binding lectin in the adhesion of extraembryonic endoderm cells. Differentiation, *21*: 175–182.

Moscona, A.A. 1960. Patterns and mechanisms of tissue reconstruction from dissociated cells. *In* D. Rudnick (ed.), *Developing Cell Systems and Their Control.* Ronald Press, New York, pp. 45–70.

Moscona, A.A., and R.E. Hausman. 1977. Biological and biochemical studies on embryonic cell-cell recognition. *In* J.W. Lash and M.M. Burger (eds.), *Cell and Tissue Interactions.* Raven Press, New York, pp. 173–185.

Müller, K., and G. Gerisch. 1978. A specific glycoprotein as the target site of adhesion blocking Fab in aggregating *Dictyostelium* cells. Nature (Lond.), *274*: 445–449.

Nicolson, G.L. 1979. Topographic display of cell surface components and their role in transmembrane signaling. Curr. Topics Dev. Biol., *13*: 305–338.

Nicolson, G.L. et al. 1977. Modifications in transformed and malignant tumor cells. *In* B.R. Brinkley and K.R. Porter (eds.), *International Cell Biology, 1976–1977.* Rockefeller University Press, New York, pp. 138–148.

Novikoff, A.B., and E. Holtzman. 1976. *Cells and Organelles,* 2nd ed. Holt, Rinehart and Winston, New York.

Pearlstein, E., L.I. Gold, and A. Garcia-Pardo. 1980. Fibronectin: A review of its structure and biological activity. Mol. Cell. Biochem., *29*: 103–128.

Pouysségur, J., and I. Pastan. 1979. The directionality of locomotion of mouse fibroblasts. Role of cell adhesiveness. Exp. Cell Res., *121*: 373–382.

Ray, J., T. Shinnick, and R. Lerner. 1979. A mutation altering the function of a carbohydrate binding protein blocks cell-cell cohesion in developing *Dictyostelium discoideum.* Nature (Lond.), *279*: 215–221.

Roberson, M.M., and P.B. Armstrong. 1980. Carbohydrate-binding component of amphibian embryo cell surfaces: Restriction to surface regions capable of cell adhesion. Proc. Natl. Acad. Sci. U.S.A., *77*: 3460–3463.

Rutishauser, U. et al. 1978a. Cell-adhesion molecules from neural tissues of the chick embryo. *In* R.A. Lerner and D. Bergsma (eds.), *Molecular Basis of Cell-Cell Interactions.* (Birth Defects: Original Article Series, Vol. 14) Alan R. Liss, New York, pp. 305–316.

Rutishauser, U. et al. 1978b. Adhesion among neural cells of the chick embryo. III. Relationship of the surface molecule CAM to cell adhesion and the development of histotypic patterns. J. Cell Biol., *79*: 371–381.

Singer, S.J., and G.L. Nicolson. 1972. The fluid mosaic model of the structure of cell membranes. Science, *175*: 720–731.

Siu, C.-H. et al. 1976. Developmentally regulated proteins of the plasma membrane of *Dictyostelium discoideum.* The carbohydrate-binding protein. J. Mol. Biol., *100*: 157–178.

Sloboda, R.D. 1980. The role of microtubules in cell structure and cell division. Am. Sci., *68*: 290–298.

Spooner, B.S., K.M. Yamada, and N.K. Wessells. 1971. Microfilaments and cell locomotion. J. Cell Biol., *49*: 595–613.

Steinberg, M.S. 1963. Reconstruction of tissues by dissociated cells. Science, *141*: 401–408.

Steinberg, M.S. 1970. Does differential adhesion govern self-assembly processes in histogenesis? Equilibrium configurations and the emergence of a hierarchy among populations of embryonic cells. J. Exp. Zool., *173*: 395–434.

Tanenbaum, S.W. (ed.). 1978. *Cytochalasins-Biochemical and Cell Biological Aspects.* Elsevier/North-Holland, Amsterdam.

Terranova, V.P., D.H. Rohrbach, and G.R. Martin. 1980. Role of laminin in the attachment of PAM 212 (epithelial) cells to basement membrane collagen. Cell, *22*: 719–726.

Thomas, W.A., and M.S. Steinberg. 1981. Two distinct adhesion mechanisms in embryonic neural retina cells. II. An immunological analysis. Dev. Biol., *81*: 106–114.

Thomas, W.A. et al. 1981. Two distinct adhesion mechanisms in embryonic neural retina cells. III. Functional specificity. Dev. Biol., *81*: 379–385.

Tilney, L.G., E.M. Bonder, and D.J. DeRosier. 1981. Actin filaments elongate from their membrane-associated ends. J. Cell Biol., *90*: 485–494.

Toole, B.P. 1981. Glycosaminoglycans in morphogenesis. *In* E.D. Hay (ed.), *Cell Biology of Extracellular Matrix.* Plenum Press, New York, pp. 259–294.

Townes, P.L., and J. Holtfreter. 1955. Directed movements and selective adhesion of embryonic amphibian cells. J. Exp. Zool., *128*: 53–120.

Trinkaus, J.P. 1976. On the mechanism of metazoan cell movements. *In* G. Poste and G.L. Nicolson (eds.), *The Cell Surface in Animal Embryogenesis and Development*. Elsevier/North Holland Biomedical Press, Amsterdam, pp. 225–329.

Virtanen, I. et al. 1982. Fibronectin in adhesion, spreading and cytoskeletal organization of cultured fibroblasts. Science, *298*: 660–663.

Weber, K., P.C. Rathke, and M. Osborn. 1978. Cytoplasmic microtubular images in glutaraldehyde-fixed tissue culture cells by electron microscopy and by immunofluorescence microscopy. Proc. Natl. Acad. Sci. U.S.A., *75*: 1820–1824.

Weeds, A. 1982. Actin-binding proteins—regulators of cell architecture and motility. Nature (Lond.), *296*: 811–816.

Wegner, A. 1976. Head-to-tail polymerization of actin. J. Mol. Biol., *108*: 139–150.

Weiss, P. 1959. Cellular dynamics. Reviews of Modern Physics, *31*: 11–20.

Wessells, N.K. et al. 1971. Microfilaments in cellular and developmental processes. Science, *171*: 135–143.

Wilson, E.B. 1896. *The Cell in Development and Inheritance*. Reprinted by Johnson Reprint Corp., New York. (1966).

Wilson, H.V. 1907. On some phenomena of coalescence and regeneration in sponges. J. Exp. Zool., *5*: 245–258.

Yamada, K.M., and K. Olden. 1978. Fibronectins—adhesive glycoproteins of cell surface and blood. Nature (Lond.), *275*: 179–184.

Yamada, K.M. et al. 1980a. Characterization of fibronectin interactions with glycosaminoglycans and identification of active proteolytic fragments. J. Biol. Chem., *255*: 6055–6063.

Yamada, K.M., K. Olden, and L.-H.E. Hahn. 1980b. Cell surface protein and cell interactions. *In* S. Subtelny and N.K. Wessells (eds.), *The Cell Surface: Mediator of Developmental Processes*. 38th Symposium of the Society for Developmental Biology. Academic Press, New York, pp. 43–77.

Yamada, K.M., K. Olden, and I. Pastan. 1978. Transformation-sensitive cell surface protein: Isolation, characterization, and role in cellular morphology and adhesion. Ann. N.Y. Acad. Sci., *312*: 256–277.

Zieve, G.W., S.R. Heidemann, and J.R. McIntosh. 1980. Isolation and partial characterization of a cage of filaments that surrounds the mammalian mitotic spindle. J. Cell Biol., *87*: 160–169.

PART THREE
GAMETOGENESIS

6 Spermatogenesis

The formation of male gametes is an extended process that begins with the reduction of chromosome number (meiosis) and proceeds through the extensive morphological changes that convert the haploid spermatid into a fully differentiated spermatozoon (**spermiogenesis**). As we discussed in Chapter 1, the germ cell population expands by mitosis of diploid spermatogonia, which serve as **stem cells.** The spermatogonial mitotic divisions may be incomplete, leaving daughter cells in continuity with one another via cytoplasmic bridges. Since clusters of interconnected cells are produced from a single spermatogonium, these cells can be considered **clones.** Meiotic divisions may also be incomplete, thus enlarging the clones, which are then composed of numerous haploid spermatids. The intercellular bridges that connect members of a clone are lost in the final stages of spermiogenesis when excess cytoplasm is sloughed from the sperm (see section 6–3).

In most species, spermatogenesis occurs while the germ cells are intimately associated with specialized somatic cells. This physical relationship has been studied extensively in the testes of mammals. Mammalian testes contain numerous **seminiferous tubules,** shown in cross section in Figure 6.1. Within the tubules there are a number of radially distributed Sertoli cells, with which the germ cells remain associated during the entire spermatogenic process. The Sertoli cells are columnar, with broad bases and narrow tips that extend into the lumen of the tubule. Spermatogonia are situated between the Sertoli cells and the underlying basal lamina (Fig. 6.2). Germ cells in meiosis and spermiogenesis are embedded in membrane-enclosed recesses in the Sertoli cells or are trapped in depressions between adjacent Sertoli cells. The germ cells are arranged in a very precise sequence. Spermatogonia remain at the bases of the Sertoli cells, whereas cells in progressive stages of

217

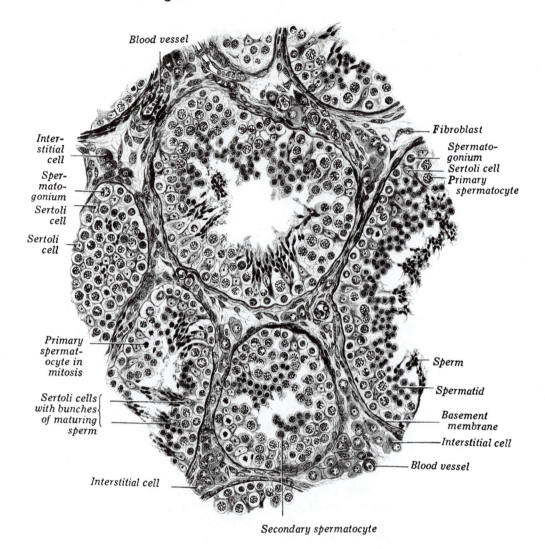

Blood vessel

Fibroblast

Spermato-
gonium

Sertoli cell

Primary
spermatocyte

Inter-
stitial
cell

Sper-
mato-
gonium

Sertoli
cell

Sertoli
cell

Primary
spermat-
ocyte in
mitosis

Sperm

Spermatid

Basement
membrane

Interstitial cell

Blood vessel

Sertoli cells
with bunches
of maturing
sperm

Interstitial cell

Secondary spermatocyte

FIGURE 6.1 Section of human testis. The transected tubules show various stages of spermatogenesis. ×170. (After A.A. Maximow. From W. Bloom and D. W. Fawcett. 1975. *A Textbook of
Histology*, 10th ed. W.B. Saunders, Philadelphia, p. 809.)

meiosis and spermiogenesis are situated at successively higher positions.
This topographical arrangement of germ cells as they proceed through
spermatogenesis is caused by their continual displacement upward from
the bases of the Sertoli cells. According to one hypothesis (Russell, 1980),
the germ cells are carried toward the tips of the Sertoli cells because they
are tightly bound to the outer surfaces of these cells, which undergo
conformational changes that displace their lateral margins (and the attached germ cells) apically. The attachments that bind the germ cells to

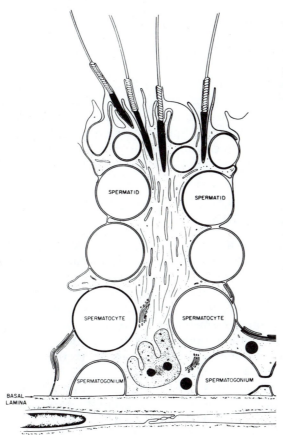

SPERMATID

SPERMATID

SPERMATOCYTE

SPERMATOCYTE

SPERMATOGONIUM

SPERMATOGONIUM

BASAL LAMINA

FIGURE 6.2 Relationship between the germ cells and a Sertoli cell. Spermatogonia occupy a basal position between the Sertoli cell and the basal lamina. Cells in progressively later stages of spermatogenesis are found at successively more apical positions. Mature spermatozoa are released into the lumen of the seminiferous tubule from the tip of the Sertoli cell. (After M. Dym and D.W. Fawcett. 1970. The blood-testis barrier in the rat and the physiological compartmentation of the seminiferous epithelium. Biol. Reprod., *3:* 324.)

the Sertoli cells are desmosome-like junctions. An alternative is that the germ cells actively move upward between the Sertoli cells. This hypothesis requires that the junctions between germ cells and Sertoli cells must break to allow germ cell movement and then reform when movement stops. Spermatozoa occupy deep recesses or crypts at the tips of the Sertoli cells and are released from there into the lumen of the seminiferous tubule.

In any particular region of a seminiferous tubule, groups of germ cells around the circumference appear to progress toward the lumen more or less in unison (Leblond and Clermont, 1952). Thus, they move to the lumen in waves, forming circumferential zones of more advanced cells inside zones of less advanced cells. The Sertoli cells are believed to be responsible for governing the rate of spermatogenesis by controlling the translocation of germ cells from the basal lamina to the lumen. Contiguous Sertoli cells must therefore have some means for coordinating their activities in translocating germ cells. Sertoli cells are also involved in metabolic regulation of germ cells. Specialized cell junctions (gap junctions) form between these cells and are thought to facilitate intercellular transport of molecules and ions (Russell, 1980). Finally, as we shall discuss in section 6–1, Sertoli cells secrete hormones that are involved in regulation of spermatogenesis.

After their differentiation in the testes, mammalian sperm are immotile. The capacity for motility is acquired as the result of a process called **maturation.** Maturation occurs in the **epididymis,** a highly convoluted duct in which the sperm are stored after their release from the testes and from which they are forcibly expelled at ejaculation. Sperm maturation will be discussed in section 6–5.

6–1. HORMONAL REGULATION OF SPERMATOGENESIS

The principal stimulus for vertebrate germ cell differentiation is the steroid hormone **testosterone** which is synthesized by specialized somatic cells, the **interstitial cells (cells of Leydig).** These steroidogenic cells are localized in the connective tissue between the seminiferous tubules. The testosterone diffuses into the tubules, where it promotes spermiogenesis. The production of testosterone by the interstitial cells is regulated by a gonadotropic hormone released by the pituitary gland into the circulation. This hormone is **luteinizing hormone (LH)**, sometimes called **interstitial cell-stimulating hormone (ICSH).**

Another pituitary gonadotropic hormone, **follicle-stimulating hormone (FSH),** is also involved in regulation of spermiogenesis. The primary sites of FSH action are the Sertoli cells. One effect of FSH on Sertoli cells in mammals is to stimulate the release of **androgen-binding protein (ABP).** This protein has a high affinity for testosterone and functions to retain the steroid within the seminiferous tubules and sustain its effects on spermiogenesis (Turner and Bagnara, 1976).

Testosterone and ABP are both found in the fluid of the epididymis. Testosterone (and possibly ABP) is necessary to promote sperm maturation (Howards et al., 1979).

6–2. SPERM STRUCTURE

Before considering the development of sperm, we shall review the end product of the process: the mature, functional sperm. Sperm morphology is highly utilitarian; each morphological aspect is designed to achieve one or both of the sperm's two major chores—movement to the vicinity of the egg and fusion with it. When we think about spermatozoa, we usually envision sleek, streamlined structures with small heads and long whiplike tails; this is indeed the form of most mammalian sperm. However, animals have evolved a variety of sperm shapes, some of which are shown in Figure 6.3. Marine and freshwater invertebrates that discharge their sperm into the water have sperm that are considered "primitive,"

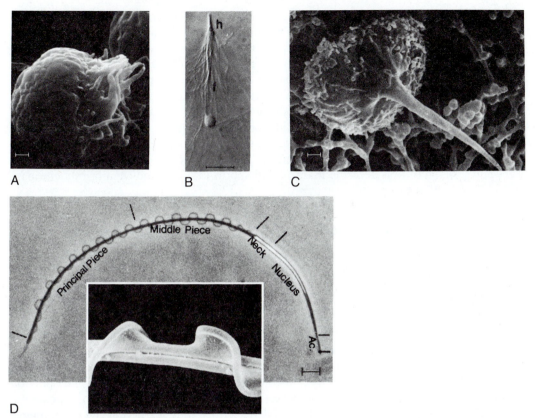

A B C

D

FIGURE 6.3 Forms of some unusual animal spermatozoa. *A,* Scanning electron micrograph of the amoeboid sperm of the round worm, *Ascaris.* Scale bar equals 1.0 μm. (From M. Abbas and G.D. Cain. 1979. In vitro activation and behavior of the amoeboid sperm of *Ascaris suum* (Nematoda). Cell Tissue Res., *200:* 276.) *B,* The multiflagellate sperm of the termite, *Mastotermes,* as seen with Nomarski optics. h: head; f: flagella. Scale bar equals 25 μm. (From B. Baccetti and R. Dallai, 1978. Reproduced from *The Journal of Cell Biology,* 1978, vol. 76, pp. 569–576 by copyright permission of the Rockefeller University Press.) *C,* Scanning electron micrograph of the tack-shaped sperm of the prawn, *Palaemonetes.* The cup-shaped basal region contains the nucleus. Scale bar equals 1.0 μm. (From L.D. Koehler. 1979. A unique case of cytodifferentiation: Spermiogenesis of the prawn, *Palaemonetes paludosus.* J. Ultrastruct. Res., *69:* 112.) *D,* Phase contrast micrograph of the sperm of the urodele amphibian, *Pleurodeles.* The sperm tail consists of a long axial fiber and a thin undulating membrane. Scale bar equals 10 μm. The inset shows details of the tail by scanning electron microscopy. Scale bar equals 0.5 μm. (From B. Picheral. 1979. Structural, comparative, and functional aspects of spermatozoa in urodeles. *In* D.W. Fawcett and J.M. Bedford [eds.], *The Spermatozoon.* Urban & Schwarzenberg, Inc., Baltimore, pp. 268, 269.)

whereas animals with internal fertilization have more elaborate sperm. Specializations found in sperm of either type are thought to be adaptations to the conditions of fertilization found in these species rather than indicative of their phylogenetic rank.

The primary components of most sperm are a **nucleus,** an **acrosome,** and a **flagellum.** The nucleus contains a highly condensed mass of chromatin, in which the individual chromosomes cannot be observed by light or electron microscopy. The acrosome exhibits a variety of morphologies and assists in penetration of the egg accessory layers and in species-specific attachment of sperm to eggs. The flagellum is the locomotor organelle found in most types of sperm. However, flagella are not universal. For example, the *Ascaris* sperm, shown in Figure 6.3 is ameboid rather than flagellate. Because of the variety of sperm morphologies, no single kind of sperm can be considered "typical." Because the sperm of mammals have been studied extensively, and because mammalian sperm are intrinsically interesting to us as a result of our taxonomic status, we shall first describe mammalian sperm in detail and then briefly compare them to primitive sperm.

Mammalian Sperm

The morphology of mammalian sperm was studied extensively by the cytologists of the nineteenth century and first half of the twentieth century, using the only instrument that was available to them, the light microscope. The general morphology of a typical mammalian sperm as reconstructed from light micrographs is shown in Figure 6.4.

The two major regions of the sperm are the **head** and the **tail**. The interior of the head consists of the nucleus. Surrounding the nucleus at the anterior end of the sperm head is the acrosome. The acrosome does not cover the nucleus entirely but forms a cap over it. The portion of the head behind the posterior margin of the acrosome is the **postacrosomal region.** The tail is subdivided into four segments: the **neck, middle piece, principal piece,** and **end piece.** The slender neck forms the articulation between the head and tail. The middle piece is characterized by a sheath of mitochondria surrounding the tail elements.

Because of its small size, which is close to the limit of resolution of the light microscope, further details of sperm structure could be resolved only by the electron microscope. This instrument became widely used in the 1950s, and as a result of extensive analysis of sperm with it, we now have a very detailed understanding of sperm ultrastructure. Figure 6.5 is a reconstruction of sperm structure that is based upon electron micrographs. The cell membrane has been removed to reveal the underlying components, and representative cross sections of the sperm are shown at various levels.

The head shows little internal organization, since the bulk is occupied by the nucleus, which contains highly compact chromatin with no resolvable detail. The acrosome cap over the nucleus also has a very simple organization. The relationship between the acrosome and nucleus is better seen in a sagittal section of a sperm head, as shown diagrammatically in Figure 6.6. The acrosome is sandwiched between the plasma membrane and the nuclear envelope. The acrosome is surrounded by its own membrane. Note that a portion of the acrosome

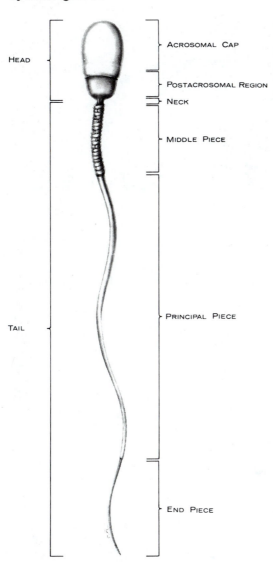

HEAD

ACROSOMAL CAP

POSTACROSOMAL REGION

NECK

MIDDLE PIECE

PRINCIPAL PIECE

TAIL

END PIECE

FIGURE 6.4 General morphology of a mammalian spermatozoon. (From W. Bloom and D.W. Fawcett. 1975. *A Textbook of Histology,* 10th ed. W.B. Saunders, Philadelphia, p. 814.)

extends anteriorly past the tip of the nucleus. This **apical segment** of the acrosome may be quite prominent in some species and often has a species-specific shape. In other species—notably the human—the apical segment is small and inconspicuous.

The amorphous-appearing contents of the acrosome include several hydrolytic enzymes. When a sperm reaches the immediate vicinity of an egg, it undergoes an **acrosome reaction,** which causes the plasma membrane and the outer acrosomal membrane to vesiculate and be shed, thus releasing the enzymes of the acrosome. These enzymes apparently assist

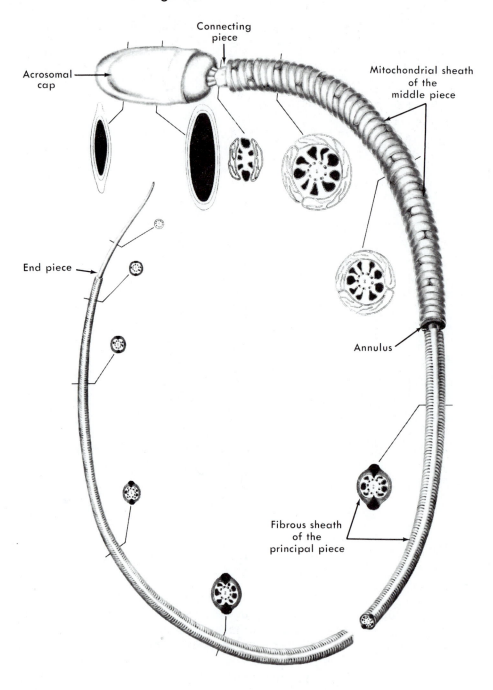

Connecting
piece

Acrosomal
cap

Mitochondrial sheath
of the
middle piece

End piece

Annulus

Fibrous sheath
of the
principal piece

FIGURE 6.5 Ultrastructure of mammalian sperm. The cell membrane has been removed. Representative cross sections are shown at several levels. (From W. Bloom and D.W. Fawcett. 1975. *A Textbook of Histology,* 10th ed. W.B. Saunders, Philadelphia, p. 821.)

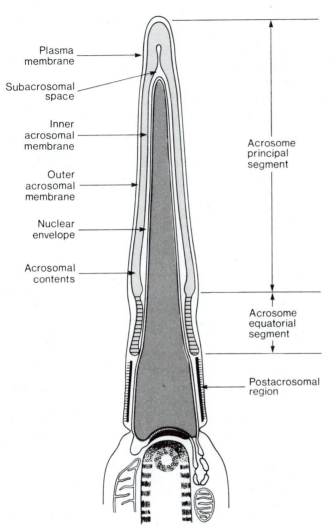

Plasma membrane

Subacrosomal space

Inner acrosomal membrane

Outer acrosomal membrane

Nuclear envelope

Acrosomal contents

Acrosome principal segment

Acrosome equatorial segment

Postacrosomal region

FIGURE 6.6 Diagrammatic representation of sagittal section through mammalian sperm head. Nucleus is in black. (After D.W. Fawcett. 1975. The mammalian spermatozoon. Dev. Biol., *44:* 400.)

the sperm in penetrating the accessory layers that surround the egg (see Chap. 9).

The posterior portion of the acrosome is narrow and may have a different density from that of the anterior portions. This region of the acrosome—the **equatorial segment**—has a unique fate during fertilization, since it is the only part of the acrosome to remain intact. The remainder is lost during the acrosome reaction. The integrity of the equatorial segment during the acrosome reaction may be due to bridges that can be seen by electron microscopy to link the inner and outer acrosomal membranes in this region. The bridges are not present in the remainder of the acrosome (Russell et al., 1980). The equatorial segment is functionally significant because it is the site of initial contact between the sperm and egg at fertilization. Beneath the plasma membrane in the postacrosomal region there is a dense layer of unknown composition.

The acrosome and the nucleus determine the shape of the sperm head, which can be quite variable (Fig. 6.7). The functional significance of such dramatic differences in sperm head shape is not known. The shape has no apparent mechanical role to play in fertilization, since the apical segment, which contributes substantially to the shape of the head, is destroyed during the acrosome reaction before the sperm makes contact with the egg.

The sperm tail is a very intricate structure that produces the flagellar movement to propel the sperm toward the egg. The motor apparatus of the sperm tail consists of two central microtubules that are surrounded by an array of nine doublet microtubules. This structure is called the **axoneme.** At high magnification (Fig. 6.8) the two subunits of the doublets are seen to have different shapes. One of them is a complete tubule and therefore circular in cross section. The second is an incomplete tubule that is C-shaped. It opens onto the wall of the cylindrical subunit. Each doublet is associated with diffuse armlike appendages that project toward the adjacent doublet.

The axoneme in Figure 6.8 has been specially stained to reveal the structure of the tubules, which are composed of minute protofilaments. The central pair of microtubules and the circular member of the outer doublets are each composed of 13 protofilaments, whereas the crescentric member of each outer doublet is composed of about ten protofilaments. These protofilaments run along the entire length of a microtubule and are composed primarily of tubulin. The arms associated with the outer doublets are composed of another protein, **dynein 1**, which possesses ATPase activity and is responsible for converting chemical energy into mechanical movement (Ogawa et al., 1977). **Radial spokes** can also be seen to extend from the outer doublets to the central pair of microtubules.

The exact mechanism involved in generating flagellar movement is unknown, but it probably involves localized sliding between adjacent doublets, much like the sliding filament mechanism of muscle contraction (Summers and Gibbons, 1971). Sliding is mediated by the dynein arms, which use the energy derived from the hydrolysis of ATP to attach to the adjacent doublet and cause one doublet to slide against the other. The central pair of microtubules interacts with the radial spokes to coordinate the sliding activities of the outer doublets in order to produce

FIGURE 6.7 Sagittal sections of sperm heads of several mammalian species, shown ▶ to permit a comparison of the sizes and shapes of the apical portion of the acrosome. (From D.W. Fawcett. 1970. A comparative view of sperm ultrastructure. Biol. Reprod. [Suppl.], 2: 95. The micrographs of the chinchilla and guinea pig sperm originally appeared in D.W. Fawcett and D.M. Phillips. 1969. Observations on the release of spermatozoa and on changes in the head during passage through epididymis. J. Reprod. Fert. [Suppl.], 6: 405–418.)

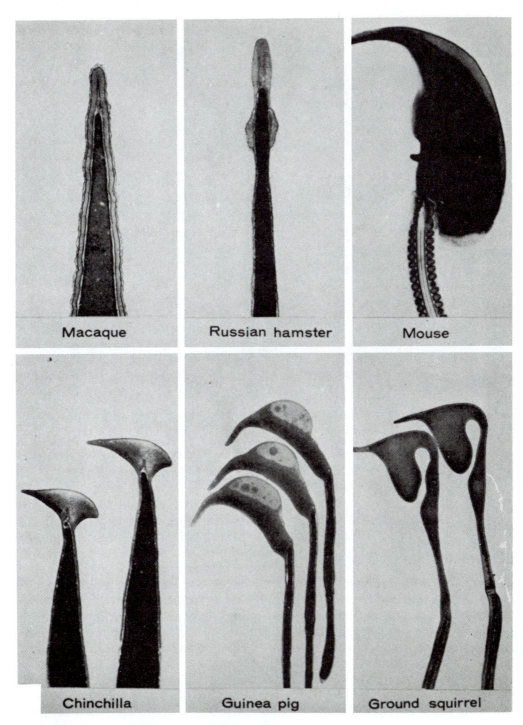

FIGURE 6.7 (See *legend on the opposite page.*)

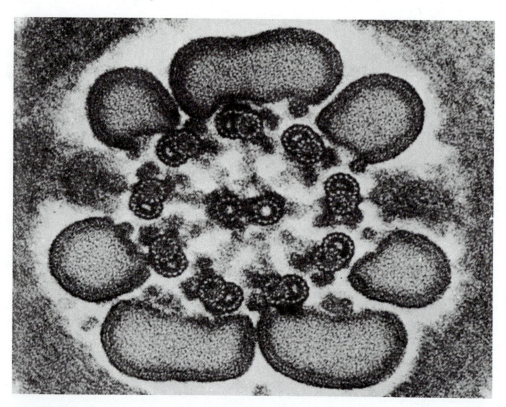

FIGURE 6.8 Axoneme and associated fibers in the principal piece of a spermatozoon after fixation in glutaraldehyde containing tannic acid. After this treatment the protofibrils in the walls of the central microtubules and of the doublets are visible in negative image. The dynein arms can be seen attached to the circular member of each doublet and projecting toward the adjacent doublet. Radial spokes can also be seen. (After D. Phillips. From D.W. Fawcett. 1975. The mammalian spermatozoon. Dev. Biol., *44:* 411.)

the bending action of the tail, which propels the sperm (Warner and Satir, 1974; Witman et al., 1978). As shown by the representative cross sections in Figure 6.5, the axoneme runs through the entire tail. A similar organization of microtubules is nearly universal in cilia and flagella throughout the plant and animal kingdoms. The evolutionary conservation of the 9 + 2 pattern suggests that this arrangement of microtubules is essential for generating ciliary and flagellar motion.

The axoneme is surrounded by a row of nine **outer dense fibers,** each of which parallels an axoneme doublet. The fibers are thickest in the proximal half of the tail and progressively decrease in diameter toward the tip. The relative thickness and extent of the fibers vary considerably among mammalian species. In some they are thick and extend nearly the whole length of the tail, whereas in others they are thin and terminate in the proximal portion of the principal piece. The dense fibers may serve to stiffen sperm tails, but their function and composition are not well understood.

The neck region forms the base of the tail. The major structural element in this region is the convex **connecting piece,** which articulates with a concave depression in the base of the sperm head. Very fine filaments in the space between the connecting piece and the sperm head are probably responsible for attachment of the head to the tail (see Fig. 6.6). Behind the articulation region the connecting piece consists of nine **segmented columns** that attach to the anterior ends of the nine outer dense fibers. The sperm of some mammalian species have a centriole (the **proximal centriole**) embedded in a depression in the connecting piece. During sperm tail development another centriole (the **distal centriole**) is also present, but it degenerates during development of the connecting piece.

The middle piece of the sperm is characterized by a sheath of elongate **mitochondria** that are wrapped end-to-end in a helical chain around the axoneme (Fig. 6.9). The mitochondria apparently provide the energy for sperm propulsion. The middle piece is terminated by a structure called the **annulus** (see Fig. 6.5). Immediately behind the annulus the axoneme is encased in a **fibrous sheath.** This region of the tail is the principal piece. The sheath consists of two longitudinal columns that are connected by a series of hemispherical ribs (Fig. 6.10). Anteriorly, the columns are attached for a short distance to two of the nine outer dense fibers. These two outer dense fibers then terminate abruptly, and along the remainder of the principal piece the columns attach directly to the two doublets that were internal to the dense fibers. As the sperm tail tapers, the columns and ribs diminish and then end abruptly a few mi-

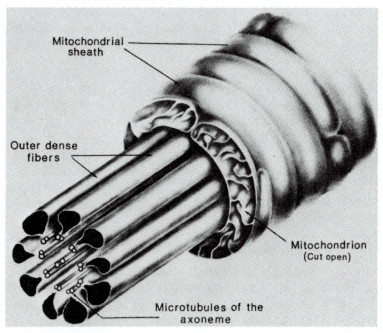

Mitochondrial sheath

Outer dense fibers

Mitochondrion (Cut open)

Microtubules of the axoneme

FIGURE 6.9 Diagrammatic representation of a segment from the middle piece of a mammalian spermatozoon. (From D.W. Fawcett. 1975. The mammalian spermatozoon. Dev. Biol., *44*: 397.)

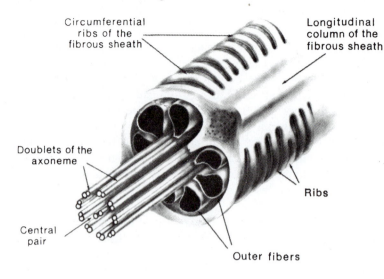

Circumferential ribs of the fibrous sheath

Longitudinal column of the fibrous sheath

Doublets of the axoneme

Central pair

Ribs

Outer fibers

FIGURE 6.10 Diagrammatic representation of a segment from the principal piece of the mammalian spermatozoon illustrating one of the two longitudinal columns of the fibrous sheath and the associated ribs. Inward projections of the longitudinal columns attaching to doublets divide the tail into two unequal compartments, one containing three outer fibers and the other containing four. (From D.W. Fawcett. 1975. The mammalian spermatozoon. Dev. Biol., *44:* 398.)

crometers from the tip of the flagellum. The termination marks the junction of the principal piece and the end piece.

Primitive Sperm

The primitive spermatozoa of marine and freshwater invertebrates are usually blunter in shape than mammalian sperm. The sperm head in these species contains a rounded or conical nucleus and is terminated by a small acrosome. Most of the variation in sperm shape among these animals is due to the acrosome, which exhibits species-specific variations in size and organization (Fig. 6.11). The changes occurring in the acrosome during the acrosome reaction differ dramatically from those during the mammalian acrosome reaction. When sperm of marine invertebrates swim into the vicinity of the egg, the acrosome everts a long, slender process (the acrosomal process) that assists in egg penetration. The surface of the acrosomal process is coated with a substance (**bindin**) that binds the sperm to the egg vitelline envelope. The acrosome reaction is an important component of fertilization, and it will be discussed extensively in Chapter 9. The middle piece of the primitive sperm is short and consists of a few spheroidal mitochondria surrounding the base of the tail. The latter structure consists of an axoneme with a typical 9 + 2 pattern of microtubules.

6–3. SPERMIOGENESIS

At the completion of meiosis, spermatids differentiate into spermatozoa. The primary events of this differentiation process are (1) modifications of the nucleus and (2) the elaboration of specialized organelles that must function during sperm transport and fertilization. These processes also

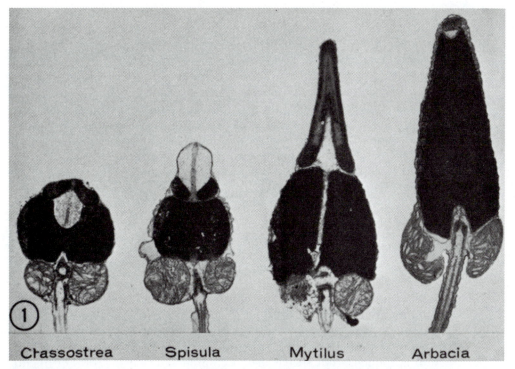

Crassostrea   Spisula   Mytilus   Arbacia

FIGURE 6.11 Examples of primitive sperm of four marine invertebrates (oyster, surf clam, mussel, and sea urchin) showing a globular or ovoid nucleus, a single ring of mitochondria, and a simple 9 + 2 flagellum. While these features are subject to little variation, the acrosomes are quite highly specialized and are characteristic of each species. (After S. Ito. From D.W. Fawcett. 1970. Comparative view of sperm ultrastructure. Biol. Reprod. [Suppl.], *2:* 91.)

produce the changes in shape that are necessary for proper sperm function.

Nuclear Modifications

The nuclear modifications involve an extensive condensation of chromatin and acquisition of morphology that are characteristic of the species. Chromatin condensation allows for a large reduction in nuclear volume and helps streamline the cell and facilitate locomotion. The tight packing of the DNA may also make it less susceptible to physical damage or mutation during storage and transport to the site of fertilization. The condensation may result from removal of chromatin-associated proteins or from the formation of unique DNA-protein complexes. The proteins that form the highly condensed sperm chromatin fall into three categories: (1) protamines, (2) histones, and (3) other sperm-specific basic proteins (Bloch, 1969).

The formation of chromatin with specialized proteins requires the replacement of the preexisting somatic histones. This process is particularly well studied in the trout and salmon. In these species the histones are enzymatically modified. This change destabilizes the DNA-histone complex and allows protamines to replace the histones (Dixon, 1972). The newly formed DNA-protamine complex has a configuration that is different from the nucleosome configuration of somatic chromatin (Dixon et al., 1977). The DNA in the newly formed DNA-protamine complex is very tightly packed and is genetically inactive. The absence of transcriptional activity in spermatids is a generalized phenomenon in the animal kingdom regardless of which proteins are complexed with DNA in the condensed chromatin. The absence of transcription following chromatin condensation means that protein synthesis, which is essential for the completion of spermiogenesis, must utilize stored messenger RNA molecules. This aspect of spermiogenesis will be discussed in more detail in section 6–4.

The changes in chromatin may also involve loss of high mobility group (HMG) proteins, at least in some species. Since certain HMG proteins are associated with transcriptionally competent chromatin (see Chap. 4), their loss may help to erase the pattern of gene expression that previously existed in the nuclei (Kennedy and Davies, 1980). This would enable the fertilized egg to establish new patterns of gene expression without interference from preexisting sperm HMG proteins.

Morphological changes in the nucleus occur concurrently with chromatin condensation. Fawcett et al. (1971) have suggested that particular nuclear shapes result from the pattern of DNA-protein interaction during condensation. It has also been proposed that the form of the nucleus is produced by pressure applied by microtubules outside the nucleus (McIntosh and Porter, 1967). Evidence to be presented later in this chapter indicates that changes in nuclear shape in *Drosophila* sperm are caused by microtubules, but there is no similar evidence for the sperm of other species. Another possibility is that the shaping of sperm nuclei and the pattern of chromatin condensation are determined by interactions between the DNA and components of the nuclear matrix (Risley et al., 1982).

Formation of Sperm Organelles

The near universal presence of the acrosome on sperm heads is a remarkable example of evolutionary conservation. The presence of hydrolytic enzymes in acrosomes has led to proposals that this structure is a modified lysosome. Like lysosomes, acrosomes are derivatives of the Golgi complex. Formation of the acrosome is illustrated in Figures 6.12

FIGURE 6.12 A series of electron micrographs of the juxtanuclear region of a guinea ▶ pig spermatocyte (*A*) and spermatids (*B* to *G*), showing the successive stages of formation of the acrosomal vesicle (*C* to *E*) and its conversion to an acrosomal cap (*F, G*). The guinea pig is selected because the large size of its acrosome makes it favorable material for study of the process of differentiation, which is qualitatively similar in all mammals. (From W. Bloom and D.W. Fawcett. 1975. *A Textbook of Histology*, 10th ed. W.B. Saunders, Philadelphia, p. 825.)

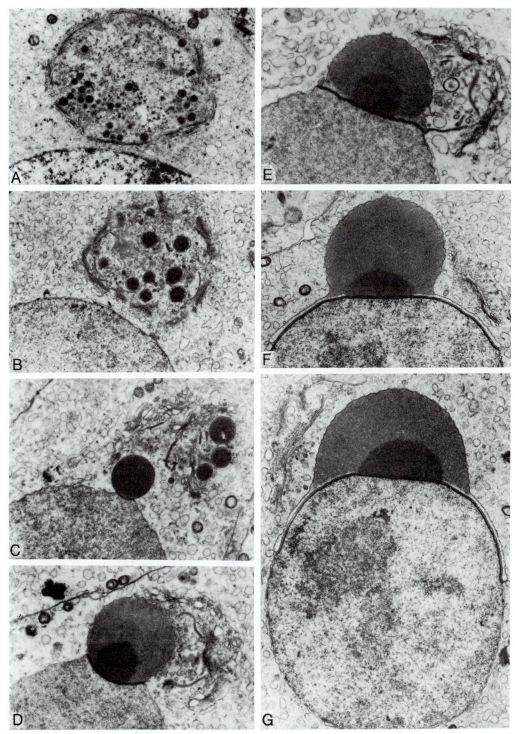

FIGURE 6.12 *(See legend on the opposite page.)*

and 6.13, which contain a series of electron micrographs showing successive stages of this process in the guinea pig. The first sign of acrosome formation is the appearance within the Golgi of numerous membrane-enclosed **proacrosomal granules** (Fig. 6.12A and B). The granules coalesce into a single large **acrosomal vesicle,** which contains a dense **acrosomal granule** (Fig. 6.12C). The acrosomal vesicle adheres to the nuclear envelope, thereby marking the future anterior tip of the sperm nucleus. The Golgi continues to form proacrosomal granules, which contribute to the enlargement of the acrosomal vesicle by fusing with it (Fig. 6.12D to G). When the acrosomal vesicle has reached its full size, the Golgi moves into the postnuclear cytoplasm, and the acrosomal vesicle becomes modified, assuming the final shape of the mature acrosome (Fig. 6.13).

Formation of the middle piece is highly variable, since the mitochondria in different species assume a number of configurations and undergo a variety of structural changes. Primitive spermatozoa have very simple middle pieces with clusters of a few enlarged mitochondria (see Fig. 6.11). During spermiogenesis there is a progressive decrease in the number of mitochondria, accompanied by an increase in size of the remaining mitochondria. It is uncertain how this happens, but it is assumed that it occurs by mitochondrial fusion (Longo and Anderson, 1974).

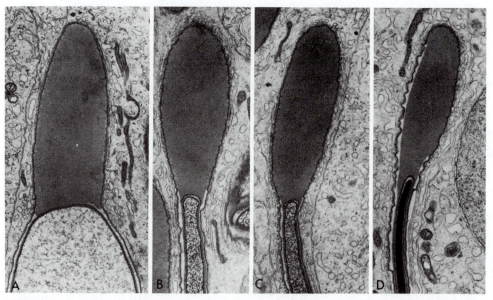

FIGURE 6.13 Further development of the acrosome of guinea pig spermatozoon. When the acrosome has attained its full size, the Golgi migrates into the postnuclear cytoplasm, and the subsequent changes in the acrosome consist of a progressive modification of its shape taking place concurrently with a condensation of the chromatin and flattening of the nucleus. (From D.W. Fawcett and D.M. Phillips. 1969. Observations on the release of spermatozoa and on changes in the head during passage through epididymis. J. Reprod. Fert. [Suppl.], *6*: 405–418.)

In mammals the mitochondria migrate to the base of the axoneme, where they associate with the outer dense fibers. This association may be mediated by molecules of the outer mitochondrial membranes that recognize and bind to the dense fibers (Phillips, 1980). Mitochondria move down the flagellum behind the annulus. The posterior movement of the annulus and mitochondria is shown in Figures 6.14A and B. Annular movement ceases when it reaches the posterior end of the middle piece. The mitochondria shown in Figure 6.14B are rounded. At a later stage (Fig. 6.14C) they have elongated. These teardrop-shaped mitochondria wrap around the axoneme to form helical chains. The mitochondria will later associate end to end to produce the helical row of mitochondria that are depicted in Figures 6.5 and 6.9.

Differentiation of the sperm tail is basically the process of axoneme formation. Obviously, the details of tail formation may vary, since various organelles are elaborated in different species. Mammalian sperm,

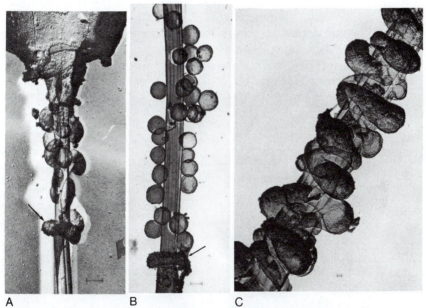

A   B   C

FIGURE 6.14 Formation of the guinea pig sperm middle piece as shown using surface replicas. In this technique the plasma membrane is removed with a detergent. A replica of the demembranated sperm is made by coating it with platinum and carbon. The replica is then examined with the electron microscope to visualize surface details. A, Replica of spermatid showing the annulus (arrow) moving posteriorly, followed by mitochondria. Scale bar equals 1.0 μm. B, Replica of mitochondria attached to the outer dense fibers of the axoneme. Annulus (arrow) has moved farther back. Scale bar equals 0.5 μm. C, Teardrop-shaped mitochondria form helical chains around the axoneme. Scale bar equals 0.1 μm. (From D.M. Phillips. 1980. Observations on mammalian spermiogenesis using surface replicas. J. Ultrastruct. Res., 72: 108–110.)

for example, must form outer dense fibers and a fibrous sheath. However, the axoneme is the fundamental organelle in all sperm flagella.

The axoneme is organized by one of the two sperm centrioles. The centrioles migrate to the end of the nucleus that is opposite to the acrosome and situate at right angles to one another. The centriole closest to the nucleus is called the proximal centriole, and the other member of the pair is the distal centriole. The distal centriole is parallel to the long axis of the cell and gives rise to the 9 + 2 axoneme (Fig. 6.15). Elongation of the flagellum continues as tubulin dimers are added to the distal ends of the growing microtubules.

The ultimate disposition of the two centrioles is quite variable. In mammals the distal centriole helps organize the connecting piece of the neck before the integrity of the centriole itself is lost. The loss of this centriole indicates that it does not serve as a basal body, or kinetosome, to initiate flagellar movement in the mature spermatozoon (Fawcett, 1972). The proximal centriole is also lost in the sperm of some mammalian species (Fawcett, 1975).

These findings have upset one of the time-honored tenets of embryology. Theodor Boveri had proposed that the fertilizing sperm provides the centrioles that organize the cleavage spindle of the zygote (Wilson, 1896). According to this theory, the centrosome of the egg is either lost or incapacitated after formation of the second polar body. The theory has persisted in the literature, and it may apply for some species. For example, in the sea urchin two centrioles enter the egg with the sperm, and the centrioles appear to participate in organizing the cleavage

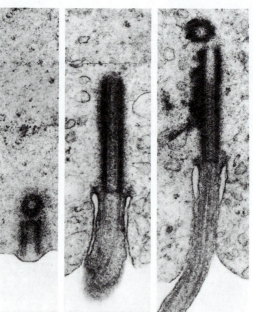

FIGURE 6.15 The earliest events in formation of the tail consist of migration of the centrioles to the cell surface in the postnuclear region, and polymerization of microtubule protein on the template provided by the distal centriole. A typical 9 + 2 axoneme is formed, and the simple flagellum elongates by accretion of microtubule subunits to its distal end. (From W. Bloom and D.W. Fawcett. 1975. *A Textbook of Histology*, 10th ed. W.B. Saunders, Philadelphia, p. 827.)

spindle (Longo and Anderson, 1968). However, as we have seen, the theory does not apply to mammals. Phillips (1970) also has shown that the centrioles disappear during insect spermiogenesis and cannot contribute to the cleavage spindle.

As we noted earlier in this chapter, large numbers of developing male germ cells are joined by intercellular bridges. The formation of syncytial clones during mammalian spermatogenesis is outlined in Figure 6.16. Incomplete cytokinesis during all but the earliest spermato-

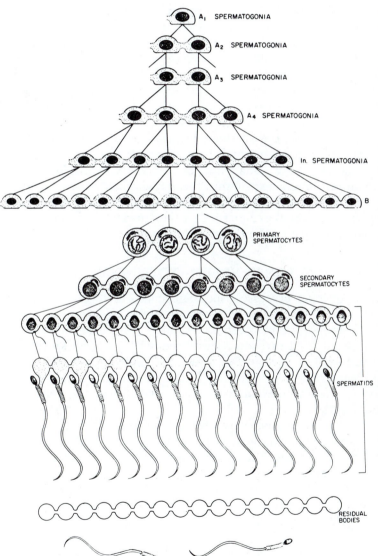

FIGURE 6.16 Diagram illustrating formation of syncytial clones of male germ cells and release of individual sperm. Once a spermatogonium is committed to differentiation, the daughter cells of all subsequent spermatogonial divisions and the two meiotic divisions remain connected by intercellular bridges. Six successive generations of spermatogonia are shown here. The earliest spermatogonial generations are A_1 to A_4. These are followed by intermediate (In.) spermatogonia, which divide to produce B spermatogonia. These in turn divide to produce the primary spermatocytes. The two meiotic divisions expand the clone. Following their differentiation, individual sperm are separated from the syncytial chain of residual bodies. The actual number of interconnected cells in a clone is much larger than shown here. (From W. Bloom and D.W. Fawcett. 1975. *A Textbook of Histology*, 10th ed. W.B. Saunders, Philadelphia, p. 824.)

gonial divisions results in groups of connected spermatogonia, and incomplete meiotic divisions expand the clones even more. The maintenance of cytoplasmic continuity during male germ cell development is widespread in the animal kingdom, which suggests that this phenomenon is of fundamental importance for spermatogenesis. Although the significance of the interconnections of these cells is not known with certainty, it is likely that interconnection is responsible for the synchrony of division and differentiation of the members of the clone and the resulting production of several mature sperm at exactly the same time.

At the completion of spermiogenesis, individual sperm are released from the syncytium by the process that removes excess cytoplasm (Fig. 6.16). The cytoplasm accumulates in the neck region of the spermatids in lobules called **residual bodies.** The cytoplasmic bridges are localized between residual bodies, while the body of each sperm is attached to its residual body by a narrow strand of cytoplasm. Constriction of these strands releases the sperm, leaving behind the syncytial chains of residual bodies. This process results in individual spermatozoa that emerge from the testes. In mammals the spermatozoa are then transported to the epididymis, the duct through which they pass prior to their forcible release at ejaculation. As we shall discuss in section 6–5, mammalian sperm undergo further changes within the epididymis that are essential to allow them to become fertile. One of these changes is the removal of a small amount of residual cytoplasm that remains attached to the neck region as a small tag called the **cytoplasmic droplet.**

6–4.   GENE FUNCTION IN SPERMATOGENESIS

Like all differentiation processes, spermatogenesis is regulated by the genome and is dependent upon coordinated expression of selected portions of the genome. Spermatogenesis is unique, however, in that the chromatin is condensed early in spermiogenesis before the sperm are completely formed. As we shall see, chromatin condensation renders the chromatin incapable of transcription. Consequently, transcription that is necessary for sperm differentiation must occur before the major differentiation events.

The loss of transcriptional capacity has been monitored in a number of species. These studies indicate that RNA is synthesized in spermatogonia and spermatocytes, but transcription cannot be detected during the final stages of differentiation. However, the exact stage at which the RNA synthetic capacity is lost is somewhat variable. In some species, notably *Drosophila*, transcription is terminated in the primary spermatocyte stage (Olivieri and Olivieri, 1965; Gould-Somero and Holland, 1974).

On the other hand, in the mouse, synthesis of messenger RNA continues after meiosis. The identification of newly synthesized postmeiotic RNA as mRNA is based upon the criteria that it is polyadenylated (Erickson et al., 1980a), is associated with polysomes (D'Agostino et al., 1978; Geremia et al., 1978), and can be translated *in vitro* (Erickson et al., 1980b). The postmeiotic synthesis of RNA ceases after nuclear elongation (and chromatin condensation) has begun (Geremia et al., 1977).

Although RNA synthesis stops, completion of sperm formation is dependent on the continued synthesis of protein (Monesi, 1965; Brink, 1968; Gould-Somero and Holland, 1974; O'Brien and Bellvé, 1980). Since there is no concurrent synthesis of RNA, protein synthesis must be supported by stable RNA produced during early stages of spermatogenesis and stored for translation during spermiogenesis. This is an example of cellular mechanisms acting at the *posttranscriptional level* to delay the ultimate expression of genes by storing transcripts until a later stage of development, when the stored transcripts are translated.

The preservation of RNA synthesized during early stages of mouse spermatogenesis has been demonstrated by labeling RNA of primary spermatocytes with ^{3}H-uridine, allowing them to develop to the spermatid stage and recovering the previously labeled RNA in the spermatids (Geremia et al., 1977). A more specific example of preservation and delayed utilization of RNA is provided by the protamine messengers of the trout.

Trout spermatogenesis is highly amenable to developmental studies. Unlike species in which spermatogenesis is continuous throughout the year, trout spermatogenesis is seasonal. At the beginning of each seasonal cycle, spermatogenesis is initiated simultaneously in a large number of cells, and they progress through the various stages of spermatogenesis more or less synchronously. The initiation of the cycle, which is regulated by hormones, can be induced experimentally by injections of pituitary extracts. Consequently, batches of cells at given stages of spermatogenesis can be obtained to follow the molecular events of spermatogenesis. Experimental methods also have been developed to separate cells according to their developmental stage, thus providing the means for eliminating unwanted contaminating cells (Louie and Dixon, 1972).

Replacement of histones by protamines occurs in trout spermatids. These protamines are synthesized in the spermatid cytoplasm and transported to the nucleus, where they displace the histones. Recently, protamine messenger has been isolated and used as a template for making protamine cDNA, which in turn has been used as a molecular probe to detect protamine messenger. In this way the life history of protamine messenger has been reconstructed (Iatrou and Dixon, 1978; Iatrou et al., 1978).

Hybridization of RNA with protamine cDNA is first detected in the primary spermatocyte stage of spermatogenesis, indicating that transcription of the protamine genes begins at this time, although there is no concurrent synthesis of protamine. Three size classes (16S, 11S, and 7.5S) of protamine RNA, which lack poly (A), are found in the spermatocyte nucleus. A fourth nuclear protamine RNA is the same size as the cytoplasmic messenger (6–6.5S), and it is polyadenylated. The cytoplasm contains polyadenylated protamine messenger combined with protein as messenger RNP complexes. No protamine messenger is associated with polysomes at this stage, which is an expected result, since no protamine synthesis occurs until the spermatid stage. The reduction in size of the nuclear protamine RNA is assumed to result from posttranscriptional processing of the initial transcript.

Following sequential size reduction, the RNA is polyadenylated and transported to the cytoplasm in the form of the RNP particles, which are stored in the germ cells until the spermatid stage. At this time the protamine messenger is first detected on polysomes, indicating that translation has been initiated. The storage of long-lived messengers for utilization during terminal cell differentiation is a common developmental strategy, and we shall encounter this phenomenon frequently as we examine differentiation of other cell types. The factors that control gene expression at the posttranscriptional level are unknown, but the unique experimental advantages of the trout protamine system may provide the means for their discovery.

Genetic Dissection of Spermatogenesis

Genetic control of spermatogenesis is elegantly demonstrated by the defective spermatogenesis that is caused by certain mutant genes. Correlation of mutant genes with specific defects in differentiation assists in explaining the genetic program for development, since the defective development implies that expression of the wild-type allele of the mutant gene is required for normal spermatogenesis. This approach is called "genetic dissection."

An excellent example of a mutation affecting spermatogenesis is one found in humans, as described by Afzelius (1976). Individuals affected by this condition produce nonmotile sperm and are therefore infertile. The syndrome apparently is inherited as an autosomal recessive mutant. Electron microscopic examination of the sperm of affected individuals reveals that the axonemes are defective. As shown in Figure 6.17, the dynein arms on the axonemal outer doublets are missing. The normal allele of this gene is apparently responsible for either the synthesis of the dynein protein or for attachment of the dynein arms to the doublets. The lack of motility underscores the importance of the dynein arms in flagellar movement. A locus (the p locus) that regulates sperm morphogenesis has also been described in mice. Microscopic examination of sperm from mice homozygous for mutant alleles at the p locus indicates that they have abnormal heads (Hollander et al., 1960a, 1960b; Wolfe et al., 1977). Thus, normal sperm head morphogenesis is apparently dependent upon expression of the wild-type allele.

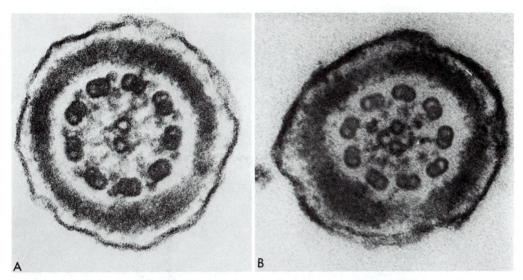

FIGURE 6.17 Electron micrographs of cross sections through the human sperm tail. *A,* The tail of an ordinary, motile spermatozoon has nine microtubular doublets; on each of the doublets there are two dynein arms. *B,* The sperm tail of a sterile male is devoid of the dynein arms. × 140,000. (From B.A. Afzelius. 1976. A human syndrome caused by immotile sperm. Science *193:* 317–319. Copyright 1976 by the American Association for the Advancement of Science.)

The genetic control of sperm differentiation is particularly well understood in *Drosophila.* A number of genes that affect spermatogenesis have been found on both the autosomes and sex chromosomes (Lindsley and Lifschytz, 1971; Kiefer, 1973; Lifschytz and Hareven, 1977; Rungger-Brändle, 1977). Sperm development in *Drosophila* is ideally suited for genetic analysis of spermatogenesis, since the details of formation of these intricate cells are well known and because of the large number of mutants that affect sperm differentiation. For example, the use of mutants provides a means for clarifying certain cause-and-effect relationships in development.

As we discussed earlier in this chapter, alteration of nuclear shape has been attributed to either the microtubules that surround the nucleus or the pattern of chromatin condensation. A mutant of *Drosophila* is known in which the nuclei fail to elongate. When spermatids of this mutant—*ms*(3)10R—are examined with the electron microscope, they are found to lack the microtubules that normally surround the nucleus during elongation (Fig. 6.18*A*). The chromatin condenses in the spermatid nuclei, but it does not undergo the highly ordered packing that occurs during elongation. An occasional spermatid with a partial complement of perinuclear microtubules is found in mutant testes (Fig. 6.18*B*). The nuclei of these spermatids have partially elongated, and their chromatin is more ordered than that in nuclei that are not associated with microtubules. The correlation between the extent of nuclear elongation and the completeness of the perinuclear microtubule layer sup-

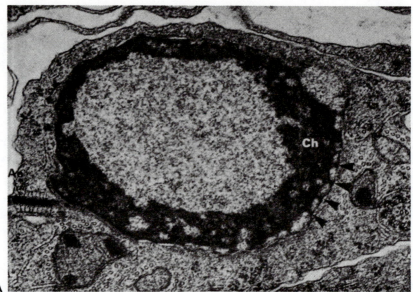

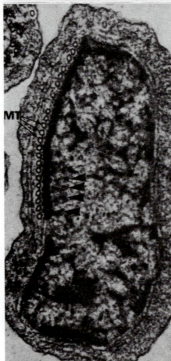

FIGURE 6.18 Spermatids of *ms*(3)10R *Drosophila*. *A,* This nucleus is devoid of surrounding microtubules, which in the wild type encircle the nucleus. Note that the chromatin (Ch) has condensed, but nuclear elongation has not occurred. Chromatin adheres to the nuclear envelope in some areas (arrows). Ac: elongate acrosome. ×42,000. *B,* A partial complement of perinuclear microtubules (MT) is present around the nucleus of this semi-elongated spermatid nucleus. The condensed chromatin is more ordered here (arrows) than in nuclei with no microtubules. ×58,600. (From R.F. Wilkinson et al. 1974. Genetic control of spermiogenesis in *Drosophila melanogaster:* The effects of abnormal cytoplasmic microtubule populations in mutant *ms*(3)10R and its colcemid-induced phenocopy. J. Ultrastruct. Res., *48:* 245.)

ports the hypothesis that the microtubules play a direct role in shaping the nucleus (at least in *Drosophila*).

The spermatozoon is an elegant cell formed under the guidance of its genome in cooperation with the Sertoli cells and in response to hormonal signals. The discovery of additional mutations affecting sperm morphology and function will greatly assist in clarifying the ways in which the genomic information is utilized in its formation.

6–5. MAMMALIAN SPERM MATURATION

Spermatozoa emerging from the testes are fertile in most animal species. In fact, laboratory fertilization of the eggs of some species (such as the frog, *Rana pipiens*) has traditionally involved macerating the testes in a small volume of water to release the sperm. However, in many mammals spermatozoa emerging from the testes are immature and, hence, incompetent to fertilize eggs. They must undergo a number of significant physiological, morphological, and biochemical changes that occur within the epididymis (Bedford, 1975, 1979). These changes, which constitute sperm maturation, are thought to be mediated by the epididymal fluids that bathe the sperm during their traverse of the epididymis. Two of the components of this fluid are testosterone and androgen-binding protein. The latter may be responsible for regulating the level of testosterone in the fluid, which in turn may play an important role in mediating the maturation changes (Howards et al., 1979).

The morphological changes of maturation include some remodeling of the acrosome and migration of the cytoplasmic droplet from the neck down the middle piece from which it is shed. Testicular spermatozoa are incapable of motility; progressive modifications occurring as the sperm traverse the epididymis allow them to develop the capacity to swim. The cell surface also undergoes certain changes, including the acquisition of new surface proteins and the modification, loss, and/or redistribution of some preexisting proteins. Changes in the antigenic properties of the sperm surface provide evidence for surface modification (Bedford and Cooper, 1978). Another technique for studying the surface changes is the use of plant lectins to identify carbohydrates on the surfaces. As we discussed in Chapter 5, carbohydrates complexed to surface proteins may mediate the interactions between the cell and its environment. Lectin-binding sites have been found to undergo changes in amount and distribution during maturation, suggesting that the cell surface glycoprotein configuration changes during this process (Nicolson and Yanagimachi, 1979). Since the cell surface is the region of the sperm that interfaces directly with components of the female reproductive tract after ejaculation, it is reasonable to assume that surface glycoproteins play an important role in the events that lead to fertilization. The identification of these molecules and clarification of their roles should greatly increase our knowledge of fertilization and, perhaps, suggest new targets for contraceptive procedures.

Through the process of maturation, spermatozoa acquire the potential ability to fertilize eggs. Realization of this potential occurs in the female reproductive tract, in which mammalian sperm must undergo even further changes in a process called **capacitation** before they can penetrate an egg (see Chap. 9).

REFERENCES

Abbas, M., and G.D. Cain. 1979. In vitro activation and behavior of the amoeboid sperm of *Ascaris suum* (Nematoda). Cell Tissue Res., *200*: 273–284.

Afzelius, B.A. 1976. A human syndrome caused by immotile sperm. Science, *193*: 317–319.

Baccetti, B., and R. Dallai. 1978. The spermatozoon of arthropoda. XXX. The multiflagellate spermatozoon in the termite Mastotermes darwiniensis. J. Cell Biol., *76*: 569–576.

Bedford, J.M. 1975. Maturation, transport, and fate of spermatozoa in the epididymis. *In* D.W. Hamilton and R.O. Greep (eds.), *Handbook of Physiology*, Vol. 5, section 7. Williams & Wilkins Co., Baltimore, pp. 303–317.

Bedford, J.M. 1979. Evolution of the sperm maturation and sperm storage functions of the epididymis. *In* D.W. Fawcett and J.M. Bedford (eds.), *The Spermatozoon*. Urban & Schwarzenberg, Baltimore-Munich, pp. 7–21.

Bedford, J.M. and G.W. Cooper. 1978. Membrane fusion events in the fertilization of vertebrate eggs. *In* G. Poste and G.L. Nicolson (eds), *Membrane Fusion. Cell Surface Reviews*, Vol. 5. Elsevier/North-Holland Biomedical Press, Amsterdam, pp. 66–125.

Bloch, D.P. 1969. A catalog of sperm histones. Genetics, *61* (Suppl.): 93–111.

Bloom, W., and D.W. Fawcett. 1975. *A Textbook of Histology*, 10th ed. W.B. Saunders Co., Philadelphia.

Brink, N.G. 1968. Protein synthesis during spermatogenesis in *Drosophila melanogaster*. Mutat. Res., *5*: 192–194.

D'Agostino, A., R. Geremia, and V. Monesi. 1978. Post-meiotic gene activity in spermatogenesis of the mouse. Cell Differ. 7: 175–183.

Dixon, G.H. 1972. The basic proteins of trout testis chromatin: Aspects of their synthesis, post-synthetic modifications and binding to DNA. *In* E. Diczfalusy (ed.), *Karolinska Symposia on Research Methods in Reproductive Endocrinology. Fifth Symposium. Gene Transcription in Reproductive Tissue.* Stockholm, pp. 130–154.

Dixon, G.H. et al. 1977. The expression of protamine genes in developing trout sperm cells. *In* P. Ts'o (ed.), *The Molecular Biology of the Mammalian Genetic Apparatus*. Elsevier/North Holland Biomedical Press, Amsterdam, pp. 355–379.

Dym, M., and D.W. Fawcett. 1970. The blood-testis barrier in the rat and the physiological compartmentation of the seminiferous epithelium. Biol. Reprod, *3*: 308–326.

Erickson, R.P. et al. 1980a. Further evidence for haploid gene expression during spermatogenesis. Heterogeneous, poly (A)-containing RNA is synthesized postmeiotically. J. Exp. Zool, *214*: 13–19.

Erickson, R.P. et al. 1980b. Quantitation of mRNAs during mouse spermatogenesis: Protamine-like histone and phosphoglycerate kinase-2 mRNAs increase after meiosis. Proc. Natl. Acad. Sci. U.S.A., 77: 6086–6090.

Fawcett, D.W. 1970. A comparative view of sperm ultrastructure. Biol. Reprod., 2 (Suppl): 90–127.

Fawcett, D.W. 1972. Observations on cell differentiation and organelle continuity in spermatogenesis. In R.A. Beatty and S. Gluecksohn-Waelsch (eds.), The Genetics of the Spermatozoan. Published by the editors, Edinburgh, pp. 37–68.

Fawcett, D.W. 1975. The mammalian spermatozoon. Dev. Biol., 44: 394–436.

Fawcett, D.W., W.A. Anderson, and D.M. Phillips. 1971. Morphogenetic factors influencing the shape of the sperm head. Dev. Biol., 26: 220–251.

Fawcett, D.W., and D.M. Phillips. 1969. Observations on the release of spermatozoa and on changes in the head during passage through the epididymis. J. Reprod. Fert., 6 (Suppl.): 405–418.

Geremia, R. et al. 1977. RNA synthesis in spermatocytes and spermatids and preservation of meiotic RNA during spermiogenesis in the mouse. Cell Differ., 5: 343–355.

Geremia, R., A. D'Agostino, and V. Monesi. 1978. Biochemical evidence of haploid gene activity in spermatogenesis of the mouse. Exp. Cell Res., 111: 23–30.

Gould-Somero, M., and L. Holland. 1974. The timing of RNA synthesis for spermiogenesis in organ cultures of Drosophila melanogaster testes. Wilhelm Roux' Archiv f. Entwicklungsmechanik, 174: 133–148.

Hollander, W.F., J.H.D. Bryan, and J.W. Gowen. 1960a. A male sterile pink-eyed mutant type in the mouse. Fertil. Steril., 11: 316–324.

Hollander, W.F., J.H.D. Bryan, and J.W. Gowen. 1960b. Pleiotropic effects of a mutant at the p-locus from X-irradiated mice. Genetics, 45: 413–418.

Howards, S., C. Lechene, and R. Vigersky. 1979. The fluid environment of the maturing spermatozoon. In D.W. Fawcett and J.M. Bedford (eds.), The Spermatozoon. Urban & Schwarzenberg, Baltimore-Munich, pp. 35–41.

Iatrou, K., and G.H. Dixon. 1978. Protamine messenger RNA: Its life history during spermatogenesis in rainbow trout. Fed. Proc., 37: 2526–2533.

Iatrou, K., A.W. Spira, and G.H. Dixon. 1978. Protamine messenger RNA: Evidence for early synthesis and accumulation during spermatogenesis in rainbow trout. Dev. Biol., 64: 82–98.

Kennedy, B.P., and P.L. Davies. 1980. Acid-soluble nuclear proteins of the testis during spermatogenesis in the winter flounder. Loss of the high mobility group proteins. J. Biol. Chem., 255: 2533–2539.

Kiefer, B.I. 1973. Genetics of sperm development in Drosophila. In F.H. Ruddle (ed.), Genetic Mechanisms of Development. Academic Press, Inc., New York, pp. 47–102.

Koehler, L.D. 1979. A unique case of cytodifferentiation: Spermiogenesis of the prawn, Palaemonetes paludosus. J. Ultrastruct. Res., 69: 109–120.

Leblond, C.P., and Y. Clermont. 1952. Definition of the stages of the cycle of the seminiferous epithelium in the rat. Ann. N.Y. Acad. Sci., 55: 548–573.

Lifschytz, E., and D. Hareven. 1977. Gene expression and the control of spermatid morphogenesis in Drosophila melanogaster. Dev. Biol., 58: 276–294.

Lindsley, D.L., and E. Lifschytz. 1971. The genetic control of spermatogenesis in Drosophila. In R.A. Beatty and S. Gluecksohn-Waelsch (eds.), The Genetics of the Spermatozoan. Published by the editors, Edinburgh, pp. 203–222.

Longo, F.J., and E. Anderson. 1968. The fine structure of pronuclear development and fusion in the sea urchin. J. Cell Biol., *39*: 339–368.

Longo, F.J., and E. Anderson. 1974. Gametogenesis. *In* J. Lash and J.R. Whittaker (eds.), *Concepts of Development.* Sinauer Associates, Inc., Sunderland, Mass., pp. 3–47.

Louie, A.J., and G.H. Dixon. 1972. Trout testis cells. I. Characterization by deoxyribonucleic acid and protein analysis of cells separated by velocity sedimentation. J. Biol. Chem., *247*: 5490–5497.

McIntosh, J.R., and K.R. Porter. 1967. Microtubules in the spermatids of the domestic fowl. J. Cell Biol., *35*: 153–173.

Monesi, V. 1965. Synthetic activities during spermatogenesis in the mouse. RNA and protein. Exp. Cell Res., *39*: 197–224.

Nicolson, G.L., and R. Yanagimachi. 1979. Cell surface changes associated with the epididymal maturation of mammalian spermatozoa. *In* D.W. Fawcett and J.M. Bedford (eds.), *The Spermatozoon.* Urban & Schwarzenberg, Baltimore-Munich, pp. 187–194.

O'Brien, D.A., and A.R. Bellvé. 1980. Protein constituents of the mouse spermatozoon. II. Temporal synthesis during spermatogenesis. Dev. Biol., *75*: 405–418.

Ogawa, K., T. Mohri, and H. Mohri. 1977. Identification of dynein as the outer arms of sea urchin sperm axonemes. Proc. Natl. Acad. Sci. U.S.A., *74*: 5006–5010.

Olivieri, G., and A. Olivieri. 1965. Autoradiographic study of nucleic acid synthesis during spermatogenesis in *Drosophila melanogaster. Mutat. Res., *2*: 366–380.

Phillips, D.M. 1970. Insect sperm: Their structure and morphogenesis. J. Cell Biol., *44*: 243–277.

Phillips, D.M. 1980. Observations on mammalian spermiogenesis using surface replicas. J. Ultrastruct. Res., *72*: 103–111.

Picheral, B. 1979. Structural, comparative, and functional aspects of spermatozoa in urodeles. *In* D.W. Fawcett and J.M. Bedford (eds.), *The Spermatozoon.* Urban & Schwarzenberg, Baltimore-Munich, pp. 267–287.

Risley, M.S. et al. 1982. Determinants of sperm nuclear shaping in the genus *Xenopus. Chromosoma, *84*: 557–570.

Rungger-Brändle, E. 1977. Abnormal microtubules in testes of the mutant l(3)pl (lethalpolyploid) of *Drosophila hydei,* cultured *in vivo.* Exp. Cell Res., *107*: 313–324.

Russell, L.D. 1980. Sertoli–germ cell interrelations: A review. Gamete Res. *3*: 179–202.

Russell, L., R.N. Peterson, and M. Freund. 1980. On the presence of bridges linking the inner and outer acrosomal membranes of boar spermatozoa. Anat. Rec., *198*: 449–459.

Summers, K.E., and I.R. Gibbons. 1971. Adenosine triphosphate-induced sliding of tubules in trypsin-treated flagella of sea-urchin sperm. Proc. Natl. Acad. Sci. U.S.A., *68*: 3092–3096.

Turner, C.D., and J.T. Bagnara. 1976. *General Endocrinology,* 6th ed. W.B. Saunders Co., Philadelphia.

Warner, F.D., and P. Satir. 1974. The structural basis of ciliary bend formation. Radial spoke positional changes accompanying microtubule sliding. J. Cell Biol., *63*: 35–63.

Wilkinson, R.F., H.P. Stanley, and J.T. Bowman. 1974. Genetic control of spermiogenesis in *Drosophila melanogaster:* The effects of abnormal cytoplasmic microtubule populations in mutant *ms*(3)10R and its colcemid-induced phenocopy. J. Ultrastruct. Res., *48:* 242–258.

Wilson, E.B. 1896. *The Cell in Development and Inheritance.* Reprinted by Johnson Reprint Corp., New York. (1966).

Witman, G.B., J. Plummer, and G. Sander. 1978. *Chlamydomonas* flagellar mutants lacking radial spokes and central tubules. Structure, composition, and function of specific axonemal components. J. Cell Biol., *76:* 729–747.

Wolfe, H.G., R.P. Erickson, and L.C. Schmidt. 1977. Effects on sperm morphology by alleles at the pink-eyed dilution locus in mice. Genetics, *85:* 303–308.

7 Oogenesis

Differentiation of the female gamete, in many ways, marks the beginning of the developmental process that produces the next generation of organisms. As we have seen, the male gamete is specialized to deliver its nuclear package to the egg, but the egg contains not only a haploid nucleus but also the materials and energy sources that are needed for construction of the embryo until it can either produce them on its own or obtain them from the environment.

The timing of oogenesis relative to meiosis is also different from that of spermatogenesis. In the male, gamete differentiation occurs *after* meiosis, but in the female, the oocyte is fully formed *before* meiosis is completed. Consequently, differentiation of the female gamete is intimately associated with meiosis. In most species the bulk of gamete differentiation occurs during prophase I of meiosis. In species producing yolky eggs, this phase is arbitrarily divided into: (1) **previtellogenesis** (before yolk deposition), (2) **vitellogenesis** (yolk deposition), and (3) **postvitellogenesis** (after yolk deposition). Most oocyte growth occurs during vitellogenesis. The resumption of meiosis (**maturation**) and ovulation produce the ripe egg, or ovum. In some organisms, such as *Xenopus laevis*, oocytes in these various stages of oogenesis can coexist in the ovary (Fig. 7.1).

Oogenesis can be quite prolonged in relation to the length of the life cycle of the organism. In the frog *Rana pipiens*, for example, oogenesis takes three years. After metamorphosis of tadpoles into juvenile frogs, and every year after that, oogonial divisions produce a new batch of oocytes. Thus, the ovary contains three generations of oocytes simultaneously, with one generation maturing each year and being replaced by a new one. In mammals, on the other hand, all of the oogonial divisions and transformations of oogonia into oocytes are completed

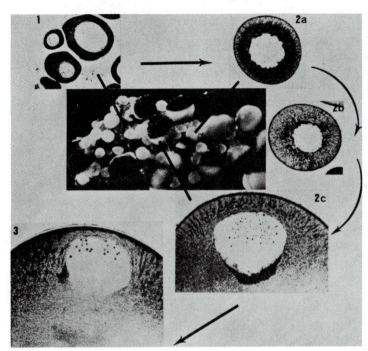

FIGURE 7.1 *Xenopus* oogenesis. In the middle: fragment of an ovary dissected from an adult female, showing oocytes at different stages of their growth. 1, 2, and 3 are oocyte sections, seen with the light microscope. 1: previtellogenic oocytes. 2a and 2b: early vitellogenic oocytes. 2c: postvitellogenic oocyte. 3: oocyte undergoing maturation (the germinal vesicle envelope is breaking down). (From J. Brachet. 1979. Oogenesis and maturation in amphibian oocytes. Endeavour, N.S. *3:* 145.)

either before or shortly after birth, producing a finite number of oocytes, each of which is retained in meiotic prophase I (Pinkerton et al., 1961; Franchi et al., 1962). Thus, the period of oogenesis covers virtually the entire life span from birth to ovulation. However, no oocyte growth occurs until puberty, after which a new group of oocytes resumes development during each cycle. Most of the growing oocytes fail to reach maturity during the cycle and degenerate. Compare this program with that of mammalian spermatogenesis, which is a continuous process. Division of spermatogonia in the sexually mature male produces a constant stream of cells that enter into meiosis and undergo spermiogenesis without interruption.

7–1. HORMONAL CONTROL OF OOGENESIS

Oogenesis is regulated by modulations in the concentrations of circulating hormones. In the vertebrates, gonadal function is regulated by gonadotropic hormones released from the pituitary gland. The secretory functions of the pituitary are, in turn, regulated by neuroendocrine factors released by the hypothalamus. The gonadotropins act on the ovary to cause egg growth and, ultimately, ovulation. These hormones also stimulate follicle cells in the ovary to synthesize steroid hormones.

The seasonal oogenesis cycle in amphibians illustrates the hormonal interactions that operate in **oviparous** species, in which a paramount function during oogenesis is the synthesis and deposition of yolk. The initiation of seasonal egg growth is triggered by environmental stimuli that cause the hypothalamus to stimulate the pituitary to secrete gonadotropins. The gonadotropins are carried to the ovaries in the circulation and cause the follicle cells to synthesize estrogen (Fig. 7.2). This hormone is then released into the circulation and stimulates the liver to synthesize vitellogenin. Next, the vitellogenin is released into the blood stream and is transported to the ovaries, where gonadotropins promote its uptake by oocytes. During this step it is incorporated into yolk-containing organelles, the **yolk platelets.** When oogenesis is complete, gonadotropins then promote meiotic maturation and ovulation. These latter processes are mediated by yet another steroid hormone, **progesterone,** which is synthesized by follicle cells under the gonadotropic influence.

A remarkably similar scheme for regulation of vitellogenesis is found in *Drosophila* (Jowett and Postlethwait, 1980; Postlethwait and Shirk, 1981). In this organism the gonadotropic hormone is the **juvenile hormone,** which is secreted by the **corpus allatum.** The juvenile hormone promotes oocyte differentiation and stimulates the ovary to produce ecdysone, which stimulates another organ, the **fat body,** to produce vitellogenin. The vitellogenin is released into the hemolymph and is sequestered into the oocytes. Juvenile hormone also stimulates the ovary itself to produce vitellogenin. Hence, there is a dual origin for yolk in *Drosophila*. It is important to realize that insects are a heterogeneous group of organisms, and not all of them follow this scheme. A detailed discussion of the various endocrine strategies for control of oogenesis can be found in Highnam and Hill (1977).

Regulation of oogenesis is perhaps most complicated in the mammals, in which there is cyclic production and ovulation of ripe eggs. The mammalian egg develops in conjunction with its surrounding follicle cells, forming a functional unit: the **follicle** (see section 7–2). Variable numbers of follicles develop during a cycle, each culminating in the expulsion of a mature egg at ovulation. A single follicle usually matures in the human ovary during each cycle, whereas in most other mammals several follicles mature simultaneously. Mammals have two gonadotropic hormones that act synergistically to regulate oogenesis. These are the follicle-stimulating hormone (FSH) and the luteinizing hormone (LH), which cause the growth of the follicle, prepare the egg for ovulation, and stimulate estrogen production by the follicle cells.

The menstrual cycle, which is found only in primates, is illustrated in Figure 7.3. The menstrual phase is considered to be the beginning of the cycle. A notable increase in FSH production by the pituitary occurs at menstruation. This hormone promotes growth and development of the oocyte and, together with LH, causes the follicle cells to release increasing amounts of estrogen, which assists in promoting follicle growth. The consequent buildup of estrogen in the blood then causes the hypothalamus to "order" the pituitary to reduce FSH production and

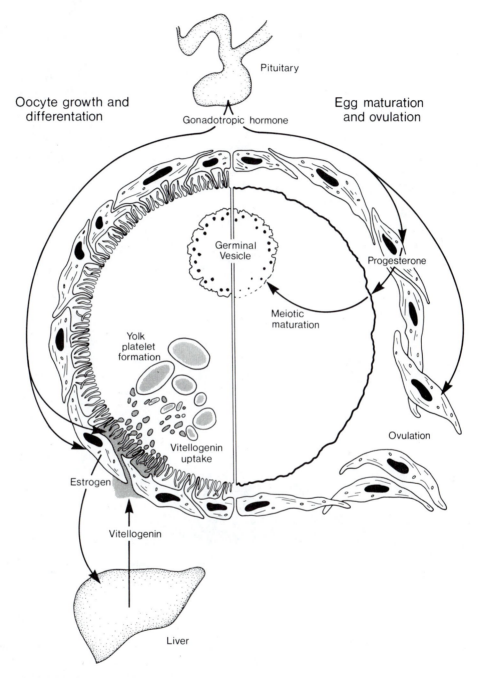

FIGURE 7.2 Hormonal regulation of amphibian oocyte growth and differentiation (left) and egg maturation and ovulation (right).

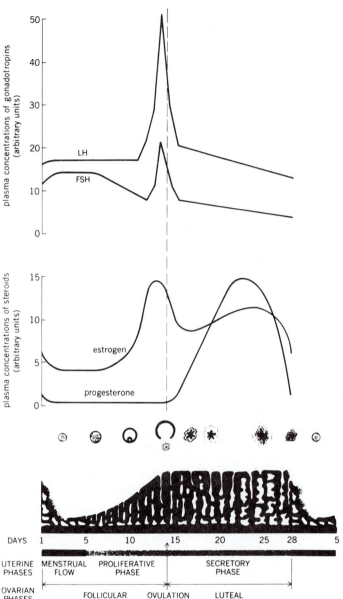

FIGURE 7.3 Diagram of the ovarian cycle showing levels of the four primary hormones (LH, FSH, estrogen, and progesterone), the condition of the follicle, and the condition of the uterus. (From A.J. Vander, J.H. Sherman, and D.S. Luciano. 1975. *Human Physiology. The Mechanisms of Body Function.* 2nd ed., p. 443. Used with permission of McGraw-Hill Book Co.)

increase LH production. A brief surge in the production of LH at about the midpoint of the cycle triggers ovulation. Although FSH production is also enhanced at this time, the significance of the FSH elevation is uncertain. After ovulation, LH promotes the conversion of the follicle—minus the oocyte—into the **corpus luteum,** which itself is an endocrine structure that produces progesterone and small amounts of estrogen. In a feedback mechanism the steroids act on the brain, causing the hypothalamus to shut down production of FSH and LH by the pituitary.

If the egg is not fertilized, the corpus luteum degenerates within eight to ten days after ovulation, and the lining of the uterus is sloughed, causing the menstrual bleeding. The destruction of the corpus luteum reduces the levels of steroid hormones in the blood. This relieves the inhibition on the pituitary, which then resumes producing gonadotropins, and the cycle repeats.

If the egg is fertilized, production of LH is continued, and the corpus luteum expands and produces even higher levels of steroids. The corpus luteum continues to produce high levels of steroids until about the fourth month, when its steroid production begins to wane. During the latter half of pregnancy, the **placenta** becomes the principal source of steroid hormone production.

Oral contraceptive development has been a direct result of research on endocrine interactions. Synthetic estrogens and progesterones are employed to take advantage of the feedback inhibition that curtails gonadotropin release by the pituitary. In the absence of the surge of LH that normally occurs at midcycle, ovulation is prevented.

7–2. OOCYTE-ACCESSORY CELL INTERACTIONS DURING OOGENESIS

Like developing male germ cells, oocytes form an intimate association with nongerm cells in the ovary. These **accessory cells** may be important in steroid hormone production, in transportation of certain essential cytoplasmic components to the oocyte, and in formation of cellular and noncellular layers that surround the fully differentiated egg. The accessory cells fall into two categories: (1) follicle cells and (2) **nurse cells.**

Follicle cells are somatic cells that form a layer surrounding the oocyte. This layer of cells is known as the **follicular epithelium** (Fig. 7.4A). The major distinction between follicle cells and nurse cells is that the former are derived from somatic cells, whereas the latter are derived from the germ cell line and remain associated with the oocyte via cytoplasmic bridges.

The Mammalian Follicle

During follicle growth in mammals the follicular epithelium proliferates to become multilayered. The cells of the expanding follicle are sometimes called **granulosa cells.** The granulosa cells and the oocyte are separated by a space that widens due to the deposition of extracellular material that is produced by the oocyte and possibly the follicle cells (see section 7–9). This material forms a continuous coat, the **zona pellucida,** around the oocyte. Examination of the zona with the electron microscope reveals that it is penetrated by numerous short microvilli from the surface of the oocyte and relatively long cytoplasmic processes from the follicle cells, which contact the oocyte surface (Fig. 7.4B and C). Desmosomes and gap junctions are found at the sites of contact between the

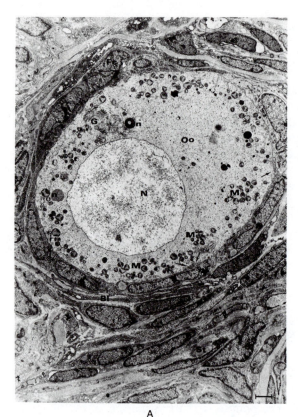

FIGURE 7.4 Relationship between mammalian oocyte and follicle cells. *A,* Electron micrograph of a primordial follicle of rabbit. The oocyte (Oo) is surrounded by a single layer of granulosa cells (Fc). The oocyte and granulosa cell membranes interdigitate extensively (arrows) and may form junctional complexes (*arrows). The most prominent feature of the oocyte is the large nucleus (N). A basal lamina (Bl) surrounds the follicle and isolates it from the stroma (St) of connective tissue. G, Golgi membranes; M, mitochondria; n, nucleolus-like bodies. Scale bar equals 2 μm. (From J. Van Blerkom and P. Motta. 1979. *The Cellular Basis of Mammalian Reproduction.* Urban & Schwarzenberg, Inc., Baltimore, p. 33.)

A

cytoplasmic processes and the oocyte membrane (Anderson and Albertini, 1976; Gilula et al., 1978). The gap junctions facilitate intercellular communication (Wassarman and Letourneau, 1976; Eppig, 1977; Gilula et al., 1978) and provide a mechanism for transport of materials into the oocyte that may be necessary for growth and differentiation (Heller et al., 1981). Prior to ovulation the cytoplasmic processes and microvilli are usually withdrawn. The zona and a few granulosa cells are retained by the egg when it is ovulated (see section 7–9).

As proliferation of granulosa cells nears completion, they are thought to secrete a fluid that accumulates in intercellular spaces. The fluid-filled spaces coalesce to form a large cavity—the **antrum.** Follicles with large antra are called **Graafian follicles** (Fig. 7.5). Formation of the antrum displaces the oocyte to one side of the follicle. The cluster of follicle cells around the oocyte is called the **cumulus oophorus.** As we shall discuss later the cumulus accompanies the egg at ovulation.

The Insect Egg Chamber

Nurse cells are found in certain invertebrates, including some coelenterates, annelids, mollusks, and insects. The formation and function of

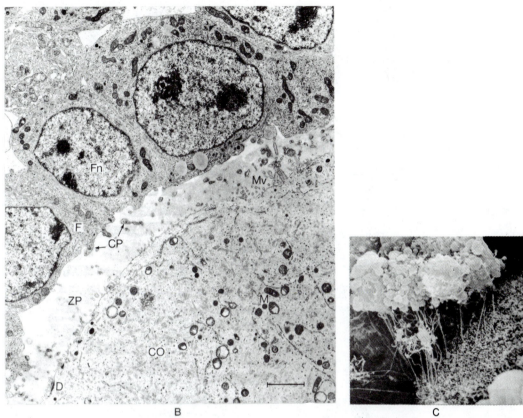

C

FIGURE 7.4 *(Continued)* B, Advanced oocyte of mouse, showing zona pellucida (ZP), microvilli (Mv) from the oocyte, and cytoplasmic processes (CP) from the follicle cells. CO, Cytoplasm of oocyte; F, follicle cell; Fn, nucleus of follicle cell; D, desmosome at point of contact of follicle cell projection and surface of oocyte; M, mitochondria. Scale bar equals 0.5 μm. (Courtesy of E. Anderson.) C, Scanning electron micrograph of the surface of a preovulatory rat oocyte. The zona pellucida has been removed enzymatically. Follicle cell processes are clearly evident. Note oocyte microvilli. × 1150. (From N. Dekel et al. 1978. Cellular association in the rat oocyte-cumulus cell complex: Morphology and ovulatory changes. Gamete Res., *1:* 53.)

insect nurse cells are particularly well understood and will serve as our example.

Oogenesis in insects with nurse cells is called **meroistic.** Much of our understanding of meroistic oogenesis is based upon light and electron microscopic studies utilizing *Drosophila melanogaster.* However, this species is certainly not representative of all insects. When *Drosophila* oogonia are undergoing mitosis, cytokinesis is incomplete, resulting in continuity of the cells via cytoplasmic bridges. Although all cells in a clone appear to be equivalent, one of them somehow becomes specialized and differentiates into the gamete; the remaining interconnected cells serve as nurse cells. The cluster of oocyte and associated nurse cells is an **egg chamber,** which is completely surrounded by follicle cells (Fig. 7.6). Nurse cells appear to provide the developing oocyte with macro-

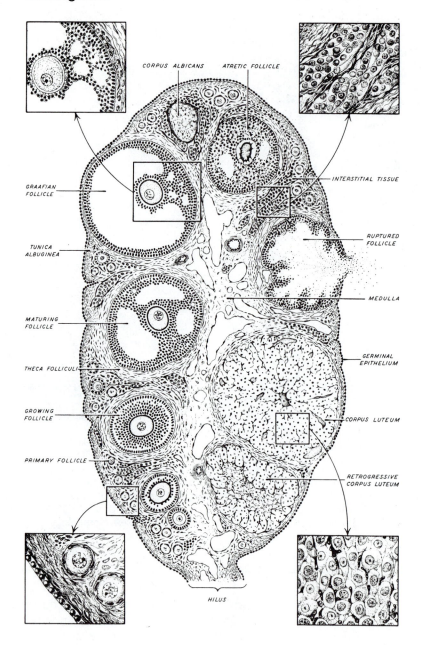

FIGURE 7.5 Diagram of a composite mammalian ovary. Progressive stages in the differentiation of a Graafian follicle are indicated on the left. The mature follicle may become atretic (top) or ovulate and undergo luteinization (right). (From C.D. Turner and J.T. Bagnara. 1976. *General Endocrinology*, 6th ed. W.B. Saunders, Philadelphia, p. 371.)

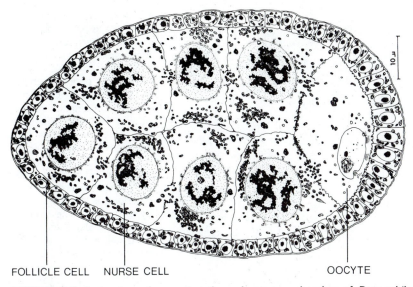

FOLLICLE CELL   NURSE CELL   OOCYTE

FIGURE 7.6 A drawing of a section through an egg chamber of *Drosophila melanogaster* as seen through the light microscope. Note the cytoplasmic continuity between cells. Organelles such as mitochondria can be seen in the intercellular bridges. The chamber is surrounded by a layer of follicle cells. The plasma membrane of the oocyte interdigitates with the membranes of the follicle cells. (From W.S. Klug, R.C. King, and J.M. Wattiaux. 1970. Oogenesis in the *suppressor*[2] of *hairy-wing* mutant of *Drosophila melanogaster*. II. Nucleolar morphology and *in vitro* studies of RNA protein synthesis. J. Exp. Zool., *174:* 129.)

molecules and even with ribosomes, which are transported from the nurse cells to the oocyte through the intercellular bridges.

Formation of the oocyte–nurse cell clone of *Drosophila* is diagrammatically represented in Figure 7.7. Mitosis of an oogonial stem cell produces two daughter cells that separate from one another. One of them continues to behave as a stem cell for the production of additional clones, whereas the other functions as a **cystoblast,** which establishes a clone by mitotic division. The daughter cells (**cystocytes**) produced by a cystoblast do not separate, thereby forming permanent **ring canals.** Mitoses involving cystocytes are also peculiar in that cell volume does not double before mitosis. Consequently, as divisions continue, the cystocytes become progressively smaller. During consecutive divisions the spindle axes of the cystocytes shift, creating a branching chain of interconnected cells. After the fourth division has produced 16 cells, two cells of the clone (1e and 2e) differ from all the rest in that they have four ring canals. Both of these cells prepare for meiosis, although the process persists to completion in only one of them. The process is aborted very early in the other.

Since only one of the cells with four ring canals develops into an oocyte, there must be an inequality between them. How do they differ

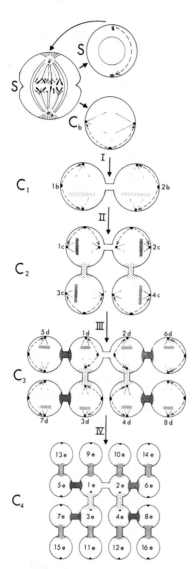

FIGURE 7.7 A diagrammatic model showing steps in the production of a cluster of 16 interconnected cystocytes. In this drawing the cells are represented by circles lying in a single plane, and the ring canals have been lengthened for clarity. The area of each circle is proportional to the volume of the cell. The stem cell (S) divides into two daughters, one of which behaves like its parent. The other differentiates into a cystoblast (C_b), which by a series of four divisions (I to IV) produces 16 interconnected cystocytes. C_1, first; C_2, second; C_3, third; and C_4, fourth generation cystocyte. The original stem cell is shown at early anaphase. Each parent-daughter pair of centrioles is attached to the plasma membrane by astral rays. The daughter centriole is drawn slightly smaller than its parent. The daughter stem cell receives one pair of centrioles. One remains in place while the other moves to the opposite pole. This movement is represented by a broken arrow. In the daughter cystoblast and all cystocytes the initial position of the original centriole pair is represented by a solid half circle, whereas their final positions are represented by solid circles. The position of the future cleavage furrow is drawn as a strip of defined texture. The canal derived from the furrow is treated in a similar fashion. (From E.A. Koch, P.A. Smith, and R.C. King. 1967. The division and differentiation of *Drosophila* cystocytes. *J. Morphol., 121:* 62.)

from one another? Two possibilities have been proposed. One is the topographical arrangement of the cystocytes. Only one of them occupies the position at the posterior tip of the egg chamber (Fig. 7.6). This cystocyte becomes the oocyte. The oocyte also has a unique association with the follicle cells that surround the egg chamber. Careful examination of Figure 7.6 will reveal that the membranes of the oocyte and the follicle cells interdigitate with one another. Perhaps the interdigitations produce a functional relationship that causes the follicle cells to transfer to the oocyte factors that promote differentiation. However, a dilemma remains. Does this cystocyte assume its position and interdigitate with follicle cells because it is inherently different from the other cells, or does it become different because it finds itself in this situation?

After determination of the oocyte, the nurse cells grow and become highly polyploid due to extensive replication of DNA without further cell division. After this **endoreplication,** *Drosophila* nurse cells may have up to 1024 times as much DNA as the haploid genome (King, 1970). The polyploid nurse cell nuclei are very active in RNA synthesis. The disposition of this RNA has been demonstrated by autoradiography of egg chambers exposed to labeled RNA precursors. The results of such an experiment are shown in Figure 7.8. In this experiment, egg chambers of the housefly, *Musca domestica,* were exposed to ^{3}H-cytidine for different periods of time. After short exposure to the label (Fig. 7.8A), the nurse cell nuclei—but not the oocyte nucleus—incorporate the precursor into RNA. The fate of this RNA is seen in Figure 7.8B, which is an autoradiograph of material that was fixed five hours after injection of ^{3}H-cytidine. Here, the labeled nurse cell RNA can be seen entering the oocyte cytoplasm via ring canals. There is still no indication of oocyte nuclear RNA synthesis by this time. Since each nurse cell has the equivalent of several genomes, the oocyte can acquire vast amounts of RNA from them to promote its differentiation. It is likely that some of this RNA is utilized in oocyte protein synthesis.

What regulates the movement of cytoplasmic constituents through the ring canals? Do they merely respond to concentration differences between the nurse cells and oocyte, or is there a means of directing particular substances in one direction or the other? There is evidence from experiments with the moth *Hyalophora cecropia* that molecular

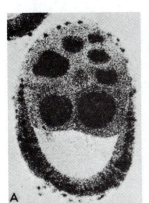
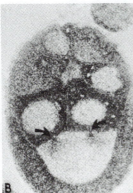

FIGURE 7.8 Synthesis of macromolecules in nurse cells of the housefly, *Musca domestica,* and their transport into the oocyte. *A,* Autoradiograph of section of an egg chamber 1 hour after injection of ^{3}H-cytidine. Intense label in nurse cell nuclei is an indication of RNA synthesis. Some labeled RNA has been transported to nurse cell cytoplasm, and labeled RNA can be seen entering the oocyte from an adjacent nurse cell (arrow). *B,* Egg chamber 5 hours after injection of label. The labeled RNA originally synthesized in the nurse cell nuclei is now mainly localized in the nurse cell cytoplasm. Transport into the oocyte via ring canals is indicated by arrows. Scale bar equals 100 μm. (From K. Bier. 1963. Autoradiographische Untersuchungen über die Leistugen des Follikelepethels und der Nährzellen bei der Dotterbildung und Eiwiessynthes im Fliegenover. Wilh. Roux' Archiv., *154*: 568. Reprinted with permission of Springer-Verlag, Heidelberg. Reproduced from B.I. Balinsky, 1975.)

flow between nurse cells and the oocyte is directed by differences in electrical potential between nurse cells and the oocyte. Electrical potential measurements reveal that the nurse cell cytoplasm is negative relative to the oocyte cytoplasm (Woodruff and Telfer, 1973). Consistent with this electrical polarity is the observation that a negatively charged fluorescent protein will diffuse from nurse cells into the oocyte (Fig. 7.9A), but not in the opposite direction (Fig. 7.9B). By contrast, a positively charged protein will move from the oocyte to the nurse cells (Fig. 7.9C), but not into the oocyte from the nurse cells (Fig. 7.9D). The be-

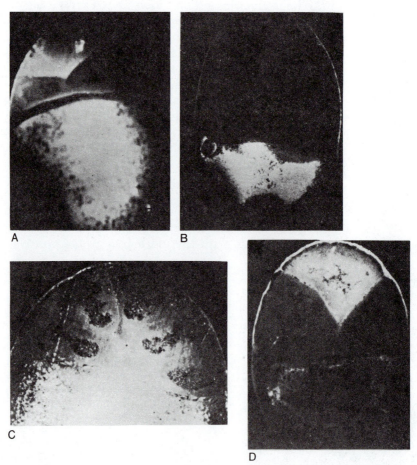

FIGURE 7.9 Fluorescent photomicrographs of *Hyalophora cecropia* follicles injected with fluorescent protein and incubated for 1 to 2 hours. *A*, Negatively charged protein injected into nurse cells. *B*, Negatively charged protein injected into oocyte. *C*, Positively charged protein injected into oocyte. *D*, Positively charged protein injected into nurse cells. The width of the nurse cell cap adjacent to the oocyte in each case is approximately 450 μm. (From R.I. Woodruff and W.H. Telfer. 1980. Reprinted by permission from *Nature* (Lond.), Vol. 286, No. 5768, pp. 84–86. Copyright © 1980 Macmillan Journals Limited.)

havior of charged proteins suggests that molecular movement between the oocyte and nurse cells is a kind of electrophoresis (see Chap. 4).

Ultimately, the nurse cells inject virtually all of their cytoplasm into the oocyte. The ring canals are then severed, and the nurse cells are sloughed from the oocyte prior to ovulation (Cummings and King, 1969).

The simultaneous support by 15 cells, each of which contains several genomes, enables the *Drosophila* oocyte to grow quite rapidly. Under optimal conditions the cytoplasmic volume can increase 90,000 times in only three days (King, 1972). Such rapid growth is a considerable achievement and testifies to the remarkable adaptive significance of this mechanism of oogenesis.

The follicle cells that circumscribe the insect egg chamber play various roles during oogenesis. As previously mentioned, they can transfer material to the oocyte. Insect follicle cells, as is the case with mammalian follicle cells, are coupled to the oocyte via gap junctions, which serve as channels for intercellular transport (Huebner, 1981). Follicle cells also play an active role in vitellogenesis; they can sequester yolk precursors from the hemolymph for transport to the oocyte, and, in some species, they may also synthesize yolk precursors (see section 7–4). The follicle cells are also involved in secretion of (initially) the vitelline envelope and (subsequently) the chorion that surround the insect egg (Cummings and King, 1969; Quattropani and Anderson, 1969; Cummings et al., 1971). The deposition of these accessory layers around the entire circumference of the oocyte is facilitated by the interposition of follicle cells between the nurse cells and oocyte in the latter stages of oogenesis so that the oocyte becomes encased by these cells except for the areas occupied by the ring canals.

7–3. ORGANIZATION OF THE EGG

In many ways an egg is an ordinary cell with typical cell organelles. However, because of its unique roles in fertilization and in providing for the maintenance of the developing embryo, the egg acquires a number of specializations during oogenesis that are not found in somatic cells. One of the most striking aspects of egg morphology is the definite spatial organization of organelles and other inclusions in the cytoplasm (often called **ooplasm**).

One critical aspect of the spatial arrangement of the egg is its **polarity**. Egg constituents may be unequally distributed along the major axis of the egg. This axis is an imaginary line connecting the two poles: the animal pole and the vegetal pole. The egg nucleus is displaced toward the animal pole, and when meiotic maturation occurs, the polar bodies are formed at this pole. Certain cytoplasmic organelles and inclusions may also be displaced toward the animal pole. In the amphibian egg, for example, ribosomes, mitochondria, and pigment granules are prevalent at the animal pole and decrease in numbers gradually toward the vegetal pole. Yolk platelets, on the other hand, are small and loosely packed in the animal hemisphere and become progressively larger and more concentrated toward the vegetal pole (Fig. 7.10). The adjective "vegetal" is applied to this pole because the yolk functions in embryonic nutrition.

Visible markers, such as pigment granules in the amphibian egg, illustrate polarity directly. The absence of such markers should not be interpreted as meaning that an oocyte lacks polarity. For example, the sea urchin egg has a predetermined animal–vegetal axis. However, it cannot always be seen, because redistribution of pigment granules, which reflect this axis, may not occur until after fertilization (Schroeder, 1980). We shall discuss the polarity of the sea urchin egg and its developmental significance in Chapter 11.

The unequal distribution of ooplasmic constituents plays an extremely important role in the establishment of regional specialization in

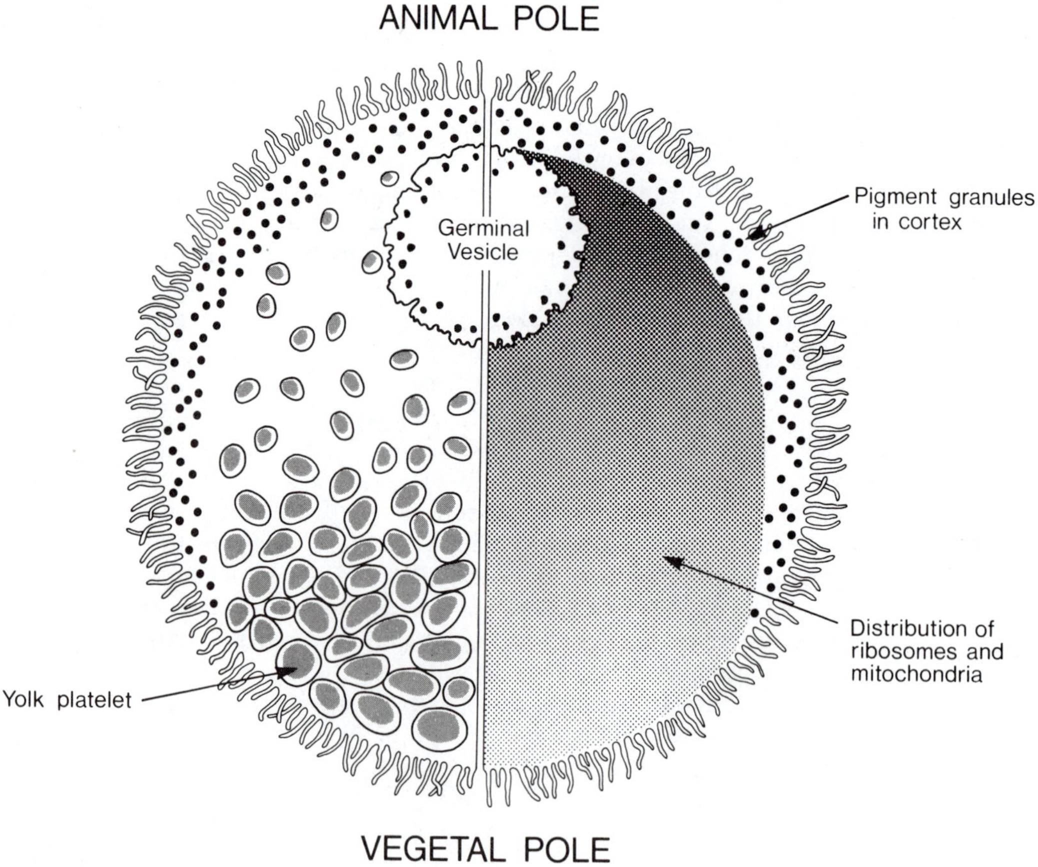

ANIMAL POLE

FIGURE 7.10 Diagrammatic representation of polarized distribution of amphibian egg components. Left: yolk distribution. Right: distribution of ribosomes and mitochondria. Note also the displacement of the germinal vesicle to the animal pole and the cortical location of pigment granules.

the embryo (see Chap. 11). Consequently, the mechanisms that establish and maintain the spatial organization of the ooplasm set the stage for orderly embryonic development. We know very little about these mechanisms. However, it is intriguing to speculate that ooplasmic polarity is a consequence of an intracellular gradient in electrical potential that governs the distribution of charged particles. We have seen that such a mechanism operates in the insect egg chamber. In Chapter 11 we shall discuss a similar mechanism that operates in the establishment of polarity in the fertilized eggs of certain brown algae. Additional research is necessary to ascertain if this is a generalized mechanism for producing polarity.

Yolk: Amount and Distribution

The amount of yolk in animal eggs is quite variable. Early investigators devised various schemes for categorizing eggs according to yolk content and distribution. Eggs with a small amount of evenly distributed yolk have been called **oligolecithal, isolecithal,** or **homolecithal.** These eggs are found in many invertebrates, such as the sea urchin and the lower chordates (e.g., *Amphioxus* and tunicates). In these eggs the only visible clue of egg polarity may be the site of polar body formation. In species with relatively large amounts of yolk the animal and vegetal hemispheres show distinctly different organization because of the concentration of the yolk in the vegetal hemisphere. In the most extreme situations (e.g., the eggs of reptiles, bony fish, birds, and some mollusks, including cephalopods and some gastropods), there is a segregation of cytoplasm and yolk, with the cytoplasm restricted to a thin layer covering the yolk. This layer thickens at the animal pole to form a **cytoplasmic cap** that contains the nucleus (Fig. 7.11A). These eggs are called **telolecithal.** The situation is not so extreme in the amphibians, in which the yolk is concentrated in the vegetal hemisphere, whereas other cytoplasmic organelles are more numerous in the animal hemisphere. Some authors place amphibian eggs in a separate category—**mesolecithal**—whereas others classify them as moderately telolecithal.

The arthropod egg, particularly the insect egg, is unique in that the yolk assumes a central position and is surrounded by a thin coat of cytoplasm. There is also an island of cytoplasm containing the nucleus in the center of the egg. These eggs are **centrolecithal.** The animal–vegetal pole relationships are irrelevant for centrolecithal eggs. There is, however, a definite polarity in insect eggs, which is reflected in the shape of the egg. The *Drosophila* egg shown in Figure 7.11B is slightly more rounded at one end. This end is destined to be the posterior portion of the embryo, whereas the opposite end will become the anterior portion. Thus, these eggs have an **anterior-posterior polarity.** The dorsal and ventral aspects of the embryo are also indicated in the egg: The convex side will be the ventral side of the embryo, whereas the concave side will be the dorsal side.

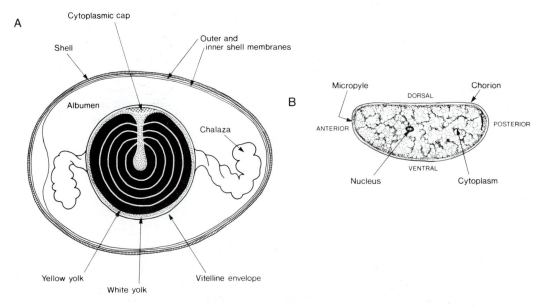

FIGURE 7.11 Representative yolky eggs (not drawn to scale). *A*, Telolecithal hen's egg. (After F.R. Lillie. 1919. Redrawn from B.I. Balinsky. 1981. *An Introduction to Embryology*, 5th ed. Saunders College Publishing, Philadelphia.) *B*, Centrolecithal insect egg. (After O.A. Johannsen and F.H. Butt. 1941. *Embryology of Insects and Myriapods*, p. 10. Reprinted with permission of McGraw-Hill Book Co. Reproduced from B.I. Balinsky, 1975.)

The amount of yolk in mammalian eggs is quite variable. The eggs of primitive mammals have a great deal of yolk, but those of placental mammals have minimal yolk reserves.

The Egg Cortex

The egg cytoplasm is organized into two definitive regions. The cytoplasmic layer just below the plasma membrane, the **cortex,** assumes physical properties that are distinct from those of the remainder of the cytoplasm. Most of the egg cytoplasm (the **endoplasm**) is in a fluid state. The cortex, however, has a higher viscosity and is a semirigid gel. These differences between the cortex and endoplasm can be demonstrated by centrifuging an egg. The components of the endoplasm are readily displaced during centrifugation, but the cortical constituents remain in place.

The cortex in the eggs of many species contains organelles that are not found in the endoplasm. The electron micrograph of the cortex of the amphibian egg (Fig. 7.12) illustrates two common types of cortical inclusions: **cortical granules** and **pigment granules.** Cortical granules are spherical structures, surrounded by a membrane and containing acid

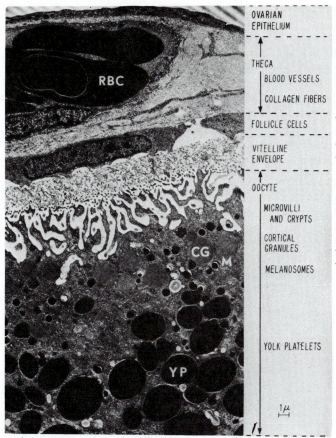

OVARIAN EPITHELIUM

THECA
BLOOD VESSELS
COLLAGEN FIBERS

FOLLICLE CELLS

VITELLINE ENVELOPE

OOCYTE

MICROVILLI AND CRYPTS

CORTICAL GRANULES

MELANOSOMES

YOLK PLATELETS

1μ

FIGURE 7.12 Electron micrograph illustrating the organization of the cortex of the *Xenopus* oocyte. Surrounding the egg are (1) the ovarian epithelium, (2) a thick thecal layer that includes blood vessels, collagen fibers, and fibroblasts, (3) a single layer of follicle cells, and (4) the vitelline envelope, which is penetrated by oocyte microvilli and follicle cell processes. In addition to microvilli, the oocyte surface is contoured with deep crypts. The cortical cytoplasm contains cortical granules (CG) and melanosomes (M). Yolk platelets (YP) appear below the cortex. (From J.N. Dumont and A.R. Brummett. 1978. Oogenesis in *Xenopus laevis* (Daudin). V. Relationships between developing oocytes and their investing follicular tissues. J. Morphol., *155:* 97.)

mucopolysaccharides and protein. These organelles function at fertilization, when their contents are extruded into the region surrounding the egg (see Chap. 9). The nature of pigment granules in the cortex is quite variable. Those in amphibian eggs contain the dark brown or black pigment, melanin, and are called **melanosomes.** Pigment granule distribution is not uniform in the amphibian egg (see Fig. 7.10). The vegetal hemisphere lacks pigment granules and is white, whereas the animal hemisphere contains many pigment granules and is darkly pigmented. Between the dark and light areas is a region of intermediate pigmentation, the **marginal zone.** The cortex persists after fertilization and has a major role to play in development: Cortical microfilaments provide the forces that constrict cells during cleavage (Chap. 10) and alter cell shape during gastrulation (Chap. 13).

The Oocyte Nucleus

The nucleus of the oocyte becomes highly modified during oogenesis and grows to an immense size. During early phases of oogenesis it occupies a nearly central position in the oocyte. As the cell increases in diameter, the nucleus becomes distinctly eccentric in location and may become situated very near the plasma membrane at the animal pole. These enlarged nuclei are given a special name—**germinal vesicles.** The germinal vesicle envelope is a highly convoluted bilaminar structure. The inner and outer membranes are joined at numerous sites to form **nuclear pores** (Fig. 7.13A), which are thought by some investigators to be involved in nucleocytoplasmic communication. The pores are not openings that provide nuclear and cytoplasmic continuity but are complexes of discrete elements. The hypothesis proposing a transport function is supported by micrographs that apparently show material in transit between nucleus and cytoplasm (Fig. 7.13B).

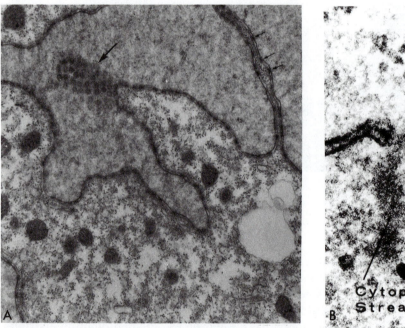
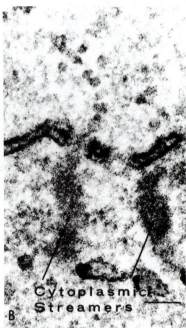

FIGURE 7.13 Nuclear pores in the nuclear envelopes of amphibian oocytes. *A,* Electron micrograph of portion of *Xenopus laevis* oocyte. Undulating nuclear envelope is penetrated by numerous regularly spaced pores (small arrows). The large arrow indicates a place where the envelope has been cut tangentially, showing cross sections of several pores. (Micrograph by Dr. G. Steinert. From J. Brachet. 1974. *Introduction to Molecular Embryology.* Springer-Verlag, New York, p. 46.) *B,* Portion of the nuclear envelope of a *Rana clamitans* oocyte. Nuclear pores are cut longitudinally, showing the apparent passage of material between cytoplasm and nucleus. Within the cytoplasm, this material appears as "cytoplasmic streamers," and within the nucleus it appears as nuclear granules. Scale bar equals 0.1 μm. (From E.M. Eddy and S. Ito, 1971. Reproduced from *The Journal of Cell Biology,* 1971, vol. 49, pp. 90–108 by copyright permission of The Rockefeller University Press.)

Cytoplasmic derivatives of the nuclear envelope, which are called **annulate lamellae,** are found in the eggs of animals of diverse phylogenetic origin. They consist of groups of parallel double membranes that contain pore complexes (Fig. 7.14). The function of annulate lamellae is not known. They are frequently accompanied by ribosomes, leading to speculation that they are involved in protein synthesis. It has also been proposed that they bear nuclear information that is dispatched to the cytoplasm, where it is stored for utilization after fertilization. Annulate lamellae have been most frequently observed in oocytes, but they have also been detected in some somatic cells. Thus, their function may not be related solely to oogenesis but may be part of normal cell physiology.

In contrast to the formation of annulate lamellae from the nuclear envelope, an alternative mode of formation has been reported in some insects. Annulate lamellae are formed within RNA-containing dense masses in the cytoplasm having no apparent connection to the nuclear envelope (Kessel and Beams, 1969; Halkka and Halkka, 1977). It is intriguing that structures with apparent ultrastructural identity can be formed in quite different ways in different organisms.

7–4. DIFFERENTIATION OF OOCYTE CONSTITUENTS

A number of specialized structures are formed during oogenesis. Since oocyte structure in different species is so variable, it is not possible to discuss the formation of every specialized oocyte inclusion. Instead, we shall discuss here the differentiation of a few of the better understood and most widespread oocyte constituents.

Vitellogenesis

Yolk is the most prominent cytoplasmic component of the egg in many species. It is a heterogeneous substance consisting of lipid, carbohydrate, and protein. The form taken by the yolk is quite variable. In some species it is present as general reserve material with no definitive ultrastructural characteristics. In others it may be present in **yolk bodies.** The amphibian yolk platelets are representative of yolk-containing organelles (see Fig. 7.12). These are flattened ovoid structures consisting primarily of two components, **phosvitin** and **lipovitellin,** in a crystalline lattice. Phosvitin is a phosphoprotein, and lipovitellin is a lipophosphoprotein.

The raw material for vitellogenesis can come from two sources: It may be synthesized within the oocyte (**autosynthesis**) or synthesized outside the oocyte and then incorporated into the oocyte during vitellogenesis (**heterosynthesis**). Organisms may use either or both of these methods to produce yolk.

Amphibians utilize both autosynthetic and heterosynthetic mechanisms of yolk production and, therefore, can serve as examples of both modes of vitellogenesis. During early vitellogenesis, yolk production oc-

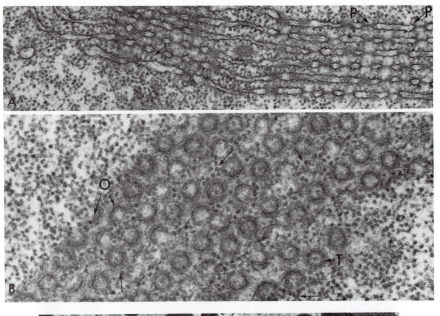

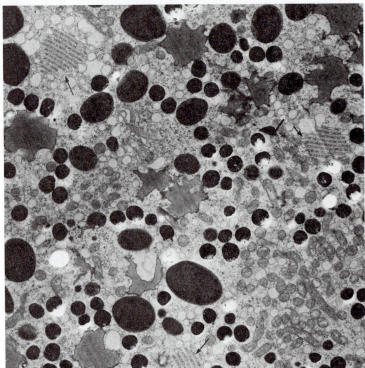

FIGURE 7.14 Examples of annulate lamellae. *A* and *B*, Oocyte of the ascidian *Ciona intestinalis.* Perpendicular (P) sections through pores are indicated in *A. B* illustrates both oblique (O) and tangential (T) sections through pores. Ribosomes appear attached to membranes of annulate lamellae at unlabeled arrows. *C*, Portion of *Rana pipiens* oocyte. Portions of several stacks of annulate lamellae are visible (unlabeled arrows). Pores are sectioned perpendicularly. *A:* ×31,300. *B:* ×35,700. *C:* ×7,500. (From R.G. Kessel. 1968. Annulate lamellae. J. Ultrastruct. Res. [Suppl.], *10:* 12, 61.)

curs primarily in the oocyte, whereas in later stages, most of the yolk is obtained by the oocyte from the blood by micropinocytosis. Endogenous yolk production occurs within structures called **multivesicular bodies** (Balinsky and Devis, 1963; Ward, 1978a). These structures (Fig. 7.15) are circumscribed by a membrane and contain a variety of membranous, fibrillar, granular, and vesicular components. The origins of multivesicular bodies are uncertain, but they are thought to arise from vesicles that originate in the Golgi complex and endoplasmic reticulum (Kress and Spornitz, 1972; Kessel and Ganion, 1980). The conversion of multivesicular bodies to yolk precursor complexes (Fig. 7.15) is marked by the accumulation of yolk protein. Growth of the yolk precursors appears to occur by fusion with additional vesicles. Ultimately, crystalline yolk protein completely fills the yolk precursors.

Some amphibians (notably the Ranidae) may utilize an unusual mechanism of endogenous yolk production during the initial phase of oogenesis: Yolklike material is produced within mitochondria (Lanzavecchia, 1961; Ward, 1962). This material appears as small crystals in the mitochondria (Fig. 7.16A), and the crystals grow to obliterate the mitochondrial internal structure (Fig. 7.16B). Preliminary information obtained from biochemical analysis of these crystals indicates that they lack lipovitellin and phosvitin (Ward, personal communication). Thus, their original designation as yolk may have been premature. Similar inclusions have also been reported for some fish and a gastropod mollusk (Yamamoto and Oota, 1967; Carasso and Favard, 1960).

During later stages of vitellogenesis, amphibian oocytes begin to acquire exogenously produced yolk. The yolk precursor (vitellogenin) is synthesized in the liver and transported via the circulation to the ovary, where it is taken up into the oocytes by micropinocytosis. This method of yolk formation, which is the major source of yolk for the amphibian oocyte, was first proposed by Roth and Porter (1964). Yolk production in

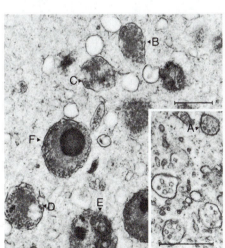

FIGURE 7.15 Endogenous yolk production within multivesicular bodies. Typical multivesicular bodies are shown in the inset. Body labeled A is acquiring an amorphous matrix. Bodies labeled B, C, and D show a progressive increase in matrix density, while E and F are later stages containing dense deposits of yolk. Scale bars equal 0.5 µm. (From R.T. Ward. 1978a. The origin of protein and fatty yolk in *Rana pipiens*. III. Intramitochondrial and primary vesicular yolk formation in frog oocytes. Tissue Cell, *10:* 522.)

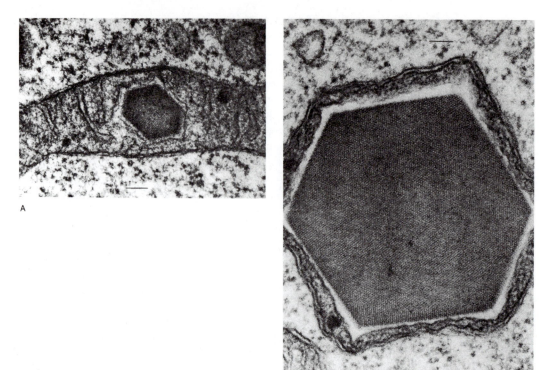

FIGURE 7.16 Intramitochondrial formation of yolk-like material in *Rana pipiens* oocytes. *A,* Small yolk-like crystal in mitochondrion. *B,* Yolk-like crystal virtually obliterates mitochondrial internal structure. Scale bars equal 0.1 μm. (From R.T. Ward, 1978a. The origin of protein and fatty yolk in *Rana pipiens.* III. Intramitochondrial and primary vesicular yolk formation in frog oocytes. Tissue Cell, *10:* 518, 519.)

the liver and its incorporation into oocytes from the blood by micropinocytosis is also typical of the reptiles, fishes, and birds.

The uptake of vitellogenin by amphibian oocytes has been extensively studied by Wallace, Dumont, and their associates. Vitellogenin reaches the follicle via a capillary network located within the thecal layer of the follicle (see Fig. 7.12). After exiting the capillaries, the vitellogenin passes through channels between the follicle cells to reach the oocyte surface (Dumont, 1978), where it is incorporated by micropinocytosis. At the bases of the microvilli that cover the oocyte surface, numerous invaginations and pits are formed. These have a so-called **bristle coat** on their convex cytoplasmic side and a fuzzy layer of **glycocalyx** on their extracellular concave surface. The glycocalyx, which may contain specific vitellogenin receptors, is thought to adsorb the vitellogenin. The invaginations and pits then pinch off from the oolemma and form **coated vesicles.** The vesicles carry the adsorbed vitellogenin into the oocyte. The vesicles lose the bristle coat and fuse to form **primordial yolk platelets**

(**PYPs**). The latter are transported to yolk platelets, which are the mature yolk bodies. This sequence of events is shown in the electron micrograph in Figure 7.17A.

Experimental confirmation of this sequence has been obtained by electron microscope autoradiography of oocytes exposed to ^{3}H-vitellogenin. As shown in Figure 7.17B, the labeled vitellogenin is initially found in small vesicles in the peripheral cytoplasm and in PYPs. After further exposure (Fig. 7.17C) the labeled vitellogenin is also found in larger PYPs and in definitive yolk platelets located deeper in the cytoplasm.

Heterosynthetic yolk production has also been discovered in insects. As discussed in section 7–1, insect vitellogenin is synthesized in the fat body. The incorporation of the blood-borne yolk precursor into oocytes has been particularly well studied in the cecropia moth where the follicle cells play an active role in removing vitellogenin from the circulation. The follicle cells are separated from one another by spaces filled with a matrix that appears to consist of sulfated mucopolysaccharide. It has been proposed that this matrix is responsible for binding the vitellogenin, thus removing it from the blood and making it available for incorporation by the micropinocytotic mechanism of the oocyte (Telfer, 1979). The matrix is preferentially produced during vitellogenesis; after vitellogenesis the matrix is lost, and tight junctions are formed between the cells, thus terminating vitellogenin uptake (Telfer, 1979; Rubenstein, 1979). The blood-borne vitellogenin is supplemented by vitellogenin produced by the follicle cells (Anderson and Telfer, 1969; Bast and Telfer, 1976). Thus, as with *Drosophila* (see section 7–1), there is a dual origin of vitellogenin in cecropia.

Mitochondria

The oocyte exaggerates the production of certain components so that the egg will have a store of these components with which to begin development. This relieves the embryo of the need to produce them during the earliest stages of development. One component that is stockpiled during oogenesis is the mitochondrion. For example, the *Xenopus* egg has 10^5 times as many mitochondria as a somatic cell.

Mitochondria are unique cytoplasmic organelles in that they possess their own self-replicating DNA. The amplification of mitochondria causes a tremendous increase in the amount of mitochondrial DNA in the cytoplasm. Whereas the ratio of nuclear DNA to mitochondrial DNA in somatic *Xenopus* cells is about 100:1, it is reversed during oogenesis and ranges from 1:1 to 1:100 in fully grown oocytes (Dawid, 1972).

During early oogenesis, mitochondria may be clustered near the germinal vesicle. These **juxtanuclear aggregates** have been variously referred to as "yolk nuclei," "mitochondrial clouds," and "Balbiani bodies." They were once thought to be involved in yolk formation, but this

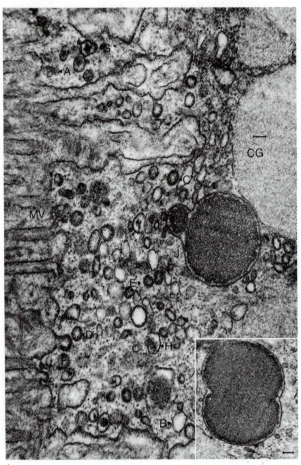

A

FIGURE 7.17 Vitellogenin uptake by amphibian oocytes. *A,* Electron micrograph of the cortical region of a *Rana pipiens* oocyte illustrating micropinocytotic uptake of vitellogenin and fusion of micropinocytotic vesicles to form yolk platelets. Bases of microvilli (MV) are to the left, and cortical granules (CG) are to the right. At A is a group of very small vesicles. At C, D, and E are coated vesicles. F shows formation of a coated vesicle. At G, H, and I smooth vesicles are seen containing dense deposits. At J a smooth vesicle is fusing with a large primordial yolk platelet. Inset: Two large vesicles have just fused. Scale bars equal 0.1 μm. (From R.T. Ward. 1978b. The origin of protein and fatty yolk in *Rana pipiens.* IV. Secondary vesicular yolk formation in frog oocytes. Tissue Cell, *10:* 527.)

idea has been discarded in recent years (Balinsky and Devis, 1963). Ultrastructural studies of oogenesis in *Xenopus laevis* have traced the formation of the mitochondrial aggregate (Fig. 7.18). The aggregate is first detected in the primordial germ cell stage. The aggregated mitochondria are associated with granular, electron-dense material. This material appears to be similar to the so-called "nuage material" associated with the nuclear pores. It is possible that the nuage material originates in the

FIGURE 7.17 *(Continued) B,* Electron microscope autoradiograph of peripheral region ▶ of oocyte cultured 20 minutes in ³H-vitellogenin. Silver grains are located over small vesicles and PYPs. × 17,800. Inset: Higher magnification showing silver grain over newly formed vesicle. × 54,500. *C,* Oocytes exposed to ³H-vitellogenin for 40 minutes. Silver grains are located over large PYPs and at the surface (arrow in inset) of yolk platelets (YP). × 18,000. Inset: × 13,000. (From A.R. Brummett and J.N. Dumont. 1977. Intracellular transport of vitellogenin in *Xenopus* oocytes: An autoradiographic study. Dev. Biol., *60:* 484, 485.)

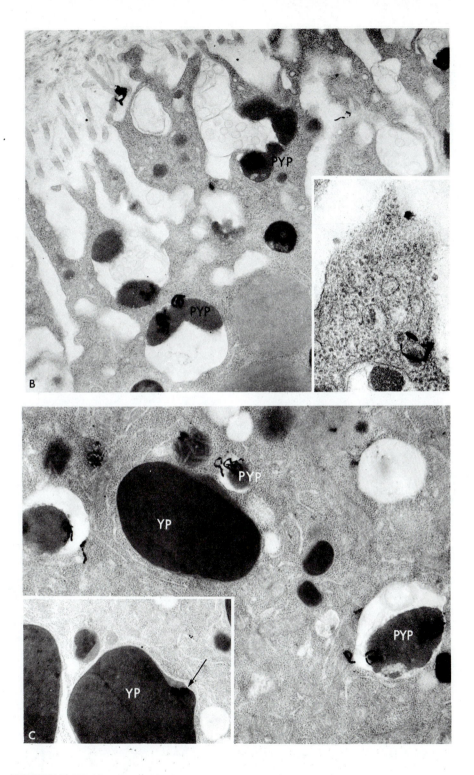

FIGURE 7.17 (*See legend on the opposite page.*)

273

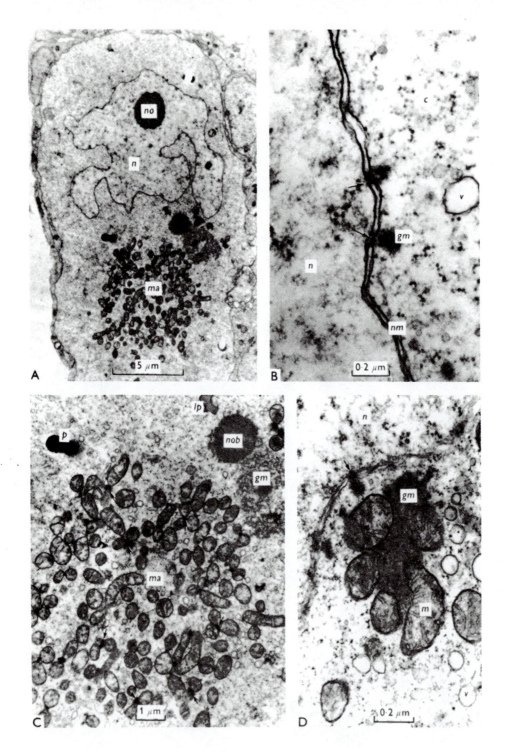

FIGURE 7.18 (*See legend on the opposite page.*)

nucleus and is transported via the pores to the cytoplasm, where it associates with the mitochondrial aggregate. On the other hand, there is presently no way of discounting the possibility that this material is being transported in the reverse direction—from cytoplasm to nucleus. The mitochondrial aggregates are also associated with structures called "nucleoluslike bodies" that have ultrastructural similarities to nucleoli, although their composition is unknown. The aggregates present in oogonia are not significantly different from those in primordial germ cells. In previtellogenic oocytes the aggregate frequently forms a cap on one side of the nucleus (Fig. 7.19A). Perhaps significantly, this site corresponds to the location of the centrioles (Tourte et al., 1981). As the oocyte enlarges, the aggregate becomes more compact and spherical and is converted into the definitive Balbiani body (Fig. 7.19B). Electron micrographs reveal that the mitochondria within the Balbiani body are long filamentous structures (Fig. 7.19C). The granular material as previously described is now present only at the periphery of the aggregate. The Balbiani body contains numerous small vesicles, which may be derived from the endoplasmic reticulum, and lipid droplets (Billett and Adam, 1976). Expansion of the Balbiani body correlates with an increase in the number of mitochondria (Balinsky and Devis, 1963; Al-Mukhtar, 1970).

Later, during vitellogenesis, the Balbiani body disappears as the mitochondria disperse. It is interesting to note that dispersal coincides with the end of the phase of rapid mitochondrial amplification (Webb and Smith, 1977). Since the Balbiani body is present during rapid mitochondrial amplification, it is possible that this configuration facilitates the rapid increase in the number of mitochondria.

Changes in the distribution of mitochondria have also been observed during oogenesis in some species of mammals. During early

◄FIGURE 7.18 Mitochondrial aggregates in primordial germ cells of *Xenopus laevis*. *A,* Electron micrograph showing the mitochondrial aggregate (ma) adjacent to the highly lobed nucleus (n) containing a large, prominent nucleolus (no). *B,* High-power electron micrograph of part of the nuclear envelope of a primordial germ cell, illustrating the electron-dense, granular material (gm) associated with the nuclear pores (arrows). *C,* The primordial germ cell organelle aggregate shown in *A* at higher magnification. Note the presence of a nucleolus-like body (nob) and electron-dense granular material (gm), some of which is associated with mitochondria within the aggregate (arrows). *D,* A high-power electron micrograph showing the association between electron-dense granular material and mitochondria (m) in the perinuclear cytoplasm of a primordial germ cell. There is a distinct similarity between the granular material in association with these mitochondria and the "nuage material" adjacent to the nuclear pores (arrows). c, cytoplasm; lp, lipid body; m, mitochondrion; p, pigment granule; v, vesicle. (From K.A.K. Al-Mukhtar and A.C. Webb. 1971. An ultrastructural study of primordial germ cells, oogonia, and early oocytes in *Xenopus laevis*. J. Embryol. Exp. Morphol., *26:* 200.)

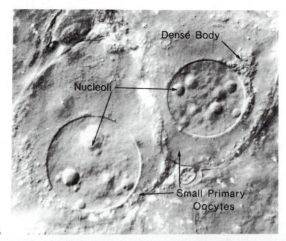

A

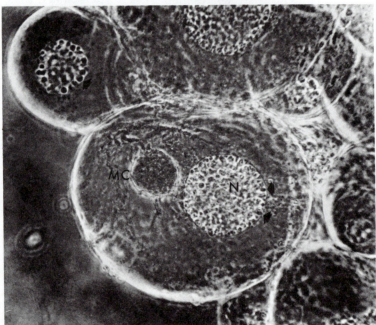

B

FIGURE 7.19 Modifications to the mitochondrial aggregate in amphibian oo-cytes. *A,* Light micrograph of living primary oocytes of *Rana clamitans* as seen with differential interference contrast optics. A large dense body is present adjacent to the nucleus of one oocyte. ×650. (From E.M. Eddy and S. Ito. 1971. Reproduced from *The Journal of Cell Biology,* 1971, vol. 49, pp. 90–108 by copyright permission of The Rockefeller University Press.) *B,* Phase contrast micrograph of living frog oocytes at a later stage of oogenesis. Mitochondria have formed a definitive Balbiani body. MC, mitochondrial cloud (Balbiani body); N, nucleus; arrows indicate nucleoli. (From B.I. Balinsky. 1981. *An Introduction to Embryology,* 5th ed. Saunders College Publishing, Philadelphia, p. 76.)

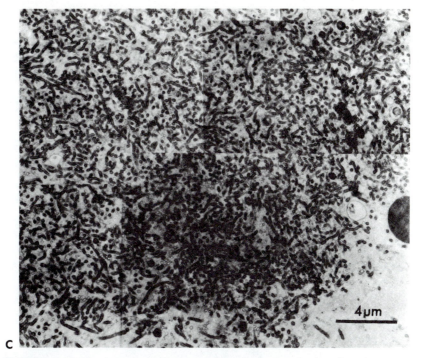

FIGURE 7.19 *(Continued) C,* Montage of electron micrographs of sections covering one-third of the diameter of the Balbiani body of the *Xenopus* oocyte. The immense number of mitochondria in the Balbiani body is obvious. (From F.S. Billett and E. Adam. 1976. The structure of the mitochondrial cloud of *Xenopus laevis* oocytes. J. Embryol. Exp. Morphol., *33*: 701.)

stages of oogenesis the mitochondria are localized to a band in the periphery of the oocytes. Thereafter, the mitochondria become dispersed throughout the ooplasm. At all times the mitochondria are associated with cisternae of the endoplasmic reticulum (Cran et al., 1980). The mitochondria of mammals may become highly modified because of the altered orientation of cristae or the deposition of material in the mitochondria. Examples of modified mitochondria are shown in Figure 7.20.

Cortical Granules

Cortical granules are formed within the endoplasm of the oocyte. Initially they are randomly distributed in the egg, but as the oocytes near completion, they migrate to the cortex. In those organisms in which their origin has been investigated, cortical granules are apparently produced by the combined efforts of the rough endoplasmic reticulum and the Golgi complex. The synthesis of cortical granule precursors apparently occurs on the ribosomes of the rough ER. The precursors are then transported within the ER to the Golgi, where they are assembled into definitive cortical granules.

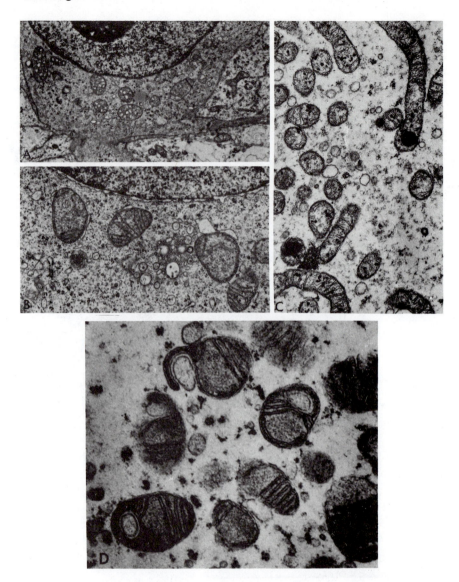

FIGURE 7.20 Examples of modified mitochondria found in the ooplasms of some species of mammals. *A,* The two components of cristae separate from one another and give the mitochondria a vesiculated appearance in human oogonia. ×5,800. *B,* Cristae are displaced peripherally in some mitochondria in rat primary oocytes, while in others they are organized in a shelf-like manner. ×12,200. *C,* In oocytes of secondary follicles of the spider monkey, electron-dense granules can be seen. ×14,000. *D,* Bovine mitochondria show the development of hood-shaped extensions. ×25,000. (*A, B, C* from D. Szöllösi. 1972. Changes of some cell organelles during oogenesis in mammals. *In* J.D. Biggers and A.W. Schuetz (eds.), *Oogenesis.* University Park Press, Baltimore, pp. 58, 61. *D* from P.L. Senger and R.G. Saacke. 1970. Reproduced from *The Journal of Cell Biology,* 1970, vol. 46, pp. 405–408 by copyright permission of The Rockefeller University Press.)

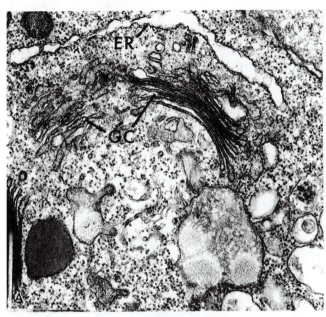

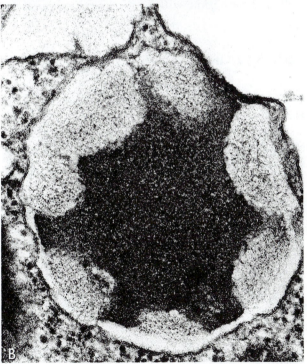

FIGURE 7.21 Formation of the cortical granules in the *Arbacia* oocyte. *A,* Stages in the initial formation of cortical granules as seen in a section through a Golgi complex (GC). Note cisternae of endoplasmic reticulum (ER). ×28,300. *B,* Section through a mature cortical granule. ×85,400. (From E. Anderson, 1968. Reproduced from *The Journal of Cell Biology,* 1968, vol. 37, pp. 514–539 by copyright permission of The Rockefeller University Press.)

In the sea urchin *Arbacia,* saccules of the Golgi containing a substance with a density similar to that of cortical granule material appear to pinch off to form membrane-enclosed vesicles. These cortical granule precursors increase in diameter, presumably by fusion of vesicles, and acquire their definitive form (Fig. 7.21). A similar pattern of cortical granule formation has been described in the golden hamster oocyte (Selman and Anderson, 1975). In this species, vesicles containing material resembling cortical granule constituents pinch off from the saccules of the Golgi complex and fuse with one another. These vesicles then fuse with others that are derived from the rough ER to form mature cortical granules.

Germ Plasm

Another significant component produced during oogenesis is the germ plasm. Although the germ plasm is produced in the oocyte, it does not function until after fertilization. It is responsible for germ cell determination in the zygote. The germ plasm becomes localized in a region of the zygote and is then segregated into particular cells during cleavage. As a consequence, these cells are the primordial germ cells, the precursors of the germ cell line. Thus, the oocyte is responsible for perpetuation of the germ cells in the next generation. The structure of the germ plasm and its role in germ cell determination will be discussed in detail in Chapter 11.

7–5.   GENE EXPRESSION DURING OOGENESIS: AMPHIBIANS

Oogenesis is a process with dual consequences: (1) the construction of a highly complex specialized cell and (2) the production of components that will be utilized during postfertilization development. Therefore, the utilization of genomic information during oogenesis has important implications for the regulation of early embryonic development. The most comprehensive studies on these two aspects of oogenesis have been conducted on the oocytes of amphibians. The ready availability of experimental material, the size of the oocytes, and the ability to regulate oogenesis by hormonal injection have contributed to the popularity of amphibians for research on oogenesis. Other important factors are the large size of the highly modified lampbrush chromosomes and the number of nucleoli produced during oogenesis. Because of this wealth of information we shall concentrate on amphibians in our analysis of gene expression during oogenesis. In section 7–6 we shall consider gene expression in other organisms.

Lampbrush Chromosomes

The process of meiosis is an integral aspect of oogenesis; consequently, the structure of oocyte chromosomes is affected by the meiotic process. The most frequently encountered situation in animal oogenesis is that in which meiosis is initiated at the beginning of oogenesis, homologous chromosomes synapse to form tetrads, and homologues begin to separate at diplonema. Then, however, meiosis is suspended, and the homologues remain attached by chiasmata. This hiatus may last for months or even years, and it is during this phase of oogenesis that the major growth and differentiation of the oocyte occur. The chromosomes at this time may become highly elongated and modified to form lampbrush chromosomes (Fig. 7.22A). Because of their extremely large size, lampbrush chromosomes of amphibians are readily amenable to experimental manipulation and have been studied in more detail than those of other groups.

To understand these chromosomes properly, we must recall that at diplonema each homologue consists of two sister chromatids arranged with their axes parallel to one another. In the lampbrush configuration the chromosomes possess a linear sequence of globular structures (chromomeres), which are composed of compacted chromatin. The chromomeres are connected by thin parallel strands of **interchromomeric chromatin.** One or more pairs of **loops** extend laterally from the chromomeres, giving the chromosomes the lampbrush appearance. The presumed relationships among the various structures seen in lampbrush chromosomes are diagrammed in the model shown in Figure 7.22B.

The axis of each homologue consists of thin sister chromatid strands that span the interchromomeric region. At intervals the chromatin becomes compacted on one side of a chromomere, extends out into loops, returns to a compacted state on the other side of the chromomere, and then extends from there through the next interchromomeric region to the next chromomere, and so on. This model assumes that, as with all chromosomes, the chromatin is continuous from one end to the other and that it undergoes local differences in compaction and folding at intervals along the length of the chromosome. Apparent confirmation of this assumption comes from experiments in which individual chromosomes are stretched. This treatment can cause partial unraveling, allowing the chromosomes to become more linear. In the experiment outlined in Figure 7.23, a chromosome has been stretched to the point where the connection between the two sides of a chromomere has been broken, and the paired loops assume the form of a "double bridge." This demonstrates the continuity between the chromosome axis and the axes of the loops.

One of the most striking aspects of lampbrush chromosome morphology is the presence of a **matrix** on the loops. The matrix morphology of sister loops is always identical. However, the matrix on different loop pairs can be quite distinctive, with the result that lampbrush chromosomes are adorned with a variety of unusual structures (Fig. 7.24). The differences in morphology of loops originating from different chromomeres provide landmarks that make it possible to construct "maps" of lampbrush chromosomes (Fig. 7.25). The structure of loops apparently has a genetic basis, since heterozygotes produced by mating individuals

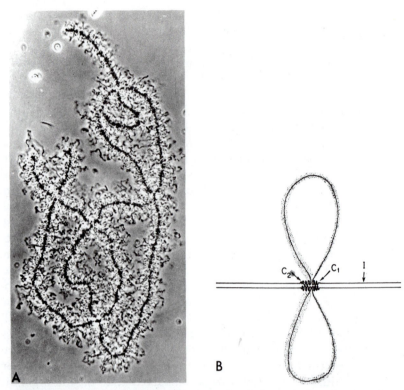

FIGURE 7.22 Lampbrush chromosome morphology. *A,* Chromosome from an oocyte nucleus of *Notophthalmus viridescens.* ×275. (From J.G. Gall. 1966. Techniques for the study of lampbrush chromosomes. *In* D.M. Prescott (ed.), *Methods in Cell Physiology,* vol. 2. Academic Press, New York, p. 38.) *B,* Diagrammatic representation of the supposed general organization of the DNA components—the 2 chromatids—that make up a single lampbrush chromosome. Each horizontal line represents a single chromatid that consists of a single DNA duplex. Each chromatid is thought to run relatively straight through the interchromomeric region (I), become locally compacted in one side of the chromomere (C_1), pass out into a loop, where it is again in a relatively extended state, return to the compacted state in the other side of the chromomere (C_2), and then proceed along the interchromomeric fibril to the next chromomere/loop complex. (From H.C. Macgregor. 1977. Lampbrush chromosomes. *In* H.J. Li and R.A. Eckhardt (eds.). *Chromatin and Chromosome Structure.* Academic Press, New York, p. 346.)

with structurally distinct loops have heterozygous lampbrush chromosomes with loop morphologies that are characteristic of each parent (Callan, 1963). This indicates that each loop includes a specific DNA region that determines the morphology of the loop. This is not to imply, however, that each loop contains a single gene. As we shall discuss later, evidence indicates that some loops may contain more than one gene.

The composition of the matrix on the lampbrush loops is indicated by the fact that it is digestible by either ribonuclease or protease (Gall,

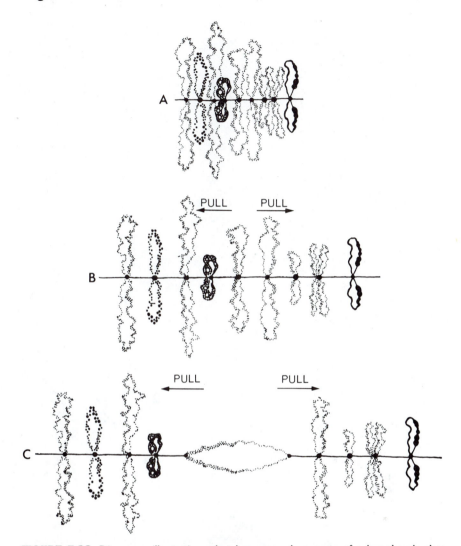

FIGURE 7.23 Diagrams illustrating what happens when parts of a lampbrush chromosome are stretched. *A*, Unstretched; solid dots represent condensed chromomeric DNA. *B*, Stretched within the elastic limit. *C*, Stretched beyond the elastic limit—one chromomere has broken and a pair of lateral loops spans the break. (After H.G. Callan. 1963. The nature of lampbrush chromosomes. Int. Rev. Cytol., *15*: 4.)

1954). Thus, the matrix consists of protein associated with RNA that is presumably being synthesized actively on the loops. The loops' activities in transcription are also indicated by autoradiographs showing incorporation of radioactive precursors into RNA on the loops (see Fig. 7.28A). An interesting aspect of the morphology of loop matrices is that they are sometimes asymmetrical; they can be thin at one end, gradually increas-

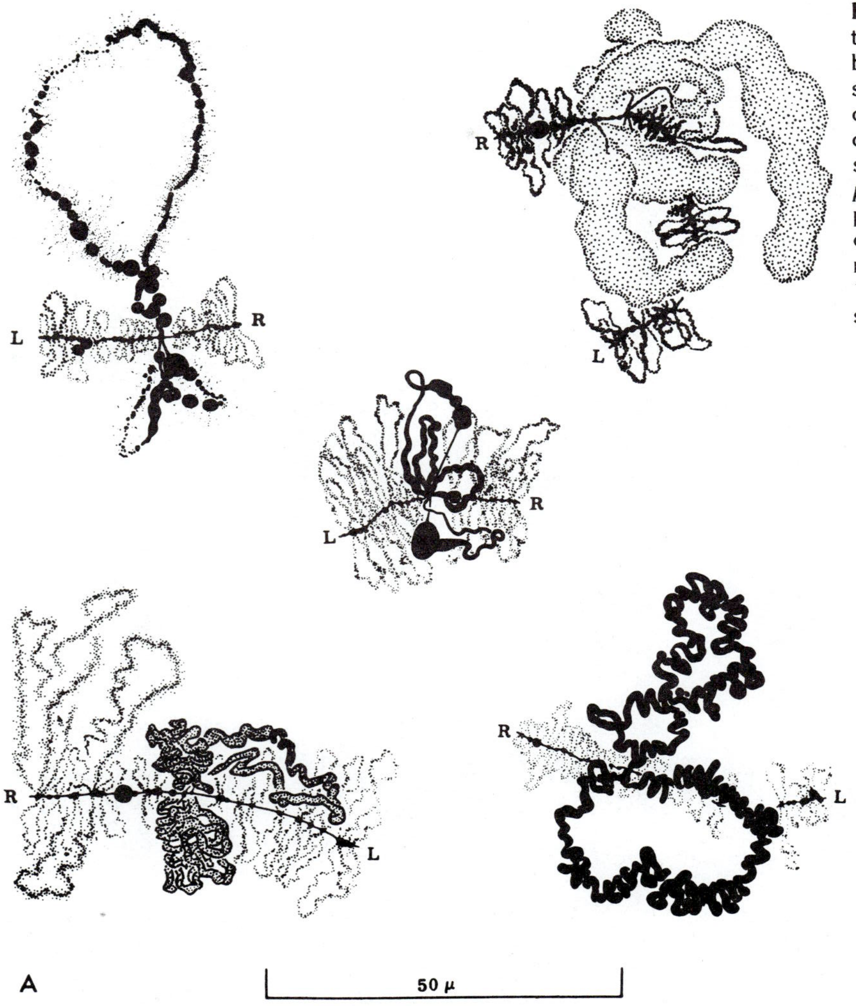

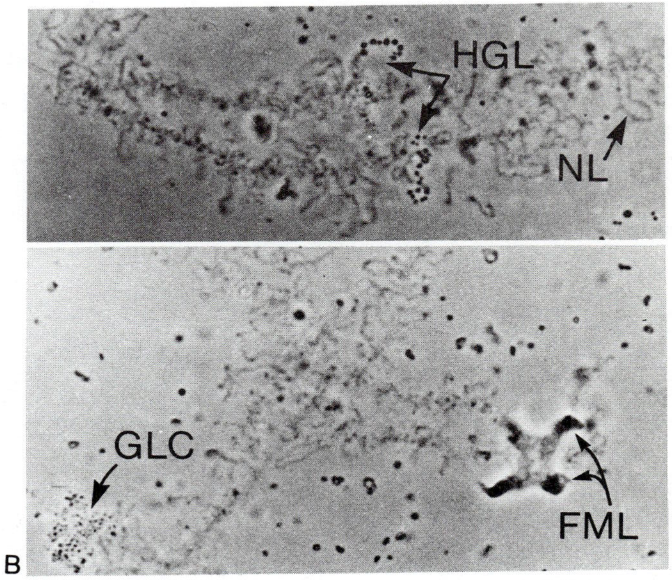

FIGURE 7.24 Examples of amphibian lampbrush loop matrix morphologies. *A*, Drawings of short segments of lampbrush chromosomes of the newt *Triturus cristatus*, illustrating some of the lampbrush loop matrix morphologies in this species. (From H.G. Callan. 1963. The nature of lampbrush chromosomes. Int. Rev. Cytol., *15*: 17.) *B*, Photomicrographs showing portions of two lampbrush chromosomes of *Rana pipiens*. Note the homologous granular loops (HGL), occupying the same site on homologues, and the granular loop cluster (GLC) and large fusing matrix loops (FML). NL, normal loop, ×570. (From R.E. Rogers and L.W. Browder. 1977. Morphological observations on cultured lampbrush-stage *Rana pipiens* oocytes. Dev. Biol., *55*: 142.)

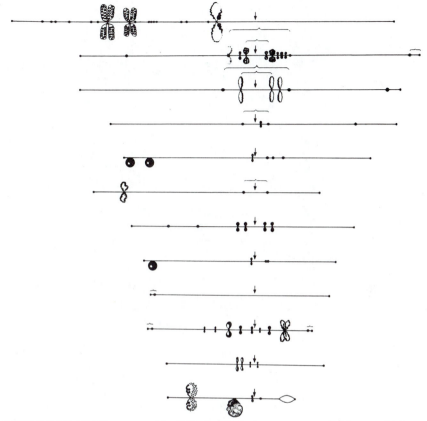

FIGURE 7.25 Working map of the 12 lampbrush chromosomes of *Triturus cristatus* (numbers 1 to 12, top to bottom, respectively). Centromere positions are indicated by the vertically aligned arrows. Structures that frequently fuse together are linked by brackets. (From H.G. Callan. 1963. The nature of lampbrush chromosomes. Int. Rev. Cytol., *15:* 14.)

ing in thickness around the loop contour (Fig. 7.26). The asymmetry indicates that transcription begins on one side of the loop and proceeds until the entire loop is transcribed. These loops would appear to correspond to single transcriptional units.

Electron micrographs of asymmetrical loops prepared by the Miller technique (see Chap. 3) beautifully demonstrate the directionality of transcription. Figure 7.27*A* is an electron micrograph showing transcription of a portion of a lampbrush loop. Individual transcripts can be seen due to dispersal of the loop matrix during sample preparation. RNA synthesis is proceeding in the direction of the arrow. Progressively longer RNA transcripts radiate from the loop axis as the distance from the loop origin increases. At the base of each transcript is a small granule, which is presumably the RNA polymerase molecule. Nucleotides are added to the base of the transcript as the polymerase moves along the loop con-

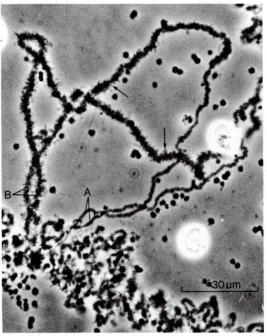

FIGURE 7.26 Phase contrast photomicrograph of a part of a lampbrush chromosome from the salamander *Plethodon cinereus* showing a pair of very long (240 μm) loops, both of which arise from the same chromomere. The loops have a fairly granular matrix of ribonucleoprotein, and in some places (arrows), the loop axis is faintly discernible. Each loop has a thin end or "insertion" (A) and grows gradually thicker toward the thick end (B), a phenomenon that is characteristic of most loops and is referred to as "loop asymmetry." (From H.C. Macgregor. 1977. Lampbrush chromosomes. *In* H.J. Li and R.A. Eckhardt (eds.). *Chromatin and Chromosome Structure.* Academic Press, New York, p. 343.)

tour, progressively increasing the length of the transcript. Newly synthesized RNA associates with protein, resulting in deposition of ribonucleoprotein (RNP) matrix on the loops. At the termination of transcription the completed transcript is released from the loop along with its associated protein.

Another class of loops apparently has more than one transcriptional unit. A portion of a loop with multiple transcriptional units is shown in Figure 7.27*B*. Portions of two transcriptional units can be seen, and they are separated by a nontranscribed spacer. This particular micrograph is especially interesting in that it shows that the two transcriptional units have opposite polarity, indicating that transcription is proceeding in different directions on the same loop. Loops with multiple transcriptional units are, in fact, quite common. On some of these loops all the transcriptional units have the same polarity, whereas in others transcription may be proceeding in different directions (Scheer et al., 1976a).

The transcriptional units shown in Figure 7.27 illustrate that the RNA polymerases are very close to one another. Compare these micrographs with those shown in Figure 3.21. The latter show nonribosomal transcriptional units in somatic cells of *Oncopeltus* and *Drosophila* and are characterized by relatively few, widely spaced transcripts. The density of polymerases on DNA is the net result of initiation of transcription by attachment of polymerases to the DNA and the rate of movement of the polymerases to the termination site. Apparently, one or both of these

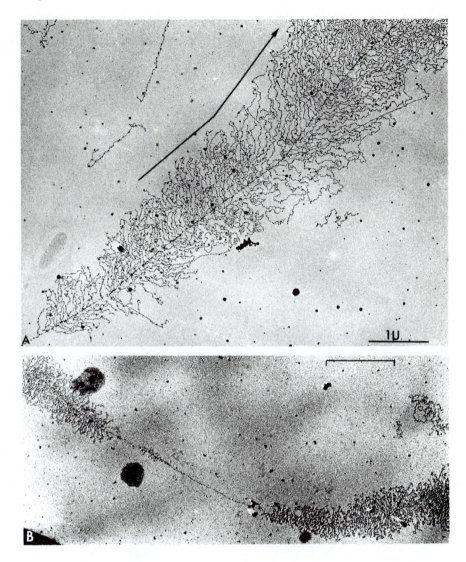

FIGURE 7.27 Electron microscopic visualization of lampbrush transcription. *A,* Micrograph of a portion of a *Notophthalmus viridescens* lampbrush chromosome loop near its thin insertion end. RNA polymerase molecules can be seen on the loop axis at the base of each nascent RNP fibril. Arrow shows direction of transcription. (From O.L. Miller, Jr., and B.R. Beatty. 1969. Portrait of a gene. J. Cell. Physiol., *74* [Suppl. 1]: 231.) *B,* Micrograph of a portion of a *Pleurodeles poireti* lampbrush chromosome loop. Two transcriptional units can be seen. Although on the same loop, the direction of transcription of the two units is opposite. Scale bar equals 2 μm. (From N. Angelier and J.C. Lacroix. 1975. Complexes de transcription d'origines nucléolaire et chromosomique d'ovocytes de *Pleurodeles waltlii* et *P. poireti* (Amphibiens, Urodèles). Chromosoma [Berl.], *51*: 331.)

functions is highly modified in lampbrush chromosomes as compared to chromosomes of somatic cells. The close packing of nascent transcripts, combined with the long duration of the lampbrush phase in most organisms, indicates that a tremendous amount of RNA synthesis is achieved during the lampbrush phase. However, the nature and fate of the RNA produced by the lampbrush chromosomes are only partially understood. Most lampbrush transcription apparently produces DNA-like, heterogeneous RNA. Evidence in support of this statement comes from an elegant experiment by Edström and Gall (1963). These investigators isolated lampbrush chromosomes and extracted RNA and DNA from them. The base compositions of the RNA and DNA were then determined by microelectrophoresis. The lampbrush RNA was found to have a base composition of 46% guanine plus cytosine (GC composition; see Chap. 3), which is quite similar to the 44% GC composition of the DNA. The large size of the nascent transcripts observed on the lampbrush loops (5 to 10×10^4 nucleotides) also indicates that they are similar to the large heterogeneous nuclear RNA found in somatic cell nuclei.

Additional evidence about the identities of lampbrush transcripts has been obtained from studies utilizing the inhibitor α-amanitin (Schultz et al., 1981). As discussed in Chapter 3, the three RNA polymerases are differentially sensitive to α-amanitin. Polymerase II, which synthesizes heterogeneous nuclear RNA, is inhibited by 0.5 μg/ml of α-amanitin. The results of an inhibitor experiment are illustrated in Figure 7.28. Newt oocytes were incubated *in vitro* in the presence of ^{3}H-UTP and ^{3}H-CTP and autoradiographs then prepared of the lampbrush chromosomes. Incorporation of label into RNA on the loops is shown in Figure 7.28A. The addition of 0.5 μg/ml of α-amanitin abolishes all RNA synthesis on lampbrush loops (Fig. 7.28B). Note that the incorporated label is confined to short stretches on the chromosomal axes. These correspond to the centromeric regions, which are the sites that contain

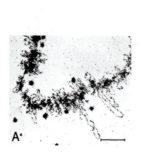

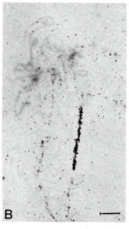

FIGURE 7.28 Autoradiographs of newt lampbrush chromosomes. *A,* Chromosomal segment showing incorporation of ^{3}H-UTP and ^{3}H-CTP. Note labeling of loops. Scale bar equals 20 μm. *B,* Short segment of chromosome after incubation in ^{3}H-UTP and ^{3}H-CTP in presence of α-amanitin. Label is found only on the chromosomal axis. Scale bar equals 10 μm. (From L.D. Schultz, B.K. Kay, and J.G. Gall. 1981. In vitro RNA synthesis in oocyte nuclei of the newt *Notophthalmus.* Chromosoma [Berl.], *82:* 178, 181.)

the 5S RNA genes. Incubation of oocytes in 200 μg/ml of α-amanitin abolishes the label over these regions. This concentration of α-amanitin is sufficient to inhibit polymerase III as well as polymerase II. As we discussed in Chapter 3, the 5S RNA genes are transcribed by polymerase III. These results suggest that 5S RNA is synthesized at certain sites on the axes of lampbrush chromosomes, whereas heterogeneous nuclear RNA is synthesized on lampbrush loops.

Participation of RNA polymerase II in transcription on lampbrush loops is also indicated by experiments in which antibodies to the polymerase are injected into oocyte germinal vesicles (Bona et al., 1981). The results of such an experiment are illustrated in Figure 7.29. A control lampbrush chromosome from a *Pleurodeles waltlii* oocyte is shown in Figure 7.29A. Within a few minutes after microinjection of anti-RNA polymerase II, all lampbrush loops retract into the chromosome axes (Fig. 7.29B), and all nascent transcripts are shed from the chromatin (Fig. 7.29C). This antibody has no effect upon transcription of ribosomal RNA genes. Thus, its effects are specific to polymerase II.

The nature of RNA transcribed by lampbrush chromosomes is also suggested by analyses of RNA in the RNP particles isolated from the nuclei of lampbrush-phase *Triturus* oocytes (Sommerville, 1973). Since nucleoplasmic RNPs are shed from the lampbrush chromosomes after transcription, the RNA in these particles should be quite similar to the

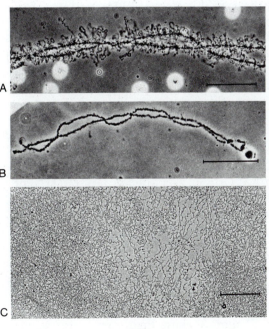

FIGURE 7.29 Effects of anti-RNA polymerase II on morphology of *Pleurodeles waltlii* lampbrush chromosomes. *A*, Control chromosome. Scale bar equals 50 μm. *B*, Loop retraction caused by injection of anti-RNA polymerase II, 5 minutes after injection. Scale bar equals 50 μm. *C*, Electron micrograph of spread chromatin from an oocyte that had been injected with anti-RNA polymerase II. Chromatin has the configuration of transcriptionally inactive chromatin. Scale bar equals 1 μm. (Reproduced with permission from M. Bona, U. Scheer, and E.K.F. Bautz. 1981. Antibodies to RNA polymerase II (B) inhibit transcription in lampbrush chromosomes after microinjection into living amphibian oocytes. Journal of Molecular Biology, *151*: 81–99. Copyright: Academic Press Inc. (London) Ltd.)

nascent RNA on the lampbrush loops. Analyses show that this RNA is very large, is heterogeneous in size, and has a GC composition of about 45%, which corresponds to the 44% GC composition of the complementary lampbrush DNA (see previous discussion). These characteristics indicate that it is DNA-like, heterogeneous RNA. Since, as we discussed in Chapter 3, there is considerable evidence that messenger RNA is derived from heterogeneous nuclear RNA, it is likely that some of the RNA produced on the lampbrush chromosomes and packaged into the RNP particles is messenger RNA precursor.

The proposed role of the RNA in nuclear RNP particles is supported by molecular hybridization studies. In these experiments, cDNA prepared from cytoplasmic polyadenylated RNA is used as a probe to detect homologies between nuclear and cytoplasmic transcripts. Lampbrush-stage oocytes accumulate polyadenylated RNA (which is thought to be messenger RNA) in cytoplasmic particles. Cytoplasmic RNP particles from *Triturus* ovaries are used as a source of polyadenylated RNA for preparation of the cDNA probe. The cDNA probe hybridizes with RNA from nuclear RNP particles, confirming the relationship between these lampbrush transcripts and the cytoplasmic RNA. These observations suggest that the lampbrush loops are transcribing large RNA molecules, which are processed in the nucleus and transported to the cytoplasm as polyadenylated messenger RNA combined with protein (Sommerville and Malcolm, 1976). This is not to imply, however, that the lampbrush configuration is essential for synthesis of high molecular weight RNA in oocytes; synthesis of this class of RNA continues unabated in postlampbrush-phase oocytes (Anderson and Smith, 1977).

The ability to clone specific DNA sequences provides the opportunity to identify transcribed genes on lampbrush loops by *in situ* hybridization by using cloned sequences (Pukkila, 1975). This technique differs from that used in the *in situ* hybridization studies discussed in Chapter 3 in one significant way: The chromosomes are not denatured. Thus, the chromosomal DNA remains double stranded and inaccessible to the DNA probe, which can be either single stranded or denatured DNA. The hybridization detected by autoradiography should, therefore, be due to annealing between the probe DNA and nascent RNA. A necessary control for these experiments is pretreatment of chromosomes with ribonuclease; the enzyme removes the nascent RNA and prevents hybridization.

Initial experiments of this type have been promising and have produced some quite unexpected results. The first loop-associated structural gene transcripts identified in this way are histone transcripts of the newt *Notophthalmus* (Gall et al., 1981). Surprisingly, satellite DNA sequences have also been localized to transcriptionally active lampbrush loops (Callan and Old, 1980; Varley et al., 1980a and b; Diaz et al., 1981). As we discussed in Chapter 3, these highly repetitive sequences are generally considered to be nontranscribed. Lampbrush chromosomes appear to be an exception to this rule. One possible explanation for transcription of satellite sequences on lampbrush chromosomes is that transcription is initiated on structural genes in the usual way, but the RNA polymerases overrun the normal termination signals. Thus, they would continue to transcribe adjacent stretches of highly repetitive DNA, which are located between adjacent histone gene clusters. The transcripts so produced would contain covalently bound satellite and structural gene transcripts, which would be separated from each other during posttranscriptional processing. It is possible that overrun of termination signals is the rule in transcription of lampbrush chromosomes, resulting in large transcriptional units and correspondingly large transcripts (Diaz et al., 1981).

The termination of the lampbrush phase is characterized by regression of the loop pairs along the length of the chromosomes and reversion of the chromosomes to their usual size. The reversibility of the lampbrush configuration indicates that it is within the normal range of morphological variability of chromosomes.

One other noteworthy aspect of lampbrush chromosomes is that they are found during oogenesis of a wide variety of animals representing nearly the entire phylogenetic spectrum (Davidson, 1976). They are not universal, however. For example, lampbrush chromosomes are not present in oocytes of meroistic insects. Thus, the lampbrush phase is not an indispensable aspect of animal oogenesis. But, it is *nearly universal*, which underscores the significance of lampbrush chromosomes, and suggests that they play a major role in information transfer during oogenesis. Additional information about lampbrush chromosomes may be found in an excellent review by Macgregor (1980).

Synthesis and Accumulation of mRNA

One of the unique processes that occurs during oogenesis is the production of messenger RNA for utilization during early development. The production of messenger during amphibian oogenesis has been analyzed by studying polyadenylated RNA and, more recently, by using cloned sequences to probe specific transcripts. The most extensive studies have used *Xenopus laevis* oocytes as experimental material.

The mature *Xenopus* oocyte contains considerable polyadenylated RNA. This RNA represents *at least a portion* of the messenger RNA of these cells, since it can be translated *in vitro* (Darnbrough and Ford, 1976). However, the majority of this RNA is not active in oocyte protein synthesis but is stored in cytoplasmic RNP particles (Rosbash and Ford, 1974). This RNA is part of the information pool with which the zygote begins its development and is the mRNA that has accumulated during oogenesis; that is, it is the net result of messenger synthesis and degradation. Therefore, two of the prime functions of oogenesis are the synthesis and accumulation of messenger RNA.

The synthesis of polyadenylated RNA begins early in oogenesis.

However, in spite of continued synthesis of this RNA, its amount does not change significantly during oogenesis (Rosbash and Ford, 1974; Dolecki and Smith, 1979). Recent experiments using recombinant DNA technology have also failed to detect any *qualitative* changes in polyadenylated RNA (Golden et al., 1980). In those experiments *Xenopus* ovarian poly (A)$^+$ RNA was used to make cDNA (see Chap. 3). The cDNA was then made double stranded, inserted into plasmids, and cloned. These cDNA clones were used to probe for complementary RNA at different stages of oogenesis. In all cases, the RNA begins to accumulate in early oogenesis, reaches a plateau in early vitellogenesis, and then remains at the same concentration throughout the remainder of oogenesis. Thus, although synthesis continues, a constant population of polyadenylated RNA is maintained, presumably by degradation of excess RNA.

Similar results have also been obtained for histone messenger RNAs, which—unlike histone messengers in somatic cells—are predominantly polyadenylated (Levenson and Marcu, 1976; Ruderman and Pardue, 1977, 1978; Ruderman et al., 1979). Cloned histone H3 genes were hybridized to RNA extracted from oocytes at different stages of oogenesis, and the level of hybridization was used as a measure of the amount of histone messenger present. As with total poly (A)$^+$ RNA, the histone messenger accumulates in early oogenesis and does not increase in amount thereafter (van Dongen et al., 1981).

The absence of quantitative and qualitative changes in the messenger RNA population raises questions about information processing during oogenesis. For example, why is the oocyte involved in the apparently wasteful exercise of continual synthesis and degradation of messenger? We obviously require a great deal more information before we have a complete understanding of messenger RNA metabolism during *Xenopus* oogenesis. However, it is clear that both transcriptional and posttranscriptional events are very important in these cells.

Synthesis of Ribosome Components

The production and accumulation of ribosomes are major chores of oogenesis. The large store of ribosomes in the egg relieves the embryo of the need to produce its own ribosomes to support protein synthesis during early embryonic development and makes it possible for the zygote to undertake immediate protein synthesis after fertilization. Therefore, the components of ribosomes must be produced in prodigious amounts during oogenesis. These components include 18S, 28S, and 5S RNA, as well as the ribosomal proteins. Except in meroistic insects, these components are apparently synthesized under the direction of the oocyte nucleus, assembled within the nucleoli, and transported to the oocyte cytoplasm—primarily as a storage product. As we shall see, the patterns of synthesis and accumulation of the individual components of ribosomes during *Xenopus* oogenesis are quite different from one another.

5S RNA

The synthesis of 5S RNA occurs rapidly during the earliest stages of oogenesis, prior to the major synthetic period for 18S and 28S RNA and ribosomal proteins. The large amounts of 5S RNA made during the early stages are stored in either 7S or 42S RNP particles and later incorporated into ribosomes when the other components become available (Ford, 1971; Ford and Southern, 1973). The rapid rate of 5S RNA production and its accumulation by the previtellogenic *Xenopus* oocyte produce large stores of this class of RNA, which account for about 45% of the total RNA of these oocytes (Ford, 1976). By early vitellogenesis, nearly half the amount of 5S RNA needed during oogenesis has accumulated, and production continues throughout oogenesis, concurrent with the synthesis of 18S and 28S RNA, but at a lower rate.

The production of large amounts of 5S RNA is facilitated by the multiplicity of 5S RNA genes. As we discussed in Chapter 3, the *Xenopus laevis* genome contains three kinds of genes that code for 5S RNA: oocyte, trace oocyte, and somatic. The oocyte 5S RNA genes are present as approximately 24,000 tandemly repeated 120 base pair coding units per haploid genome (Brown and Sugimoto, 1973). Since the chromosomes are in meiosis during oogenesis, each oocyte nucleus contains four times this many 5S RNA genes, or 96,000 of them.

Initiation of 5S RNA gene transcription is regulated by an interaction between an oocyte transcription factor and the promoter site, which is located within each coding sequence (see Chaps. 3 and 4). The transcription factor also binds with the 5S RNA after its synthesis and becomes a component of the 7S storage particle. By binding to its own transcription factor, 5S RNA can repress further transcription of the 5S RNA genes (Pelham and Brown, 1980; Honda and Roeder, 1980). This system of feedback inhibition allows us to hypothesize a mechanism for regulation of 5S RNA synthesis during oogenesis in *Xenopus:* In early stages of oogenesis, massive synthesis of transcription factor would result in accumulation of large amounts of 5S RNA. As the RNA accumulates, it binds the factor and causes transcription to level off (Pelham and Brown, 1980). Synthesis of 5S RNA could continue so long as unbound transcription factor is available to the 5S RNA genes.

Following oocyte maturation, 5S RNA synthesis is no longer detectable. The amount of transcription factor has also been drastically reduced (Pelham et al., 1981). In fact, egg extracts will only support transcription of 5S RNA genes if exogenous transcription factor is added (Fig. 7.30). By contrast, oocyte extracts will transcribe 5S RNA genes without added factor. These results suggest that the reduction in the

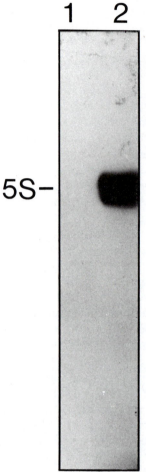

1   2

5S−

FIGURE 7.30 Assay for 5S RNA gene transcription factor in *Xenopus* eggs. The figure shows an autoradiograph of a polyacrylamide gel analysis of labeled RNA produced *in vitro*. The position of 5S RNA is noted on the figure. In the absence of added transcription factor (Lane 1) an egg extract will not promote synthesis of 5S RNA. This class of RNA is synthesized if transcription factor is added to the egg extract (Lane 2). (From B.M. Honda and R.G. Roeder. 1980. Association of 5S gene transcription factor with 5S RNA and altered levels of the factor during cell differentiation. Cell, *22:* 121. Copyright © Massachusetts Institute of Technology; published by the MIT Press.)

amount of transcription factor after maturation leads to a shutdown of 5S RNA synthesis (Honda and Roeder, 1980). After fertilization, the embryo reinitiates 5S RNA synthesis, but only from the somatic genes (see Chap. 12).

RIBOSOMAL RNA

Ribosomal RNA (i.e., 18S and 28S RNA) is the major component of oocyte RNA. Synthesis of large amounts of ribosomal RNA in many organisms is facilitated by a mechanism that is unique to oocytes: **amplification of ribosomal RNA genes.** This phenomenon occurs in many animals and has been demonstrated in some species of fish, insects, mollusks, and amphibians, although this discussion will be limited to *Xenopus*. One striking aspect of amplification is that it can be seen readily with the light microscope. The genetic locus that contains the ribosomal

RNA genes is marked in somatic cells by the presence of the nucleolus. Somatic nuclei of *Xenopus* contain two nucleoli, one for each haploid chromosome set. Thus, the tetraploid oocyte nucleus would be expected to have four nucleoli. However, it contains literally hundreds of small nucleoli, which are detached from the chromosomes and are predominantly located lining the inner membrane of the nuclear envelope. The amplification process is readily monitored by incorporation of ^{3}H-thymidine into oocyte DNA. All of the thymidine incorporation is due to replication of the nucleolar DNA, since there is no other nuclear DNA synthesis occurring in oocytes. As is shown in the autoradiographs in Figure 7.31, the label is incorporated into the extrachromosomal DNA that forms a cap of fibrillar material on one side of the nucleus. The major incorporation occurs during the prelampbrush pachytene stage.

This fibrillar material consists of numerous circular replicates of the nucleolar organizer region. During late pachynema and in diplonema, miniature nucleoli form in association with these rings of extrachromosomal DNA. The nucleoli then become more or less evenly distributed by spreading over the inner surface of the nuclear envelope. As many as 1500 nucleoli are formed during diplonema in *Xenopus*. This is about 375 times the tetraploid nucleolar number. The amount of ribosomal DNA, as measured by RNA-DNA hybridization, is increased by a comparable amount (Brown and Dawid, 1968). These multiple nucleoli provide the oocyte with numerous ribosomal coding units for the prodigious production of 18S and 28S RNA that occurs during oogenesis. It has been calculated that if amplification did not occur, it would take more than 400 years to produce the amount of ribosomal RNA found in the mature oocyte (Perkowska et al., 1968).

Synthesis of ribosomal RNA occurs at very low rates during pachynema when amplification is in progress, and the same rate prevails in the interval between amplification and the beginning of vitellogenesis, when it then undergoes a dramatic increase. There is evidence that the rate of ribosomal RNA synthesis is reduced somewhat in postvitellogenic oocytes, but it still occurs at rates exceeding those found in previtellogenesis (Davidson, 1976).

As we noted previously, maximal transcriptional rates are characterized by very close packing of polymerases on the chromatin. The proximity of these polymerases to one another, which can be monitored by observing transcription with the electron microscope, depends upon the frequency of initiation of transcription on a transcriptional unit and the rate of polymerase movement. Scheer et al. (1976b) have observed nucleolar transcription of *Triturus* oocytes, which have a ribosomal transcriptional pattern similar to that of *Xenopus*. They have found that the packing of polymerases on the ribosomal RNA genes reflects the rate of transcription. As shown in Figure 7.32, maximum packing is found in nucleoli from vitellogenic oocytes, whereas previtellogenic and postvitellogenic oocytes have fewer polymerases per unit length of nucleolar

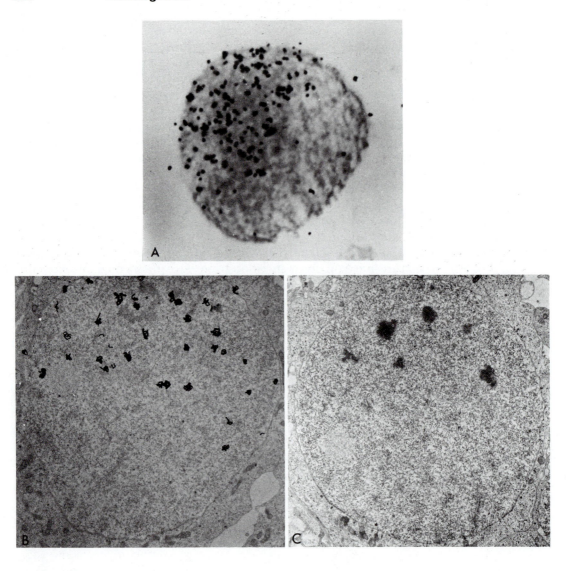

FIGURE 7.31 Nucleolar amplification. *A,* Light microscope autoradiograph, showing incorporation of ³H-thymidine into nucleus of pachytene oocyte. Label is over nuclear cap. × 1500. (From L.W. Coggins and J.G. Gall. 1972. Reproduced from *The Journal of Cell Biology,* 1972, vol. 52, pp. 569–576 by copyright permission of The Rockefeller University Press. *B,* EM autoradiograph, showing incorporation of ³H-thymidine into fibrillar cap of pachytene nucleus. × 4,100. *C,* Electron micrograph of late pachytene oocyte, showing formation of several small nucleoli in nuclear cap. × 5,700. (*B, C* from L.W. Coggins. 1973. An ultrastructural and radioautographic study of early oogenesis in the toad, *Xenopus laevis.* J. Cell Sci., *12:* 91, 93.)

chromatin. These results clearly indicate that amplification itself is not sufficient to cause maximal rates of ribosomal RNA synthesis, since amplified nucleoli may have low rates of transcription. It is apparent that transcriptional-level controls are operating to modulate the rate of rRNA production by the nucleoli.

RIBOSOMAL PROTEINS

The increased ribosomal RNA synthetic activity at vitellogenesis corresponds with the appearance of ribosomes in the cytoplasm. Ribosome assembly is thought to occur in the nucleoli, where the newly available 18S and 28S RNA form complexes with the 5S RNA and ribosomal proteins. Much of the 5S RNA for the ribosomes is synthesized in earlier stages and stored in RNP particles for use during vitellogenesis. But what about the ribosomal proteins? Does their synthesis correlate with that of the 5S RNA or with that of 18S and 28S RNA? The available evidence suggests that ribosomal protein synthesis is occurring at higher rates during vitellogenesis than during previtellogenesis (Hallberg and Smith, 1975), which indicates that their synthesis is correlated with the period when 18S and 28S RNA are synthesized at high rates and when concurrent ribosome assembly is taking place. Because of the requirements for assembly of large numbers of ribosomes during oogenesis, ribosomal proteins are made in very large amounts in oocytes and account for a large percentage of the protein synthesis. For example, during midvitellogenesis 20 to 30% of the protein synthesized in the oocyte is ribosomal protein. Obviously, large amounts of the messengers for ribosomal protein must be made during oogenesis. If the synthesis and accumulation of these messengers could be monitored, we could gain some greatly needed insight into structural gene expression during oogenesis.

Transfer RNA Synthesis

The production of transfer RNA appears to parallel that of 5S RNA; that is, it is synthesized in large amounts during previtellogenesis and accumulates in the 42S storage particles in which 5S RNA is also stored (Ford, 1976). Each 42S particle contains 12 molecules of tRNA and four 5S RNA molecules (Picard et al., 1980).

It is curious that 5S RNA can be stored in two different particles. The relationship between the 7S and 42S particles is obscure; although both contain 5S RNA, only the 7S particle contains the 5S regulatory protein (Pelham and Brown, 1980). Each category of storage particle contains approximately one half the cell's 5S RNA, whereas the 42S particles contain approximately 90% of the tRNA (Picard et al., 1980). Both of these classes of RNA are extremely long lived in the oocyte, having a lifetime of several months (Mairy and Denis, 1972).

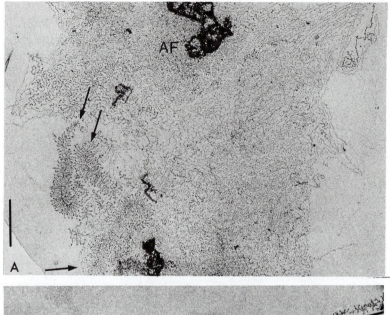

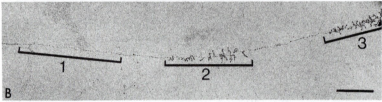

FIGURE 7.32 Nucleolar transcription in *Triturus* oocytes. *A* and *B*, Previtellogenic pattern. A large proportion of the chromatin is transcriptionally silent and tends to form fibrillar aggregates (AF). Regions of low transcriptional activity and a few maximally active matrix units (arrows) are also seen in *A* Scale bar equals 1.0 μm. *B* shows three adjacent rRNA coding units from spread chromatin. Regions 2 and 3 show sparse packing of polymerases, while region 1 is inactive. Scale bar equals 1.0 μm.

Like 5S RNA, the accumulated tRNA accounts for about 45% of the total RNA of previtellogenic oocytes, and together they account for the vast majority of the RNA in these oocytes. The stored tRNA is released into the cytoplasm at vitellogenesis. The 42S storage particles are no longer detectable during this stage (Denis and Mairy, 1972).

Oocyte Protein Synthesis

The utilization of oocyte messengers for protein synthesis during oogenesis is a topic that has received far too little attention from investigators. As a result, we have only very sketchy information. Much of the information we do have has been obtained from work on *Xenopus laevis*. Since we concentrated on *Xenopus* during the preceding discussions on

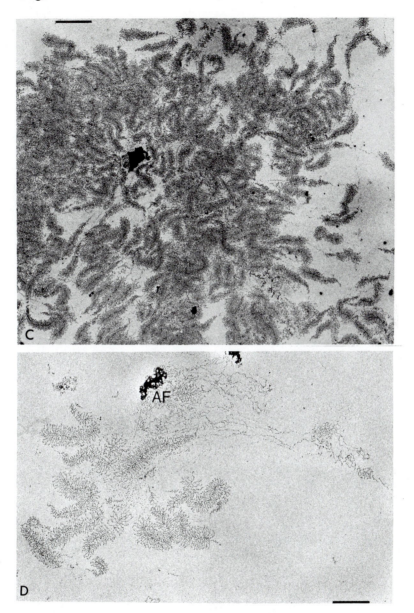

FIGURE 7.32 *(Continued)* C, Vitellogenic pattern. A typical nucleolar transcriptional pattern is seen. Virtually all of the chromatin has maximally active
matrix units alternating with untranscribed spacers. Scale bar equals 2.0 μm.
D, Postvitellogenic pattern. A large part of the chromatin is transcriptionally
silent and has a tendency to form fibrillar aggregates. Some regions of low
transcriptional activity and maximally active matrix units are seen. Scale bar
equals 1.0 μm. (From U. Scheer, M.F. Trendelenburg, and W.W. Franke.
1976b. Reproduced from *The Journal of Cell Biology,* 1976, vol. 69, pp.
465–489 by copyright permission of The Rockefeller University Press.)

RNA synthesis, we shall also center this discussion on *Xenopus*, which will help round out our understanding of the utilization of genomic information during oogenesis in this species.

As differentiation of the *Xenopus* oocytes proceeds, the pattern of protein synthesis changes. These patterns have been determined by electrophoretic analyses of the proteins synthesized at various stages. As shown in Figure 7.33, the major changes occur during vitellogenesis, with little apparent qualitative change afterward. The dynamic pattern of protein synthesis contrasts sharply with the apparently constant messenger RNA population. On the surface, these results suggest some sort of post-transcriptional regulation of gene expression. However, much more work needs to be done to characterize the messenger population more completely and to attempt to correlate synthesis and accumulation of *specific* messengers with their translation and the accumulation of translational products. One difficulty in obtaining these kinds of data is that oocytes make a wide variety of proteins in small amounts. These undoubtedly include typical metabolic enzymes and structural proteins.

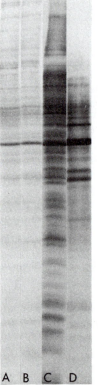

FIGURE 7.33 Products of oocyte protein synthesis. Fragments of ovary were incubated in culture medium containing ^{35}S-methionine. Oocytes of different stages were prepared and the proteins analyzed by polyacrylamide electrophoresis. The gel was then prepared for autoradiography. The autoradiographs are shown here. *A*, Postvitellogenic oocytes. *B* and *C*, mid- and early vitellogenic oocytes, respectively. *D*, Previtellogenic oocytes. (From C. Darnbrough and P.J. Ford. 1976. Cell-free translation of messenger RNA from oocytes of *Xenopus laevis*. Dev. Biol., *50:* 296.)

A   B   C   D

Although oocytes, unlike differentiating reticulocytes or oviducal gland cells, are not committed to the production of a few characteristic proteins, some proteins are made in relatively large amounts (e.g., ribosomal proteins, 5S RNA regulatory protein, and histones). These proteins could be valuable tools for studying gene expression during oogenesis, particularly if transcription of their structural genes, accumulation of their messengers, and translation of the messengers into the proteins are traced during the various stages of oogenesis. This type of analysis has yielded some information on histone gene expression, as we have previously discussed.

7–6. GENE EXPRESSION DURING OOGENESIS: NONAMPHIBIAN SPECIES

In this section we shall expand the discussion of gene expression during oogenesis by examining this process in nonamphibian species. We shall not attempt a general survey of oogenesis in the animal kingdom. Instead, we shall concentrate on two examples: meroistic insects and sea urchins. The meroistic insects have been chosen to illustrate the role of nurse cells in macromolecular synthesis. The sea urchins are significant because of the rapidly expanding literature on the utilization of oogenic transcripts during embryonic development.

Meroistic Insects

As we discussed previously, oocyte chromosomes of meroistic insects do not go through a lampbrush phase. In fact, the oocyte nucleus is of diminished functional significance in these insects, and the nurse cell nuclei assume a major role in production of oocyte cytoplasmic RNA. Most of the RNA transported to the oocyte from its nurse cells is ribosomal RNA that enters the oocyte in the form of ribosomes (Davidson, 1976). Recently obtained evidence from work on giant silkmoths indicates that polyadenylated RNA (i.e., putative mRNA) and 4S RNA are also synthesized in the nurse cells and transported into the oocytes (Paglia et al., 1976).

In contrast to those organisms with lampbrush-type oogenesis, the whole process of oogenesis in meroistic insects is extremely rapid. For example, in the cricket (which has lampbrush chromosomes) oogenesis lasts 100 days, whereas in *Drosophila* (which has nurse cells) oogenesis takes only about eight days. Since the time allotted for RNA synthesis in meroistic insects is so short, the requirement for extensive RNA synthesis during oogenesis appears to be met by a dramatic increase in ploidy (see section 7–2). Since there are 15 nurse cells associated with each oocyte, the equivalent of several thousand genomes apparently satisfies the needs of the oocyte cytoplasm for RNA in a very short period of time.

Sea Urchins

A great deal of information about the messenger RNA population of sea urchin eggs and embryos has recently been obtained by studies with RNA-DNA hybridization technology. Single copy DNA is used as a probe for mRNA, since most messengers are transcribed from this DNA. These hybridization studies yield valuable qualitative information about sequence complexity of RNA. They also allow comparisons to be made between RNA populations. Thus, changes in complexity can also be examined.

The sequence complexity of unfertilized sea urchin eggs is approximately 3.7×10^7 nucleotides, which is equivalent to 20,000 to 30,000 different structural gene transcripts (Galau et al., 1976; Hough-Evans et al., 1977). To determine whether sequence complexity changes during oogenesis, the single copy DNA from egg RNA-DNA hybrids has been used as a probe for the presence of homologous transcripts in the cytoplasmic RNA from previtellogenic oocytes (Hough-Evans et al., 1979). Reaction of the previtellogenic oocyte RNA with this DNA measures the fraction of the egg single copy gene set that is also represented in the cytoplasm of immature oocytes. Less than half (44%) of the egg single copy gene set is represented in previtellogenic oocyte cytoplasmic RNA. Thus, 56% of the messenger sequences are added to the cytoplasm during vitellogenesis. These results contrast with those obtained with *Xenopus* oocytes, in which messenger complexity does not increase during vitellogenesis. Thus, it would appear that the patterns of structural gene expression during oogenesis in the sea urchin and *Xenopus* differ dramatically.

Investigators have recently begun to analyze the synthesis of specific kinds of RNA during sea urchin oogenesis and to follow the utilization of this RNA by the cells. The synthesis of histone messengers has been detected by polyacrylamide gel analysis of ^{3}H-labeled oocyte RNA (Ruderman and Schmidt, 1981). Interestingly, the sea urchin oocyte

A   B   C

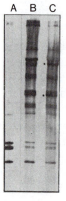

FIGURE 7.34 Polyacrylamide gel autoradiographs showing newly synthesized RNAs from sea urchin oocytes. RNA is labeled with ^{3}H-uridine. *A*, Marker histone messengers. *B*, Polysomal RNA from oocytes. *C*, Nonribosome–associated RNA from oocytes. In addition to histone messengers oocytes contain heterogeneous, high molecular weight RNA of undetermined nature and mitochondrial RNA (dots). (From J.V. Ruderman and M.R. Schmidt. 1981. RNA transcription and translation in sea urchin oocytes and eggs. Dev. Biol., *81*: 226.)

histone messengers are nonpolyadenylated, a result that contrasts with that obtained for *Xenopus* histone messengers (see section 7–5). The fate of the newly synthesized histone mRNA has been studied by comparing polysomal and nonribosome-associated labeled RNA from oocyte cytoplasm: Histone messengers are found in both fractions (Fig. 7.34). Thus, a portion of the messenger appears to be actively translated in oocytes. This result supports earlier experiments in which histone synthesis was detected in sea urchin oocytes (Cognetti et al., 1974, 1977).

7–7. UTILIZATION OF GENOMIC INFORMATION IN OOGENESIS: CONCLUSIONS

Like any differentiation process, oogenesis involves differential utilization of genomic information. The completed oocyte is composed of numerous specialized components that enable it to function in fertilization, as well as components that will be utilized during postfertilization development. Consequently, gene expression during oogenesis has both immediate and long-term implications for reproduction. A portion of the transcribed information directs the synthesis of oocyte protein, whereas the remainder is retained after fertilization and has profound effects upon early postfertilization development (see Chap. 12). Because of this dichotomy, the full-grown oocyte can be seen as the *result* of a complex differentiation process or an *interim stage* in a continuing process.

The oocyte is unusual in another fundamental way: Its proteins may be either synthesized endogenously or incorporated within the oocyte from exogenous sources. In some cases (as in meroistic insects), even the oocyte organelles may be synthesized exogenously. Therefore, the oocyte may be a product of the coordinated expression of genes in a number of cells; the overall control of these processes is exerted by the hormones that control the reproductive cycle.

7–8. CONVERSION OF THE OOCYTE TO THE EGG— MATURATION AND OVULATION

Differentiation of the female gamete occurs while the cell nucleus remains in a prolonged meiotic prophase. The completion of meiosis is essential for the production of the haploid female pronucleus that associates with the male pronucleus at fertilization. The timing of meiotic completion is quite variable among animals (Table 7–1). The most familiar pattern (because it occurs in humans and nearly all other vertebrates) is the one in which resumption of meiosis is initiated at ovulation and proceeds as far as metaphase of meiosis II. The egg remains at this stage until fertilization, which triggers the completion of meiosis II.

The phase between the initial resumption of meiosis and metaphase II is the period of oocyte maturation. The most conspicuous events of maturation are the dissolution of the germinal vesicle (which indicates

TABLE 7–1
Stage of Egg Maturation at which Sperm Penetration Occurs in Different Animals

Young primary oocyte	Fully grown primary oocyte	First metaphase	Second metaphase	Female pronucleus
The annulate worm *Dinophilus*	The round worm *Ascaris*	The nemertine worm *Cerebratulus*	The lancelet *Amphioxus*	Coelenterates, e.g., anemones
The polychaete worm *Histriobdella*	The mesozoan *Dicyema*	The polychaete worm *Chaetopterus*	The amphibians *Siredon, Rana, Xenopus*	Echinoids, e.g., sea urchins
The flatworm *Otomesostoma*	The sponge *Grantia*	The mollusk *Dentalium*	Most mammals	
The onychophoran *Peripatopsis*	The polychaete worm *Myzostoma*	The cone worm *Pectinaria*		
The annulate worm *Saccocirrus*	The clam worm *Nereis*	Many insects		
	The clam *Spisula*			
	The echiuran worm *Thalassema*			
	Dog and fox			

Modified from Austin, C.R. *Fertilization.* © 1965, p. 87. Reprinted by permission of Prentice-Hall, Inc. Englewood Cliffs, N.J.

the end of prophase I), condensation of the chromosomes, formation of a spindle, and production of the first polar body (which marks the completion of meiosis I). Thereafter, the chromosomes realign on the second metaphase spindle, where they remain until fertilization. However, maturation is much more than merely the reinitiation of meiosis; it includes numerous *physiological changes* that are necessary for fertilization to occur.

Regulation of Maturation and Ovulation

The most extensive investigations of the events of vertebrate oocyte maturation have been conducted with the amphibians *Rana pipiens* and *Xenopus laevis*, primarily because maturation and ovulation in these species can be readily induced in the laboratory and because the large size of their eggs facilitates experimental investigation. It has long been the practice to induce ovulation in amphibians by injecting gonadotropins into gravid females. Since gonadotropins also cause oocyte maturation, it has been recognized for some time that both of these events are under hormonal control. Steroid hormones are also effective inducers of maturation, indicating some sort of interaction between the gonadotropins and steroids.

Detailed investigations of the precise hormonal regulation of maturation became possible when it was discovered that maturation can be induced *in vitro* by adding either gonadotropin or the steroid hormone progesterone to culture medium containing ovarian fragments. However, gonadotropin is ineffective in inducing maturation if the oocytes are separated from their follicles, whereas addition of either ovarian tissue fragments or progesterone to the culture medium will promote maturation. These observations led to the conclusions that the follicle cells are the targets of gonadotropin action and that they produce progesterone, which stimulates the oocytes to undergo maturation (Masui, 1967; Schuetz, 1967; Smith et al., 1968).

The effects of progesterone on the oocyte in bringing about maturation have been carefully documented. One surprising result is the finding that progesterone apparently acts on the oocyte surface to trigger maturation. This conclusion was drawn from attempts to induce maturation by microinjecting progesterone directly into oocytes, a treatment that invariably fails to promote maturation, although these oocytes can subsequently be induced to mature by bathing them in the hormone (Smith and Ecker, 1971; Masui and Markert, 1971). The surface action of progesterone is also supported by the observation that oocytes will mature when exposed to a progesterone analog that is covalently linked to a large polymer that does not enter the cells (Godeau et al., 1978). The interaction of the hormone with the oocyte surface differs from the usual interaction between steroid hormones and cytoplasmic receptors (see Chap. 4).

There is another significant difference between the action of progesterone in inducing oocyte maturation and the classical mechanism of steroid hormone action on somatic cells. In the latter situation the effects of steroid hormones are mediated at the level of transcription. However, in oocytes, inhibitors of RNA synthesis fail to prevent the progesterone effect. On the other hand, maturation is prevented by inhibition of protein synthesis, an indication that the induction of maturation is mediated by the cytoplasm rather than by the nucleus—apparently by components synthesized during oogenesis, stored in the cytoplasm, and activated in response to the hormone. These components must include messenger RNA, ribosomes, and other elements of the protein synthetic machinery (Wasserman and Smith, 1978).

The unique mode of progesterone action on oocytes has focused attention on possible changes in the oocyte surface that could mediate the progesterone effect. Measurements of membrane potentials indicate that progesterone triggers an electrical depolarization of the oocyte membrane (Moreau et al., 1976; Wallace and Steinhardt, 1977). Changes in potential could result from changes in oocyte membrane permeability to ions (O'Connor et al., 1977, Morrill and Ziegler, 1980).

We shall now examine the roles of a number of cellular factors in maturation in an attempt to reconstruct the events that are triggered by exposure of oocytes to progesterone.

THE ROLE OF CALCIUM

The effects of progesterone appear to be mediated by ionic calcium. A role for calcium in maturation is suggested by numerous experiments in which calcium reduction in the medium leads to inhibition of maturation, or in which elevated calcium initiates maturation (for review, see Masui and Clarke [1979]; Maller and Krebs [1980]). Evidence has been accumulating that a transient increase in *ionic* calcium is a necessary event in the promotion of maturation by progesterone.

Direct evidence for the increase in Ca^{++} has been made possible through a technique that utilizes aequorin, a protein extracted from jellyfish, which luminesces in the presence of Ca^{++}. The aequorin must be injected into the oocyte, where it can interact with liberated Ca^{++}. Ordinary *Xenopus* oocytes are pigmented. Hence, the aequorin luminescence is partially masked by the pigment. Wasserman et al. (1980) and Moreau et al. (1980) utilized albino *Xenopus* oocytes, in which aequorin luminescence could be more easily detected and electronically measured. Figure 7.35 shows the measured luminescence of a progesterone-treated oocyte. The progesterone causes a rapid increase in luminescence. Light emission returns to the resting value within a few minutes. This result must reflect the release of bound calcium and a subsequent sequestering of the ion into the bound form once again. Internal free Ca^{++} levels have also been monitored by using Ca^{++}-sensitive electrodes, with similar results: Free Ca^{++} increases for a time and then declines.

Recent evidence suggests that the increase in calcium activity may cause maturation via the mediation of the calcium-binding protein, cal-

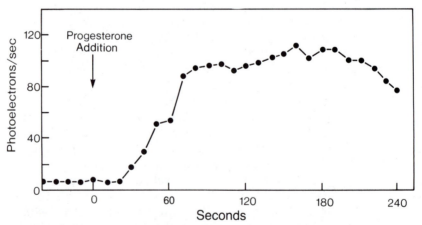

FIGURE 7.35 Aequorin luminescence from progesterone-stimulated albino *Xenopus laevis* oocyte. Oocyte was injected with aequorin prior to progesterone addition. (After W.J. Wasserman et al. 1980. Progesterone induces a rapid increase in $[Ca^{2+}]_{in}$ of *Xenopus laevis* oocytes. Proc. Natl. Acad. Sci. U.S.A., *77*: 1535.)

modulin. This protein is a ubiquitous calcium-dependent regulator that is found in virtually all eukaryotic cells. Calcium causes conformational changes in calmodulin, which can alter its biological function. Oocyte calmodulin bound to calcium will trigger maturation when injected into oocytes, whereas calmodulin or Ca^{++} alone has no effect (Maller and Krebs, 1980; Wasserman and Smith, 1981). Thus, the Ca^{++}-calmodulin complex can substitute for progesterone. These results suggest that the free calcium released under progesterone stimulation binds to the calmodulin and stimulates it to cause maturation.

THE ROLE OF CYCLIC AMP

Cyclic AMP is known to be a "second messenger" that mediates the effects of surface-acting hormones in many cell types. Cyclic AMP activates **protein kinase** enzymes, which in turn phosphorylate proteins. The phosphorylated proteins then perform particular cell functions that cannot be performed by their nonphosphorylated precursors. Evidence concerning a role for cyclic AMP in maturation comes from experiments in which maturation is blocked by inhibitors of **phosphodiesterase,** the enzyme responsible for breaking down cyclic AMP (O'Connor and Smith, 1976; Bravo et al., 1978). Thus, prolonging the life of cyclic AMP in the oocyte prevents maturation. Cyclic AMP would appear to be a *negative* effector of maturation; that is, a reduction in cyclic AMP levels is a prerequisite for maturation.

Recent studies have confirmed that cyclic AMP levels are reduced as a result of progesterone exposure. Figure 7.36 shows the effects of progesterone on cyclic AMP levels in *Xenopus* oocytes. Note that cyclic AMP levels drop rapidly within the first minute after progesterone exposure. This rapid drop is followed by a return to the initial, or basal, level of cyclic AMP. The decrease in cyclic AMP levels is due to an imbalance between the synthesis and degradation of cyclic AMP; that is,

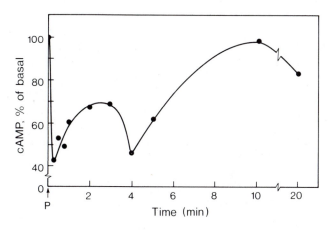

FIGURE 7.36 Effect of progesterone on cyclic AMP levels in *Xenopus laevis* oocytes. The time of addition of progesterone is indicated by an arrow. Cyclic AMP levels are represented as percentages of the initial, or basal, level. (After J.L. Maller, F.R. Butcher, and E.G. Krebs. 1979. Early effect of progesterone on levels of cyclic adenosine 3':5'-monophosphate in *Xenopus* oocytes. J. Biol. Chem., *254:* 580.)

degradation exceeds synthesis. Cyclic AMP is synthesized from ATP in a reaction catalyzed by the enzyme **adenylate cyclase.** Recent evidence indicates that progesterone causes a transient decline in adenylate cyclase activity (Baltus et al., 1981). Another means for reducing cyclic AMP levels is an increase in the activity of phosphodiesterase. Additional research is necessary to determine the relative importance of the synthetic and degradative processes in causing the decline in cyclic AMP levels that leads to maturation.

The correlation between the drop in cyclic AMP levels and the initiation of maturation implies that the prophase arrest that prevails in the immature oocyte is maintained by phosphoproteins that are produced by the cyclic AMP-dependent protein kinase. The decrease in cyclic AMP levels would reduce protein kinase activity and end the prophase arrest.

Because of the importance of the protein kinase in this scheme, we should examine this enzyme and its relationship with cyclic AMP. Protein kinase has two kinds of subunits: regulatory and catalytic. When the subunits are associated as the holoenzyme complex it is enzymatically inactive. As its name implies, the catalytic subunit is responsible for promoting phosphorylation. Cyclic AMP binds to the regulatory subunit of the complex, causing it to dissociate from the catalytic subunit, which then becomes enzymatically active. If purified regulatory subunits are injected into oocytes, they initiate maturation in the absence of progesterone (Fig. 7.37A). Presumably they bind to free catalytic subunits, thus rendering them inactive. The resulting reduction in phosphorylation would allow maturation to begin. The same effect has been obtained by injecting a naturally occurring inhibitor of catalytic subunits. Conversely, progesterone-induced maturation can be inhibited by injection of catalytic subunits (Fig. 7.37B). Apparently, the phosphorylation caused by the injected subunits is sufficient to maintain the prophase block.

As we discussed earlier, protein synthesis is necessary for progesterone induction of maturation. Synthesis of oocyte proteins is inhibited by injection of the catalytic subunit of protein kinase. This result suggests that the phosphorylation of proteins by the cyclic AMP-dependent kinase inhibits protein synthesis, perhaps by directly affecting the protein synthetic machinery of the oocyte (Maller and Krebs, 1980). Likewise, release of the kinase-mediated inhibition may be necessary to allow protein synthesis to proceed. This interpretation is also supported by the following experiment. Cycloheximide (an inhibitor of protein synthesis—see Chap. 3) added before injection of the inhibitor of protein kinase catalytic subunits blocks the ability of the inhibitor to induce maturation (Fig. 7.37C). Thus, protein synthesis must normally occur concurrently with, or subsequent to, kinase inhibition.

An apparent contradiction of the foregoing discussion on cyclic AMP is the observation that progesterone exposure causes an increase in

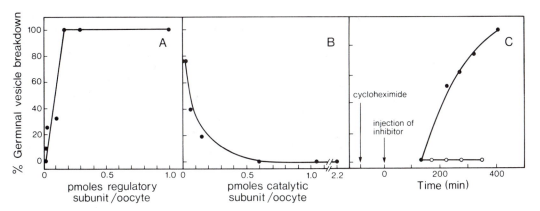

FIGURE 7.37 Experiments illustrating the role of protein kinase in maturation. *A*, Induction of maturation by protein kinase regulatory subunit complexed to cyclic AMP. The regulatory subunit-cyclic AMP complex was diluted as indicated and injected into the oocytes. Maturation was scored by germinal vesicle breakdown. *B*, Inhibition of maturation by catalytic subunit of protein kinase. *C*, Effect of cycloheximide on maturation induced by injection of inhibitor of protein kinase. Cycloheximide was added prior to injection of the inhibitor. Solid circles indicate percentage of germinal vesicle breakdown induced by inhibitor alone. Open circles indicate response in oocytes pretreated with cycloheximide. (*A, B* after J.L. Maller and E.G. Krebs. 1977. Progesterone-stimulated meiotic cell division in *Xenopus* oocytes. Induction by regulatory subunit and inhibition by catalytic subunit of adenosine 3':5'-monophosphate-dependent protein kinase. J. Biol. Chem., *252:* 1714, 1715. *C* after J.L. Maller and E.G. Krebs. 1980. Regulation of oocyte maturation. Curr. Topics Cell. Reg., *16:* 289.)

phosphorylation of proteins (Morrill and Murphy, 1972; Wallace, 1974; Maller et al., 1977). However, current evidence suggests that this phosphorylation is mediated by a cyclic AMP-*independent* protein kinase. Interestingly, this kinase is dependent upon the presence of ionic calcium and calmodulin for maximal activity (Wasserman and Smith, 1981). The activation of this kinase by Ca^{++}-calmodulin may be an essential consequence of the release of ionic calcium that we discussed in the preceding section.

Phosphorylation of proteins is a rather late event, occurring several hours after progesterone exposure in *Xenopus* and shortly before the breakdown of the germinal vesicle (Maller et al., 1977). This burst of phosphorylation is thought to be related to the function of the **maturation promotion factor,** which we shall discuss next.

THE ROLE OF MATURATION PROMOTION FACTOR

Progesterone treatment of oocytes leads to the appearance in the cytoplasm of a "factor" that mediates the hormonal effects on the oocyte. This maturation promotion factor (MPF) can be demonstrated by injection of cytoplasm from progesterone-treated eggs into untreated recipients, a procedure that causes the recipient oocytes to undergo maturation (Fig. 7.38). Repeated serial transfers of cytoplasm from maturing recipi-

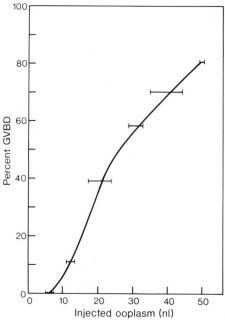

FIGURE 7.38 The graph reveals that the frequency of oocyte maturation is proportional to the volume of injected cytoplasm. The cytoplasm was taken from oocytes 18 to 20 hours after progesterone treatment. Abscissa: volume of injected cytoplasm. Ordinate: frequency of germinal vesicle breakdown (GVBD). (After Y. Masui and C.L. Markert. 1971. Cytoplasmic control of nuclear behavior during meiotic maturation of frog oocytes. J. Exp. Zool., *177:* 134.)

ent oocytes will continue to induce maturation in sequential recipients even though only the original donor oocyte was exposed to progesterone. This result indicates that injection of MPF triggers the production of still more MPF in an autocatalytic reaction.

The formation of MPF is entirely independent of the nucleus, since it can be produced in enucleated oocytes (Masui and Markert, 1971; Reynhout and Smith, 1974). This can be demonstrated by treating enucleated oocytes with progesterone and injecting their cytoplasm into nucleated nonprogesterone-treated oocytes; the injected cells mature, indicating the formation of MPF in the enucleated oocytes (Table 7–2).

Protein synthesis does appear to be necessary for the initial production of MPF but is not required for MPF action in promoting maturation. These conclusions stem from the observations that maturation of progesterone-treated oocytes can be inhibited with cycloheximide until the time of first appearance of MPF, but after this event, maturation occurs even in the presence of the inhibitor. Furthermore, cycloheximide fails to prevent the autocatalytic amplification of MPF (Wasserman and Masui, 1975).

The initial appearance of MPF and the development of resistance to cycloheximide occur at about two thirds of the normal time interval between application of progesterone and germinal vesicle breakdown. Thus, appearance of MPF is a relatively late event, occurring subsequent to the release of ionic calcium and the reduction in cyclic AMP levels, which causes a reduction in protein kinase activity, and is dependent upon these early events.

TABLE 7–2
Injection of Cytoplasm from Enucleated Oocytes Treated with Progesterone*

Volume (nl) injected	Enucleated donor		Nucleated donor	
	Number	G.V.B.D. (%)†	Number	G.V.B.D. (%)
5–6	18	0	12	0
11–13	34	18	54	11
19–24	49	20	54	32
35–36	40	58	45	76
50	18	67	21	71

*Injection was carried out 18 to 24 hours after progesterone treatment of donors.
†G.V.B.D.: Percentage of injected oocytes showing germinal vesicle breakdown.
From Masui, Y. and C.L. Markert. 1971. Cytoplasmic control of nuclear behavior during meiotic maturation of frog oocytes. J. Exp. Zool., *17*: 135.

The injection of cytoplasm from progesterone-treated oocytes into oocytes that are not exposed to the hormone provides an assay for the effects of MPF and may provide the means for discovering the mode of action of MPF in promoting maturation. A major effect of an injection of cytoplasm containing MPF is an immediate burst of protein phosphorylation in the recipient oocyte, which is mediated by a cyclic AMP-independent protein kinase (see preceding discussion). Like other aspects of MPF function, the phosphorylation is not inhibited by cycloheximide and is, therefore, independent of protein synthesis (Maller et al., 1977).

The nature of MPF is suggested by findings that MPF activity is terminated by heat, high levels of ionic calcium, or treatment with protease enzyme, and that the activity is dependent upon magnesium ions. These results indicate that MPF activity depends upon a protein whose activity is regulated by divalent cations (Wasserman and Masui, 1976). Recently, some success has been reported in attempts to purify and more precisely characterize MPF (Wu and Gerhart, 1980). One of the most interesting observations made on the MPF preparations is that their activity is enhanced more than twofold by ATP and that they have phosphorylating activity. One possible interpretation of these results is that MPF exists in the oocyte in an inactive nonphosphorylated form and that phosphorylation activates it. Thus, the burst of phosphorylation that follows MPF injection may reflect this step. Furthermore, phosphorylated MPF may itself be a kinase, which can phosphorylate even more MPF and, consequently, autocatalytically amplify its own activity.

Molecular Events in Amphibian Oocyte Maturation

The meiotic events of maturation are among a number of significant oocyte responses to hormonal stimulation, which include changes in the patterns of macromolecular synthesis. Although nuclear RNA synthesis continues in the interval between hormonal stimulation and germinal vesicle breakdown, this RNA is apparently of no immediate consequence to the maturation process. This conclusion is based on the finding that maturation occurs in oocytes treated with inhibitors of RNA synthesis and in oocytes that are enucleated prior to exposure to progesterone. The latter produce MPF and undergo an activation response similar to that of nucleated eggs (see Chap. 9), but they do not undergo normal cleavage. Conversely, inhibition of protein synthesis prevents maturation, an indication that the induction of maturation is mediated by the cytoplasm— apparently by components synthesized during oogenesis, stored in the cytoplasm, and activated by the hormone.

Because of the significance of protein synthesis in maturation, a number of studies have attempted to characterize it. Hormonal stimulation apparently results in both quantitative and qualitative changes in protein synthesis. There is a generalized twofold to fourfold increase in the rate of protein synthesis after hormonal stimulation. Furthermore, this increase occurs in enucleated oocytes, underscoring the fact that protein synthesis during maturation is independent of nuclear regulation. Qualitative changes in protein synthesis following hormonal stimulation are poorly characterized. However, there is some evidence of definite changes in the *kinds* of protein synthesized after progesterone exposure. Electrophoretic studies have shown that a small number of new proteins are synthesized by *Xenopus* oocytes in response to progesterone treatment, but the identity of these proteins is not known (Wasserman and Smith, 1978). We have already discussed the evidence that protein synthesis is essential for maturation. Obviously, synthesis of one or more specific proteins is essential for this activity, but again the identity of the proteins is not known.

One specific effect of progesterone exposure on protein synthesis in *Xenopus* oocytes is known—a dramatic preferential stimulation of histone synthesis. This increase averages sixteenfold or more, as compared to the twofold to fourfold increase in general protein synthesis. Furthermore, an increase in histone synthesis also occurs in eggs enucleated prior to progesterone exposure (Adamson and Woodland, 1977). It is clear from these results that the hormone treatment promotes a preferential translation of certain kinds of stored messenger (including histone messenger). This histone synthesis apparently enlarges the store of histones in the egg, which are to be used for the formation of chromatin during the rapid mitotic divisions that follow fertilization.

Histone synthesis by mature *Xenopus* eggs occurs on histone messengers that are *nonpolyadenylated* (Ruderman et al., 1979). This contrasts with histone synthesis in oocytes, which occurs predominantly on polyadenylated histone messengers (see section 7–5). Since egg histone synthesis utilizes stored messengers, we must conclude that either the oocyte polyadenylated RNA is preferentially degraded or it is deadenylated during maturation.

The Consequences of Germinal Vesicle Breakdown

The significance of germinal vesicle breakdown and the resultant mixing of nucleoplasm and cytoplasm is apparent from experiments in which diploid blastula nuclei are transplanted into *Rana pipiens* eggs after progesterone treatment. When blastula nuclei are transplanted into eggs enucleated *after germinal vesicle breakdown*, a high percentage of the recipients will cleave and continue development. However, if blastula nuclei are transplanted into eggs enucleated *before progesterone treatment*, none of them will undergo genuine cleavage after activation. Injection of germinal vesicle material into enucleated eggs prior to nuclear transfer can restore the ability of recipient eggs to cleave. Therefore, the cleavage restriction results solely from the absence of germinal vesicle components in the cytoplasm. Furthermore, since the germinal vesicle material used to restore the cleavage capability comes from unstimulated oocytes, the "cleavage factor" apparently preexists in the nucleus and is not synthesized as a result of hormonal stimulation (Smith and Ecker, 1969).

The events of maturation provide the egg with the physiological capability to become fertilized and initiate development. The acquisition of physiological maturity involves the utilization of information stored in the oocyte, by either translation or shifting components from nucleus to cytoplasm. As we shall see in Chapter 12, the tapping of this informational store during maturation samples only a small amount of a very large pool of information, since the mature egg contains the complete program and machinery for cleavage—poised for utilization. Only the stimulus of fertilization is required to activate the program and utilize the machinery.

Ovulation

The mechanism of oocyte release from the mature follicle is poorly understood. One possibility is that certain follicle cells contract, causing the follicle to burst and release the oocyte. Another possibility is that the follicle ruptures because of degenerative changes that weaken its wall. The weakened wall ruptures, and the egg passively flows out of the follicle. In mammals the weakening of the follicle is facilitated by a buildup of fluid just prior to ovulation. The fluid distends the follicle, causing it to rupture. Further work is necessary to determine whether the egg is passively or actively expelled from the follicle at ovulation. It is entirely possible that ovulation is active in some species and passive in others. Some species may use a combination of these mechanisms: Degenerative changes may first weaken the follicle wall, which ruptures and allows the egg to be expelled by cellular contractions.

7–9. EGG ENVELOPES

Ovulated eggs are normally covered with some sort of noncellular egg envelopes. Egg envelopes are divided into three categories, based on their source: (1) **primary envelopes,** which are produced by the oocyte itself during oogenesis, (2) **secondary envelopes,** which are produced by the follicle cells that surround the oocyte, and (3) **tertiary envelopes,** which are added as the egg passes through the reproductive tract after ovulation. Unfortunately, the specific names given to egg envelopes cause a great deal of confusion, since the same names are often given to envelopes of entirely different origin.

Examples of primary envelopes include the chorion of the fish egg, the jelly coat of the echinoderm egg, and the vitelline envelope of the amphibian egg. The latter consists of glycoproteins (Wolf et al., 1976) and is formed as an acellular layer between the oocyte and the follicle cells during oogenesis. Because the oocyte needs to maintain functional contact with the follicle cells, numerous processes from the follicle cells make contact with the oocyte surface. During maturation, the cytoplasmic processes and oocyte microvilli are withdrawn, and the vitelline envelope becomes a layer of constant thickness around the egg. The withdrawal of the follicle cell processes from the vitelline envelope is preparatory to ovulation, when the egg, with the vitelline envelope intact, is released from the surrounding follicle cells (Schuetz, 1972).

The morphology of the zona pellucida of the mammalian oocyte is very similar to the vitelline envelope of the amphibian oocyte (compare Figs. 7.4 and 7.12). However, the origin of the mammalian egg envelope is not certain. Some investigators consider it to be a primary envelope, whereas others assume that it is a secondary envelope. Recent evidence favors an oocyte origin for the zona material (Bleil and Wassarman, 1980; Bousquet et al., 1981). Withdrawal of cellular processes during mammalian maturation precedes ovulation. However, unlike in the amphibian, some of the surrounding follicle cells do not detach from the zona during maturation but remain associated with the egg to form a structure called the cumulus oophorus. This consists of an outer region containing a loose aggregate of cells and a single layer of elongated cells with fine processes that radiate toward the egg, which is called the **corona radiata,** lining the outer surface of the zona pellucida (Fig. 7.39). The cumulus cells are eventually sloughed from the zona, but in many species they are still present at fertilization.

Examples of secondary envelopes are the vitelline envelope and the chorion that surround insect eggs. Tertiary envelopes include such structures as the amphibian jelly coat and the albumin and shells of reptilian and avian eggs.

Egg envelopes serve various functions, such as protection and nutrition. They often provide formidable barriers through which sperm must penetrate. As we learned in Chapter 6, some types of sperm have

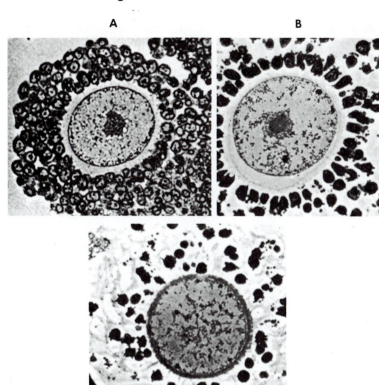

FIGURE 7.39 Rat cumulus-oocyte complex at various stages. *A*, Preovulatory complex in a follicle. *B*, Preovulatory complex at a later stage. The cumulus is reduced to a single layered corona radiata. *C*, Postovulatory ovum. Some corona cells remain attached to the zona, but most have dissociated. ×420. (From N.B. Gilula, M.L. Epstein, and W.H. Beers. 1978. Reproduced from *The Journal of Cell Biology*, 1978, vol. 78, pp. 58–75 by copyright permission of The Rockefeller University Press.)

evolved specializations that allow them to penetrate the envelopes and make contact with the egg at fertilization. Besides being barriers to sperm entry, the envelopes may also serve positive functions in fertilization (see Chap. 9).

7–10. THE FERTILIZABLE EGG—JUST THE BEGINNING

The mature egg is the product of one of the most intricate developmental processes known. The constituents of the egg have both immediate and long-term implications for reproduction. As a differentiated cell, the egg contains specialized proteins and organelles that determine its functional capabilities in fertilization. But, unlike other differentiated cells, fulfillment of its unique role is not a terminal function, but the initial step that leads to formation of a complete new individual.

The roles of egg constituents in development are far reaching. Stored yolk fulfills the nutritional requirements of the embryo until it develops its own mechanism for acquiring nutrients. The egg cytoplasm contains vast amounts of messenger RNA as well as ribosomes, tRNA, and all the other elements that are necessary for protein synthesis. These constituents not only enable the embryo to begin immediate protein synthesis once development is activated but even determine *which* proteins will be made, until the embryonic nuclei assume control over development. Furthermore, the *organization of egg constituents* has significant implications for the *organization of the embryo.* The egg is a highly ordered cell, and the location of individual components may determine the sites of specific embryonic regions. The implications of egg constituents for development are immense and will become clear in Part Four, when we discuss early embryonic development in detail.

REFERENCES

Adamson, E.D., and H.R. Woodland. 1977. Changes in the rate of histone synthesis during oocyte maturation and very early development of *Xenopus laevis.* Dev. Biol., *57:* 136–149.

Al-Mukhtar, K.A.K. 1970. Oogenesis in amphibia with special reference to the formation, replication and derivatives of mitochondria. Ph.D. thesis. Southhampton University, England.

Al-Mukhtar, K.A.K., and A.C. Webb. 1971. An ultrastructural study of primordial germ cells, oogonia and early oocytes in *Xenopus laevis.* J. Embryol. Exp. Morphol., *26:* 195–217.

Anderson, D.M., and L.D. Smith. 1977. Synthesis of heterogeneous nuclear RNA in full-grown oocytes of *Xenopus laevis* (Daudin). Cell, *11:* 663–671.

Anderson, E. 1968. Oocyte differentiation in the sea urchin, *Arbacia punctulata,* with particular reference to the origin of cortical granules and their participation in the cortical reaction. J. Cell Biol., *37:* 514–539.

Anderson, E., and D.F. Albertini. 1976. Gap junctions between the oocyte and companion follicle cells in the mammalian ovary. J. Cell Biol., *71:* 680–686.

Anderson, L.M., and W.H. Telfer. 1969. A follicle cell contribution to the yolk spheres of moth oocytes. Tissue Cell, *1:* 633–644.

Angelier, N., and J.C. Lacroix. 1975. Complexes de transcription d'origines nucléolaire et chromosomique d'ovocytes de *Pleurodeles waltlii* et *P. poireti* (Amphibiens, Urodèles). Chromosoma, *51:* 323–335.

Austin, C.R. 1965. *Fertilization.* Prentice-Hall, Inc. Englewood Cliffs, N.J.

Balinsky, B.I. 1975. *An Introduction to Embryology,* 4th ed. W.B. Saunders Co., Philadelphia.

Balinsky, B.I. 1981. *An Introduction to Embryology,* 5th ed. Saunders College Publishing, Philadelphia.

Balinsky, B.I., and R.J. Devis. 1963. Origin and differentiation of cytoplasmic structures in the oocytes of *Xenopus laevis.* Acta Embryol. Morphol. Exp., 6:5–108.

Baltus, E., J. Hanocq-Quertier, and M. Guyaux. 1981. Adenylate cyclase and cyclic AMP-phosphodiesterase activities during the early phase of maturation in *Xenopus laevis* oocytes. FEBS Lett., *123:* 37–40.

Bast, R.E., and W.H. Telfer. 1976. Follicle cell protein synthesis and its contribution to the yolk of the *Cecropia* moth oocyte. Dev. Biol., *52:* 83–97.

Bier, K. 1963. Autoradiographische Unterschungen über die Leistugen des Follikelepethels und der Nährzellen bei der Dotterbildung und Eiwiessynthes im Fliegenover. Wilhelm Roux' Archiv f. Entwicklungsmechanik, *154*: 552–575.

Billett, F.S., and E. Adam. 1976. The structure of the mitochondrial cloud of *Xenopus laevis* oocytes. J. Embryol. Exp. Morphol., *33*: 697–710.

Bleil, J.D., and P.M. Wassarman. 1980. Synthesis of zona pellucida proteins by denuded and follicle-enclosed mouse oocytes during culture *in vitro.* Proc. Natl. Acad. Sci. U.S.A., *77*: 1029–1033.

Bona, M., U. Scheer, and E.K.F. Bautz. 1981. Antibodies to RNA polymerase II (B) inhibit transcription in lampbrush chromosomes after microinjection into living amphibian oocytes. J. Mol. Biol., *151*: 81–99.

Bousquet, D. et al. 1981. The cellular origin of the zona pellucida antigen in the human and hamster. J. Exp. Zool., *215*: 215–218.

Brachet, J. 1974. *Introduction to Molecular Embryology.* Springer-Verlag, New York.

Brachet, J. 1979. Oogenesis and maturation in amphibian oocytes. Endeavour, N.S., *3*: 144–149.

Bravo, R. et al. 1978. Amphibian oocyte maturation and protein synthesis: Related inhibition by cyclic AMP, theophylline, and papaverine. Proc. Natl. Acad. Sci. U.S.A., *75*: 1242–1246.

Brown, D.D., and I.B. Dawid. 1968. Specific gene amplification in oocytes. Science, *160*: 272–280.

Brown, D.D., and K. Sugimoto. 1973. 5S DNAs of *Xenopus laevis* and *Xenopus mulleri:* Evolution of a gene family. J. Mol. Biol., *48*: 397–415.

Brummett, A.R., and J.N. Dumont. 1977. Intracellular transport of vitellogenin in *Xenopus* oocytes: An autoradiographic study. Dev. Biol., *60*: 482–486.

Callan, H.G. 1963. The nature of lampbrush chromosomes. Int. Rev. Cytol., *15*: 1–34.

Callan, H.G., and R.W. Old. 1980. *In situ* hybridization to lampbrush chromosomes: A potential source of error exposed. J. Cell Sci., *41*: 115–123.

Carasso, N., and P. Favard. 1960. Vitellogenèse de la planorbe. Ultrastructure des plaquettes vitellines. *Fourth International Conference on Electron Microscopy.* Berlin, Vol. II. Berlin, Göttingen, Heidelberg, pp. 431–435.

Coggins, L.W. 1973. An ultrastructural and radioautographic study of early oogenesis in the toad *Xenopus laevis.* J. Cell Sci., *12*: 71–93.

Coggins, L.W., and J.G. Gall. 1972. The timing of meiosis and DNA synthesis during early oogenesis in the toad, *Xenopus laevis.* J. Cell Biol., *52*: 569–576.

Cognetti, G., G. Spinelli, and A. Vivoli. 1974. Synthesis of histones during sea urchin oogenesis. Biochim. Biophys. Acta., *349*: 447–455.

Cognetti, G. et al. 1977. Studies on protein synthesis during sea urchin oogenesis. Cell Differ., *5*: 283–291.

Cran, D.G., R.M. Moor, and M.F. Hay. 1980. Fine structure of the sheep oocyte during antral follicle development. J. Reprod. Fert., *59*: 125–132.

Cummings, M.R., and R.C. King. 1969. The cytology of the vitellogenic stages of oogenesis in *Drosophila melanogaster.* I. General staging characteristics. J. Morphol., *128*: 427–442.

Cummings, M.R., N.M. Brown, and R.C. King. 1971. The cytology of the vitellogenic stages of oogenesis in *Drosophila melanogaster.* III. Formation of the vitelline membrane. Zeitschrift f. Zellforsch., *118*: 482–492.

Darnbrough, C., and P.J. Ford. 1976. Cell-free translation of messenger RNA from oocytes of *Xenopus laevis*. Dev. Biol., *50*: 285–301.

Davidson, E.H. 1976. *Gene Activity in Early Development*, 2nd ed. Academic Press, Inc., New York.

Dawid, I.B. 1972. Cytoplasmic DNA. *In* J.D. Biggers and A.W. Schuetz (eds.), *Oogenesis*. University Park Press, Baltimore, pp. 215–226.

Dekel, N. et al. 1978. Cellular associations in the rat oocyte-cumulus cell complex: Morphology and ovulatory changes. Gamete Res., *1*:47–57.

Denis, H., and M. Mairy. 1972. Recherches biochimiques sur l'oogenèse. 1. Distribution intracellulaire du RNA dans les petits oocytes de *Xenopus laevis*. Eur. J. Biochem., *25*: 524–534.

Diaz, M.O. et al. 1981. Transcripts from both strands of a satellite DNA occur on lampbrush chromosome loops of the newt Notophthalmus. Cell, *24*: 649–659.

Dolecki, G.J., and L.D. Smith. 1979. Poly(A)$^+$ RNA metabolism during oogenesis in *Xenopus laevis*. Dev. Biol., *69*: 217–236.

Dumont, J.N. 1978. Oogenesis in *Xenopus laevis* (Daudin). VI. The route of injected tracer transport in the follicle and developing oocyte. J. Exp. Zool., *204*: 193–218.

Dumont, J.N., and A.R. Brummett. 1978. Oogenesis in *Xenopus laevis* (Daudin). V. Relationships between developing oocytes and their investing follicular tissues. J. Morphol., *155*: 73–97.

Eddy, E.M., and S. Ito. 1971. Fine structural and radioautographic observations on dense perinuclear cytoplasmic material in tadpole oocytes. J. Cell Biol., *49*: 90–108.

Edström, J.-E., and J.G. Gall. 1963. The base composition of ribonucleic acid in lampbrush chromosomes, nucleoli, nuclear sap, and cytoplasm of *Triturus* oocytes. J. Cell Biol., *19*: 279–284.

Eppig, J.J. 1977. Mouse oocyte development *in vitro* with various culture systems. Dev. Biol., *60*: 371–388.

Ford, P.J. 1971. Non-coordinated accumulation and synthesis of 5S ribonucleic acid by ovaries of *Xenopus laevis*. Nature (Lond.), *233*: 561–564.

Ford, P.J. 1976. Control of gene expression during differentiation and development. *In* C.F. Grahan and P.F. Wareing (eds.), *The Developmental Biology of Plants and Animals*. W.B. Saunders Co., Philadelphia. pp. 302–345.

Ford, P.J., and E.M. Southern. 1973. Different sequences for 5S RNA in kidney cells and ovaries of *Xenopus laevis*. Nature (New Biol.), *241*: 7–12.

Franchi, L.L., A.M. Mandl, and S. Zuckerman. 1962. The development of the ovary and the process of oogenesis. *In* S. Zuckerman, A.M. Mandl, and P. Eckstein (eds.), *The Ovary*, Vol. 1. Academic Press, Inc., New York, pp. 1–88.

Galau, G.A. et al. 1976. Structural gene sets active in embryos and adult tissues of the sea urchin. Cell, *7*: 487–505.

Gall, J.G. 1954. Lampbrush chromosomes from oocyte nuclei of the newt. J. Morphol., *94*: 283–351.

Gall, J.G. 1966. Techniques for the study of lampbrush chromosomes. *In* D.M. Prescott (ed.), *Methods in Cell Physiology*, Vol. 2. Academic Press, Inc., New York, pp. 37–60.

Gall, J.G. et al. 1981. Histone genes are located at the sphere loci of newt lampbrush chromosomes. Chromosoma, *84*: 159–171.

Gilula, N.B., M.L. Epstein, and W.H. Beers. 1978. Cell-to-cell communication and ovulation. A study of the cumulus-oocyte complex. J. Cell Biol., 78:58–75.

Godeau, J.F. et al. 1978. Induction of maturation in *Xenopus laevis* oocytes by a steroid linked to a polymer. Proc. Natl. Acad. Sci. U.S.A., *75*: 2353–2357.

Golden, L., U. Schafer, and M. Rosbash. 1980. Accumulation of individual pA$^+$ RNAs during oogenesis of Xenopus laevis. Cell, *22*: 835–844.

Halkka, L., and O. Halkka. 1977. Accumulation of gene products in the oocytes of the dragonfly *Cordulia aenea*. II. Induction of annulate lamellae within dense masses during diapause. J. Cell Sci., *26*: 217–228.

Hallberg, R.L., and D.C. Smith. 1975. Ribosomal protein synthesis in *Xenopus laevis* oocytes. Dev. Biol., *42*: 40–52.

Heller, D.T., D.M. Cahill, and R.M. Schultz. 1981. Biochemical studies of mammalian oogenesis: Metabolic cooperativity between granulosa cells and growing mouse oocytes. Dev. Biol., *84*: 455–464.

Highnam, K.C., and L. Hill. 1977. *The Comparative Endocrinology of the Invertebrates*, 2nd ed. Edward Arnold (Publishers) Ltd., London.

Honda, B.M., and R.G. Roeder. 1980. Association of a 5S gene transcription factor with 5S RNA and altered levels of the factor during cell differentiation. Cell, *22*: 119–126.

Hough-Evans, B. R. et al. 1977. Appearance and persistence of maternal RNA sequences in sea urchin development. Dev. Biol., *60*: 258–277.

Hough-Evans, B.R. et al. 1979. RNA complexity in developing sea urchin oocytes. Dev. Biol., *69*: 258-269.

Huebner, E. 1981. Oocyte-follicle cell interaction during normal oogenesis and atresia in an insect. J. Ultrastruct. Res., *74*: 95–104.

Johannsen, O.A., and F.H. Butt. 1941. *Embryology of Insects and Myriapods*. McGraw-Hill Book Co., New York.

Jowett, T., and J.H. Postlethwait. 1980. The regulation of yolk polypeptide synthesis in *Drosophila* ovaries and fat body by 20-hydroxyecdysone and a juvenile hormone analog. Dev. Biol., *80*: 225–234.

Kessel, R.G. 1968. Annulate lamellae. J. Ultrastruct. Res. (Suppl.), *10*: 1–82.

Kessel, R.G., and H.W. Beams. 1969. Annulate lamellae and "yolk nuclei" in oocytes of the dragonfly, *Libellula pulchella*. J. Cell Biol., *42*: 185–201.

Kessel, R.G., and L.R. Ganion. 1980. Electron microscopic and autoradiographic studies on vitellogenesis in *Necturus maculosus*. J. Morph., *164*: 215–233.

King, R.C. 1970. *Ovarian Development in* Drosophila melanogaster. Academic Press, Inc., New York.

King, R.C. 1972. *Drosophila* oogenesis and its genetic control. *In* J.D. Biggers and A.W. Scheutz (eds.), *Oogenesis*. University Park Press, Baltimore, pp. 253–275.

Klug, W.S., R.C. King, and J.M. Wattiaux. 1970. Oogenesis in the *suppressor*^{2} of *hairy-wing* mutant of *Drosophila melanogaster*. II. Nucleolar morphology and *in vitro* studies of RNA protein synthesis. J. Exp. Zool., *174*: 125–140.

Koch, E.A., P.A. Smith, and R.C. King. 1967. The division and differentiation of *Drosophila* cystocytes. J. Morphol., *121*: 55–70.

Kress, A., and U.M. Spornitz. 1972. Ultrastructural studies of oogenesis in some European amphibians. I. *Rana esculenta* and *Rana temporaria*. Zeitschrift f. Zellforsch., *128*: 438–456.

Lanzavecchia, G. 1961. The formation of yolk in frog oocytes. Proceedings of

Second European Regional Conference on Electron Microscopy, Delft, 1960, *2*: 746–749.

Levenson, R.G., and K.B. Marcu. 1976. On the existence of polyadenylated histone mRNA in Xenopus laevis oocytes. Cell, *9*: 311–322.

Lillie, F.R. 1919. *The Development of the Chick*, 2nd ed. Henry Holt, New York.

Macgregor, H.C. 1977. Lampbrush chromosomes. *In* H.J. Li and R.A. Eckhardt (eds.), *Chromatin and Chromosome Structure.* Academic Press, Inc., New York, pp. 339–357.

Macgregor, H.C. 1980. Recent developments in the study of lampbrush chromosomes. Heredity, *44*: 3–35.

Mairy, M., and H. Denis. 1972. Recherches biochimiques sur l'oogenèse. 2. Assemblage des ribosomes pendant le grand accroissement des oocytes de *Xenopus laevis.* Eur. J. Biochem., *25*: 535–543.

Maller, J.L., F.R. Butcher, and E.G. Krebs. 1979. Early effect of progesterone on levels of cyclic adenosine $3':5'$-monophosphate in *Xenopus* oocytes. J. Biol. Chem., *254*: 579–582.

Maller, J.L., and E.G. Krebs. 1977. Progesterone-stimulated meiotic cell division in *Xenopus* oocytes. Induction by regulatory subunit and inhibition by catalytic subunit of adenosine $3':5'$-monophosphate-dependent protein kinase. J. Biol. Chem., *252*: 1712–1718.

Maller, J.L., and E.G. Krebs. 1980. Regulation of oocyte maturation. Curr. Top. Cell. Regul., *16*: 271–311.

Maller, J., M. Wu, and J.C. Gerhart. 1977. Changes in protein phosphorylation accompanying maturation of *Xenopus laevis* oocytes. Dev. Biol., *58*: 295–312.

Masui, Y. 1967. Relative roles of the pituitary, follicle cells, and progesterone in the induction of oocyte maturation in *Rana pipiens.* J. Exp. Zool., *166*: 365–376.

Masui, Y., and H.J. Clarke. 1979. Oocyte maturation. Int. Rev. Cytol., *57*: 185–282.

Masui, Y., and C.L. Markert. 1971. Cytoplasmic control of nuclear behavior during meiotic maturation of frog oocytes. J. Exp. Zool., *177*: 129–146.

Miller, O.L., Jr., and A.H. Bakken. 1972. Morphological studies of transcription. In E. Diczfalusy (ed.), *Karolinska Symposia on Research Methods in Reproductive Endocrinology. Fifth Symposium. Gene Transcription in Reproductive Tissue,* Stockholm, pp. 155–177.

Miller, O.L., Jr., and B.R. Beatty. 1969. Portrait of a gene. J. Cell. Physiol., *74* (Suppl. 1): 225–232.

Moreau, M., P. Guerrier, and M. Dorée. 1976. Modifications précoces des propriétés électrique de la membrane plasmique des ovocytes de *Xenopus laevis* au cours de la réinition méiotique induite par la progestérone, le parachloromercuribenzoate (pCMB) ou l'ionophore A 23187. C.R. Acad. Sci. (D) (Paris), *282*: 1309–1312.

Moreau, M., J.P. Vilain, and P. Guerrier. 1980. Free calcium changes associated with hormone action in amphibian oocytes. Dev. Biol., *78*: 201–214.

Morrill, G.A., and J.B. Murphy. 1972. Role for protein phosphorylation in meiosis and in early cleavage phase of amphibian embryonic development. Nature (Lond.), *238*: 282–284.

Morrill, G.A., and D. Ziegler. 1980. Na^+ and K^+ uptake and exchange by the amphibian oocyte during the first meiotic division. Dev. Biol., *74*: 216–223.

O'Connor, C.M., K.R. Robinson, and L.D. Smith. 1977. Calcium, potassium, and sodium exchange by full-grown and maturing *Xenopus laevis* oocytes. Dev. Biol., *61*: 28–40.

O'Connor, C.M., and L.D. Smith. 1976. Inhibition of oocyte maturation by theophylline: Possible mechanism of action. Dev. Biol., *52*: 318–322.

Paglia, L.M., S.J. Berry, and W.H. Kastern. 1976. Messenger RNA synthesis, transport, and storage in silkmoth ovarian follicles. Dev. Biol., *51*: 173–181.

Pelham, H.R.B., and D.D. Brown. 1980. A specific transcription factor that can bind either the 5S RNA gene or 5S RNA. Proc. Natl. Acad. Sci. U.S.A., 77: 4170–4174.

Pelham, H.R.B., M.W. Wormington, and D.D. Brown. 1981. Related 5S RNA transcription factors in *Xenopus* oocytes and somatic cells. Proc. Natl. Acad. Sci. U.S.A., 78: 1760–1764.

Perkowska, E., H.C. Macgregor, and M.L. Birnstiel. 1968. Gene amplification in the oocyte nucleus of mutant and wild-type *Xenopus laevis.* Nature (Lond.), *217*: 649–650.

Picard, B. et al. 1980. Biochemical research on oogenesis. Composition of the 42S storage particles of *Xenopus laevis* oocytes. Eur. J. Biochem., *109*: 359–368.

Pinkerton, J.H.M. et al. 1961. Development of the human ovary—A study using histochemical technics. Obstet. Gynecol., *18*: 152–181.

Postlethwait, J.H., and P.D. Shirk. 1981. Genetic and endocrine regulation of vitellogenesis in *Drosophila.* Am. Zool., *21*: 687–700.

Pukkila, P.J. 1975. Identification of the lampbrush chromosome loops which transcribe 5S ribosomal RNA in *Notophthalmus (Triturus) viridescens.* Chromosoma, *53*: 71–89.

Quattropani, S.L., and E. Anderson. 1969. The origin and structure of the secondary coat of the egg of *Drosophila melanogaster.* Zeitschrift f. Zellforsch., *95*: 495–510.

Reynhout, J.K., and L.D. Smith. 1974. Studies on the appearance and nature of a maturation-inducing factor in the cytoplasm of amphibian oocytes exposed to progesterone. Dev. Biol., *38*: 394–400.

Rogers, R.E., and L.W. Browder. 1977. Morphological observations on cultured lampbrush-stage *Rana pipiens* oocytes. Dev. Biol., *55*: 135–147.

Rosbash, M., and P.J. Ford. 1974. Polyadenylic acid-containing RNA in *Xenopus laevis* oocytes. J. Mol. Biol., *85*: 87–101.

Roth, T.F., and K.R. Porter. 1964. Yolk protein uptake in the oocyte of the mosquito *Aedes aegypti* L. J. Cell Biol., *20*: 313–332.

Rubenstein, E.C. 1979. The role of an epithelial occlusion zone in the termination of vitellogenesis in *Hyalophora cecropia* ovarian follicles. Dev. Biol., *71*: 115–127.

Ruderman, J.V., and M.L. Pardue. 1977. Cell-free translation analysis of messenger RNA in echinoderm and amphibian early development. Dev. Biol., *60*: 48–68.

Ruderman, J.V., and M.L. Pardue. 1978. A portion of all major classes of histone messenger RNA in amphibian oocytes is polyadenylated. J. Biol. Chem., *253*: 2018–2025.

Ruderman, J.V., and M.R. Schmidt. 1981. RNA transcription and translation in sea urchin oocytes and eggs. Dev. Biol., *81*: 220–228.

Ruderman, J.V., H.R. Woodland, and E.A. Sturgess. 1979. Modulations of histone messenger RNA during the early development of *Xenopus laevis.* Dev. Biol., *71*: 71–82.

Scheer, U. et al. 1976a. Classification of loops of lampbrush chromosomes according to the arrangement of transcriptional complexes. J. Cell Sci., *22*: 503–519.

Scheer, U., M.F. Trendelenburg, and W.W. Franke. 1976b. Regulation of transcription of genes of ribosomal RNA during amphibian oogenesis. A biochemical and morphological study. J. Cell Biol. *69*: 465–489.

Schroeder, T.E. 1980. Expressions of the prefertilization polar axis in sea urchin eggs. Dev. Biol., *79*: 428–443.

Schuetz, A.W. 1967. Action of hormones on germinal vesicle breakdown in frog (*Rana pipiens*) oocytes. J. Exp. Zool., *166*: 347–354.

Schuetz, A.W. 1972. Hormones and follicular functions. *In* J.D. Biggers and A.W. Schuetz (eds.), *Oogenesis.* University Park Press, Baltimore, pp. 479–511.

Schultz, L.D., B.K. Kay, and J.G. Gall. 1981. In vitro RNA synthesis in oocyte nuclei of the newt *Notophthalmus. Chromosoma, 82*: 171–187.

Selman, K., and E. Anderson. 1975. The formation and cytochemical characterization of cortical granules in ovarian oocytes of the golden hamster (*Mesocricetus auratus*). J. Morphol., *147*: 251–274.

Senger, P.L., and R.G. Saacke. 1970. Unusual mitochondria of the bovine oocyte. J. Cell Biol., *46*: 405–408.

Smith, L.D., and R.E. Ecker. 1969. Role of the oocyte nucleus in physiological maturation in *Rana pipiens.* Dev. Biol., *19*: 281–309.

Smith, L.D., and R.E. Ecker. 1971. The interaction of steroids with *Rana pipiens* oocytes in the induction of maturation. Dev. Biol., *25*: 232–247.

Smith, L.D., R.E. Ecker, and S. Subtelny. 1968. *In vitro* induction of physiological maturation in *Rana pipiens* oocytes removed from their ovarian follicles. Dev. Biol., *17*: 627–643.

Sommerville, J. 1973. Ribonucleoprotein particles derived from the lampbrush chromosomes of newt oocytes. J. Mol. Biol., *78*: 487–503.

Sommerville, J., and D.B. Malcolm. 1976. Transcription of genetic information in amphibian oocytes. Chromosoma, *55*: 183–208.

Szöllösi, D. 1972. Changes of some cell organelles during oogenesis in mammals. *In* J.D. Biggers and A.W. Schuetz (eds.), *Oogenesis.* University Park Press, Baltimore, pp. 47–64.

Telfer, W.H. 1979. Sulfate and glucosamine labelling of the intercellular matrix in vitellogenic follicles of a moth. Wilhelm Roux's Archiv. Dev. Biol., *185*: 347–362.

Tourte, M., F. Mignotte, and J.-C. Mounolou. 1981. Organization and replication activity of the mitochondrial mass of oogonia and previtellogenic oocytes in *Xenopus laevis.* Dev., Growth Differ., *23*: 9–21.

Turner, C.D., and J.T. Bagnara. 1976. *General Endocrinology,* 6th ed. W.B. Saunders Co., Philadelphia.

Van Blerkom, J., and P. Motta. 1979. *The Cellular Basis of Mammalian Reproduction.* Urban & Schwarzenberg, Baltimore-Munich.

Vander, A.J., J.H. Sherman, and D.S. Luciano. 1975. *Human Physiology. The Mechanisms of Body Function,* 2nd ed. McGraw-Hill Book Co., New York.

van Dongen, W. et al. 1981. Quantitation of the accumulation of histone messenger RNA during oogenesis in *Xenopus laevis.* Dev. Biol., *86*: 303–314.

Varley, J.M., H.C. Macgregor, and H.P. Erba. 1980a. Satellite DNA is transcribed on lampbrush chromosomes. Nature (Lond.), *283*: 686–688.

Varley, J.M. et al. 1980b. Cytological evidence of transcription of highly repeated DNA sequences during the lampbrush stage in *Triturus cristatus carnifex.* Chromosoma, *80*: 289–307.

Wallace, R.A. 1974. Protein phosphorylation during oocyte maturation. Nature (Lond.), *252*: 510–511.

Wallace, R.A., and R.A. Steinhardt. 1977. Maturation of *Xenopus* oocytes. II. Observations on membrane potential. Dev. Biol., *57*: 305–316.

Ward, R.T. 1962. The origin of protein and fatty yolk in *Rana pipiens*. II. Electron microscopical and cytochemical observations of young and mature oocytes. J. Cell Biol., *14*: 309–341.

Ward, R.T. 1978a. The origin of protein and fatty yolk in *Rana pipiens*. III. Intramitochondrial and primary vesicular yolk formation in frog oocytes. Tissue Cell, *10*: 515–524.

Ward, R.T. 1978b. The origin of protein and fatty yolk in *Rana pipiens*. IV. Secondary vesicular yolk formation in frog oocytes. Tissue Cell, *10*: 525–534.

Wassarman, P.M., and G.E. Letourneau. 1976. RNA synthesis in fully grown mouse oocytes. Nature (Lond.), *261*: 73–74.

Wasserman, W.J., and Y. Masui. 1975. Effects of cycloheximide on a cytoplasmic factor initiating meiotic maturation in *Xenopus* oocytes. Exp. Cell Res., *91*: 381–388.

Wasserman, W.J., and Y. Masui. 1976. A cytoplasmic factor promoting oocyte maturation: Its extraction and preliminary characterization. Science, *191*: 1266–1268.

Wasserman, W.J., et al. 1980. Progesterone induces a rapid increase in $[Ca^{2+}]_{in}$ of *Xenopus laevis* oocytes. Proc. Natl. Acad. Sci. U.S.A., *77*: 1534–1536.

Wasserman, W.J., and L.D. Smith. 1978. Oocyte maturation: Non-mammalian vertebrates. *In* R.E. Jones (ed.), *The Vertebrate Ovary*. Plenum Publishing Corp., New York.

Wasserman, W.J., and L.D. Smith. 1981. Calmodulin triggers the resumption of meiosis in amphibian oocytes. J. Cell Biol., *89*: 389–394.

Webb, A.C., and L.D. Smith. 1977. Accumulation of mitochondrial DNA during oogenesis in *Xenopus laevis*. Dev. Biol., *56*: 219–225.

Wolf, D.P. et al. 1976. Isolation, physicochemical properties, and the macromolecular composition of the vitelline and fertilization envelopes from *Xenopus laevis* eggs. Biochemistry, *15*: 3671–3675.

Woodruff, R.I., and W.H. Telfer. 1973. Polarized intercellular bridges in ovarian follicles of the cecropia moth. J. Cell Biol., *58*: 172–188.

Woodruff, R.I., and W.H. Telfer. 1980. Electrophoresis of proteins in intercellular bridges. Nature (Lond.), *286*: 84–86.

Wu, M., and J.C. Gerhart. 1980. Partial purification and characterization of the maturation-promoting factor from eggs of *Xenopus laevis*. Dev. Biol., *79*: 465–477.

Yamamoto, K., and I. Oota. 1967. An electron microscope study of the formation of the yolk globule in the oocyte of zebrafish, *Brachydanio rerio*. Bull. Fac. Fish. Hokkaido University, *17*: 165–174.

8 Plant Gametogenesis

As originally proposed by August Weismann, animal gametes are derivatives of a distinct germ cell line that is established early in embryonic development. Unlike animals, plants do not segregate a germ cell line from somatic cells. This difference between plants and animals is related to more basic differences in overall growth and development strategies. During animal development, cells can undergo considerable rearrangement and migration, and growth can occur by enlargement of internal body regions and by mitosis of selected stem cell populations. However, plant cells are surrounded by rigid cellulose walls that prevent any large-scale movement or migration of cells. In vascular plants, two growth centers are established during early development, and they elaborate the cells from which organs are produced. The aerial portions of the plant are produced by mitosis in the **shoot apical meristem,** and the root system is produced by the **root apical meristem.** Organs are formed *in situ* from cells produced by the meristems. The same cells may give rise to various structures under different conditions. For example, in flowering plants the reproductive organs (flowers) develop from meristems that, under different conditions, would produce leaves. Hormones control whether a flower will develop from the meristem, and local conditions within the developing primordium determine which cells will form the gametes.

Plant life cycles are more complicated than those of animals. Higher plants, as well as some algae and fungi, exhibit an **alternation of generations.** The diploid zygote divides mitotically and develops into a plant called a **sporophyte.** Cells within the reproductive organs of adult sporophytes undergo meiosis to form haploid **meiospores,** which divide by mitosis to form a second plant generation called the **gameotophyte.** The latter matures to produce gametes, and union of male and female gametes produces the diploid zygote, thus completing the cycle. Although in this chapter we shall concentrate on the formation of gametes, the process is necessarily linked to the formation of the gametophyte. Consequently, we shall also discuss gametophyte formation where relevant.

Reproductive strategies in the plant kingdom are so varied that it is not feasible either to survey the formation of all the different kinds of plant gametes or to discuss a "typical" pattern of male or female gametogenesis. Instead, we shall discuss gametogenesis in organisms that represent two highly contrasting patterns of reproduction. One group includes primitive plants that shed eggs and motile sperm that appear very similar to those of animals. The sperm of some primitive plants are represented in Figure 8.1. In contrast to animal sperm, most plant sperm have two or more *anteriorly* directed flagella. The other group consists of the flowering plants, or **angiosperms.** In these organisms the egg is embedded deep within plant tissue that is in aerial portions of the plant. Consequently, swimming male gametes are impractical and have been replaced in evolution by pollen-borne gametes. The pollen grains are

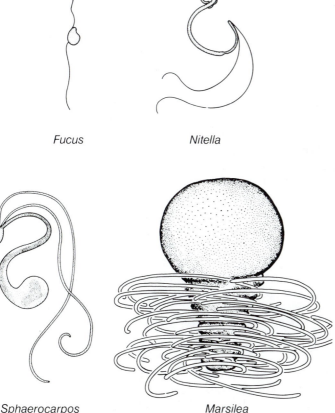

Fucus   *Nitella*

Sphaerocarpos   *Marsilea*

FIGURE 8.1 Forms of some plant sperm. (*Fucus* after E.G. Pollock. 1970. Fertilization in *Fucus*. Planta [Berl.], *92:* 90; *Nitella* after F.R. Turner. 1968. Modified from *The Journal of Cell Biology*, 1968, vol. 37, pp. 370–393 by copyright permission of The Rockefeller University Press. *Sphaerocarpos* after L. Diers. 1967. Der Feinbau des Spermatozoids von *Sphaerocarpos donnellii* Aust. (*Hepaticae*). Planta [Berl.], *72:* 122; *Marsilea* after D.G. Myles and P.K. Hepler. 1977. Spermiogenesis in the fern *Marsilea*: Microtubules, nuclear shaping, and cytomorphogenesis. J. Cell Sci., *23:* 57–83.)

dispersed in the air or are carried by animals (e.g., insects). In addition, angiosperms have evolved a mechanism to deliver the male gamete to the egg, cloistered within the plant tissue.

Because of the small size of gametes, the study of their morphogenesis must be based upon electron microscopy. As we have seen in the previous two chapters, this kind of analysis has been very successful in amassing considerable knowledge about formation of animal gametes. The usual approach is to follow the formation of specific gamete organelles that are functionally important in reproduction. In studying plants, one finds this approach is complicated by the difficulties encountered in preparing cells for electron microscopy. These difficulties are primarily due to the presence of cell walls and vacuoles.

Studies of gene expression during gametogenesis, which are important in understanding how morphogenesis is regulated, have been facilitated in animal research by the ready accessibility of large numbers of developing gametes for biochemical and molecular analysis. There has been some success in applying these approaches to the study of plants, particularly for male gametogenesis. Unfortunately, very little information of this type is available for female gametogenesis. The major difficulty is due to the location of female gametes deep within the vegetative tissues. Poor accessibility makes biochemical studies very difficult.

Despite the technical difficulties associated with plant gametogenesis, progress has been made in recent years. It is only a beginning, however, and considerable work remains to be done.

8–1. GAMETOGENESIS IN PRIMITIVE PLANTS

The sexual reproduction of primitive plants is highly variable, and no species is typical of this vague category of diverse organisms. In this and subsequent chapters, we shall use two organisms as examples of primitive plants. These are the marine brown alga *Fucus* and the water fern *Marsilea*. These organisms have been chosen because they have been successfully exploited as experimental organisms and because their reproduction is reminiscent of that of marine invertebrate animals. Not all aspects of gametogenesis are equally understood for these two organisms. For example, egg differentiation is better understood for *Fucus*, but descriptions of spermiogenesis are more complete for *Marsilea*. We can only hope that complete descriptions of gametogenesis in both of these species will soon be available.

Fucus

The life cycle of *Fucus* is somewhat atypical of plants and is reminiscent of an animal life cycle. The vegetative *Fucus* plant is diploid and contains reproductive organs in which gametes are formed directly without the intervention of a gametophyte stage. Meiosis produces haploid cells (meiospores) that undergo one or more mitotic divisions. The meiospores directly differentiate into gametes that are released into the open sea, where fertilization occurs much like in marine invertebrate animals. *Fucus* zygotes have been extensively exploited by investigators for developmental studies, particularly those concentrating on the formation of regional differences in the zygote that affect the pattern of development. These investigations will be discussed in Chapter 11.

The sperm of *Fucus* develop within structures called **antheridia.** Each antheridium produces 64 sperm by meiosis followed by four mitotic divisions. The biflagellated sperm of *Fucus* is represented in Figure 8.1. Additional details of *Fucus* sperm structure are described in Chapter 9.

Fucus eggs develop within a structure called an **oogonium.** Each oogonium begins as a chamber containing a single diploid cell—the **oogonial mother cell.** The nucleus of this cell undergoes meiosis to produce four haploid nuclei; each of these undergoes one mitotic division, resulting in the eight-nucleate stage. This brief stage is terminated by the formation of cell membranes between the nuclei, producing eight eggs (McCully, 1968).

Eggs of *Fucus* are characterized by a highly convoluted nucleus, numerous cytoplasmic vesicles, and the absence of a cell wall. The egg nucleus contains nucleoli, and the chromatin is diffuse, with little apparent heterochromatin (Fig. 8.2A). The nuclear envelope is very similar to that of animal eggs and is perforated by numerous pores. The perinuclear region contains a considerable amount of electron-dense material, which is frequently found within indentations of the nuclear envelope. This material appears to be analogous to the nuage material observed in transit between the nucleus and cytoplasm in animal oocytes (see Fig. 7.13). The direction of transit, if any, has not been established. However, after fertilization this material becomes surrounded by the nuclear envelope, suggesting that it may be incorporated into the nucleus from the cytoplasm (Fig. 8.2B).

Unfertilized *Fucus* eggs contain two types of vesicles: V_1 and V_2. The former are produced during egg differentiation from the smooth endoplasmic reticulum (Fig. 8.2C). These vesicles are translucent before fertilization but acquire a fibrillar component after fertilization.. V_2 vesicles, which contain fibrillar material, are also present before fertilization. Their origin is unknown. After fertilization the Golgi bodies become highly active and produce a third type of vesicle: V_3.

A cell wall is deposited around the zygote soon after the egg is fertilized. The components for construction of the wall are present in vesicles that release their contents over the egg surface. Most of the cell wall components are deposited by V_3 vesicles, with lesser amounts coming from V_2 vesicles (Brawley et al., 1976).

The absence of a cell wall is a general property of plant gametes and an adaptation that facilitates fusion of gametes. Because of it, one of the first chores of the zygote after fertilization is the construction of a cell wall. The ready accessibility of *Fucus* zygotes, which are free of surrounding somatic tissue, makes them an excellent model system for the study of cell wall deposition. This topic will be discussed in more detail in Chapter 9.

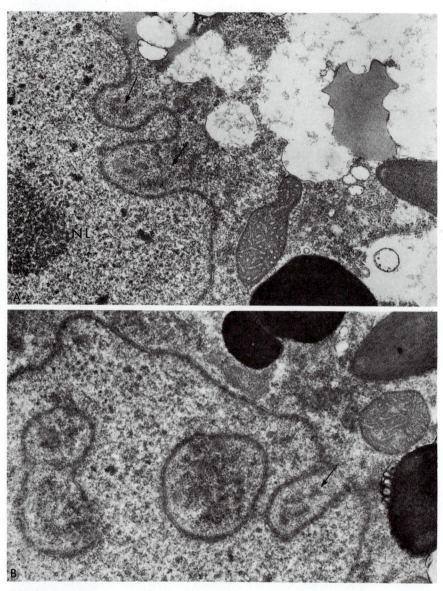

FIGURE 8.2 Electron micrographs of the *Fucus* egg. *A*, Nucleus and perinuclear cytoplasm. Note nucleolus (NL) and pores in nuclear envelope. Indentations of the nuclear envelope contain dense material (arrows). ×36,000. *B*, After fertilization, dense material is found both in the perinuclear region (arrow) and surrounded by the nuclear envelope. ×36,000.

Illustration continued on the opposite page

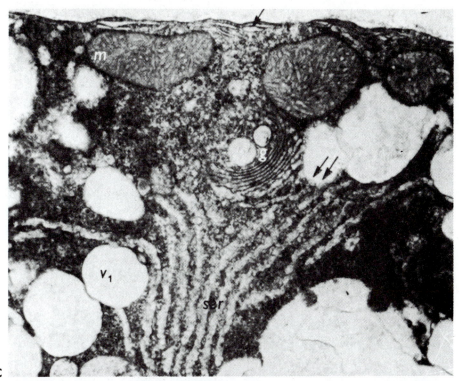

FIGURE 8.2 (*Continued*) *C*, Formation of egg vesicles (V_1) by smooth endoplasmic reticulum (double arrows). The Golgi complex appears to be inactive. The smooth endoplasmic reticulum (single arrow) near the plasmalemma is not distended. ser: smooth endoplasmic reticulum; m: mitochondrion. ×40,500. (From S.H. Brawley, R. Wetherbee, and R.S. Quatrano. 1976. Fine-structural studies of the gametes and embryo of *Fucus vesiculosis* L. (Phaeophyta). II. The cytoplasm of the egg and young zygote. J. Cell Sci., *20:* 263, 267.)

Marsilea

The *Marsilea* life cycle is quite different from that of *Fucus* in that there is a haploid gametophyte stage that alternates with the diploid sporophyte. The sporophyte (Fig. 8.3) forms a spore-bearing appendage called the **sporocarp.** Within the sporocarp there are **microsporangia** and **megasporangia.** The microsporangia produce **microspores** by meiosis. These haploid cells are released from the sporocarp into water and develop into **microgametophytes,** in which the sperm differentiate. Megasporangia undergo meiosis to produce **megaspores,** which are released from the sporocarp along with the microspores. The haploid megaspores undergo numerous mitotic divisions and develop into tiny, multicellular **megagametophytes.** Each megagametophyte forms a single, large egg located in a terminal structure called an **archegonium.** During fertilization, sperm must enter the archegonium to make contact with the egg (see Chap. 9).

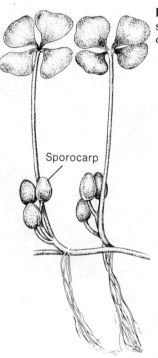

FIGURE 8.3 The water fern *Marsilea*. (From K. Norstog and R.W. Long. 1976. *Plant Biology*. W.B. Saunders, Philadelphia, p. 362.)

Sporocarp

Recent ultrastructural studies on *Marsilea* sperm differentiation constitute one of the most extensive analyses of the morphogenesis of any single cell type—plant or animal. For this reason, we shall discuss its structure and morphogenesis in detail. These elaborate cells are pear-shaped and have a spiral coil at the anterior end that consists of about ten turns (see Fig. 8.1). The coil begins at the tapered anterior end and extends part way into the bulbous portion. The coil bears between 100 and 150 flagella distributed along its length and has a very complex internal structure (Fig. 8.4). A single mitochondrion extends along the entire ten gyres of the coil. A ribbon of microtubules also extends the length of the coil and forms a cytoskeleton. At the anterior tip of the coil, the microtubular ribbon is associated with the **multilayered structure** (**MLS**). The latter extends for less than one gyre of the coil and is composed of a few layers of repeating units (Fig. 8.5*A*). The flagella arise from rows of basal bodies embedded in a layer of dense material called the **flagellated band** (Fig. 8.5*B*). The spiral nucleus conforms to the shape of the coil, but it is confined to the posteriormost four or five gyres. It has the appearance of an elongated rod of very condensed chromatin. The interrelationships of the various organelles in the coil are shown in Figure 8.5*C*. A head-on view of *Marsilea* sperm is shown in Figure 8.5*D*, which reveals that the coil is a left-handed spiral. This configuration causes the sperm to spiral in a counterclockwise direction, propelled by flagella on the coil.

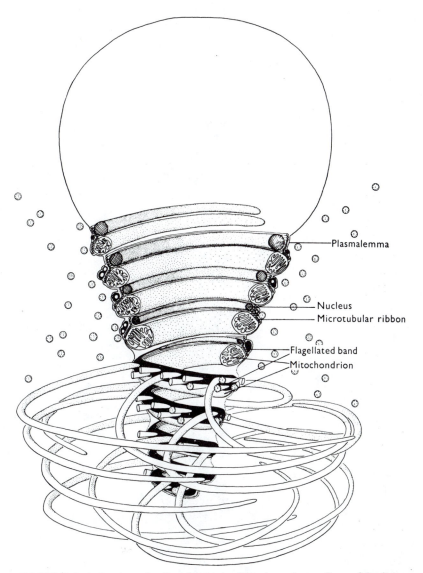

FIGURE 8.4 A drawing of a *Marsilea* sperm based on observations of the living sperm and on electron micrographs. The anterior region is represented as it would be seen intact with some of the flagella cut away. The posterior region is shown with half of the sperm removed, revealing the cut ends of the organelles of the coil. (From D.G. Myles and P.R. Bell. 1975. An ultrastructural study of the spermatozoid of the fern, *Marsilea vestita*. J. Cell Sci., *17*: 635.)

As previously mentioned, *Marsilea* sperm are formed within the microgametophytes, which are formed by repeated mitotic division of the microspore to produce 32 spermatids and 7 sterile cells. After mitosis the spermatids differentiate into sperm. Both the mitotic and differen-

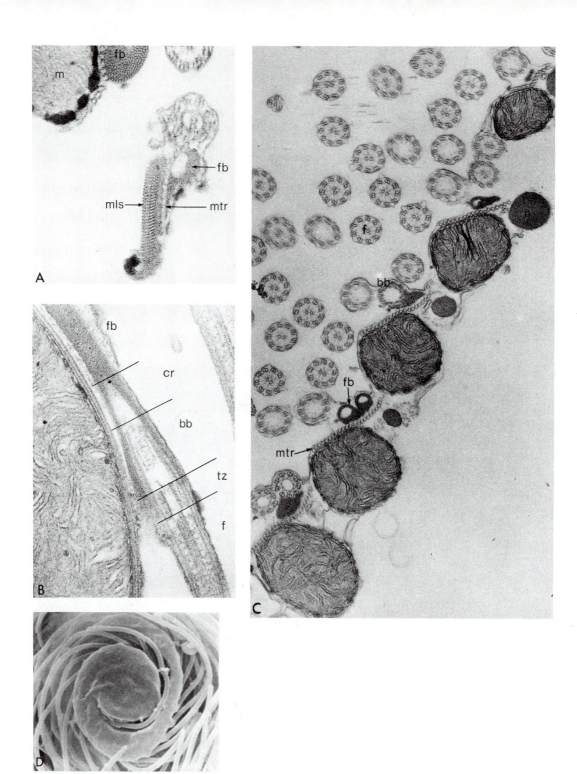

A

B

C

D

FIGURE 8.5 (*See legend on the opposite page.*)

◀ **FIGURE 8.5** Ultrastructure of *Marsilea* sperm. *A*, Multilayered structure (mls) at tip of anterior coil. The microtubule ribbon (mtr) extends posteriorly from the multilayered structure. A portion of the flagellated band (fb) can also be seen. m: mitochondrion. ×72,000. *B*, This section illustrates the attachment of flagellum and basal body to the flagellated band. Proceeding from the proximal to the distal region is the flagellated band (fb), flaring into an open cylindrical region (cr), basal body (bb), transition zone (tz), and flagellum (f). ×49,400. *C*, Cross-section through five of the posterior gyres of the helix. The helical structures include a single large mitochondrion (m), a nucleus (n) containing highly condensed chromatin, and a densely stained flagellated band (fb). A basal body (bb) and flagella (f) are also seen in cross section. ×37,200. *D*, The anterior coil observed head on by scanning electron microscopy reveals the left-handed nature of the spiral. ×10,800. (*A, B, C* from D.G. Myles and P.R. Bell. 1975. An ultrastructural study of the spermatozoid of the fern, *Marsilea vestita*. J. Cell Sci., *17:* 643, 645; *D* from P.K. Hepler and D.G. Myles. 1977. Spermatogenesis in Marsilea: An example of male gamete development in plants. *In* B.R. Brinkley and K.R. Porter (eds.), *International Cell Biology. 1976–1977.* The Rockefeller University Press, New York, p. 576. © The Rockefeller University Press.)

tiation phases are synchronous and occur at predictable rates at given temperatures (Hepler and Myles, 1977). This means that the development of numerous cells can be followed simultaneously. Furthermore, development of the sperm occurs within microspores, which are independent of the sporophyte and can thrive in distilled water. These properties make *Marsilea* spermatogenesis an excellent experimental system, with much promise for biochemical and molecular analyses.

After mitosis, the nondescript spermatids are packed in tetrad clusters within a spherical compartment, and their external shape is determined by the packing arrangement. However, important changes have already occurred in the internal organization of these cells in preparation for formation of the anterior coil. In telophase of the penultimate division, flocculent material aggregates in an indentation of the nucleus (Fig. 8.6*A*). Within this flocculent material, two plaques form, each of which develops into a **blepharoplast** (Fig. 8.6*B* and *C*). As the daughter cells enter prophase of the final mitotic division, the blepharoplasts separate from one another (Fig. 8.6*D*) and move to opposite poles of the spindle apparatus. The microtubules of the spindle appear to radiate from the blepharoplasts (Fig. 8.6*E*). During metaphase and anaphase of this division, each blepharoplast begins to form the 100 to 150 basal bodies for the sperm flagella. An early stage in this process is shown in Figure 8.7*A*, which illustrates the formation of procentrioles by a blepharoplast. The procentrioles mature into basal bodies during telophase.

Also during telophase the multilayered structure forms in association with the basal bodies and a mitochondrion (Fig. 8.7*B*). It seems to elaborate and direct the positioning of microtubules that participate in shaping the mature sperm. The components of the MLS include a plaque that overlies the mitochondrion, numerous thin partitions, and a ribbon of microtubules. A layer of flocculent material is present between the microtubule ribbon and the basal bodies. This material may be involved in the continued growth and maturation of the basal bodies.

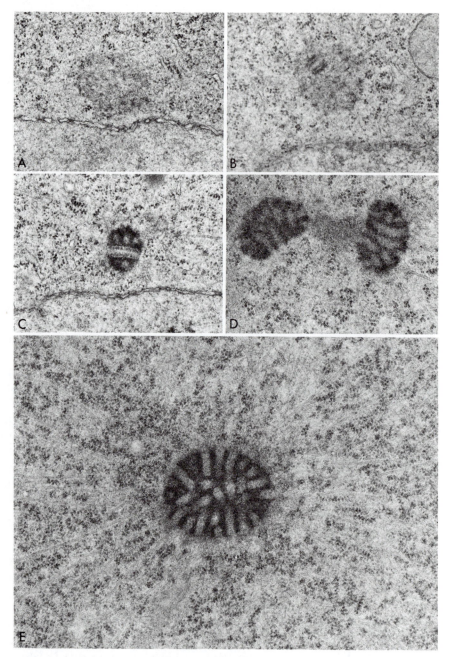

FIGURE 8.6 (*See legend on the opposite page.*)

◄ **FIGURE 8.6** Sequential stages of blepharoplast formation. *A*, The blepharoplast arises in a small indentation in the nucleus during telophase of the next-to-last division. It first appears as a sphere of flocculent material. ×32,200. *B*, Two parallel plaques emerge from the flocculent material. Each plaque is triple layered and is separated from its partner by 0.045 μm. The distal layer of each plaque becomes more densely stained during subsequent stages of development. ×32,200. *C*, Continued condensation of material onto the distal surfaces of each plaque produces a pair of young hemispherical blepharoplasts. ×32,200. *D*, The two enlarging blepharoplasts begin to separate from each other as the cell enters prophase of the last division. Some residual material still connects the pair. ×32,200. *E*, The blepharoplast is the focal point for spindle microtubules during prophase of the final division. A mature blepharoplast is 0.5 to 1.0 μm in diameter and is interpenetrated by numerous lightly stained channels. ×27,200. (*A* to *D* from P.K. Hepler. 1976. The blepharoplast of *Marsilea:* Its *de novo* formation and spindle association. *J. Cell Sci., 21:* 376; *E* from P.K. Hepler and D.G. Myles. 1977. Spermatogenesis in Marsilea: An example of male gamete development in plants. *In* B.R. Brinkley and K.R. Porter (eds.), *International Cell Biology. 1976–1977.* The Rockefeller University Press, New York, p. 571. © The Rockefeller University Press.)

The shaping of the sperm is mainly a matter of forming the anterior coil, with the concurrent production of new organelles (such as the flagella) and modifications to existing ones, including the nucleus and mitochondria. The nucleus becomes elongate, and its chromatin becomes condensed, while mitochondria fuse together and elongate. The multilayered structure appears to play a key role in these processes. Early in spermiogenesis the MLS, its accompanying mitochondrion, and the basal bodies associate with a lateral edge of the nucleus. The microtubule ribbon extends posteriorly from the MLS and overlies the nuclear envelope. The nucleus begins to elongate in a direction that is parallel to the ribbon of microtubules. An early stage in nuclear elongation is shown in Figure 8.8. The organelle spiral coil develops as an anteriorly growing projection from the cell body, with the MLS constantly at its anterior tip. The microtubule ribbon trails behind the MLS, and the nucleus elongates along the inner surface of this ribbon, extending forward for four or five gyres. The elongating nucleus becomes surrounded by mitochondria, which probably aggregate and fuse with each other and the MLS mitochondrion to form a single mitochondrion, which later extends along the entire ten gyres of the anterior coil.

The association of microtubules with the outer membrane of the nuclear envelope during nuclear elongation is reminiscent of the situation during nuclear shaping in animal spermiogenesis (see Chap. 6). The role of the microtubules in nuclear shaping in *Marsilea* spermiogenesis has been studied by treatment of developing spermatids with colchicine (Myles and Hepler, 1982). In the presence of this drug, which inhibits microtubule polymerization, nuclear elongation occurs, but the nucleus does not develop its normal spiral shape. Therefore, the mechanical force for elongation is still present, but the normal process of shape determination is absent. The mechanical force is thought to be associated with the nuclear envelope, perhaps in the form of a nuclear proteinaceous network. The microtubules may be necessary either: (1) to organize this

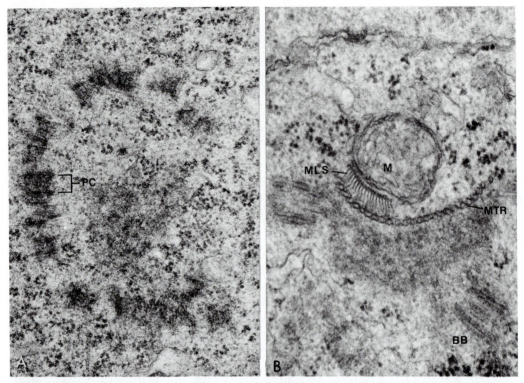

FIGURE 8.7 Formation of basal bodies and multilayered structure. *A*, The spherical blepharoplast enlarges and transforms into numerous procentrioles (PC). These will subsequently mature into basal bodies with the typical nine triplet tubule structure. ×42,000. *B*, The multilayered structure (MLS) arises at telophase of the last division. It is composed of a plaque that overlies a mitochondrion (M), numerous thin partitions, and a ribbon of closely aligned microtubules (MTR). Flocculent material occurs on the distal surface of the microtubule ribbon and may contribute to the continued growth and maturation of the basal bodies (BB). ×63,000. (*A* from P.K. Hepler. 1976. The blepharoplast of *Marsilea:* Its *de novo* formation and spindle association. *J. Cell Sci., 21:* 384; *B* from P.K. Hepler and D.G. Myles. 1977. Spermatogenesis in Marsilea: An example of male gamete development in plants. *In* B.R. Brinkley and K.R. Porter (eds.), *International Cell Biology. 1976–1977.* The Rockefeller University Press, New York, p. 572. © The Rockefeller University Press.)

system into a form that determines the direction of coiling or (2) to restrict or guide the nucleus externally into the correct form as it elongates (Myles and Hepler, 1977, 1982).

It is intriguing to note that the edge of the nucleus associated with the microtubule ribbon is flattened and is devoid of nuclear pores, which are distributed over the remainder of the nuclear envelope (Myles et al., 1978). At the time of nuclear elongation along the inner surface of the ribbon of microtubules, the basal bodies spread out on the outer surface of the ribbon and are sandwiched between it and the plasmalemma (Fig. 8.9A). During the middle stages of spermiogenesis the basal bodies become organized in two rows, with a flagellum emanating from each basal body and forming the flagellar band. These two rows are found along nearly the entire length of the coil (Fig. 8.9*B* and *C*). The formation of the anterior coil is summarized in the three drawings of Figure 8.10, which show successive stages in spermiogenesis.

Besides elongation, the other major change in the nucleus is chromatin condensation. Some condensation occurs during the coiling process, but most occurs very late in spermiogenesis, predominantly near the pore-free region of the nuclear envelope associated with the microtubules (Fig. 8.11). The morphological relationship implicates the microtubles in condensation, and this role for the microtubules has been confirmed by treatment with colchicine: Chromatin condensation occurs in patches scattered throughout the nucleus, rather than following the orderly pattern as previously described. Presumably, the microtubules are necessary to organize the nuclear factors that initiate chromatin condensation (Myles and Hepler, 1982).

The molecular events of chromatin condensation are uncertain. Some plant sperm that acquire elongated, highly condensed nuclei undergo a replacement of somatic histones by protaminelike basic proteins, similar to those of animal sperm (Reynolds and Wolfe, 1978). Perhaps a similar mechanism occurs in *Marsilea*.

One of the final events in spermiogenesis is one that is analogous to a process in late animal spermiogenesis—sloughing of excess cytoplasm. This process in *Marsilea* is illustrated in Figure 8.12. A ring of cytoplasm grows down over the anterior coil, completely encircling it and creating an internal, but extracellular, cavity. Organelles such as ribosomes, mitochondria, and endoplasmic reticulum accumulate in the **cytoplasmic bridge,** which is eventually shed from the sperm, taking the nuclear envelope with it, when the sperm are released from the microgametophyte.

Formation of *Marsilea* sperm is an elegant process that produces very complex single cells. Fortunately, *Marsilea* spermatids are readily accessible for developmental studies. The microspores develop in water, providing a natural equivalent of the *in vitro* situation that has been so successfully exploited for experimental analysis of differentiation of many somatic cell types. Similar exploitation of this system should yield important information about the processes involved in cell morphogenesis, particularly the ways that microtubules participate in cellular shaping.

8–2. GAMETOGENESIS IN ANGIOSPERMS

The flowers of angiosperm sporophytes are the sites of production of the male and female gametophytes. Although the male gametophytes are released from the flower, the female gametophytes are located deep within flower tissue, where eggs they contain are fertilized. After fertil-

Text continued on page 343.

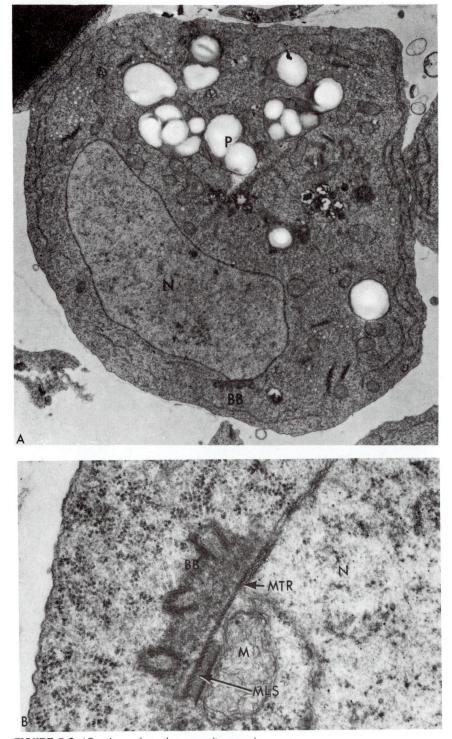

FIGURE 8.8 (*See legend on the opposite page.*)

◀FIGURE 8.8 Ultrastructure of *Marsilea* spermatid at beginning of nuclear elongation. *A,* Low-power micrograph of spermatid sectioned longitudinally. Basal bodies (BB) are distributed along one edge of the nucleus (N). At this stage, the cell has a marked polarity, with the nucleus occupying the future anterior end of the sperm and plastids (P) restricted to the posterior region. ×8900. *B,* A higher magnification showing a portion of the nucleus (N), microtubule ribbon (MTR), and multilayered structure (MLS) with its associated mitochondrion (M). Basal bodies (BB) lie outside the ribbon and are embedded in a dense material. ×43,500. (From D.G. Myles and P.K. Hepler. 1977. Spermiogenesis in the fern, *Marsilea:* Microtubules, nuclear shaping, and cytomorphogenesis. J. Cell Sci., *23:* 62.)

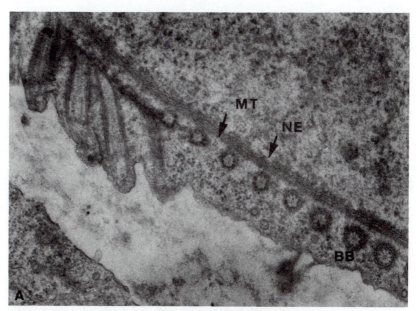

FIGURE 8.9 Ultrastructure of *Marsilea* spermatid during formation of the anterior coil. *A,* Cross section through a spermatid in a middle developmental stage. Adjacent to the nuclear envelope (NE) is a longitudinally-cut microtubule (MT) of the ribbon. Basal bodies (BB) are cut in cross section at successive levels and their proximal-to-distal structure is seen from left to right. ×37,000. *B,* Longitudinal section through a spermatid in a late developmental stage. The organelle coil now makes more than four gyres, and successive cross sections of it are numbered from the anterior end. The nucleus is attenuated in the anterior region. The cell is beginning to acquire its final pear shape. ×9200. Inset: Differential interference micrograph at about the same stage in development. ×1200. *C,* Higher magnification of *B,* showing organelle coil cross sections 2, 4, and 6. Coil mitochondrion (M), microtubule ribbon (MTR), flagellar band (FB), and nucleus (N) are cut in cross section. ×36,400. (From D.G. Myles and P.K. Hepler. 1977. Spermiogenesis in the fern, *Marsilea:* Microtubules, nuclear shaping, and cytomorphogenesis. J. Cell Sci., *23:* 57–83.)

Illustration continued on page 340.

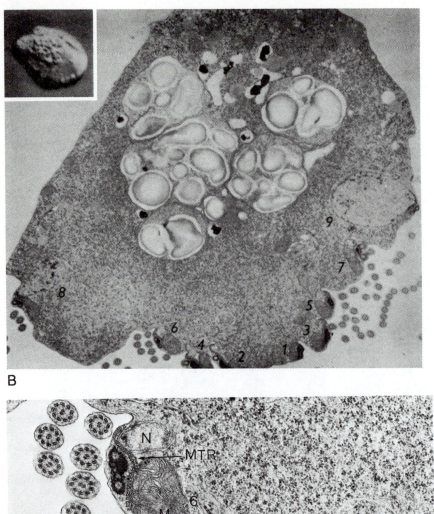

B

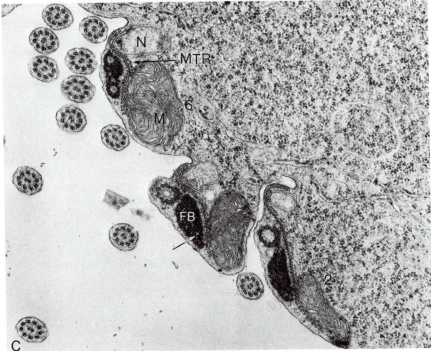

C

FIGURE 8.9 (Continued) See legend on page 339.

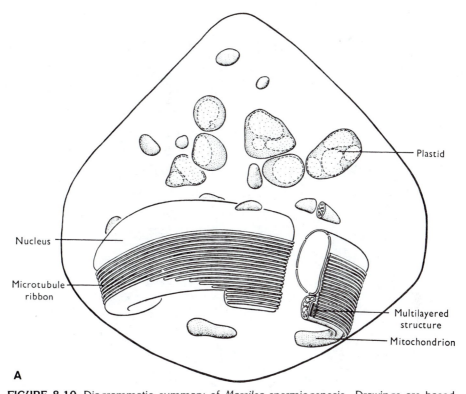

A

FIGURE 8.10 Diagrammatic summary of *Marsilea* spermiogenesis. Drawings are based upon light and electron microscope observations. *A,* A spermatid during early development. The nucleus is beginning to coil around the anterior edge of the cell. A ribbon of micro-tubules runs along the exterior edge of the nucleus and the coil mitochondrion. Between the microtubule ribbon and mitochondrion is the multilayered structure. Other mitochondria are clustered near the nuclear envelope. Plastids are restricted to the posterior region of the cell. The basal bodies have been omitted to reveal the other organelles of the coil. *B,* A spermatid during the middle stage of development. The nucleus is coiled slightly more than one complete gyre. A coil mitochondrion extends a short distance in front of the nucleus and is growing backwards toward the posterior end of the nucleus. The microtubule ribbon is continuous along the length of the coil except for a short distance in the posterior end of the nucleus. The multilayered structure makes about one gyre at this stage. Mitochondria cluster around the nucleus, and plastids are restricted to the posterior region of the cell. The flagellar band and flagella have been omitted to reveal the other organelles of the coil. *C,* A spermatid during late development. The nucleus now makes about four gyres and is attenuated in the anterior end. The nucleus has almost reached its final shape, but the mitochondrion will continue to elongate in an anterior direction until it makes about nine gyres. The microtubule ribbon increases to its maximum number of microtubules (not all drawn in this illustration) shortly behind the anterior tip and then gradually reduces in number toward the posterior end of the coil. The flagellar band and flagella are omitted to reveal the other organelles of the coil. (From D.G. Myles and P.K. Hepler. 1977. Spermiogenesis in the fern, *Marsilea:* Microtubules, nuclear shaping, and cytomorphogenesis. J. Cell Sci., *23:* 57–83.)

Illustration continued on page 342.

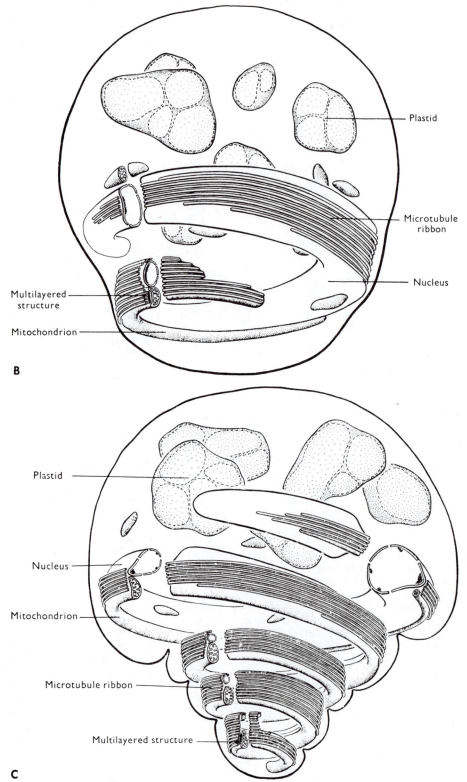

B

Plastid

Microtubule
ribbon

Nucleus

Multilayered
structure

Mitochondrion

Plastid

Nucleus

Mitochondrion

Microtubule ribbon

Multilayered structure

C

FIGURE 8.10 (*Continued*) *See legend on page 341.*

FIGURE 8.11 Cross section through three gyres of the organelle coil during a late stage of spermiogenesis. The nucleus (N) is close to its final shape, and the chromatin (C) is in its final stages of condensation. ×30,500. (From D.G. Myles and P.K. Hepler. 1977. Spermiogenesis in the fern, *Marsilea:* Microtubules, nuclear shaping, and cytomorphogenesis. J. Cell Sci., *23:* 77.)

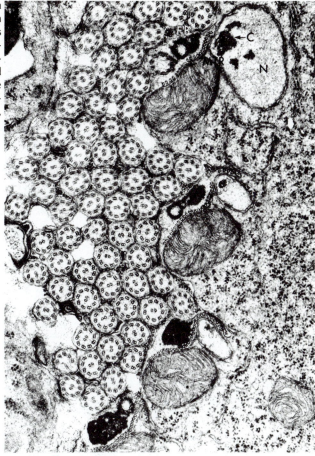

ization the zygote is retained by the sporophyte, in which it undergoes embryonic development. After its development has been completed, the embryo becomes metabolically quiescent; then, encased in a seed, it is released from the parental plant to embark upon an independent existence. In this section we shall discuss the formation of the gametophytes and the gametes they enclose. In subsequent chapters we shall resume discussion of later phases of plant reproduction. To set the stage for detailed analyses of plant gametogenesis and development, we shall first review some of the mechanics of plant sexual reproduction and introduce some of the terminology that will be used in subsequent discussions.

Sexual Reproduction of Angiosperms

The structure of a typical flower is diagramed in Figure 8.13. The male reproductive organs of the flower are the **stamens,** and the female reproductive organ is the **pistil.** The major parts of a stamen are the **anther** (pollen sac), the **filament** supporting the anther, and the **connective,** which

FIGURE 8.12 Illustration, based on light and electron microscopic observations, showing the cytomorphogenesis that results in the shedding of excess cytoplasm. *A,* A ring of cytoplasm begins to grow down around the anterior coil of the developing spermatid. *B,* The ring has almost completely enclosed the anterior end, forming a veil of two thicknesses of plasmalemma and a thin layer of cytoplasm over the flagella and anterior coil. *C,* The ring fuses with itself, enclosing the flagella and anterior region of the sperm in an internal but extracellular space; the anterior bridge begins to fill with cytoplasm. *D,* Upon release from the microspore, the sperm pinches off the anterior bridge of excess cytoplasm, the cytoplasm rounds up into a sphere, and the sperm swims away. (From D.G. Myles and P.K. Hepler. 1977. Spermiogenesis in the fern, *Marsilea:* Microtubules, nuclear shaping, and cytomorphogenesis. J. Cell Sci., *23:* 76.)

joins the filament and the anther. The formation of pollen, which occurs within the anther, is outlined in Figure 8.14. Initially the anther contains a number of diploid microspore mother cells. These undergo meiosis, each producing four haploid microspores, which are associated in a cluster called a **tetrad.** The microspores usually separate from one another, and the cell walls thicken. Concurrently, the single haploid nucleus divides mitotically to yield a **generative cell** and a **vegetative cell,** which are separated by a thin wall. This two-celled structure is the male gametophyte, or pollen grain. After their differentiation is complete, pollen grains are released from the anther.

The pistil consists of three regions: The enlarged tip is the **stigma,** the stalk is the **style,** and the enlarged base is the **ovary,** which is composed of one or more **ovules.** The eggs develop within the ovules. An immature ovule contains two distinct cell layers: the central **nucellus** and the outer **integuments.** The latter encircle the nucellus except for a minute opening near the base, the **micropyle.** One of the nucellar cells enlarges to become the megaspore mother cell, which undergoes meiosis to produce four haploid megaspores (Fig. 8.15). Only one of these cells survives; it is nourished by the cells of the nucellus and enlarges to become the **embryo sac,** which forms the **female gametophyte.**

The nucleus of the embryo sac divides mitotically three times to produce eight nuclei. Three nuclei migrate to either end of the embryo sac, acquire their own cytoplasm, and become partitioned off as individual cells. The three at the top of the embryo sac (the **chalazal end**) are called **antipodals.** Of the three at the bottom (the **micropylar end**), the middle one is the egg, which is flanked by two **synergids.** The two remaining nuclei (**polar nuclei**) migrate to the center of the embryo sac, where they become the nuclei of the large **central cell.** The mature gametophyte thus consists of seven cells embedded within the tissues of the ovary.

Sexual reproduction in flowering plants is initiated when pollen grains are transferred to a stigma (Fig. 8.16). After landing on the stigma, the pollen grains begin germination. This process involves mitotic division of the generative cell to form the two **sperm cells,** which together constitute the male gamete, and the formation of the **pollen tube,** which grows down the style and provides a means for the sperm cells to reach the female gametophyte. In some species, mitosis of the generative cell occurs in the pollen grain before its release from the anther.

As the pollen tube grows down the style, the sperm follow the vegetative nucleus down. The pollen tube enters the ovule via the micropyle, and the two sperm cells are forcefully discharged into the female gametophyte. One of the sperm nuclei enters the egg and fuses with the egg nucleus to form the diploid zygote, whereas the other one fuses with the two polar nuclei of the central cell to form a triploid **fusion nucleus.** The endosperm, which develops from the central cell, is utilized as an energy source by the developing embryo. The type of fertilization found in flowering plants is called **double fertilization.**

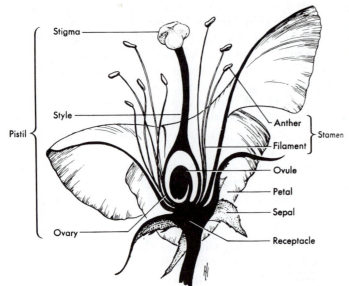

Stigma

Style

Pistil

Ovary

Anther

Filament

Stamen

Ovule

Petal

Sepal

Receptacle

FIGURE 8.13 Section of mature flower. (Reprinted with permission of Macmillan Publishing Co., Inc. from *Botany: A Functional Approach,* 3rd Edition by W.H. Muller. Copyright © 1974 by Walter H. Muller.)

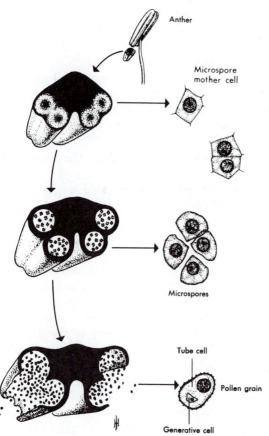

Anther

Microspore mother cell

Microspores

Tube cell

Pollen grain

Generative cell

FIGURE 8.14 Diagrammatic representation of development of mature pollen grains. (Reprinted with permission of Macmillan Publishing Co., Inc. from *Botany: A Functional Approach,* 3rd Edition by W.H. Muller. Copyright © 1974 by Walter H. Muller.)

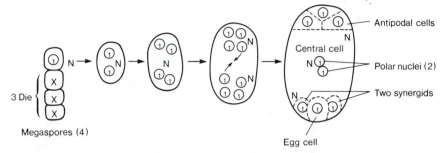

FIGURE 8.15 Diagrammatic outline of embryo sac development. (After K. Norstog and R.W. Long. 1976. *Plant Biology.* W.B. Saunders, Philadelphia, p. 438.)

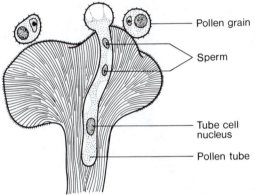

FIGURE 8.16 Pollen tube development. (Redrawn with permission of Macmillan Publishing Co., Inc. from *Botany: A Functional Approach,* 4th Edition by W.H. Muller. Copyright © 1979 by Walter H. Muller.)

Formation of Angiosperm Male Gametes

Although the angiosperm male gamete is reduced to two small cells in the pollen tube, the process of gametogenesis is complex and involves the intermediate formation of pollen grains. These serve as the delivery system for the gametes and provide protection during their development and transport to the stigma. There the pollen grains interact with the stigma and germinate, forming the pollen tubes that deliver the gametes to the female gametophyte. To fulfill these functions, the pollen grains must acquire a number of morphological and physiological specializations during their differentiation. The internal structure of a pollen grain is shown in the electron micrograph of Figure 8.17*A*. The pollen grain contains two cells of unequal size: the vegetative cell and the generative cell.

The surface of pollen is covered by an intricately patterned wall, which may give it a very bizarre appearance (Fig. 8.17*B* to *D*). The wall is quite significant functionally, since it protects the pollen and is responsible for the interaction with the stigma. The wall is a double-layered structure consisting of an outer **exine** and an inner **intine.** The exine is covered with pores and furrows that determine the appearance

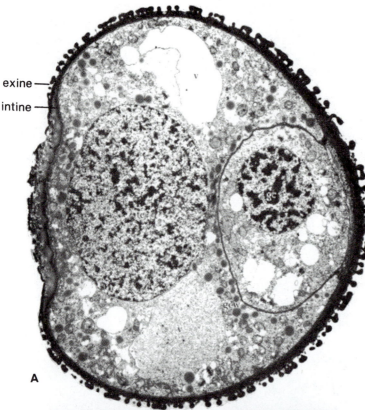

exine —

intine —

A

FIGURE 8.17 Pollen grain structure. *A,* Transmission electron micrograph of pollen grain of blood lily. tc, nucleus of vegetative cell; gc, nucleus of generative cell; gcw, generative cell wall; v, vacuole. (From J.M. Sanger and W.T. Jackson. 1971. Fine structure study of pollen development in *Haemanthus katherinae* Baker. I. Formation of vegetative and generative cells. J. Cell. Sci., *8:* 299.) *B* to *D,* Scanning electron micrographs of pollen of various angiosperm species. *B, Eranthemum.* ×1000. (From K. Norstog and R.W. Long. 1976. *Plant Biology.* W.B. Saunders, Philadelphia, p. 429.)

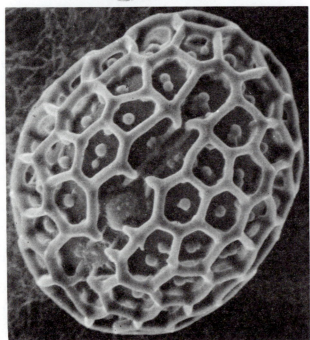

B

Illustration continued on the opposite page.

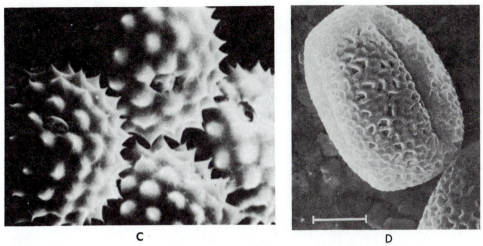

C · D

FIGURE 8.17. (*Continued*) *C, Ambrosia trifida* (ragweed). ×3400. (From R.B. Knox. 1976. Cell recognition and pattern formation in plants. *In* C.F. Graham and P.F. Wareing (eds.), *The Developmental Biology of Plants and Animals.* W.B. Saunders, Philadelphia, p. 145.) *D, Olea europaea.* Scale bar equals 5.0 μm. (From E. Pacini and C.G. Vosa. 1979. Scanning electron microscopy analysis of exine patterns in cultivars of olive [*Olea europaea* L.]. Ann. Bot., *44:* 745–748.)

of the grain. The exine has one or more **germinal apertures,** through which the pollen tube emerges during germination. The discontinuity in the exine wall at the germinal aperture exposes a portion of the intine wall to the surface.

The composition of the two layers is quite different. The exine is comprised of **sporopollenin,** which is a carotenoid-containing polymer. The intine has a pectin-cellulose composition that is similar to that of primary walls of somatic cells. The pollen wall also contains a number of proteins. These proteins apparently function as **recognition substances** that determine whether the stigma will accept the pollen and allow germination to occur. Some of these proteins are the **allergens** that are responsible for hay fever caused by inhalation of certain pollen grains. Pollen wall proteins are located in both the exine and intine. Exine proteins fill the numerous cavities present in this layer, whereas intine proteins are integral intine wall components. Although the intine wall is similar to the primary cell wall of somatic cells, the presence of protein makes this wall unique. Thus, neither of the pollen walls has a close counterpart in somatic tissues (Heslop-Harrison, 1971).

Male gametogenesis is initiated in the anther, where a germ cell line of **sporogenous cells** is established. This line forms the microspore mother cells (or **microsporocytes**) that undergo meiosis to produce the haploid microspores that subsequently differentiate into pollen grains. Since these processes occur within the anther, they are obviously influenced by the surrounding tissue. The innermost layer surrounding the microsporocytes is the **tapetum** (Fig. 8.18) The tapetal cells have been

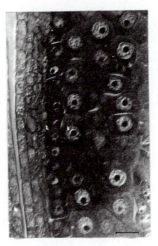

FIGURE 8.18 Longitudinal section of an anther of *Hyoscyamus niger* at the microsporocyte stage. The tapetal cells (T) form a single layer surrounding the microsporocytes (MS). Photographed with differential interference optics. Scale bar equals 20 μm. (From V. Raghavan. 1981. A transient accumulation of poly (A)-containing RNA in the tapetum of *Hyoscyamus niger* during microsporogenesis. Dev. Biol., *81:* 344.)

implicated as being directly involved in microsporogenesis, particularly in providing nutrients, in temporal regulation, and in formation of the pollen wall. Thus, as in animal gametogenesis, pollen grain development is dependent upon somatic cells.

Young microsporocytes are enclosed by primary cell walls but remain interconnected by narrow plasmodesmata. As they enter into prophase of the first meiotic division, a secondary wall is deposited around the cells; this wall is composed of a substance known as **callose.** The callose walls do not sever the cytoplasmic connections between microsporocytes. On the contrary, the cytoplasmic continuity increases with the formation of fewer, but much wider, channels (Fig. 8.19) that allow for interchange of cytoplasmic constituents and are even wide enough for the passage of organelles such as mitochondria. The channels cause cytoplasmic continuity among large numbers of cells, making them interdependent and apparently contributing to their developmental synchrony, since they proceed through meiosis in unison. The channels may also provide a means for rapid distribution of nutrients among the interconnected microsporocytes. Whereas somatic cells are surrounded by cellulose walls that allow intercellular movement of solutes by diffusion, the callose wall is relatively impermeable. Hence, the capability for interchange via diffusion is lost once the callose wall is formed, and the channels may provide a substitute mechanism for intercellular transport.

At the completion of prophase the channels are sealed by callose, and each microsporocyte is completely isolated within an impermeable callose wall. It continues meiosis as an independent entity (Heslop-Harrison, 1965; Knox and Heslop-Harrison, 1970). Meiosis of a microsporocyte results in the formation of four haploid microspores, which remain encased in the original callose wall to form a tetrad. The cells of the tetrad are also separated from one another by callose (Fig. 8.20*A* and *B*).

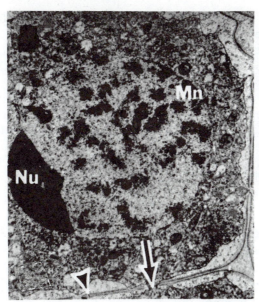

FIGURE 8.19 *Pisum* microsporo-cyte at diplonema of meiosis I. Broad channel connecting this cell with adjacent microsporocyte is in-dicated by the arrow. A severed plasmodesm is indicated by the ar-rowhead. Callose walls are evident between cells. An electron-dense nucleolus (Nu) and a micronucleo-lus (Mn) are evident in the nucleus. ×2785. (From J.A. Biddle. 1979. Anther and pollen development in garden pea and cultivated lentil. Can. J. Bot., *57:* 1889.)

POLLEN MORPHOGENESIS

Pollen wall formation is initiated in the tetrad stage (Knox, 1976). Initial elements of the wall are deposited between the microspores and the sur-rounding callose. The sporopollenin exine wall is the first to be laid down. Exine components are synthesized within the microspores. The exine material is deposited on the microspore surfaces in intricate spe-cies-specific patterns that determine the ultimate appearance of the pol-len grain. The generation of these patterns is an intriguing example of cell morphogenesis.

Pollen grains are released from the tetrad into the anther cavity before wall formation is complete. Pollen release is caused by the diges-tion of the callose (Fig. 8.20C) by a callase enzyme produced by the so-matic cells of the anther (Stieglitz and Stern, 1973). This event begins the final stage of differentiation—**pollen maturation.**

Completion of the pollen wall involves deposition of protein and additional sporopollenin in the exine wall and formation of the intine wall. The sporopollenin is produced by both the pollen and the tapetal cells. The latter also produce the exine proteins. The transfer of tapetal material to the pollen grains is facilitated by dissolution of the tapetal cell walls, leaving naked protoplasts that release vesicles containing ex-ine components (Heslop-Harrison, 1975). Tapetal material may also be transferred to the exines by direct contact between the tapetal protoplasts and pollen grains (Pacini and Juniper, 1979b). This is illustrated in Figure 8.21. Another mechanism involves the secretion of fibrous material by tapetal cells, which fills the cavities of the exine (Fig. 8.22). Because of their origin from tapetal cells, which are sporophyte tissue, and their role in the pollen-stigma interaction, exine proteins are called **sporophytic recognition substances.**

The intine wall components are synthesized in the pollen cytoplasm and deposited on the inner surface of the exine. The proteins in this wall are called **gametophytic recognition substances.** In preparation for formation of the intine, the plasmalemma withdraws irregularly from the inner layer of the exine, leaving an undulating outline below the exine (Fig. 8.23A). Fibrillar precursors of the pectin-cellulose intine wall are deposited in the space between the plasmalemma and exine. Golgi vesicles just below the plasmalemma are presumed to be involved in formation of the polysaccharide intine components (Fig. 8.23B).

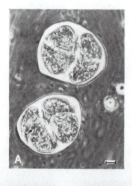

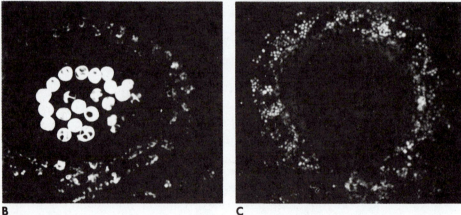

B   C

FIGURE 8.20 Callose wall of *Lilium* microspores. *A,* Light micrograph of two young tetrads. Each microspore is surrounded by a callose wall. Scale bar equals 10 μm. (From J. Heslop-Harrison. 1968b. Wall development within the microspore tetrad of *Lilium longiflorum.* Can. J. Bot., *46:* 1192.) *B* and *C,* Anther sections stained with aqueous aniline blue and photographed under ultraviolet light. Callose shows intense fluorescence, and starch grains in the cells of the anther wall also show some activity. *B,* Tetrads surrounded by callose. *C,* Young pollen grains. The callose wall has been dispersed. The pollen grains no longer show intense fluorescence. (From J. Heslop-Harrison. 1965. Cytoplasmic continuities during spore formation in flowering plants. Endeavour, *25:* 67.)

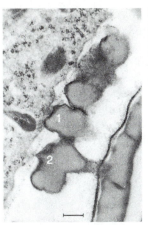

FIGURE 8.21 Contact transfer of tapetal material to pollen grain of *Olea europaea*. The tips of the exine are in contact with the tapetal cell plasmalemma at 1. When the exine detaches (2), a thin layer derived from the tapetal cell is present on the surface. Scale bar equals 1.0 μm. (Courtesy of E. Pacini.)

FIGURE 8.22 Transfer of fibrillar material from tapetal cells to pollen exine in *Olea europaea*. *A*, Fibrillar material is seen between two tapetal cells and in space surrounding pollen exine. Scale bar equals 1.0 μm. (Courtesy of E. Pacini). *B*, Fibrillar material from tapetal cells is present in exine cavities. Scale bar equals 1.0 μm. (From E. Pacini and B.E. Juniper. 1979a. The ultrastructure of pollen-grain development in the olive (*Olea europaea*). 1. Proteins in the pore. New Phytol., *83:* 157–163.)

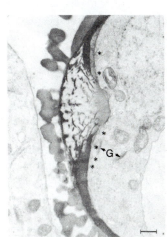

FIGURE 8.23 Formation of the intine in *Olea europaea* pollen. The plasmalemma is retracting from the exine (asterisks). Golgi bodies (G) are thought to be involved in formation of the intine in the space between the plasmalemma and exine. Scale bar equals 1.0 μm. (From E. Pacini and B.E. Juniper. 1979a. The ultrastructure of pollen-grain development in the olive (*Olea europaea*). 1. Proteins in the pore. New Phytol., *83:* 157–163.)

As the polysaccharide components are being deposited, derivatives of the plasmalemma are incorporated into the intine. In some species these structures are most abundant in the aperture region. They may be in the form of microvillilike extensions of the plasmalemma or successive delaminations from the plasmalemma, separated from one another by polysaccharide (Fig. 8.24). These structures contain the intine proteins. Eventually, a polysaccharide layer cuts these proteinaceous components from the remainder of the pollen plasmalemma. The intine proteins are presumed to be synthesized on the ribosomes of the rough endoplasmic reticulum just below the cell surface (Heslop-Harrison, 1975).

In addition to the morphological events, pollen maturation also entails two mitotic divisions. The first of these is decidedly unequal, producing the large vegetative cell and the smaller generative cell. The generative cell is often completely surrounded by the vegetative cell. The second division involves only the generative cell, which divides to produce the two gametes. In some species this division occurs in pollen grains; in others it may occur after pollen germination.

POLLEN GERMINATION

Mature pollen grains are released by dehiscence of the anther and are passively transported to a stigma. On the surface of the stigma, the pollen grain takes up water and begins germination, during which the intine bulges through the germinal aperture, and the pollen tube emerges (Fig. 8.25). Once initiated, the pollen tube continues growing down the style, delivering the sperm cells to the female gametophyte.

Since the possibility exists that pollen grains of numerous species will land on the stigma, plants have evolved a mechanism of pollen recognition to ensure that the egg is fertilized by the appropriate gametes. This recognition process depends upon an interaction between pollen wall proteins and the stigma surface. If the pollen proteins and the stigma are compatible, germination is allowed to begin and proceed to completion. The nature of the pollen-stigma interaction is not fully understood. However, there is evidence that the interaction is similar to that involved in cell–cell adhesion in animals (see Chap. 5).

The lectin concanavalin A (Con A) has been found to bind specifically to the stigma surface, indicating the presence of carbohydrate-containing surface receptors for this lectin (Heslop-Harrison, 1976; Knox et al., 1976). Furthermore, Con A reduces the capacity of the stigma to bind pollen proteins (Clarke et al., 1979). One interpretation of the latter result is that the binding of pollen protein to the stigma is prevented because the Con A itself is attached to the pollen protein-binding sites. Thus, the pollen proteins are lectins that recognize these carbohydrate-containing receptors. Alternatively, it is possible that the Con A causes a nonspecific interference with pollen protein binding. Additional experiments are necessary to distinguish between these possibilities.

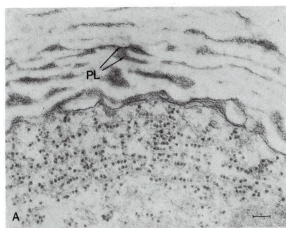

FIGURE 8.24 Incorporation of proteinaceous plasmalemma derivatives into the intine of pollen grains. *A*, Protein lamellae (PL) in the intine of developing *Cosmos bipinnatus* pollen grain. Scale bar equals 0.1 μm. *B*, Microvilli-like extensions of the plasmalemma (M) in the intine of developing *Malvaviscus arboreus* pollen grains. Scale bar equals 1.0 μm. (From J. Heslop-Harrison. 1975. The physiology of the pollen grain surface. Proc. R. Soc. Lond. Ser. B, *190:* 275–299.)

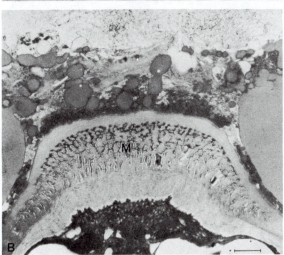

 The emergence of the pollen tube at germination begins the final phase in transport of the male gametes to the female gametophyte. Elongation of the pollen tube occurs at the tip by fusion of Golgi-derived vesicles to the plasmalemma (Fig. 8.26). The Golgi vesicles provide additional membrane for cellular expansion, and their contents, which are expelled at fusion, contribute to remodeling of the cell wall to facilitate this expansion (Picton and Steer, 1981).

 The Golgi vesicles are produced in a zone located several micrometers behind the tip. A zone of mitochondria and endoplasmic reticulum is located between the tip and the Golgi zone. The Golgi vesicles must be transported through the mitochondrial zone to the tip. Transportation of the Golgi vesicles, and hence, tube growth are inhibited by cytochalasin B, suggesting that microfilaments are responsible for vesicle transport (Mascarenhas and La Fountain, 1972).

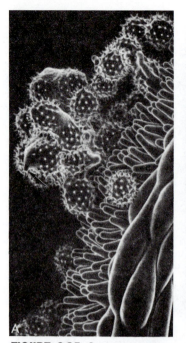
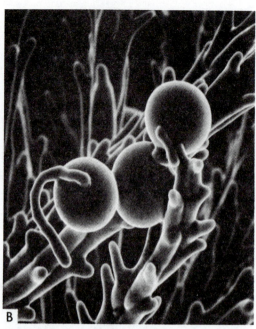

FIGURE 8.25 Scanning electron micrographs of pollen germination. *A*, Clump of *Cosmos bippinatus* pollen grains on stigma soon after pollination. Pollen tubes are emerging (arrow). ×250. (From R.B. Knox. 1973. Pollen wall proteins: Pollen-stigma interactions in ragweed and *Cosmos* [Compositae]. J. Cell Sci., *12:* 435.) *B, Phalaris* pollen, showing emerging pollen tube. (From R.B. Knox. 1976. Cell recognition and pattern formation in plants. *In* C.F. Graham and P.F. Wareing (eds.), *The Developmental Biology of Plants and Animals.* W.B. Saunders, Philadelphia, p. 147.)

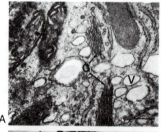
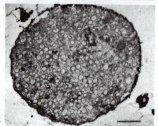

FIGURE 8.26 Mechanism of pollen tube elongation in *Tradescantia virginiana. A,* Golgi zone showing formation of vesicles (V) from Golgi bodies (G). Scale bar equals 0.25 μm. *B,* Apex of pollen tube. Note the densely packed secretory vesicles. Scale bar equals 1.0 μm. (Courtesy of J.M. Picton.)

Within the pollen tube the vegetative cell nucleus and the sperm nuclei migrate toward the female gametophyte. Nuclear migration may be facilitated by contractile microfilaments (Cresti et al., 1976). The germination phase ends with the entrance of the gametes into the female gametophyte, where double fertilization occurs. Our discussion of fertilization will continue in the next chapter.

GENE EXPRESSION DURING MALE GAMETOGENESIS

Differentiation of the male gametes and their vehicle—the pollen grain— involves the coordinated expression of genes of the germ cell line and of the surrounding sporophytic cells, the tapetal cells. Both contribute components to the pollen grain, and both contribute to the temporal regulation of pollen differentiation.

Transcriptional activity in the developing germ cells has been found to occur in distinct phases. Furthermore, posttranscriptional mechanisms that regulate the utilization of the RNA produced in these phases have been identified. Microsporocyte transcription occurs mainly during the premeiotic period and during early prophase (Stanley and Linskens, 1974). Heslop-Harrison (1971) has demonstrated that the genes determining the pattern of the exine wall of the pollen grain are transcribed in this stage. Either these transcripts or the proteins they encode are retained in the cytoplasms of the four haploid microspores, which each microsporocyte produces. The microspores then elaborate the exine walls. This is an example of delayed gene expression that is reminiscent of the delayed expression of genes that are transcribed during animal spermatogenesis. In both cases the expression of genes transcribed prior to meiosis is delayed until the haploid phase, when meiosis has been completed.

Not all products of this early transcriptional phase are retained, however. Many of them appear to be lost from the microsporocyte cytoplasm during midprophase of meiosis I, when there is a dramatic drop in the amount of RNA and in the number of ribosomes in the cytoplasm. This change is apparently part of a massive reorganization of the cytoplasm that may involve selective elimination of cytoplasmic components. It has been interpreted as a mechanism whereby the cell can eliminate certain constituents produced by the diploid nucleus, such as long-lived, ribosome-associated mRNA, before the initiation of transcription in the haploid microspore nuclei (Dickinson and Heslop-Harrison, 1977). Ribosomes are restored in the cytoplasm in late stages of meiosis. They are apparently formed in supernumerary nucleoli that are reminiscent of the multiple nucleoli in the amphibian oocyte nucleus (Dickinson and Heslop-Harrison, 1977).

After meiosis, the genetic composition of individual haploid microspores differs from that of their sister microspores and of the premeiotic diploid nucleus because of segregation and recombination.

Hence, elimination of messengers transcribed by the diploid nucleus would ensure, for example, that microspore genes coding for proteins that function as gametophytic recognition substances could be expressed in each individual pollen grain and not be masked by prior expression of the diploid genome. The callose wall that is completed at the end of prophase I also helps to maintain the genetic integrity of the individual microspores, since it prevents diffusion of components from other members of the tetrad (Heslop-Harrison, 1968a). If this interpretation of cytoplasmic reorganization is correct, it would be one of the most unusual examples of posttranscriptional regulation that has been encountered.

The postmeiotic (i.e., haploid) microspores are also active in RNA synthesis. Genes coding for gametophytic pollen recognition substances must be expressed at this time, since these substances result from expression of the haploid, not the diploid, genome. However, portions of the transcriptional products that accumulate in the microspore cytoplasm are not immediately utilized. Instead, they are stored for utilization during pollen germination, which follows a period of metabolic quiescence that begins at the completion of pollen development.

Metabolic activity is abruptly reinitiated at germination, and the pollen tube may emerge within minutes. The rapid emergence of the pollen tube involves an equally rapid production of new components. In *Tradescantia*, for example, protein synthesis begins within two minutes of placing pollen grains in a germination medium. For such a rapid response the cellular protein synthetic machinery must be poised to begin synthesis when called upon. Otherwise, the cell would be forced first to synthesize components such as ribosomes before the synthesis of pollen tube constituents could commence. Indeed, no *de novo* synthesis of RNA is required for germination, since the pollen tube can be initiated in the presence of actinomycin D, which inhibits RNA synthesis. Although some growth of the pollen tube occurs in the presence of this inhibitor, elongation of the pollen tube stops prematurely. Cycloheximide, which inhibits protein synthesis, allows germination, but tube growth ceases very early—before actinomycin-induced inhibition (Mascarenhas, 1975, 1978). These various results lead to the following conclusions:

1. Proteins essential for the earliest germination events preexist in the pollen cytoplasm.
2. Completion of the pollen tube requires *de novo* protein synthesis.
3. RNA to support the initiation of protein synthesis preexists in the pollen cytoplasm.
4. Protein synthesis in the final phases of pollen tube elongation is supported by RNA that is synthesized after germination has begun.

Direct evidence for stable preexisting messenger RNA in *Tradescantia* pollen has been obtained by isolation and *in vitro* translation of polyadenylated RNA from ungerminated pollen grains. Several of the proteins synthesized *in vitro* comigrate in electrophoresis gels with proteins made in the germinating pollen grains (Frankis and Mascarenhas, 1980).

Formation of Angiosperm Female Gametes

Differentiation of the angiosperm egg is the culmination of the series of events that stem from the formation of the megaspore mother cell within the nucellus of the ovule. The megaspore mother cell enters into meiosis to produce four haploid megaspores. During prophase in many species, the cytoplasm undergoes a reorganization that is similar to that of the microsporocytes, during which considerable organelle degeneration occurs (Fig. 8.27). A reduction in the number of ribosomes is particularly striking. Restoration of ribosomes usually occurs later in meiosis. In most angiosperms only one of the haploid megaspores survives to form the embryo sac (Fig. 8.28). The mechanisms operating to select the surviving megaspore are uncertain but presumably reflect an inherent polarity within the ovule.

Development of the female gametophyte from the surviving megaspore, which results in formation of the egg, is quite rapid. We shall discuss here the eggs of *Capsella bursa-pastoris* (the shepherd's purse) and *Gossypium* (cotton). We have chosen to use *Capsella* not because it is necessarily a typical angiosperm but because it has been used as a model for angiosperm development for more than 80 years by numerous authors and investigators. In recent years, development of the *Capsella* embryo has been examined with the electron microscope, vastly increasing our knowledge of the embryology of this species. We shall describe the embryogenesis of *Capsella* in Chapter 10. *Gossypium* is included primarily because it has been exploited as an experimental organism. Our discussion of angiosperm fertilization in Chapter 9 will be based mainly on *Gossypium*.

One of the most striking characteristics of the *Capsella* egg is its polarity. The basal two thirds of the egg contains a single, large vacuole, whereas the cytoplasm and the nucleus are in the apical (i.e., chalazal) third of the cell (see Fig. 9.42). This polarity somewhat parallels that observed in yolky animal eggs, where the nucleus is restricted to the animal hemisphere. As we shall see in later chapters, the polarity that is established during formation of the egg influences the developmental pattern of the embryo. Therefore, the mechanism that establishes egg polarity has a long-lasting effect upon development.

The shape of the large egg nucleus (Fig. 8.29A) differs dramatically from that of *Fucus*. Whereas the nuclear contour in the latter species is highly convoluted, the *Capsella* nucleus is quite smooth. The nucleus contains a single, highly prominent nucleolus that has a unique morphology and consists of two distinct regions: a dense granular outer layer and a loosely arranged interior composed of an amorphous substance

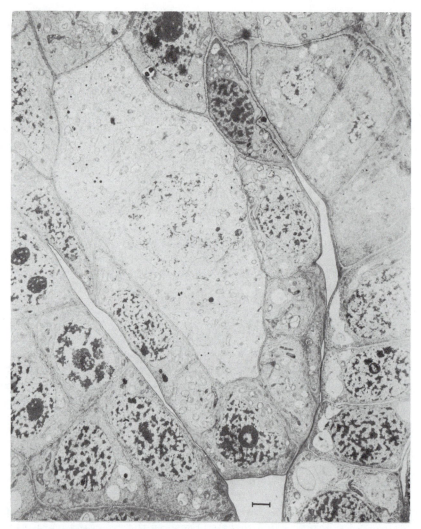

FIGURE 8.27 Megaspore mother cell cytoplasm during meiosis. This megaspore mother cell of *Crepis tectorum* shows organelle degeneration in the cytoplasm. The cell has a centrally located nucleus. Nucellar cells surround the meiotic cell. Scale bar equals 1.0 μm. (From J.-C. Godineau. 1973. Le sac embryonaire des angiospermes. Morphogénèse et infrastructure. Bulletin de la Societe Botanique de France Mémoires Coll. Morphologie, *1973:* 25–54.)

intertwined with fibrous material. The nuclear envelope is perforated by numerous pores that are similar to those of the nuclear envelopes of animal oocytes and the *Fucus* egg.

The egg cytoplasm is notable for its ultrastructural simplicity: It contains very little endoplasmic reticulum and only a few kinds of organelles and inclusions, notably the starch-containing **plastids** and mitochondria. The plastids apparently surround the nucleus, whereas the numerous mitochondria are spread throughout the cytoplasm. The cytoplasm is literally packed with ribosomes, which appear to be randomly distributed rather than in polysomal clusters (Fig. 8.29B). As we shall see in Chapter 9, this condition changes dramatically after fertilization, when the ribosomes appear to be in polysomes. Although there are no substantiating molecular data, the ultrastructural appearance of the cytoplasm suggests that there is considerable synthesis of ribosomes during egg development, storage of these ribosomes in the mature egg, and utilization of the ribosomes in forming polysomes after fertilization. This is an obvious analogy to the ribosome synthesis and accumulation pattern in animal oogenesis, but it remains to be confirmed by molecular data. The scarcity and poor development of Golgi bodies (often called **dictyosomes** by botanists) in the cytoplasm are also consistent with a low level of synthetic activity in the mature egg cytoplasm.

Unlike the *Fucus* egg, the angiosperm egg has a cell wall. However, the wall between the *Capsella* egg and the central cell is incomplete, causing the cell membranes of the two cells to lie close together in numerous places. The cell walls between the egg and the synergids contain

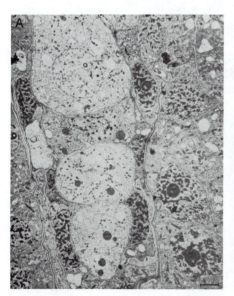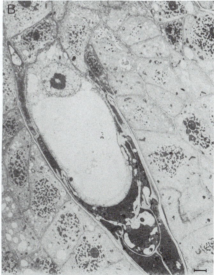

FIGURE 8.28 Megaspores of *Crepis tectorum*. *A*, Linear tetrad of megaspores. The cell at the chalazal end is largest. The cell below it is showing signs of degeneration. *B*, Surviving megaspore at the chalazal end has acquired a large vacuole. The three megaspores toward the micropyle have degenerated. Scale bars equal 2.0 μm. (From J.-C. Godineau. 1973. Le sac embryonaire des angiospermes. Morphogénèse et infrastructure. Bulletin de la Societe Botanique de France Mémoires Coll. Morphologie, *1973*: 25–54.)

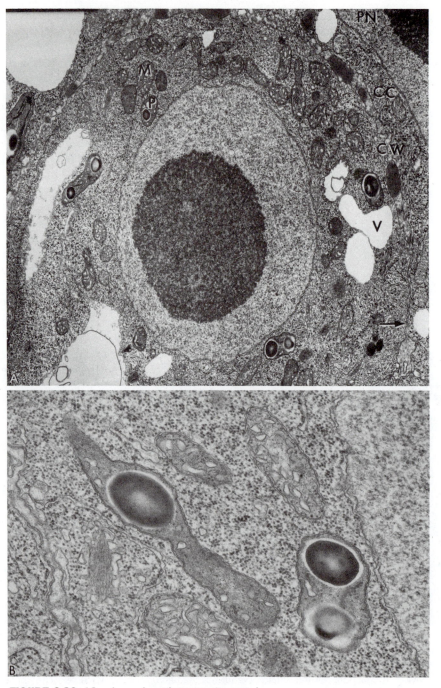

FIGURE 8.29 (*See legend on the opposite page.*)

◀FIGURE 8.29 Ultrastructure of the egg of *Capsella*. *A*, Nucleus and surrounding cytoplasm. The nucleus has a smooth contour and contains a single, large nucleolus. The surrounding cytoplasm contains plastids (P), mitochondria (M), and vacuoles (V). The plasma membranes of the egg and central cell (CC) are separated by a partial cell wall (CW), except at points where the two appear to be in contact (arrow). Note also a portion of one polar nucleus. ×11,000. *B*, Egg cytoplasm containing plastids, mitochondria, and numerous randomly oriented ribosomes. No ribosomal clusters are evident. ×32,500. (From R. Schulz and W.A. Jensen. 1968. *Capsella* embryogenesis: The egg, zygote, and the young embryo. Am. J. Bot., *55*: 810, 811.)

numerous plasmodesmata, which are sites of cytoplasmic continuity that penetrate the walls. It has been proposed that the plasmodesmata allow material for construction of the egg cell to be transported from the synergids. If this mechanism does occur, the synergids may be analogous to the nurse cells of meroistic insects (Jensen, 1965). The basal cell walls of the synergids are modified to form fingerlike extensions, which are called the **filiform apparatus.** This is the site of sperm entry during fertilization.

The *Gossypium* egg has a strikingly different appearance from that of *Capsella* mainly because of the difference in the relative volumes of cytoplasm and vacuole (see the electron micrograph of the *Gossypium* egg in Figure 9.41). As we discussed previously, the vacuole of the *Capsella* egg occupies the basal two thirds of the cell. By contrast, in *Gossypium* the vacuole fills nearly the entire cell, and the cytoplasm is spread in a thin layer between the vacuole and the plasma membrane (Jensen, 1968). Like the *Capsella* egg, the nucleus is in the chalazal third of the *Gossypium* egg; the greatest concentration of cytoplasm surrounds the nucleus. Thus, the *Gossypium* egg also exhibits polarity. The egg is only partially surrounded by a wall; there is no wall between the egg and synergids or central cell. As in the *Capsella* egg, the ribosomes are plentiful and exist predominantly as monosomes rather than as polysomes (see Fig. 9.43*A*). The dramatic changes that occur in the cytoplasm after fertilization will be discussed in the next chapter.

REFERENCES

Biddle, J.A. 1979. Anther and pollen development in garden pea and cultivated lentil. Can. J. Bot., *57*: 1883–1900.

Brawley, S.H., R. Wetherbee, and R.S. Quatrano. 1976. Fine-structural studies of the gametes and embryo of *Fucus vesiculosus* L. (Phaeophyta). II. The cytoplasm of the egg and young zygote. J. Cell Sci., *20*: 255–271.

Clarke, A. et al. 1979. Pollen-stigma interactions: Identification and characterization of surface components with recognition potential. Proc. Natl. Acad. Sci. U.S.A., *76*: 3358–3362.

Cresti, M.M. et al. 1976. Fibrous masses and cell and nucleus movement in the pollen tube of *Petunia hybrida*. Acta. Bot. Neerl., *25*: 381–383.

Dickinson, H.G., and J. Heslop-Harrison, 1977. Ribosomes, membranes, and organelles during meiosis in angiosperms. Phil. Trans. R. Soc. Lond. [Series B], *277*: 327–342.

Diers, L. 1967. Der Feinbau des Spermatozoids von *Sphaerocarpos donnellii* Aust. (*Hepaticae*). Planta (Berl.), *72*: 119–145.

Frankis, R., and J.P. Mascarenhas. 1980. Messenger RNA in the ungerminated pollen grain: A direct demonstration of its presence. Ann. Bot., *45*: 595–599.

Godineau, J.-C. 1973. Le sac embryonaire des angiospermes. Morphogénèse et infrastructure. Bulletin de la Societe Botanique de France Mémoires Coll. Morphologie, 1973: 25–54.

Hepler, P.K. 1976. The blepharoplast of *Marsilea:* Its *de novo* formation and spindle association. J. Cell Sci., *21*: 361–390.

Hepler, P.K., and D.G. Myles. 1977. Spermatogenesis in Marsilea: An example of male gamete development in plants. *In* B.R. Brinkley and K.R. Porter (eds.), *International Cell Biology. 1976–1977.* The Rockefeller University Press, New York, pp. 569–579.

Heslop-Harrison, J. 1965. Cytoplasmic continuities during spore formation in flowering plants. Endeavour, *25*:65–72.

Heslop-Harrison, J. 1968a. Pollen wall development. Science, *161*: 230–237.

Heslop-Harrison, J. 1968b. Wall development within the microspore tetrad of *Lilium longiflorum*. Can. J. Bot., *46*: 1185–1192.

Heslop-Harrison, J. 1971. Wall pattern formation in angiosperm sporogenesis. Symp. Soc. Exp. Biol. *25*: 277–300.

Heslop-Harrison, J. 1975. The physiology of the pollen grain surface. Proc. Roy. Soc. Lond. Ser. B, *190*: 275–299.

Heslop-Harrison, Y. 1976. Localisation of concanavalin A binding sites on the stigma surface of a grass species. Micron, *7*: 33–36.

Jensen, W.A. 1965. The ultrastructure and histochemistry of the synergids of cotton. Am. J. Bot., *52*: 238–256.

Jensen, W.A. 1968. Cotton embryogenesis: The zygote. Planta (Berl.), *79*: 346–366.

Knox, R.B. 1973. Pollen wall proteins: Pollen-stigma interactions in ragweed and *Cosmos* (Compositae). J. Cell Sci., *12*: 421–443.

Knox, R.B. 1976. Cell recognition and pattern formation in plants. *In* C.F. Graham and P.F. Wareing (eds.), *The Developmental Biology of Plants and Animals.* W.B. Saunders Co., Philadelphia.

Knox, R.B., and J. Heslop-Harrison. 1970. Direct demonstration of the low permeability of the angiosperm meiotic tetrad using a fluorogenic ester. Z. Pflanzenphysiol., *62*: 451–459.

Knox, R.B. et al., 1976. Cell recognition in plants: Determinants of the stigma surface and their pollen interactions. Proc. Natl. Acad. Sci. U.S.A., *73*: 2788–2792.

Mascarenhas, J.P. 1975. The biochemistry of angiosperm pollen development. Bot. Rev., *41*: 259–314.

Mascarenhas, J.P. 1978. Ribonucleic acids and proteins in pollen germination. Proc. IV. Int. Palynol. Conf., Lucknow (1976–1977), *1*: 400–406.

Mascarenhas, J.P., and J. La Fountain. 1972. Protoplasmic streaming, cytochalasin B, and growth of the pollen tube. Tissue Cell, *4*: 11–14.

McCully, M.E. 1968. Histological studies on the genus *Fucus*. II. Histology of the reproductive tissues. Protoplasma, *66*: 205–230.

Muller, W.H. 1974. *Botany: A Functional Approach*, 3rd ed. Macmillan, New York.

Muller, W.H. 1979. *Botany: A Functional Approach*, 4th ed. Macmillan, New York.

Myles, D.G., and P.R. Bell. 1975. An ultrastructural study of the spermatozoid of the fern, *Marsilea vestita*. J. Cell Sci., *17*: 633–645.

Myles, D.G., and P.K. Hepler. 1977. Spermiogenesis in the fern *Marsilea*: Microtubules, nuclear shaping, and cytomorphogenesis. J. Cell Sci., *23*: 57–83.

Myles, D.G., and P.K. Hepler. 1982. Shaping of the sperm nucleus in *Marsilea*: A distinction between factors responsible for shape generation and shape determination. Dev. Biol., *90*: 238–252.

Myles, D.G., D. Southworth, and P.K. Hepler. 1978. A freeze-fracture study of the nuclear envelope during spermiogenesis in *Marsilea*. Formation of a pore-free zone associated with the microtubule ribbon. Protoplasma, *93*: 419–431.

Norstog, K., and R.W. Long. 1976. *Plant Biology*. W.B. Saunders Co., Philadelphia.

Pacini, E., and B.E. Juniper. 1979a. The ultrastructure of pollen-grain development in the olive (*Olea europaea*). 1. Proteins in the pore. New Phytol., *83*: 157–163.

Pacini, E., and B.E. Juniper. 1979b. The ultrastructure of pollen-grain development in the olive (*Olea europaea*). 2. Secretion by the tapetal cells. New Phytol., *83*: 165–174.

Pacini, E., and C.G. Vosa. 1979. Scanning electron microscopy analysis of exine patterns in cultivars of olive (*Olea europaea* L.). Ann. Bot., *44*: 745–748.

Picton, J.M., and M.W. Steer. 1981. Determination of secretory vesicle production rates by dictyosomes in pollen tubes of *Tradescantia* using cytochalasin D. J. Cell Sci., *49*: 261–272.

Pollock, E.G. 1970. Fertilization in *Fucus*. Planta (Berl.), *92*: 85–99.

Raghavan, V. 1981. A transient accumulation of poly (A)-containing RNA in the tapetum of *Hyoscyamus niger* during microsporogenesis. Dev. Biol., *81*: 342–348.

Reynolds, W.F., and S.L. Wolfe. 1978. Changes in basic proteins during sperm maturation in a plant, *Marchantia polymorpha*. Exp. Cell Res., *116*: 269–273.

Sanger, J.M., and W.T. Jackson. 1971. Fine structure study of pollen development in *Haemanthus katherinae* Baker. I. Formation of vegetative and generative cells. J. Cell Sci., *8*: 289–301.

Schulz, R., and W.A. Jensen. 1968. *Capsella* embryogenesis: The egg, zygote, and the young embryo. Am. J. Bot., *55*: 807–819.

Stanley, R.G., and H.F. Linskens. 1974. *Pollen. Biology, Biochemistry, Management*. Springer-Verlag, Berlin.

Stieglitz, H., and H. Stern. 1973. Regulation of β-1,3-glucanase activity in developing anthers of *Lilium*. Dev. Biol., *34*: 169–173.

Turner, F.R. 1968. An ultrastructural study of plant spermatogenesis. Spermatogenesis in *Nitella*. J. Cell Biol., *37*:370–393.

PART FOUR
FROM EGG
TO EMBRYO

9 Fertilization

Mature gametes are products of complex differentiation processes that prepare them to participate in establishing a new generation of organisms. During their differentiation, male and female gametes acquire distinct characteristics that define their separate roles in fertilization. The male gametes normally are motile and must travel to the egg. To facilitate their mobility, sperm acquire specialized locomotory organelles, and they divest themselves of excess cytoplasm, thus easing the burden on the transport mechanism. They must also have the means to attach to the egg surface, penetrate it, and deliver the haploid nucleus to the egg interior. The egg, on the other hand, need not have elaborate locomotory organelles, but it must have large amounts of cytoplasm, from which the embryo is formed. In addition, the egg possesses the physiological and morphological capacity to be fertilized by the sperm.

Successful fertilization is the culmination of **mating,** which is a behavioral phenomenon adapted to the conditions of fertilization for each species. The essential aspects of mating are that the sperm and egg encounter one another and that the zygote be in an environment favorable for development. Each sex must form sufficient gametes to ensure that enough zygotes will be produced to perpetuate the species. The enormity of this task is illustrated by marine invertebrates, which release their gametes into the open sea, where they are rapidly dispersed. For this reason, vast numbers of gametes may be produced to maintain a sufficient gamete concentration in the water. A single sea urchin, for example, may release up to 400 million eggs or 100 billion sperm during a single breeding season lasting but a few months (Epel, 1977). An improvement in the efficiency of reproduction is clearly advantageous to any species. Any adaptation that facilitates successful fertilization reduces the energy cost of reproduction. This is especially true in the female, since the production of eggs requires the synthesis of considerable cytoplasm.

Various mechanisms have evolved to facilitate successful reproduction. One kind of mechanism involves the *timing of mating*. If mating occurs when the gametes of both sexes are mature and when the zygotes are most likely to thrive, fewer gametes will be "wasted." Most species have a restricted breeding season that coincides with gamete maturity. The timing of the breeding season is usually controlled by climatic conditions. The estrous cycle of mammals is an example of close synchrony between mating and gamete release. Another type of mechanism that facilitates fertilization is *morphological adaptation*, which increases the likelihood of sperm-egg contact. An example is internal fertilization in mammals in which the gametes of both sexes are deposited in the female reproductive tract. Containment of the gametes prevents dispersal such as occurs in water. Furthermore, fluid movements within the reproductive tract assist in transporting the gametes to the site of fertilization.

Sperm movement toward the egg may either be random or the consequence of directed movement caused by an "attractive force" associated with the egg. The latter mechanism, called chemotaxis, typifies a number of plant species, including both plants we have been discussing as examples of primitive plants—*Fucus* and *Marsilea*. As we shall discuss in section 9-5, plant chemotactic substances are released by the immotile egg or by the surrounding archegonium to guide the free-swimming sperm to the immediate vicinity of the egg, where the gametes may fuse. Recent systematic searches for chemotaxis in animals by Miller have revealed that the mechanism is widespread. Chemotaxis has been demonstrated in certain hydrozoan cnidarians, ascidians, mollusks, and echinoderms (Miller, 1977a, b, 1981; Metz, 1978).

Mechanisms that facilitate sperm–egg encounter ensure that sufficient sperm will arrive in the vicinity of the eggs. The sequence of events after the encounter culminates in the development of a diploid zygote from the haploid gametes. These events are critical to the survival of the species. For this reason, a number of mechanisms have evolved to ensure (1) that the sperm can penetrate the egg accessory layers and fuse with the egg and (2) that only a single sperm of the same species fertilizes the egg. We shall first discuss how gametes make contact with one another and then consider the response of the egg and the formation of a diploid zygote nucleus.

9–1. SPERM–EGG ASSOCIATION

Embryologists have been fascinated by the relationships between sperm and eggs since fertilization was first observed a century ago. What enables these cells to recognize one another? How do eggs discriminate between sperm of their own species and unrelated sperm? Why does only a single sperm fuse with an egg? These questions have led to considerable speculation during the past 100 years, but the groundwork for the resolution of these problems has been laid only in recent years. Detailed experimental analyses of fertilization were first conducted with sea urchins and other echinoid echinoderms; these species have also been the targets of recent investigations that have greatly clarified how sperm–egg association occurs. It is therefore appropriate that we begin this section with the echinoids.

Echinoids

These common marine organisms have many advantages for the study of fertilization. They can be readily induced to shed large numbers of gametes simply by injecting the mature adult with a solution of potassium chloride. The gametes can be mixed by the investigator, who can follow the sequence of fertilization events in precise chronological order.

The echinoid egg (Fig. 9.1) is surrounded by an outer jelly coat composed of a polysaccharide-glycoprotein complex and by a vitelline envelope that is composed of a network of glycoprotein fibers. The vitelline envelope is attached to the egg surface by a series of short processes called **vitelline posts.** Initial sperm–egg contact involves the jelly coat. A great deal of controversy about the jelly coat has arisen over the years, beginning with an observation made by F. R. Lillie (1912). He described the response of spermatozoa to seawater in which unfertilized eggs had been suspended. This "egg water" causes sperm to form clusters consisting of aggregates of sperm oriented head to head. Lillie proposed that this **agglutination** of sperm is due to a substance that he called **fertilizin**, which diffuses from the eggs into the seawater. It was later demonstrated that the fertilizin is derived from the jelly coat. The agglutination was described as a cross-linking of sperm heads by the fertilizin, much like the agglutination of cells by antibodies (Tyler, 1948). A receptor molecule (**antifertilizin**) was proposed to exist in sperm heads and to combine with fertilizin to cause the agglutination. When sperm and egg jelly from different species were combined, it was reported that egg jelly agglutinated sperm of its own species more readily than unrelated sperm. These observations led to the widespread belief that the fertilizin functions as the species-specific sperm receptor during fertilization (see Metz, 1967, for review).

The fertilizin–antifertilizin system has dominated thought on fertilization for several decades, and many attempts have been made to establish it as a universal system for sperm–egg interaction in the animal kingdom. However, in recent years the fertilizin concept has come under attack. It has been shown that agglutination is not necessarily species-specific. Furthermore, it has recently been proposed that the clusters of sperm that form in response to jelly are caused by an increase in sperm motility induced by the jelly. The rapidly moving sperm swarm together, forming a large cluster. The sperm are not physically held together, but are moving within the cluster. According to this hypothesis, the clustering is not actually an agglutination (Loeb, 1914; Collins, 1976).

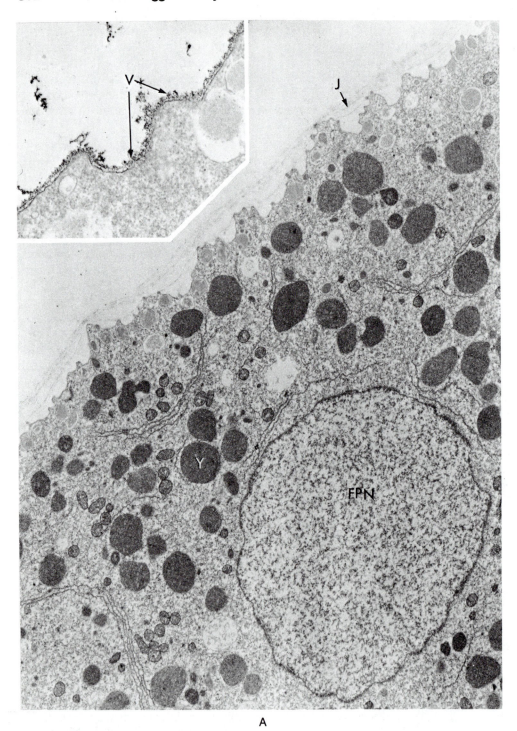

A

FIGURE 9.1 (*See legend on the opposite page.*)

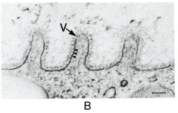

B

FIGURE 9.1 The echinoid egg surface. *A*, Electron micrograph of the cortex of the sand dollar egg. The female pronucleus (FPN) lies in the cortex. The cytoplasm contains many yolk bodies (Y). The outer jelly layer (J) has shrunken during preparation. × 7500. Inset: The vitelline envelope (V) is closely applied to the outer surface of the oolemma. × 32,500. (From R.G. Summers and B.L. Hylander. 1974. An ultrastructural analysis of early fertilization in the sand dollar, *Echinarachnius parma*. Cell Tissue Res., *150:* 352. Reprinted with permission of Springer-Verlag, Heidelberg.) *B*, Electron micrograph of the cortex of the sea urchin egg. Vitelline envelope is attached to the egg surface by a series of vitelline posts (arrowheads). Scale bar equals 0.1 μm. (From D.E. Chandler and J. Heuser. 1980. Reproduced from *The Journal of Cell Biology*, 1980, vol. 84, pp. 618–632 by copyright permission of The Rockefeller University Press.)

The responses of sperm to dissolved egg jelly, discussed earlier, are artifacts that are observed in the laboratory and are not normal events in fertilization. However, they do reflect specific interactions between egg jelly and sperm that are of undoubted significance in fertilization. One of the most important effects of the egg jelly is to trigger the release of the **acrosomal process,** the sperm organelle that attaches and fuses with the egg surface. Release of this process is the acrosome reaction. The formation of the acrosomal process in the sand dollar, *Echinarachnius parma*, is shown in Figure 9.2. The anterior tip of the intact sperm prior to the acrosome reaction is shown in Figure 9.2*A*. The major components of the acrosome are the spherical, membrane-enclosed acrosomal vesicle, which contains the acrosomal granule, and the surrounding **periacrosomal material.** The acrosomal vesicle rests in a cup-shaped depression in the apex of the nucleus, the **subacrosomal fossa.** Most of the periacrosomal material is located in the fossa. Note the close association between the acrosomal vesicle membrane and the sperm plasma membrane in the region anterior to the two arrows. During the acrosome reaction these two membranes fuse along a line that circumscribes the acrosome at the "rim of dehiscence," which is marked by the arrows. As a consequence, the entire anterior half of the acrosomal vesicle membrane and the overlying plasma membrane are shed (Fig. 9.2*B* and *C*). The basal half of the acrosomal vesicle membrane is everted to form the acrosomal process. The growing process is coated with the contents of the acrosomal vesicle. The acrosomal process contains numerous microfilaments that are organized from the periacrosomal material of the subacrosomal fossa. The polymerization of the actin subunits to form microfilaments has been studied in the sea cucumber, *Thyone briareus*. In sperm of this species, polymerization is initiated in an organelle called the **actomere,** which

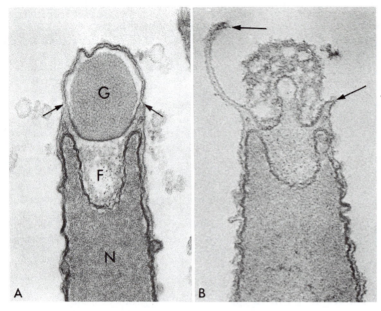

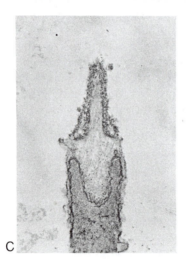

FIGURE 9.2 Acrosome reaction in sand dollar sperm. A, Intact sperm head. An acrosomal granule (G) lies within the acrosomal vesicle, which is completely surrounded by a membrane. Periacrosomal material is particularly prevalent in the subacrosomal fossa (F), which is a depression in the tip of the nucleus (N). The prospective rim of dehiscence is indicated by arrows. ×77,600. B, Initial stage of reaction in which the sperm plasmalemma and the acrosomal vesicle membrane have partially fused. The fused edges are indicated by arrows. The acrosomal process has begun to elongate by eversion of the remaining basal portion of the acrosomal vesicle membrane. Acrosomal vesicle contents have been altered in their morphology. ×70,400. C, Later stage of reaction in which the process has undergone primary elongation. Note that the remaining contents of the vesicle now form a coating on that portion of membrane that is of acrosomal vesicle origin. Microfilaments are clearly observable within the process at this stage of its formation. ×58,800. (From R.G. Summers and B.L. Hylander. 1974. An ultrastructural analysis of early fertilization in the sand dollar, *Echinarachnius parma.* Cell Tissue Res., *150:* 348, 354. Reprinted with permission of Springer-Verlag, Heidelberg.)

appears to contain a small number of preformed microfilaments embedded in a dense matrix (Tilney, 1978). It will be interesting to learn whether this—or a similar—organelle also initiates microfilament formation in the echinoid acrosome reaction.

The belief that the acrosome reaction is triggered by the jelly coat is based upon observations that the sperm of some echinoid species undergo the acrosome reaction while traversing the jelly coat (Dan, 1967) and that a solution of dissolved jelly coat material can cause sperm to undergo an acrosome reaction (Dan et al., 1964; Decker et al., 1976). The component of the egg jelly that induces the acrosome reaction has been identified as a sulfated fucose polymer (SeGall and Lennarz, 1979). Treatment of sperm with univalent fragments of antibodies (Fab fragments; see Chap. 5) to a sperm surface glycoprotein prevents the sperm from responding to egg jelly (Fig. 9.3). This result indicates that a component of the egg jelly, possibly the fucose polymer, interacts with a sperm surface glycoprotein to induce the acrosome reaction.

The sperm of some sea urchin species are induced to undergo the acrosome reaction only with jelly coat from the same species, whereas others will respond equally well to either a homologous or heterologous jelly coat. Since only acrosome-reacted sperm can fertilize an egg, the sperm–egg jelly interaction in the former group appears to play a role in the species specificity of fertilization. The latter group may rely upon a

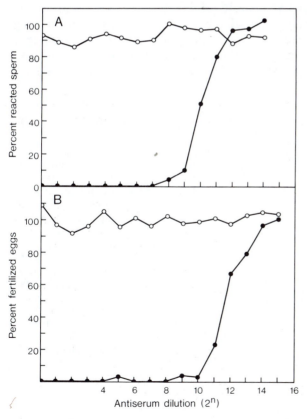

FIGURE 9.3 Effects of sperm surface glycoprotein Fab on the jelly-induced acrosome reaction (*A*) and on fertilization (*B*). Lines with open circles represent responses to preimmune Fab and serve as controls. Lines with closed circles represent responses to immune Fab. (After A.C. Lopo and V.D. Vacquier. 1980. Antibody to a sperm surface glycoprotein inhibits the egg jelly-induced acrosome reaction of sea urchin sperm. Dev. Biol., *79*: 331, 332.)

block to cross-fertilization, which occurs at a subsequent step (see following discussion).

It has been known for some time that induction of the acrosome reaction by egg jelly requires the presence of calcium ions in the medium (Dan, 1954). Dan interpreted this to mean that the acrosome reaction is triggered by an increased permeability to Ca^{++} caused by egg jelly (Dan et al., 1964). In recent years a simple tool has become available for approaching this problem. This tool belongs to a group of antibiotics called **ionophores,** which are lipid soluble molecules that selectively bind to certain cations and transport them across membranes. One ionophore— A23187—is specific for transport of divalent cations such as Ca^{++} and Mg^{++}. Thus, A23187 can be used to raise the levels of Ca^{++} in the sperm. Indeed, the acrosome reaction is triggered by the ionophore, but only when Ca^{++} is present in the medium (Fig. 9.4). The Ca^{++} dependency indicates that it, not Mg^{++}, is responsible for mediating the effects of the ionophore.

Additional evidence implicating Ca^{++} uptake as a trigger of the acrosome reaction is provided by experiments in which uptake is prevented by the use of drugs. For example, two drugs that specifically block Ca^{++} transport (D600 and verapamil) inhibit the acrosome reaction (Fig. 9.5). Finally, $^{45}Ca^{++}$ has been found to be incorporated into sperm during the acrosome reaction (Schackmann et al., 1978). The Ca^{++} is thought to be necessary for the fusion of the acrosomal vesicle membrane and the sperm plasma membrane, which exposes the contents of the acrosomal vesicle (Tilney et al., 1978).

The acrosome reaction also involves uptake of sodium ions and a discharge of protons. A role for sodium in the acrosome reaction is sug-

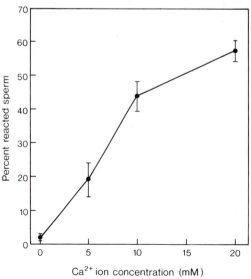

FIGURE 9.4 Calcium ion dependency of ionophore induction of the sea urchin acrosome reaction. Sperm were added to artificial sea water containing A23187 and the indicated amounts of Ca^{++}. After 2.5 minutes the sperm were fixed and examined with the electron microscope to determine percentage of reacted sperm. (After G.L. Decker, D.B. Joseph, and W.J. Lennarz. 1976. A study of factors involved in induction of the acrosomal reaction in sperm of the sea urchin, *Arbacia punctulata*. Dev. Biol., *53:* 121.)

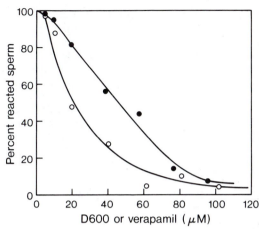

FIGURE 9.5 Inhibition of the acrosome reaction by D600 and verapamil. Sperm were added to artificial sea water containing egg jelly, $CaCl_2$, and the indicated amounts of D600 (line with closed circles) or verapamil (line with open circles). (After R.W. Schackmann, E.M. Eddy, and B.M. Shapiro. 1978. The acrosome reaction of *Strongylocentrotus purpuratus* sperm. Ion requirements and movements. Dev. Biol., *65:* 490.)

gested by this evidence: (1) The reaction fails to occur in Na^+-free seawater; (2) $^{22}Na^+$ is incorporated into sperm during the acrosome reaction (Schackmann and Shapiro, 1981); (3) The acrosome reaction can be induced in the absence of jelly by the monovalent cation ionophore gramicidin S (Schackmann et al., 1978). This action of gramicidin S is dependent upon the presence of sodium ions in the medium, indicating that it causes a Na^+ influx.

The proton release occurs very rapidly upon exposure of sperm to jelly and is reflected by an acidification of the medium during the acrosome reaction (Fig. 9.6). The Na^+ uptake and the H^+ release both occur within 15 seconds of sperm exposure to egg jelly, which is approximately the time of the appearance of the acrosomal process. Furthermore, roughly equal molar amounts of Na^+ and H^+ are transported in opposite directions. These results suggest that the Na^+ is exchanged for the H^+, which is then transported out of the cell. The efflux of H^+ raises the sperm intracellular pH (Schackmann et al., 1981). The rise in pH triggers the formation of the acrosomal process, possibly by inducing the polymerization of actin, which is necessary for eversion of the process (Tilney et al., 1978).

An increase in the levels of the monovalent cation potassium in the medium has been shown to *inhibit* the acrosome reaction (Schackmann et al., 1978). The drug tetraethylammonium chloride, a potent inhibitor of K^+ transport, also inhibits the acrosome reaction. These results suggest that K^+ must be transported out of the sperm to allow the acrosome reaction to proceed. The movement of K^+ has been traced by using $^{42}K^+$ and K^+-selective electrodes. These studies have confirmed that an efflux of K^+ occurs during the acrosome reaction (Schackmann and Shapiro, 1981). A consequence of K^+ efflux could be a change in the sperm plasma membrane potential. In fact, a decrease in K^+-dependent membrane po-

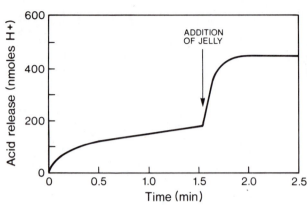

FIGURE 9.6 Acid release from sperm induced by egg jelly. Sperm were added to artificial sea water and were allowed to approach a steady-state level of acid release. At this time jelly was added, and the resulting pH shift was followed. (After R.W. Schackmann, E.M. Eddy, and B.M. Shapiro. 1978. The acrosome reaction of *Strongylocentrotus purpuratus* sperm. Ion requirements and movements. Dev. Biol., *65:* 490.)

tential upon exposure of sperm to egg jelly has been observed (Schackmann et al., 1981).

The extensive ionic permeability changes we have discussed appear to play key roles in mediating the induction of the acrosome reaction by egg jelly. These events have a close parallel at a later stage in fertilization. As we shall learn in section 9–2, the activation of the sea urchin egg by the sperm is also mediated by ionic fluxes.

After traversing the jelly coat, sperm encounter the vitelline envelope, to which they attach in large numbers by their acrosomal processes (Figs. 9.7 and 9.8). The acrosomal processes appear to bind to the vitelline envelope by means of the acrosomal material that coats the processes. This material has been called bindin because of its functional role in fertilization (Vacquier and Moy, 1977). The attachment between the acrosomal process and vitelline envelope is species-specific (Summers and Hylander, 1975); the specificity is due to a lectinlike interaction between the sperm bindin (which is a protein) and a specific polysaccharide sperm receptor on the vitelline envelope (Glabe et al., 1982). The specificity of this interaction appears to be a significant block to cross-fertilization between species (Glabe and Vacquier, 1978).

Further clarification of the sperm-binding mechanism could be obtained if the receptor molecule were isolated and its role in sperm binding were studied in detail. Recently, partial purification of the putative receptor has been reported (Rossignol et al., 1981). The substance, which has a glycoprotein component, has been obtained from isolated membranes of the eggs of *Arbacia punctulata* and *Strongylocentrotus purpuratus*. The partially purified receptors will bind to sperm of their own species but will not bind sperm of the other species. If these substances are the actual vitelline envelope sperm receptors, they should compete with eggs for sperm binding and therefore inhibit fertilization in a *species-specific manner*. In fact, fertilization of *S. purpuratus* eggs by *S. purpuratus* sperm is inhibited by the homologous receptor, but *A. punctulata* receptor has no effect on *S. purpuratus* fertilization. Thus, these

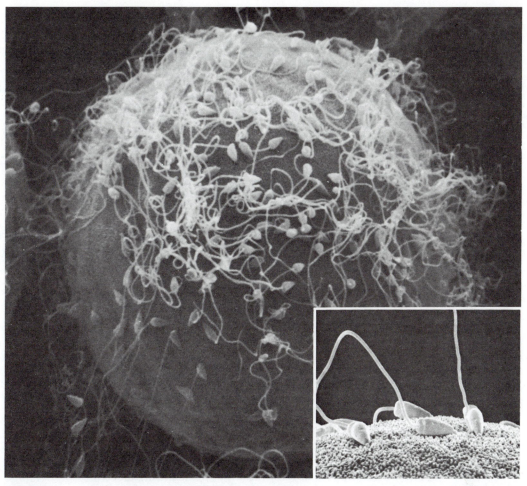

FIGURE 9.7 Scanning electron micrograph showing sperm bound to the vitelline envelope of the sea urchin egg. ×1900. Inset shows a close-up of the vitelline envelope with several sperm bound by their acrosomal processes. The vitelline envelope follows the contours of the egg surface, which is covered with microvilli. ×4800. (From D. Epel. 1977. The program of fertilization. Sci. Am., *237*(5): 128, 129.)

substances have properties that are consistent with those of the actual receptors for their respective sperm. Further study of these substances should greatly increase our understanding of sperm–egg binding.

In spite of the large numbers of sperm that can bind to the vitelline envelope, only a single sperm normally penetrates this layer and fuses with the egg plasma membrane. Penetration of the vitelline envelope may be due to a chymotrypsinlike protease enzyme associated with the acrosome process. A role for the enzyme in vitelline envelope penetration is indicated by experiments in which an inhibitor of chymotrypsin-like enzymes was shown to prevent specifically sperm penetration of the

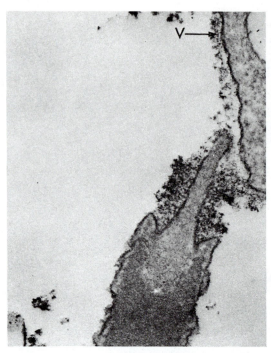

FIGURE 9.8 Binding of the sand dollar sperm to the egg vitelline envelope (V). The material coating the acrosomal process forms a bond with the envelope. ×66,000. (From R.G. Summers and B.L. Hylander. 1974. An ultrastructural analysis of early fertilization in the sand dollar, *Echinarachnius parma.* Cell Tissue Res., *150:* 359.)

vitelline envelope (Green and Summers, 1982). Fusion of the sperm with the egg plasma membrane initially involves the tip of the acrosomal process (Fig. 9.9) and results in the plasma membranes of the two cells becoming continuous and forming a cytoplasmic bridge that eventually enlarges to allow the entire sperm (minus its plasma membrane) to enter the egg. The egg appears to be actively involved in internalizing the sperm. This process is illustrated by the scanning electron micrographs in Figure 9.10. After fusion, microvilli adjacent to the sperm begin to elongate and cluster around the sperm to form a **fertilization cone,** which engulfs the sperm. This interpretation of sperm incorporation is supported by experiments in which both fertilization cone formation and sperm incorporation are inhibited by cytochalasin B (Schatten and Schatten, 1980; Longo, 1980). These results suggest that microfilaments in the egg cortex are responsible for actively drawing the sperm into the egg and that the fertilization cone is a manifestation of this activity. The subsequent behavior of the sperm head in the egg interior will be discussed in detail in section 9–3.

Mammals

The study of mammalian fertilization is much more than an academic exercise, because of the social and economic consequences of reproduction of humans and of domesticated animals. Human overpopulation

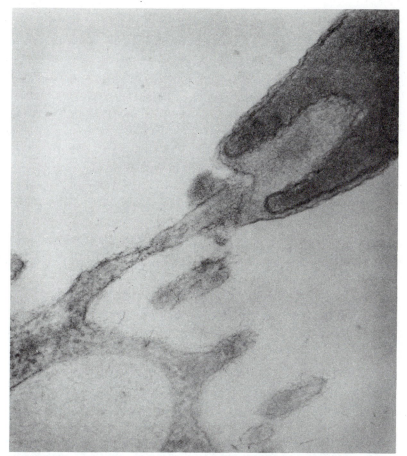

FIGURE 9.9 Moment of fusion between the tip of the sea urchin sperm acrosomal process and an egg microvillus is captured in this transmission electron micrograph. The fusion results in the formation of a cytoplasmic bridge through which the sperm cell enters the egg. ×54,600. (From D. Epel. 1977. The program of fertilization. Sci. Am., *237*(5): 131.)

can be a major social problem, and infertility can be the source of much personal anxiety. An understanding of gamete interaction might assist in solving these problems by suggesting new means of birth control or ways of improving fertility, when desired (see box, p. 383). Furthermore, the economic implications of reproduction of domesticated mammals are of mammoth proportions, and proper management of livestock reproduction is important in optimizing the food supply in any economy.

Mammals present special technical problems for the study of fertilization. Since gamete fusion occurs internally, it is difficult to observe the initial stages of gamete interaction. Investigators prefer to observe these events *in vitro*. Early attempts at *in vitro* fertilization were unsuccessful, since sperm collected directly from the male are unable to

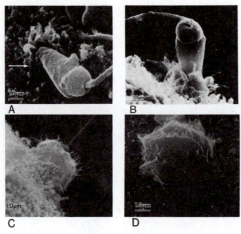

FIGURE 9.10 Scanning electron microscopic study of sperm incorporation by the sea urchin egg. *A*, Microvilli (arrow) elongate and cluster around the sperm head. *B*, Microvilli have elongated and surround the sperm to form the fertilization cone. *C*, As the sperm head and midpiece enter the egg, the tail projects from the fertilization cone. *D*, A patch of wrinkled membrane on the egg surface marks the point of complete sperm entry. (From H. Schatten and G. Schatten. 1980. Surface activity at the egg plasma membrane during sperm incorporation and its cytochalasin B sensitivity. Scanning electron microscopy and time-lapse video microscopy during fertilization of the sea urchin, *Lytechinus variegatus*. Dev. Biol., *78:* 435–449.)

penetrate the egg accessory layers and make contact with the egg surface. This ability is normally acquired by sperm only after they have resided in the female reproductive tract for a number of hours. This acquisition of fertilizing competence by sperm in the female reproductive tract is called capacitation (Austin, 1952). The changes undergone by the sperm include morphological modifications to the plasma membrane (Friend et al., 1977) and modification, redistribution, or loss of surface glycoprotein molecules (Kinsey and Koehler, 1978). These alterations are assumed to prepare the membrane to participate in the acrosome reaction. Sperm also become more motile after exposure to the female reproductive tract (Yanagimachi, 1970). The enhancement of motility may also be a component of capacitation. Increased motility enables the sperm to move farther up the female tract where they ultimately make contact with the egg (Johnson et al., 1981). *In vitro* techniques for inducing sperm capacitation are now available, and capacitated sperm are routinely used for *in vitro* fertilization of mammalian eggs.

Before encountering the egg, sperm must penetrate the egg accessory layers. Penetration is usually attributed to the enzymes released from the acrosome during the acrosome reaction (McRorie and Williams, 1974), which occurs in sperm during their approach to the zona shortly before making contact with the cumulus oophorus. The enzyme **hyaluronidase** is responsible for the penetration of the cumulus oophorus. Hyaluronidase digests the hyaluronic acid that binds the loosely organized cumulus cells. Another enzyme—**corona-penetrating enzyme**—is apparently responsible for allowing sperm to penetrate between the tightly bound corona radiata cells. As we shall discuss subsequently, penetration of the zona pellucida may also be due to an acrosomal enzyme.

The mammalian acrosome reaction differs from that of the sea urchin in that no acrosomal process is formed. The dehiscence of the acrosomal membrane occurs in much the same way, however. The plasmalemma and the outer acrosomal membrane fuse intermittently over the tip of the sperm head, causing the formation of numerous vesicles. This process produces gaps through which the acrosomal contents are released and leaves the inner membrane of the acrosome as the outer surface of the sperm head. The equatorial segment of the acrosome is spared from this vesiculation process and remains intact (Fig. 9.11). Like the sea urchin acrosome reaction, the mammalian reaction requires calcium ions (Yanagimachi and Usui, 1974).

Regular text continued on page 385

ADVANCES IN HUMAN FERTILITY REGULATION: IMMUNOCONTRACEPTION AND *IN VITRO* FERTILIZATION

The steroid contraceptive pill, which inhibits ovulation (see Chap. 7), and the diaphragm, which is a barrier to sperm–egg contact, have been the mainstays of fertility control for several decades. However, recent research has focused on the possibility of perfecting immunological means of contraception. Considerable attention has also been directed in recent years toward finding means to allow women with fertility disorders to conceive and bear children. Spectacular results in this regard have been reported using *in vitro* fertilization techniques. In this section we shall discuss recent progress in these two exciting new areas of human reproductive biology research.

Fertilization is dependent upon successful binding of sperm to the zona pellucida, which is mediated by complementary receptor sites on the sperm plasma membrane and the zona surface. As shown in the following figures, *in vitro* exposure of unfertilized mouse eggs to antibodies directed at the zona will inhibit the binding of sperm to the zona (Aitken and Richardson, 1981). This kind of result has led to the possibility that the zona might be an appropriate target for an immunological approach to contraception. In fact, some naturally occurring fertility disorders in humans may be due to the presence of zona antibodies (Shivers and Dunbar, 1977; Mori et al., 1978). Thus, immunization of a woman with zona antigens might elicit antibodies that would be directed at her zonae and prevent fertilization. Widespread use of this procedure would be facilitated if zona immunity were reversible.

Preliminary results from immunization studies with experimental animals have been encouraging. Two approaches have been used—passive and active immunization. Passive immunization involves injection of serum containing antibodies. For

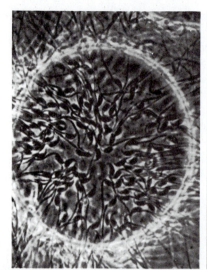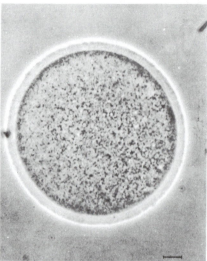

Sperm binding to the zona pellucida of an unfertilized mouse egg (left) and inhibition of binding after incubation of unfertilized eggs in presence of anti-zona antibody (right). Scale bar equals 10 µm. (Courtesy of R.J. Aitken.)

example, passive immunization of mice with rabbit antisera against zona pellucida antigens induces temporary infertility (Tsunoda and Chang, 1978). Active immunization involves injection of zona antigens into the target female, who would then produce antibodies. This procedure is most effective when the zona antigen is from another species. For example, mice immunized with solubilized hamster zonae become temporarily infertile; after sufficient reduction of the antibody titer, the formerly infertile mice become capable of delivering normal young (Gwatkin et al., 1977). Similar results for both passive and active immunization have been obtained in a variety of mammalian species (Aitken et al., 1981). Although the assumption inherent in these experiments has been that the antibodies inhibit fertilization, recent evidence suggests that the antibodies may be affecting ovarian function rather than (or in addition to) fertilization (Wood et al., 1981). Further investigation is necessary to establish the mode of reproductive dysfunction in immunized females.

Another possibility for immunocontraception is to induce immunity to the sperm. Recent *in vitro* studies have shown that Fab fragments of rabbit antihamster sperm antibodies will inhibit hamster sperm binding to and passage through the zona, thus interfering with fertilization (Tzartos, 1979). This and similar results with antisperm antibodies (Lopo and Vacquier, 1980) suggest the possibility of passive immunization with such antisera or of active immunization against sperm as effective strategies for contraception.

Successful application of the techniques of *in vitro* fertilization and implantation of the embryo in the mother's uterus has made the "test tube human baby" a reality (Edwards, 1981). These procedures were perfected by Edwards and Steptoe in the United Kingdom and have since been introduced into other countries where fertility clinics have been established for the primary purpose of helping women with occluded oviducts to conceive and bear children. Preovulatory oocytes are aspirated from their follicles by the technique of **laparoscopy.** For the success of this procedure it is important to predict with a high degree of accuracy the presence of preovulatory oocytes in the ovary. In the natural menstrual cycle, the approach of ovulation can be monitored by measuring estrogen levels, which assesses follicle growth, and the onset of the luteinizing hormone (LH) surge, which indicates that ovulation is imminent (see Chap. 7). Alternatively, gonadotropin or the drug clomiphene can be used to regulate follicular growth and ovulation. This approach has the advantage that several follicles will develop during the cycle. Oocytes can be collected simultaneously from each of these follicles.

After collection, the oocytes are cultured to allow oocyte maturation to conclude. Mature eggs are then fertilized in droplets of medium under oil, which prevents evaporation, or in small culture tubes. A small amount of semen is introduced into the medium containing the egg, and the zygote is allowed to develop to the hatched blastocyst stage in a culture tube before it is implanted through the cervix into the uterus with a catheter. Current research is underway to improve the success of implantation, which is the most unpredictable step in this whole procedure.

Regular text continued from page 383

After penetrating the cellular layers surrounding the egg, the head of the sperm encounters the zona pellucida, a rather thick envelope composed of glycoprotein. Sperm binding occurs between the newly exposed inner acrosomal membrane and the surface of the zona (Hartmann and Gwatkin, 1971). This binding is species-specific in some (but not all) species and is analogous to the binding of the sea urchin sperm to the egg vitelline envelope. In the mammal, the sperm bind to species-compatible sperm receptors on the zona. This receptor system has been the target of recent attempts to develop new contraceptive techniques (see box, p. 383).

Penetration of the zona (Fig. 9.12) has been ascribed to the action of an acrosomal proteolytic enzyme, **acrosin,** which has been proposed to digest a path through the zona in advance of the sperm (Stambaugh and Buckley, 1969; Stambaugh et al., 1969; Gwatkin, 1977). Recent evidence, however, suggests that the proteolytic activity may be necessary for sperm binding to the zona, rather than penetration (Saling, 1981). Further

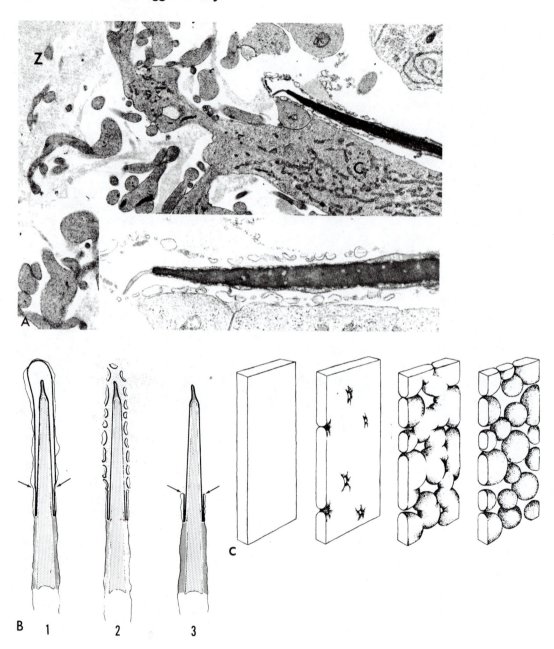

FIGURE 9.11 (*See legend on the opposite page.*)

◄ **FIGURE 9.11** The mammalian (rabbit) acrosome reaction. *A*, A rabbit sperm is approaching the zona pellucida (Z) between corona cells (C) and corona cell processes. The acrosomal cap has been transformed into a series of vesicles that still cover the rostral region of the sperm head. ×10,600. Inset: Higher magnification to show the arrangement of the vesiculated acrosome. ×24,400. *B*, Diagrammatic representation of the acrosome reaction in rabbit spermatozoa. (1) Intact sperm head as seen in samples of ejaculate, epididymal, and capacitated uterine spermatozoa. (2) The outer membrane of the acrosomal cap fuses at several points, progressively, with the overlying plasma membrane to form a series of vesicles around the rostral part of the sperm head; the acrosomal content escapes through the ports that appear between the vesiculated membranes. (3) The vesicles are shed, leaving the sharply defined inner acrosomal membrane as the limiting border of the anterior sperm head. Continuity of the sperm surface is ensured by fusion at the anterior border of the persistent "equatorial" segment between the acrosome and plasma membranes that remain (arrowed). *C*, Diagrams illustrating inferred stages in the process of vesiculation between two apposed cellular membranes. From the available evidence, it is not certain whether the fourth stage is fully achieved in the sperm acrosome reaction. (*A* from J.M. Bedford. 1968. Ultrastructural changes in the sperm head during fertilization in the rabbit. Am. J. Anat., *123:* 339; *B* courtesy of Dr. J.M. Bedford; *C* from C. Barros et al. 1967. Reproduced from *The Journal of Cell Biology*, 1967, vol. 34, pp. C1–C5 by copyright permission of The Rockefeller University Press.)

research is necessary to establish the function of the proteolytic enzyme and to determine whether penetration is enzymatic or mechanical. Factors released at the point of contact between sperm and the zona may block penetration by supernumerary sperm and provide an initial block to polyspermy (Hartmann and Hutchison, 1981). Subsequent polyspermy blocks are also operative and will be discussed in the next section.

Passage of the sperm through the zona is tangential to the egg surface. As a consequence, the tip of the sperm does not make the initial contact with the surface of the egg. Instead, the equatorial segment of the sperm head contacts the egg (Fig. 9.13*A*). A process of egg cytoplasm flows around the midposterior part of the sperm head, superficially internalizing it, while the tip of the head and the tail project into the perivitelline space. Microvilli from the egg surface seem to draw the anterior portion of the sperm head into the egg by a process similar to phagocytosis (Fig. 9.13*B*). After the head has been incorporated, the tail follows and is stripped of its plasma membrane in the process. The sequence of events in sperm entry is summarized in Figure 9.14.

9–2. THE FERTILIZATION RESPONSE: EGG ACTIVATION

Sperm–egg interaction elicits far-ranging changes in the egg. These responses, collectively called **egg activation,** include a block to the entry of additional sperm ("block to polyspermy") and characteristic morphological and physiological changes, which initiate embryonic development.

Block to Polyspermy

Since polyspermy usually leads to abnormal development, it is beneficial for the egg to prevent it. In most organisms, polyspermy is prevented by

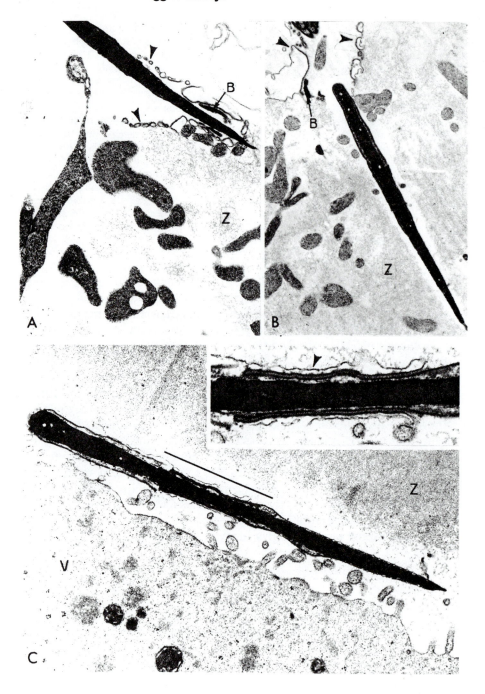

FIGURE 9.12 (*See legend on the opposite page.*)

◀ **FIGURE 9.12** Sperm penetration of the rabbit egg zona. *A*, Sperm head at the zonal surface (Z) of a penetrated egg recovered 12.5 hours after coitus. The acrosome reaction has already occurred. The vesiculated products of the reaction, indicated by heavy arrowheads, are not necessarily of uniform size, and, in the rabbit, are often accompanied by a dense body (B), sometimes seen within the acrosome of ejaculated sperm. × 15,200. *B*, Sperm head penetrating the zona pellucida of an activated egg collected 12.5 hours after coitus. The vesiculated acrosomal remnants discarded by this sperm head at the zonal surface testify strongly to the functional competence of sperm that exhibit this vesiculation reaction. Note that the equatorial region is intact, and that the inner membrane of the acrosome is devoid of visible acrosomal content. × 15,200. *C*, Sperm lying in the perivitelline space of a reacted egg collected 14 hours after coitus. The parasagittal section shown here is a typical example that demonstrates the intact equatorial segment overlain by plasma membrane (see inset) and the clearly defined inner membrane of the acrosome, devoid of acrosomal remnants. V: vitellus. × 23,100; inset, × 47,400. (From J.M. Bedford. 1972. An electron microscopic study of sperm penetration into the rabbit egg after natural mating. Am. J. Anat., *133:* 235, 237.)

mechanisms that bar additional sperm–egg fusions after the initial sperm has interacted with the egg. In the sea urchin two sequential blocking mechanisms have been proposed. The first is transient and occurs within seconds of sperm–egg contact, whereas the other is permanent and takes longer to develop.

The fast block, which was first demonstrated by Rothschild and Swann (1952), is caused by electrical depolarization of the egg plasma membrane from a negative value to a positive value that accompanies the entry of sperm (Jaffe, 1976). This depolarization is the result of an influx of sodium ions that is due to an opening of Na^+ channels, triggered by the fertilizing sperm (Gould-Somero et al., 1979; Jaffe et al., 1979; Jaffe, 1980). The positive charge on the membrane renders it unreceptive for fusion with additional sperm. A similar fast block to polyspermy has been reported for anuran amphibians (Cross and Elinson, 1980; Grey et al., 1982), starfish (Miyazaki and Hirai, 1979), and the echiuran worm *Urechis* (Gould-Somero et al., 1979).

The slower block to polyspermy results from the **cortical reaction** (Fig. 9.15). This change in the cortex begins at the site of sperm entry and propagates around the egg circumference. During the cortical reaction the cortical granules rupture due to fusion of their membranes with the egg plasma membrane, which causes vesiculations (Anderson, 1968). As a consequence of the cortical reaction, the vitelline envelope dissociates from the egg plasma membrane. This dissociation results from breakage of the vitelline posts that attach the vitelline envelope to the membrane (Chandler and Heuser, 1980). The breakage of the vitelline posts is thought to be caused by a protease enzyme that is released from the ruptured cortical granules (Carroll and Epel, 1975).

Separation of the vitelline envelope from the egg surface creates a space into which the contents of the cortical granules spill. This results in elevation of the vitelline envelope from the egg surface. It is then called the **fertilization envelope,** or **activation calyx** (Fig. 9.16). The area between the membranes is the **perivitelline space.**

An early stage in the propagation of the cortical reaction is shown in Figure 9.16. Gamete contact has occurred at some point to the right of this section. Release of cortical granule contents has caused elevation of the fertilization envelope on the right. However, elevation is only beginning on the left. A portion of an intact cortical granule is shown at the extreme left, which is the limit of the cortical reaction. A stop-action micrograph of cortical granule release is shown in Figure 9.17. The contents of recently ruptured cortical granules can be seen in the perivitelline space. Note the effect that the cortical reaction has upon

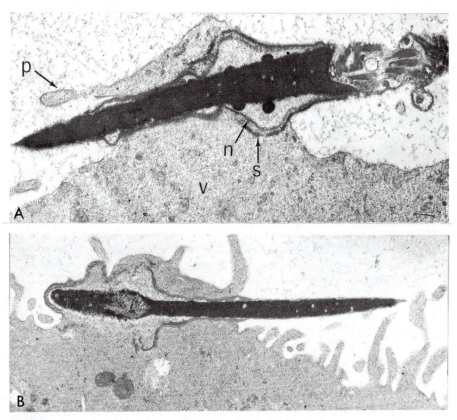

FIGURE 9.13 Sperm entry in the rabbitt egg. *A,* Section through the head of the fertilizing sperm taken soon after its fusion with the vitelline surface of an egg recovered 11.5 hours after coitus. The ooplasm has come to envelop the midposterior region of the sperm head; two separate components (n and s) of the postacrosomal region are evident—it is likely that "n" is the nuclear envelope. Note the vitelline process (p) beginning to envelop the acrosomal region of the head. v, vitellus. ×27,500. *B,* Parasagittal section of the fertilizing sperm head in the early stages of fusion with an egg recovered 11.5 hours after coitus. The anterior portion of the sperm head has been internalized. ×24,600. (From J.M. Bedford. 1972. An electron microscopic study of sperm penetration into the rabbit egg after natural mating. Am. J. Anat., *133:* 239, 241.)

the egg plasma membrane, which becomes a mosaic composed of the original membrane plus the membranes that formerly enclosed the cortical granules. This results in a tremendous increase in surface area. The excess membrane is accommodated by the formation of elongated and highly branched microvilli (Fig. 9.18) and by the formation of endocytotic

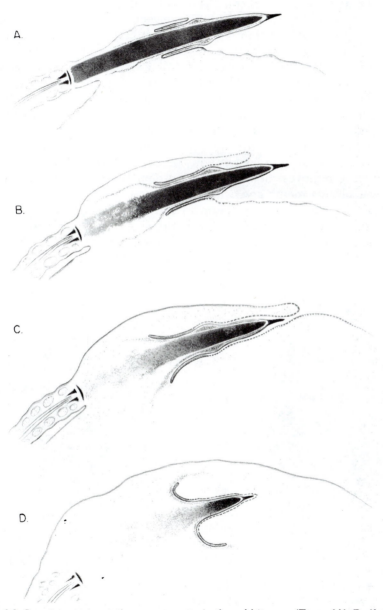

FIGURE 9.14 Drawings representing sperm entry in the rabbit egg. (From J.M. Bedford. 1972. An electron microscopic study of sperm penetration into the rabbit egg after natural mating. Am. J. Anat., *133:* 220, 221.)

vesicles, which may be the vehicles for resorption of excess membrane into the egg (Chandler and Heuser, 1979).

   The cortical reaction results in formation of hydrogen peroxide (H_2O_2), release of enzymes that are involved in establishing blocks to polyspermy, and release of material for the formation of the **hyaline layer.** The H_2O_2 is thought to inactivate supernumerary sperm after entry of the first sperm. This conclusion is based upon the observations that (1) treatment of sperm with H_2O_2 at concentrations similar to those released by eggs at fertilization reduces the fertilizing capacity of sperm and (2) addition of catalase (which decomposes H_2O_2 to H_2O and O_2) to the seawater causes 100% polyspermy (Coburn et al., 1981; Boldt et al., 1981). Thus this polyspermy block is only effective if H_2O_2 is present. One of the enzymes released from cortical granules is a protease that breaks the bonds attaching the vitelline envelope to the egg plasma membrane (see preceeding discussion). Another protease destroys the glycoprotein sperm receptors on the vitelline envelope. Both of these enzymatic activities assist in the polyspermy block. Elevation of the fertilization envelope removes the sites of sperm binding from their proximity to the egg plasma membrane, whereas destruction of the receptors themselves causes supernumerary sperm to become detached from the membrane and prevents the binding of additional sperm (Carroll and Epel, 1975). The clear, viscous hyaline layer is formed from material that spills out

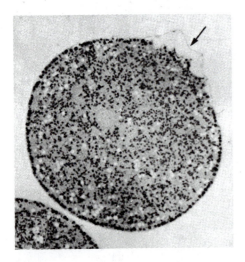 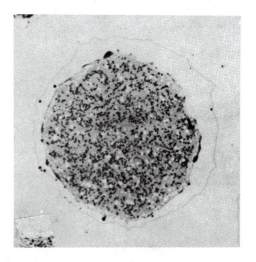

A   **B**

FIGURE 9.15 The cortical reaction of the fertilized sea urchin egg. *A*, Cortical granule rupture is initiated at the fertilization site, causing localized elevation of the vitelline envelope (arrow). *B*, The cortical reaction is complete as indicated by a fully elevated fertilization envelope and absence of cortical granules. Scale bar equals 10 μm. (From D.E. Chandler and J. Heuser. 1979. Reproduced from *The Journal of Cell Biology*, 1979, vol. 83, pp. 91–108 by copyright permission of The Rockefeller University Press.)

of the cortical granules and covers the egg surface. This layer has an important role to play during morphogenesis and will be discussed further in Chapters 10 and 13.

Upon elevation, the fertilization envelope undergoes changes in its chemical and physical properties; it becomes hardened and resistant to solubilization and proteolytic digestion. Hardening of the envelope is caused by its interaction with a peroxidase enzyme released from the cortical granules, that utilizes H_2O_2 released from the eggs to promote cross-linking of the individual molecules that compose the envelope, which is thus converted to a polymer (Foerder and Shapiro, 1977; Hall, 1978). The hardened envelope is resistant to sperm entry and surrounds and protects the embryo until it is dissolved by the hatching enzyme released from the embryo during early development.

A cortical reaction to fertilization is not unique to sea urchins but is found in a wide variety of organisms, including vertebrates. Anuran amphibians have a cortical reaction (Fig. 9.19A) that is remarkably similar to that of sea urchins; that is, the release of cortical granule components causes the elevation of the vitelline envelope and its conversion

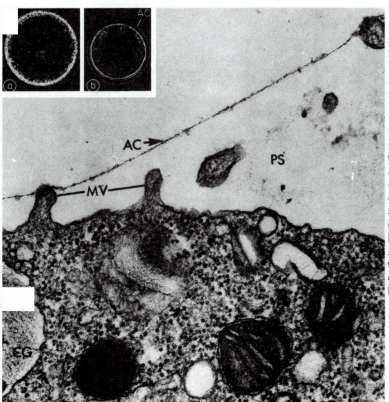

FIGURE 9.16 A section of an activated egg of the sea urchin *Arbacia* showing the fertilization envelope or activation calyx (AC), perivitelline space (PS), microvilli (MV), and a portion of a nonactivated cortical granule (CG). ×60,000. Inset *a* is a phase-contrast photomicrograph of a living mature egg. ×460. Inset *b* is a phase-contrast photomicrograph of a living fertilized egg showing the complete activation calyx (AC). ×400. (From E. Anderson. 1968. Reproduced from *The Journal of Cell Biology*, 1968, vol. 37, pp. 514–539 by copyright permission of The Rockefeller University Press.)

to a fertilization envelope that is impenetrable to sperm and has a reduced binding affinity for sperm (Grey et al., 1976).

Likewise, in some mammals, release of cortical granule constituents (Fig. 9.19*B*) results in modifications of the zona pellucida (called the **zona reaction**), which in turn cause the loss of sperm-binding sites on the zona (Barros and Yanagimachi, 1971). As in sea urchins, sperm detachment apparently results from the action of a protease enzyme (Gwatkin et al., 1973). The zona also undergoes hardening, which is analogous to hardening of the sea urchin fertilization envelope. Recent evidence suggests that the mechanism of hardening of the zona is also similar to the hardening mechanism in sea urchins; that is, peroxidase released during the cortical reaction catalyzes cross-linking of molecules of the zona (Schmell and Gulyas, 1980). The similarity between the polyspermy blocks in sea urchins and mammals is remarkable in view of the evolutionary distance between them.

An additional block to polyspermy occurs at the mammalian egg surface (Wolf, 1978). In a few mammals, such as the rabbit, a zona reaction is lacking. In these species the egg surface block is apparently sufficient to prevent polyspermy.

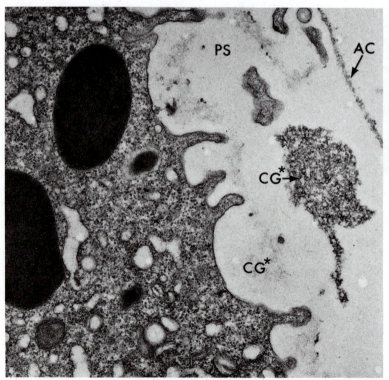

FIGURE 9.17 A section through the surface of a fertilized egg. AC, activation calyx; PS, perivitelline space; CG*, dense and less dense portions of discharged cortical granules. ×21,000. (From E. Anderson. 1968. Reproduced from *The Journal of Cell Biology*, 1968, vol. 37, pp. 514–539 by copyright permission of The Rockefeller University Press.)

The Initiation of Development

Fertilization of the egg initiates the developmental program. Considerable evidence has recently been obtained, indicating that activation is mediated by changes in the ionic composition of the egg. We shall now examine this evidence, beginning with a consideration of ionic calcium, which plays a central role in activation. Early evidence that calcium undergoes dynamic changes after fertilization was provided by Mazia (1937), who demonstrated that the proportion of free calcium increases

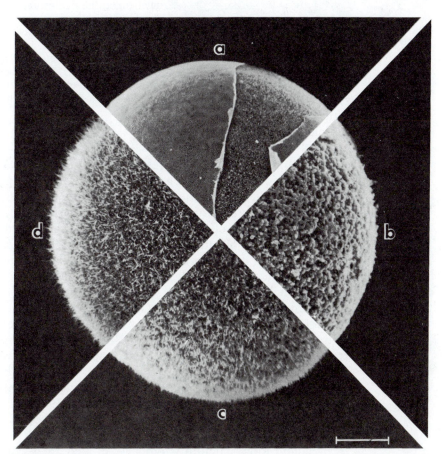

FIGURE 9.18 A composite of scanning electron micrographs of plasma membrane surfaces of sea urchin eggs at selected stages: (a) before fertilization, with the plasma membrane revealed where the vitelline envelope is torn away; (b) 1 minute after fertilization, when the surface is obscured by globules released from ruptured cortical granules; (c) 5 minutes after fertilization, following removal of the hyaline layer precursor; and (d) 13 minutes after fertilization, when the presence of a few long microvilli gives a distinctive "fuzzy" appearance. Scale bar equals 10 μm. (From T.E. Schroeder. 1979. Surface area change at fertilization: Resorption of the mosaic membrane. Dev. Biol., *70:* 311.)

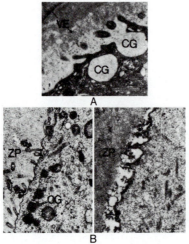

FIGURE 9.19 Cortical reactions in *Xenopus laevis* egg *(A)* and mouse eggs *(B)*. *A* is a section of a *Xenopus* egg fixed 5 minutes after fertilization. Note the ruptured cortical granule (CG) on the right and the intact granule on the left. The vitelline envelope (VE) is being converted into the fertilization envelope, which is refractory to sperm. Scale bar equals 1.0 μm. (From R.D. Grey et al. 1982. An electrical block is required to prevent polyspermy in eggs fertilized by natural mating of *Xenopus laevis*. Dev. Biol., *89:* 480.) *B* illustrates mouse eggs before (left) and after (right) the cortical reaction. ZP: zona pellucida. Scale bar equals 0.5 μm. (Figure on the left from B.J. Gulyas. 1980. Cortical granules of mammalian eggs. Int. Rev. Cytol., *63:* 357–392; figure on the right from B.J. Gulyas and E.D. Schmell. 1980. Ovoperoxidase activity in ionophore treated mouse eggs. I. Electron microscopic localization. Gamete Res., *3:* 273.)

after fertilization of the sea urchin egg, apparently as a result of the release of calcium ions from a bound state within the egg. Mazia's observations could be explained in one of two ways. Either free Ca^{++} increases *as a result* of activation, or it is a *primary cause* of activation.

If it is a cause of activation, then activation should be dependent upon an elevation of intracellular Ca^{++}. This dependence is illustrated by the fact that parthenogenic activation of fish or amphibian eggs by pricking them with a needle is only possible if Ca^{++} is present in the external medium. Presumably, the calcium ions enter the eggs at the wound and trigger development (Gilkey et al., 1978). Recently, the ionophore A23187 has been used as a means of raising internal free Ca^{++} in unfertilized eggs. For example, bathing sea urchin eggs in seawater containing A23187 will activate them. Surprisingly, the ionophore will also activate eggs in media lacking Ca^{++} and Mg^{++}. This observation led to the proposal that A23187 causes the release of internally bound calcium or magnesium (Steinhardt and Epel, 1974; Chambers et al., 1974). Most calcium in unfertilized eggs is in a bound form, whereas magnesium is unbound; thus, calcium is presumably the ion involved. In its bound form, calcium is probably in a complex with one or more proteins. It is assumed that the ionophore mimics fertilization and that the release of bound Ca^{++} is the normal response to sperm contact. Ionophore activation has been demonstrated for a wide variety of organisms, including starfish, mollusks, amphibians, fish, and mammals (Steinhardt et al., 1974; Belanger and Schuetz, 1975), suggesting that release of bound Ca^{++} is a generalized mechanism for egg activation in the animal kingdom.

Indirect evidence in support of the central role of Ca^{++} in egg activation is provided by an experiment in which the calcium chelator EGTA was injected into unfertilized sea urchin eggs (Zucker and Steinhardt, 1978). Upon fertilization these eggs failed to activate. Since reducing calcium prevents activation, it is likely that the elevation in ionic calcium is essential for activation.

Direct observation of Ca^{++} release is now possible through a technique that utilizes aequorin, a protein extracted from jellyfish, which luminesces in the presence of Ca^{++}. The aequorin must be injected into the egg where it can interact with liberated Ca^{++}. The initial aequorin experiments were done with eggs of the fish *Oryzias latipes* (commonly called "medaka"), since they are relatively easy to microinject and since they are transparent, allowing the luminescence to be seen and measured. Fertilization or ionophore treatment causes a transient burst of luminescence that develops rapidly and dissipates more slowly (Fig. 9.20). Aequorin injection has now been accomplished with sea urchin eggs as well, with similar results (Steinhardt et al., 1977). Clearly, bound calcium is liberated after fertilization. Furthermore, since the aequorin experiments indicate that this effect is transient, it would appear that the Ca^{++} again becomes bound, possibly at a new site in the cell.

THE CORTICAL REACTION

One of the consequences of the elevation in free Ca^{++} in the sea urchin egg is triggering of the cortical reaction. This was first demonstrated by Vacquier (1975), using isolated sea urchin egg plasma membranes that possess intact cortical granules. Vacquier developed an ingenious method for isolating the membranes so that the granules can be observed. A substrate (e.g., plastic, glass) is coated with a solution of protamine sulfate. Egg vitelline envelopes adhere to the protamine-coated surface. Treatment of the attached eggs with calcium-free seawater containing a chelating agent causes the eggs to lyse, and the egg membranes spread out on the surface "inside-out," that is, with the inner surface and attached cortical granules exposed (Fig. 9.21A). A cortical reaction can be induced on these exposed membranes by treating them with Ca^{++}. The response is an instantaneous swelling of granules and a discharge of their contents, which form a gel network on the exposed surface of the membrane (Fig. 9.21B).

A means by which Ca^{++} may be responsible for triggering the cortical reaction is suggested by recent experiments of Steinhardt and Alderton (1982). These investigators prepared antibodies to the calcium-binding protein calmodulin (see Chap. 7). When isolated sea urchin plasma membranes are treated with the antibodies, the ability of Ca^{++} to induce the cortical reaction is lost. This result suggests that calmodulin is responsible for mediating the effects of Ca^{++} on the cortical reaction after binding to the ionic calcium. Steinhardt and Alderton have

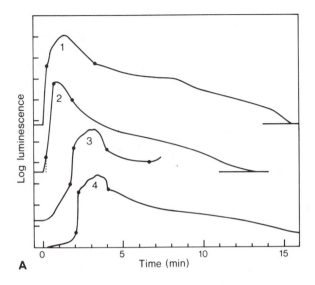

A

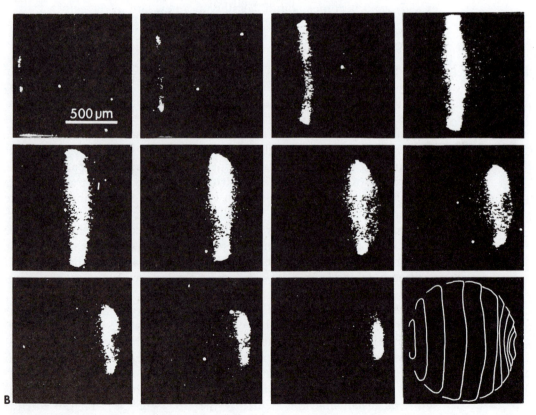

B

FIGURE 9.20 (*See legend on the opposite page.*)

FIGURE 9.20 Calcium release in medaka eggs as monitored with aequorin. A, Semilogarithmic plots of activation responses, showing rapid rise in luminescence followed by slower drop to resting level. Eggs 1 and 2 were sperm-activated, while 3 and 4 were ionophore-activated. B, A free calcium wave propagating across a sperm-activated egg. Successive photographs are 10 seconds apart. Last frame is a tracing showing the leading edges of the 11 wave fronts. Egg is oriented with micropyle (site of sperm entry) to the left. (A after E.B. Ridgway, J.C. Gilkey, and L.F. Jaffe. 1977. Free calcium increases explosively in activating medaka eggs. Proc. Natl. Acad. Sci. U.S.A., 74: 626; B from J.C. Gilkey et al. 1978. Reproduced from The Journal of Cell Biology, 1978, vol. 76, pp. 448–466 by copyright permission of The Rockefeller University Press.)

demonstrated by indirect immunofluorescence (see Chap. 5) that calmodulin is present on the inner plasma membrane surface where it could perform this function.

The effect of Ca^{++} on the cortical reaction has also been investigated in the larger amphibian eggs, which are of sufficient size that Ca^{++} can be directly micro-injected into them. As with the isolated sea urchin membranes, Ca^{++} injections induce a cortical reaction (Hollinger and Schuetz, 1976; Hollinger et al., 1979). Recently, Ca^{++} has been injected into mouse eggs, causing not only cortical granule breakdown but the initiation of cell division as well. This result suggests that the postfertilization increase of intracellular free Ca^{++} activates the normal sequence of events that initiate mammalian development (Fulton and Whittingham, 1978).

METABOLIC ACTIVATION

Fertilization is often considered a "trigger" that switches on the metabolically hypoactive unfertilized egg so that a predetermined program of metabolic processes can unfold. Since experimentally induced elevations of ionic calcium will mimic the effects of the sperm, we assume that the transient rise of ionic calcium that follows sperm entry mediates the effects of the sperm in activating this program. In the sea urchin this program includes increases in particular enzymatic activities, an increase in the rate of respiration, initiation of macromolecular synthesis, and changes in the transport characteristics of the egg plasma membrane (Epel et al., 1969). As we shall discuss in Chapter 12, the initial developmental program is entirely cytoplasmic and does not rely upon concurrent nuclear input.

When is this program initiated, and how is the transient rise in ionic calcium involved in its initiation? The earliest metabolic changes include activation of respiratory enzymes, followed by a large burst in the respiratory rate. These changes are initiated within seconds of the cortical reaction. Activation of protein synthesis is a "late response," beginning five minutes after fertilization (Fig. 9.22). Amino acid, phosphate, and nucleoside transport mechanisms and DNA synthesis are also activated as part of the late response (Epel, 1975). Calcium can activate both the early and late responses. However, since the calcium release is transient and precedes the late responses by several minutes, there presumably are intervening events that mediate the effects of the calcium release. In fact, the late events can be experimentally uncoupled from a

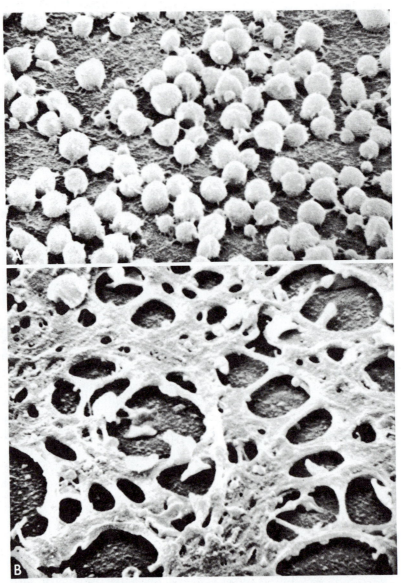

FIGURE 9.21 Scanning electron micrographs of cortical granules bound to the inner surface of the plasma membrane. *A,* Intact cortical granules. ×6400. *B,* Fused cortical granule material after cortical granule discharge. ×7500. (From V.D. Vacquier. 1975. The isolation of intact cortical granules from sea urchin eggs: Calcium ions trigger granule discharge. Dev. Biol., *43:* 65, 66.)

dependency upon Ca^{++} release. This can be achieved by activating eggs with ammonia. The late events of activation occur, but the calcium release, the cortical reaction, and the respiratory changes do not occur (Steinhardt and Mazia, 1973; Epel et al., 1974). In contrast, the late events can be prevented by transferring eggs to sodium-free seawater after fertilization.

These effects of ammonia and sodium are both related to another event that follows fertilization—an elevation of intracellular pH (Fig. 9.23). This pH rise is reflected in the release of acid from the eggs into the surrounding medium. This "fertilization acid" is due to an efflux of H^+ from the eggs, which begins at the start of the cortical reaction and continues for four minutes. Release of H^+ is prevented if eggs are transferred to sodium-free seawater after fertilization. However, when Na^+ is added to the water, H^+ efflux occurs, and its rate is linearly dependent upon Na^+ concentration (Fig. 9.24). Likewise, the intracellular pH rise is dependent upon the presence of extracellular Na^+ (Shen and Steinhardt, 1979). Thus, the failure of eggs to develop in sodium-free seawater is due to the inhibition of the pH rise.

The parthenogenic activation of eggs by ammonia is due to its effects on pH; that is, it mimics the sodium-dependent pH elevation (Fig. 9.25). The pH rise produced by ammonia is caused by the penetration of the cell by the uncharged base NH_3, which subsequently picks up a hydrogen ion (forming NH_4^+) and directly raises the pH of the cytoplasm (Winkler and Grainger, 1978). Thus, although their mechanisms are different, both sodium and ammonia produce the same result, that is, an increase in pH.

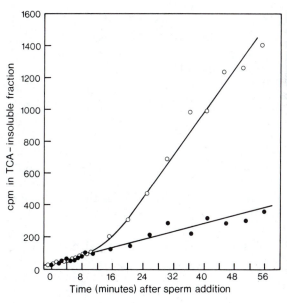

FIGURE 9.22 Protein synthesis by unfertilized eggs (dots) and fertilized eggs (circles) as measured by incorporation of ^{14}C-leucine. (After D. Epel. 1967. Protein synthesis in sea urchin eggs: A "late" response to fertilization. Proc. Natl. Acad. Sci. U.S.A., *57:* 901.)

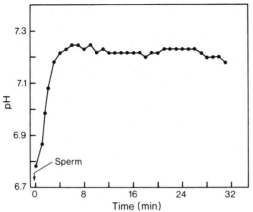

FIGURE 9.23 Continuous recording of intracellular pH during fertilization. (After S.S. Shen and R.A. Steinhardt. 1978. Reprinted by permission from *Nature*, Vol. 272, No. 5650, pp. 253–254. Copyright © 1978 Macmillan Journals Limited.)

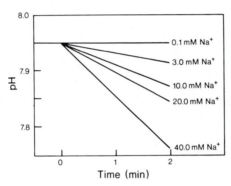

FIGURE 9.24 Acid efflux from sea urchin eggs as a function of sodium concentration. Acid release is indicated by a drop in the pH in the surrounding sea water. (After J.D. Johnson, D. Epel, and M. Paul. 1976. Reprinted by permission from *Nature*, Vol. 262, No. 5570, pp. 661–664. Copyright © 1976 Macmillan Journals Limited.)

What is the significance of the sodium-dependent pH shift? Johnson et al. (1976) have proposed that the low cytoplasmic pH of the unfertilized egg maintains metabolic inhibition, whereas the rise in pH relieves the inhibition and allows development to proceed. In this connection, Grainger et al. (1979) have recently demonstrated that the increase and maintenance of protein synthesis at fertilization can be largely accounted for by the increase in intracellular pH. The molecular mechanisms involved in the activation of protein synthesis will be discussed in Chapter 12.

As we previously discussed, metabolic activation is mediated by the release of bound Ca^{++}. Calcium release, in turn, triggers the sodium-dependent increase in intracellular pH. The dual roles of calcium and sodium in the pH rise are illustrated in Figure 9.26. In this experiment eggs were activated by the calcium ionophore A23187, which normally causes an elevation in pH. However, the pH remains acidic if sodium is absent from the medium; addition of sodium allows alkalization of the eggs and activation of development. Clearly, an understanding of the interrelationships among Ca^{++}, Na^{+}, and the pH rise is critical to our

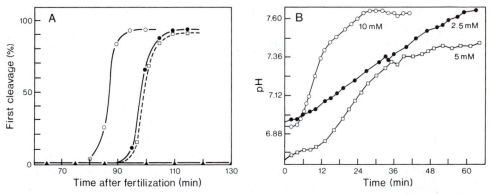

FIGURE 9.25 The ammonia effect. *A,* Ammonia can substitute for sodium in initiating development. At 1 minute after fertilization, eggs were washed in sodium-free sea water to prevent activation. Transfer of eggs to sea water will initiate development (see solid line with open circles on graph). Likewise, transfer of eggs to sea water containing either 50 mM sodium (solid line with solid circles) or 5 mM NH_4Cl (dashed line with open squares) will initiate development. Control eggs transferred to sodium-free sea water (solid line with solid triangles) will not cleave. (After J.D. Johnson, D. Epel, and M. Paul. 1976. Reprinted by permission from *Nature,* Vol. 262, No. 5570, pp. 661–664. Copyright © 1976 Macmillan Journals Limited.) *B,* Continuous recordings of intracellular pH during activation by NH_4Cl at 2.5, 5, and 10 mM. (After S.S. Shen and R.A. Steinhardt. 1978. Reprinted by permission from *Nature,* Vol. 272, No. 5650, pp. 253–254. Copyright © 1978 Macmillan Journals Limited.)

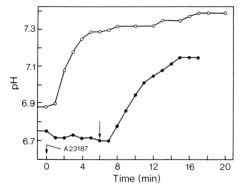

FIGURE 9.26 Dual roles of Ca^{++} and Na^+ in alkalization after activation of the sea urchin egg. Eggs were activated with A23187, and pH was monitored with an electrode. Line with open circles represents intracellular pH in eggs activated in artificial sea water. Line with closed circles represents intracellular pH in eggs activated in sodium-free sea water. At the arrow the external sodium concentration was raised to 80 mM. (After S.S. Shen and R.A. Steinhardt. 1979. Reprinted by permission from *Nature,* Vol. 282, No. 5734, pp. 87–89. Copyright © 1979 Macmillan Journals Limited.)

understanding of metabolic activation. Additional research should help clarify this relationship.

How generalized are the results obtained with sea urchins? Does a similar pattern of metabolic activation under ionic control pertain to other organisms? We have only partial answers to these questions. We do know that release of intracellular Ca^{++} activates development of the eggs of a variety of organisms. However, there is reason to believe that the events after Ca^{++} release are variable. For example, the sea urchin is in a different metabolic state at fertilization from that of the vertebrates and other organisms in which meiotic maturation is under hormonal control. In the sea urchin, meiotic maturation is completed before ovulation, and the eggs become metabolically hypoactive until fertilization. In contrast, maturation of the amphibian egg (and that of most other vertebrates) is initiated at ovulation and proceeds only as far as metaphase II. Fertilization triggers completion of meiosis so that pronuclear fusion can occur. The metaphase block in unfertilized eggs is thought to be maintained by an endogenous inhibitor that is inactivated by Ca^{++} release at fertilization (Meyerhof and Masui, 1977). The hormonal stimulus that initiates maturation also activates protein synthesis and a number of other metabolic changes (Ecker and Smith, 1968; Wasserman and Smith, 1978). Thus, the unfertilized amphibian egg, unlike that of the sea urchin, is metabolically very active. Fertilization causes further metabolic changes, including alterations in certain enzymatic activities and an increase in the number of polysomes in the cytoplasm (Woodland, 1974). An increase in polysomes after fertilization suggests that fertilization causes a further elevation in the rate of protein synthesis. Taken together, these results indicate that metabolic *activation* in the amphibian is initiated by the hormonal stimulus, and sperm entry triggers metabolic *adjustments*. The studies on metabolic activation in other species are too meager to draw a generalized picture of the mechanisms of metabolic activation. Clearly, considerable research remains to be done on this subject.

EGG REARRANGEMENTS

Fertilization may elicit substantial reorganization of cytoplasmic components. Some of these components are **morphogenic determinants** that become segregated into particular cells during cleavage and impart specific developmental properties on the cells by affecting their patterns of gene expression. Because of their developmental significance, rearrangement of determinants is a vital fertilization response. We shall discuss examples of rearrangement of morphogenic determinants and the roles of these rearranged components on cell determination in Chapter 11.

9–3. FORMATION OF THE DIPLOID ZYGOTE NUCLEUS

Both male and female gamete nuclei must undergo a certain amount of "processing" in the egg before they are capable of forming a diploid zygote nucleus. Furthermore, they must migrate to a common point where they can make contact. The sperm nucleus has a particularly complex history in the egg. It is delivered to the egg as a streamlined structure with highly condensed chromatin. It must enlarge, and its chromatin must decondense before fusing with the female pronucleus. This process has been thoroughly documented in sea urchins, and this discussion will be based primarily upon the results obtained for that group.

As discussed earlier in this chapter, the sea urchin sperm is initially incorporated into an elevated fertilization cone that forms in response to sperm–egg contact. An early stage in this incorporation is shown in Figure 9.27A. Shortly after its incorporation the nuclear envelope surrounding the sperm nucleus begins to vesiculate and dissociate (Fig. 9.27B). The dissociation is not complete, however, since the apical and basal portions of the sperm nuclear envelope remain intact and are later incorporated into the male pronuclear envelope. Chromatin dispersal begins at the periphery of the nucleus after nuclear envelope dispersion and proceeds toward the center of the chromatin mass. Before dispersion

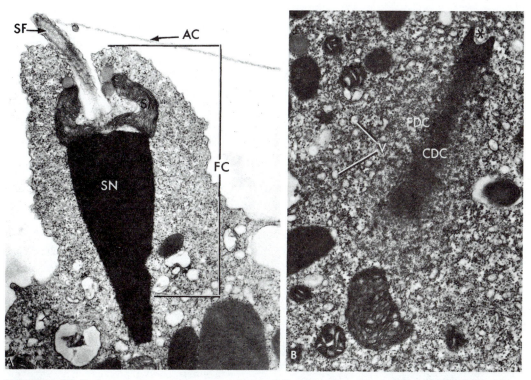

FIGURE 9.27 Early stages in sperm incorporation in *Arbacia*. *A*, Sperm in fertilization cone (FC). Identifiable sperm organelles include sperm nucleus (SN), sperm mitochondria (SM), and sperm flagellum (SF). The activation calyx (AC) is visible above the egg surface. ×16,300. (From F.J. Longo. 1973. Fertilization: A comparative ultrastructural review. Biol. Reprod., *9:* 154.) *B*, Dissociation of sperm nuclear envelope and dispersal of chromatin. Envelope dispersion is complete except at the tip (asterisk) and the base (the latter not seen in this section) of the nucleus. The center of the chromatin mass (CDC) remains condensed at this stage, while the peripheral chromatin is finely dispersed (FDC). Smooth vesicles (V) are evident at the periphery of the finely dispersed chromatin. The pronuclear envelope will be constructed by an aggregation of these vesicles. ×15,000. (From F.J. Longo and E. Anderson. 1968. Reproduced from *The Journal of Cell Biology*, 1968, vol. 39, pp. 339–368 by copyright permission of The Rockefeller University Press.)

is complete, membranous vesicles begin to aggregate along the periphery of the chromatin. These vesicles fuse with one another and with the persisting apical and basal portions of the original sperm nuclear envelope to produce the male pronuclear envelope.

In addition to undergoing internal morphological and physiological changes, the pronuclei must also move toward the site in the egg where they will encounter one another. In sea urchins, movement of the male pronucleus is caused by the sperm aster, which radiates from the pronucleus and pushes against the egg cortex, causing the pronucleus to be displaced toward the center of the egg (Chambers, 1939; Hamaguchi and Hiramoto, 1980). The astral fibers contact the female pronucleus, interconnecting the pronuclei and abruptly drawing the female pronucleus to the male pronucleus. Continuing action of the aster seems to displace the two adjacent pronuclei to the egg center, where they fuse to form a diploid zygote nucleus. The migration and fusion of the pronuclei are shown in the series of photographs in Figure 9.28. The role of astral fibers (which are microtubules) in pronuclear movement is supported by experiments in which pronuclear movement is prevented by inhibitors of microtubules (Zimmerman and Zimmerman, 1967; Schatten and Schatten, 1981).

The eggs of some animal species do not demonstrate pronuclear fusion. In these eggs, the pronuclei approximate one another and remain closely apposed until the dissolution of the pronuclear envelopes and alignment of the chromosomes on a common metaphase plate. The rabbit zygote provides an example of pronuclear approximation. The pronuclei migrate to a central position in the egg, where they flatten and become highly convoluted on the sides facing one another (Fig. 9.29). These surfaces then interdigitate and become tightly apposed (Fig. 9.30*A*). After apposition, chromatin condensation occurs, indicating that the chromatin has begun preparing for the first cleavage division. Next, the pronuclear envelopes begin to break down by the familiar process of fusion of inner and outer layers followed by vesiculation (Fig. 9.30*B*).

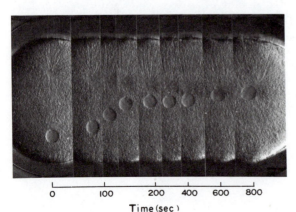

O 100 200 400 600 800

Time (sec)

FIGURE 9.28 Interference contrast micrographs showing migration of the male (top) and female (bottom) pronuclei in an egg of the sea urchin *Clypeaster japonicus*. Note the aster radiating from the small male pronucleus. Arrows indicate time when the photographs were taken. Scale bar equals 20 μm. (From M.S. Hamaguchi and Y. Hiramoto. 1980. Fertilization process in the heart-urchin, *Clypeaster japonicus* observed with a differential interference microscope. Develop., Growth & Differ., *22*: 524.)

After disruption of the pronuclear envelopes the condensing chromosomes from the two pronuclei intermix and orient on the first mitotic spindle (Fig. 9.30C). Thus, in contrast to the sea urchin, a zygote nucleus is not formed until the two-cell stage, when the maternal and paternal genomes are enclosed within a common nucleus for the first time. The absence of pronuclear fusion has also been reported for the mouse and golden hamster zygotes (Zamboni et al., 1972; Kunkle and Longo, 1975). It will be interesting to discover whether this is a general feature of mammalian zygotes.

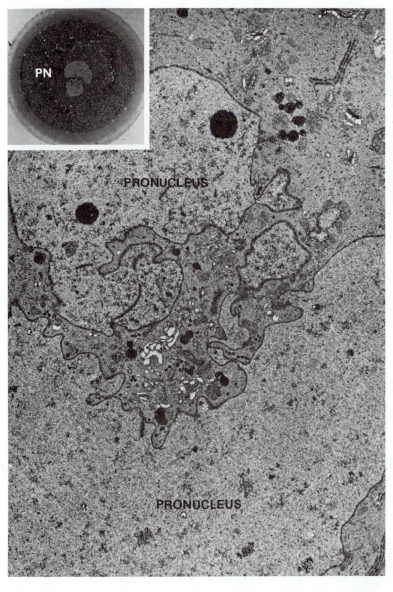

FIGURE 9.29 Light (inset) and electron micrographs of closely associated pronuclei (PN) of the rabbit zygote. Note the convolutions on the surfaces facing one another. ×7200; inset, ×300. (From F.J. Longo and E. Anderson. 1969. Cytological events leading to the formation of the two-cell stage in the rabbit: Association of the maternally and paternally derived genomes. J. Ultrastruct. Res., *29*: 95.)

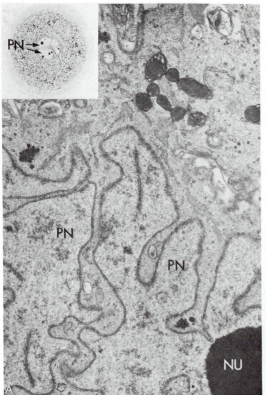

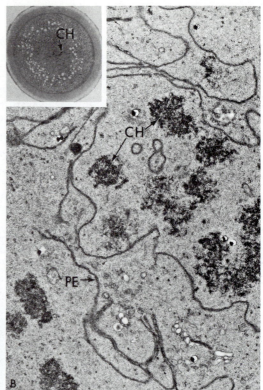

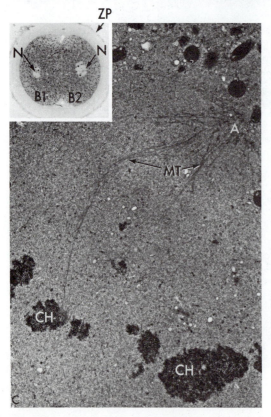

FIGURE 9.30 Changes following pronuclear apposition in the rabbit zygote. *A,* Light (inset) and electron micrographs of interdigitating pronuclei (PN). NU: nucleolus-like body. ×11,900; inset, ×200. *B,* Light (inset) and electron micrographs displaying the breakdown of the pronuclear envelopes (PE) and the condensation of chromatin (CH). ×12,440; inset, ×200. *C,* Portion of the metaphase figure demonstrating the maternally and paternally derived chromosomes (CH). MT: microtubules; A: aster. ×8750. Inset, photomicrograph of the two-cell stage of the rabbit showing the two blastomeres (B$_1$ and B$_2$) and their nuclei (N). ZP: zona pellucida. ×200. (From F.J. Longo and E. Anderson. 1969. Cytological events leading to the formation of the two-cell stage in the rabbit: Association of the maternally and paternally derived genomes. J. Ultrastruct. Res., *29:* 103, 105, 113.)

9–4. PARTHENOGENESIS

Fertilization is the usual method by which embryonic development begins. However, it is by no means the only method, since a number of species have evolved mechanisms for initiation of development of the egg by **parthenogenesis,** that is, initiation that does not require sperm. Parthenogenesis is part of the normal life cycle of many organisms. For example, in aphids a parthenogenic phase alternates with a sexual phase. Females that hatch in the spring reproduce parthenogenically throughout the summer, and at the end of summer a single sexual generation develops. Eggs produced by the sexual females are fertilized, and these are the eggs that over-winter and produce the female nymphs that emerge the following spring. In some species of flatworms, rotifers, arthropods, and lizards, every individual arises parthenogenically. In these cases of **complete parthenogenesis** the populations consist entirely of females—the males have been dispensed with.

Parthenogenic females (sometimes called **parthenogens**) must overcome two significant problems that are normally solved by sperm entry. One is the activation of development. The parthenogenic activation mechanisms are, in most cases, poorly understood. In the case of the wasp, *Habrobracon*, activation apparently results from a mechanical stimulus received by the egg as it is squeezed through the ovipositor when it is laid. Eggs that do not receive this stimulus do not develop. Once the egg is activated, the nuclei of the embryo must have the somatic chromosome number, just like the female parent. This constitutes the second significant problem. Adjustment of chromosome number often results from modified meiotic mechanisms. For example, in the parthenogenic lizard, *Cnemidophorus uniparens*, the germ cell chromosome number is doubled before meiosis, but two meiotic divisions reduce the chromosome number in the eggs back to the somatic level. These eggs go on to develop as normal embryos (Cuellar, 1971). Instead of modified meiosis, some parthenogenic species will undergo normal meiosis, and diploidy is then reconstituted by fusion of pairs of cleavage nuclei.

Parthenogenesis illustrates the marvelous plasticity of nature, with basic biological mechanisms being modified to suit a species' particular needs. It also underscores the significance of oogenesis, since the egg alone can possess every quality required for development.

9–5. FERTILIZATION IN PLANTS

Fertilization mechanisms in plants are extremely diverse. Primitive plants that produce flagellated sperm and immotile eggs resemble animals in their mode of fertilization. On the other hand, angiosperms, in which the sperm cells are delivered to the egg in the pollen tube, have evolved quite a different fertilization program. Although these are not the only types of fertilization mechanisms in the plant kingdom, they are among the best studied, and an analysis of fertilization in these forms will complement our earlier discussion of gametogenesis and our upcoming discussions of embryogenesis.

Primitive Plants

As in the previous chapter on gametogenesis, we shall discuss two examples of fertilization in primitive plants: the brown alga *Fucus* and the water fern *Marsilea*. *Fucus* has a fertilization program that is remarkably similar to that of marine invertebrate animals: Sperm and eggs are shed into the sea, and the sperm moves to the vicinity of the egg by means of flagellar movement. It then fuses with the egg to produce a free-living zygote that undergoes cleavage and differentiation to produce an adult. As discussed earlier in this chapter, *Fucus* sperm are attracted to the eggs by a substance released into the sea (Fig. 9.31*A*). The attractant can be obtained for experimental purposes by placing large numbers of eggs in buffered seawater. The attractant is released into this "egg water," and the supernatant, which contains the substance, is collected. The sperm-attracting properties of the supernatant can be demonstrated by placing a capillary tube containing it in a sperm suspension on a microscope stage. Sperm are attracted to the capillary in large numbers, producing a "sperm cloud" at the capillary tip (Fig. 9.31*B*). The attractant of *Fucus serratus* has been purified and identified as a volatile hydrocarbon, and

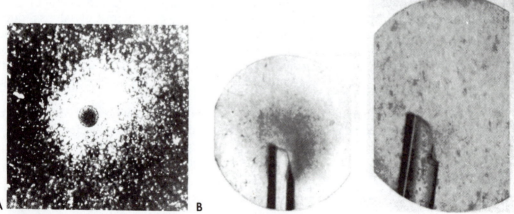

A   B

FIGURE 9.31 Chemotaxis in *Fucus. A.* Egg attracting a halo of sperm. Darkfield. The egg is about 0.1 mm at its widest. *B,* Demonstration of sperm-attracting properties of supernatant collected from an egg suspension. Left: capillary containing supernatant attracts sperm; right: capillary containing sea water fails to attract sperm. (*A* from D.G. Müller and K. Seferiadis. 1977. Specificity of sexual chemotaxis in *Fucus serratus* and *Fucus vesiculosus* (Phaeophyceae). Z. Pflanzenphysiol., *84:* 93. *B* from A.H. Cook and J.A. Elvidge. 1951. Fertilization in the Fucaceae: Investigations on the nature of the chemotactic substance produced by eggs of *Fucus serratus* and *F. vesiculosus.* Proc. Roy. Soc. Lond. (Biol.), *138:* 114.)

it has been designated **fucoserraten** (Müller and Jaenicke, 1973; Müller and Gassmann, 1978).

Sperm attracted to the vicinity of eggs have a swelling at the tip of one of the flagella. This flagellum probes the egg surface and establishes contact between the gametes. It is therefore analogous to the acrosome of animal sperm. Numerous *Fucus* sperm can attach to the surface of each egg. Sperm binding, which is highly species-specific, is mediated by an association between carbohydrate-containing ligands on the egg surface and lectinlike proteins on the sperm surface (Bolwell et al., 1979, 1980). This mechanism is reminiscent of sperm binding to the sea urchin vitelline envelope or to the mammalian zona pellucida (see section 9–1). The free flagella of the numerous attached sperm continue to beat and, as a result, cause the egg to spin rapidly in the water. However, after penetration by one sperm, the other attached sperm and those within the immediate vicinity of the fertilized egg become immobilized. This observation suggests that the fertilized egg releases a sperm-inactivating substance that helps to prevent polyspermy (Pollock, 1970).

An early stage in sperm–egg contact is shown in Figure 9.32. Here, one of the sperm flagella can be seen in contact with the egg plasmalemma. This micrograph also reveals ultrastructural features of the *Fucus* sperm. The nucleus contains highly condensed chromatin, similar to that of animal sperm. It is surrounded by an envelope that lacks nuclear pores. Sperm mitochondria are modified, having electron-dense material in the intracristal space and longitudinally oriented cristae. The sperm also contains an eyespot adjacent to the mitochondria. The eyespot is thought to be involved in sperm phototaxis.

After sperm–egg fusion, the sperm is drawn into the egg. Sperm entry triggers an egg response that is quite similar to the cortical reaction in sea urchins. In *Fucus* fertilization acts on cytoplasmic vesicles lying beneath the plasma membrane and triggers the discharge of their contents, which assemble the cell wall of the zygote (Fig. 9.33). Release of cytoplasmic vesicle contents is thought to occur by an exocytotic mechanism like that described for the breakdown of cortical granules by the sea urchin (Peng and Jaffe, 1976). The rigid cell wall that is formed serves as a block to the entry of additional sperm into the egg (Callow et al., 1978).

Fertilization of the *Marsilea* egg is complicated by the internal location of the egg within an archegonium. The free-swimming sperm must enter the archegonium in order to make contact with the egg. The archegonium is shaped like a round-bottom flask. The neck consists of a group of cells that form a channel to the egg, which is located in the bulbous base of the archegonium. Sperm swimming in the vicinity of the archegonial opening are trapped by a gelatinous substance that surrounds the opening (Fig. 9.34).

Once trapped, the sperm at first swim about at random but later are oriented toward the channel as if following a chemical track. This behavior suggests that a chemotactic substance is released from the archegonium to draw the trapped sperm into the channel and toward the egg. Movement of the sperm through the channel is thought to be facilitated by the spiral symmetry of the sperm as well as by the action of the multiple flagella. Several sperm can be observed in the cavity of the archegonium above the egg (Fig. 9.35). However, the egg is overlain by a thick wall containing a hole that is only slightly larger in diameter than the sperm. This narrow opening allows only a single sperm to penetrate and make contact with the egg plasma membrane via the anteriormost tip of the sperm (Fig. 9.36). After sperm–egg contact, the egg cytoplasm flows around the sperm, creating a fertilization cone that incorporates the entire sperm.

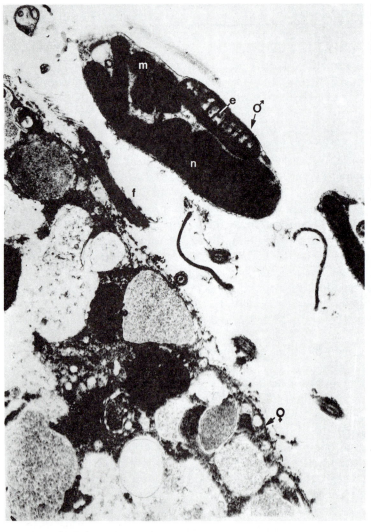

FIGURE 9.32 The *Fucus* sperm (♂) and egg (♀) prior to fertilization. One of the sperm flagella (f) is against the egg plasmalemma. e: eyespot; m: mitochondria; n: nucleus. ×19,800. (From S.H. Brawley, R. Wetherbee, and R.S. Quatrano. 1976. Fine-structural studies of the gametes and embryo of *Fucus vesiculosus* L. (Phaeophyta). I. Fertilization and pronuclear fusion. J. Cell Sci., *20:* 241.)

Shortly after sperm penetration a new extracellular layer appears over the egg (Figs. 9.35*C* and 9.36*B*). This layer apparently blocks the entry of additional sperm, preventing polyspermy (Myles, 1978). Inside the egg the sperm nucleus separates from the spiral band and other or-

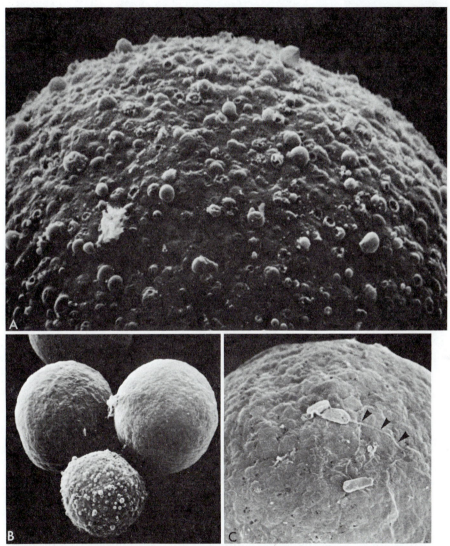

FIGURE 9.33 Scanning electron microscopy of fertilization in *Fucus*. *A*, Surface of an unfertilized egg showing the irregularity due to the protrusion of cytoplasmic vesicles. ×1900. *B*, Group of cells observed 10 minutes after mixing eggs and sperm. Note the smooth surfaces of the two fertilized eggs and the rough surface of the unfertilized egg (lower). ×400. *C*, High-power SEM of surface of fertilized egg. Note the smooth surface. The fertilizing sperm is seen with the tip of the anterior flagellum embedded in secreted wall material and the long posterior flagellum (arrowheads) lying on the surface. ×970. (From M.E. Callow et al. 1978. Fertilization in brown algae. I. SEM and other observations on *Fucus serratus*. J. Cell Sci., *32:* 48, 50, 52.)

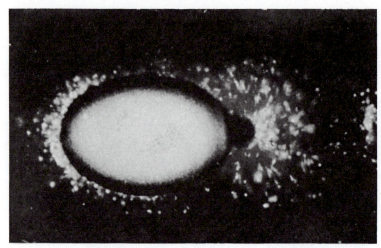

FIGURE 9.34 A megagametophyte of *Marsilea* with numerous sperm trapped in the mucus surrounding it. An archegonium is apparent at its tip. ×450. (From D.G. Myles. 1975. Reproduced with permission from *The Biology of the Male Gamete*. J.G. Duckett and P.A. Racey, Eds. Copyright by The Linnean Society of London.)

ganelles and is converted into a pronucleus as it moves toward the egg pronucleus. As described in the preceding chapter, the sperm nuclear envelope is shed before fertilization. Hence, the chromatin of the fertilizing sperm is not envelope-enclosed. During the formation of the male pronucleus a nuclear envelope is formed with the involvement of the egg endoplasmic reticulum.

Flowering Plants

Delivery of sperm nuclei to the embryo sac is accomplished by the pollen tube that grows down the style after the pollen lands on the stigma. In most angiosperm species the pollen tube enters the embryo sac through the micropyle, which is an opening in the integument layer that surrounds the embryo sac. After passing through the micropyle, there are three possible modes of entry of the pollen tube into the embryo sac: (1) between the egg and one synergid; (2) between the embryo sac wall and one synergid; and (3) directly into one of the synergids. We shall discuss the latter mechanism, which is found in *Capsella*, the organism that we have chosen to illustrate embryo sac development (Chap. 8) and embryonic development (Chap. 10) in angiosperms. This mechanism is also found in *Gossypium* (cotton), where it has been studied in detail. We shall use *Gossypium* to illustrate angiosperm fertilization. With few exceptions, *Capsella* and *Gossypium* have very similar patterns of fertilization.

As we discussed previously, the mature embryo sac contains two synergids flanking the egg. If the flower is not pollinated, this situation remains unchanged. However, after pollination (but before the pollen tube reaches the embryo sac) one of the synergids begins to degenerate (Fig. 9.37); this is the synergid that the pollen tube enters (Jensen and

Fisher, 1968). The discovery that one of the synergids begins degeneration before the entry of the pollen tube upsets a long-standing belief that the synergid is destroyed by the invading pollen tube. The nature of the signal relayed to the embryo sac to cause one of the synergids to degenerate is completely unknown but very intriguing. Since the pollen tube enters this synergid, it is likely that it releases a chemotactic substance that attracts the pollen tube to it.

The pollen tube enters the degenerating synergid through the filiform apparatus (Fig. 9.38). The entrance of the pollen tube is followed by more rapid degeneration of the synergid. The tube grows for a short distance within the synergid and then ceases to grow. After the pollen

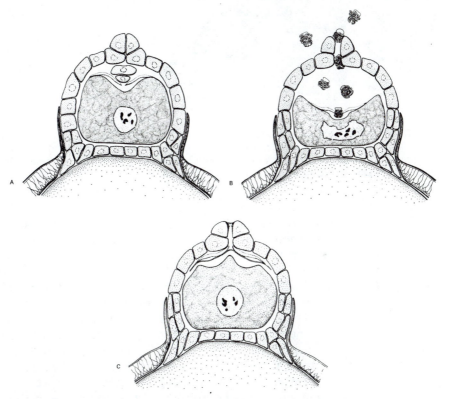

FIGURE 9.35 Diagram of the archegonium at three different stages in development. *A*, The spheroidal egg with ventral canal and neck canal cells still intact. A thick wall with a hole in its center separates the egg and the ventral canal cell. The neck cells (four of eight are shown) are still close together so that the neck is closed. *B*, The egg and its nucleus have become cup-shaped. The two canal cells have disintegrated or been extruded through the open neck. The neck cells have moved apart and spermatozoids are able to move through the open neck and into the archegonial cavity. One spermatozoid is passing through the hole in the thick wall. *C*, The zygote after it has regained the spheroidal shape of the egg. A new extracellular layer is found across the top of the cell underneath the original thick wall. (From D.G. Myles. 1978. The fine structure of fertilization in the fern *Marsilea vestita*. J. Cell Sci., *30:* 267.)

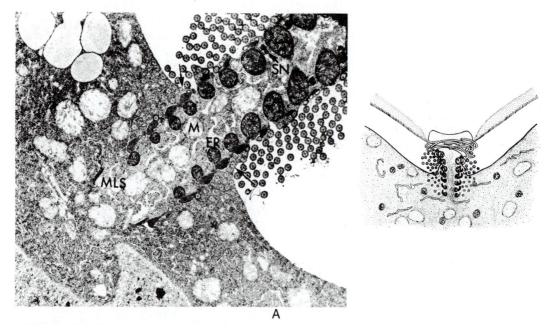

FIGURE 9.36 Electron micrographs (left) and drawings (right) of sperm entry in *Marsilea. A*, A sperm fusing with the egg. The two plasma membranes have fused (arrows), and the egg cytoplasm has penetrated into the region between the sperm coils. ER: egg endoplasmic reticulum; M: egg mitochondrion; MLS: sperm multilayered structure; SN: sperm nucleus. ×10,400.

Illustration continued on the opposite page.

tube has reached its maximum length, it discharges some cytoplasm and the two sperm cells. Discharge is through a pore that is located subterminally in the pollen tube. After discharge, synergid degeneration is completed; almost all synergid organelles become unrecognizable, and the plasmalemma disappears. Following their discharge from the pollen tube, one sperm makes contact with the egg cell membrane, whereas the other contacts the central cell. These contacts are facilitated by the loss of the synergid plasmalemma during degeneration. The contacts are also facilitated by the absence of a cell wall in this portion of the synergid. By contrast, the *Capsella* synergid is surrounded by a cell wall (see Chap. 8). It is uncertain how the sperm nuclei enter the egg and central cell of *Capsella*, but it has been proposed that the cell wall ruptures near the open end of the pollen tube, allowing contact to be made (Schulz and Jensen, 1968a).

It is assumed that the sperm nuclei alone enter the egg and central cell, while their cytoplasms remain behind. The remnants of these cells may be the X-bodies (Fig. 9.39), which are found in the synergid after pollen tube discharge (Jensen and Fisher, 1968). The sperm nucleus that enters the egg cytoplasm approaches the egg nucleus, apparently moving by cytoplasmic streaming. The nuclear envelopes fuse, and the contents of the two nuclei coalesce. The other sperm nucleus enters the central

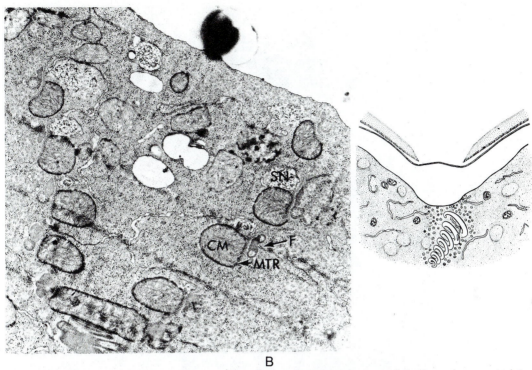

B

FIGURE 9.36 *(Continued) B,* A sperm in the egg interior. A new extracellular layer (represented in drawing) has appeared over the egg, preventing the entry of additional sperm. CM: coil mitochondrion; F: flagellar band; MTR: microtubule ribbon; SN: sperm nucleus. ×14,100. (From D.G. Myles. 1978. The fine structure of fertilization in the fern *Marsilea vestita.* J. Cell Sci., *30:* 270, 271, 272.)

cell cytoplasm and approaches the polar nuclei, fusing with them to form the triploid endosperm. The endosperm nucleus begins dividing soon after its formation. The endosperm cytoplasm remains syncytial in some plants and becomes cellularized in others (including *Gossypium*). The endosperm provides the nutrients that are utilized by the growing embryo at some stages of development.

The persistent synergid remains intact for a short time after pollen tube discharge, but then it, too, degenerates.

As in animal fertilization, the fusion of gametes triggers a number of responses by the egg that ultimately result in cleavage of the zygote into a multicellular embryo (Jensen, 1974). The precleavage phase can last for a very long time in angiosperms. For example, the *Gossypium* zygote remains undivided for approximately two and one half days following fertilization. The visible changes in the zygote nucleus involve the nucleoli, which appear to undergo reorganization. In *Gossypium*, a number of small nucleoli are seen in the sperm end of the zygote nucleus; these fuse to form a single large nucleolus so that the zygote nucleus has

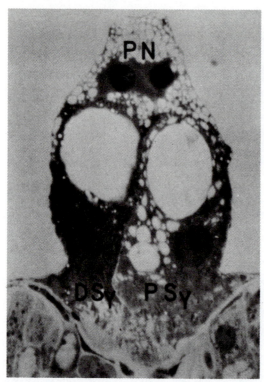

FIGURE 9.37 Degenerating synergid (DSy) and persistent synergid (PSy) in a *Gossypium* embryo sac not yet penetrated by the pollen tube. The fusing polar nuclei (PN) are visible in the central cell. Light micrograph. ×900. (From W.A. Jensen and D.B. Fisher. 1968. Cotton embryogenesis: The entrance and discharge of the pollen tube in the embryo sac. Planta [Berl.], *78:* 160.)

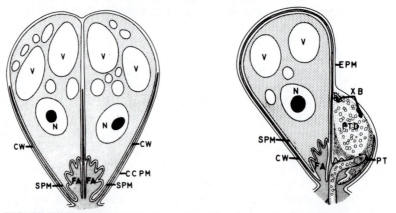

FIGURE 9.38 Diagrammatic summary of the changes in the synergids before and after discharge of the pollen tube. Left diagram shows synergids in unpollinated flower. Right diagram shows synergids after pollen tube discharge. N: nucleus; V: vacuoles; CW: cell wall; FA: filiform apparatus; SPM: synergid plasma membrane; CCPM: central cell plasma membrane; PT: pollen tube; XB: X bodies; PTD: pollen tube discharge; EPM: endosperm plasma membrane. (From W.A. Jensen and D.B. Fisher. 1968. Cotton embryogenesis: The entrance and discharge of the pollen tube in the embryo sac. Planta [Berl.], *78:* 164.)

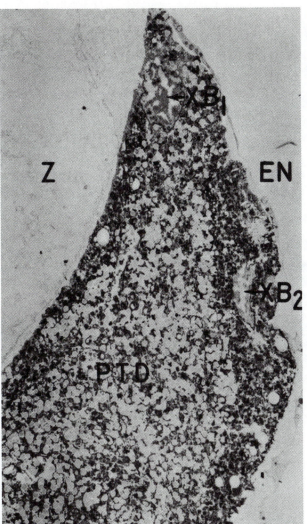

FIGURE 9.39 Degenerating synergid 24 hours after pollination (10 to 12 hours after discharge) with pollen tube discharge (PTD) and two X-bodies. One of the X-bodies (XB_1) is near the zygote (Z) while the second (XB_2) is near the endosperm (EN). ×2400. (From W.A. Jensen and D.B. Fisher. 1968. Cotton embryogenesis: The entrance and discharge of the pollen tube in the embryo sac. Planta [Berl.], *78:* 174.)

two nucleoli—one per haploid gene set. In *Capsella*, on the other hand, the sperm nucleolus fuses with that of the egg, producing a single nucleolus in the zygote nucleus.

Quite dramatic changes occur in the cytoplasm of the zygote prior to cleavage. One of the most spectacular changes involves the size of the zygote and density of the cytoplasm. In cotton the zygote shrinks to one half the volume of the egg, mainly because of a reduction of the large central vacuole of the egg (Fig. 9.40). After vacuole shrinkage, the cyto-

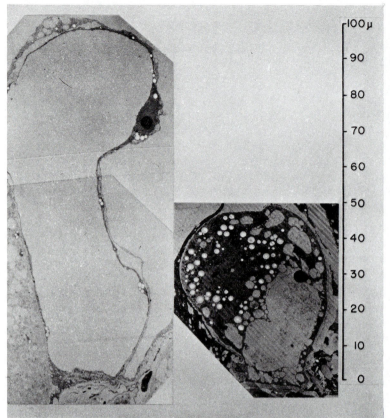

FIGURE 9.40 Median sections of the cotton egg (left) and a 3-day-old zygote (right). The position of the egg nucleus in the upper right third of the egg is characteristic although the egg has collapsed slightly. The zygote nucleus is in division. The micropyle is down. Both sections at the same magnification. (From W.A. Jensen. 1968. Cotton embryogenesis: The zygote. Planta [Berl.], *79:* 350.)

plasm—which was formerly spread thinly at the periphery of the chalazal end of the egg—occupies a large area in the chalazal end, packed around the zygote nucleus. The latter is surrounded by mitochondria and plastids. Zygote shrinkage also occurs in some other angiosperms, but it is by no means a universal phenomenon. In *Capsella,* where the egg vacuole is much smaller than in *Gossypium,* the zygote actually increases in size somewhat. As in cotton, the vacuole is at the micropylar end, whereas the bulk of the cytoplasm, which contains the nucleus, is at the chalazal end (Fig. 9.41).

The ultrastructural appearance of the zygote cytoplasm suggests that fertilization activates considerable metabolic activity. Before fertilization, ribosomes appear to be randomly scattered in the cytoplasm,

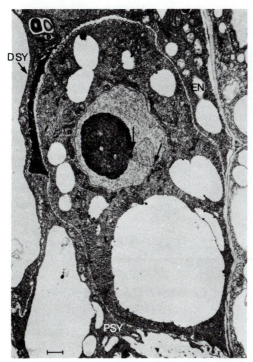

FIGURE 9.41 Electron micrograph of the *Capsella* zygote. Nuclear bridges are still present in the zygote nucleus (arrows), indicating the incomplete fusion of the male and female nuclei. PSY: persistent synergid; DSY: degenerating synergid; EN: endosperm. Scale bar equals 1.0 μm. (From R. Schulz and W.A. Jensen. 1968b. *Capsella* embryogenesis: The egg, zygote, and the young embryo. Am. J. Bot., *55*: 813.)

primarily as monosomes. However, after formation of the zygote nucleus, large polysome complexes appear, indicating that intense protein synthesis is under way. Polysome complexes have been observed in the zygote cytoplasm of a number of angiosperm species, but they are particularly dramatic in cotton, where each complex may contain as many as 30 to 40 ribosomes (Fig. 9.42). These polysomes must be translating extremely large proteins. Somewhat later, a new population of ribosomes appears in the cytoplasm; they associate as small polysomes (Fig. 9.43). Presumably smaller proteins are produced by these polysomes.

Various other ultrastructural changes in the cytoplasm are also indicative of enhanced metabolic activity in the zygote. For example, starch begins to accumulate in the plastids surrounding the nucleus (Fig. 9.42*B*), and a new cell wall is being laid down around the zygote. The Golgi bodies appear to be involved in the latter process. These organelles, which seem quiescent in the egg (Fig. 9.42*A*), appear in the zygote to bud off vesicles that may contain cell wall material.

A

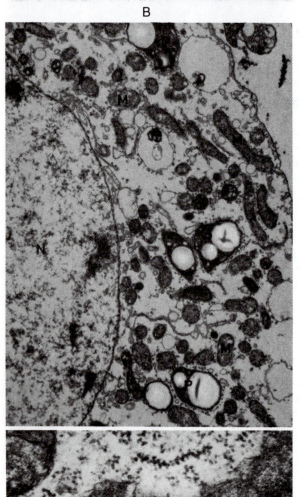

B

FIGURE 9.42 Ultrastructural changes in the cotton egg cytoplasm, resulting from fertilization. *A*, Portion of the egg and central cell (CC) prior to fertilization. Note the random arrangement of ribosomes and the quiescent appearance of the Golgi apparatus (D). P: plastid. ×37,000. *B*, Portion of a zygote. The plastids (P) and mitochondria (M) form a shell around the nucleus (N) with the ribosomes, aggregated as polysomes, forming shells around the plastids and mitochondria. ×7300. Inset: A large helical polysome in a zygote. ×33,400. (From W.A. Jensen. 1968. Cotton embryogenesis: The zygote. Planta [Berl.], *79:* 351, 354.)

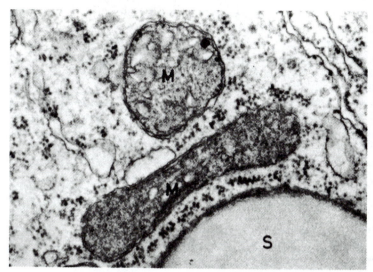

FIGURE 9.43 Cytoplasm of a cotton zygote three days after pollination. Both helical polysomes as well as second-generation ribosomes and small polysomes can be seen. ×42,250. (From W.A. Jensen. 1968. Cotton embryogenesis: The zygote. Planta [Berl.], *79:* 363.)

REFERENCES

Aitken, R.J., and D.W. Richardson. 1981. Measurement of the sperm binding capacity of the mouse zona pellucida and its use in the estimation of anti-zona antibody titres. J. Reprod. Fert., *63:* 295–307.

Aitken, R.J. et al. 1981. The influence of anti-zona and anti-sperm antibodies on sperm-egg interactions. J. Reprod. Fert., *62:* 597–606.

Anderson, E. 1968. Oocyte differentiation in the sea urchin *Arbacia punctulata,* with particular reference to the origin of cortical granules and their participation in the cortical reaction. J. Cell Biol., *37:* 514–539.

Austin, C.R. 1952. The "capacitation" of the mammalian spermatozoa. Nature (Lond.), *170:* 326.

Barros, C. et al. 1967. Membrane vesiculation as a feature of the mammalian acrosome reaction. J. Cell Biol., *34:* C1–C5.

Barros, C., and R. Yanagimachi. 1971. Induction of zona reaction in golden hamster eggs by cortical granule material. Nature (Lond.), *233:* 268–269.

Bedford, J.M. 1968. Ultrastructural changes in the sperm head during fertilization in the rabbit. Am. J. Anat., *123:* 329–357.

Bedford, J.M. 1972. An electron microscopic study of sperm penetration into the rabbit egg after natural mating. Am. J. Anat., *133:* 213–254.

Belanger, A.M., and A.W. Schuetz. 1975. Precocious induction of activation responses in amphibian oocytes by divalent ionophore A23187. Dev. Biol., *45:* 378–381.

Boldt, J. et al. 1981. Reaction of sperm with egg-derived hydrogen peroxide helps prevent polyspermy during fertilization in the sea urchin. Gamete Res., *4:* 365–377.

Bolwell, G.P. et al. 1979. Fertilization in brown algae. II. Evidence for lectin-sensitive complementary receptors involved in gamete recognition in *Fucus serratus.* J. Cell Sci., *36:* 19–30.

Bolwell, G.P., J.A. Callow, and L.V. Evans. 1980. Fertilization in brown algae. III. Preliminary characterization of putative gamete receptors from eggs and sperm of *Fucus serratus*. J. Cell Sci., *43:* 209–224.

Brawley, S.H., R. Wetherbee, and R.S. Quatrano. 1976. Fine-structural studies of the gametes and embryo of *Fucus vesiculosus* L. (Phaeophyta). I. Fertilization and pronuclear fusion. J. Cell Sci., *20:* 233–254.

Callow, M.E. et al. 1978. Fertilization in brown algae. I. SEM and other observations on *Fucus serratus*. J. Cell Sci., *32:* 45–54.

Carroll, E.J., and D. Epel. 1975. Isolation and biological activity of the proteases released by sea urchin eggs following fertilization. Dev. Biol., *44:* 22–32.

Chambers, E.L. 1939. The movement of the egg nucleus in relation to the sperm aster in the echinoderm egg. J. Exp. Biol., *16:* 409–424.

Chambers, E.L., B.C. Pressman, and B. Rose. 1974. The activation of sea urchin eggs by the divalent ionophores A23187 and X-537A. Biochem. Biophys. Res. Commun., *60:* 126–132.

Chandler, D.E., and J. Heuser. 1979. Membrane fusion during secretion. Cortical granule exocytosis in sea urchin eggs as studied by quick-freezing and freeze-fracture. J. Cell Biol., *83:* 91–108.

Chandler, D.E., and J. Heuser. 1980. The vitelline layer of the sea urchin egg and its modification during fertilization. J. Cell Biol., *84:* 618–632.

Coburn, M., H. Schuel, and W. Troll. 1981. A hydrogen peroxide block to polyspermy in the sea urchin *Arbacia punctulata*. Dev. Biol., *84:* 235–238.

Collins, F. 1976. A reevaluation of the fertilizin hypothesis of sperm agglutination and the description of a novel form of sperm adhesion. Dev. Biol., *49:* 381–394.

Cook, A.H., and J.A. Elvidge. 1951. Fertilization in the Fucaceae: Investigations on the nature of the chemotactic substance produced by eggs of *Fucus serratus* and *F. vesiculosus*. Proc. Roy. Soc. Lond. (Biol.), *138:* 97–114.

Cross, N.L., and R.P. Elinson. 1980. A fast block to polyspermy in frogs mediated by changes in the membrane potential. Dev. Biol., *75:* 187–198.

Cuellar, O. 1971. Reproduction and the mechanism of meiotic restitution in the parthenogenetic lizard *Cnemidophorus uniparens*. J. Morphol., *133:* 139–165.

Dan, J.C. 1954. Studies on the acrosome. III. Effect of calcium deficiency. Biol. Bull., *107:* 335–349.

Dan, J.C. 1967. Acrosome reaction and lysins. *In* C.B. Metz and A. Monroy (eds.), *Fertilization*, Vol. 1. Academic Press, New York, pp. 237–367.

Dan, J.C., Y. Ohori, and H. Kushida. 1964. Studies on the acrosome. VII. Formation of the acrosomal process in sea urchin spermatozoa. J. Ultrastruct. Res., *11:* 508–524.

Decker, G.L., D.B. Joseph, and W.J. Lennarz. 1976. A study of factors involved in induction of the acrosomal reaction in sperm of the sea urchin, *Arbacia punctulata*. Dev. Biol., *53:* 115–125.

Ecker, R.E., and L.D. Smith. 1968. Protein synthesis in amphibian oocytes and early embryos. Dev. Biol., *18:* 232–249.

Edwards, R.G. 1981. Test-tube babies, 1981. Nature (Lond.), *293:* 253–256.

Epel, D. 1967. Protein synthesis in sea urchin eggs: A "late" response to fertilization. Proc. Natl. Acad. Sci. U.S.A., *57:* 899–906.

Epel, D. 1975. The program of and mechanisms of fertilization in the echinoderm egg. Am. Zool., *15:* 507–522.

Epel, D. 1977. The program of fertilization. Sci. Am., 237(5): 128–138.

Epel, D. et al. 1969. The program of structural and metabolic changes following fertilization of sea urchin eggs. *In* G. Padilla, G.L. Whitson, and I. Cameron (eds.), *The Cell Cycle: Gene-Enzyme Relationships*. Academic Press, New York, pp. 279–298.

Epel, D. et al. 1974. An analysis of the partial metabolic derepression of sea urchin eggs by ammonia; the existence of independent pathways. Dev. Biol., 40: 245–255.

Foerder, C.A., and B.M. Shapiro. 1977. Release of ovoperoxidase from sea urchin eggs hardens the fertilization membrane with tyrosine crosslinks. Proc. Natl. Acad. Sci. U.S.A., 74: 4214–4218.

Friend, D.S. et al. 1977. Membrane particle changes attending the acrosome reaction in guinea pig spermatozoa. J. Cell Biol., 74: 561–577.

Fulton, B.P., and D.G. Whittingham. 1978. Activation of mammalian oocytes by intracellular injection of calcium. Nature (Lond.), 273: 149–151.

Gilkey, J.C. et al. 1978. A free calcium wave traverses the activating egg of the medaka, *Oryzias latipes*. J. Cell Biol., 76: 448–466.

Glabe, C.G., and V.D. Vacquier. 1978. Egg surface glycoprotein receptor for sea urchin sperm binding. Proc. Natl. Acad. Sci. U.S.A., 75: 881–885.

Glabe, C.G. et al. 1982. Carbohydrate specificity of sea urchin sperm binding: a cell surface lectin mediating sperm-egg adhesion. J. Cell Biol., 94: 123–128.

Gould-Somero, M., L.A. Jaffe, and L.Z. Holland. 1979. Electrically mediated fast polyspermy block in eggs of the marine worm, *Urechis caupo*. J. Cell Biol., 82: 426–440.

Grainger, J.L. et al. 1979. Intracellular pH controls protein synthesis rate in the sea urchin egg and early embryo. Dev. Biol., 68: 396–406.

Green, J.D., and R.G. Summers. 1982. Effects of protease inhibitors on sperm-related events in sea urchin fertilization. Dev. Biol., 93: 139–144.

Grey, R.D., P.K. Working, and J.L. Hedrick. 1976. Evidence that the fertilization envelope blocks sperm entry in eggs of *Xenopus laevis*: Interaction of sperm with isolated envelopes. Dev. Biol., 54: 52–60.

Grey, R.D. et al. 1982. An electrical block is required to prevent polyspermy in eggs fertilized by natural mating of *Xenopus laevis*. Dev. Biol., 89: 475–484.

Gulyas, B.J. 1980. Cortical granules of mammalian eggs. Int. Rev. Cytol., 63: 357–392.

Gulyas, B.J., and E.D. Schmell. 1980. Ovoperoxidase activity in ionophore treated mouse eggs. I. Electron microscopic localization. Gamete Res., 3: 267–277.

Gwatkin, R.B.L. 1977. *Fertilization Mechanisms in Man and Mammals*. Plenum Press, New York.

Gwatkin, R.B.L. et al. 1973. The zona reaction of hamster and mouse eggs: Production *in vitro* by a trypsin-like protease from cortical granules. J. Reprod. Fertil., 32: 259–265.

Gwatkin, R.B.L., D.T. Williams, and D.J. Carlo. 1977. Immunization of mice with heat-solubilized hamster zonae: Production of anti-zona antibody and inhibition of fertility. Fertil. Steril., 28: 871–877.

Hall, H.G. 1978. Hardening of the sea urchin fertilization envelope by peroxidase-catalyzed phenolic coupling of tyrosines. Cell, 15: 343–355.

Hamaguchi, M.S., and Y. Hiramoto. 1980. Fertilization processes in the heart-urchin, *Clypeaster japonicus* observed with a differential interference microscope. Dev., Growth Differ., 22: 517–530.

Hartmann, J.F., and R.B.L. Gwatkin. 1971. Alteration of sites on the mammalian sperm surface following capacitation. Nature (Lond.), *234*: 479–481.

Hartmann, J.F., and C.F. Hutchison. 1981. Modulation of fertilization *in vitro* by peptides released during hamster sperm-zona pellucida interaction. Proc. Natl. Acad. Sci. U.S.A., *78*: 1690–1694.

Hollinger, T.G., J.N. Dumont, and R.A. Wallace. 1979. Calcium-induced dehiscence of cortical granules in *Xenopus laevis* oocytes. J. Exp. Zool., *210*: 107–116.

Hollinger, T.G., and A.W. Schuetz. 1976. "Cleavage" and cortical granule breakdown in *Rana pipiens* oocytes induced by direct micro-injection of calcium. J. Cell Biol., *71*: 395–401.

Jaffe, L.A. 1976. Fast block to polyspermy in sea urchin eggs is electrically mediated. Nature (Lond.), *261*: 68–71.

Jaffe, L.A. 1980. Electrical polyspermy block in sea urchins: Nicotine and low sodium experiments. Dev., Growth Differ., *22*: 503–507.

Jaffe, L.A., M. Gould-Somero, and L. Holland. 1979. Ionic mechanism of the fertilization potential of the marine worm, *Urechis caupo* (Echiura). J. Gen. Physiol., *73*: 469–492.

Jensen, W.A. 1968. Cotton embryogenesis: The zygote. Planta (Berl.), *79*: 346–366.

Jensen, W.A. 1974. Reproduction in flowering plants. *In* A.W. Robards (ed.), *Dynamic Aspects of Plant Ultrastructure*. McGraw-Hill Book Co., London, pp. 481–503.

Jensen, W.A., and D.B. Fisher. 1968. Cotton embryogenesis: The entrance and discharge of the pollen tube in the embryo sac. Planta (Berl.), *78*: 158–183.

Johnson, J.D., D. Epel, and M. Paul. 1976. Intracellular pH and activation of sea urchin eggs after fertilization. Nature (Lond.), *262*: 661–664.

Johnson, L.L., D.F. Katz, and J.W. Overstreet. 1981. The movement characteristics of rabbit spermatozoa before and after activation. Gamete Res., *4*: 275–282.

Kinsey, W.H., and J.K. Koehler. 1978. Cell surface changes associated with *in vitro* capacitation of hamster sperm. J. Ultrastruct. Res., *64*: 1–13.

Kunkle, M., and F.J. Longo. 1975. Cytological events leading to the cleavage of golden hamster zygotes. J. Morphol., *146*: 197–214.

Lillie, F.R. 1912. The production of sperm iso-agglutinins by ova. Science, *36*: 527–530.

Loeb, J. 1914. Cluster formation of spermatozoa caused by specific substances from eggs. J. Exp. Zool., *17*: 123–140.

Longo, F.J. 1973. Fertilization: A comparative ultrastructural review. Biol. Reprod., *9*: 149–215.

Longo, F.J. 1980. Organization of microfilaments in sea urchin (*Arbacia punctulata*) eggs at fertilization: Effects of cytochalasin B. Dev. Biol., *74*: 422–433.

Longo, F.J., and E. Anderson. 1968. The fine structure of pronuclear development and fusion in the sea urchin, *Arbacia punctulata*. J. Cell Biol., *39*: 339–368.

Longo, F.J., and E. Anderson. 1969. Cytological events leading to the formation of the two-cell stage in the rabbit: Association of the maternally and paternally derived genomes. J. Ultrastruct. Res., *29*: 86–118.

Lopo, A.C., and V.D. Vacquier. 1980. Antibody to a sperm surface glycoprotein inhibits the egg jelly-induced acrosome reaction of sea urchin sperm. Dev. Biol., *79*: 325–333.

Mazia, D. 1937. The release of calcium in *Arbacia* eggs on fertilization. J. Cell. Comp. Physiol., *10*: 291–308.

McRorie, R.A., and W.L. Williams. 1974. Biochemistry of mammalian fertilization. Ann. Rev. Biochem., *43*: 777–803.

Metz, C.B. 1967. Gamete surface components and their role in fertilization. *In* C.B. Metz and A. Monroy (eds.), *Fertilization*, Vol. 1. Academic Press, New York. pp. 163–236.

Metz, C.B. 1978. Sperm and egg receptors involved in fertilization. *In* A.A. Moscona and A. Monroy (eds.), *Current Topics in Developmental Biology*, Vol. 12. Academic Press, New York, pp. 107–147.

Meyerhof, P.G., and Y. Masui. 1977. Ca and Mg control of cytostatic factors from *Rana pipiens* oocytes which cause metaphase and cleavage arrest. Dev. Biol., *61*: 214–229.

Miller, R.L. 1977a. Chemotactic behavior of the sperm of chitons (Mollusca: Polyplacophora). J. Exp. Zool., *202*: 203–212.

Miller, R.L. 1977b. Distribution of sperm chemotaxis in the animal kingdom. *In* K.G. Adiyodi and R.G. Adiyodi (eds.), *Advances in Invertebrate Reproduction*, Vol. 1. Peralam-Kenoth, Kerala, India, pp. 99–119.

Miller, R.L. 1981. Sperm chemotaxis occurs in echinoderms. Am. Zool., *21*: 985.

Miyazaki, S., and S. Hirai. 1979. Fast polyspermy block and activation potential. Correlated changes during oocyte maturation of a starfish. Dev. Biol., *70*: 327–340.

Mori, T. et al. 1978. Possible presence of autoantibodies to zona pellucida in infertile women. Experientia, *34*: 797–799.

Müller, D.G., and G. Gassmann. 1978. Identification of the sex attractant in the marine brown alga *Fucus vesiculosus*. Naturwissenschaften, *65*: 389.

Müller, D.G., and L. Jaenicke. 1973. Fucoserraten, the female sex attractant of *Fucus serratus* L. (Phaeophyta). FEBS Lett., *30*: 137–139.

Müller, D.G., and K. Seferiadis. 1977. Specificity of sexual chemotaxis in *Fucus serratus* and *Fucus vesiculosus* (Phaeophyceae). Z. Pflphysiol.,*84*: 85–94.

Myles, D.G. 1975. Structural changes in the sperm of *Marsilea vestita* before and after fertilization. *In* J.G. Duckett and P.A. Racey (eds.), *The Biology of the Male Gamete: Linnean Society Supplement No. 1 to the Biological Journal*. Vol. 7. Academic Press, New York, pp. 129–134.

Myles, D.G. 1978. The fine structure of fertilization in the fern *Marsilea vestita*. J. Cell Sci., *30*: 265–281.

Peng, H.B., and L.F. Jaffe. 1976. Cell-wall formation in *Pelvetia* embryos. A freeze-fracture study. Planta (Berl.), *133*: 57–71.

Pollock, E.G. 1970. Fertilization in *Fucus*. Planta (Berl.), *92*: 85–99.

Ridgway, E.B., J. C. Gilkey, and L.F. Jaffe. 1977. Free calcium increases explosively in activating medaka eggs. Proc. Natl. Acad. Sci. U.S.A., *74*: 623–627.

Rossignol, D.P., A.J. Roschelle, and W.J. Lennarz. 1981. Sperm-egg binding: Identification of a species-specific sperm receptor from eggs of Strongylocentrotus purpuratus. J. Supramol. Struct. Cell. Biochem., *15*: 347–358.

Rothschild, Lord, and M.M. Swann. 1952. The fertilization reaction in the sea-urchin. The block to polyspermy. J. Exp. Biol., *29*: 469–483.

Saling, P.M. 1981. Involvement of trypsin-like activity in binding of mouse spermatozoa to zonae pellucidae. Proc. Natl. Acad. Sci. U.S.A., *78*: 6231–6235.

Schackmann, R.W., R. Christen, and B.M. Shapiro. 1981. Membrane potential depolarization and increased intracellular pH accompany the acrosome reaction of sea urchin sperm. Proc. Natl. Acad. Sci. U.S.A., *78*: 6066–6070.

Schackmann, R.W., E.M. Eddy, and B.M. Shapiro. 1978. The acrosome reaction of *Strongylocentrotus purpuratus* sperm. Ion requirements and movements. Dev. Biol., *65*: 483–495.

Schackmann, R.W., and B.M. Shapiro. 1981. A partial sequence of ionic changes associated with the acrosome reaction of *Strongylocentrotus purpuratus*. Dev. Biol., *81*: 145–154.

Schatten, G., and H. Schatten. 1981. Effects of motility inhibitors during sea urchin fertilization. Microfilament inhibitors prevent sperm incorporation and restructuring of fertilized egg cortex, whereas microtubule inhibitors prevent pronuclear migrations. Exp. Cell Res., *135*: 311–330.

Schatten, H., and G. Schatten. 1980. Surface activity at the egg plasma membrane during sperm incorporation and its cytochalasin B sensitivity. Scanning electron microscopy and time-lapse video microscopy during fertilization of the sea urchin *Lytechinus variegatus*. Dev. Biol., *78*: 435–449.

Schmell, E.D., and B.J. Gulyas. 1980. Ovoperoxidase activity in ionophore treated mouse eggs. II. Evidence for the enzyme's role in hardening the zona pellucida. Gamete Res., *3*: 279–290.

Schroeder, T.E. 1979. Surface area change at fertilization: Resorption of the mosaic membrane. Dev. Biol., *70*: 306–326.

Schulz, R., and W.A. Jensen. 1968a. *Capsella* embryogenesis: The synergids before and after fertilization. Am. J. Bot., *55*: 541–552.

Schulz, R., and W.A. Jensen. 1968b. *Capsella* embryogenesis: The egg, zygote, and young embryo. Am. J. Bot., *55*: 807–819.

SeGall, G.K., and W.J. Lennarz. 1979. Chemical characterization of the component of the jelly coat from sea urchin eggs responsible for induction of the acrosome reaction. Dev. Biol., *71*: 33–48.

Shen, S.S., and R.A. Steinhardt. 1978. Direct measurement of intracellular pH during metabolic derepression of the sea urchin egg. Nature (Lond.), *272*: 253–254.

Shen, S.S., and R.A. Steinhardt. 1979. Intracellular *p*H and the sodium requirement at fertilization. Nature (Lond.), *282*: 87–89.

Shivers, C.A., and B.S. Dunbar. 1977. Autoantibodies to zona pellucida: A possible cause for infertility in women. Science, *197*: 1082–1084.

Stambaugh, R., R.G. Brackett, and L. Mastroianni. 1969. Inhibition of *in vitro* fertilization of rabbit ova by trypsin inhibitors. Biol. Reprod., *1*: 223–227.

Stambaugh, R., and J. Buckley. 1969. Identification and subcellular localization of the enzymes effecting penetration of the zona pellucida by rabbit spermatozoa. J. Reprod. Fertil., *19*: 423–432.

Steinhardt, R.A., and J.M. Alderton. 1982. Calmodulin confers calcium sensitivity on secretory exocytosis. Nature (Lond.), *295*: 154–155.

Steinhardt, R.A., and D. Epel. 1974. Activation of sea-urchin eggs by a calcium ionophore. Proc. Natl. Acad. Sci. U.S.A., *71*: 1915–1919.

Steinhardt, R., and D. Mazia. 1973. Development of K^+-conductance and membrane potentials in unfertilized sea urchin eggs after exposure to NH_4OH. Nature (Lond.), *241*: 400–401.

Steinhardt, R.A. et al. 1974. Is calcium ionophore a universal activator for unfertilized eggs? Nature (Lond.), *252*: 41–43.

Steinhardt, R., R. Zucker, and G. Schatten. 1977. Intracellular calcium release at fertilization in the sea urchin egg. Dev. Biol., *58*: 185–196.

Summers, R.G., and B.L. Hylander. 1974. An ultrastructural analysis of early fertilization in the sand dollar, *Echinarchnius parma.* Cell Tissue Res., *150*: 343–368.

Summers, R.G., and B.L. Hylander. 1975. Species-specificity of acrosome reaction and primary gamete binding in echinoids. Exp. Cell Res., *96*: 63–68.

Tilney, L.G. 1978. Polymerization of actin. V. A new organelle, the actomere, that initiates the assembly of actin filaments in Thyone sperm. J. Cell Biol., *77*: 551–564.

Tilney, L.G. et al. 1978. The polymerization of actin. IV. The role of Ca^{++} and H^+ in the assembly of actin and in membrane fusion in the acrosomal reaction of echinoderm sperm. J. Cell Biol., *77*: 536–550.

Tsunoda, Y., and M.C. Chang. 1978. Effect of antisera against eggs and zonae pellucidae on fertilization and development of mouse eggs *in vivo* and in culture. J. Reprod. Fert., *54*: 233–237.

Tyler, A. 1948. Fertilization and immunity. Physiol. Rev., *28*: 180–219.

Tzartos, S.J. 1979. Inhibition of in-vitro fertilization of intact and denuded hamster eggs by univalent anti-sperm antibodies. J. Reprod. Fert., *55*: 447–455.

Vacquier, V.D. 1975. The isolation of intact cortical granules from sea urchin eggs: Calcium ions trigger granule discharge. Dev. Biol., *43*: 62–74.

Vacquier, V.D., and G.W. Moy. 1977. Isolation of bindin: the protein responsible for adhesion of sperm to sea urchin eggs. Proc. Natl. Acad. Sci. U.S.A., *74*: 2456–2460.

Wasserman, W.J., and L.D. Smith. 1978. Oocyte maturation: Nonmammalian vertebrates. *In* R.E. Jones (ed.), *The Vertebrate Ovary.* Plenum Press, New York. pp. 443–468.

Winkler, W.M., and J.L. Grainger. 1978. Mechanism of action of NH_4Cl and other weak bases in the activation of sea urchin eggs. Nature (Lond.), *273*: 536–538.

Wolf, D.P. 1978. The block to sperm penetration in zona-free mouse eggs. Dev. Biol., *64*: 1–10.

Wood, D.M., C. Liu, and B.S. Dunbar. 1981. Effect of alloimmunization and heteroimmunization with zonae pellucidae on fertility in rabbits. Biol. Reprod., *25*: 439–450.

Woodland, H.R. 1974. Changes in the polysome content of developing *Xenopus laevis* embryos. Dev. Biol., *40*: 90–101.

Yanagimachi, R. 1970. The movement of golden hamster spermatozoa before and after capacitation. J. Reprod. Fertil., *23*: 193–196.

Yanagimachi, R., and N. Usui. 1974. Calcium dependence of the acrosome reaction and activation of guinea pig spermatozoa. Exp. Cell Res., *89*: 161–174.

Zamboni, L., J. Chakraborty, and D.M. Smith. 1972. First cleavage division of the mouse zygote. Biol. Reprod., *7*: 170–193.

Zimmerman, A.M., and S. Zimmerman. 1967. Action of colcemid in sea urchin eggs. J. Cell Biol., *34*: 483–488.

Zucker, R.S., and R.A. Steinhardt. 1978. Prevention of the cortical reaction in fertilized sea urchin eggs by injection of calcium-chelating ligands. Biochim. Biophys. Acta, *541*: 459–466.

10 Becoming Multicellular

After fertilization the zygote is converted into an aggregate of cells, each of which occupies a specific location and differentiates to acquire a specialized function. How does an embryo become multicellular? How do cells locate in their assigned positions? How do they acquire their specialized physiological properties so that they can contribute to total organismic function? These questions are all interrelated, since, as we shall see, these processes are occurring simultaneously, and the consequences of each bear upon the others. In this chapter we shall focus upon the division of zygotes into multicellular embryos. Although cell division is the predominant visible change in the embryo at this stage, the results of the division process are more significant than a mere increase in the number of cells composing the embryo. As we shall see in Chapter 11, much of the groundwork is being laid for regional specialization of the embryo, which will occur later in development.

10–1. CLEAVAGE OF THE ANIMAL ZYGOTE

Fusion of the sperm and egg to form the diploid zygote is followed by cleavage of the zygote into numerous small cells. The major characteristic distinguishing cell division during cleavage from ordinary cell division is the absence of cell growth. Ordinarily, mitosis is followed by a rather protracted interphase, during which the daughter cells increase in size to equal that of the parental cell. However, during zygotic cleavage the interphase is relatively short, and the growth phase is apparently bypassed. Without interphase growth the cytoplasmic volume relative to that of the nucleus is progressively reduced until it approaches that of ordinary somatic cells. For example, the volume of the cytoplasm of the mature sea urchin egg is 550 times that of its nucleus, but by the end of cleavage the cytoplasmic volume is only six times that of the nucleus (Brachet, 1950). Once this nucleocytoplasmic volume ratio is established, the rate of cell division decreases markedly. The transition to a slower rate of division at the end of cleavage is quite abrupt. This is shown for the frog embryo in Figure 10.1. Here, the logarithm of the number of cells per embryo is plotted against developmental time. During cleavage there is a rapid, constant rate of increase in cell number. But at about 40 hours, at the end of the cleavage stage, the increase in cell number is considerably slower.

The rate of cleavage is dramatically affected by temperature. This effect can be easily demonstrated by exposing zygotes to different environmental temperatures; at higher temperatures the embryonic cells cleave more rapidly. However, the paramount control over the division rate is exerted by the genome. This is illustrated very nicely by comparing the cleavage rate of the frog embryo, reared at 15°C, with that of a mammalian embryo, which develops at 37°C. Even though the frog embryo is maintained at a much lower temperature, it consists of approximately 32,000 cells at 43 hours (which corresponds to the early gastrula stage), whereas the rabbit embryo, which is at a much warmer temperature, consists of only 32 cells at a comparable time. The mouse embryo, which also develops at 37°C, consists of only 8 cells at that time.

The mechanism of control over division rate is unknown. Although regulated by the genome, cleavage rate is apparently mediated by the egg cytoplasm. This conclusion can be drawn from the following experiment. If eggs are hybridized with sperm from a species with a different cleavage rate, the maternal rate predominates *even if the egg has been enucleated prior to fertilization.* Clearly, the egg cytoplasm has been preprogrammed by the genome before fertilization to divide at a certain rate. Since the division rate decreases when the somatic nucleocytoplasmic volume ratio is attained (see previous discussion), it is tempting to speculate that the nucleocytoplasmic volume ratio of the zygote determines the division rate during cleavage. However, this seems unlikely, since when a sea urchin egg is divided in half before fertilization,

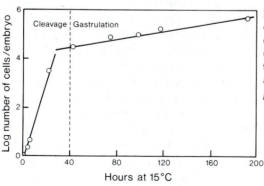

FIGURE 10.1 Increase in number of cells during early development of the egg of the frog (*Rana*). (After L.C. Sze. 1953. Changes in the amount of desoxyribonucleic acid in the development of *Rana pipiens*. J. Exp. Zool., *122*: 594.)

both halves cleave at the same rate as normal, full-sized eggs (Rustad et al., 1970).

The Initiation of Cleavage

The first cleavage division is the culmination of the activation reaction initiated by the fertilizing sperm. The preparation for cleavage must include a number of metabolic and morphological events in both the nucleus and cytoplasm. Both gamete nuclei must be transformed into pronuclei, the maternal and paternal DNA must be replicated, and the chromosomes must condense. The centrioles must position themselves on opposite sides of the zygote, and the mitotic apparatus, including the spindle and the asters that radiate from the centrioles, must be constructed. Finally, the structural components that are responsible for separating the zygote into daughter cells (called blastomeres) must become functional. For cleavage to proceed, numerous metabolic prerequisites must also be met, including mobilization of cytoplasmic reserves to generate the energy and synthesize the structural components that are necessary for both division and construction of new blastomere constituents. As we shall see in Chapter 12, the genetic control over most of these synthetic processes is mediated via the egg cytoplasm and is a delayed manifestation of transcription during oogenesis. The utilization of the cytoplasmic program for cleavage unfolds as a consequence of activation, which we now know is under ionic regulation. We do not know, however, how the ionic shifts direct the zygote to initiate cleavage.

The cleavage program itself has been the subject of intense investigation for nearly a century. Cleavage consists of two distinct processes—nuclear division (**karyokinesis**) and cytoplasmic division (**cytokinesis**). Karyokinesis is a function of the mitotic apparatus, whereas cytokinesis is caused by the formation of a cleavage furrow, which forms perpendicular to the spindle axis and constricts the zygote into two blastomeres. The constriction of the zygote is caused by structural elements located in the cortex. The cortex thickens in the furrow region to form a **contractile ring** that separates the cells. The nature and function of the contractile ring will be discussed later in this chapter.

Karyokinesis and cytokinesis are normally coupled so that daughter nuclei are distributed to separate cells. However, these processes can be experimentally uncoupled so that the functional relationship between them can be examined. For example, if the mitotic apparatus of the sea urchin zygote is removed at late metaphase, a stage when the zygote is still spherical, division will occur by formation of a normal-appearing cleavage furrow that bisects the zygote into two enucleate blastomeres (Fig. 10.2). Likewise, any treatment that disrupts or displaces the mitotic figure at anaphase has no effect upon cleavage. The independence of cytokinesis after such treatment is shown in Figure 10.3. In 10.3*A* the mitotic apparatus has been destroyed with an injection of seawater, and

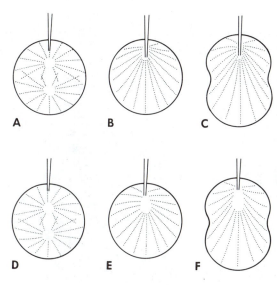

FIGURE 10.2 Effects of displacement of mitotic apparatus of sea urchin zygote on cleavage. A to C, Successive stages of cleavage when the spindle is removed. A, Before removal of the spindle. B, After removal. C, Furrow appears at a predetermined position. D to F, Successive stages of cleavage when the spindle is displaced by removal of part of the zygote cytoplasm. D, before displacement of the spindle. E, After displacement. F, Furrow appears at a predetermined position. In both series, cleavage planes are independent of shifted position of asters. (After Y. Hiramoto. 1956. Cell division without mitotic apparatus in sea urchin eggs. Exp. Cell Res., *11:* 631).

in 10.3*B* it has been displaced with a large oil droplet. These drastic treatments fail to prevent furrowing. The cytoplasm in the region constricted by the cleavage furrow can also be disrupted by stirring it vigorously with a needle without interfering with cleavage (Fig. 10.4). Thus, furrowing is independent of structural elements in the central cytoplasm; it is entirely a function of the cortex.

The independence of the cortex during the latter phases of mitosis implies that some earlier event has established the site of furrow formation in the cortex. It has been known for some time that furrowing occurs at a position on the surface that is equidistant from the two centrioles and their asters. Early theories of cleavage assumed that the asters play an active role in inducing the furrow. These theories were based primarily upon observations that furrowing would not occur in eggs lacking a *pair* of asters located near the surface. Furthermore, if there are three asters, three furrows will form equidistant from each aster pair; if there are four asters, four furrows will form. The importance of the asters (rather than the nucleus or spindle) in furrow induction has also been demonstrated by the cleavage of enucleate sea urchin eggs after parthenogenic activation. In spite of the absence of chromosomes and a spindle, asters developed, and furrows formed between them, cleaving the cytoplasm (Harvey, 1936).

Experiments designed to examine the relationship between the asters and the surface have been conducted by Rappaport and his associates. Rappaport (1961) assumed that if the asters induce furrow formation, there must be a period during early mitosis when the furrow site has not yet been established. Hence, manipulation of cell geometry at this time to alter the normal relationship between the asters and the surface should affect furrow formation. These experiments have confirmed that the location of the asters during a critical period of early anaphase determines

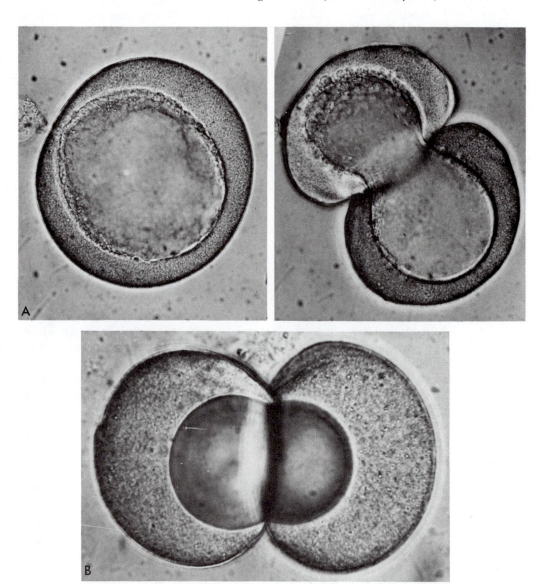

FIGURE 10.3 Disruption of the mitotic apparatus in the sea urchin embryo during late stages of mitosis has no effect on cleavage. *A,* Sea water is injected into the zygote to destroy the mitotic apparatus. The zygote on the left was injected before the onset of cleavage; cleavage is prevented. The zygote on the right was injected during cleavage; cytokinesis continues in spite of total destruction of the mitotic apparatus. *B,* This zygote was injected with an oil droplet at anaphase. Although the mitotic apparatus has been competely disrupted, cytokinesis occurs normally, compressing the oil droplet. (From Y. Hiramoto. 1965. Reproduced from *The Journal of Cell Biology,* 1965, vol. 25, pp. 161–167 by copyright permission of The Rockefeller University Press.)

the furrow site and that once the site is determined, cytokinesis proceeds independently of the mitotic apparatus. One such experiment is outlined in Figure 10.5. Here, a small glass ball is used to displace the asters to one side of the zygote. A cleavage furrow is formed between these two asters only on this side of the zygote, producing a binucleate horseshoe-shaped structure. Furrowing is not initiated on the opposite side because the asters have been displaced from their normal position near this surface. At the next division, three furrows occur simultaneously. Two spindles form in each arm of the horseshoe, and cytokinesis occurs between the asters of each spindle. The third furrow is initiated between the two asters in the bend of the horseshoe, *even though no spindle apparatus is present.* Thus, a furrow is produced between asters if they are close enough to the surface for the interaction to occur.

The interaction is completed in a remarkably short time. This has been demonstrated for the sand dollar egg (Fig. 10.6) by Rappaport and Ebstein (1965). They displaced the mitotic apparatus to one side of the egg prior to furrow establishment (late prophase or early metaphase). After the furrow developed on the near surface, they pushed the distal

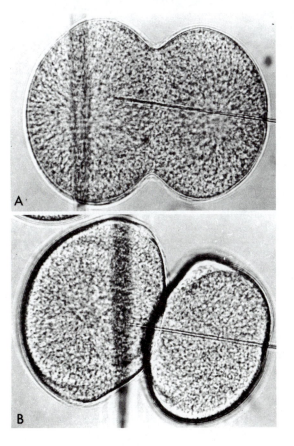

A

B

FIGURE 10.4 Cleavage in a sand dollar egg with a moving needle inserted through the cleavage plane. The needle was swept back and forth during the period between the photographs. (From R. Rappaport. 1966. Experiments concerning the cleavage furrow in invertebrate eggs. J. Exp. Zool., *161:* 6.)

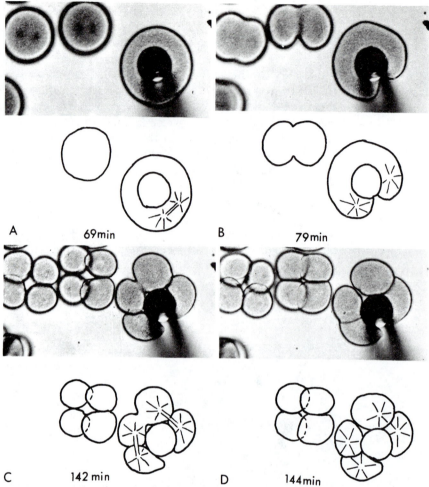

FIGURE 10.5 Cleavage of a torus-shaped sand dollar zygote. Condition of the mitotic apparatus is shown in line drawings. Note synchrony with controls. Timing begins at fertilization. *A,* Immediately before furrowing. *B,* First cleavage completed, producing a binucleate cell. *C,* Second cleavage. Two cells have divided from the free ends of the horseshoe and the binucleate cell, and the binucleate cell is dividing between the polar regions of the aster of the second division. *D,* Division completed. Each cell contains one nucleus. (From R. Rappaport. 1961. Experiments concerning the cleavage stimulus in sand dollar eggs. J. Exp. Zool., *148:* 83.)

surface toward the mitotic apparatus and held it there to allow it to interact with the asters. The minimum exposure necessary to initiate furrow formation was one minute; when the distal surface was released before one minute had elapsed, a furrow would not form. However, when the distal surface was held in for one minute, a furrow would develop two and one half minutes after release. Hence, after the interaction has occurred, a short latent period ensues before the appearance of a furrow.

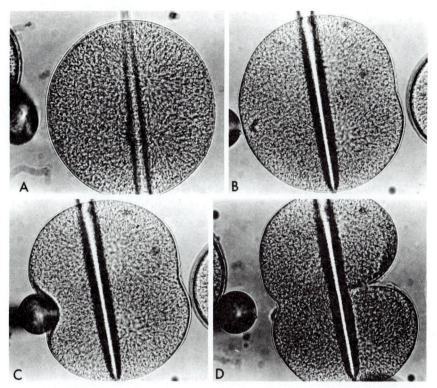

FIGURE 10.6 Experimental exposure of distal surface of sand dollar egg to displaced mitotic apparatus. *A,* Standard initial conditions. The egg is flattened with a needle that crosses the geometric center, displacing the astral centers to the right. The furrow will be unilateral and cut inward from the right. *B* to *D,* A furrow is forming on the right. Meanwhile, a glass ball is used to push the distal surface toward the mitotic apparatus. When the surface is held in for a minute or more and then released, a furrow forms three and one-half minutes after surface and mitotic apparatus were first brought together. (From R. Rappaport and R.P. Ebstein. 1965. Duration of stimulus and latent periods preceding furrow formation in sand dollar eggs. J. Exp. Zool., *158:* 377, 378.)

The Mechanism of Cytokinesis

The nature of the interaction between asters and the cortex that results in cytokinesis is unknown. Recent evidence casts doubt on the possibility of any physical interaction between the mitotic apparatus and the cortex in the equatorial region. Asnes and Schroeder (1979) examined the equatorial cortex of sea urchin eggs by electron microscopy to test the hypothesis that astral rays from the two centrioles intersect at the equator, and thus mediate the effects of the asters on the cortex. They found that microtubules fail to penetrate the equatorial cortex during the period from metaphase to midanaphase, which includes the period (early anaphase) when furrow determination occurs. Therefore, microtubules are not responsible for stimulating the equatorial cortex. Rappaport (1978) has subjected the cytoplasm between the asters and equatorial cortex of sand dollar and sea urchin eggs to mechanical agitation with needles before furrow establishment. Furrows appeared in spite of continual agitation. These results suggest that the furrow may not be a direct consequence of interaction of asters with the equatorial cortex.

Cleavage results from tension in the cortex in the furrow region that causes separation of the egg into two halves. Schroeder (1981a) has demonstrated that this tension is not induced at the time of furrow establishment. On the contrary, the entire egg cortex is under tension well before this time. Since the tensile forces are evenly distributed, the egg remains spherical. At the onset of cleavage the tension is greatest at the equator and lowest at the two poles. Schroeder has reintroduced Wolpert's (1960) proposal that the cleavage stimulus from the asters causes *relaxation of the cortex at the poles.* Thus, the asters directly affect the cortex in closest proximity to them. Between the asters the equatorial cortex retains its contractile properties. Since the balance of tension on the surface has been upset, cleavage can ensue.

The generation of contractile forces in the cleavage furrow results from a circumferentially oriented ring of microfilaments, ranging from 3 to 7 nanometers in diameter (Fig. 10.7). This **contractile ring of microfilaments,** which is localized in the cortex at the base of the cleavage furrow, is thought to constrict the zygote like an "intracellular drawstring" (Schroeder, 1973). In those zygotes with one-sided cleavage furrows, the organelle is a *band* of microfilaments that knifes its way through the cytoplasm from the outer surface (Fig. 10.8). A partial list of organisms in which microfilaments have been observed in cleavage furrows is presented in Table 10.1.

The presence of microfilaments in the contractile ring is an indication that these structures generate the force that causes furrowing. This conclusion is supported by experiments on sea urchin zygotes with the drug cytochalasin B. If cytochalasin is applied before furrow formation, the contractile ring is not formed, and cleavage does not occur. If the drug is applied after the furrow has been initiated, the microfilaments of the contractile ring are disorganized, and cytokinesis is arrested. These results imply that furrowing is dependent upon microfilament organization (Schroeder, 1972). However, as in other experiments utilizing drugs, possible nonspecific effects on cellular metabolism might be the primary influence of cytochalasin, and the inhibition of cytokinesis could be a secondary effect. Thus, these experiments should be viewed with caution.

As we discussed in Chapter 5, microfilaments function in intracellular contraction in a variety of cell types. They are composed of actin, which is one of the contractile proteins in muscle. The other major contractile protein of muscle—myosin—has also been extracted from the sea urchin egg cortex (Mabuchi, 1973). Since muscle contraction depends upon an interaction between actin and myosin, it is possible that a similar interaction between the actin-containing microfilaments and myosin causes furrowing. Recently, myosin has been identified within the furrow itself with the use of a fluorescent antibody against myosin (Fujiwara and Pollard, 1976). This observation places myosin in the correct location for interaction with the microfilaments. Furthermore, injected

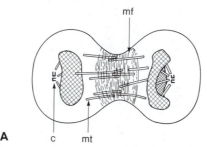

A

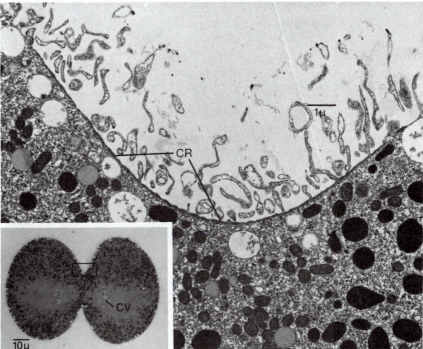

B

FIGURE 10.7 Ultrastructure of the cleavage furrow in the dividing sea urchin zygote. A, Drawing to illustrate the relationships among organelles and orientation of contractile ring. Contractile ring of microfilaments (mf) is intimately involved in equatorial constriction. Microtubules (mt) belonging to the spindle extend through the equatorial region between the daughter cells. C: centrioles. Cross-hatching represents nuclei. (After T.E. Schroeder, 1973. Cell constriction: Contractile role of microfilaments in division and development. Am. Zool., *13:* 950.) B, Meridional sections of an *Arbacia* zygote in first cleavage. The electron micrograph corresponds to the boxed area of the inset, which is a light micrograph of a zygote at a comparable stage. Clusters of chromosomal vesicles (CV) characterize the reforming nuclei. F: cleavage furrow. The contractile ring (CR) is seen below the plasmalemma in the furrow region.

Illustration continued on following page

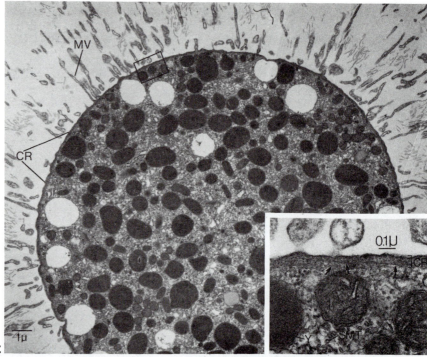

FIGURE 10.7 *(Continued)* C, Equatorial section of cleaving *Arbacia* zygote. The contractile ring (CR) completely encircles the zygote. Box frames region shown at higher magnification in inset. Individual microfilaments are indicated by arrows. MV: microvilli; Y: yolk granule; SMT: spindle microtubules. (*B* and *C* from T.E. Schroeder. 1972. Reproduced from *The Journal of Cell Biology*, 1972, vol. 53, pp. 419–434 by copyright permission of The Rockefeller University Press.)

myosin antibody inhibits cytokinesis of starfish blastomeres (Mabuchi and Okuno, 1977). These observations strongly suggest that an interaction between actin and myosin causes furrowing.

Although there is no evidence concerning the effects of the asters on microfilaments, Schroeder (1981a) has speculated that the asters may disrupt the contractile microfilaments in the polar region. The remaining microfilament system in the equatorial cortex could generate the contractile force and would become the contractile ring. The microfilaments in the equator might also be augmented by displacement of the polar microfilaments.

The cleavage of the zygote requires a substantial increase in cell surface area. This increase could occur by a stretching of the original membrane or by the assembly of new membrane. In eggs with many relatively long microvilli, such as the sea urchin egg (see Figs. 9.18 and 10.7), the microvilli increase the surface area of the membrane of the spherical egg considerably. This redundant membrane may satisfy most, if not all, of the surface requirements of daughter cells at cleavage (Schroeder, 1981b). In amphibian cleavage the additional surface area

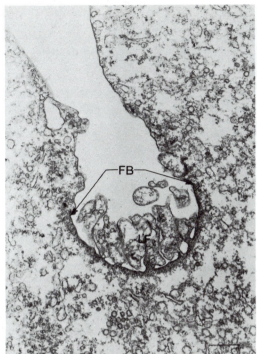

FIGURE 10.8 Cleavage of the squid zygote. Cleavage furrow progresses inward from one side. Furrowing is caused by formation of a band of microfilaments (FB), seen here in cross section just below the plasmalemma on the lateral margins of the furrow. LF: longitudinal folds at base of furrow. Scale bar equals 0.5 μm. (Courtesy of J.M. Arnold.)

results, to a large extent, from construction of new membrane in the cleavage furrow. This new membrane is readily observed, since the cortex below it is unpigmented, whereas the cortex below preexisting membrane contains pigment granules (Fig. 10.9). The initiation of furrowing in the amphibian zygote is evident by the invagination of the pigmented surface at the animal pole to form a shallow groove. Then, as the furrow deepens, new membrane is formed. At the completion of furrowing, the

TABLE 10–1
Occurrence of Microfilaments in Cleavage Furrows*

Species	References
Aequorea (cnidarian)	Szollosi (1970)
Armandia (polychaete)	Szollosi (1970)
Loligo (squid)	Arnold (1969)
Arbacia (sea urchin)	Schroeder (1969); Tilney and Marsland (1969)
Xenopus (frog)	Kalt (1971); Singal and Sanders (1974)
Triturus (newt)	Selman and Perry (1970)
Ambystoma (axolotl)	Bluemink (1970)
Gallus (chicken)	Gipson (1974)
Oryctolagus (rabbit)	Gulyas (1973)

*This is a partial list of species in which a contractile ring or contractile band has been observed by electron microscopy.

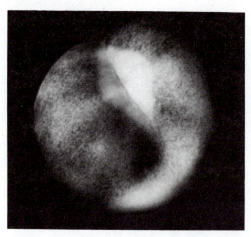

FIGURE 10.9 New membrane formation in the cleavage furrow of the newt zygote. New surface is white, whereas pre-existing surface is pigmented. (From G. Selman and M. Perry. 1970. This micrograph first appeared in New Scientist, London, the weekly review of science and technology.)

apposing walls of the blastomeres are composed of new membrane.

It has been proposed that the new membrane material is synthesized either in the Golgi apparatus or in association with the endoplasmic reticulum. The mode of insertion of the new membrane into the preexisting membrane is uncertain. Various insertion mechanisms, which may not be mutually exclusive, have been proposed. These include: fusion of vesicles of new membrane with the existing membrane (Singal and Sanders, 1974), direct insertion of newly synthesized material (Bluemink and De Laat, 1973), and insertion of new membrane as microvilli, which flatten by stretching (Denis-Donini et al., 1976).

Cleavage Patterns

The determination of cleavage furrow sites by the mitotic apparatus means that the location and orientation of the zygote nucleus in the egg can affect the pattern of cleavage. Theoretically, the nucleus could orient at random so that the first cleavage division would have no particular orientation. Indeed, in some coelenterate species, cleavage is irregular, with furrows forming at random. However, in most organisms, cleavage is highly ordered, and the orientation of not only the first but also successive furrows occurs in a definite pattern. The location and orientation of the mitotic apparatus, and hence cleavage furrow locations, are dependent upon the inherent polarity of the egg (see Chap. 7) and, in certain species, upon the site of sperm entry. As we shall learn in Chapter 11, sperm entry can trigger a cytoplasmic reorganization in those species, which establishes the symmetry of the embryo.

The mitotic figure tends to be positioned in the center of the cell's protoplasm, and the axis of the spindle parallels the long axis of the cytoplasmic mass; the cleavage furrow cuts this axis transversely, equalizing the amount of protoplasm distributed to the daughter cells. Thus, cell division tends to split a cell into equal parts (Wilson, 1925). The simplest form of cleavage occurs in echinoderms, which have spherical eggs with sparse and evenly distributed yolk. The cleavage of the sea cucumber *Synapta digitata* is illustrated in Figure 10.10. The plane of the first division is through the animal–vegetal pole axis (**meridional** or vertical) and divides the zygote into right and left halves. Since the echinoderm nucleus is centrally located, the furrow appears simultaneously at all points in the cleavage plane, causing symmetrical cleavage.

At the completion of this division the centrioles in each blastomere divide, and the daughter centrioles orient parallel to the long axis of the blastomere. Since each blastomere is a hemisphere, it is longer in the plane of division than in any other plane. Therefore, the daughter centrioles orient parallel to it, resulting in the formation of the second cleavage furrow, which is perpendicular to the first furrow and passes through the animal–vegetal pole axis. After this division the embryo consists of four cells, each as long as the egg while no more than one half its diameter. The next division is, therefore, perpendicular to the long axis and parallel to the equator of the egg. The embryo now consists of two quartets of cells; one quartet makes up the animal hemisphere, and the other, the vegetal hemisphere. The fourth cleavage is again meridional, producing two tiers of eight cells each, and the fifth is parallel to the equator, dividing each tier into an upper and a lower layer.

The blastomeres in the animal half of the echinoderm embryo lie directly above the corresponding blastomeres in the vegetal half. Since any plane passing through the major embryonic axis divides the embryo into symmetrical halves, these embryos have radial symmetry; this pattern of cleavage is accordingly called **radial.** The regularity of cleavage in *Synapta* is especially long-lasting, even for echinoderms. Other echinoderm species may show deviation from this regular pattern after the third division. A modification that is particularly noteworthy is that found in the sea urchin embryo (Fig. 10.11). At the fourth cleavage, each of the

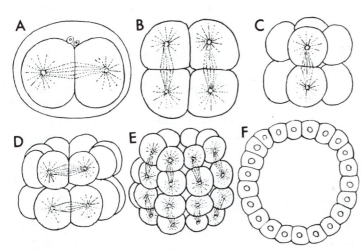

FIGURE 10.10 Radial cleavage with almost equal size of blastomeres in sea cucumber *Synapta digitata. A,* Two-cell stage; *B,* four-cell stage (viewed from animal pole); *C,* eight-cell stage, lateral view; *D,* 16-cell stage; *E,* 32-cell stage; *F,* blastula, vertical section. (After Selenka, from E. Korschelt, 1936. Vergleichende Entwicklungsgeschichte der Tiere. Band 1. Gustav Fischer Verlag, Jena, p. 66. Reproduced from B.I. Balinsky. 1975.)

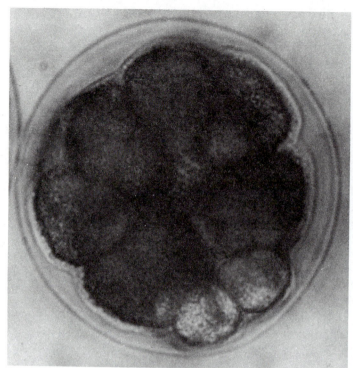

FIGURE 10.11 Photomicrograph of a living *Arbacia* embryo at the 16-cell stage. The diameter of the embryo (including fertilization envelope) is about 70 μm. Micromeres are the small cells with relatively clear cytoplasm at the lower right—two are visible of a total of four. Four of the eight mesomeres are at the upper left, which is the animal pole, while two of the four large macromeres can be seen between mesomeres and micromeres. All cells are surrounded by the hyaline layer, and the whole embryo is enclosed by the fertilization envelope. (Photomicrograph by R. Arceci. From D.R. Senger and P.R. Gross. 1978. Macromolecule synthesis and determination in sea urchin blastomeres at the sixteen-cell stage. Dev. Biol., *65*: 405.)

blastomeres in the animal hemisphere undergoes meridional cleavage to form an octet of equal-sized cells, called **mesomeres.** In the vegetal hemisphere the cleavage is unequal, and the cleavage plane is shifted so that it is parallel to the equator. This produces four small cells, the **micromeres,** clustered at the vegetal pole and four large cells, the **macromeres.** Thus, the sea urchin embryo at this stage consists of three tiers of cells of unequal size. However, the radial symmetry is maintained in spite of the cell size differential.

The idealized pattern of cleavage shown by the echinoderms is not typical of most other groups of organisms. Most embryos do, however, maintain a cleavage pattern that is similar to that of the echinoderms until at least the third division. Variations from this pattern are due to factors that cause displacement of the mitotic apparatus or that interfere with the progress of cleavage furrows. In addition to the radial pattern of cleavage, two other cleavage patterns are generally recognized: **spiral cleavage** and **bilateral cleavage.**

Spiral cleavage is found in some flatworms, the nemerteans, the mollusks (except cephalopods), and some annelids. The spiral pattern is apparent at the eight-cell stage, when the animal quartet cells are displaced, and instead of lying directly above the cells of the vegetal quartet, each lies at the interface between two underlying vegetal blastomeres. As cleavage continues, cells become stacked in an alternating spiral pattern (Fig. 10.12). This cell arrangement causes cells to interlock with one

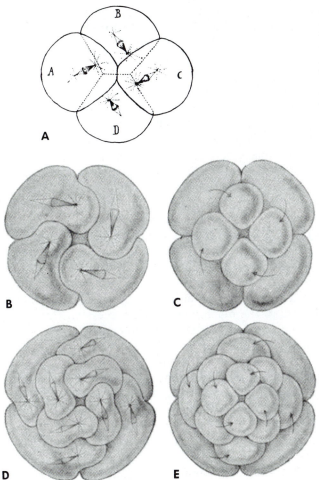

FIGURE 10.12 Spiral cleavage. *A* and *B*, Four-cell stage in preparation for third division. *C*, Eight-cell stage. *D*, Preparation for fourth division. *E*, 16-cell stage. (*A* after Robert, from E. Korschelt. 1936. Vergleichende Entwicklungsgeschichte der Tiere. Band 1. Gustav Fischer Verlag, Jena, p. 864. Reproduced from B.I. Balinsky, 1975; *B* to *E* also from E. Korschelt, 1936, p. 68.)

another to form polyhedral shapes. A similar pattern is formed by masses of spheres crowded together by pressure or capillary action and gives the greatest economy of space for the mass. This is dramatically illustrated by the form assumed by soap bubbles, which reflects almost exactly the configuration of cells in a spirally cleaving embryo (Fig. 10.13).

The arrangement of blastomeres in a spirally cleaving embryo is caused not by rearrangement of cells after cleavage but rather by the oblique orientation of spindles *before cleavage.* The first division is typical: it passes through the animal–vegetal pole axis, although it may be inclined somewhat. The spindles at the second division are perpendicular to the first; however, unlike radial cleavage in which they would also be perpendicular to the main axis, these spindles are inclined and in opposite directions to one another. This causes one daughter blastomere from each cleavage pair to lie slightly above the other, whereas cells opposite one another are at the same level. To trace the cells during cleavage, one traditionally designates the two higher cells as A and C, and the two lower cells as B and D (Fig. 10.12*A*). The former meet at the animal pole but diverge away from the vegetal pole. Conversely, B and D meet at the vegetal pole but are inclined away from the animal pole. The identities of these cells are easy to ascertain, since the polar bodies are located at the site where A and C are in contact.

The third cleavage is equatorial (i.e., horizontal) as in radial cleavage, but the spindles are oblique rather than parallel to the major axis. Consequently, the cleavage planes are shifted, causing the cells of the animal quartet to lie in the spaces between the lower cells. In most cases of spiral cleavage the spindles of the third cleavage are shifted to the right when the embryo is observed from the animal pole. This causes clockwise displacement of the upper tier of cells and is known as **dextral** cleavage. Anticlockwise, or **sinistral,** cleavage, is rare. As in radial cleavage, successive cleavage planes are perpendicular to one another, causing an alternation of the directions of the planes. Therefore, all odd-numbered furrows (cleavages 3, 5, and so forth) are dextral, whereas all even-numbered furrows are sinistral. The third cleavage is also frequently *unequal*, so that the cells of the upper tier (the micromeres) are smaller than those of the lower tier (the macromeres). At successive cleavages the micromeres divide equally, whereas the macromeres continue to divide unequally, producing additional quartets of micromeres, which are arranged in tiers stacked below preexisting ones.

The rigid pattern of spiral cleavage produces a highly ordered embryo, with each cell occupying a specific position. As we shall see in Chapter 11, each cell and its progeny have a particular role to play in forming parts of the embryo and, subsequently, the larva. Hence, in these embryos, the *pattern of division* is involved in developmental organization.

Bilateral cleavage is found, for example, in ascidians and cephalopod mollusks. In these forms the first cleavage passes through the plane of symmetry, which is in the median plane of the resulting embryo, and subsequent cleavages are symmetrical to it. Hence, one side of the embryo is a mirror image of the other. This is in contrast to radial sym-

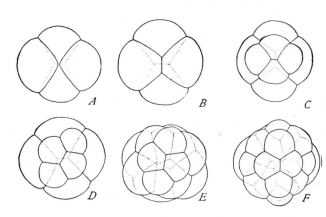

FIGURE 10.13 Soap-bubble models of cleavage figures in *Trochus. A* and *B,* Models of four-cell stages; *C* and *D,* Eight-cell models; *E,* 12-cell model; *F,* 16-cell model. (From E.B. Wilson. 1925. Reprinted with permission of Macmillan Publishing Co., Inc. From *The Cell in Development and Heredity,* 3rd edition, by E.B. Wilson. Copyright 1925 by Macmillan Publishing Co., Inc., renewed 1953 by Anna M.K. Wilson.)

metry, in which any plane passing through the animal–vegetal pole axis divides the embryo into symmetrical halves.

The early cleavage stages of an ascidian zygote are shown in Figure 10.14. The first cleavage plane passes through the animal and vegetal poles, bisecting zones of distinct cytoplasmic constituents (see p. 484). The second cleavage plane is perpendicular to the first. The resultant four-cell–stage embryo has two cells on either side of the median plane, each cell is mirrored by an identical cell on the side opposite. The embryo now has recognizable right and left sides, as well as anterior and posterior regions. Each subsequent cleavage on the left side is mirrored by an identical cleavage on the right, adding to the bilateral symmetry. This is shown in Figure 10.14F, which is a ventral view of the 16-cell–stage embryo. The orientation of cleavage furrows at the previous division is indicated by short lines that intersect daughter blastomeres.

The cephalopod egg also has bilateral cleavage. However, because of the presence of large amounts of dense yolk in the egg, these cleavages are superficial. We shall deal with the effects of yolk on cleavage in the following section of this chapter. The cephalopod cleavage pattern will be described there as well.

As in spirally cleaving eggs, the pattern of bilateral cleavage is important in embryonic organization. This is evident as early as the first cleavage, which parcels the presumptive right and left halves of the embryo into daughter blastomeres. The site of the first cleavage furrow occurs at a predictable location, which is established by the organization of the egg itself or by the site of sperm entry. The effect of sperm entry on establishment of the first cleavage furrow, and hence the plane of bilateral symmetry of the ascidian embryo, will be discussed in Chapter 11.

In addition to the cleavage patterns we have discussed earlier, the unusual nature of mammalian cleavage should also be noted. The first cleavage plane is typical and is in the usual animal–vegetal pole axis (Fig. 10.15). However, the arrangement of blastomeres after the next division is unique (Lewis and Hartman, 1933; Gulyas, 1975); one pair of blasto-

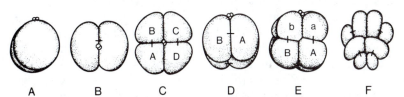

A   B   C   D   E   F

FIGURE 10.14 Bilateral cleavage of an ascidian. *A*, Right view of two-cell stage. *B*, Animal pole view of two-cell stage. *C*, Animal pole view of four-cell stage. *D*, Right view of four-cell stage. *E*, Right view of eight-cell stage. *F*, Vegetal pole view of 16-cell stage. (After E.G. Conklin. 1905. *J. Acad. Nat. Sci. Philad. Ser. 2, 13:* 1–119. Redrawn from P. Grant. 1978. *Biology of Developing Systems.* Holt, Rinehart and Winston, New York.)

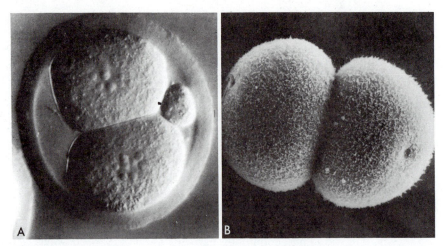

FIGURE 10.15 First cleavage of the mouse zygote. *A*, Photographed using Nomarski differential interference optics. The nuclei contain several round nucleoli. Note polar body on the right (animal pole). ×250. *B*, Scanning electron micrograph after removal of the zona pellucida. ×1200. (Micrographs by Dr. Patricia G. Calarco. From C.J. Epstein. 1975. Gene expression and macromolecular synthesis during preimplantation embryonic development. Biol. Reprod., *12:* 83.)

meres is arranged perpendicular to the other pair (Fig. 10.16*A* and *B*). This arrangement is due to an unusual pattern of cleavage at the second division. At this division the cleavage plane in one of the two blastomeres is rotated 90° with respect to the animal–vegetal pole axis. Hence, this pattern of cleavage is called **rotational** (Gulyas, 1975). The orientation of the first two cleavages is illustrated in the drawing of Figure 10.16*D*, where it is compared with sea urchin cleavage. As cleavage continues, blastomeres divide at quite different rates, and mammalian embryos may consist of uneven numbers of cells at various times.

The Influence of Yolk on Cleavage

The amount and distribution of yolk have a profound effect on cleavage. These effects are illustrated by the amphibian egg (Fig. 10.17). Yolk accumulates in the vegetal hemisphere of the egg, while the nucleus and the bulk of the cytoplasm are displaced toward the animal pole. Because of the eccentric location of the zygote nucleus, the first and second cleavage furrows are initiated at the animal pole and continue around the egg toward the vegetal pole. The location of the nuclei in the blastomeres at the four-cell stage influences the third cleavage. This is the equatorial (horizontal) cleavage, which in echinoderms divides the embryo into two tiers of equal-sized cells. However, since the nucleus of the amphibian egg is displaced toward the animal pole, the plane of the third cleavage is shifted toward the animal pole. This produces an animal quartet of cells that are considerably smaller than the cells of the vegetal quartet.

This shift of cleavage plane causes a size asymmetry that is retained throughout cleavage; animal hemisphere cells remain smaller than the vegetal hemisphere cells.

Because of the influence of the yolk on cleavage, eggs are classified according to the amount and distribution of yolk. Eggs with a limited amount of yolk that is evenly distributed are called isolecithal (sometimes referred to as oligolecithal or homolecithal). Eggs of this type are found in many invertebrates (such as echinoderms), lower chordates (such as *Amphioxus*), and mammals. Cleavage in these forms is unaffected by yolk, and it is complete (or **holoblastic**) with the first cleavage furrow forming simultaneously at all points around the circumference of the egg in the animal–vegetal plane. Holoblastic cleavage is usually equal, producing daughter blastomeres of equivalent size. However, some isolecithal eggs (such as those of certain mollusks) may cleave unequally. Cleavage asymmetry in these eggs reflects other aspects of cytoplasmic organization.

Eggs with larger amounts of yolk are called telolecithal. The yolk tends to concentrate in the vegetal hemisphere in these eggs, whereas the cytoplasm in the animal hemisphere contains very little yolk. Moderately telolecithal eggs (sometimes called mesolecithal), such as those of amphibians, have holoblastic cleavage that is also unequal. O. Hertwig in 1898 demonstrated that if a frog's egg is centrifuged so that the yolk is densely compacted at the vegetal pole, cleavage furrows are unable to penetrate the yolk, and cleavage is then incomplete. A comparable situation occurs naturally in certain yolk-laden eggs, such as the avian egg. Cleavage in these eggs is restricted to the cytoplasm near the animal pole, and the yolk-rich vegetal portion of the egg remains uncleaved. This type of cleavage is incomplete (or **meroblastic**).

In the avian egg the cytoplasm is restricted to a thin peripheral layer at the animal pole, forming a structure called the **blastodisc.** The first few cleavages of the hen's egg are shown in Figures 10.18 and 10.19. The first cleavage is initiated near the center of the blastodisc, but it proceeds only as far as a short slit in the surface before a second slit, the second cleavage furrow, forms at right angles to it. The four blastomeres remain continuous with the yolk both below and at their outer margins. As cleavage continues, cells in the center are separated from those at the edge by intervening cleavage furrows, but the central cells still retain their continuity with the underlying yolk. At the periphery, cleavage furrows radiate for a short distance in all directions; these blastomeres are continuous with both the yolk and the peripheral cytoplasm (Fig. 10.19, stages I and II). Complete cells, surrounded on all sides by a plasma membrane, are first formed in the center of the blastodisc (Fig. 10.19, stage III). These blastomeres can separate from the yolk in one of two ways—either by convergence and fusion of the original furrows below the nucleated portion of a blastomere or by cell division with horizontal cleavage furrows. In the latter case the upper blastomere would be com-

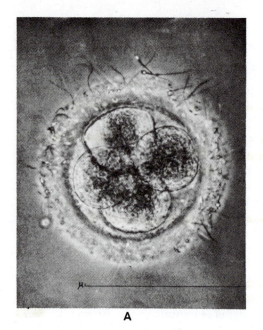

A

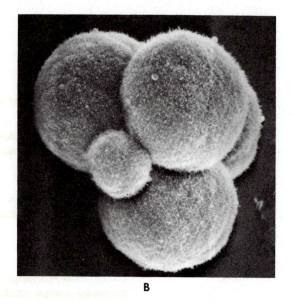

B

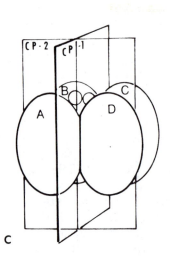

C

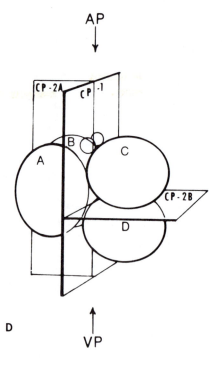

D

FIGURE 10.16   (*See legend on the opposite page.*)

◄ **FIGURE 10.16** The second cleavage in mammalian development. *A,* Human 4-cell stage. Note numerous spermatozoa outside of zona pellucida. (From R.G. Edwards and R.E. Fowler. 1970. Human embryos in the laboratory. Sci. Am., *223*(6): 44. Reproduced from B.I. Balinsky, 1975.) *B,* Scanning electron micrograph of the 4-cell mouse embryo. Second polar body at lower right. × 1700. (From P.G. Calarco. 1975. Cleavage [mouse]. *In* E.S.E. Hafez (ed.), *Scanning Electron Microscopic Atlas of Mammalian Reproduction.* Igaku-Shoin, Ltd., p. 309.) *C* and *D,* Comparison of cleavage planes during the first two divisions of the sea urchin and rabbit zygotes. *C,* Diagrammatic presentation of cleavage planes in the sea urchin egg. The first cleavage plane (CP-1) is meridional (vertical), and it passes through the animal and vegetal poles. The second cleavage plane (CP-2) is also meridional through both blastomeres and passes through the two poles, but bisects CP-1 at 90°. Observing from the animal pole, the blastomeres at the 4-cell stage shall be designated as A, B, C, and D, going clockwise. *D,* Diagrammatic summary of the formation of the first and second cleavage planes in the rabbit egg. The first cleavage plane (CP-1) passes through the polar axis of the egg. The plane of second cleavage (CP-2A) through the "first blastomere" of the two-cell embryo passes through the polar axis and is perpendicular to CP-1. The plane of second cleavage (CP-2B) through the "second blastomere" is perpendicular to both CP-1 and CP-2A. The view of an embryo, along the plane of first cleavage, with CP-2A to the left and CP-2B to the right of CP-1 (as diagrammed here), shall be defined as a *frontal view.* A view from above shall be referred to as an *animal pole view* (AP). A view from below will be called a *vegetal pole view.* In this illustration the zona pellucida and the mucin coat surrounding the embryo are omitted for the sake of clarity. (*C* and *D* from B.J. Gulyas. 1975. A re-examination of cleavage patterns in eutherian mammalian eggs: Rotation of blastomere pairs during second cleavage in the rabbit. J. Exp. Zool., *193:* 237, 244.)

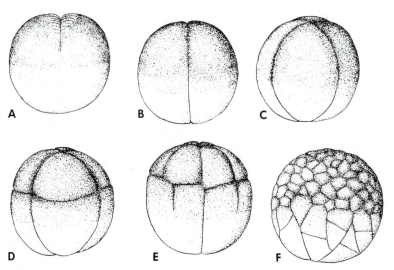

FIGURE 10.17 Cleavage in the frog, semidiagrammatic. (From B.I. Balinsky. 1981. *An Introduction to Embryology,* 5th ed. Saunders College Publishing, Philadelphia, p. 110.)

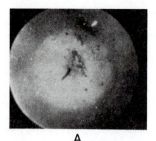

A   B

FIGURE 10.18 Early cleavage of the avian zygote. *A*, First cleavage. *B*, Second cleavage. (From M.W. Olsen. 1942. Maturation, fertilization, and early cleavage in the hen's egg. J. Morphol., *70:* 533.)

plete, whereas the lower one would retain its continuity with the yolk. The marginal cells eventually lose the partial membranes that intervene between nuclei; they form a syncytium called the **periblast.** The periblast may function to break down the yolk into usable nutrients for the embryo. The central mass of complete blastomeres eventually is separated from the underlying yolk by a fluid-filled cavity called the **subgerminal cavity,** which forms by a coalescence of small cavities that form under the blastomeres in the center of the blastodisc (Fig. 10.19, stages II to IV).

Fish eggs are also telolecithal, but the distribution and relative amounts of yolk are quite variable in this diverse group of organisms. In the teleost fishes the cytoplasm is segregated from the yolk and is restricted to a thin layer that surrounds a compact internal mass of yolk. The cytoplasm thickens at the animal pole to form a cytoplasmic cap (blastodisc) in which the nucleus resides. Cleavage is restricted to this cap. The first cleavages are meroblastic, since the cleavage furrows are unable to traverse the yolk. The pattern of early cleavage of the zebrafish egg is shown by the scanning electron micrographs of Figure 10.20. Note that the embryo during the first stages of cleavage has bilateral symmetry, and all of the blastomeres lie in rows. By the 32-cell stage the cells begin to layer as horizontal cleavages produce the first complete cells, which are surrounded on all sides by a plasma membrane. Since the blastodisc is restricted to a small area and is prevented from moving laterally, continued cleavage causes the buildup of cells, which form an elevated embryo perched above the yolk mass (Fig. 10.20*H*).

The eggs of cephalopod mollusks are also highly telolecithal, dividing by meroblastic cleavage. The early cleavage pattern results in bilateral symmetry. Figure 10.21 shows the pattern of early cleavage for the egg of the squid.

The eggs of some insects have an unusual organization in which the nucleus is in the center of the egg and surrounded by a small island of cytoplasm, which in turn is surrounded by the yolk. The remaining cytoplasm is located in the spaces surrounding the yolk particles and in a thin, peripheral layer (called the **periplasm**) just below the plasmalemma. The peripheral layer of cytoplasm is formed soon after fertilization as the result of a reorganization of egg constituents. Since the yolk

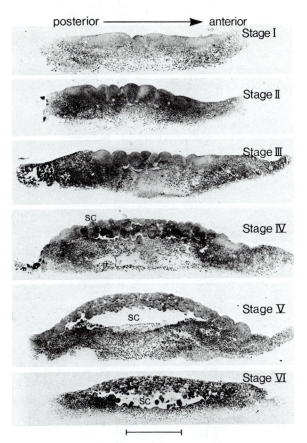

FIGURE 10.19 Median sagittal sections of cleaving chick blastodiscs. Stages indicated are based upon the staging system of Eyal-Giladi and Kochav (1976). The first horizontal cleavages, which create a multilayered blastodisc, appear at stage III. Small cavities under blastomeres in the center of the blastodisc fuse to form the subgerminal cavity (SC). By stage VI the blastodisc is four or five cells thick. Scale bar equals 0.5 mm. (From S. Kochav, M. Ginsburg, and H. Eyal-Giladi. 1980. From cleavage to primitive streak formation: A complementary normal table and a new look at the first stages of the development of the chick. II. Miroscopic anatomy and cell population dynamics. Dev. Biol., *79*: 298.)

is centrally located and surrounded by cytoplasm, this type of egg is classified as centrolecithal. The *Drosophila* egg is centrolecithal, and we shall use it to describe the unique cleavage in this type of egg.

The initial mitotic divisions in the centrolecithal zygote are not accompanied by cytokinesis. As a result, the zygote becomes a multinucleate structure, with nuclei scattered throughout the yolk (Fig. 10.22). Each daughter nucleus is surrounded by cytoplasm that is apparently sequestered from that distributed around the yolk particles. Each of these cytoplasmic islands containing a nucleus is called an **energid.** The postfertilization mitoses result in massive reorganization of egg constituents as the yolk and cytoplasm become progressively segregated from one another. Each nuclear division is also paralleled by rhythmic movements of periplasm to and from the poles of the egg. It is likely that cytoplasm is being sequestered from the interior into the periplasm during these rhythmic movements, since the periplasm becomes notably thicker as

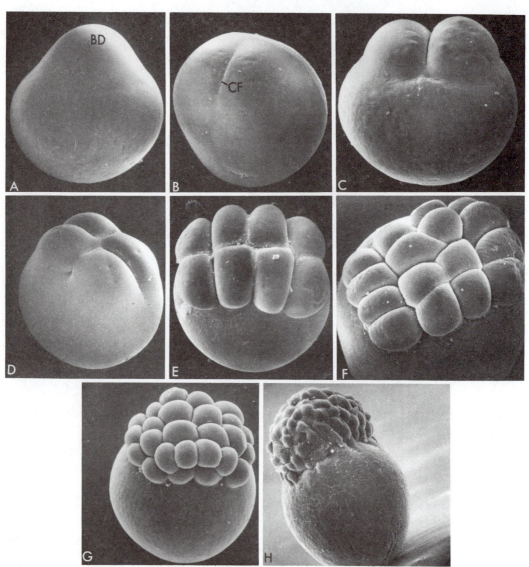

FIGURE 10.20 Scanning electron micrographs showing stages in the development of a zebrafish zygote through the blastula stage. *A*, The blastodisc (BD) is clearly distinguishable on this zygote. ×120. *B*, The first cleavage furrow (CF) begins to form. ×120. *C*, Two-cell stage. ×120. *D*, Four-cell stage. ×120. *E*, Eight-cell stage. ×120. *F*, 16-cell stage. ×160. *G*, 32-cell stage. ×120. *H*, Blastula with many dividing blastomeres. ×130. (From H.W. Beams and R.G. Kessel. 1976. Cytokinesis: A comparative study of cytoplasmic division in animal cells. Am. Sci., *64:* 280, 281, 284.)

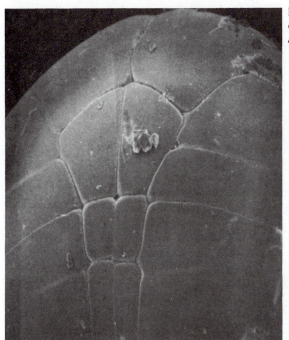

FIGURE 10.21 Scanning electron micrograph of cleaving squid embryo. (Courtesy of Dr. J.M. Arnold.)

nuclear division proceeds. After several divisions some of the energids move into the periplasm, populating it with nuclei. Since the nuclei are accompanied by cytoplasm, the surfacing of energids contributes to the thickness of the periplasm. Some nuclei do not surface but remain in the egg interior to form a syncytial cytoplasmic layer that surrounds the yolk, presumably to assist in resorption of the yolk so that its energy can be utilized for the development of the embryo. The entire complex of the central yolk and the yolk nuclei in their cytoplasmic matrix constitutes the **yolk sac.**

After the nuclei arrive in the periplasm, this zone is called the blastoderm. The nuclei of the blastoderm continue a few rounds of division without cytokinesis. This phase of development is called the **syncytial blastoderm stage.** After each nuclear division a partial cleavage furrow is formed between adjacent nuclei. Thus, each nucleus in the syncytial blastoderm is surmounted by a hillock of cytoplasm, giving the surface of the embryo a blebbed appearance. The partial furrows are short-lived, disappearing soon after their formation. They reform after the next mitosis, creating blebs that are more numerous and smaller than before (Fig. 10.23).

An exception to this pattern is found at the posterior pole of the embryo. Immediately after nuclei reach the posterior pole periplasm, they become pinched off into definitive cells, surrounded by membranes (Fig. 10.24). These cells are called **pole cells,** which function as primordial germ cells. This unique behavior at the posterior pole is due to a specialized cytoplasm that is localized there. We shall discuss this fascinating case of cytoplasmic localization in Chapter 11.

Cellularization of the syncytial blastoderm is a dramatic event that occurs simultaneously over the entire surface. In *Drosophila* it occurs

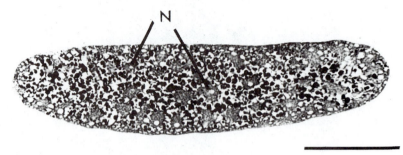

A

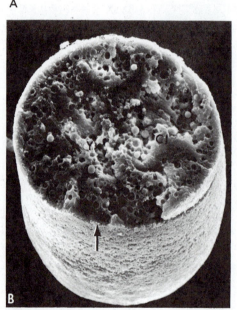

B

FIGURE 10.22 *Drosophila* embryos with numerous nuclei scattered in the center of the egg. *A,* Longitudinal section at about the time of the eighth mitosis. Nuclei (N) in cytoplasmic islands (energids) are fairly evenly spaced throughout the egg interior. Scale bar equals 0.1 mm. (Histological preparation by J.C. Wilson. Reproduced with permission from S.L. Fullove and A.G. Jacobson. 1978. *The Genetics and Biology of Drosophila,* Vol. 2c, M. Ashburner and T.R.F. Wright (eds.). Copyright by Academic Press Inc. [London] Ltd.) *B,* Scanning electron micrograph of transverse fracture through an embryo at a stage comparable to that of A. Note the cytoplasmic islands (CI) distributed throughout the yolk (Y). There is only a thin layer of periplasm (arrow) at the surface of the embryo at this stage. ×410. (From F.R. Turner and A.P. Mahowald. 1976. Scanning electron microscopy of *Drosophila* embryogenesis. I. The structure of the egg envelopes and the formation of the cellular blastoderm. Dev. Biol., *50:* 99.)

following the twelfth or thirteenth nuclear division. After this division the spherical nuclei begin to enlarge, elongating in a plane perpendicular to the plasmalemma. As shown in Figure 10.25A, bundles of microtubules surround the elongating nuclei. Fullilove and Jacobson (1971) have suggested that the microtubules are either actively involved in elonga-

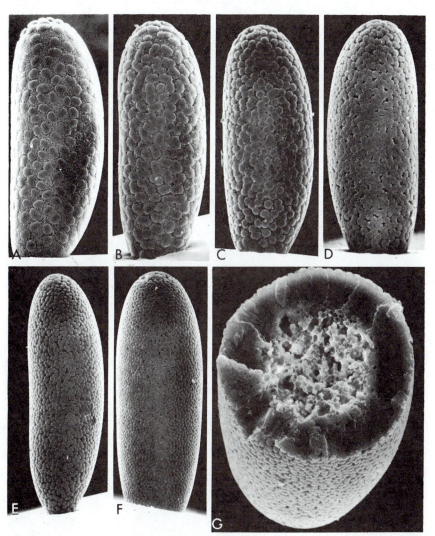

FIGURE 10.23 A series of SEM pictures illustrating the appearance of the *Drosophila* embryo during the syncytial blastoderm stage. *A* to *F*, External views. Note the reduction in size and increase in the number of surface bulges above the nuclei. *G*, Transverse fracture through an embryo similar to that shown in *F*. Note the superficial nature of the surface bulges. *A*, ×180; *B*, ×170; *C*, ×180; *D*, ×200; *E*, ×155; *F*, ×170; *G*, ×425. (From F.R. Turner and A.P. Mahowald. 1976. Scanning electron microscopy of *Drosophila* embryogenesis. I. The structure of the egg envelopes and the formation of the cellular blastoderm. Dev. Biol., *50*: 101, 105.)

tion or serve to constrain and orient the nuclei as they enlarge, so that they lie with their long axes directed inward toward the yolk. As the nuclei elongate, cleavage furrows form between them, pushing in from the surface and segregating the nuclei into distinct cells (Fig. 10.25*B*). The membranes at the bases of the cleavage furrows do not fuse below the nuclei to encircle these cells completely. Instead, the cells remain connected with the yolk sac by a cytoplasmic stalk. The membrane connecting these stalks is the **yolk membrane.** The continuity between the cellular membranes and the yolk membrane is shown in Figure 10.25*B* and *C*. The embryo now consists of a single layer of columnar epithelial cells (the **cellular blastoderm**), a syncytial core of yolk surrounded by a yolk membrane, and a cluster of pole cells at the posterior end.

The process of cellularization is summarized in the drawings of Figure 10.26. These drawings also indicate the orientation of the asters. Note that furrows form both at the midpoint of spindles and also between adjacent asters belonging to different spindles. Presumably, the furrowing sites are determined by the interaction between asters and the cortex in a manner similar to that demonstrated for the echinoid zygote.

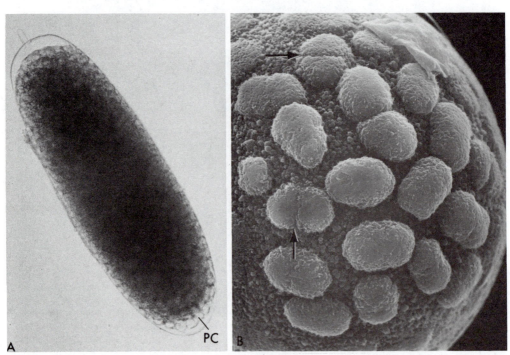

FIGURE 10.24 *Drosophila* pole cells. *A,* Light micrograph of *Drosophila* embryo shortly after cellularization of pole cells (PC) at the posterior pole. ×230. Micrograph by R.E. Rogers from the author's laboratory. *B,* High-power SEM of posterior pole of an embryo, showing numerous pole cells. Note the apparent division planes (arrows) in several of the pole cells. ×860. (*B* from F.R. Turner and A.P. Mahowald. 1976. Scanning electron microscopy of *Drosophila* embryogenesis. I. The structure of the egg envelopes and the formation of the cellular blastoderm. Dev. Biol., *50:* 103.)

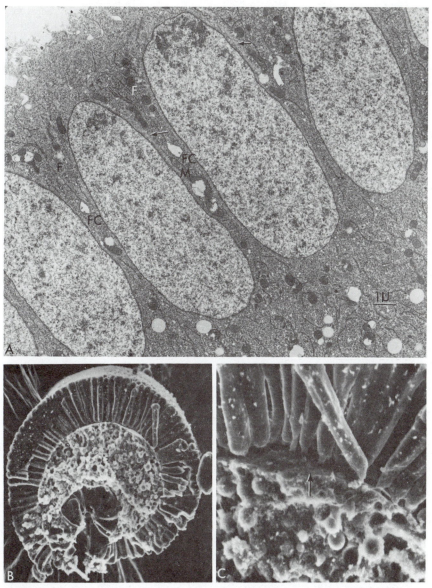

FIGURE 10.25 Nuclear elongation and cellularization of the *Drosophila* embryonic blastoderm. *A*, Nuclear elongation. Bundles of microtubules run parallel to the nuclei at arrows. F: cleavage furrows; M: mitochondrion; FC: furrow canal at base of furrow. Bar equals 1 μm. (From S.L. Fullilove and A.G. Jacobson. 1971. Nuclear elongation and cytokinesis in *Drosophila montana*. Dev. Biol., *26:* 564.) *B*, Scanning electron micrograph of transverse fracture through an embryo in which cellularization is complete. Compare with Figure 10.23 *G.* ×400. *C*, High-power resolution of a portion of *B* showing that blastoderm cells retain cytoplasmic connections with the yolk sac. In the area where cells have been broken off, circular depressions in the yolk sac membrane can be seen (arrow). ×1030. (*B* and *C* from W.L. Rickoll. 1976. Cytoplasmic continuity between embryonic cells and the primitive yolk sac during early gastrulation in *Drosophila melanogaster*. Dev. Biol., *49:* 308).

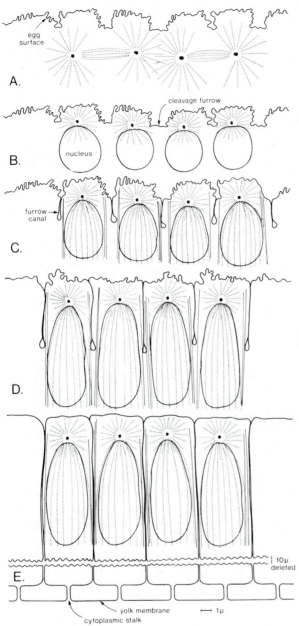

FIGURE 10.26 Schematic representation of nuclear elongation and cellularization of the blastoderm of *Drosophila*. (After S.L. Fullilove and A.G. Jacbson. 1971. Nuclear elongation and cytokinesis in *Drosophila montana*. Dev. Biol., *26*: 575.)

The Morula and Blastula

The three-dimensional relationships of cells during cleavage are quite variable among different animal species. During the first few cleavages, embryos with holoblastic cleavage may appear as a solid cluster of cells. This stage of development is sometimes called the **morula** because of its superficial resemblance to a mulberry (*morula* is the Latin word for mulberry). No comparable form is found in embryos with meroblastic cleavage. Very few cleaving embryos are actually solid structures packed with cells. Instead, a fluid-filled cavity (the blastocoele) is present, and it gradually increases in size with successive divisions. The cells of the embryo surround this cavity to form an epithelial layer, the blastoderm. The embryo at this stage is called a blastula. The simplest blastula is found in radially symmetrical embryos with complete cleavage, including the echinoids (Fig. 10.27). Here, the blastocoele increases in size as the cells increase in number, but it decreases in thickness to form a progressively thinner blastoderm.

One distinguishing feature of this blastula is that every cell has an outer surface that is in intimate contact with the extracellular hyaline layer (see Chap. 9) and an inner surface that is in contact with the blastocoele fluid. As cell number increases during cleavage, this orientation is retained by the daughter cells; the cell layer expands and becomes thinner. The outer surfaces of cells of the sea urchin blastoderm develop cilia (Fig. 10.27*F*), which beat and cause the embryo to rotate within the fertilization envelope. The embryo later secretes a hatching enzyme that digests the fertilization envelope, allowing the embryo to escape from confinement into the sea. This process is called **hatching.** The hatched blastula (sometimes called the swimming blastula) develops an **apical tuft** of long, stiff cilia at the animal pole and retains its spherical shape for several hours. Then the vegetal pole begins to flatten, forming the **vegetal plate,** and the embryo becomes pear-shaped. The central cells dissociate from the vegetal plate and enter into the blastocoele (Fig. 10.27*D* and *E*; see Chap. 13 for a discussion of this process). These cells, which are derivatives of the micromeres, are called the **primary mesenchyme.** They later form the triradiate spicules and eventually give rise to the skeleton of the pluteus larva. During the formation of the primary mesenchyme, the embryo is designated a **mesenchyme blastula.** This stage is followed immediately by gastrulation.

One consequence of the sea urchin type of blastula formation is the polarity produced in the cells themselves. The outer and inner surfaces are in contact with quite different environments: The outer surface abuts the hyaline layer, whereas the inner surface is bathed by the blastocoele fluid. These surfaces are structurally quite different from one another, with a cilium forming on the outer surface. As we shall see in Chapter 11, polarity of this sort can have profound effects on development.

In telolecithal eggs the presence of yolk leads to modifications in the form of the blastula. For example, in the moderately telolecithal frog's egg the blastocoele is displaced to the animal hemisphere (see Fig. 1.9). Hence, the blastoderm is covered on top with a thin layer of small cells and is underlain by a thick floor of yolky vegetal hemisphere cells. The roof of the blastocoele is arched, whereas the floor is relatively flat. Further modification is found in highly telolecithal eggs that exhibit meroblastic cleavage. In these species the blastoderm is a disc rather

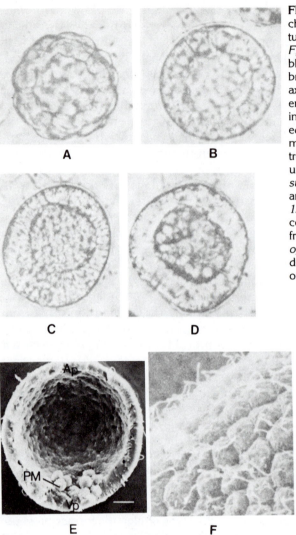

FIGURE 10.27 The blastula stage of sea urchin development. *A* and *B,* Prehatching blastulae. *C* and *D,* Mesenchyme blastulae. *E* and *F,* Scanning electron micrographs of hatched blastulae. *E,* Mesenchyme blastula-stage embryo bisected along the animal-vegetal pole axis (Ap and Vp, respectively). Note the presence of primary mesenchyme cells (PM) on the inner surface of the vegetal plate. Scale bar equals 10 μm. (*E* from K. Akasaka, S. Amemiya, and H. Terayama. 1980. Scanning electron microscopical study of the inside of sea urchin embryos (*Pseudocentrotus depressus*). Effects of aryl β-xyloside, tunicamycin and deprivation of sulfate ions. Exp. Cell Res., *129:* 3.) *F,* Small area of external surface, each cell bearing a single cilium. ×2000. (*A* to *D* from G. Giudice. 1973. *Developmental Biology of the Sea Urchin Embryo,* 1st ed. Academic Press, New York, pp. 11, 12. *F* courtesy of Dr. W.J. Humphreys.)

than a sphere. In the avian egg, for example, the blastoderm is initially a disc of cells lying atop the subgerminal cavity.

Before the formation of a three-dimensional embryo from the avian blastoderm, the central area of the blastoderm must be modified from a stratified epithelium, several cell layers thick, to a single-layered epithelium. This process has been described in various ways. Recent studies, however, appear to clarify the events that are occurring at this time (Eyal-Giladi and Kochav, 1976; Kochav et al., 1980). Cells become detached from the ventral surface of the blastoderm and collect in the subgerminal cavity. The loss of these cells causes a thinning of the blastoderm (Fig. 10.28). The thinning first occurs at the future posterior end of the embryo and proceeds to the future anterior end. This thin, circular region is now called the **area pellucida.** The marginal region remains thick and is called the **area opaca.** The cells shed from the blastoderm fall to the bottom of the subgerminal cavity. Any further role in development for these cells is uncertain. Some of these cells can be seen in Figure 10.28, stages IX and X.

The initial stage in forming a three-dimensional avian embryo is the restructuring of the blastoderm to form a two-layered structure. A lower layer is formed by cells from two sources. One source is small cells that ingress at numerous sites (**polyingression**) from the area pellucida epithelium and adhere to similarly derived cells to form isolated cell clusters (Fig. 10.28, stage X). The other source is cells from the posterior margin of the area pellucida. Cells stream forward from this region (see Chap. 13) and adhere with the cell clusters to form a continuous disc-shaped sheet of cells (the **hypoblast**) in the center of the area pellucida (Figs. 10.29 and 10.30). The overlying or "surface" layer is the **epiblast.**

The arrangement of cells of the cleaving mammalian embryo is unique. Embryos consisting of approximately 16 cells are considered to be at the morula stage. The exact number of cells is variable, since cleavage is asynchronous. At this time the embryo is a spherical cluster of cells (Fig. 10.31). Fluid-filled cavities form between cells, and they subsequently coalesce to form a single cavity that is displaced to one end of the embryo. The embryo at this stage is called a **blastocyst** (Fig. 10.32). The blastocyst cavity usually becomes evident at approximately the 32-cell stage.

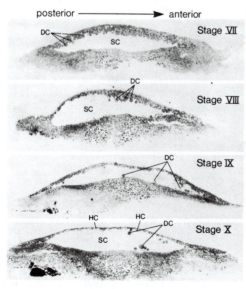

posterior ———▶ anterior

Stage VII

Stage VIII

Stage IX

Stage X

FIGURE 10.28 Formation of the area pellucida in the chick embryo. These median sagittal sections are oriented with the anteior end on the right. Thinning of the blastoderm begins at the posterior end and spreads anteriorly. Detached cells (DC) accumulate at the bottom of the subgerminal cavity (SC). A few isolated clusters of hypoblast cells (HC) are seen below the blastoderm at stage X. Scale bar equals 0.5 mm. (From S. Kochav, M. Ginsburg, and H. Eyal-Giladi. 1980. From cleavage to primitive streak formation: A complementary normal table and a new look at the first stages of the development of the chick. II. Microscopic anatomy and cell population dynamics. Dev. Biol., *79*: 299.)

Stage XII   Stage XIII

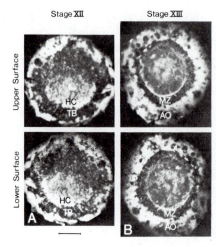

Upper Surface

Lower Surface

FIGURE 10.29 Upper and lower surfaces of isolated chick blastoderms showing formation of the hypoblast. *A,* Hypoblast cells (HC) are clustered in a horseshoe-shaped region on the lower surface of the posterior region of the area pellucida. The posterior-most region of the area pellucida remains single-layered and therefore appears as a transparent band (TB). *B,* The hypoblast becomes a disc-shaped sheet in the center of the area pellucida. The transparent band extends completely around the central zone containing the hypoblast. This transparent region of the area pellucida is the marginal zone (MZ), which in turn is surrounded by the area opaca (AO). Scale bar equals 1.0 mm. (From H. Eyal-Giladi and S. Kochav. 1976. From cleavage to primitive streak formation: A complementary normal table and a new look at the first stages of development of the chick. I. General morphology. Dev. Biol., *49:* 330, 332.)

posterior ——————→ anterior

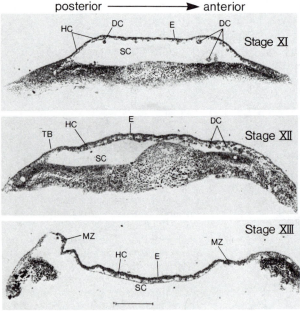

Stage XI

Stage XII

Stage XIII

FIGURE 10.30 Median sagittal sections of chick blastoderms showing formation of the hypoblast. At stage XI several fragmentary sheets of hypoblast cells (HC) are seen below the epiblast (E) in the posterior area pellucida. Cells (DC) that detached during thinning of the blastodisc are seen in the subgerminal cavity (SC). At stage XII the hypoblast is more extensive. The transparent band (TB) is seen at the posterior-most end of the area pellucida as a region lacking hypoblast cells. At stage XIII the hypoblast is complete and is surrounded by the marginal zone (MZ). Scale bar equals 0.5 mm. (From S. Kochav, M. Ginsburg, and H. Eyal-Giladi. 1980. From cleavage to primitive streak formation: A complementary normal table and a new look at the first stages of the development of the chick. II. Microscopic anatomy and cell population dynamics. Dev. Biol., *79:* 300.)

Two distinct populations of cells are identifiable in the blastocyst—those that are internal and an outer layer of epithelial cells. The internal cells cluster at one end (the **embryonic pole**) of the blastocyst (with the cavity at the opposite end); this cluster of cells is the **inner cell mass** or **embryoblast.** The outer cells are the **trophoblastic cells** or **trophectoderm cells.** The fates of these cells during development are quite different. The

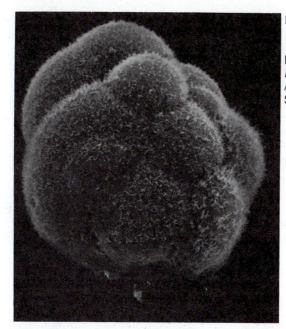

FIGURE 10.31 Mouse morula. ×1540. (From P.G. Calarco. 1975. Cleavage [mouse]. *In* E.S.E. Hafez (ed.), *Scanning Electron Microscopic Atlas of Mammalian Reproduction.* Igaku-Shoin, Ltd., p. 309.)

inner cell mass gives rise to the embryo proper, as well as extraembryonic structures, such as the amnion and yolk sac, whereas the trophoblastic cells (collectively called the **trophoblast** or **trophectoderm**) are responsible for the attachment of the blastocyst to the uterine wall and invasion of the uterine tissue (**implantation**). The trophoblast later forms the major portion of the placenta.

During the initial blastocyst stages the embryo is encased by the zona pellucida. However, the zona must be shed so the embryo can implant in the uterine wall. Shedding of the zona occurs by a "hatching" process whereby the embryo escapes through a crack in the zona (Fig. 10.32*C*). The zona-free blastocyst is shown in Figure 10.32*D*. Note the microvilli that cover the surfaces of the trophoblastic cells. The microvilli presumably assist in implantation.

At the time of implantation (Fig. 10.33*A*) the trophoblast consists of a single layer of cells, and the inner cell mass is an amorphous cluster of cells. As implantation proceeds (Fig. 10.33*B*), the trophoblast differentiates into two layers, the inner **cytotrophoblast** and the outer **syncytiotrophoblast.** The latter is a multinucleate cytoplasmic mass with fingerlike processes that penetrate the uterine lining and invade the underlying stroma. Meanwhile, a layer of cells appears on the surface of the inner cell mass adjacent to the blastocyst cavity. This cell layer is the hypoblast. Small spaces appear within the inner cell mass, which coalesce to produce the **amniotic cavity** (Fig. 10.33*C*). This cavity is overlain by a thin epithelial covering, the **amnion,** which is a derivative of the inner cell mass (Luckett, 1975).

Concurrently, the inner cell mass is converted into a bilayered **embryonic disc** composed of the epiblast, which forms the floor of the amniotic cavity, and the hypoblast above the blastocyst cavity. The embryonic disc bears a resemblance to the bilayered chick embryo, discussed previously.

The hypoblast expands laterally to spread over the inner surface of the cytotrophoblast. This layer of extraembryonic endoderm forms the

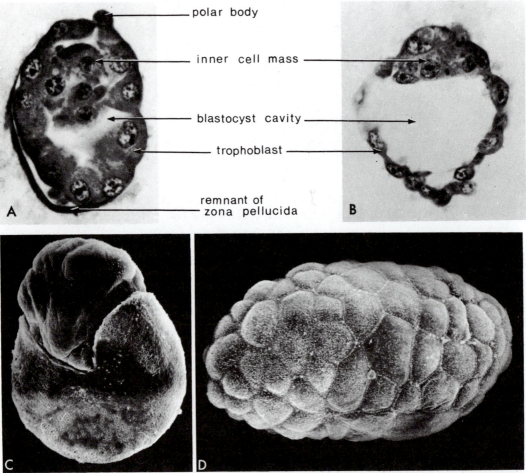

polar body

inner cell mass

blastocyst cavity

trophoblast

remnant of
zona pellucida

FIGURE 10.32 Mammalian blastocysts. *A* and *B,* Sections of human blastocyst recovered from the uterine cavity. ×600. *A,* At 4 days of development, the blastocyst cavity is just beginning to form. *B,* At 4½ days, the blastocyst cavity has enlarged, and the inner cell mass and trophoblast are clearly defined. (*A* and *B* from A.T. Hertig, J. Rock, and E.C. Adams. 1956. A description of 34 human ova within the first 17 days of development. Am. J. Anat., *98:* 461.) *C* and *D,* Scanning electron micrographs of mouse blastocysts. *C,* Blastocyst "hatching" through crack in zona. ×575. *D,* Zona-free blastocyst. ×2170. (*C* and *D* from S. Bergström, 1971. *Surface Ultrastructure of Mouse Blastocysts Before and At Implantation.* Uppsala Offset Center, Uppsala, Sweden, p. 39.)

primitive (or **primary**) **yolk sac** below the embryonic disc (Fig. 10.33*D* and *E*). The space between the primitive yolk sac and cytotrophoblast is filled with a meshwork of extraembryonic endoderm (Luckett, 1978).

As with the chick embryo, a primitive streak later forms at the posterior margin of the epiblast, through which mesoderm and endoderm presumably immigrate during gastrulation to form the three-layered embryo. Note that the interpretation of early mammalian development presented here is based upon the recent work of Luckett (1975, 1978), who has reinterpreted the origins of certain blastocyst tissues through a reevaluation of histological preparations of human embryos. The origins of these tissues remain controversial. However, Luckett's interpretation is consistent with the results of experimental work on the origins of embryonic tissues in blastocysts of other mammals.

As we have seen, there is tremendous variability in the shapes of animal blastulae and in the processes that form them. The next stage of development involves massive rearrangements of the cells of the blastula to form a three-layered embryo. We shall return to the formative processes of animal embryogenesis in Chapter 13.

10–2. PLANT EMBRYOGENESIS

After fertilization, the plant zygote, like the animal zygote, undergoes cleavage, which distributes the cytoplasm of the egg among numerous smaller cells. A major distinction between animal and plant cell division is the presence of cellulose walls around plant cells. As we learned in the previous chapter, the egg either lacks a cell wall or has an incomplete wall, allowing for contact and fusion of gamete membranes. The formation of a complete cell wall occurs in the zygote in response to fertilization, and cellular encasement is maintained by the laying down of new cell wall material after each mitosis as development begins. The need for cell walls in the cleavage plane dictates that the cleavage mechanism for plants be different from that for animals, in which daughter cells separate from one another by furrowing (see Chap. 5).

The rigidity with which plant cells are maintained in position after mitosis precludes the dramatic rearrangement of cells that characterizes animal development. Hence, cleavage of the plant zygote is *deliberate*; each division plane occurs in a precise site so as to fix the location of every cell. These cells may in turn divide and differentiate to produce distinct regions of the plant body. Consequently, cleavage is directly linked to the formation of body regions.

Plant body regions are arranged along a definite axis: Structures with highly contrasting form and function are localized at the opposite ends of this axis, and plant growth may be directed in opposite directions at either end. Thus, we say that the plant body has polarity. The rigid positioning of plant cells means that polarity must be accomplished by division patterns rather than by rearrangement of cells after mitosis. The

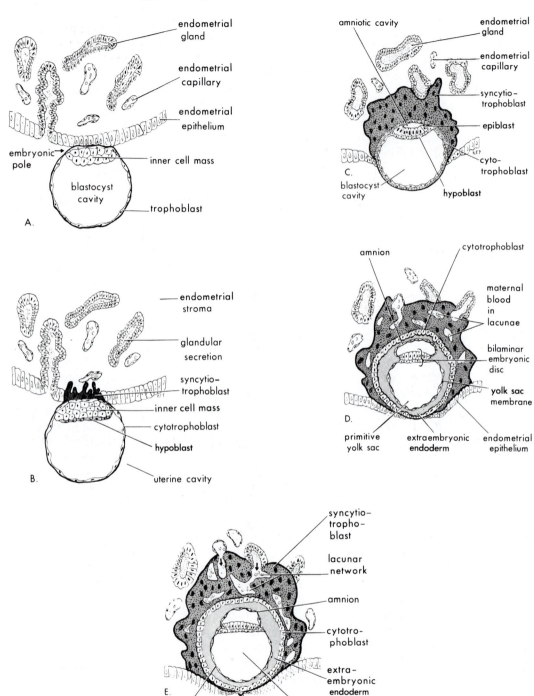

FIGURE 10.33 (*See legend on the opposite page.*)

◄ **FIGURE 10.33** Drawings of sections through implanting human embryos. *A*, At six days, the trophoblast attaches to the uterine wall at the embryonic pole of the blastocyst. *B*, At seven days, the syncytiotrophoblast has penetrated the uterine epithelium and started to invade the stroma. The hypoblast has formed. *C*, At eight days, the blastocyst is partially implanted. The amniotic cavity has formed, and the embryo is a bilayered disc. *D*, At nine days, the embryo has implanted. The yolk sac membrane and the extraembryonic endoderm have formed. *E*, At ten days, the uterine epithelium completely covers the implanted embryo. (From Moore, K.L.: *The Developing Human. Clinically Oriented Embryology.* 3rd Ed. Courtesy W.B. Saunders Company. 1982. *C, D* redrawn after A.T. Hertig and J. Rock. 1945. Two human ova of the previllous stage, having a developmental age of about seven and nine days respectively. Contr. Embryol. Carneg. Instn., *31:* 65–84 by copyright permission of the Carnegie Institution of Washington. *E* redrawn after A.T. Hertig and J. Rock. 1941. Two human ova of the previllous stage, having a developmental age of about eleven and twelve days respectively. Contr. Embryol. Carneg. Instn., *29:* 127–156 by copyright permission of the Carnegie Institution of Washington.)

establishment of polarity is one of the first orders of business during development, and in many developmental strategies the very first division establishes the axis. In such a situation, one of the two daughter cells will be the precursor of one end of the plant, and the other will form the opposite end.

This is beautifully illustrated by the brown alga *Fucus.* Eggs of *Fucus* are fertilized in the open sea, and the zygotes then settle onto a solid substratum, such as a rock. The fucoid zygote is radially symmetrical, with the nucleus in the center, and has no signs of regional differences. However, it then undergoes dramatic changes that result in the plant becoming highly polarized (Fig. 10.34). A protuberance forms from one side of the formerly spherical zygote, causing it to become pear-shaped. The first cell division occurs at right angles to the protrusion, forming two unequal cells. The protuberance is located entirely in one daughter cell (the **rhizoid cell**), and the spherical portion is in the other (the **thallus cell**). The elongate rhizoid cell continues to divide and differentiates into the **holdfast,** which attaches the embryo to a rock, anchoring it in the surf. The globular thallus cell divides and forms the bulk of the plant, which consists primarily of the branched thallus that bears the reproductive organs. The polarization of the spherical zygote is initially established by the point of sperm entry. However, polarity determination by sperm entry is quite labile and can be readily shifted by a variety of environmental factors. This lability has been exploited extensively by investigators, who have devised experimental means to polarize fucoid zygotes in order to probe the polarization mechanism. These studies have established the fucoid zygote as an important experimental system for the analysis of polarity. The results of these studies will be discussed in detail in Chapter 11.

Polarity is also evident in early stages of angiosperm development. In the preceding chapter the ultrastructure of the angiosperm zygote was described, and it was pointed out that the large vacuole is located in the micropylar end of the egg, whereas the majority of the cytoplasm—which contains the nucleus—is in the chalazal end. The first division plane is

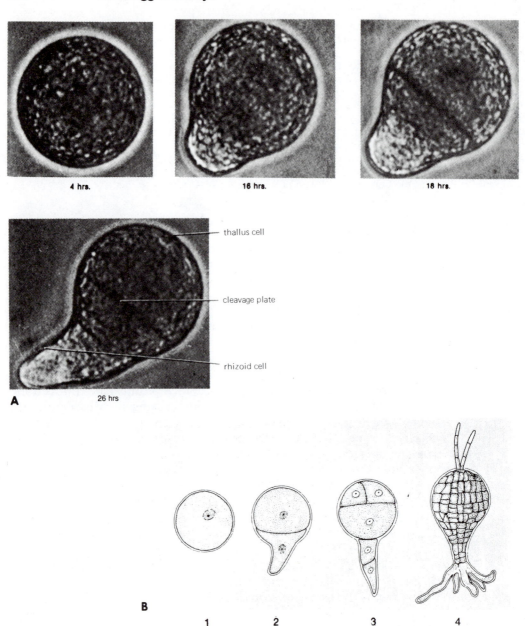

FIGURE 10.34 Development of *Fucus*. *A*, Zygotes at 4, 16, 18, and 26 hours after fertilization, seen by phase contrast, showing establishment of primary polarity, formation of rhizoid, and first cleavage. (Photographs by G.B. Bouck. From L.F. Jaffe. 1968. Localization in the developing *Fucus* egg and the general role of localizing currents. Adv. Morphogen., *7:* 298.) *B*, Diagrammatic representation of development. 1. Zygote. 2. After first division. 3. The cell formed by the protuberance forms the rhizoid, while the other cell forms the thallus. 4. Young embryo about 12 days old, with a well-developed rhizoid and apical hairs on the thallus. Shown at reduced magnificaion. (Reprinted with permission of Macmillan Publishing Co., Inc. from *Development in Flowering Plants* by John G. Torrey. Copyright © 1967 by John G. Torrey.)

transverse to the long axis of the egg. Because of the eccentric location of the nucleus, this division is unequal. It produces a small **terminal cell** at the chalazal end of the zygote, which is predominantly cytoplasmic, and a large **basal cell** at the micropylar end, which contains the large vacuole. The fates of these two cells vary somewhat in different species. In *Capsella*—the species we have chosen to demonstrate angiosperm development—the terminal cell divides to form the embryo proper (see subsequent discussion), whereas the basal cell divides twice in the transverse direction to produce two **suspensor cells,** which are interspersed between the terminal cell and basal cell. The suspensor cells and their descendants undergo repeated transverse divisions to produce a long filament, or **suspensor,** that connects the embryo to the large basal cell.

Figure 10.35 is a montage of electron micrographs showing the developing *Capsella* embryo attached to the elongating suspensor, which is terminated at the micropylar end by the basal cell. The embryo, suspensor cells, and basal cell are joined by numerous plasmodesmata, which are penetrated by endoplasmic reticulum, indicating that they are functionally interconnected. The lower suspensor cells fuse with the wall of the embryo sac, and the endosperm wall extends fingerlike projections into these cells. The basal cell also interdigitates with the surrounding integuments. These extensions resemble the filiform apparatus of the synergid and may function to absorb the rich starch reserves from the cells of the integuments. The plasmodesmata between the basal cell and the cells of the suspensor may serve as a system to transport these reserves from the integuments into the embryo. However, the mechanism of nutrient incorporation by the embryo remains unclear. The suspensor cells eventually lose their cytoplasm and become crushed by the growing embryo.

Division of the terminal cell to produce the multicellular embryo proceeds while the suspensor is being formed. The drawings of Figure 10.36 illustrate the division patterns of these two regions. Note that the first division of the terminal cell is longitudinal. Both of the resulting cells also divide longitudinally at right angles to the previous division to form the quadrant stage; each of the quadrant cells then divides transversely to form the octant stage. The latter is shown in Figure 10.35. The transverse divisions are significant in that they segregate the presumptive shoot apex and cotyledons (four apical cells) from the presumptive hypocotyl (four basal cells). However, other than their positions in the embryo, no ultrastructural or histochemical distinction can be made between these two groups of cells. The next division is also an important morphogenic event. This division is **periclinal,** and each cell produces an inner cell and an outer cell. The resulting 16-cell embryo consists of an outer layer of eight cells (collectively called the **protoderm**) and an inner group of eight cells that will give rise to the **procambium** and **ground meristem** (Fig. 10.37). These three tissues will produce the principal tissues of the plant: The protoderm forms the epidermis; the procambium forms the xylem, phloem, and cambium of the vascular bundles; and the ground meristem forms the parenchyma of the pith and cortex. As before, these cells with quite different developmental fates do not as yet show any ultrastructural differentiation.

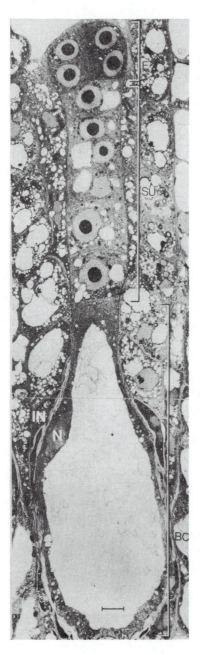

FIGURE 10.35 A montage of the octant embryo (E), suspensor (SU), and basal cell (BC). The nucleus (N) is located in the peripheral cytoplasm of the basal cell. The expanding basal cell has crushed the cells of the inner integument (IN). The lower suspensor cells are larger and less dense than the upper suspensor cells and will not divide again. Scale bar equals 5.0 μm. (From P. Schulz and W.A. Jensen. 1969. *Capsella* embryogenesis: The suspensor and the basal cell. Protoplasma, *67:* 140.)

The upper suspensor cell also contributes to the formation of the embryo. This cell, which is called the **hypophysis,** is also shown in Figure 10.37A. The hypophysis divides to produce cells that initiate the cortex of the embryonic root and form the central portion of the root cap. The

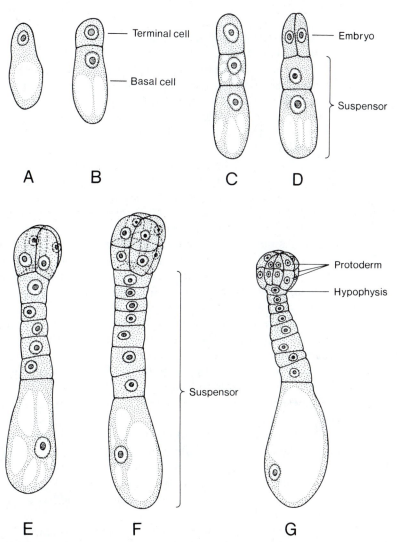

FIGURE 10.36 Early embryonic development of *Capsella. A* and *B,* Division of zygote into terminal and basal cells. *C,* Basal cell divides transversely. *D,* Longitudinal division of the teminal cell produces the two-celled embryo. *E,* Quadrant stage. Five suspensor cells connect the embryo to the basal cell. *F,* Octant stage. *G,* Sixteen-cell stage. Periclinal division has produced eight outer protoderm cells and eight inner cells. The upper suspensor cell is designated the hypophysis. (Adapted from a Turtox Wall Chart; courtesy Science Kit and Boreal Laboratories.)

hypophysis is the only cell contributing to formation of the embryo that shows any specialization in its cytoplasmic ultrastructure; it has a low ribosome density and is highly vacuolated. Embryonic cell division continues in the globular embryo, increasing the number of protoderm cells by **anticlinal** division and producing more inner cells by **horizontal** division (Fig. 10.37*B*).

After the globular stage, the embryo begins to undergo a shape change as the cotyledons emerge as two ridges on the apical side of the embryo. The altered shape of the embryo at this stage has led to its designation as the **heart stage** (Fig. 10.38). Now, for the first time, there are cytological differences in the cells of different embryonic regions. As shown in Figure 10.38, the cells of the procambium and ground meristem appear to be more highly vacuolated than those of the protoderm, and the cells at the tips of the cotyledons stain for proteins and nucleic acids more intensely than other cells of the embryo.

Between the bases of the cotyledons there is an undifferentiated group of cells that is the presumptive shoot apical meristem. At the opposite end of the procambial system, near the suspensor, is a group of cells that will form the root apical meristem. The meristems will function after germination to produce the shoot and root systems, respectively. As embryogenesis continues past the heart stage, the two meristems are progressively separated from one another by elongation of the embryonic axis, or **hypocotyl.** Continued elongation of the hypocotyl and cotyledons bends the cotyledons so that they will fit inside the enveloping ovule wall (Fig. 10.39). The rate of embryo growth gradually declines and eventually ceases as the embryo becomes quiescent. The quiescence is brought about by dehydration of the embryo. The entire process of embryonic development of *Capsella* from zygote to mature quiescent embryo takes about ten days. Meanwhile, the ovule wall is thickening and hardening to form the *seed coat.* The seed coat protects the quiescent embryo until development is reinitiated by seed germination.

The pattern of development we have described here is typical of dicotyledonous plants (dicots). These plants have two cotyledons, whereas the other major group of angiosperms, the monocots, have a single cotyledon. These two groups also differ in the organization of tissues in the primary plant body, although the basic tissues themselves are quite similar.

Angiosperm embryonic development is sustained by nutrients derived from the endosperm. Like the embryo, the endosperm is formed at fertilization (see Chap. 9). The endosperm initially develops more rapidly than the embryo and may undergo several mitotic divisions before the zygote divides. In some plants, endosperm nuclear division is accompanied by cell wall formation, producing cellular endosperm. In others, such as *Capsella*, the first nuclear divisions do not produce cells. The resultant **nuclear endosperm** is a syncytium with a fluid consistency (Fig. 10.39*A*). When cell wall formation begins, the entire endosperm may become cellular (Fig. 10.39*B*), or—in some species—a portion of the en-

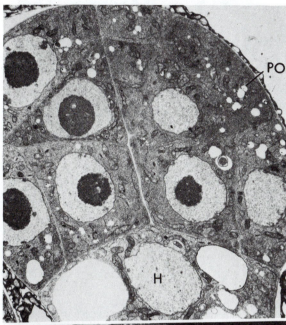

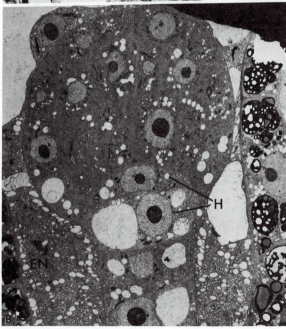

FIGURE 10.37 Cell division patterns of the apical cell derivatives in *Capsella*. *A*, Origin of the protoderm (PO) (16-celled embryo). There are no apparent ultrastructural differences between the cells of the embryo at this stage. The upper suspensor cell (hypophysis, H) has a lower ribosome density and is more vacuolate than the cells of the embryo and will contribute to the formation of the embryonic root. ×6150. *B*, The globular embryo showing the fine structural similarity of the cells of the protoderm, future procambium, and ground meristem. The two suspensor cells next to the embryo are derived from the hypophysis (H). The lenticule-shaped cell has a greater ribosome density and is less vacuolate than its sister cell. The endosperm (EN) closely invests the embryo at this stage. ×1500. (From R. Schulz and W.A. Jensen. 1968. *Capsella* embryogennesis: The early embryo. *J. Ultrastruct. Res., 22:* 384, 385.)

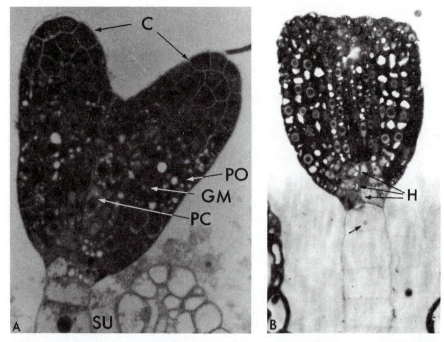

FIGURE 10.38 The "heart stage" of *Capsella* development. Embryos stained for protein (*A*) or nucleic acids (*B*). *A*, The heart-shaped embryo showing structural and histochemical differences between the cells of the protoderm (PO), procambium (PC), and ground meristem (GM). The cells at the tips of the cotyledons (C) stain very intensely for protein, and the cells of the procambium and ground meristem are more vacuolate than the cells of the protoderm. The cells in the region of the hypophysis stain with less intensity than those of the embryo, and the suspensor (SU) stains very lightly. Aniline blue black. ×740. *B*, Side view of the heart-shaped embryo showing the intense staining of the embryo for nucleic acids. The region of the hypophysis (H) stains less intensely than the embryo, and the suspensor stains faintly except for the nucleoli, which give a strong reaction for RNA (arrow). Azure B. ×550. (From R. Schulz and W.A. Jensen. 1968. *Capsella* embryogenesis: The early embryo. J. Ultrastruct. Res., *22:* 388.)

dosperm may remain as a fluid. The latter situation occurs in the coconut, in which the liquid endosperm is the "milk."

Quiescence terminates the embryonic phase of angiosperm development. At this time the only presumptive adult plant structures are the two meristems. The remaining parts of the plant body must be produced by division of the meristems, followed by differentiation of the cells produced. Consequently, the major developmental events in the formation of the mature plant body are *postembryonic,* and since the meristems are permanent sources of new cells, development is extended throughout the adult phase of the life of the plant. We shall discuss the formation of the mature plant body in Chapter 13.

B

A

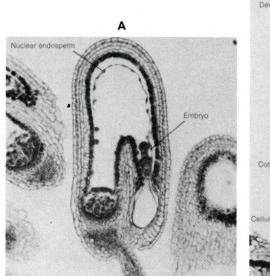
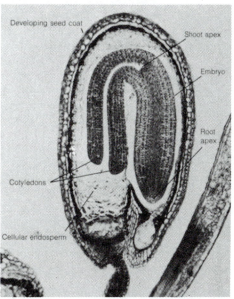

FIGURE 10.39 Longitudinal sections of developing seeds in *Capsella*. *A,* Early embryo. *B,* Nearly mature embryo. (From *Comparative Morphology of Vascular Plants,* Second Edition, by Adriance S. Foster and Ernest M. Gifford, Jr. Copyright © 1974 by W.H. Freeman and Company. All rights reserved.)

REFERENCES

Akasaka, K., S. Amemiya, and H. Terayama. 1980. Scanning electron microsopical study of the inside of sea urchin embryos (*Pseudocentrotus depressus*). Effects of aryl β-xyloside, tunicamycin and deprivation of sulfate ions. Exp. Cell Res., *129*: 1–13.

Arnold, J.M. 1969. Cleavage furrow formation in a telolecithal egg (*Loligo pealii*). J. Cell Biol., *41*: 894–904.

Asnes, C.F., and T.E. Schroeder. 1979. Cell cleavage. Ultrastructural evidence against equatorial stimulation by aster microtubules. Exp. Cell Res., *122*: 327–338.

Balinsky, B.I. 1975. *An Introduction to Embryology,* 4th ed. W.B. Saunders Co., Philadelphia.

Balinsky, B.I. 1981. *An Introduction to Embryology,* 5th ed. Saunders College Publishing, Philadelphia.

Beams, H.W., and R.G. Kessel. 1976. Cytokinesis: A comparative study of cytoplasmic division in animal cells. Am. Sci., *64*: 279–290.

Bergström, S. 1971. *Surface Ultrastructure of Mouse Blastocysts before and at Implantation.* Uppsala Offset Center, Uppsala, Sweden.

Berrill, N.J., and G. Karp. 1976. *Development.* McGraw-Hill Book Co., New York.

Bluemink, J.G. 1970. The first cleavage of the amphibian egg. An electron microscope study of the onset of cytokinesis in the egg of *Ambystoma mexicanum*. J. Ultrastruct. Res., *32*: 142–166.

Bluemink, J.G., and S.W. De Laat. 1973. New membrane formation during cytokinesis in normal and cytochalasin B-treated eggs of *Xenopus laevis*. I. Electron microscope observations. J. Cell Biol., *59*: 89–108.

Brachet, J. 1950. *Chemical Embryology*, L.G. Barth (transl.). Facsimile edition published by Hafner Press (Macmillan), New York. (1968).

Calarco, P.G. 1975. Cleavage (mouse). *In* E.S.E. Hafez (ed.), *Scanning Electron Microscopic Atlas of Mammalian Reproduction*. Igaku Shoin Ltd., Tokyo. pp. 306–317.

Conklin, E.G. 1905. The orientation and cell-lineage of the ascidian egg. J. Acad. Nat. Sci. Philad. Ser. 2, *13*: 1–119.

Denis-Donini, S., B. Baccetti, and A. Monroy. 1976. Morphological changes of the surface of the eggs of *Xenopus laevis* in the course of development. 2. Cytokinesis and early cleavage. J. Ultrastruct. Res., *57*: 104–112.

Edwards, R.G., and R.E. Fowler. 1970. Human embryos in the laboratory. Sci. Am., *223*(6): 44–54.

Epstein, C.J. 1975. Gene expression and macromolecular synthesis during preimplantation embryonic development. Biol. Reprod., *12*: 82–105.

Eyal-Giladi, H., and S. Kochav. 1976. From cleavage to primitive streak formation: A complementary normal table and a new look at the first stages of development of the chick. I. General morphology. Dev. Biol., *49*: 321–337.

Foster, A.S., and E.M. Gifford, Jr. 1974. *Comparative Morphology of Vascular Plants*, 2nd ed. W.H. Freeman and Co., San Francisco.

Fujiwara, K., and T.D. Pollard. 1976. Fluorescent antibody localization of myosin in the cytoplasm, cleavage furrow, and mitotic spindle of human cells. J. Cell Biol., *71*: 848–875.

Fullilove, S.L., and A.G. Jacobson. 1971. Nuclear elongation and cytokinesis in *Drosophila montana*. Dev. Biol., *26*: 560–577.

Fullilove, S.L., and A.G. Jacobson. 1978. Embryonic development: Descriptive. *In* M. Ashburner and T.R.F. Wright (eds.), *The Genetics and Biology of Drosophila*, Vol. 2c. Academic Press, London, pp. 106–143.

Gipson, I. 1974. Electron microscopy of early cleavage furrows in the chick blastodisc. J. Ultrastruct. Res., *49*: 331–347.

Giudice, G. 1973. *Developmental Biology of the Sea Urchin Embryo*. Academic Press, New York.

Grant, P. 1978. *Biology of Developing Systems*. Holt, Rinehart and Winston, Inc., New York.

Gulyas, B.J. 1973. Cytokinesis in the rabbit zygote: Fine-structural study of the contractile ring and the mid-body. Anat. Rec., *177*: 195–208.

Gulyas, B.J. 1975. A reexamination of cleavage patterns in eutherian mammalian eggs: Rotation of blastomere pairs during second cleavage in the rabbit. J. Exp. Zool., *193*: 235–248.

Harvey, E.B. 1936. Parthenogenetic merogony or cleavage without nuclei in *Arbacia punctulata*. Biol. Bull., *71*: 101–121.

Hertig, A.T., and J. Rock. 1941. Two human ova of the previllous stage, having a developmental age of about eleven and twelve days respectively. Contrib. Embryol. Carneg. Instn., *29*: 127–156.

Hertig, A.T., and J. Rock. 1945. Two human ova of the previllous stage, having a developmental age of about seven and nine days respectively. Contrib. Embryol. Carneg. Instn., *31*: 65–84.

Hertig, A.T., J. Rock, and E.C. Adams. 1956. A description of 34 human ova within the first 17 days of development. Am. J. Anat., *98*: 435–493.

Hiramoto, Y. 1956. Cell division without mitotic apparatus in sea urchin eggs. Exp. Cell Res., *11*: 630–636.

Hiramoto, Y. 1965. Further studies on cell division without mitotic apparatus in sea urchin eggs. J. Cell Biol., *25*: 161–167.

Jaffe, L.F. 1968. Localization in the developing *Fucus* egg and the general role of localizing currents. Adv. Morphogen., *7*: 295–328.

Kalt, M.R. 1971. The relationship between cleavage and blastocoel formation in *Xenopus laevis*. J. Embryol. Exp. Morphol., *26*: 51–66.

Kochav, S., M. Ginsburg, and H. Eyal-Giladi. 1980. From cleavage to primitive streak formation: A complementary normal table and a new look at the first stages of the development of the chick. II. Microscopic anatomy and cell population dynamics. Dev. Biol., *79*: 296–308.

Korschelt, E. 1936. Vergleichende Entwicklungsgeschichte der Tiere. Band I. Gustav Fischer Verlag, Jena.

Lewis, W.H., and C.G. Hartman. 1933. Early cleavage stages of the egg of the monkey (*Macacus rhesus*). Contrib. Embryol. Carneg. Instn., *24*: 187–201.

Luckett, W.P. 1975. The development of primordial and definitive amniotic cavities in early rhesus monkey and human embryos. Am. J. Anat., *144*: 149–168.

Luckett, W.P. 1978. Origin and differentiation of the yolk sac and extraembryonic mesoderm in presomite human and rhesus monkey embryos. Am. J. Anat., *152*: 59–98.

Mabuchi, I. 1973. A myosin-like protein in the cortical layer of the sea urchin egg. J. Cell Biol., *59*: 542–547.

Mabuchi, I., and M. Okuno. 1977. The effect of myosin antibody on the division of starfish blastomeres. J. Cell Biol., *74*: 251–263.

Moore, K.L. 1977. *The Developing Human*, 2nd ed. W.B. Saunders Co., Philadelphia.

Moore, K.L. 1982. *The Developing Human. Clinically Oriented Embryology*, 3rd ed. W.B. Saunders, Co., Philadelphia.

Olsen, M.W. 1942. Maturation, fertilization, and early cleavage in the hen's egg. J. Morphol., *70*: 513–533.

Rappaport, R. 1961. Experiments concerning the cleavage stimulus in sand dollar eggs. J. Exp. Zool., *148*: 81–89.

Rappaport, R. 1966. Experiments concerning the cleavage furrow in invertebrate eggs. J. Exp. Zool., *161*: 1–8.

Rappaport, R. 1978. Effects of continual mechanical agitation prior to cleavage in echinoderm eggs. J. Exp. Zool., *206*: 1–12.

Rappaport, R., and R.P. Ebstein. 1965. Duration of stimulus and latent periods preceding furrow formation in sand dollar eggs. J. Exp. Zool., *158*: 373–382.

Rickoll, W.L. 1976. Cytoplasmic continuity between embryonic cells and the primitive yolk sac during early gastrulation in *Drosophila melanogaster*. Dev. Biol., *49*: 304–310.

Rustad, R.C., S. Yuyama, and L.C. Rustad. 1970. Nuclear cytoplasmic relations in the mitosis of sea urchin eggs. II. The division times of whole eggs and haploid and diploid half-eggs. Biol. Bull., *138*: 184–193.

Schroeder, T.E. 1969. The role of "contractile ring" filaments in dividing *Arbacia* eggs. Biol. Bull., *137*: 413–414.

Schroeder, T.E. 1972. The contractile ring. II. Determining its brief existence, volumetric changes, and vital role in cleaving *Arbacia* eggs. J. Cell Biol., *53*: 419–434.

Schroeder, T.E. 1973. Cell constriction: Contractile role of microfilaments in division and development. Am. Zool., *13*: 949–960.

Schroeder, T.E. 1981a. The origin of cleavage forces in dividing eggs. A mechanism in two steps. Exp. Cell Res., *134*: 231–240.

Schroeder, T.E. 1981b. Interrelations between the cell surface and the cytoskeleton in cleaving sea urchin eggs. *In* G. Poste and G.L. Nicolson (eds.), *Cytoskeletal Elements and Plasma Membrane Organization.* Elsevier/North-Holland Biomedical Press, Amsterdam, pp. 169–216.

Schulz, R., and W.A. Jensen. 1968. *Capsella* embryogenesis: The early embryo. J. Ultrastruct. Res., *22*: 376–392.

Schulz, P., and W.A. Jensen. 1969. *Capsella* embryogenesis: The suspensor and the basal cell. Protoplasma, *67*: 139–163.

Selman, G. and M. Perry. 1970. How cells cleave. New Sci., *46*(2 April): 12–14.

Senger, D.R., and P.R. Gross, 1978. Macromolecule synthesis and determination in sea urchin blastomeres at the sixteen-cell stage. Dev. Biol., *65*: 404–415.

Singal, P.K., and E.J. Sanders. 1974. An ultrastructural study of the first cleavage of *Xenopus* embryos. J. Ultrastruct. Res., *47*: 433–451.

Sze, L.C. 1953. Changes in the amount of desoxyribonucleic acid in the development of *Rana pipiens.* J. Exp. Zool., *122*: 577–601.

Szollosi, D. 1970. Cortical cytoplasmic filaments of cleaving eggs: A structural element corresponding to the contractile ring. J. Cell Biol., *44*: 192–209.

Tilney, L.G., and D. Marsland. 1969. A fine structural analysis of cleavage induction and furrowing in the eggs of *Arbacia punctulata.* J. Cell Biol., *42*: 170–184.

Torrey, J.G. 1967. *Development in Flowering Plants.* Macmillan, New York.

Turner, F.R., and A.P. Mahowald. 1976. Scanning electron microscopy of *Drosophila* embryogenesis. I. The structure of the egg envelopes and the formation of the cellular blastoderm. Dev. Biol., *50*: 95–108.

Wilson, E.B. 1925. *The Cell in Development and Heredity,* 3rd ed. Macmillan, New York.

Wolpert, L. 1960. The mechanics and mechanism of cleavage. Int. Rev. Cytol., *10*: 163–216.

11 The Developmental Consequences of Cleavage

During cleavage the zygote is divided into numerous smaller cells that subsequently become specialized entities in the emerging multicellular organism. The characteristics of each of these specialized cell types are specified by the genome of the newly formed zygote nucleus. As we learned in Chapter 2, the genome is distributed equally among the daughter cells at each cleavage division. Since the nuclei are equivalent, how can specialized cells emerge in which different portions of the genome are utilized?

The answer to this puzzle is that nuclei in different regions of the embryo receive unique stimuli from their surroundings; these stimuli initiate the chain of events that establish different populations of cells in different embryonic regions. Two generalized sources of such stimuli have been identified: those *intrinsic to the cells* in which the nuclei reside and those *extrinsic to the cells*. Intrinsic differences in the cytoplasms of blastomeres are established as a direct result of cleavage, which segregates unique ooplasmic constituents into specific blastomeres. Cells may also come under distinct extrinsic influences that variably influence differentiation. For example, a zygote lying at the bottom of a body of water during daylight is exposed to light from above, but its lower part is shaded. The light gradient could establish localized cytoplasmic differences that would cause the derivatives of the illuminated half of the zygote to follow a different developmental path from the shaded half. Cells may also be influenced by their position within the embryo. In a morula, for example, there are "inside" cells and "outside" cells. The inside cells are completely surrounded by the outside cells, whereas the latter are usually in contact with the egg accessory layers.

The initial establishment of unequal influences on nuclei is one of the most critical events during development. Numerous experiments

481

have demonstrated that in the absence of a proper pattern of differential nuclear environments, development will not proceed normally. Thus, cleavage, which is responsible for organizing the three-dimensional arrangement of cells in the embryo, has a pivotal role to play in the initiation of regional differentiation. We shall now consider examples of the two general mechanisms for establishing differential nuclear environments during cleavage.

11–1. CYTOPLASMIC LOCALIZATION AS A MECHANISM REGULATING DIFFERENTIATION

The redistribution of localized cytoplasmic constituents during cleavage has been recognized for a very long time. Various investigators have traced the ultimate fates of these constituents and have frequently noted a correlation between their presence and the ultimate developmental fates of cells that receive them. Once such a correlation is established, the next step is to manipulate the egg experimentally to determine whether the cytoplasmic constituents are *essential* for the development of particular embryonic structures. More than a century ago, investigators began documenting a number of clear-cut cases in which specific egg cytoplasmic constituents are responsible for directing specific pathways of development. These constituents are collectively known as **morphogenic determinants,** which are defined as substances that (1) are heterogeneously distributed in the zygote and (2) dictate a specific pathway of cellular differentiation.

Determinants are produced during oogenesis. During the activation response that follows fertilization, they frequently undergo major reorganization, becoming arranged along distinct axes of symmetry. Cleavage then fixes the positions of the determinants. Consequently, the sites of cleavage furrows may be correlated with the interfaces between adjacent determinants. Cleavage assumes enormous significance in the overall scheme of development, for it is not merely a passive process of increasing the number of cells from which the embryo is then formed, but a mechanism by which the initial differences between cells of the embryo may be established.

A few organisms have traditionally been used as examples of cytoplasmic localization of determinants. These organisms have continued to serve as experimental material over the years, and a large body of knowledge about them has accumulated. Thus, although we know little about this phenomenon in the vast majority of species, it has been well studied in a few cases, and we shall discuss the results of some of those studies here.

Cytoplasmic Localization in Ascidians

The ascidian egg provides one of the most clear-cut examples of cytoplasmic localization of morphogenic determinants. The eggs of certain ascidians contain cytoplasmic regions (ooplasms) that can be readily identified by their distinct pigmentation. The pigmentation serves as a marker to trace the developmental fates of the various regions. Three different ooplasms are visible in these eggs: (1) the transparent **ectoplasm;** (2) the yolky **endoplasm;** and (3) the **myoplasm.** In the genus *Styela* the myoplasm contains yellow pigment granules, and in *Boltenia* the granules are orange or red. The distribution of these ooplasms is shown in Figure 11.1*A*.

The ectoplasm, which is located in the animal hemisphere, is derived primarily from the substance liberated from the germinal vesicle when it breaks down during oocyte maturation (Fig. 11.1*B*). The endoplasm is located predominantly in the vegetal hemisphere, whereas the myoplasm occupies the entire egg cortex. Fertilization elicits a massive reorganization of the ooplasms in a process known as **ooplasmic segregation** (Fig. 11.1*C–I*). Fertilization (Fig. 11.1*C*) occurs near the vegetal pole and triggers a streaming of the cortical myoplasm into the vegetal hemisphere where it localizes as a highly pigmented cap at the vegetal pole. The ectoplasm follows the myoplasm to the vegetal hemisphere. During its movement the ectoplasm separates into a number of separate aggregates (Fig. 11.1*F*) and finally collects just above the myoplasmic cap (Fig. 11.1*G*). Concurrent with the movement of the myoplasm and ectoplasm (and possibly as a consequence of their movement) the endoplasm is shifted into the animal hemisphere. The ooplasms are now stratified.

Ooplasmic segregation is a convulsive event that occurs with such force that the egg is temporarily distorted (Fig. 11.1*E–G*; Fig. 11.2). The distortion takes the form of a constriction that spreads from the animal pole to the vegetal pole. As the constriction moves downward, the egg changes shape from pear-shaped to dumbbell-shaped and to pear-shaped again before reforming the spherical shape. Both ooplasmic segregation and the contraction are inhibited by cytochalasin B (Sawada and Osani, 1981). These results and the combined electron microscopic and biochemical studies of Jeffery and Meier (1983) indicate that contraction of an actin microfilament network in the cortex is responsible for ooplasmic segregation.

The first phase of ooplasmic segregation is complete with the stratification of ooplasms. The stratification is temporary, however, and is followed by rearrangements that establish bilateral symmetry in the egg. The male pronucleus is a major factor in the latter phase. It migrates from the fertilization site near the vegetal pole, moving upward on the future posterior side of the egg as it travels toward the animal hemisphere to contact the female pronucleus. The myoplasm and ectoplasm accompany the nucleus, streaming upward on this side of the egg. The myoplasm localizes in a subequatorial position on the posterior side where it forms a crescent-shaped structure (Fig. 11.1*I*). In *Styela* this is known as the **yellow crescent** (Conklin, 1905). The ectoplasm accompanies the male pronucleus into the animal hemisphere where the pronuclei contact one another. Meanwhile, the endoplasm returns to the vegetal hemisphere, completing the ooplasmic segregation.

In his detailed cell-lineage studies E. G. Conklin (1905) described how intervening cleavage furrows segregate the ooplasms into specific blastomeres that are the precursors for definitive regions of the larva. The first cleavage division passes through the plane of bilateral symmetry, bisecting the yellow crescent (Fig. 11.1*J*). Each resultant blastomere is a mirror image of the other and receives an equal amount of each cytoplasmic region. The second division is perpendicular to the first, oriented parallel to the animal–vegetal axis. Hence, the two posterior blastomeres receive all the yellow crescent material (Fig. 11.1 *K–L*). The next cleavage further segregates the cytoplasmic regions. This division is equatorial; the ectoplasm goes primarily into the animal hemisphere blastomeres, and much of the endoplasm enters the anterior pair of vegetal hemisphere blastomeres. The sorting out of cytoplasm continues as cleavage proceeds. By following the ultimate fates of the various regions, Conklin determined that the cells into which yellow crescent material is segregated become the muscle and mesenchyme of the tail, cells with the endoplasm material become the notochord, neural tube, and digestive

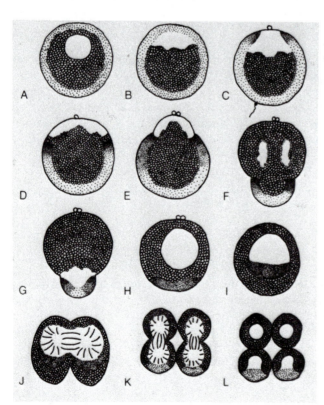

FIGURE 11.1 Ooplasmic segregation in the egg of *Styela plicata. A.* Meridional section through a fully grown oocyte with an intact germinal vesicle (center). *B.* At maturation the transparent ectoplasm (clear area) is released from the germinal vesicle. The cortical myoplasm is represented with fine granulation, and the endoplasm is shown as a coarse granular area, which is predominantly in the vegetal hemisphere. *C–H.* Sperm entry (shown in *C*) triggers ooplasmic segregation. *I.* The yellow crescent forms on the future posterior side of the embryo. *J.* The first cleavage furrow bisects the yellow (myoplasmic) crescent. *K–L,* Subequatorial sections through the myoplasmic crescent during and after the second cleavage division. (From W.R. Jeffery. 1983. Messenger RNA localization and cytoskeletal domains in ascidian embryos. *In* W.R. Jeffery and R.A. Raff (eds.), *Time. Space and Pattern in Embryonic Development.* Alan R. Liss, New York, p. 243.)

system, and the cells containing ectoplasm develop into the epidermis that covers the surface of the larva.

These observations do not indicate that the cytoplasmic regions play a *causative role* in determination of larval structures. Conklin investigated this possibility by destruction of particular blastomeres. He found that the remaining blastomeres produced an incomplete larva containing only those structures that they would have produced in normal development; those regions corresponding to the damaged blastomeres were missing. In a later study he applied the technique of displacement of cytoplasm by centrifugal force, which stratified the contents of the zygotes. Cleavage fixed the stratified cytoplasm in its abnormal configuration, and the effects on development were monitored. The resulting larvae developed abnormally, with the positions of tissues and organs correlated to the sites of the corresponding cytoplasms. Conklin (1931) concluded that the ". . . different areas of cytoplasm are specific factors in the development of particular tissues and organs, and in this sense they are 'organ-forming substances.' "

Conklin's conclusion has been confirmed by cell deletion experiments performed on eight-cell–stage embryos of a related ascidian, *Ciona intestinalis* (Whittaker et al., 1977). *Ciona* is quite similar to *Styela* in its pattern of development and in the organization of the egg cytoplasm, except that the *Ciona* cytoplasmic regions are less distinct. These ex-

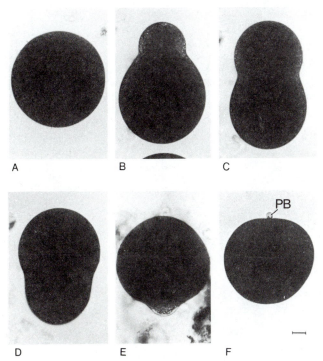

A   B   C

D   E   F

PB

FIGURE 11.2 Modification of egg shape during ooplasmic segregation in living *Ciona intestinalis* eggs. *A,* Unfertilized egg. *B–E,* Successive shape changes of fertilized eggs, culminating in formation of the myoplasmic cap at the vegetal pole. *F,* The spherical shape is reassumed. The animal pole is at the top in each photomicrograph. PB: polar body. Scale bar equals 20 μm. (From T. Sawada and K. Osani. 1981. The cortical contraction related to the ooplasmic segregation in *Ciona intestinalis* eggs. Wilh. Roux' Arch., *190:* 209. Reprinted with permission of Springer-Verlag, Heidelberg.)

periments have also utilized a highly specific histochemical stain that identifies an enzyme that occurs only in larval muscle. Consequently, the appearance of this enzyme—acetylcholinesterase (AChE)—is a rapid means of detecting whether muscle cells have differentiated. At the eight-cell stage the muscle determinants are segregated into the two lower posterior cells. Removal of these cells with a fine glass needle (Fig. 11.3) results in a developing larva that does not stain for AChE (Fig. 11.4B). Furthermore, the two isolated blastomeres divide to form a cell aggregate that does stain positively for AChE (Fig. 11.4C). Although these experiments are more sophisticated in their execution than those by Conklin, they merely confirm his conclusions that the yellow crescent cytoplasm contains determinants that direct the cells in which it is localized to differentiate as muscle.

The histochemical technique has also been combined with a cleavage-inhibiting treatment to confirm the role of cleavage in segregation of determinants. Embryos treated with cytochalasin B cease cytokinesis, but nuclear division continues. When untreated controls show a positive stain for AChE, these cleavage-arrested embryos also stain for the enzyme. Thus, acquisition of AChE activity occurs even though cell division is prevented. Whereas in normal development the muscle determinants would ultimately be segregated into a small number of cells that form muscle, inhibition of cleavage stops the segregation process and freezes in place the pattern of determinants. The presence of these determinants at the time of inhibition is then assessed by staining for the

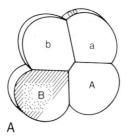

A

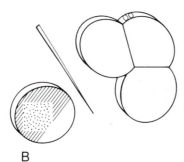

B

FIGURE 11.3 Diagram illustrating *(A)* the lateral orientation of the *Ciona intestinalis* embryo at the eight-cell stage and *(B)* the surgical separation of the B cell pair. Mesodermal regions fated to become muscle (area of diagonal lines) and mesenchyme (dotted area) are indicated. (After J.R. Whittaker, G. Ortolani, and N. Farinella-Ferruzza. 1977. Autonomy of acetylcholinesterase differentiation in muscle-lineage cells of ascidian embryos. Dev. Biol., *55:* 197.)

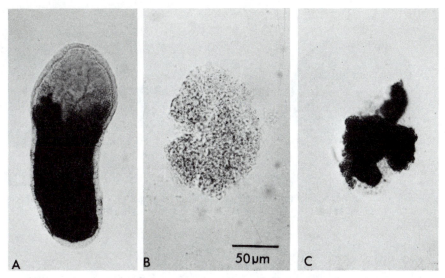

FIGURE 11.4 Histochemical localization of AChE activity in developing larvae of *Ciona intestinalis*. *A*, Control. *B*, Developing larva from which both B cells were removed at the eight-cell stage. No AChE activity is detected. *C*, Cell mass from B cells removed at the eight-cell stage. Positive AChE stain. (From J.R. Whittaker, G. Ortolani, and N. Farinella-Ferruzza. 1977. Autonomy of acetylcholinesterase differentiation in muscle lineage cells of ascidian embryos. Dev. Biol., *55:* 198.)

muscle-specific enzyme. The series of photomicrographs of stained embryos in Figure 11.5 illustrates this point. The embryos were treated with cytochalasin B at progressively later cleavage stages and allowed to continue developing. They were then fixed and stained for AChE activity 12 to 14 hours after fertilization. In Figure 11.5*A*, cleavage was inhibited at the two-cell stage. Both blastomeres stain for AChE, confirming the presence of muscle determinants in both blastomeres. In Figure 11.5*B*, cleavage was inhibited at the four-cell stage. However, only two blastomeres—those corresponding to the two posterior blastomeres—stain positively. Likewise, at the eight-cell stage (Fig. 11.5*C*), only two blastomeres stain positively. These correspond to the two blastomeres that receive the muscle determinants after the third cleavage (Whittaker, 1973).

A combination of the AChE histochemical technique and experimental alteration of the cleavage pattern has demonstrated that a modification in segregation can alter the developmental fates of blastomeres. Eggs of *Styela plicata* were compressed during the third cleavage, which caused this cleavage to be meridional instead of equatorial (Fig. 11.6). As a consequence, the compressed embryos had four cells with myoplasm instead of two as in normal cleavage (compare Fig. 11.6*D* and *F*; see also Fig. 11.7). The developmental fate of the myoplasm-containing blastomeres is demonstrated by treatment with cytochalasin B to inhibit cleav-

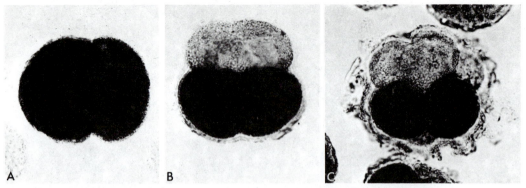

FIGURE 11.5 AChE activity in cleavage-arrested *Ciona intestinalis* embryos. *A*, Inhibited at the two-cell stage. *B*, Inhibited at the four-cell stage. *C*, Inhibited at the eight-cell stage. ×235. (Courtesy of Dr. J.R. Whittaker.)

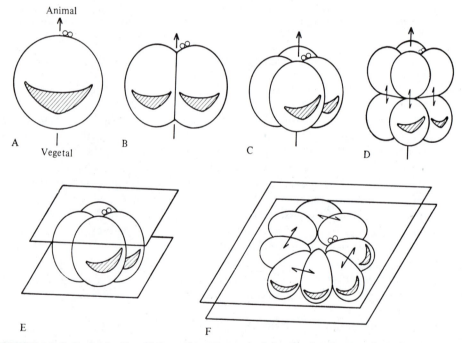

FIGURE 11.6 Redistribution of the yellow crescent of the *Styela plicata* embryo by compression. *A–D*, Normal development through the eight-cell stage. The animal-vegetal axis is indicated with an arrow. Polar bodies are shown at the animal pole. The yellow crescent is indicated by cross-hatching. Sister blastomeres formed at the third cleavage are indicated by a barbed line. *E*, Four-cell stage embryo is compressed as shown. *F*, A compressed embryo divides to form a flat plate of cells lying in the same plane. Sister blastomeres are indicated as in *D*. Four blastomeres now contain the yellow crescent material. (From J.R. Whittaker. 1980. Acetyl-cholinesterase development in extra cells caused by changing the distribution of myoplasm in ascidian embryos. J. Embryol. Exp. Morphol., *55:* 347.)

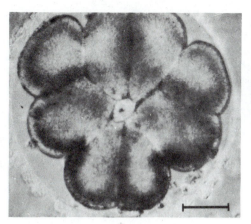

FIGURE 11.7 *Styela plicata* embryo that had been compressed at the four-cell stage and photographed after the third cleavage. The granular myoplasm can be seen in the four lower blastomeres. Scale bar equals 50 μm. (From J.R. Whittaker. 1980. Acetycholinesterase development in extra cells caused by changing the distribution of myoplasm in ascidian embryos. J. Embryol. Exp. Morphol., *55:* 348.)

age, followed by staining for AChE activity. As shown in Figure 11.8*A*, the normal eight-cell–stage embryo that had been cleavage-arrested shows AChE activity in two blastomeres. However, the compressed eight-cell–stage cleavage-arrested embryos frequently show AChE activity in the four blastomeres that received the myoplasm (Fig. 11.8*B*). These experiments demonstrate that one can confer myogenic properties (i.e., the ability to develop into muscle) on blastomeres by causing them to receive the myoplasm.

The implications of these observations are enormous. They imply that the zygote cytoplasm contains specific morphogenic components and that these components are precisely distributed during cleavage into the correct location in the embryo. After segregation the determinants selectively allow expression of those genes that are necessary for differentiation of specific cell types. Conversely, although the information for the differentiation of these cells exists in the genomes of every cell, it will not be utilized without the stimulation received from this cytoplasm. Consequently, the spatial specification of specific gene expression is a *direct result of cleavage.*

Current speculation as to the identities of the morphogenic determinants of the myoplasm centers on messenger RNA molecules. In Chapter 12 we shall discuss evidence that messenger RNA is heterogeneously distributed in ascidian eggs and may therefore function in determination of developmental fate.

Localization in Spirally Cleaving Zygotes

The pattern of spiral cleavage in annelids and mollusks is very precise and highly predictable. Cell lineage studies have demonstrated that each blastomere has a definite fate in development. The importance of the spatial organization of blastomeres is demonstrated by numerous observations that any disturbance of blastomere organization causes abnormal

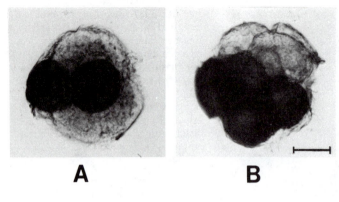

A   B

FIGURE 11.8 Comparison of normal *(A)* and compressed *(B)* eight-cell–stage embryos that had been cleavage-arrested with cytochalasin B and then reacted for acetylcholinesterase activity. The normal embryo shows AChE activity in two blastomeres, while the compressed embryo shows activity in four blastomeres. Scale bar equals 50 μm. (From J.R. Whittaker. 1980. Acetylcholinesterase development in extra cells caused by changing the distribution of myoplasm in ascidian embryos. J. Embryol. Exp. Morphol., *55:* 348.)

development. Furthermore, isolated blastomeres develop as they would in the intact organism, and blastomeres at progressively later cleavages have more limited fates. These experiments confirm that specific developmental potentialities are restricted to particular blastomeres by cytoplasmic segregation during spiral cleavage.

A structure found in certain molluscan and annelid embryos is particularly useful for demonstrating localization. This is the **polar lobe,** which segregates a specialized cytoplasm during cleavage by forming a structure that is especially accessible to experimentation. The formation of the polar lobe during development of the scaphopod mollusk *Dentalium* is shown in Figure 11.9. In the *Dentalium* zygote, a layer of cytoplasm at the vegetal pole protrudes into the bulbous polar lobe that appears prior to the first cleavage. The formation of the polar lobe is analogous to a rubber band constricting a balloon below the equator, changing its shape from that of a sphere to that of an asymmetrical dumbbell (Conrad and Williams, 1974). The first cleavage is initiated at the animal pole and constricts the zygote asymmetrically, leaving the polar lobe attached to the CD blastomere. This stage has been designated the **trefoil stage** because of its three-lobed appearance (Fig. 11.10). After cytokinesis the lobe is resorbed by the CD blastomere, restoring the spherical shape, in a process that is analogous to the relaxation of a rubber band. The CD blastomere receives the cytoplasm contained in the polar lobe, whereas the AB blastomere is deficient in polar lobe cytoplasm. A polar lobe is also formed before the second cleavage. When this division is concluded, the lobe remains attached to the D blastomere only, and C (like A and B) is deficient in polar lobe cytoplasm. Once again, the polar lobe constriction relaxes, restoring the spherical shape of the D blastomere. Cleavage continues in a typical spiral pattern with no further evidence of polar lobe formation, and the embryo develops into a **trochophore larva.**

The developmental significance of the polar lobe was demonstrated by E. B. Wilson (1904), who removed the polar lobe at the trefoil stage.

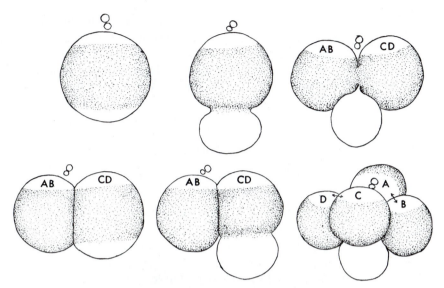

FIGURE 11.9 Cleavage of the mollusk *Dentalium*. (After E.B. Wilson. 1904. J. Exp. Zool., *1:* 1–72. From B.I. Balinsky. 1981. *An Introduction to Embryology.* 5th ed. Saunders College Publishing, Philadelphia.)

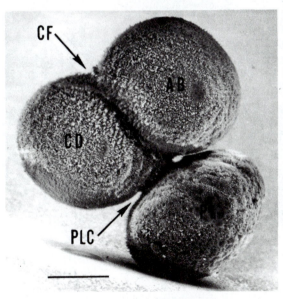

FIGURE 11.10 Scanning electron micrograph of an *Ilyanassa* embryo at the trefoil stage. The AB and CD blastomeres are indicated, as are the polar lobe (PL), the first cleavage furrow (CF), and the polar lobe constriction (PLC). Scale bar equals 50 μm. (From G.W. Conrad, D.C. Williams, F.R. Turner, K.M. Newrock, and R.A. Raff. 1973. Reproduced from *The Journal of Cell Biology*, 1973, vol. 59, pp. 228–233 by copyright permission of The Rockefeller University Press.)

These lobeless embryos continued their spiral cleavage and formed trochophore larvae that were deficient in mesodermal derivatives. An identical type of defect was found in cell separation experiments, in which AB-half embryos developed with a mesodermal deficiency, whereas CD halves developed into complete larvae with mesodermal derivatives.

Wilson also separated blastomeres at the four-cell stage. Only the D blastomere developed into a complete trochophore; the remainder developed into incomplete larvae lacking mesoderm. Wilson concluded that these defects are due to the lack of a *particular kind of cytoplasmic constituent* that is localized in the polar lobe, segregated into the D blastomere, and subsequently parceled out to the cells of the D quadrant. Only if the polar lobe material is received will mesoderm develop from this quadrant.

The most detailed studies on the significance of the polar lobe have been conducted in recent years on another mollusk, the marine snail *Ilyanassa obsoleta.* The *Ilyanassa* embryo develops into a larva called a **veliger,** which includes a shell, digestive tract, foot, otocyst, heart, eyes, and velum. The latter is a ciliated larval structure that serves for motility and food gathering. Lobe-deficient larvae possess cilia, an endodermal mass, pigment cells, nervous system, muscle, and stomach but lack a velum, heart, intestine, otocyst, and eyes (Fig. 11.11); that is, these larvae lack the derivatives of the D quadrant but possess those structures that do not depend upon the polar lobe cytoplasm.

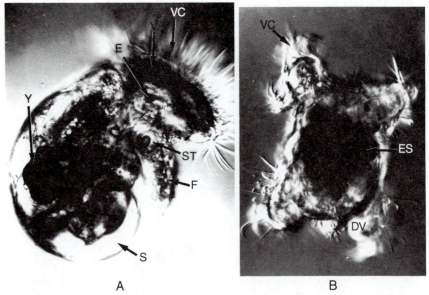

A   B

FIGURE 11.11 Normal and lobeless veliger larvae of *Ilyanassa*. A, Normal veliger larva. B, Veliger larva showing the typical lobeless syndrome. Photographs of living embryos were taken with Nomarski differential interference optics. Note in A the foot (F), the eye (E), the velum (V) with long velar cilia (VC), the shell (S), the otocyst (ST), and the residual yolk (Y). Note in B the disorganized velum (DV) and velar cilia (VC). Out of the plane of this photomicrograph is the everted stomodeum (ES) typical of lobeless larvae. (From K.M. Newrock and R.A. Raff. 1975. Polar lobe specific regulation of translation in embryos of *Ilyanassa obsoleta*. Dev. Biol., *42*: 243.)

Once segregated into the D blastomere, different determinants are progressively parceled out to its mitotic derivatives. As discussed in Chapter 10, subsequent cleavages in spiralians are unequal, producing stacks of micromeres lying above the basal quartet of macromeres. The distribution of the polar lobe determinants into the D micromeres has been studied by a series of cell deletion experiments conducted by Clement (1962). Clement destroyed the D macromere following successive cleavages. By observing which structures developed in the resulting larvae, he was able to ascertain which determinants are passed into the micromeres at each division. The first quartet of micromeres is produced at the third cleavage division (Fig. 11.12). In the D quadrant the blastomeres are designated 1d (micromere) and 1D (macromere). Destruction of 1D leaves the ABC descendants and 1d intact. However, the resultant larva has the same defects as the lobeless larva, indicating that no determinants are segregated from D at this division. However, after the next two divisions, macromere deletion has less drastic implications for development. Destruction of 2D leaves ABC descendants, 1d, and 2d intact, and the resulting larva develops a shell. The most dramatic improvement is produced after the next division. With the addition of 3d the larva contains the velum, eyes, shell, and foot, but the heart and intestine are lacking. After the next division, destruction of 4D yields a complete larva, indicating that the remaining determinants are passed into the 4d micromere. The latter is the precursor of the mesodermal derivatives. The 4D macromere itself is of little developmental significance.

These results indicate that each micromere of the D quadrant (except 1d) receives cytoplasm that contributes to the development of specific larval organs. In some cases this cytoplasm allows D micromeres to interact with micromeres of the other quadrants to induce them to differentiate. For instance, the eyes develop from cells of the A and C quadrants after an interaction with D micromeres. The most clear-cut case of determinants parceled into a micromere to cause them to undergo a specific pathway of development is the formation of 4d. This cell receives from D the capability to form the mesodermal derivatives. At the same time, the D macromere loses this potentiality.

The demonstration of determinants that promote specific kinds of differentiation provides a very graphic illustration of how gene expression can be selectively modulated. Hence, it is very important to identify the determinants and to obtain specific determinants in amounts that are sufficient to examine their characteristics and their mode of action. However, the search for polar lobe determinants has proved to be very difficult. Clement (1968) has demonstrated by centrifugation experiments that the determinants may in fact be in the cortex rather than in the cytoplasm. When *Ilyanassa* eggs are inverted during centrifugation, the cytoplasmic components normally found in the vegetal region are driven to the animal hemisphere. The eggs fragment in the centrifugal field to form animal halves (with vegetal cytoplasm) and vegetal halves. In spite of the displacement of the normal cytoplasm out of the vegetal region, the vegetal halves form polar lobes and ultimately produce com-

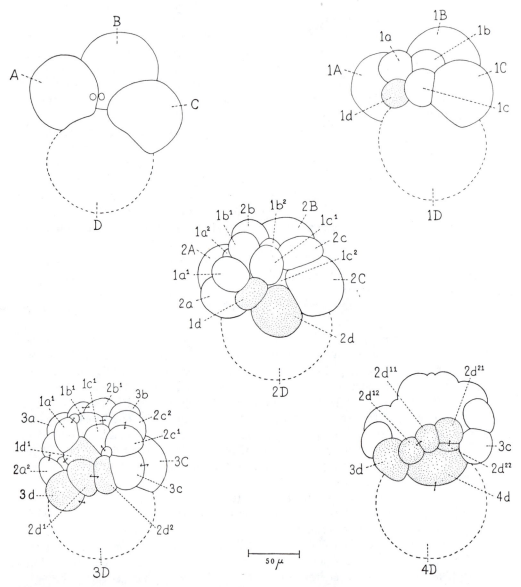

FIGURE 11.12 Stages of removal of D quadrant macromere of *Ilyanassa*. The cell removed is shown in broken outline. All D quadrant micromere derivatives are shaded. *A*, D blastomere removed. ABC combination. *B*, 1D macromere removed. ABC + 1d combination. *C*, 2D macromere removed. ABC + 1d + 2d combination. *D*, 3D macromere removed. ABC + 1d + 2d + 3d combination. *E*, 4D macromere removed. ABC + 1d + 2d + 3d + 4d combination. (From A.C. Clement. 1962. Development of *Ilyanassa* following removal of the D macromere at successive cleavage stages. J. Exp. Zool., *149*: 194.)

plete larvae. Conversely, the animal halves fail to form polar lobes and produce larvae lacking lobe-dependent structures. The failure to displace the determinants by centrifugal force indicates that they are tightly bound to the vegetal pole plasmalemma, presumably as integral components of the cortex.

The search for the polar lobe determinants has also been conducted by electron microscopy. Thus far, no structural correlate of polar lobe determinants has been found in *Ilyanassa*. However, the small polar lobe of the freshwater snail *Bithynia tentaculata* does contain a structure that may be involved in morphogenic determination. This structure, called the **vegetal body,** can be observed with either light or electron microscopy. It is closely associated with the egg surface at the vegetal pole of the egg and becomes segregated into the polar lobe (Fig. 11.13). These are properties that would be expected of the active polar lobe constituents.

The vegetal body is a cup-shaped mass of small vesicles, most of which are filled with a dark-staining substance. The contents of these vesicles have not been positively identified. However, some preliminary experiments indicate that the vesicles may contain RNA. The possibility that the vegetal body contains the polar lobe morphogenic determinants is exciting, since it opens up the possibility that identifiable determinants can be isolated for characterization and analysis of their mode of action in regulating differentiation.

Localization in Ctenophores

Ctenophores are simple, biradially symmetrical animals. The embryonic phase is followed by a larva that possesses two particularly striking kinds of structures that have proved very useful for developmental studies. These are the **comb plates** and the **photocytes.** The comb plates are rows of large cilia, which are locomotory organs. The photocytes are cells specialized to produce light. Cell deletion experiments have established that the comb plate determinants segregate from the photocyte determinants at the third cleavage. Prior to this, isolated blastomeres will produce both structures. After this cleavage (the eight-cell stage), the embryo consists of four inner cells called M blastomeres, and four outer E blastomeres (Fig. 11.14). An isolated E blastomere will continue to cleave as if it were still part of an intact embryo, and comb plates will differentiate from the partial embryo that is formed, whereas photocytes will not. Conversely, a partial embryo produced from cleavage of an isolated M blastomere will produce photocytes but not comb plates.

In contrast to the localization mechanism discussed for ascidians, discrete regions containing cytoplasmic determinants for these potentialities do not preexist in the portions of the zygote that will become segregated into the E or M cells. On the contrary, determinants become localized into those segments *during cleavage* (Freeman, 1976). Cleavage in ctenophores is therefore a period of active rearrangement of determinants, and the orientation of the cleavage planes ensures that the determinants are fixed in the proper spatial arrangement within the body plan of the organism. These observations confirm that cleavage is a critical phase in the establishment of the future pattern of cell differentiation.

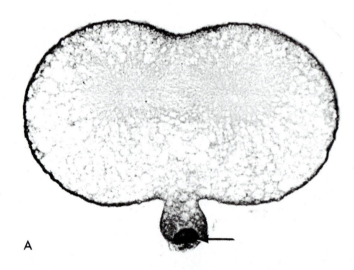

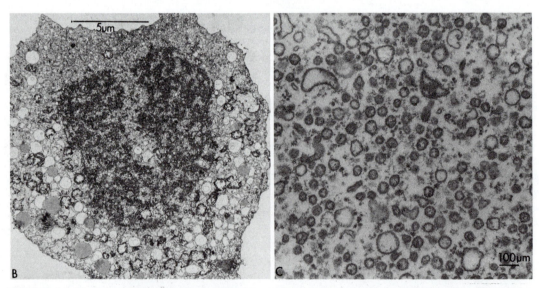

FIGURE 11.13 Polar lobe of *Bithynia*. *A*, Light micrograph of a zygote at first cleavage. The arrow indicates the darkly staining vegetal body in the polar lobe. ×450. *B*, Electron micrograph of polar lobe with vegetal body. *C*, Detail of the vegetal body, showing its small vesicles. Most vesicles are filled with a darkly staining substance that may be RNA. (From M.R. Dohmen and N.H. Verdonk. 1974. The structure of morphogenetic cytoplasm present in the polar lobe of *Bithynia tentaculata* (Gastropoda, Prosobranchia). *J. Embryol. Exp. Morphol., 31:* 425, 426, 428.)

Localization of Germ Cell Determinants

One of the most universal localizations in the animal kingdom is that which determines the germ cell line. Germ cells are determined by a specialized egg cytoplasmic inclusion called the germ plasm, which becomes segregated during cleavage into the primordial germ cells—the precursors of the adult gametes. The concept of a germ plasm can be traced to the nineteenth century. August Weismann proposed that a substance responsible for maintaining germ cell continuity from generation to generation is transmitted from parent germ cells to progeny germ cells. He called this substance *Keimplasma* (germ plasm). As we discussed in Chapter 1, Weismann believed that the germ plasm was the hereditary material itself. It was thought that the germ cell line retained the complete heritage of the species, but the somatic cells lost hereditary qualities by qualitative division of the germ plasm, retaining only those qualities needed for differentiation of specific cell types.

These ideas were the basis for Wilhelm Roux's theory of preformation, which has subsequently been discredited. However, the concept of the germ plasm has persisted—albeit in a highly modified form. Numerous investigators reported in the 1890s and early 1900s that the primordial germ cells (PGCs) are, indeed, set aside from the somatic cells during early development. The PGCs could often be traced through development by their distinctive cytology. These studies also revealed that the PGCs may contain a specialized cytoplasm that is initially present in a portion of the zygote cytoplasm and becomes segregated into the germ cell line. Subsequent experimental work has established that this specialized cytoplasm—the germ plasm—is responsible for determination of the germ cells.

CHROMOSOME DIMINUTION IN *ASCARIS*

The existence of a germ plasm was convincingly demonstrated by a series of descriptive and experimental studies on germ cell origin in *Ascaris megalocephala* by Theodor Boveri. The origin of the germ cell line was traced back to the two-cell stage of cleavage. As shown in Figure 11.15,

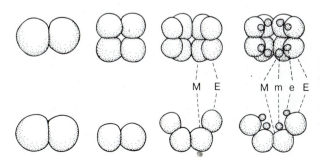

FIGURE 11.14 Diagrammatic representation of the 2-, 4-, 8-, and 16-cell stages of the ctenophore embryo. The top part of the figure depicts these stages from above. The bottom part of the figure shows the same stages from the side. M and E identify the macromeres, while m and e identify the micromeres. (After G. Freeman. 1976. The role of cleavage in the localization of developmental potential in the ctenophore *Mnemiopsis leidyi*. Dev. Biol., *49:* 147.)

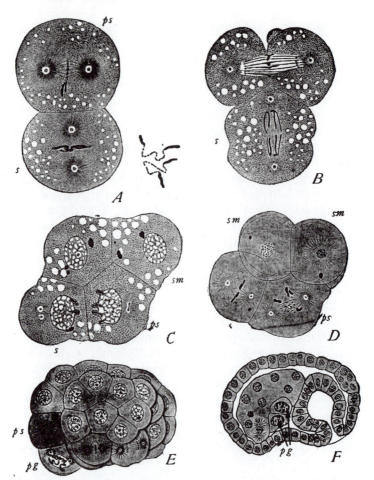

FIGURE 11.15 Early development of *Ascaris megalocephala,* showing chromosome diminution. *A,* Second division in progress. Diminution is occurring in the upper blastomere, which is the primordial somatic cell (ps). Conventional division occurs in the lower blastomere, the stem cell (s). The line drawing next to *A* is a polar view of chromosomes of the upper blastomere showing diminution in progress. *B,* Later stage of second division. Eliminated chromatin remains at the equator of the upper spindle. *C,* At the completion of the second division, the blastomeres undergo a reorganization. The vegetal-most blastomere is labeled s. It will be protected from diminution at the next division, whereas its sister blastomere (ps) will undergo diminution. Note the eliminated chromatin in the cytoplasm of the upper blastomeres (sm). *D,* Third cleavage in progress. Diminution occurs in ps, while s produces another s cell and another ps. Another diminution occurs before the 16-cell stage. At the conclusion of the fourth division (not shown), two cells remain with undiminished nuclei. As the embryo prepares for the fifth division, the final diminution occurs in one cell, and the primordial germ cell divides to produce two primordial germ cells. *E,* An embryo in the process of the fifth division. Most cells have in fact completed cleavage. However, the final diminution is in progress in the ps cell, whereas the primordial germ cell (pg) is dividing normally. *F,* The two pg cells sink into the interior of the embryo, where they will populate the gonad and multiply to form the germ cells. (After Boveri. From E.B. Wilson. 1925. Reprinted with permission of Macmillan Publishing Company from *The Cell in Development and Heredity,* 3rd Edition by E.B. Wilson. Copyright 1925 by Macmillan Publishing Company, renewed 1953 by Anna M.K. Wilson.)

the configuration of the two-cell *Ascaris* embryo is unusual, since the first cleavage division is equatorial (i.e., perpendicular to the animal–vegetal axis). Each of these blastomeres initially has a nucleus identical to that of the zygote, with two long chromosomes. However, as the animal blastomere prepares for second cleavage, the thickened, heterochromatic ends of the chromosomes are shed into the cytoplasm, where they degenerate, and the central euchromatic portions segment into numerous small chromosomes that divide and are distributed to the daughter cells. As a result of this process of chromosome diminution, the two animal blastomeres lose a portion of the genome. They are destined to develop as somatic cells, each of which has a nucleus with numerous small chromosomes.

In contrast, the nucleus of the vegetal blastomere behaves in a conventional manner: The two intact chromosomes divide, and two chromosomes are distributed to each of the daughter cells. Before the third division, one of these vegetal blastomeres undergoes diminution, whereas the other remains undiminished. Thus, the vegetal blastomere has acted as a stem cell, producing one somatic cell and another stem cell. An identical process occurs before the fourth division as well. As a result, at the 16-cell stage, two cells are left with undiminished nuclei, one of which is the progenitor of a cell line that undergoes no further diminution. This cell is the **first primordial germ cell.** At the 32-cell stage it divides into two cells, which multiply to form all the germ cells of the adult. As a result of this process, the germ cells are the only cells that contain the complete genome. Chromosome diminution seems to support Weismann's concepts of germ plasm continuity. However, diminution is an exceptional situation that occurs in only a few isolated species.

Boveri assumed that the behavior of *Ascaris* blastomere nuclei is determined by the cytoplasm in which they lie. Thus, the vegetal cytoplasm protects nuclei from diminution. As it is progressively segregated during cleavage, this cytoplasm becomes restricted to particular blastomeres. Its presence in these blastomeres has two consequences for the cells—it protects their nuclei from diminution, and thus it determines their fate as progenitors of the germ cell line. Therefore, any nucleus would be capable of either diminution or retention, depending upon the cytoplasmic composition; if the cytoplasm were properly manipulated, nuclei that would otherwise undergo diminution could be induced to retain their complete genomes (and develop as germ cells). To test his hypothesis, Boveri centrifuged *Ascaris* eggs before cleavage, thereby shifting the orientation of the spindle relative to cytoplasmic regions. For example, if the spindle is displaced 90° (Fig. 11.16), both of the first two blastomeres would be exposed to the "vegetal cytoplasm." As a result, neither nucleus undergoes diminution during the subsequent division. As the nuclei of the four-cell–stage embryo prepare to divide, diminution occurs in two blastomeres, but two nuclei remain undiminished. The resulting embryo thus has twice the normal number of germ cells. These

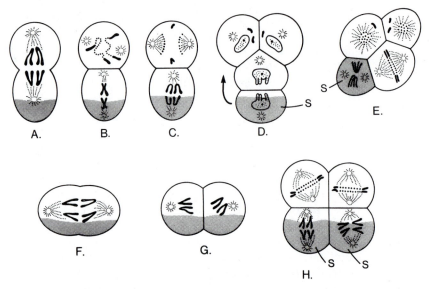

FIGURE 11.16 Distribution of germ plasm (shaded region) during cleavage of normal (*A–E*) and centrifuged (*F–H*) *Ascaris* zygotes. In the former, the distribution of germ plasm to the vegetal blastomere during the first and second cleavages protects them from diminution as the next divisions begin. At the four-cell stage, the normal embryo has one stem cell (S). After centrifugation (*F*), the spindle has been displaced 90°, and cleavage is vertical, causing germ plasm to be distributed to both blastomeres at the first division. Thus, neither cell undergoes diminution. After the second cleavage (*H*), both vegetal cells retain germ plasm and function as stem cells. (After C.H. Waddington. 1966. Reprinted with permission of Macmillan Publishing Company from *Principles of Development and Differentiation* by C.H. Waddington. Copyright © 1966 by C.H. Waddington.)

results confirm that the vegetal cytoplasm contains a "germ plasm" that protects nuclei from diminution.

INSECT POLAR PLASM

The germ plasm of *Ascaris* possesses no obvious physical characteristics that make it visible. However, the germ plasm in many species is marked by the presence of distinct cytoplasmic inclusions. One of the most striking examples is found in insect embryos. A number of nineteenth-century embryologists traced the origin of the germ cells of insects back in early development to the pole cells. As we discussed in the preceding chapter, the pole cells form at the posterior end of the embryo prior to cellularization of the blastoderm. The pole cells contain specially staining granules. Various investigators observed that the granules are present in the egg before the appearance of the pole cells. Then, during pole cell formation, the granules enter them, and hence become segregated from the somatic portion of the embryo. Hegner (1911) suggested that this "polar plasm" contains the determinants of the germ cells, the granules themselves being the visible evidence of a specialized cytoplasm that controls production of primordial germ cells.

To test his hypothesis, Hegner cauterized the posterior end of insect embryos before pole cell formation. As shown in the drawings of Figure 11.17, the cauterized embryos formed a blastoderm and continued to develop, eventually hatching. However, pole cells did not form, and the adults lacked germ cells. Therefore, destruction of the polar plasm removes the germ cell determinants. The early segregation of these determinants into distinct cells has made the insect egg a favorite experimental subject for investigators.

Much of the work on the polar plasm has been conducted on *Drosophila*, which has the added advantage that genetic manipulation can also be used to study germ cell determination. The proximity of the polar plasm of *Drosophila* to the surface of the posterior tip of the egg makes it readily accessible to experimental manipulation, allowing for convincing demonstration of the precise relationship between the polar plasm and germ cell differentiation. Geigy (1931) demonstrated that the germ cell determinants can be destroyed by irradiating the posterior tips of *Drosophila* eggs with ultraviolet light. The irradiated eggs develop into sterile flies. More recently, Okada et al. (1974) have found that

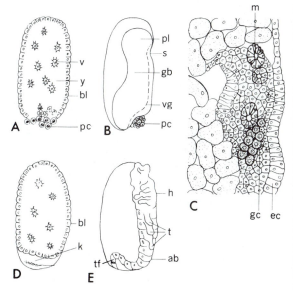

FIGURE 11.17 Experimental test of the proposed function of the polar plasm. *A* to *C*, Normal development of the beetle, *Leptinotarsa decemlineata*. *A*, Longitudinal section of blastoderm stage embryo. Note pole cells (pc). *B*, Superficial view of embryo at later stage, again showing the presence of pole cells. *C*, Longitudinal section through an embryo after pole cells have internalized. Recognizable germ cells are present in the interior. *D* and *E*, Embryos of the same species after cauterization of the posterior end before pole cell formation. *D*, Longitudinal section of cauterized embryo. Blastoderm forms, but pole cells do not appear. *E*, Side view of a later stage. Development is normal with the exception that germ cells are lacking. ab: abdomen; bl: blastoderm; ec: ectoderm; gb: germ band; gc: germ cell; h: head; k: portion of egg killed; m: malpighian tubules; pl: prothoracic lobes; s: stomodeum; t: thoracic appendages; tf: tailfold; v: vitellophage; vg: ventral groove; y: yolk. (After R.W. Hegner. 1911. Biol. Bull., *20:* 237–251. From E.H. Davidson. 1976. *Gene Activity in Early Development*, 2nd ed. Academic Press, New York, p. 283.)

sterility can be prevented in irradiated embryos by microinjection of polar plasm from nonirradiated eggs. The procedure for these experiments is diagramed in Figure 11.18. A section of the posterior end of a normal unirradiated embryo is shown in Figure 11.19A. The comparable region of an irradiated embryo is shown in Figure 11.19B; pole cells are missing. Compare this section with Figure 11.19C, which was prepared from an embryo that was irradiated and subsequently injected with polar plasm. A significant proportion (42%) of the injected embryos developed into fertile adults.

A more elaborate demonstration of the germ cell-determining properties of the posterior polar plasm is summarized in the experiment outlined in Figure 11.20. In this experiment by Illmensee and Mahowald (1974), polar plasm was transplanted into the anterior end of recipient preblastoderm embryos. Normally, the anterior end will not give rise to germ cells. However, the recipient embryos produced cells that had the morphological and ultrastructural appearance (see discussion of ultrastructure that follows) of pole cells.

Since the presence of pole cell-like structures is insufficient evidence that they are in fact primordial germ cells, Illmensee and Mahowald devised an ingenious scheme to demonstrate that the cells have the capacity to develop into functional germ cells. Since it is unlikely that anterior pole cells could reach the gonad, Illmensee and Mahowald transplanted them into the posterior region of genetically different blastoderm-stage embryos. If the transplanted cells developed into functional germ cells, the gonads of the recipients should be mosaics (i.e., they

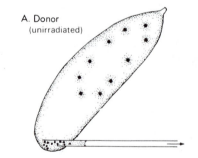

A. Donor
(unirradiated)

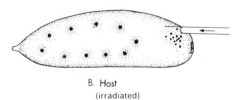

B. Host
(irradiated)

FIGURE 11.18 Diagrams showing technique of polar plasm transplantation. A, The needle is inserted into the donor egg to remove polar plasm. B, The polar plasm is then injected, just off-center, into the posterior pole of the irradiated host egg. (From M. Okada, I.A. Kleinman, and H.A. Schneiderman. 1974. Restoration of fertility in sterilized *Drosophila* eggs by transplantation of polar cytoplasm. Dev. Biol., *37*: 46.)

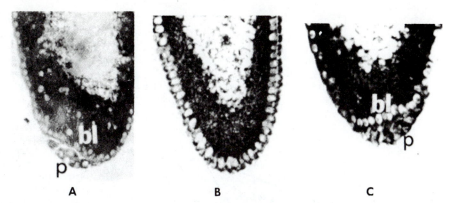

FIGURE 11.19 Longitudinal sections of posterior part of *Drosophila* embryos fixed at the blastoderm stage. *A,* Normal unirradiated embryo. Complete blastoderm (bl) and pole cells (p) are evident. ×300. *B,* Embryo was irradiated during preblastoderm stage. Blastoderm has formed over the entire embryo, but no pole cells are found. ×300. *C,* Embryo was irradiated during preblastoderm stage and subsequently injected with polar plasm. Blastoderm and pole cells appear to be similar to those of normal embryos. ×450. (From M. Okada, I.A. Kleinman, and H.A. Schneiderman. 1974. Restoration of fertility in sterilized *Drosophila* eggs by transplantation of polar cytoplasm. Dev. Biol., *37:* 48).

would have both donor and host germ cells). The mosaicism could then be tested by mating the flies after they reached adulthood.

The embryos that served as recipients for injected polar plasm were from a mutant stock homozygous for two third chromosome mutants—multiple wing hair (*mwh*) and ebony body (*e*). When the nuclei entered the injected polar plasm, they formed pole cells having the *mwh e* genotype. These cells were then transplanted to the posterior end of embryos from a stock with the X chromosome mutant genes—yellow body (*y*), white eyes (*w*), and singed bristles (*sn^3*). After reaching adulthood, these *y w sn^3* flies (which might also contain *mwh e* germ cells) were mated to *y w sn^3* flies. Since the latter flies have the dominant wild-type alleles to *mwh e*, and the *mwh e* germ cells would have the dominant wild-type alleles to *y w sn^3*, the presence of *mwh e* germ cells would be indicated by the production of progeny that are heterozygous for all mutant alleles, and therefore phenotypically wild-type. As shown in Figure 11.20, four recipients of putative pole cells (4% of all cases) produced some wild-type progeny, indicating that the anterior "pole cells" were indeed functional primordial germ cells.

The granular nature of the *Drosophila* polar plasm is due to the presence of structures called **polar granules.** These are dense nonmembranous organelles. Cytochemical studies indicate that the polar granules consist of both RNA and protein. Large polar granules closely associated with mitochondria are found in mature eggs. They retain this configuration for a short time after fertilization (Fig. 11.21*A*), but before pole cell formation they detach from mitochondria and fragment. The

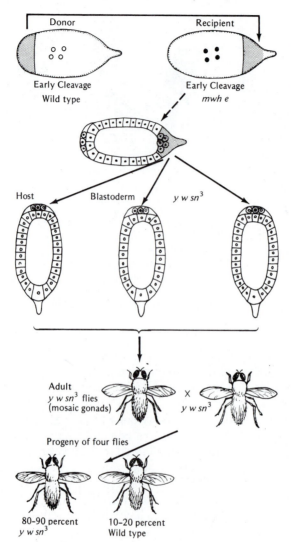

FIGURE 11.20 Experiment demonstrating the determination of germ cells by transplantation of polar plasm in *Drosophila*. Polar plasm from a wild-type fly is withdrawn in a micropipette and injected into the anterior pole of eggs of mutant flies (*mwh e*). The anterior cells formed in this region during cleavage and blastoderm formation are removed and transplanted into the posterior pole of cleaving embryos of another mutant, *y w sn³*. All these host embryos developed into mutant *y w sn³* adults and were mated to other *y w sn³* flies. Of the 92 fertile host flies mated, 88 flies gave only *y w sn³* progeny, indicating they contained no germ cells derived from the transplanted cells. Four flies, however, gave the progeny shown. The high percentages of wild-type progeny obtained from these flies indicated that some of the germ cells came from the transplanted cells and must have had the *mwh e* genotype to account for the appearance of the wild-type flies. (After K. Illmensee and A.P. Mahowald. 1974. Proc. Natl. Acad. Sci. U.S.A., *71:* 1016–1020. From Biology of Developing Systems by Philip Grant. Copyright © 1978 by Holt, Rinehart and Winston. Reprinted by permission of Holt, Rinehart and Winston.)

smaller structures disperse in the polar plasm and become surrounded by polysomes (Fig. 11.21*B*). After pole cell formation the granules reaggregate to form larger structures that appear to have lost their RNA component: They no longer stain for RNA, and the polysomes are no longer present (Mahowald, 1971b). Then, when pole cells migrate to the site of gonad formation, the polar granules again fragment and apparently become replaced by fibrillar derivatives that associate with the nuclear envelope (Fig. 11.22). The fibrillar structures retain this association in the primordial germ cell and oogonial stages. Later, as the definitive

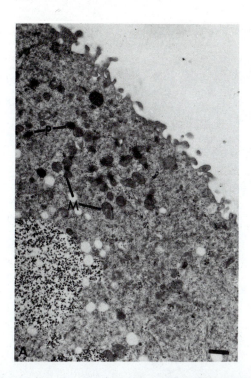

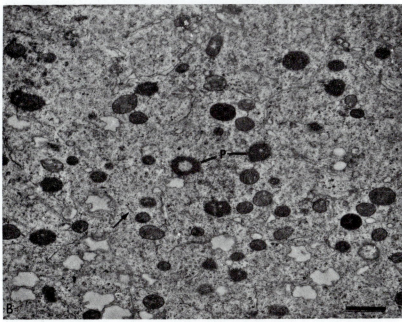

FIGURE 11.21 *Drosophila* polar granules. *A*, Posterior tip of a *D. melanogaster* embryo less than 30 minutes old. Polar granules (P) are still attached to mitochondria (M) and are located within a few micrometers of the surface of the embryo. *B*, Posterior tip of *D. willistoni* embryo before nuclei have reached the polar region. Polar granules are no longer attached to mitochondria. Helical polysomes (arrow) accumulate near the polar granules. Scale bars equal 1.0 μm. (From A.P. Mahowald. 1968. Polar granules of *Drosophila*. II. Ultrastructural changes during early embryogenesis. J. Exp. Zool., *167*: 247.)

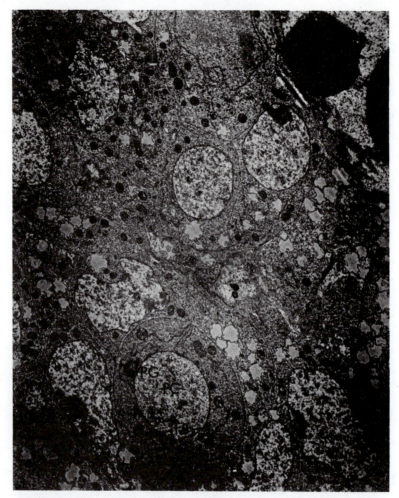

FIGURE 11.22 Low-magnification electron micrograph of a pole cell (PC) of *D. immigrans* immersed in the embryonic tissue of the posterior midgut invagination and surrounding cells. The pole cell is oval in shape, while the adjoining cells have irregular shapes. The distinctive features of the cell are the polar granules (PG) and derivatives of the polar granules (FB) located adjacent to the nuclear envelope. ×5100. (From A.P. Mahowald. 1971a. Polar granules of *Drosophila*. III. The continuity of polar granules during the life cycle of *Drosophila*. J. Exp. Zool., *176*: 331.)

gametes are differentiating, typical polar granules reappear. The polar plasm from these oocytes is already competent to promote germ cell differentiation. This has been demonstrated by Illmensee et al. (1976), who microinjected oocyte polar plasm into the anterior ends of *Drosophila* embryos. Anterior pole cells were induced and were shown to be capable of forming functional gametes.

It is generally thought that the polar granules are the germ cell determinants whose presence promotes cellularization of pole cells and their ultimate differentiation into gametes. Functional polar granules are initially produced during oogenesis and segregated into pole cells, which they induce to form gametes. One of the consequences of gamete differentiation is the formation of new polar granules that will be the germ cell determinants for the next generation, completing the cycle. If this interpretation is correct, the RNA and/or protein constituents of the polar granules may be the active agents that determine germ cell differentiation. Recently, Waring et al. (1978) have isolated polar granules from *Drosophila* embryos. Initial characterizations of the isolated polar granules reveal the presence of a basic protein, which has been tentatively identified as a polar granule–specific protein. The ability to isolate polar granules is a major breakthrough, and further studies on them may establish the functional significance of this protein and other components, such as the RNA.

AMPHIBIAN GERM PLASM

Germ plasm has also been demonstrated in vertebrates, primarily as the result of work on various anuran amphibians. The initial report of a germ plasmlike substance in anurans was the observation by Bounoure (1934) that a material with staining properties similar to those of insect egg polar plasm is present in the vegetal hemisphere of uncleaved frog zygotes. Since Bounoure's original discovery, numerous investigators have reported observing germ plasm in a wide variety of anurans. The fate of the germ plasm has been traced in several of these species, and although details vary, a generalized pattern has emerged (Fig. 11.23). The germ plasm is located in the subcortical cytoplasm at the vegetal pole of the zygote. As cleavage ensues, the substance becomes restricted to cells within the endoderm. Eventually, these cells enter the mesodermal regions from which the gonads will be formed (the **genital ridges**). There they populate the gonads, proliferate as gonia, and differentiate into gametes.

The endodermal origin of anuran germ cells has been confirmed by the experiment with *Xenopus laevis* outlined in Figure 11.24. The endoderm of an embryo from a strain heterozygous for a deletion of the nucleolus was transplanted into the ventral region of an embryo from a wild-type strain. The use of the nucleolar marker gives an unequivocal means of identifying donor and host cells (see Chap. 12 for a detailed discussion of this mutation). These heterozygotes have nuclei with one nucleolus, whereas wild-types have nuclei with two nucleoli. The donor embryos developed into sterile adults—an indication that the source of

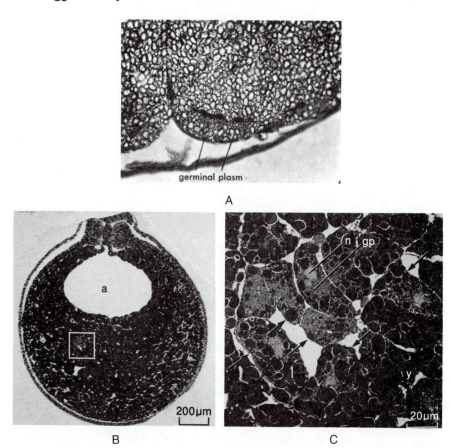

FIGURE 11.23 The germ plasm in the anuran (*Xenopus*) embryo. *A*, Germ plasm (darkly staining material) near the vegetal pole in the two-cell stage. (Courtesy of A.W. Blackler and M. Fischberg. From B.I. Balinsky. 1981. *An Introduction to Embryology*, 5th ed. Saunders College Publishing, Philadelphia, p. 430.) *B*, Transverse section of neurula. Location of primordial germ cells within the endoderm is indicated by the square. a: archenteron. *C*, Higher magnification of region marked in *B*. Germ plasm (gp) is seen in six cells (arrows). n: nucleus; y: yolk platelet. (*B* and *C* from M. Kamimura et al. 1976. Observations on the migration and proliferation of gonocytes in *Xenopus laevis*. J. Embryol. Exp. Morphol., *36:* 199.)

their germ cells has been eliminated. The host embryos developed into fertile adults whose gonads contained gametes derived from the donor strain. When mated to wild-type adults, they produced some heterozygous progeny (i.e., progeny with only one nucleolus). These progeny could only have been derived from germ cells originating from the donor endoderm.

The evidence that the amphibian germ plasm is the actual germ cell determinant is similar to that for *Drosophila*. Bounoure (1937) reported that the germ plasm is sensitive to ultraviolet (UV) radiation, providing a correlation between it and the insect polar plasm. Subsequently, Smith (1966) reported that microinjection of nonirradiated vegetal pole cytoplasm into the vegetal hemisphere of UV-irradiated eggs restores fertility to the recipients. The usual interpretation of irradiation experiments is that the UV destroys the ability of the germ plasm to determine germ cells. However, long-term studies of embryos and tadpoles that had been irradiated during cleavage have revealed that germ cells are present in the endoderm of the embryos, but the migration of these germ cells to the genital ridges of the tadpoles is delayed or inhibited (Züst and Dixon, 1977; Ikenishi and Kotani, 1979). Hence, the role of the germ plasm (which is modified by UV irradiation) may be related to the ability of germ cells to migrate from the endoderm to the genital ridges.

In addition to the proposed functional similarities, the amphibian germ plasm also bears a striking ultrastructural resemblance to insect polar granules. In the unfertilized egg and during early embryogenesis it

A

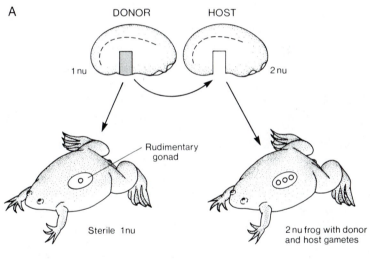

B

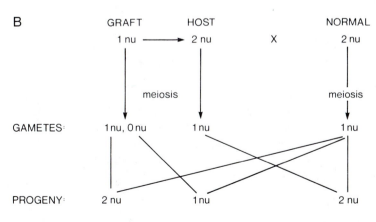

FIGURE 11.24 Experiment to demonstrate the presence of primordial germ cells in the endoderm. *A*, A block of ventral tissue is grafted into an embryo of the same stage. The donor embryo is from a *Xenopus* strain with a single nucleolus, and the host embryo is from a wild-type strain (2 nucleoli). The host metamorphoses into a fertile adult. (After P. Grant. 1978. *Biology of Developing Systems*. Holt, Rinehart and Winston, New York, p. 260.) *B*, The host is mated to a 2-nucleolate frog. It is assumed that some host sex cells remain, resulting in a chimeric gonad. All 1-nucleolate progeny must be derived from the donor endoderm. (After A.W. Blackler and M. Fischberg. 1961. Transfer of primordial germ cells in *Xenopus laevis*. J. Embryol. Exp. Morphol., *9*: 635, and A.W. Blackler. 1966. Embryonic sex cells of amphibia. Adv. Reprod. Physiol., *1*: 17.)

has the appearance of fibrogranular masses associated with mitochondria and surrounded by ribosomes (Fig. 11.25). By the time the primordial germ cells enter the genital ridges, the granular germ plasm can no longer be detected. Instead, these cells have fibrous material attached to mitochondria located adjacent to the nuclear envelope (Fig. 11.26). This so-called **nuage material** is also present in oogonia and persists through the early stages of oogenesis (see Chap. 7) but can no longer be found in postvitellogenic oocytes. The nuage is also similar to the material associated with the nuclei of oogonia and young oocytes of *Drosophila*. It has been proposed that the nuage material is derived from the germ plasm, but no direct evidence of such a transition is available. Typical germ plasm reappears in the vegetal cytoplasm during hormone-induced maturation. The available evidence suggests that the germ plasm—like the *Drosophila* polar granules—is produced during oogenesis (assuming its granular appearance and locating at the vegetal pole at maturation) and is localized into endodermal cells that will, as a result, form the germ cell line. The primordial germ cells migrate into the genital ridges to form the gonia that proliferate and differentiate into functional germ cells. After the germ cells reach the genital ridges, the germ plasm is no longer present; it has presumably been transformed into nuage material associated with the nucleus. Finally, when the oocytes undergo maturation, germ plasm reappears, and the cycle has been completed.

One difficulty with the preceding scheme is the lack of certainty that the germ plasm is transformed into the nuage. Until the transition of germ plasm into nuage has been documented, we must consider this interpretation to be speculative. The association of nuage with the nuclear envelope is intriguing. It is tempting to speculate that this association has functional significance in germ cell differentiation. Equally intriguing is the fact that nuage is present in germ cells of several divergent groups of organisms, including mammals (Eddy, 1975). It would therefore appear that the nuage has a nearly universal role to play in gametogenesis. Its retention during evolution indicates that this role is a vital one. Therefore, establishment of this role is an important priority in developmental biology research.

Localization in "Regulatory" Embryos

Developmental schemes in which the pattern of cleavage segregates localized morphogenic determinants into specific blastomeres have classically been called **mosaic development.** Cleavage in mosaic embryos is said to be **determinate,** since the intervention of cleavage furrows determines the fates of different regions of the zygote. Typical mosaic developmental patterns are found in ascidians and in spirally cleaving embryos (e.g., mollusks, annelids). A consequence of mosaicism is that the destruction or rearrangement of blastomeres will result in the loss or displacement of the correlative body regions.

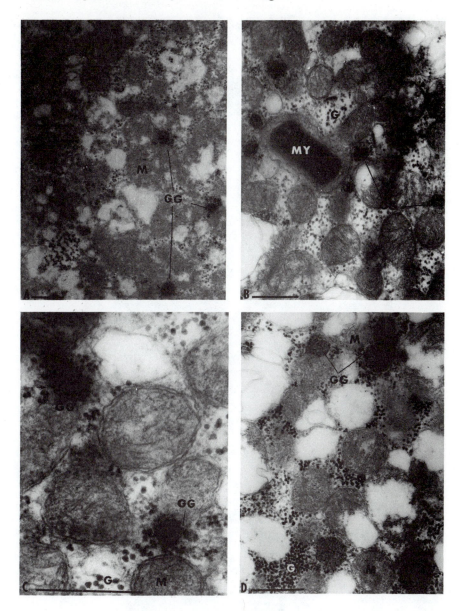

FIGURE 11.25 Ultrastructure of germinal plasm region in fertilized eggs. *A*, 1.5 hours after fertilization; *B*, two-cell stage; *C*, four-cell stage; *D*, 16-cell stage. These areas contain germinal granules (GG), mitochondria (M), mitochondria-containing yolk (MY), and glycogen (G). Arrow in *C* points to fibril connecting germinal granule to one of the ribosomes that surround it. Scale bar equals 0.5 μm. (From M.A. Williams and L.D. Smith. 1971. Ultrastructure of the "germinal plasm" during maturation and early cleavage in *Rana pipiens*. Dev. Biol., *25:* 573.)

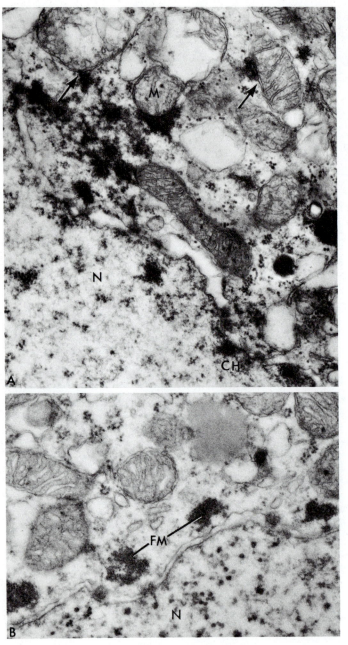

FIGURE 11.26 When the primordial germ cells reach the genital ridges, the cells lack the granular germinal plasm. Instead, fibrous material (FM) is seen adjacent to the nuclear envelope. In A, some of this material (arrows) is intimately associated with mitochondria (M). In B, the mitochondrial association is not seen. The fibrous material is not associated with clusters of ribosomes. CH: Chromatin; N: nucleus. A: ×26,800. B: ×39,000. (From A.P. Mahowald and S. Hennen. 1971. Ultrastructure of the "germ plasm" in eggs and embryos of Rana pipiens. Dev. Biol., 24: 46.)

By way of contrast, cleavage in echinoderm and amphibian embryos is frequently described as being **indeterminate,** since these embryos can be subjected to considerable experimental abuse and still regulate to form normal organisms. This pattern of development is thus called **regulatory.** As an example of regulation, isolated blastomeres of two-cell amphibian or sea urchin embryos will develop normally (Chap. 1). Clearly, the prospective right and left halves have the potential to form their opposite number. Does this mean that there is no localization of morphogenic determinants in so-called regulatory embryos? It does not, and as we shall see, the use of the terms "mosaic" and "regulatory" overstates the differences between these developmental strategies and is probably misleading. For example, we have already discussed the germ plasm in the amphibian zygote, which is an obvious case of localization of specialized cytoplasm. But in addition, there are even more fundamental localizations in these embryos that can be traced to events occurring immediately after fertilization. At this time a pigmentation change is elicited, reflecting a major reorganization in the egg that establishes the axial organization of the embryo.

As described in Chapter 7 and depicted in Figure 11.27A, the unfertilized amphibian egg appears dark in the animal hemisphere and light in the vegetal hemisphere. This difference in coloration is due to the presence of melanin pigment granules in the cortical cytoplasm of the animal hemisphere and their absence in the vegetal hemisphere. The unfertilized egg is radially symmetrical about its animal–vegetal axis. Fertilization occurs in the animal hemisphere. The sperm entry point in fertilized *Xenopus laevis* eggs is marked by a local accumulation of pigment. Sperm entry elicits an asymmetrical shift of the cortex toward the fertilization site that draws pigment granules toward this site (Palaček et al., 1978), leaving a crescent-shaped region of reduced pigmentation on the side opposite (Fig. 11.27B). This depigmented region is called the gray crescent. Although gray crescent formation is a generalized response to fertilization in amphibian eggs, the details of the cortical rearrangements that result in its formation may vary.

The zygote is now bilaterally symmetrical, and the gray crescent is an external manifestation of the symmetrization. However, as we shall discuss subsequently, basic internal changes have also occurred in egg organization during the cortical shift. The bilateral symmetry of the zygote is manifested by the relationship of body organization to the region marked by the gray crescent. In frog embryos such as *Xenopus laevis* (Fig. 11.27C–F) the first cleavage division bisects the gray crescent and passes through the sperm entry point; the right half of the embryo will be derived from one of the two blastomeres, whereas the other gives rise to the left half of the embryo. The embryo also has dorso-ventral organization; the blastopore will be formed at the lower limit of the gray crescent region, and the dorsal aspect of the embryo will originate from the gray crescent side of the embryo, whereas the opposite side forms ventral structures. As we shall discuss in Chapter 13, the dorsal lip of the blastopore determines the developmental fates of cells that come under its influence during gastrulation, thus establishing the basic axial organization of the developing embryo. Consequently, a fateful chain of events is set in motion by the entry of the sperm.

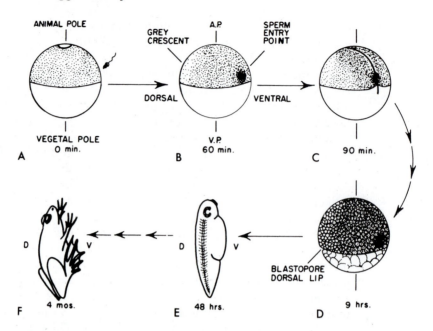

FIGURE 11.27 Formation of the dorso-ventral axis in the *Xenopus* zygote. *A*, A sperm is approaching the unfertilized egg. At this time the egg is radially symmetrical. *B*, Bilateral symmetry has been established as evidenced by the appearance of the gray crescent. *C*, The first cleavage furrow passes through the gray crescent and the sperm entry point. *D*, At gastrulation the dorsal lip of the blastopore forms at the lower limit of the gray crescent. *E* and *F* show the tadpole and frog, respectively, which retain the dorsoventral orientation established in the fertilized egg. (From M. Kirschner et al. 1980. Initiation of the cell cycle and establishment of bilateral symmetry in *Xenopus* eggs. *In* S. Subtelny and N.K. Wessells (eds.), *The Cell Surface: Mediator of Developmental Processes.* Academic Press, New York, p. 198.)

As we discussed earlier the amphibian embryo has been described as being regulatory in the sense that isolated blastomeres of two-cell embryos will develop normally. However, an experiment by Spemann demonstrates the importance of the plane of the first division in achieving this result (Fig. 11.28). In newts the plane of the first cleavage may be either perpendicular to the gray crescent or parallel to it. If perpendicular, the gray crescent is bisected and both blastomeres receive a portion of it. Ligation of embryos in this plane produces two normal embryos. However, if the first cleavage is parallel to the gray crescent, one blastomere receives all of it. If embryos are ligated in this plane, only the half receiving the crescent is capable of normal development. The other half forms only structures that are typical of the ventral portion of an embryo. Thus, the prospective dorsal region of the embryo must be present for formation of a complete embryo. The difference between regulatory and mosaic embryos is therefore a matter of degree. Each

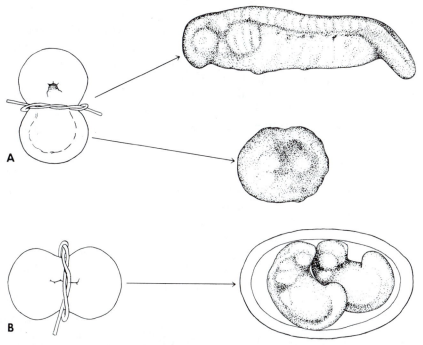

FIGURE 11.28 Spemann's ligation experiment. The plane of ligation of the newt zygote affects the developmental capacities of the separated halves. In *A*, ligation is parallel to the gray crescent, whereas in *B* the gray crescent is bisected by the ligature. After constriction in *A*, the half containing the gray crescent develops normally, but the half lacking the crescent fails to form a complete embryo. However, constriction in *B* produces two complete embryos. (After H. Spemann. 1938. *Embryonic Development and Induction.* Yale University Press, New Haven, Conn. Copyright © 1938, Yale University Press. From B.I. Balinsky. 1981. *An Introduction to Embryology,* 5th ed. Saunders College Publishing, Philadelphia, p. 126.)

blastomere of a two-cell amphibian embryo can regulate to form a complete embryo, but only if each receives elements of the prospective dorsal region.

We shall now examine in more detail the possible basis for the rearrangements at fertilization that produce the indispensable prospective dorsal region. Concurrent with formation of the gray crescent the egg cytoplasm is modified from a radial symmetry to a bilateral symmetry. One result of the cytoplasmic reorganization is the segregation of a specialized yolk-free cytoplasm to the dorsal region (Herkovits and Ubbels, 1979). Presumably, the cortical shift that produces the gray crescent also redistributes cytoplasmic contents. The sperm aster may be responsible for triggering these events (Gerhart et al., 1981). Gerhart et al. have proposed that the cytoplasmic organization, rather than the cortical region marked by the gray crescent, is responsible for establishing the dorso-ventral organization of the embryo. They base their conclusions partially upon experiments in which the blastopore site and hence the prospective dorsal side of the embryo appear on the opposite side of the egg from their normal position after the eggs are rotated, which allows gravity to reorganize the cytoplasm. They suggest that gravity has relocated the cytoplasmic region that is responsible for specifying the dorsal side of the egg. In normal development, then, the fertilizing sperm sets the following chain of events into motion that leads to axial organization of the egg: (1) Sperm entry causes a cortical contraction that produces the gray crescent and reorganizes the cytoplasm, segregating a specialized cytoplasm to the prospective dorsal side. (2) The blastopore forms on the dorsal side of the embryo, serving as the site for involution of cells during gastrulation. During gastrulation, an influence from the dorsal lip of the blastopore specifies the developmental fates of cells as the axial organization of the embryo is established. In Chapter 13, we shall return to these latter events and discuss them in more detail.

Isolated sea urchin blastomeres will also develop normally if separation occurs after the first cleavage or even after the second cleavage. However, the third cleavage, which is perpendicular to the animal–vegetal axis, may segregate regions with different developmental properties. If an eight-cell–stage embryo is ligated through the animal–vegetal axis, each group of four cells develops into a normal embryo. However, if the embryo is divided so that the upper quartet of blastomeres is separated from the lower quartet, neither the animal half nor the vegetal half can produce a normal embryo. The animal half forms primarily ectodermal derivatives, whereas the vegetal half forms primarily endodermal structures. Apparently the animal and vegetal hemispheres contain different determinants, which are first segregated from one another at the eight-cell stage. Further segregation of developmental potential occurs at the fourth cleavage division. This division is unequal: eight mesomeres are produced in the animal hemisphere, whereas the vegetal quartet divides to produce four micromeres at the vegetal pole and four macromeres between the micromeres and mesomeres (see Chap. 10). The developmental fates of the descendants of these three cell types differ, and isolates of the cell types behave individually much like they would in an intact embryo. For example, micromeres, which form skeletal structures during normal larval development, produce these same structures when isolated and maintained in vitro (Okazaki, 1975a, b).

These observations suggest that distinct morphogenic determinants are unequally distributed during sea urchin cleavage. As is the case with the amphibian embryo, specification of blastomere fate by cytoplasmic heterogeneity is not so extensive in echinoderms as in typical mosaic embryos, but cytoplasmic heterogeneities do play an important role in their development. Furthermore, cleavage is involved in segregating these different regions from one another.

11–2. EXTRINSIC DETERMINATION OF BLASTOMERE FATE

In contrast to the determination of blastomere fate by inclusion of specific cytoplasmic constituents, determination is also induced by blastomere response to external influences. In this mode of development the fate of a blastomere or a region of a zygote depends upon its position relative to the external influence. We shall discuss here two examples of this mode of determination: establishment of polarity in fucoid algae and determination of blastomeres in the mammalian embryo.

Polarity in Fucus and Related Brown Algae

Polarity is an essential property of developing systems. In animal zygotes, polarity is usually established by the structural arrangement of egg organelles, which occurs during oogenesis or after fertilization. Likewise, in many plants, polarity has already been established in the egg before fertilization, and the site of the first division is entirely predictable. Conversely, the *Fucus* egg is radially symmetrical and has no obvious indication of regional differences. The latter do not manifest themselves until several hours after fertilization, when formation of the rhizoid is initiated. The presumptive rhizoid cytoplasm sequesters certain organelles, including Golgi bodies and mitochondria (Quatrano, 1972). Thus, when the first cell division occurs, two quite dissimilar cells are formed, and the nuclei of these cells come under entirely different sets of influence. These different influences are apparently responsible for establishing the two contrasting patterns of cell differentiation in the derivatives of these two cells (i.e., one is the precursor of thallus cells, the other is the precursor of cells composing the holdfast).

The site of rhizoid formation is initially determined by the site of sperm entry. However, this determination is extremely labile and can be overidden by a variety of environmental factors (Torrey, 1967; Quatrano, 1978; Robinson and Cone, 1980):

1. In a population of closely grouped zygotes, the rhizoids all grow toward the center. This phenomenon is known as the **group effect.**
2. The rhizoid always forms toward the direction of lower pH in a pH gradient.
3. The rhizoid forms on the warmer side in a temperature gradient.
4. When illuminated from one side, the rhizoid always develops on the dark side.
5. The rhizoid forms on the positive side of a voltage gradient.
6. In a Ca^{++} or K^+ gradient, the rhizoid appears on the more concentrated side.

The light effect is particularly easy to control and has provided a simple means for inducing polarity for experimental analysis of the polarization phenomenon. The ultimate goal of this research is to discover how the zygote responds to the outside influence to form an axis that in turn determines the developmental fates of the derivatives of regions of the zygote. The results of these investigations may also clarify the polarization that occurs during gametogenesis of animals and higher plants.

Unilateral illumination of the *Fucus* zygote can determine polarity as early as four hours after fertilization, and the zygote remains sensitive to light for several hours. Light sensitivity reaches a peak at eight hours and declines thereafter. The long-term lability of polarity means that if an initial light treatment is followed by a second exposure rotated 180° from the original source, the zygotes will be oriented by the second illumination. After polarity has been fixed, there is a lag period while the zygote prepares to form the rhizoid.

The mechanism of response to polarizing influences is suggested by a series of experiments by Lionel Jaffe and his associates. By placing a number of *Fucus* zygotes in a capillary tube and illuminating them all from the same side, Jaffe was able to produce a series of zygotes polarized in the same direction (Fig. 11.29). He then measured the electrical currents in the tube and discovered that at the time of fixation of polarity,

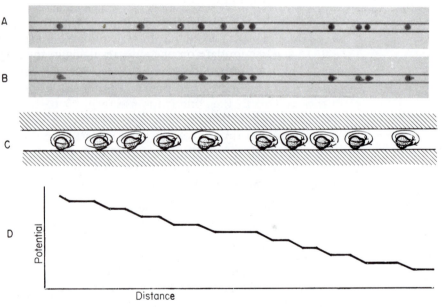

FIGURE 11.29 Detection of current flow through polarizing *Fucus* zygotes. *A,* Zygotes in a capillary tube are illuminated from the left side. *B,* Several hours later, rhizoids appear on the dark (right) side of all zygotes. *C,* Highly schematic view of inferred current pattern in the tube. *D,* Schematic graph of inferred change of potential along the tube. (From L.F. Jaffe. 1966. Electrical currents through the developing *Fucus* egg. Proc. Natl. Acad. Sci. U.S.A., *56:* 1103.)

the rhizoid end of the tube becomes electronegative. Apparently a minute current passes through each zygote. Since the zygotes are arranged in series in the tube, the current increases as it is transmitted down the tube and becomes strong enough to be detected. The current continues to flow after light exposure, indicating that it is self-generated during polarization. Because the amount of current generated by each egg is so small, it is difficult to detect the current of each individual zygote. However, highly sensitive electrodes have recently been used to confirm the current flow through individual zygotes (Nuccitelli and Jaffe, 1974). This current appears to enter the rhizoid end and leave through the opposite end; the current then reenters the rhizoid end, completing the loop.

Current flow in a living cell is dependent upon movement of ions across the plasma membrane. Thus, in *Fucus*, positive ions apparently enter the zygote on the prospective rhizoid side and carry the current through the cell. A study conducted by Robinson and Jaffe (1975) suggests that Ca^{++} ions are reponsible for carrying at least a portion of the current. By using radioactive calcium ions ($^{45}Ca^{++}$), they found that Ca^{++} enters the prospective rhizoid side of a polarizing zygote and effluxes through the prospective thallus side. Therefore, the critical event in polarization may be the localization of sites permeable to Ca^{++} (Ca^{++} channels) on the membrane. Recent evidence (Nuccitelli, 1978) indicates that the Ca^{++} channels in the membrane are mobile, and the effect of a polarizing influence such as light is to cause these channels to redistribute and accumulate on one side, resulting in membrane asymmetry and an electrical potential gradient across the cell.

After the current is established in the cell, how can the regionalization be established? Two possible mechanisms have been proposed. The entrance of a large excess of Ca^{++} could cause the accumulation of negatively charged secretory vesicles at the site of influx by a process of electrophoresis. In fact, both light and electron microscopic studies do confirm that vesicles accumulate at the site of rhizoid formation. After accumulating at this site, these vesicles secrete their contents, which are apparently necessary to soften existing cell wall and build the expanding cell wall of the growing rhizoid tip. Another possible mechanism for localization of vesicles is that the current orients microfilaments that produce mechanical force to move the vesicles to the presumptive rhizoid site. Studies in progress may determine whether either of these mechanisms is involved. But regardless of the mechanism of cytoplasmic rearrangement, it seem to be well established that a reorganization of ion permeability sites causes polarization in response to external stimuli. Virtually nothing is known about the major result of polarization, that is, the differential control of gene expression in the derivatives of the two cytoplasmic regions. However, now that the mechanism of polarization is apparently known, investigation of this problem should be much easier.

Determination of Mammalian Blastomeres

The initial diversification of cell types in mammalian development occurs with the formation of trophoblastic cells and cells of the inner cell mass (ICM). These two groups of cells develop into entirely different tissues; that is, the ICM forms the embryo proper and the trophoblast contributes to the placenta. Does their determination depend upon intrinsic or extrinsic factors?

One approach to this question is the traditional blastomere isolation experiment. Using this approach, various investigators have shown that individual blastomeres of the two-cell, four-cell, and eight-cell stages have the capacity to form both trophoblastic and ICM cells (see Herbert and Graham [1974] for review). These experiments are technically difficult, since isolated blastomeres have a limited ability to survive. A more successful approach has been to combine blastomeres of two or more embryos to determine whether the resultant **chimera** can regulate to form a normal embryo. This approach is facilitated in experiments with mice by the availability of genetic markers that allow the investigator to distinguish between cells derived from the different donor embryos. According to Mintz (1965), when two or more embryos are combined, cells of the resultant chimeras do not sort out (i.e., presumptive inside and presumptive outside cells do not migrate to their normal positions during formation of the chimera). Instead, the cells remain in place, and those that happen to be positioned on the outside form the trophoblast of the chimera, whereas those on the inside become ICM cells.

In a different experimental approach Stern (1972) disrupted mouse embryos by separating the cells from one another. She then allowed the cells to reaggregate and cultured the reaggregated structures. In spite of complete disorganization, embryos up to the early blastocyst stage could reassemble and subsequently form normal blastocysts with an ICM and trophoblast. Since cells apparently reaggregated without regard to their original position in the embryo, it is clear that the determination of ICM and trophoblastic cells is not due to preexisting cytoplasmic determinants but is imposed upon the cells by some condition that exists *during development.*

Tarkowski and Wroblewska (1967) proposed that the major factor that determines whether a cell will become either a trophoblastic cell or an ICM cell is its relative position in the embryo. Thus, "inside" cells become part of the ICM and "outside" cells become trophoblastic cells. This proposal has been subjected to direct experimental test, and the results support their hypothesis. In these experiments, cleavage stage embryos are dissociated, and isolated blastomeres are placed either on the outside or the inside of a host embryo. As expected, blastomeres placed on the outside contribute to the trophoblast with high frequency,

STEP 1   STEP 2   STEP 3   STEP 4   STEP 5

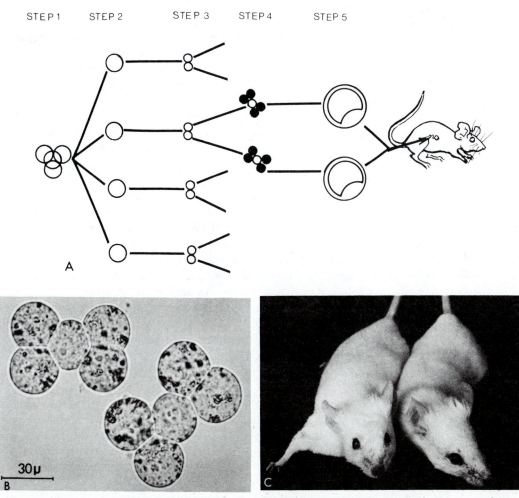

A

B   30µ

C

FIGURE 11.30 Experiment to test the "inside-outside" hypothesis. *A,* Experimental design. The donor embryo is from an albino strain (designated "A"), and the host blastomeres are from a strain with black pigmentation (designated "B"). Step 1. The donor, "A"-type embryo is obtained at the four-cell stage and the zona is removed. Step 2. The blastomeres are dissociated. Step 3. They divide to the eight-cell stage, giving four "octet pairs" of the blastomeres. Step 4. Each pair of blastomeres is dissociated and each individual blastomere is combined with four eight-cell-stage carrier, "B"-type blastomeres. Step 5. The pairs of composite embryos are cultured to the blastocyst stage. Each pair of blastocysts is transferred to one of the uterine horns of a pseudopregnant recipient. *B,* An "octet pair" of composite embryos. The "A" type donor blastomeres are each surrounded by four "B" type blastomeres. *C,* Two mice recovered from this experiment. Each mouse was derived from one of the eight-cell-stage blastomeres of a particular "A"-type embryo, combined with four eight-cell-stage "B"-type blastomeres. (From S.J. Kelly. 1977. Studies of the developmental potential of 4- and 8-cell stage mouse blastomeres. *J. Exp. Zool., 200:* 373–375.)

whereas those placed inside usually contribute to the embryo proper (Hillman et al., 1972; Kelly, 1977).

A representative experiment of the latter type is outlined in Figure 11.30. Here, four-cell–stage embryos of an albino mouse strain are dissociated, and the individual blastomeres are allowed to divide once more. The blastomere pairs are isolated, and each individual blastomere (corresponding to the eight-cell stage) is placed in the center of four blastomeres obtained from an eight-cell–stage embryo of a black mouse strain (Fig. 11.30B). The albino and black strains also have different electrophoretic variants of the enzyme glucose phosphate isomerase. Therefore, samples of tissues can be taken to determine whether they are derived from the inner blastomere or the outer blastomeres. A total of 32 embryos were recovered and analyzed electrophoretically, and in all but one case the inner blastomere made a contribution to the embryo proper. Six embryos were allowed to develop to term. Of these, all but one showed albino characteristics. Two of these mice are pictured in Figure 11.30C. Their coat color is clearly albino, confirming that the inner cell contributed to formation of the embryo proper. The results of this type of experiment indicate that the blastomeres of the cleavage-stage mammalian embryo are undetermined and that their developmental fate depends upon their relative position within the multicellular mass.

The impressive regulatory capacity of mammalian embryos has recently been exploited by researchers analyzing the developmental potential of nonembryonic cells. For example, malignant mouse teratocarcinoma cells have been injected into mouse host blastocysts. Using the differences in genetic composition as markers, investigators have definitively demonstrated that the malignant cells can contribute to the normal development of the chimeric embryos (Mintz and Illmensee, 1975; Brinster, 1976). The implications of these experiments are immense, for they demonstrate that the genome of the donor teratocarcinoma cell can be reprogrammed so that its expression of malignant properties is replaced by the controlled pattern of gene expression that leads to normal cell differentiation.

An even more complex experiment was reported by Illmensee et al. (1978). They fused mouse teratocarcinoma cells with human cells to obtain hybrids with human chromosomes in their nuclei. Injection of the hybrid cells into mouse blastocysts produced chimeric mice in which the hybrid cells contributed to a number of tissues. One of these mice is shown in Figure 11.31. The host blastocyst was from a black strain of mice. The coat of this mouse has discrete white patches that are due to the injected hybrid cell. It is uncertain whether the human genes in the hybrid cells were also expressed. However, this experiment opens up the possibility of examining human gene expression during mouse development.

FIGURE 11.31 Coat mosaicism in chimeric mouse developed from a blastocyst of a black mouse strain after micro-injection of a human-mouse hybrid cell. In addition to the expected black coat phenotype of the recipient, this mouse also exhibited several white patches on the back, tail, and feet and a large white clone extending from the midlateral to the ventral side, all of which derived from the injected hybrid cell and comprised about 20% of the total coat. (From K. Illmensee, P.C. Hoppe, and C.M. Croce. 1978. Chimeric mice derived from human-mouse hybrid cells. Proc. Natl. Acad. Sci. U.S.A., *75:* 1916.)

REFERENCES

Balinsky, B.I. 1981. *An Introduction to Embryology,* 5th ed. Saunders College Publishing, Philadelphia.

Blackler, A.W. 1966. Embryonic sex cells of amphibia. Adv. Reprod. Physiol., *1:* 9–28.

Blackler, A.W., and M. Fischberg. 1961. Transfer of primordial germ-cells in *Xenopus laevis.* J. Embryol. Exp. Morph., *9:* 634–641.

Bounoure, L. 1934. Recherches sur la lignée germinale chez la grenouille rousse aux premier stades au développement. Ann. Sci. Natur. Zool., 10ᵉ Ser., *17:* 67–248.

Bounoure, L. 1937. Le sort de la lignée germinale chez la grenouille rousse aprés l'action des rayons ultraviolets sur le pôle inferieur de l'oeuf. C. R. Acad. Sci. Paris, *204:* 1837–1839.

Brinster, R.L. 1976. Participation of teratocarcinoma cells in mouse embryo development. Cancer Res., *36:* 3412–3414.

Clement, A.C. 1962. Development of *Ilyanassa* following removal of the D macromere at successive cleavage stages. J. Exp. Zool., *149:* 193–215.

Clement, A.C. 1968. Development of the vegetal half of the *Ilyanassa* egg after removal of most of the yolk by centrifugal force compared with the development of animal halves of similar visible composition. Dev. Biol., *17:* 165–186.

Conklin, E.G. 1905. The organization and cell-lineage of the ascidian egg. J. Acad. Nat. Sci. Philadelphia, Ser. 2, *13:* 1–119.

Conklin, E.G. 1931. The development of centrifuged eggs of ascidians. J. Exp. Zool., *60*: 1–119.

Conrad, G.W., et al. 1973. Microfilaments in the polar lobe constriction of fertilized eggs of *Ilyanassa obsoleta.* J. Cell Biol., *59*: 228–233.

Conrad, G.W., and D.C. Williams. 1974. Polar lobe formation and cytokinesis in fertilized eggs of *Ilyanassa obsoleta.* I. Ultrastructure and effects of cytochalasin B and colchicine. J. Cell Biol., *36*: 363–378.

Davidson, E.H. 1976. *Gene Activity in Early Development,* 2nd ed. Academic Press, New York.

Dohmen, M.R., and N.H. Verdonk. 1974. The structure of a morphogenetic cytoplasm, present in the polar lobe of *Bithynia tentaculata* (Gastropoda, Prosobranchia). J. Embryol. Exp. Morphol., *31*: 423–433.

Eddy, E.M. 1975. Germ plasm and the differentiation of the germ cell line. Int. Rev. Cytol., *43*: 229–280.

Freeman, G. 1976. The role of cleavage in the localization of developmental potential in the ctenophore *Mnemiopsis leidyi.* Dev. Biol., *49*: 143–177.

Geigy, R. 1931. Action de l'ultra-violet sur le pole germinal dans l'oeuf de *Drosophila melanogaster* (Castration et mutabilité). Rev. Suisse Zool., *38*: 187–288.

Gerhart, J. et al. 1981. A reinvestigation of the role of the grey crescent in axis formation in *Xenopus laevis.* Nature (Lond.) *292*: 511–516.

Grant, P. 1978. *Biology of Developing Systems.* Holt, Rinehart and Winston, New York.

Hegner, R.W. 1911. Experiments with chrysomelid beetles. III. The effects of killing parts of the eggs of *Leptinotarsa decemlineata.* Biol. Bull., *20*: 237–251.

Herbert, M.C., and C.F. Graham. 1974. Cell determination and biochemical differentiation of the early mammalian embryo. *In* A.A. Moscona and A. Monroy (eds.), *Current Topics in Developmental Biology,* Vol. 8. Academic Press, New York, pp. 151–178.

Herkovits, J., and G.A. Ubbels. 1979. The ultrastructure of the dorsal yolk-free cytoplasm and the immediately surrounding cytoplasm in the symmetrized egg of *Xenopus laevis.* J. Embryol. Exp. Morph., *51*: 155–164.

Hillman, N., M.I. Sherman, and C. Graham. 1972. The effect of spatial arrangement on cell determination during mouse development. J. Embryol. Exp. Morphol., *28*: 263–278.

Ikenishi, K., and M. Kotani. 1979. Ultraviolet effects on presumptive primordial germ cells (pPGCs) in *Xenopus laevis* after the cleavage stage. Dev. Biol., *69*: 237–246.

Illmensee, K., P.C. Hoppe, and C.M. Croce. 1978. Chimeric mice derived from human–mouse hybrid cells. Proc. Natl. Acad. Sci. U.S.A., *75*: 1914–1918.

Illmensee, K., and A.P. Mahowald. 1974. Transplantation of posterior polar plasm in *Drosophila.* Induction of germ cells at the anterior pole of the egg. Proc. Natl. Acad. Sci. U.S.A., *71*: 1016–1020.

Illmensee, K., A.P. Mahowald, and M.R. Loomis. 1976. The ontogeny of germ plasm during oogenesis in *Drosophila.* Dev. Biol., *49*: 40–65.

Jaffe, L.F. 1966. Electrical currents through the developing *Fucus* egg. Proc. Natl. Acad. Sci. U.S.A., *56*: 1102–1109.

Jeffery, W.R. 1983. Messenger RNA localization and cytoskeletal domains in ascidian embryos. *In* W.R. Jeffery and R.A. Raff (eds.), *Time, Space and Pattern in Embryonic Development.* Alan R. Liss, New York, pp. 241–259.

Jeffery, W.R., and S. Meier. 1983. A yellow crescent cytoskeletal domain in ascidian eggs and its role in early development. Dev. Biol., *96:* 125–143.

Kamimura, M. et al. 1976. Observations on the migration and proliferation of gonocytes in *Xenopus laevis.* J. Embryol. Exp. Morphol., *36:* 197–207.

Kelly, S.J. 1977. Studies of the developmental potential of 4- and 8-cell stage mouse blastomeres. J. Exp. Zool., *200:* 365–376.

Kirschner, M. et al. 1980. Initiation of the cell cycle and establishment of bilateral symmetry in *Xenopus* eggs. In S. Subtelny and N.K. Wessells (eds.), *The Cell Surface: Mediator of Developmental Processes.* 38th Symposium of the Society for Developmental Biology. Academic Press, New York, pp. 187–215.

Mahowald, A.P. 1968. Polar granules of *Drosophila.* II. Ultrastructural changes during early embryogenesis. J. Exp. Zool., *167:* 237–262.

Mahowald, A.P. 1971a. Polar granules of *Drosophila.* III. The continuity of polar granules during the life cycle of *Drosophila.* J. Exp. Zool., *176:* 329–344.

Mahowald, A.P. 1971b. Polar granules of *Drosophila.* IV. Cytochemical studies showing loss of RNA from polar granules during early stages of embryogenesis. J. Exp. Zool., *176:* 345–352.

Mahowald, A.P., and S. Hennen. 1971. Ultrastructure of the "germ plasm" in eggs and embryos of *Rana pipiens.* Dev. Biol., *24:* 37–53.

Mintz, B. 1965. Experimental genetic mosaicism in the mouse. *In* G.E. Wolstenholme and M. O'Connor (eds), *Preimplantation Stages of Pregnancy.* J. and A. Churchill, London, pp. 194–207.

Mintz, B., and K. Illmensee. 1975. Normal genetically mosaic mice produced from malignant teratocarcinoma cells. Proc. Natl. Acad. Sci. U.S.A., *72:* 3585–3589.

Newrock, K.M., and R.A. Raff. 1975. Polar lobe specific regulation of translation in embryos of *Ilyanassa obsoleta.* Dev. Biol., *42:* 242–261.

Nuccitelli, R. 1978. Ooplasmic segregation and secretion in the *Pelvetia* egg is accompanied by a membrane-generated electrical current. Dev. Biol., *62:* 13–33.

Nuccitelli, R., and L.F. Jaffe. 1974. Spontaneous current pulses through developing fucoid eggs. Proc. Natl. Acad. Sci. U.S.A., *71:* 4855–4859.

Okada, M., I.A. Kleinman, and H.A. Schneiderman. 1974. Restoration of fertility in sterilized *Drosophila* eggs by transplantation of polar cytoplasm. Dev. Biol., *37:* 43–54.

Okazaki, K. 1975a. Spicule formation by isolated micromeres of the sea urchin embryo. Am. Zool., *15:* 567–581.

Okazaki, K. 1975b. Spicule formation by micromeres isolated from embryos of *Arbacia punctulata.* Biol. Bull., *149:* 439–440.

Palaček, J., G.A. Ubbels, and K. Rzehak. 1978. Changes of the external and internal pigment pattern upon fertilization in the egg of *Xenopus laevis.* J. Embryol. Exp. Morph., *45:* 203–214.

Quatrano, R.S. 1972. An ultrastructural study of the determined site of rhizoid formation in *Fucus* zygotes. Exp. Cell Res., *70:* 1–12.

Quatrano, R.S. 1978. Development of cell polarity. Ann. Rev. Plant Physiol., *29*: 487–510.

Robinson, K.R., and R. Cone. 1980. Polarization of fucoid eggs by a calcium ionophore gradient. Science, *207*: 77–78.

Robinson, K.R., and L.F. Jaffe. 1975. Polarizing fucoid eggs drive a calcium current through themselves. Science, *187*: 70–72.

Sawada, T., and K. Osani. 1981. The cortical contraction related to the ooplasmic segregation in *Ciona intestinalis* eggs. Wilh. Roux's Arch., *190*: 208–214.

Smith, L.D. 1966. The role of a "germinal plasm" in the formation of primordial germ cells in *Rana pipiens*. Dev. Biol., *14*: 330–347.

Spemann, H. 1938. *Embryonic Development and Induction.* Yale University Press, New Haven, Conn.

Stern, M.S. 1972. Experimental studies on the organization of the preimplantation mouse embryo. II. Reaggregation of disaggregated embryos. J. Embryol. Exp. Morphol., *28*: 255–261.

Tarkowski, A.K., and J. Wroblewska. 1967. Development of blastomeres of mouse eggs isolated at the 4- and 8-cell stage. J. Embryol. Exp. Morphol., *18*: 155–180.

Torrey, J.G. 1967. *Development in Flowering Plants.* Macmillan, New York.

Waddington, C.H. 1966. *Principles of Development and Differentiation.* Macmillan, New York.

Waring, G.L., C.D. Allis, and A.P. Mahowald. 1978. Isolation of polar granules and the identification of polar granule–specific protein. Dev. Biol., *66*: 197–206.

Whittaker, J.R. 1973. Segregation during ascidian embryogenesis of egg cytoplasmic information for tissue-specific enzyme development. Proc. Natl. Acad. Sci. U.S.A., *70*: 2096–2100.

Whittaker, J.R. 1980. Acetylcholinesterase development in extra cells caused by changing the distribution of myoplasm in ascidian embryos. J. Embryol. Exp. Morph., *55*: 343–354.

Whittaker, J.R., G. Ortolani, and N. Farinella-Ferruzza. 1977. Autonomy of acetylcholinesterase differentiation in muscle-lineage cells of ascidian embryos. Dev. Biol., *55*: 196–200.

Williams, M.A., and L.D. Smith. 1971. Ultrastructure of the "germinal plasm" during maturation and early cleavage in *Rana pipiens*. Dev. Biol., *25*: 568–580.

Wilson, E.B. 1904. Experimental studies on germinal localization. I. The germ regions in the egg of *Dentalium*. II. Experiments on the cleavage-mosaic in *Patella* and *Dentalium*. J. Exp. Zool., 1: 1–72.

Wilson, E.B. 1925. *The Cell in Development and Heredity,* 3rd ed. Macmillan, New York.

Züst, B., and K.E. Dixon. 1977. Events in the germ cell lineage after entry of the primordial germ cells into the genital ridges in normal and UV-irradiated *Xenopus laevis*. J. Embryol. Exp. Morphol., *41*: 33–46.

12 Gene Expression During the Initiation of Development

The union of the haploid genomes of the egg and sperm at fertilization produces the diploid zygote genome that directs the development of the embryo and specifies the phenotypic characteristics of the resulting adult organism. The impact of the zygotic genome on development is not detected in many animal species, however, until after a considerable delay. Instead, the developmental program is initiated by constituents of the egg cytoplasm. Experimental demonstration of this role of the egg cytoplasm was first provided during the late 1800s and early 1900s by echinoid species hybridization studies. In these experiments, eggs of echinoids were fertilized in the laboratory with sperm from different species (or even different genera), and the characteristics of the hybrid embryos were monitored. In such crosses the morphological characteristics of cleavage-stage embryos (such as the form and rate of cleavage) are found to be entirely maternal, whereas paternal characteristics are not evident until later stages of development. These experiments are interpreted as showing that the cleavage process is directed entirely by the cytoplasm. The embryonic nucleus—which has both maternal and paternal genes—is excluded from participation in the earliest developmental events and does not influence development until somewhat later. The exact timing of the intervention by the embryonic nucleus will be discussed in section 12–2.

The absence of embryonic nuclear involvement in the regulation of cleavage is dramatically demonstrated by a series of experiments conducted by E.B. Harvey (1936, 1940), who developed a procedure to produce enucleated fragments of sea urchin eggs, which could then be parthenogenically activated to initiate development. The behavior of these fragments (called **merogones**) is entirely attributable to the components of the cytoplasm. In Harvey's procedure, unfertilized eggs are centrifuged

527

in a sucrose solution of the same density as the eggs (Fig. 12.1). The eggs remain suspended during centrifugation, and their contents stratify as the eggs elongate and assume a dumbbell shape before the centrifugal forces separate them into two fragments. The light half always contains the nucleus, and the heavy fragment is enucleated. Parthenogenic activation of these enucleated fragments can be achieved by treating them with a solution of hypotonic seawater. These **parthenogenic merogones** go on to cleave and form abnormal blastulae (Fig. 12.2). The enucleated blastulae superficially resemble nucleated blastulae and can even hatch from the fertilization envelope. However, they lack a blastocoel and have a reduced number of cilia, which are short and irregularly distributed. No development beyond the blastula stage is observed.

Cleavage in the absence of a functional nucleus has also been demonstrated in the frog, *Rana pipiens*. Achromosomal embryos are prepared by a two-step operation. First, sperm are irradiated with x-rays to destroy the chromatin; then the irradiated sperm are used to activate eggs. After activation the haploid egg nuclei are removed manually (see Chap. 2). These embryos, which contain only functionally inactive chromatin derived from the irradiated sperm nucleus, undergo somewhat abnormal cleavage and form partial blastulae (Fig. 12.3A). Histological sections of the partial blastulae (Fig. 12.3B and C) confirm that most cells lack chro-

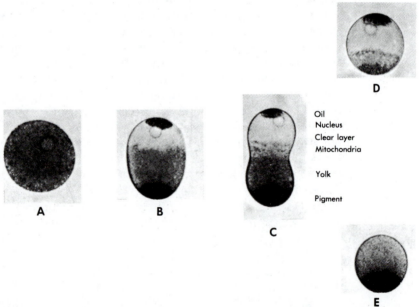

FIGURE 12.1 Production of a sea urchin merogone by centrifugation in a sucrose solution. *A*, Normal egg. *B*, Partly stratified. *C*, Fully stratified. *D*, Nucleated half. *E*, Anucleate half. (From E.B. Harvey. 1940. A comparison of the development of nucleate and non-nucleate eggs of *Arbacia punctulata*. Biol. Bull., *79:* 167.)

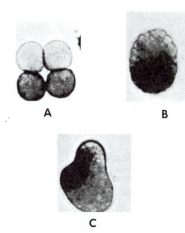

A

B

C

FIGURE 12.2 Development of activated sea urchin merogones. *A,* Perfect four-cell stage. Note asters, which form even in the absence of a nucleus. *B,* "Blastula." *C,* "Blastula" in process of hatching from fertilization envelope. (From E.B. Harvey. 1940. A comparison of the development of nucleate and non-nucleate eggs of *Arbacia punctulata.* Biol. Bull., *79:* 179.)

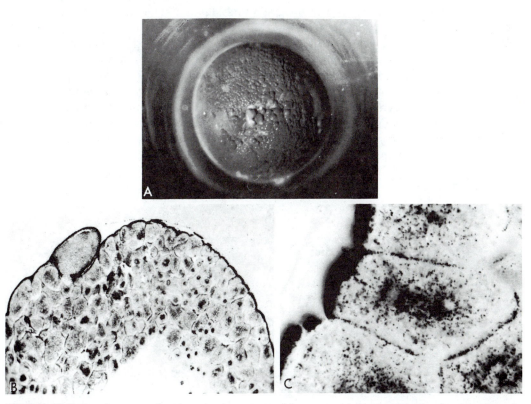

FIGURE 12.3 Development of activated achromosomal frog eggs. *A,* Partial blastula, showing cleaved area on animal hemisphere. *B,* Section through cleaved animal hemisphere of partial blastula. ×75. *C,* Higher magnification of portion of *B.* Note the astral figure. ×570. (From R. Briggs, E.U. Green, and T.J. King. 1951. An investigation of the capacity for cleavage and differentiation in *Rana pipiens* eggs lacking "functional" chromosomes. J. Exp. Zool., *116:* 497, 499.)

matin. Irradiated sperm chromatin is found in a small proportion of cells, but these cells cleave to the same extent as those without chromatin. Thus, the presence of this chromatin is not correlated with the ability to cleave.

An alternative to removal of the nucleus is the inhibition of nuclear function by using actinomycin D or α-amanitin, both of which interfere with RNA synthesis (see Chap. 3). The use of these inhibitors mimics some aspects of physical enucleation, since the nucleus is prevented from carrying on transcriptional activities. Because these transcriptional inhibitors also have other toxic side effects on cells, interpretation of experiments in which they are used must be made with caution. Although often inconclusive by themselves, experiments employing inhibitors can, however, provide valuable corroborative evidence for data obtained by other techniques. Inhibitors were first used to assess the necessity of nuclear function during early sea urchin development (Gross and Cousineau, 1964). Although treatment of these embryos with actinomycin D severely curtails RNA synthesis, cleavage still occurs, and the embryos proceed as far as the mesenchyme blastula stage before development is halted (Fig. 12.4). Similar results have also been obtained with animals from every level of the evolutionary scale with the exception of mammals. A partial list of inhibitor studies on nonmammalian embryos is presented in Table 12–1. These inhibitor studies reinforce the concept that initial embryonic development is independent of *de novo* (i.e., new) RNA synthesis. However, this generalization cannot be applied to mammalian embryos without qualification. The evidence from similar studies with mammalian embryos indicates a greater reliance on *de novo* transcription in that group. We shall deal with mammalian embryos in section 12–3.

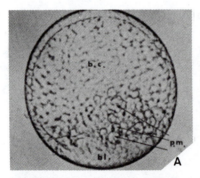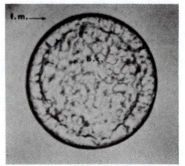

FIGURE 12.4 Effects of actinomycin D on sea urchin (*Lytechinus*) development. *A*, Control mesenchyme blastula. *B*, Actinomycin-treated blastula. b.c.: blastocoele; p.m.: primary mesenchyme; f.m.: fertilization envelope. (From P.R. Gross, L.I. Malkin, and W.A. Moyer. 1964. Templates for the first proteins of embryonic development. Proc. Natl. Acad. Sci. U.S.A., *51:* 412.)

TABLE 12–1
Effects of Inhibitors of RNA Synthesis on Early Development

Organism	Observations	References
Coleopteran insects	Arrest at blastoderm stage with either actinomycin D or α-amanitin	Lockshin, 1966; Maisonhaute, 1977
Ilyanassa obsoleta (marine snail)	Arrest at gastrulation with actinomycin D	Collier, 1966; Feigenbaum and Goldberg, 1965
Sea urchin	Arrest at mesenchyme blastula stage with actinomycin D	Gross and Cousineau, 1964; Gross et al., 1964; Summers, 1970
Ascidia nigra (ascidean)	Arrest at blastula stage with either actinomycin D or α-amanitin	Brachet et al., 1964; Brachet et al., 1972
Rana pipiens	Arrest at blastula stage with actinomycin D	Wallace and Elsdale, 1963
Fundulus heteroclitus (marine teleost)	Arrest at blastula stage with actinomycin D	Wilde and Crawford, 1966

12–1. MATERNAL REGULATION OVER EARLY DEVELOPMENT

The ability of the egg cytoplasm to support initial development without embryonic nuclear involvement indicates that preparations for cleavage are made during oogenesis, and that the cleavage mechanism itself must be a product of maternal gene expression rather than of zygotic gene expression. Thus, much of the gene transcription that occurs during oogenesis is in preparation for cleavage, and the effects of this gene activity are seen in the zygote rather than in the oocyte. Genetic evidence for this interpretation has been obtained from a number of genetic mutants with so-called **maternal effects,** which are specific phenotypic traits in the embryo that are determined by the maternal rather than the embryonic genotype. Maternal effect mutants provide some of the most convincing evidence that early development depends to a large extent on substances that preexist in the egg.

The best known example of a maternal effect is that of shell coiling in the snail *Limnaea peregra* (Boycott et al., 1930). Coiling is usually *dextral* (opening of the shell is to the right). In rare individuals the coiling is reversed, or *sinistral* (opening of the shell is to the left). The direction of coiling is determined by genes at a single locus, with the wild-type dextral allele (+) acting as a dominant to the sinistral allele (s). However, the direction of coiling is determined not by the genotype of the developing snail but by the genotype of its mother. As shown in Figure 12.5A, a sinistral (s/s) mother will produce only sinistral offspring, even though the progeny may receive the dominant dextral gene (+) from their father.

Note that the shells of the progeny are not necessarily like the mother's shell but are *determined by her genotype.* For example, a sinistral $+/s$ mother has only dextral progeny, and a dextral s/s mother has only sinistral progeny.

Although shell coiling is a trait seen in adult life, coiling is actually imposed upon the adult by the spiral mode of cleavage, and the direction of coiling depends upon the orientation of the cleavage spindle (Fig. 12.5B; see also pages 444–446 for a discussion of spiral cleavage). Thus, the spindle of sinistral embryos is tipped to the left, whereas that of dextral embryos is tipped to the right. However, the orientation of the spindle is determined *during oogenesis* and is under the control of the maternal genome. Consequently, the maternal genome regulates the direction of coiling in her progeny *regardless of their genotypes.*

A recent experiment suggests that the direction of cleavage in these embryos is regulated by a substance present in egg cytoplasm. Freeman and Lundelius (1982) have reported that cytoplasm from dextral eggs injected into sinistral eggs causes the recipients to cleave in a dextral pattern. The reciprocal cytoplasmic transfer is ineffective, however: Sinistral cytoplasm will not reverse the cleavage pattern of dextral embryos. Therefore, expression of the dextral gene during oogenesis produces a cytoplasmic substance that causes dextral cleavage after fertilization. In the absence of this substance, cleavage will be sinistral.

Injection of dextral cytoplasm is only effective in "curing" the sinistral condition if done before second polar body formation. Thus, the direction of cleavage must be determined at this early stage. This provides a probable explanation for the failure of dextral sperm to influence the cleavage pattern (and subsequently, the direction of shell coiling) of sinistral eggs: The dextral genes of fertilizing sperm nuclei are not yet capable at this early stage of directing the synthesis of the cytoplasmic product (remember that the pronuclei have not yet fused). Therefore, the direction of cleavage is governed entirely by the preexisting egg cytoplasmic components, which were produced under the control of the maternal genotype. The nature of the dextral substance is unknown, but the microinjection technique provides an ideal assay system to test substances extracted from dextral egg cytoplasm.

A very large number of maternal effect mutants are known in *Drosophila,* providing evidence for the extent of the maternal influence over the embryo. For example, *every gene* that is known to affect the for-

FIGURE 12.5 Shell coiling in the snail, *Limnaea peregra. A,* Pattern of inheritance. (Redrawn with ▶ permission from Sinott, E., L.C. Dunn, and T. Dobzhansky. 1958. *Principles of Genetics.* 5th Edition. McGraw-Hill Book Co, New York, p. 361.) *B,* Relationship between cleavage and adult shell type. Polar bodies mark the animal pole. The first (I) and second (II) cleavage furrows of the two types of embryos are mirror images of each other. The orientations of these furrows result from the angles of the spindle. The dextral pattern results from tilting of the spindle to the right, while the sinistral pattern results from tilting of the spindle to the left. (After A.H. Sturtevant and G.W. Beadle. 1962. *An Introduction to Genetics.* Dover Publications, Inc., New York, p. 329.)

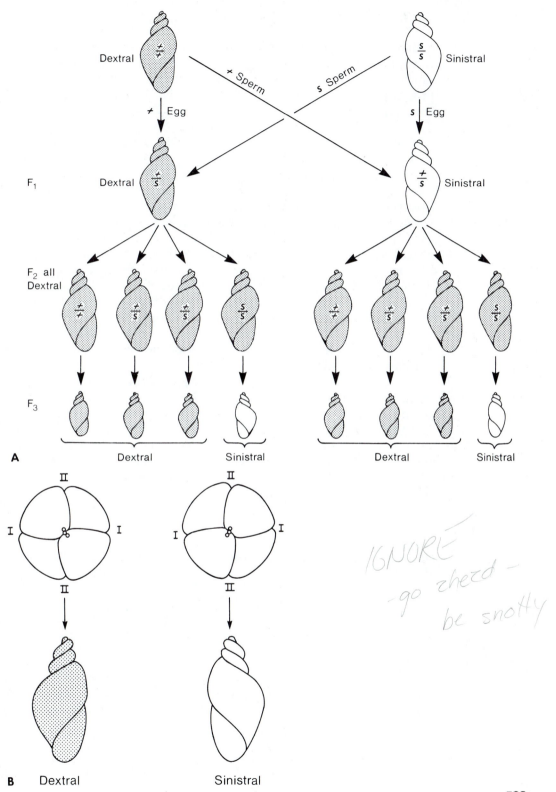

A

F₁

F₂ all
Dextral

F₃

Dextral   Sinistral   Dextral   Sinistral

B   Dextral   Sinistral

mation of the cellular blastoderm is a maternal effect mutant (Rice and Garen, 1975; Bakken, 1973; Fullilove and Woodruff, 1974). Thus, genetic control over this vital stage in early development is apparently mediated through egg constituents that are produced by the transcription of maternal genes during oogenesis.

One of the best studied maternal effect mutants in *Drosophila* is *deep orange* (*dor*), a sex-linked eye color mutant. In addition to its effects on eye color, the gene is also involved in early development. The crosses outlined in Table 12–2 illustrate the latter. When +/*dor* females are mated to *dor*/Y males, their *deep orange* progeny develop normally. However, *dor*/*dor* females do not produce viable *deep orange* progeny. Apparently, the wild-type allele of *deep orange* produces a substance that is essential for normal development. This substance is incorporated into all eggs produced by +/*dor* females, and these eggs are capable of normal development *regardless of the genotypes of the sperm that fertilize them.* On the other hand, eggs of *dor*/*dor* females are deficient in this substance. These eggs will not develop if fertilized by *dor*-bearing sperm. However, if the sperm carry the wild-type allele, about half of the +/*dor* daughters survive. This result is interpreted to mean that expression of the + gene after fertilization may be sufficient to provide the missing substance that is necessary for development.

Further evidence that there is indeed a cytoplasmic substance produced by expression of the wild-type allele of *deep orange* comes from experiments by Garen and Gehring (1972), who injected cytoplasm of wild-type eggs into *dor* embryos at the syncytial blastoderm stage. About one third of the injected embryos developed to the stage of embryonic movement, a stage never attained by uninjected *deep orange* embryos. As expected, the cytoplasm of *dor* eggs is not effective in repairing the *dor* developmental defect.

TABLE 12–2
Effects of the *Dor* Mutation on Embryonic Development of *Drosophila* Melanogaster

Genotypes of parents	Genotypes of progeny	Phenotypes of progeny
+/*dor* ♀ × *dor*/Y ♂	+/*dor*	Wild-type ♀
	dor/*dor*	Deep orange ♀
	+/Y	Wild-type ♂
	dor/Y	Deep orange ♂
dor/*dor* ♀ × *dor*/Y ♂	*dor*/*dor*	Embryonic lethal
	dor/Y	Embryonic lethal
dor/*dor* ♀ × +/Y ♂	+/*dor*	Wild-type ♀ (50% lethal)
	dor/Y	Embryonic lethal

Maternal effect mutants have also been discovered among higher organisms, particularly the amphibians. An example is *ova deficient* (*o*) in the Mexican axolotl, *Ambystoma mexicanum*. Progeny of *o/o* females die during gastrulation, regardless of the genotypes of the sperm that fertilize the eggs. Cleavage is normal until the late blastula stage, when the cleavage rate becomes reduced. Gastrulation is initiated, but it is abortive. This block can be corrected experimentally by injecting the contents of the germinal vesicle of a normal (+/*o* or +/+) ovarian oocyte into *o/o* eggs following fertilization. The recipients will develop well beyond the stage of the uninjected eggs, and in some cases develop into swimming larvae (Briggs and Cassens, 1966). Correction of the defect can also be obtained by injection of normal egg cytoplasm after germinal vesicle breakdown, but injection of cytoplasm of ovarian oocytes is only slightly effective. Apparently the normal germinal vesicle contains a factor (the product of the wild-type allele) that accumulates in the germinal vesicle and is released into the cytoplasm at germinal vesicle breakdown. This factor (called o^+ substance) is essential for normal development. Unlike *dor*, the o^+ substance is only produced during oogenesis and cannot be supplied by expression of the paternally derived wild-type allele of the embryonic genome. Further characterization of the o^+ substance indicates that it is a protein, as shown by its heat lability, its inactivation by the protease trypsin, and its precipitation by ammonium sulfate (Briggs and Justus, 1968).

The Oogenic Messenger Hypothesis

The overwhelming evidence that early development is regulated by components of the egg cytoplasm has stimulated a search for the identity of these substances and for their modes of action in mediating the influence of the maternal genome. This search has been a key area of investigation during the past decade and is one of the major success stories in developmental biology. Most of this research has been conducted on sea urchin and amphibian embryos, but the principles derived from work on these organisms are nearly universal.

As with all cells, the functional properties of the fertilized egg and the blastomeres of the cleaving embryo depend upon the proteins they contain. These cells are specialized for rapid mitotic division, utilizing stored energy in the form of yolk. Accordingly, they require structural proteins for the production of mitotic spindles and a vast new membrane system, including a plasmalemma, nuclear envelope, and endoplasmic reticulum for each blastomere. Likewise, they must produce the enzymes for their major metabolic functions, including energy utilization and DNA synthesis. Some proteins are incorporated into the oocyte cytoplasm during oogenesis. However, these components are *not sufficient* to sustain the embryo through the rapid cleavages of initial development without additional protein synthesis. This conclusion is suggested by the results obtained when embryos are treated with puromycin. This inhibitor, which prevents protein synthesis, causes a rapid inhibition of embryonic development (Hultin, 1961; Ecker and Smith, 1971). Thus, protein synthesis is apparently of immediate consequence to the embryo in the maintenance of developmental functions. As we discussed in Chapter 9, protein synthesis increases within minutes of fertilization of the sea urchin egg, as a component of the activation response. In amphibians, protein synthesis increases at meiotic maturation in anticipation of fertilization and undergoes an additional increase after sperm entry (Shih et al., 1978). The puromycin studies suggest that this protein synthesis *is essential* for development to occur.

A parodox becomes immediately apparent. Protein synthesis requires messenger RNA, transfer RNA, and ribosomes, all of nuclear origin. And yet we have learned that nuclear function—indeed, the very presence of a nucleus—is *not essential* during early development. We must therefore conclude that the cytoplasm of the egg contains the complete protein synthetic machinery (i.e., messenger RNA, transfer RNA, and ribosomes) in an underutilized form, and activation of development allows utilization of this preexisting store of information. Striking evidence supporting this conclusion comes from experiments in which protein synthesis is demonstrated in the absence of nuclear function. For example, protein synthesis occurs at near normal levels in enucleated eggs of *Rana pipiens* (Smith and Ecker, 1965), in activated sea urchin merogones (Denny and Tyler, 1964), and in sea urchin zygotes treated with actinomycin D (Gross and Cousineau, 1964). Not only is protein synthesis maintained in the absence of nuclear function, but the *qualitative pattern* of protein synthesis is apparently normal in enucleated *Rana pipiens* eggs (Ecker and Smith, 1971). These results suggest that the egg cytoplasm is preprogrammed to direct the protein synthesis that is essential to guide the development of the embryo through cleavage and blastulation.

The hypothesis that messenger RNA is stored in the oocyte cytoplasm for utilization after fertilization has stimulated considerable research to demonstrate its existence more directly. Furthermore, since these **oogenic messengers** are hypothesized to direct initial embryonic development, investigators have sought to identify specific messengers and determine the roles they play in initial development.

One means of demonstrating the existence of oogenic messengers is to extract RNA from unfertilized eggs and use it to synthesize protein *in vitro*. The *in vitro* synthesized polypeptides can then be analyzed following electrophoresis. In this way not only can the *existence* of a pool of stored mRNA be demonstrated, but also the *identities* of some specific messengers can be inferred by examining the polypeptides they encode.

The *in vitro* translation procedure demonstrates the presence of the most prevalent messenger RNA molecules (i.e., messengers, each of which is present in very large numbers). The remaining messengers (which constitute the vast majority of the RNA complexity) are present in such small numbers that their translational products cannot be detected readily by current technology. The prevalent messengers, by producing a large amount of protein when they are translated, present a technical advantage, since the proteins they encode can be analyzed. Furthermore, the messengers themselves may be analyzed more readily than the less prevalent transcripts.

Infante and Heilmann (1981) analyzed the *in vitro* translation products of polyadenylated RNA isolated from unfertilized sea urchin eggs by two-dimensional gel electrophoresis (see box, p. 88, for discussion of this technique). As we discussed in Chapter 3, the presence of poly (A) tracts on most messenger RNA molecules facilitates their isolation. More than 400 polypeptides are synthesized on the oogenic templates (Fig. 12.6). Some of the spots on this gel have been identified. For example, actin is located in the region labeled A, and tubulin is located in the region labeled T. Consequently, the messengers for these proteins are present in the unfertilized egg.

One of the most prevalent classes of oogenic messengers in sea urchin eggs is the set coding for histone proteins. These messengers are readily separated from the remainder of oogenic messengers by sucrose density gradient centrifugation (see box, p. 88), since the RNP particles containing the histone messengers complexed with protein sediment at 20S, whereas the remaining messengers are in particles that are larger and more variable in size. The presence of histone messengers in these particles is demonstrated by *in vitro* translation experiments.

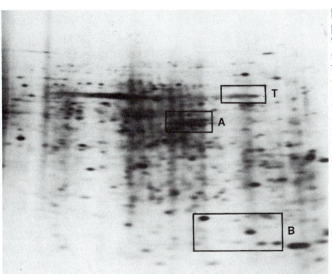

FIGURE 12.6 Cell-free translation products of poly (A)$^+$ RNA from unfertilized sea urchin eggs. This autoradiograph of the two-dimensional gel illustrates the large number of polypeptides encoded by polyadenylated egg RNA. Boxes T and A refer to tubulin and actin, respectively. Box B is a region shown in enlargement in Figure 12.10. (Reprinted with permission from A.A. Infante and L.J. Heilmann. 1981. Copyright 1981 American Chemical Society.)

The *utilization* of oogenic messengers in the first hours after fertilization in sea urchin zygotes is indicated by the observation that RNA extracted from the *polysomes* of cleavage-stage sea urchin embryos produces exactly the same proteins as the RNP-associated RNA isolated from unfertilized eggs when translated *in vitro* (Infante and Heilmann, 1981). The presence of these RNAs on polysomes indicates that they are functioning as messengers. As we shall discuss later in this chapter, the vast majority of these messenger molecules are of oogenic origin. The similarity between the *in vitro* protein synthetic patterns suggests that there is no qualitative selection of these messengers in the first few hours after fertilization; each kind of preexisting messenger is translated.

The utilization of oogenic messengers after fertilization is also indicated by the fact that synthesis of proteins known to be encoded by these messengers occurs in the absence of *de novo* RNA synthesis. For example, tubulin is synthesized by sea urchin merogones that have been parthenogenically activated (Raff et al., 1972). The synthesis of all five classes of histones occurs in cleavage- and blastula-stage sea urchin embryos in the presence of actinomycin D, which drastically inhibits *de novo* synthesis of RNA (Ruderman and Gross, 1974).

Evaluation of the oogenic messenger population by *in vitro* translation or by the use of inhibitors is indirect. The use of cloned probes permits direct identification of specific messengers and provides a sensitive means of monitoring for the presence of these transcripts in unfertilized eggs and their utilization after fertilization. Actin (Scheller et al., 1981; Crain et al., 1981) and histone (Woods and Fitschen, 1978) transcripts have been identified in this way. The most detailed studies have focused on the histone messengers. The egg of *Strongylocentrotus purpuratus* contains messengers for each of the five classes of histones. Together these messengers constitute approximately 5 to 10% of the total quantity of oogenic messenger RNA (Davidson et al., 1982). These messengers code for modified histones that are only synthesized during initial development and are assembled into the chromatin of cleavage- and (in some cases) blastula-stage embryos. We shall discuss the modified histones of early sea urchin embryos later in this chapter.

The prevalent messengers account for most of the total mass of mRNA and, as we have seen, present major advantages for analyzing oogenic messenger RNA. However, they account for only a very small minority of the total number of *different* messengers that are present (i.e., the messenger complexity; see Chap. 4), since a very large number of messengers are present in only a few copies each. The latter are collectively called the **complex class of messenger RNA,** as distinguished from the former, which constitute the **prevalent class of messenger RNA** (Davidson, 1976).

A measure of the complexity of sea urchin oogenic messenger can be obtained by determining the level of hybridization of RNA to total single copy DNA, the class of DNA that contains most structural genes (see Chap. 3). As we discussed in Chapter 7, the complexity of sea urchin egg messenger RNA is approximately 37×10^6 nucleotides, which is equivalent to approximately 20,000 to 30,000 different mRNA sequences (Hough-Evans et al., 1977). As we shall discuss subsequently (see p. 548), the vast majority of these sequences are being translated at the 16-cell stage. Thus, most of these sequences are indeed oogenic messengers for utilization during development.

Oogenic messengers have been demonstrated in amphibian eggs in much the same way as for sea urchin eggs. As we discussed in Chapter 7, most of the polyadenylated RNA in fully grown *Xenopus* oocytes is in stored RNP particles (Rosbash and Ford, 1974). The potential utilization of this RNA as messenger is indicated by its ability to be translated *in vitro* (Darnbrough and Ford, 1976). Two-dimensional gel electrophoresis of the proteins synthesized *in vitro* by RNA extracted from unfertilized *Xenopus* eggs reveals an array of proteins, including actin (Ballantine et al., 1979) and histones (Ruderman et al., 1979). The messengers for the latter are nonpolyadenylated, whereas histone messengers in oocytes are mostly polyadenylated (see Chap. 7). Complexity measurements on messenger RNA in *Xenopus* oocytes reveal that they contain approximately 20,000 structural gene transcripts (Perlman and Rosbash, 1978). This estimate is remarkably similar to the estimate obtained for sea urchin egg messenger RNA (see previous discussion).

The stored RNA in the unfertilized egg cytoplasm is the informational pool that the embryo utilizes to support the protein synthesis that is necessary for the initial phase of postfertilization development. The amount of stored oogenic mRNA in eggs has been estimated by various investigators. Davidson (1976), who reviewed the literature on this subject, has concluded that oogenic mRNA constitutes approximately 1.5 to 3% of the total RNA of the sea urchin egg and about 1 to 2% of the total RNA in the fully grown *Xenopus* oocyte.

Ribosomal and Transfer RNA for Initial Development

The rapid initiation of protein synthesis that characterizes the beginning of embryonic development requires that the translational machinery be poised for immediate utilization by the oogenic messenger RNA when it is mobilized. Protein synthesis in the absence of RNA synthesis (either in transcriptionally inhibited embryos or in activated enucleated eggs) demonstrates that the cytoplasm of the mature egg contains the complete protein synthetic machinery, including ribosomes and transfer RNA. The vast majority of the RNA in an unfertilized egg is 18S and 28S ribosomal RNA, which are integral components of the cytoplasmic ribosomes. We have previously discussed the mechanisms by which the synthesis of these large amounts of rRNA is facilitated during oogenesis (see Chap. 7).

The magnitude of the ribosomal store in the egg is beautifully illustrated by the anucleolate mutant of *Xenopus laevis.* Wild-type *X.
laevis* have two nucleoli (2-*nu*) in each diploid cell—one per haploid
chromosome set. Elsdale et al. (1958) described a heterozygote mutant
with one nucleolus per cell (1-*nu*). The heterozygotes lack a nucleolar
organizer on one of two homologous chromosomes. This deficiency is
inherited in typical Mendelian fashion. Thus, when two heterozygotes
are mated, the progeny fall into three classes having two, one, or no
nucleoli per diploid cell in a 1:2:1 ratio, respectively. The progeny lacking nucleoli (0-*nu* or anucleolates) appear perfectly normal as embryos
but are retarded after hatching and die as swimming tadpoles before feeding stages.

Since the nucleolar organizer contains the genes that code for ribosomal RNA (see Chap. 3), the anucleolate mutant embryos lack the
ability to synthesize this RNA. A comparison of RNA synthesis of anucleolate and control embryos is shown in Figure 12.7. In this experiment,
anucleolate (0-*nu*) and control (1-*nu* and 2-*nu*) embryos were allowed to

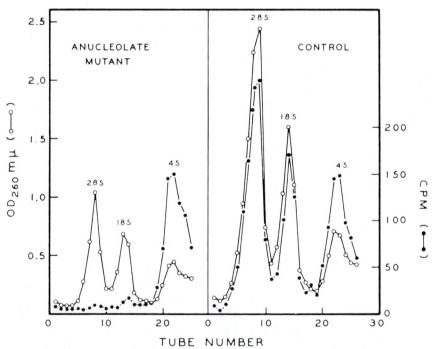

FIGURE 12.7 Sucrose density gradient centrifugation of RNA isolated from anucleolate (0-*nu*) and control embryos. Total RNA is represented by optical density measurements (line with open circles), while RNA synthesized by the embryos is represented by the radioactivity measurements (line with solid circles). (From D.D. Brown
and J.B. Gurdon. 1964. Absence of ribosomal RNA synthesis in the anucleolate mutant of *Xenopus laevis.* Proc. Natl. Acad. Sci. U.S.A., *51:* 141.)

incorporate radioactive precursor (labeled with ^{14}C) into RNA. The latter was extracted and fractionated by sucrose density gradient centrifugation. Total RNA (i.e., preexisting and newly synthesized) has been determined by optical density measurements, whereas newly synthesized RNA is indicated by radioactivity determinations. The control embryos have synthesized all the classes of RNA that are resolved by this procedure (28S and 18S rRNA and 4S RNA). In contrast, the anucleolates have synthesized 4S RNA, but the ribosomal RNA is entirely preexisting; no new 18S and 28S RNA has been produced. Since the 0-*nu* progeny are derived from heterozygous mothers, the embryos are able to survive until the swimming tadpole stage on the ribosomes made during oogenesis and stored in the egg cytoplasm.

In Chapter 7 we discussed the interesting mechanism that facilitates the accumulation of considerable amounts of 4S and 5S RNA during *Xenopus* oogenesis. Ford (1971) has determined that the late vitellogenic *Xenopus* oocyte has more than five 4S RNA molecules and more than

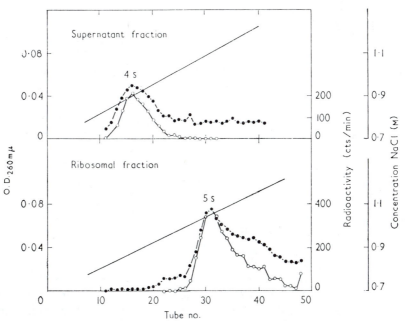

FIGURE 12.8 Column chromatography of 4S and 5S RNA from mature *Xenopus* eggs. Eggs were obtained from a female who had been injected with ^{3}H-uridine. RNA synthesized during oogenesis is radioactive. RNA was isolated from egg supernatant or ribosomal fraction, and low molecular weight RNA from these fractions was chromatographed on protamine-Celite columns. The RNA was eluted from the columns by a NaCl gradient. Molarity of NaCl: solid line. Optical density: line with open circles. Radioactivity: line with solid circles. (Reproduced with permission from Brown, D.D. and E. Littna. 1966. Journal of Molecular Biology, *20:* 95–112. Copyright by Academic Press Inc. [London] Ltd.)

seven 5S RNA molecules for each 18S and 28S RNA molecule. The retention of 4S and 5S RNA after oogenesis has been shown by labeling them during oogenesis and recovering the labeled molecules in eggs (Fig. 12.8).

Why Are Stored Messengers Inactive in Protein Synthesis?

The low level of protein synthesis before fertilization (in the sea urchin) or maturation (in the amphibian) must result from limitations on the components of the protein-synthesizing system. It has been proposed that the messenger synthesized during oogenesis is stored in the egg cytoplasm in an inactive form and that after fertilization it is mobilized for translation. This is the **masked oogenic messenger** hypothesis. According to this hypothesis, the messenger is masked in the sense that it is in a nontranslated form and protected from degradation by nucleases, making it extremely long-lived. Although the masked oogenic messenger concept has dominated thought about limitations on protein synthesis in eggs for some time, recent evidence suggests that this concept must be reexamined.

Laskey et al. (1977) injected large amounts of mRNA into *Xenopus* oocytes, which synthesize protein from endogenous messengers at a very low level. The exogenous messenger fails to increase the level of protein synthesis. Although the exogenous mRNA is translated, its translation is at the expense of endogenous messenger utilization. Thus, they are both competing for a limiting component. These results are totally incompatible with the masked oogenic messenger hypothesis and suggest that other factors are responsible for limiting the rate of protein synthesis.

Another approach to this problem is to isolate the messenger RNP particles from unfertilized eggs and determine whether they are readily translated *in vitro* or whether their utilization is limited by their innate properties. Messenger RNP particles can be isolated from homogenates of unfertilized sea urchin eggs by sucrose density gradient centrifugation. The messenger-containing particles sediment more slowly than ribosomes and are found in a fraction that has been called the **subribosomal fraction.** Jenkins et al. (1978) have reported that when these particles are isolated under conditions that preserve their stability, the RNA remains nontranslatable. However, under conditions that do not preserve RNP stability, the RNA can be readily translated *in vitro*. These results suggest that an alteration in the RNP particles after fertilization may allow translation of stored messengers *in vivo*. More recently, Moon et al. (1982) have reported that the RNA in egg RNP particles *is translatable in vitro*. They conclude that the RNA is *not masked*. The results of *in vitro* translation experiments are heavily dependent upon the techniques used in RNP isolation and translation. Thus, this experimental approach may be inappropriate to determine whether oogenic messengers are "masked."

In the absence of definitive evidence in favor of masking—at least in sea urchins and amphibians—we should consider other alternatives as explanations for the inactivity of stored messengers. One mechanism for limiting access to the translational machinery could be physical sequestration of messengers that would restrict their access to the translational machinery. Perhaps stored messengers are localized to cytoplasmic structures such as annulate lamellae (see Chap. 7). An unusual form of sequestration has been discovered for a subset of the oogenic histone messengers in the sea urchin egg; they are sequestered in the female pronucleus (see p. 547). Another possibility involves the cytoskeleton. Moon et al. (1983) have demonstrated that actively translated messengers in sea urchin embryos are associated with the cytoskeleton, whereas most messengers in eggs are not associated with it. Perhaps the translation of messengers in the egg is limited by a mechanism that restricts access of messengers to the cytoskeleton.

Finally, it is possible that egg messengers are fully translatable with unrestricted access to ribosomes, but the ribosomes are unable to translate them, perhaps because initiation factors are limiting. We should, however, keep in mind that a variety of mechanisms to limit egg protein synthesis may operate in different organisms and even for different kinds of messenger RNA in the same organism. Investigation of the mechanisms responsible for restricting the utilization of stored messengers is an active, exciting area of contemporary research. The literature in this area should yield much valuable information in the near future.

Embryonic Processing of Oogenic Messenger

After fertilization, the translation of oogenic messengers contributes to an acceleration in the rate of protein synthesis. The mobilization of stored oogenic mRNA and ribosomes to function in protein synthesis is called **recruitment.** As a consequence of this mobilization, ribosomes bind to messenger RNA molecules to form polysomes. Thus, the messengers shift from the subribosomal fraction to the ribosomal fraction. This shift has been detected in sea urchin eggs by tracing the subcellular location of polyadenylated RNA before and after fertilization. As shown in Figure 12.9, polyadenylated RNA shifts from the subribosomal to the ribosomal fraction within 30 minutes after fertilization.

The active translation of oogenic messengers commences with recruitment and the formation of polysome complexes. Humphreys (1971) has demonstrated that most of the RNA being translated on polysomes during the early cleavage stages of sea urchins is of oogenic origin. The increase in polysome content in sea urchin embryos after fertilization (Fig. 12.10) is reflected in the increase in protein synthesis during the

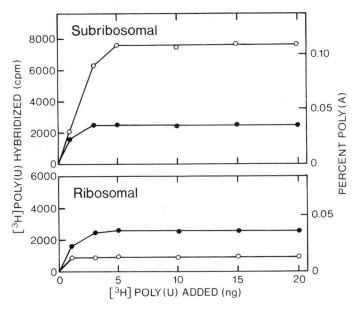

FIGURE 12.9 Saturation hybridization between ^{3}H-poly (U) and either subribosomal- or ribosomal-associated RNA from sea urchin eggs before fertilization (line with open circles) and 30 minutes after fertilization (line with solid circles). (After I. Slater, D. Gillespie, and D.W. Slater. 1973. Cytoplasmic adenylation and processing of maternal RNA. Proc. Natl. Acad. Sci. U.S.A., 70: 407.)

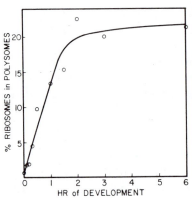

FIGURE 12.10 The increase in ribosomes in polysomes after fertilization of the sea urchin egg. The proportion of ribosomes was measured at various times after fertilization. (After T. Humphreys. 1971. Measurements of messenger RNA entering polysomes upon fertilization of sea urchin eggs. Dev. Biol., 26: 203.)

same period (see Fig. 9.23 for measurements of protein synthesis). In *Xenopus laevis* the initial increase in polysome content begins at maturation, and it rises even more after fertilization (Woodland, 1974).

In sea urchins the increase in protein synthesis is an integral part of the release of metabolic inhibition after fertilization. As we learned in Chapter 9, the release of inhibition is thought to be triggered by the increase in intracellular pH that follows fertilization. The dependency of the rate of protein synthesis on pH is illustrated by experiments in which manipulation of intracellular pH of eggs is shown to affect the rate of protein synthesis. Sea urchin eggs can be activated with NH_4Cl, which generates ammonia that enters the eggs and causes intracellular pH to rise (see Fig. 9.26). As shown in Figure 12.11, the NH_4Cl treatment

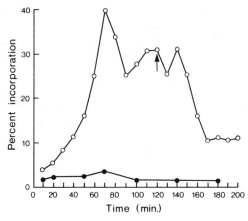

FIGURE 12.11 The rate of protein synthesis in NH$_4$Cl-activated sea urchin eggs. Unfertilized eggs were placed in either sea water (line with solid circles) or sea water containing NH$_4$Cl (line with open circles), and the rate of protein synthesis was determined at 10-minute intervals. The rate of protein synthesis is indicated by percent incorporation of ^{3}H-valine. This was calculated by dividing the amount of isotope incorporated into macromolecules (which are precipitated by trichloroacetic acid) by the total amount of isotope accumulated in the eggs. At 120 minutes (arrow) the eggs were removed from NH$_4$Cl and resuspended in sea water. Note the subsequent drop in the rate of protein synthesis. (After J.L. Grainger et al. 1979. Intracellular pH controls protein synthesis rate in the sea urchin egg and early embryo. Dev. Biol., *68*: 400.)

causes an increase in the rate of protein synthesis. Transfer of eggs to seawater lacking NH$_4$Cl (arrow) causes a drop in intracellular pH and a concomitant fall in the rate of protein synthesis. The relationship between intracellular pH and the rate of protein synthesis is also shown in Figure 12.12. In this experiment sea urchin eggs of *Lytechinus pictus* were fertilized, and the internal pH was regulated by (1) adding seawater containing sodium acetate, which enters the eggs and lowers intracellular pH, followed by (2) removal of the acetate stimulus, which allows pH to increase once again. The effects of sodium acetate on intracellular pH are shown in Figure 12.12A; within 25 minutes of adding seawater containing acetate the intracellular pH drops to the unfertilized egg level. These pH changes have profound effects on the rate of protein synthesis (Fig. 12.12B). Addition of sodium acetate stops the rise in the rate of protein synthesis, and after the intracellular pH has returned to the unfertilized egg level, the rate of protein synthesis drops off linearly until it, too, reaches the unfertilized egg rate. This effect of reduced pH on protein synthesis is reversible, since resuspending the eggs in seawater lacking sodium acetate allows protein synthesis to increase once again. Clearly, the activation and maintenance of a high rate of protein synthesis are dependent upon the maintenance of an alkaline pH in the egg cytoplasm.

Although the elevation of intracellular pH with NH$_4$Cl mimics the effects of fertilization on protein synthesis, this response is not the same as the response to sperm. Grainger et al. (1979) observed that the rate of protein synthesis in ammonia-activated eggs lags behind that of fertilized eggs. This lag can be completely overcome if the ionophore A23187 (see

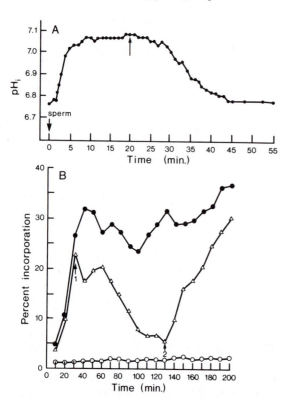

FIGURE 12.12 The relationship between intracellular pH and the rate of protein synthesis in fertilized sea urchin eggs. *A,* Eggs were fertilized and maintained in sea water. Internal pH (pH_i) was measured with a microelectrode. At 20 minutes after fertilization (arrow) eggs were exposed to sodium acetate. Internal pH declined, dropping to the unfertilized egg level within 25 minutes. *B,* The effects of sodium acetate on the rate of protein synthesis (line with open triangles). At 30 minutes after fertilization (arrow no. 1) eggs were resuspended in sea water containing sodium acetate. The rate of protein synthesis dropped to the unfertilized egg level. At 130 minutes (arrow no. 2) the eggs were returned to sea water lacking sodium acetate. Protein synthesis accelerated. For comparison, the rates of protein synthesis for unfertilized eggs (line with open circles) and fertilized eggs maintained in sea water (line with solid circles) are shown. (After J.L. Grainger, et al. 1979. Intracellular pH controls protein synthesis rate in the sea urchin egg and early embryo. Dev. Biol., *68:* 401.)

Chap. 9) is used to trigger a release of intracellular Ca^{++} in NH_4Cl-activated eggs (Winkler et al., 1980). This result suggests that both the Ca^{++} release and the increase in intracellular pH are necessary for maximal stimulation of protein synthesis.

The acceleration of protein synthesis involves at least two changes in the protein synthetic apparatus. These are the mobilization of messengers and changes in the utilization of messengers by the ribosomes. The mobilization of messengers in sea urchin zygotes is apparently a direct consequence of the pH increase. How this effect is mediated remains uncertain. The relatively low pH that exists before fertilization may maintain the messenger RNP particles in an untranslatable state, and the pH increase may release the inhibition. Their behavior may be related to a number of specialized proteins associated with the RNA, which are different from those found in RNPs from late postfertilization embryos (Moon et al., 1980). However, as we have previously discussed (see p. 541), masking of RNA in RNPs is now uncertain. One alternative is that mobilization of messengers could be facilitated by factors that must bind to the mRNPs before they can be translated. The pH increase could either release or activate these factors (for more details, see Moon et al., 1982). Another alternative is implicated by Danilchik and Hille

(1981). They challenged ribosomes of unfertilized eggs and embryos with exogenous messenger in an *in vitro* protein-synthesizing system and have demonstrated that the unfertilized egg ribosomes are less active in protein synthesis than ribosomes from embryos. This deficiency in unfertilized egg ribosomes is partially corrected by an increase in pH. Thus, the pH increase may be necessary to activate the ribosomes so that they may associate with the messengers.

Utilization of messengers by the translational machinery changes after fertilization in a second way: The rate of chain elongation of polypeptides increases (Brandis and Raff, 1978; Hille and Albers, 1979). Recently, Brandis and Raff (1979) have shown that this function is not affected by the pH increase. The temporal release of Ca^{++} at fertilization may be responsible for this change (Winkler et al., 1980).

Danilchik and Hille (1981) suggest that egg activation may involve the response of a number of components of the translational machinery to the altered chemistry of the cytoplasm to bring the rate of protein synthesis up to the level required to initiate development. Recently, Winkler and Steinhardt (1981) have developed a cell-free protein synthetic system from sea urchin eggs that may provide the assay system that will lead to a resolution of the problem of control of protein synthesis after fertilization.

An important consideration about oogenic messengers concerns possible regulation over their utilization: Are they mobilized en masse, or is there regulation over when individual transcripts will be utilized or, indeed, if they will be utilized at all? If there were no such regulation, the entire oogenic RNA population would be represented on polysomes throughout initial development. The utilization of prevalent messengers can be inferred by comparing the *in vitro* translation products of polysomal RNA with the products of transcripts stored as RNP. As we discussed earlier, RNA extracted from the polysomes of cleavage-stage sea urchin embryos produces the same kinds of proteins as the RNA isolated from unfertilized eggs when translated *in vitro*. However, careful comparison of gel autoradiographs of proteins synthesized *in vitro* from polysomal and free messengers from cleavage-stage embryos reveals certain quantitative differences (Fig. 12.13). Thus, there is a translational control mechanism that regulates the number of certain transcripts that are mobilized (Infante and Heilmann, 1981).

Histone messengers in the sea urchin zygote are also subject to translational control. This is demonstrated by the delayed recruitment of a subset of these messengers as compared to the bulk of oogenic messenger RNA (Goustin, 1981; Wells et al., 1981). Delayed recruitment of histone messengers has been attributed to their unusual mode of storage in the egg: They are stored in the egg pronucleus, rather than in the cytoplasm (Venezky et al., 1981; Showman et al., 1982; De Leon et al., 1983). As we discussed in Chapter 7, meiosis is completed before the sea urchin oocyte has ovulated. Thus, the unfertilized egg contains a haploid

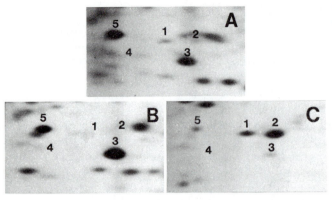

FIGURE 12.13 Comparison of proteins synthesized *in vitro* from (*A*) sea urchin total egg RNA, (*B*) cleavage-stage polysomal RNA, and (*C*) cleavage-stage free (i.e., nonpolysomal) RNA. Shown here are small portions of two-dimensional gels, which correspond to the region labeled B in Figure 12.6. (Reprinted with permission from A.A. Infante and L.J. Heilmann, 1981. Copyright 1981 American Chemical Society.)

pronucleus that fuses with the male pronucleus after fertilization. In preparation for the first cleavage division the envelope surrounding the fusion nucleus breaks down before metaphase. At this time the sequestered histone messengers are released from the nucleus and can be translated.

The utilization of the complex class of messengers is demonstrated by RNA-DNA hybridization technology. A specific probe for oogenic messengers is made by hybridizing egg RNA with single copy DNA and recovering the DNA from the hybrids. This DNA is then used to detect homologous sequences after fertilization. This DNA has been designated **oDNA** (i.e., **oocyte DNA**). The percent of hybridization between the oDNA and embryonic RNA populations is a measure of their relatedness to the original oogenic messenger set. RNA extracted from polysomes at the 16-cell stage is, by definition, the RNA that is being actively translated at that stage, and as shown by Humphreys (1971), it is predominantly of oogenic origin. The oDNA hybridizes with this polysomal RNA to 73% of the reaction with oogenic RNA. In other words, approximately 73% of the oogenic RNA sequences are being translated at this time. Interestingly, oDNA hybridizes with *total* cytoplasmic 16-cell–stage RNA to 100% of the reaction with oogenic RNA, an indication that virtually all of the oogenic sequences are present in the cytoplasm, although they are not all being concurrently translated (Hough-Evans et al., 1977). One possible interpretation of these experiments is that a portion of the complex class RNA is not functional as messengers.

The possibility that oogenic RNA contains transcripts that are not translated is also suggested by the observation that most of the total mass of the cytoplasmic complex class transcripts of sea urchin eggs contains repetitive sequences that are interspersed (i.e., covalently linked) with the single copy sequences (Costantini et al., 1980). This suggests that adjoining single copy and repetitive sequences of DNA are transcribed during oogenesis, and the repetitive sequences are not removed from the transcripts before they are transported out of the oocyte nucleus (see Chap. 3 for discussion of RNA processing). Perhaps the incompletely processed transcripts are not messengers but perform some as yet undiscovered function during development. Alternatively, they may be translatable if they undergo processing in the cytoplasm to remove the interspersed regions. However, no cytoplasmic mechanism of this kind has been described. It is interesting and perhaps significant that a major fraction of *Xenopus* oocyte polyadenylated RNA also has this same interspersed sequence configuration (Anderson et al., 1982). The presence of these transcripts in these two evolutionarily distant groups suggests that this may be a common feature of oogenic RNA with an important function in initial development.

Localization Of Oogenic Messenger

Since oogenic messenger RNA plays a critical role in directing initial development, it is logical to inquire whether it is equally distributed in the cytoplasm or if there is a pattern of distribution that may play a role in directing the differential development of regions of the embryo. The most logical embryos in which to search for mRNA distributional patterns are those that contain recognizable cytoplasmic localizations of morphogenic determinants. A coincidence of messenger localization with regions containing morphogenic determinants would suggest a role for the messenger in directing regional development. Ascidian eggs, with their three identifiable ooplasms (ectoplasm, endoplasm, and myoplasm; see Chap. 11), are excellent material in which to search for messenger localization. Recent experiments with these embryos do reveal a coincidence of messenger RNA distribution with certain ooplasms.

Jeffery and Capco (1978) used ^{3}H-poly (U) to localize polyadenylated RNA in sectioned *Styela* embryos by *in situ* hybridization (see Chap. 3 for discussion of this technique). Their results indicate that polyadenylated messenger RNA is enriched in the ectoplasm (Fig. 12.14), whereas the endoplasm has an intermediate concentration of this RNA, and the myoplasm has very low concentrations. Their experiments also show that the characteristic concentrations of poly (A)$^+$ RNA are retained while the ooplasms are shifted during ooplasmic segregation. This observation suggests that the RNA molecules have an affinity for specific ooplasmic constituents. Recent experiments indicate that the mRNA is bound to the cytoskeleton (Jeffery, 1983).

The distribution of *specific* mRNA sequences has also been studied by *in situ* hybridization with cloned probes, which were labeled with ^{125}I (Jeffery, 1983). Two probes were used in this study: a histone gene probe and an actin gene probe. Results with the histone probe indicate that histone oogenic messengers are evenly distributed in the egg and embryo. However, actin mRNA distribution is distinctly asymmetrical. As shown in Figure 12.15, the highest concentrations of actin mRNA molecules are detected in the ectoplasm and myoplasm, with very low con-

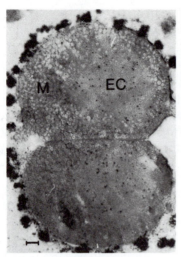

FIGURE 12.14 Distribution of polyadenylated RNA in a meridional section of a two-cell–stage *Styela* embryo. Poly (A)$^+$ RNA is localized by *in situ* hybridization with ^{3}H-poly (U), followed by autoradiography. M, myoplasm; EC, ectoplasm. Scale bar equals 10 μm. (From W.R. Jeffery and D.G. Capco. 1978. Differential accumulation and localization of maternal poly (A)-containing RNA during early development of the ascidian, *Styela*. Dev. Biol., *67:* 162.)

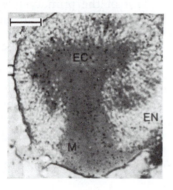

FIGURE 12.15 Distribution of actin mRNA in *Styela* embryo. One of the blastomeres of a two-cell embryo is shown here. Actin mRNA is localized by *in situ* hybridization with ^{125}I-labeled actin DNA, followed by autoradiography. Silver grains are prevalent over the ectoplasm (EC) and myoplasm (M) but not the endoplasm (EN). Scale bar equals 10 μm. (From W.R. Jeffery. 1983. Messenger RNA localization and cytoskeletal domains in ascidian embryos. *In* W.R. Jeffery and R.A. Raff (eds.), *Time, Space and Pattern in Embryonic Development.* Alan R. Liss, New York, p. 248.)

centrations in the endoplasm. The high concentrations of actin messenger in the myoplasm are particularly striking, since very low concentrations of messenger (i.e., poly (A)$^+$ RNA) are found there. Hence, actin messengers must account for a substantial percentage of the messenger RNA in the myoplasm. The significance of the distribution pattern of actin mRNA is uncertain. However, since the myoplasm contains muscle determinants, it is tempting to speculate that the presence of the messenger RNA for a muscle protein is related to muscle differentiation. Further experimentation is necessary to establish such a relationship.

12–2. EXPRESSION OF THE EMBRYONIC GENOME

The results of experiments involving species hybridization and enucleation indicate that the initial stages of embryonic development are extensively influenced by the maternal genome, whereas the newly formed embryonic genome seems to have little impact on development at this time. As development proceeds, the oogenic influence wanes, and the embryonic nuclei assume preeminence. The assumption of embryonic nuclear control actually involves two stages. The first is the *synthesis of RNA* by the nuclei of embryonic cells, and the second is the *utilization of the newly synthesized RNA* to produce functional proteins that impart on the embryonic cells a new identity that reflects the new combination of paternal and maternal genes.

Onset of Embryonic Nuclear Control

The concept of delayed nuclear influence over development originated with the echinoid species hybridization experiments conducted at the turn of the century. These experiments pinpointed the mesenchyme blastula stage as the time of onset of embryonic nuclear influence. Before this stage, phenotypic characteristics are entirely maternal; afterward paternal or hybrid characteristics are detected. This shift of responsibilities is illustrated by the behavior of the primary mesenchyme cells. These cells detach from the blastoderm during the late blastula stage and populate the blastocoele (see Chap. 13). Later, during midgastrulation, they begin to form triradiate spicules and eventually produce the branched skeleton of the pluteus larva. The formation of the primary mesenchyme cells appears to be a responsibility of the maternal genome, whereas the differentiation of these cells to form the skeleton is regulated by the embryonic genome.

A cross between echinoids that differ in the origin and subsequent differentiation of the mesenchyme illustrates this point (Table 12–3). *Cidaris tribuloides* is a primitive echinoid that differs from the more advanced sea urchins in certain aspects of primary mesenchyme formation. In *Cidaris* the mesenchyme forms from the tip of the archenteron

TABLE 12–3
Development of *Cidaris* (♀) × *Lytechinus* (♂) Hybrids

	Archenteron invagination (hours)	Primary mesenchyme formation (hours)	Site of origin of primary mesenchyme cells
Cidaris (♀)	20–33	23–26 (follows invagination)	Archenteron tip
Lytechinus (♂)	9	8 (precedes invagination)	Archenteron base and sides
Cidaris × *Lytechinus*	20	24 (follows invagination)	Archenteron base and sides

From Davidson, E.H. 1976. *Gene Activity in Early Development*, 2nd ed. Academic Press, New York. Collated from the data of Tennent, D.H. 1914. The early influence of the spermatozoan upon the characters of echinoid larvae. Carnegie Inst. Wash. Publ., *182*: 129–138.

after its invagination. Conversely, in *Lytechinus variegatus*, a typical urchin, the primary mesenchyme is formed *before invagination* of the archenteron, and the cells originate from the base and sides of the presumptive archenteron. When *Cidaris* females are mated with *Lytechinus* males, the characteristics of the cleavage- and blastula-stage embryos are entirely maternal; the primary mesenchyme cells originate after gastrulation has begun, albeit somewhat earlier than in *Cidaris* embryos. However, the site of origin of these cells is at the base and sides of the archenteron—a paternal characteristic. Thus, the timing of their formation is primarily maternally determined, but the mode of their formation is determined by embryonic genes. The results of other hybridization experiments indicate that the skeletal elements of the pluteus, which are formed from the primary mesenchyme cells, assume hybrid characteristics, an indication of embryonic gene expression during their differentiation. Treatment of sea urchin zygotes with actinomycin D confirms the conclusions of these experiments: Primary mesenchyme is formed in the presence of the inhibitor, but the mesenchyme cells do not differentiate to form skeletal elements.

A more precise means of detecting the assumption of control of the embryonic genome over development is to monitor the participation of the embryonic genome in RNA and protein synthesis. We shall now examine the initiation of transcription in sea urchin development and the appearance of these newly synthesized transcripts on polysomes. Where possible, we shall analyze the synthesis of *specific transcripts* and their utilization in protein synthesis.

As we shall see, transcription of the embryonic genome begins during early cleavage. However, as we learned earlier in this chapter, the majority of the messenger translated by cleavage-stage embryos is of oogenic origin. Thus, the expression of embryonic genes can make little impact on protein synthesis during cleavage. One reason for this is that the small number of nuclei during early cleavage is able to produce only a small amount of RNA in comparison to the large amount of oogenic messenger stored in the cytoplasm.

The impact of embryonic gene expression increases as oogenic messengers are replaced by embryonic transcripts. By the blastula stage the polysomes contain predominantly *newly synthesized* messengers (Brandhorst and Humphreys, 1972; Galau et al., 1977). This replacement is dependent upon the degradation of preexisting messengers and the accumulation of newly synthesized messengers. This does not happen all at once but is phased in over a period of time and varies, depending on the particular RNA.

One way to monitor the accumulation of newly synthesized messengers is to measure the level of messenger abundance as development proceeds. An increase in abundance of a particular messenger indicates *de novo* transcription of its gene. Changes in messenger abundance during initial development can be determined by hybridization analysis between the RNA and cloned DNA probes. Actin messenger abundance has been determined in this way. The procedure used in these experiments is a modification of the Southern blotting technique discussed in Chapter 3. In this procedure (called **RNA blot analysis** or **Northern blotting**) the RNA is transferred by blotting to a filter, on which it is hybridized to labeled single-strand DNA. The results of RNA blot analysis with RNA from *Strongylocentrotus purpuratus* embryos show that two size classes of actin mRNA are present at relatively low levels in the egg and remain at low levels until the blastula stage, when they have increased substantially due to *de novo* synthesis (Crain et al., 1981). An RNA blot analysis for cleavage through blastula stages is shown in Figure 12.16. An interesting variation on this pattern is seen in the related species, *Strongylocentrotus drobachiensis,* in which the actin messengers are more abundant in eggs than in 16-cell embryos and then begin to accumulate in the blastula and gastrula stages. The oogenic actin messenger is apparently degraded shortly after fertilization and is replaced by embryonic messengers at the later stages (Bushman and Crain, 1983). Clearly, there are species-specific variations in the longevity of oogenic messengers.

Histone messengers present a unique opportunity to monitor the utilization of oogenic messengers and the expression of embryonic genes, since specialized histone gene families are expressed at specific times during early development. Each family contains genes for the five types of histones. Genes in one family differ in nucleotide sequence from the homologous genes in the other families. Messengers produced by one family are distinguishable from messengers transcribed from another family by their distinct electrophoretic mobility, the specificity of hybridization to cloned histone genes, and the thermal stability of RNA-DNA hybrids. The latter property is used to distinguish messengers that

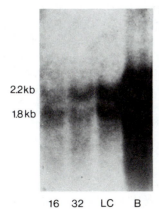

2.2kb

1.8kb

16   32   LC   B

Stage of
development

FIGURE 12.16 RNA blot analysis of actin mRNA from *Strongylocentrotus purpuratus* embryos. Total cytoplasmic RNA was electrophoresed, blotted, and hybridized to labeled actin DNA. Two size classes of actin mRNA (1.8 kilobases and 2.2 kilobases) are detected. Stages of development (16-cell, 32-cell, late cleavage, and blastula) are shown for each RNA sample. A dramatic increase in actin mRNA occurs between the late cleavage stage (12 hours post-fertilization) and the blastula stage (18 hours post-fertilization). (From W.R. Crain, Jr., D.S. Durica, and K. Van Doren. 1981. Actin gene expression in developing sea urchin embryos. Mol. Cell. Biol., *1:* 718.)

incompletely hybridize with a DNA probe, since hybrids of this sort have reduced stability. For example, messengers transcribed from the late histone gene family (see following discussion) form hybrids having reduced stability with genes of the early histone gene family (Childs et al., 1979).

In some cases the differences in nucleotide sequence among families are sufficient to produce electrophoretically distinct histone proteins. This is most pronounced for the H1, H2A, and H2B histones (Newrock et al., 1978). The messengers for these histones from different families can be identified by the specific histones they encode *in vitro*.

During very early cleavage stages the histones that are synthesized are called the **cleavage-stage (CS) histone subtypes.** The messengers for CS histones are present in the oogenic messenger population and are also synthesized for a short time after fertilization until the genes encoding them are silenced.

By the eight-cell stage, synthesis of the so-called **early histone subtypes** (sometimes called **α variants**) is initiated. These histones continue to be made until the blastula stage. Like the CS histone messengers, the messengers for the early histones are also present in the oogenic messenger population. However, these two types of messengers are stored in separate compartments in the egg. The CS histone messengers are stored in the cytoplasm along with the messengers for other proteins (Showman et al., 1983), whereas the early histone messengers are stored in the female pronucleus. We have discussed previously this unusual mode of messenger storage and its implications for messenger utilization (see p. 547). Synthesis and accumulation in the embryo of transcripts of the early histone genes begin in early cleavage but persist longer than for the CS histone messengers. The messengers transcribed from the early genes reach a maximum level of accumulation by the 200-cell stage, after which they rapidly decline in amount (Mauron et al, 1982; Maxson and Wilt, 1982).

A third category of histones, called the **late histone subtypes,** is synthesized from the 16-cell stage through the pluteus stage (Maxson et al., 1983). By the midgastrula stage these are the only histones being synthesized. Transcription of the genes encoding the late embryonic histones occurs as the levels of the early histone messengers begin to decline. As a consequence, late histone messenger accumulation coincides with the disappearance of the early histone messengers (Childs et al., 1979; Hieter et al., 1979; Maxson and Wilt, 1982; Maxson et al, 1983). The changing levels of the early and late forms of histone H2B messenger RNA are shown in Figure 12.17. The transition to late histone messenger accumulation is concomitant with the reduction in the rate of cell division that follows cleavage.

The stage-specific synthesis of histones provides an excellent system to study the utilization of oogenic transcripts, initiation of transcription, and the regulation of messenger accumulation and utilization during early embryonic development.

Hours

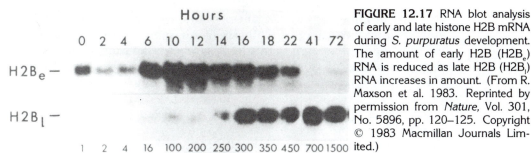

FIGURE 12.17 RNA blot analysis of early and late histone H2B mRNA during *S. purpuratus* development. The amount of early H2B (H2B$_e$) RNA is reduced as late H2B (H2B$_l$) RNA increases in amount. (From R. Maxson et al. 1983. Reprinted by permission from *Nature*, Vol. 301, No. 5896, pp. 120–125. Copyright © 1983 Macmillan Journals Limited.)

The transition from utilization of oogenic messengers to embryonic messengers shifts the responsibility for protein synthesis to the embryonic genome. As we have seen, this can result in qualitative changes in the histone messenger population. Is there a generalized qualitative change in messengers as a consequence of this transition, or does the repertoire of the messenger population remain much the same in spite of the shift in the source of the messengers? A variety of studies have revealed that a great many sequences utilized in embryonic protein synthesis are represented in both the oogenic and embryonic messenger RNA, whereas others are supplied entirely as oogenic messengers, and still others are supplied entirely by transcription after fertilization (for additional discussion, see Crain et al. [1981]; Davidson et al. [1982]).

As development continues past the blastula stage, the messenger RNA complexity is reduced, reflecting a decrease in the number of different mRNA sequences (Galau et al., 1976; Shepherd and Nemer, 1980). This is illustrated by preparation of cDNA from the prevalent polyadenylated RNA isolated from polysomes of blastula-stage sea urchin embryos and the use of this cDNA to detect homologous sequences on gastrula polysomes (Shepherd and Nemer, 1980). Of the prevalent blastula messengers, only about 16% remain prevalent on gastrula polysomes, another 16% are present at reduced levels, whereas the remaining two thirds are absent. However, when *nuclear* RNA from gastrulae is hybridized with the cDNA, it hybridizes to the same extent as blastula mRNA. Thus, all of the abundant blastula messenger species are present in the gastrula nucleus, despite the deficiency of two thirds of them from the polysomes. Clearly, the nuclei of gastrula cells continue to synthesize the abundant blastula messenger sequences, but posttranscriptional mechanisms regulate their utilization as messengers in the cytoplasm. Similar results have been reported for complex-class mRNA sequences (Wold et al., 1978), indicating that the reduction of complexity of both prevalent and complex-class messengers is due to posttranscriptional mechanisms.

In addition to synthesizing messenger RNA, the nuclei of sea urchin embryos are active in the synthesis of ribosomal RNA, thus supplementing the large amount of this RNA synthesized during oogenesis and stored in the egg ribosomes. However, the rate of synthesis of rRNA is quite low in comparison to the synthetic rate during oogenesis (Griffith et al., 1981). Electron micrographs of sea urchin embryo chromatin prepared by the Miller technique (as we discussed in Chap. 3, chromatin prepared by this procedure retains RNA in the process of synthesis) reveal that the small number of ribosomal RNA genes that are being transcribed are highly active (Busby and Bakken, 1980). The difference in synthetic rates in oocytes and embryos is probably due to a difference in the number of rRNA genes that are synthetically active. The vast majority of rRNA genes remain inactive during embryogenesis. Whether the genes that are active during embryonic development are a small subset of those genes that are active during oogenesis, or whether there is a distinct large set of these genes that are active during oogenesis and a much smaller set that is active after fertilization is not known. The rate of rRNA synthesis is increased at the pluteus larval stage after feeding has begun (Humphreys, 1973).

For comparative purposes we shall now examine the onset of embryonic nuclear control over development in amphibians. Most of the research on the molecular biology of early amphibian development has been conducted on *Xenopus laevis*. Hence, this discussion is based upon data obtained with this species.

Transcription is demonstrated in early cleavage-stage *Xenopus* embryos by incorporation of radioactive precursors into RNA. The type of RNA synthesized by these embryos is determined by sucrose density gradient centrifugation and base composition analysis. This RNA is heterogeneous in size, has DNA-like base composition, and is primarily nuclear (Davidson, 1976). Thus, it is primarily HnRNA (heterogeneous nuclear RNA). Not all of the newly synthesized RNA is restricted to the nucleus, however, since it has also been detected on polysomes (Brown and Gurdon, 1966). Some of it must therefore be messenger RNA. Little mRNA is synthesized during amphibian cleavage, and this RNA decays very rapidly. The minute amounts of newly synthesized messenger RNA that accumulate have little impact on the RNA content of these embryos because the preexisting oogenic RNA is present in overwhelming amounts. In the blastula, messenger RNA accumulation increases markedly because of an apparent stimulation of transcriptional activity (Shiokawa et al., 1981b).

As we discussed previously, the decay of oogenic messengers and accumulation of newly synthesized mRNA shifts control over development to products of embryonic gene transcription. Hybridization analyses with cDNA probes representative of individual oogenic polyadenylated mRNA sequences have monitored changes in these sequences during early development (Colot and Rosbash, 1982). Most (80%) of these sequences increase in amount during early development and are therefore produced by embryonic transcription as well as by oocyte transcription. One sequence largely disappears at the late blastula or early gastrula stage (Fig. 12.18). Embryonic synthesis of this sequence is thus of little or no consequence. This study has examined a small handful of literally thousands of oogenic sequences. It is, consequently, premature to generalize on the findings. However, the results suggest that the vast majority of oogenic transcripts are resynthesized during development.

In spite of the resynthesis of oogenic messenger sequences by embryonic gene transcription, *Xenopus* embryos also acquire messenger sequences that do not preexist as oogenic messenger. Schafer et al. (1982) have monitored for the appearance and accumulation of such sequences during development, using cDNA clones prepared from polyadenylated RNAs isolated from *Xenopus* somatic cells. Clones that do not hybridize detectably to oocyte RNA were isolated. RNA blot analyses reveal that messengers that are complementary to these probes appear at various times *after gastrulation*. Presumably, the appearance of each sequence is strategically timed to coincide with a particular developmental event that requires the messenger. More data of this type should provide a detailed chronology of the first appearance of a large number of specific messengers during development and provide a reconstruction of the process of assumption of control over development by the embryonic genome.

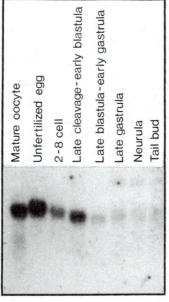

FIGURE 12.18 RNA blot analysis showing the loss of an oogenic RNA sequence during early *Xenopus* development. Polyadenylated RNA from oocytes, eggs, and embryos was electrophoresed, blotted, and hybridized to a cloned *Xenopus* probe. (From H.V. Colot and M. Rosbash. 1982. Behavior of individual maternal pA⁺ RNAs during embryogenesis of *Xenopus laevis. Dev. Biol., 94*: 82.)

We shall now consider the *de novo* transcription of the other major classes of RNA; that is, tRNA, 5S RNA, and 18S and 28S RNA. Unlike messenger RNA, no new synthesis of these types of transcripts has been detected during cleavage. Their synthesis is first detected at the blastula stage, coincident with the acceleration in messenger RNA synthesis (Table 12–4). As we discussed in Chapter 7, transcription of both oocyte and somatic 5S RNA genes occurs during oogenesis; this transcription is terminated after oocyte maturation. However, in the embryo only the somatic 5S RNA genes are transcribed, whereas the oocyte 5S RNA genes remain inactive. The differential activation of 5S RNA genes is an excellent system for the analysis of the regulation of gene transcription.

The activation of transcription of tRNA, 5S RNA, and 18S and 28S RNA genes and the acceleration in synthesis of messenger RNA during the blastula stage of *Xenopus* development indicate that the nuclei have undergone a major transition at this stage that results in transcriptional stimulation. In addition to the stimulation of transcription, a number of other parameters change at this time: The rate of cleavage slows, the synchrony of cell division is lost, and cells become motile for the first time. This constellation of events marks the **midblastula transition** (Gerhart, 1980; Newport and Kirschner, 1982a). Newport and Kirschner have demonstrated that the midblastula transition is not due to an endogenous counting mechanism based on either elapsed time, the number of rounds of DNA synthesis, or the number of cleavages. Instead, they proposed that the transition occurs when the ratio of nuclear and cytoplasmic volumes has reached a critical level (see also pp. 430–431).

As the number of nuclei increases during cleavage, the total cytoplasmic volume remains constant. Therefore, the nucleo-cytoplasmic volume ratio increases. Newport and Kirschner (1982b) have proposed that a substance present in excess in the egg cytoplasm (and not synthesized after fertilization) is removed from the cytoplasm by the nuclei. This substance is proposed to maintain the high rate of mitosis found during cleavage. After the nuclei have reached a critical number the substance is reduced to a threshold level, below which the high cleavage

TABLE 12–4
Initiation of RNA Synthesis During Development of *Xenopus Laevis*

Class of RNA	Stage of initial synthesis	References
mRNA	Cleavage	Brown and Littna, 1964; Shiokawa et al., 1981b
tRNA	Blastula	Bachvarova et al., 1966; Brown and Littna, 1966
5S RNA	Blastula	Miller, 1974
18S and 28S RNA	Blastula	Shiokawa et al., 1981a

rate can no longer be maintained; the slower division rate that ensues allows for the expression of previously suppressed functions such as transcription. The nuclear component may be DNA, which would bind the cytoplasmic substance in stoichiometric amounts (i.e., binding would be directly proportional to the total amount of DNA).

To test these hypotheses, Newport and Kirschner injected plasmids containing a yeast tRNA gene into fertilized eggs, which were prevented from cleavage by briefly centrifuging them. These eggs (called **coenocytic eggs**) continue to undergo nuclear division even though cytoplasmic division does not occur. Activation of transcription in these coenocytic eggs occurs at the same time as in control embryos. The plasmids injected into these eggs become transcriptionally inactive shortly after injection and remain inactive until the midblastula transition—the time when endogenous gene transcription is also accelerated.

Newport and Kirschner reasoned that if the amount of DNA is the critical factor that determines when transcription is activated, transcription of the plasmids might be activated prematurely if the amount of DNA were increased experimentally. Thus, they injected large amounts of an unrelated plasmid into the eggs to increase the total amount of DNA. Since these coenocytic eggs have not cleaved, this increases the ratio of DNA to total cytoplasm substantially. Transcription of the yeast tRNA genes was reactivated after this injection—well before the midblastula transition. This result suggests that activation of transcription at the midblastula transition is, indeed, dependent upon the accumulation of a critical amount of DNA.

The large size of the *Xenopus* egg facilitates microinjection, which makes it possible to conduct the kinds of experiments we have just discussed. The ability to monitor transcription of exogenous genes in the living embryo provides a powerful tool for testing hypotheses concerning regulation of transcription. Additional experiments of this type may eventually lead to the identification of the various components that exert regulatory control over gene transcription during early development.

12–3. GENE EXPRESSION DURING INITIAL DEVELOPMENT OF MAMMALS

The eggs of most animals are macroscopic and contain relatively large amounts of yolk and cytoplasm, which sustain the embryos until they have reached the stage where they can feed for themselves. The mammalian egg, on the other hand, is small and relatively poor in yolk and cytoplasm. During development the embryo receives nutrients directly from the mother, thereby reducing the embryo's reliance upon its own stored nutrients. These unique aspects of mammalian development require that it be discussed separately.

In this discussion we are utilizing data gathered primarily from studies on rabbit and mouse embryos. This approach produces an oversimplification of mammalian development, since mammals are a very diverse phylogenetic group. However, these are the only mammals for which significant information is available.

The choice of these two organisms by the majority of investigators has not been accidental. Both are common laboratory animals that are easy to maintain, and techniques are available for obtaining large numbers of embryos by hormonal superovulation and for culturing embryos *in vitro* for experimentation. The mouse offers the added advantage that a great deal of information on its genetic characteristics has been obtained over the years. The rabbit embryo presents an advantage over the mouse embryo for many kinds of biochemical and molecular studies: Its diameter is approximately twice that of the mouse embryo, and in addition, it undergoes a tremendous increase in size during the blastocyst stage. At the beginning of this stage the rabbit embryo consists of 100 to 130 cells, but it expands to 80,000 cells before implantation. This contrasts with the mouse blastocyst, which consists of only about 60 cells at the time of implantation. The increase in cell number in rabbit development is true growth, in contrast to the increase in cell number during cleavage of nonmammalian embryos, in which the increases in cell number are at the expense of cell size. Blastocyst expansion provides investigators with more material for conducting biochemical and molecular studies. These and other differences between the two embryos indicate that they utilize quite different developmental programs. Thus, we cannot extract a concept of typical mammalian development from either species. The final picture of mammalian development is likely to be a composite of several different modes of differentiation, and it remains to be seen which differences are significant and which are trivial (Manes, 1975).

Gene expression during early mammalian development is of more immediate consequence than it is for the nonmammalian embryos that were discussed earlier in this chapter. This reliance of the early mammalian embryo on expression of embryonic genes is evident when fertilized eggs are treated with inhibitors of RNA synthesis. Although actinomycin D has been used in a number of these studies, it has several side effects on embryos that are unrelated to its inhibitory effects on transcription (Manes, 1975). Accordingly, α-amanitin is considered to be a more specific inhibitor of transcription. Treatment of mouse and rabbit embryos with this inhibitor causes development to be halted during cleavage (Golbus et al., 1973; Manes, 1973). Cleavage arrest is not immediate, however, since a limited amount of cleavage occurs in the presence of α-amanitin. This delay is particularly pronounced in the rabbit embryo; zygotes will develop to the eight-cell stage before development is arrested. The concentration of α-amanitin used in these experiments (10^{-4}M) is sufficient to virtually abolish RNA synthesis immediately, but the effects on protein synthesis are delayed (Table 12–5).

These results lead to two conclusions. First, the delayed effect of the inhibitor on protein synthesis and cleavage indicates that some oo-

TABLE 12–5
Effect of α-Amanitin on Uridine and Amino Acid Incorporation by Rabbit Embryos *In Vitro*

Label	Embryonic age	Counts per minute per embryo	
		Control	α-Amanitin
^{3}H-Uridine	Day 1	198 ± 21	1.4 ± 1.2
^{3}H-Amino Acids	Day 1	191 ± 5	193 ± 7
	Day 2	202 ± 21	98 ± 12

From Manes, C. 1973. The participation of the embryonic genome during early cleavage in the rabbit. Dev. Biol., *32*: 453–459.

genic messenger is translated during the initial cleavages. Second, embryonic gene expression is essential for the continuation of cleavage. Thus, cleavage is maintained by stored oogenic transcripts as well as by new transcripts synthesized from the genomes of embryonic cells. This points out a major difference between mammals and nonmammalian species: Cleavage in the latter group does not depend upon new transcripts but can occur in the absence of *de novo* transcription. However, it is necessary to point out that although the oogenic transcripts in the rabbit egg are of little consequence in terms of the *extent* of development they will support, they are certainly significant in the *length of time* during which they are capable of exerting an influence on the embryo. Oogenic RNA will support development of the mouse zygote to the two-cell stage and the rabbit zygote to the eight-cell stage, which is a period of 24 hours in both cases. This is a relatively long time in developmental terms.

The results of inhibitor studies are supported by the results of experiments with manually enucleated mouse eggs. In these experiments unfertilized and fertilized eggs were enucleated and cultured in a medium containing ^{35}S-methionine. The pattern of protein synthesis in the enucleated eggs was compared to the pattern in control embryos by examining the array of radioactive polypeptides on two-dimensional electrophoresis gels. The enucleated fertilized eggs synthesized a number of polypeptides that are characteristically synthesized at the early two-cell stage of embryonic development (Fig. 12.19). However, unfertilized enucleated eggs failed to synthesize these characteristic polypeptides. These results suggest the presence of oogenic messenger in mouse eggs. The difference in the protein synthetic patterns of unfertilized and fertilized enucleated eggs indicates that utilization of certain of these transcripts can occur only after fertilization.

The continuation of cleavage and specific protein synthesis in the absence of *de novo* RNA synthesis implies that the unfertilized mammalian egg contains a store of ribosomes, tRNA, and messengers. The presence of oogenic messenger RNA in unfertilized rabbit and mouse

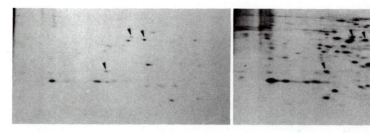

Fertilized eggs, enucleated + 2 days   2-cell embryos, controls

FIGURE 12.19 Autoradiogram of two-dimensional gel of proteins synthesized by (left) enucleated fertilized mouse eggs cultured for two days and (right) two-cell embryos. The arrowheads indicate polypeptides typical of two-cell embryos. Some of these polypeptides are also synthesized by the enucleated fertilized eggs. (From U. Petzoldt, P.C. Hoppe, and K. Illmensee. 1980. Protein synthesis in enucleated fertilized and unfertilized mouse eggs. Wilh. Roux' Archiv., *189:* 218. Reprinted with permission of Springer-Verlag, Heidelberg.)

eggs is also suggested by the detection of polyadenylated RNA in them (Schultz, 1975; Levey et al., 1978; Bachvarova and De Leon, 1980). Although not absolute proof of their function as messengers, the presence of poly (A) tracts on the RNA molecules is compatible with this interpretation. A large proportion of this polyadenylated RNA is found associated with ribosomes in unfertilized eggs, indicating that it is functioning as messenger RNA. The majority of the ribosomes in unfertilized mouse eggs are, however, not active in protein synthesis and are presumably stored to support postfertilization protein synthesis (Bachvarova and De Leon, 1977).

The presence of oogenic messengers in unfertilized eggs is confirmed by *in vitro* translation of mRNA extracted from unfertilized mouse eggs and by RNA blot analysis, which demonstrates the presence of *specific* messenger sequences. Braude et al. (1979) translated messenger RNA isolated from unfertilized mouse eggs. As shown in Figure 12.20B, a large number of polypeptides are encoded by these messengers. Figures 12.20A and C show the polypeptides that are synthesized by *intact* unfertilized eggs and early two-cell–stage embryos, respectively. Note the polypeptides indicated by the three arrows. These three polypeptides are synthesized in embryos but not in eggs. Braude et al. (1979) have shown that they are also synthesized in α-amanitin–treated embryos, which implies that they are produced from oogenic messengers. As shown in Figure 12.20B, the messengers for these polypeptides are, indeed, present in the oogenic messenger RNA. Their translation only after fertilization indicates that their utilization is under posttranscriptional control during development.

RNA blot analysis is a powerful tool for the demonstration of specific RNA sequences. Giebelhaus et al. (1983) have used this technique to demonstrate the presence of both actin and histone H3 mRNA in unfertilized mouse eggs (see the next section). As additional probes become available, RNA blot analyses with them should eventually result in the emergence of a catalog of sequences in the mammalian oogenic messenger population.

Transition Of Control Over Development From Oogenic To Embryonic Transcripts

The necessity of nuclear intervention during early mammalian development, which is indicated by the inhibitor studies discussed earlier, is supported by a growing body of genetic evidence obtained from the study

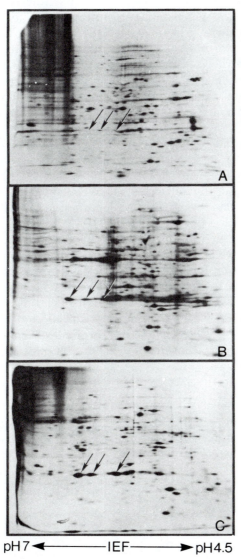

FIGURE 12.20 Comparison of the polypeptides synthesized by (A) intact, unfertilized mouse eggs, (B) mRNA from unfertilized mouse eggs translated *in vitro*, and (C) intact, early two-cell embryos. Arrows indicate polypeptides synthesized in embryos but not in eggs. However, these polypeptides are synthesized *in vitro* from egg mRNA. (From P. Braude et al. 1979. Reprinted by permission from *Nature*, Vol. 282, No. 5734, pp. 102–105. Copyright © 1979 Macmillan Journals Limited.)

pH 7 ◄──────IEF──────► pH 4.5

of mouse mutants. In these experiments, mutants that affect the primary structure of proteins have been used to estimate the time of onset of embryonic gene expression during mouse development. Males and females that differ genetically are mated, and embryonic gene expression is indicated by detection of paternal polypeptides, usually by electrophoresis. The earliest detection of a paternal polypeptide in such an experiment with mouse mutants is the synthesis of β-microglobulin at the two-cell stage (Sawicki et al., 1981). Clearly, embryonic gene expression occurs very early in the mammalian developmental program.

Analyses of RNA synthesis during mammalian development indicate that transcription occurs from the very beginning of development. Mintz (1964) was the first to demonstrate RNA synthesis in early mammalian cleavage stages. Figure 12.21 is Mintz's autoradiograph of a two-cell mouse embryo, which has been labeled with ³H-uridine. Incorporation of the label into RNA is clearly evident. Characterization of the classes of RNA synthesized during mouse development indicates that DNA-like heterogeneous RNA is synthesized as early as the two-cell stage (Woodland and Graham, 1969). This RNA is likely messenger RNA, an interpretation that is reinforced by the detection of newly synthesized polyadenylated RNA in two-celled mouse embryos (Levey et al., 1978). Ribosomal and transfer RNA synthesis are apparently delayed somewhat. There is an indication of rRNA synthesis at the two-cell stage (Knowland and Graham, 1972), and its synthesis has been unequivocally demonstrated at the four-cell stage (Woodland and Graham, 1969). The four-cell stage also marks the first appearance of newly synthesized 4S RNA (Woodland and Graham, 1969).

The cleavage-stage mouse embryo possesses transcripts of dual origin. However, a sustained impact of oogenic transcripts on development is tempered by degradation of these transcripts. The loss of oogenic transcripts is particularly striking during the two-cell stage when the amount of polyadenylated RNA falls dramatically. This is followed by a buildup of this category of RNA between the late two-cell and early blastocyst stages (Pikó and Clegg, 1982). The latter, of course, results from embryonic gene transcription. The shift in the source of transcripts apparently results in modifications to the messenger population that are sufficient to produce changes in the pattern of protein synthesis. This conclusion

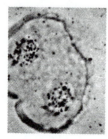

FIGURE 12.21 Autoradiograph of two-cell mouse embryo labeled with ³H-uridine. (From B. Mintz, 1964. Synthetic processes and early development in the mammalian egg. J. Exp. Zool., *157*: 89.)

is based upon the observation that the appearance of a distinct set of proteins at the late two-cell stage is prevented by treatment of the embryos with α-amanitin (Flach et al., 1982). Before this stage, of course, α-amanitin does not have a significant effect on the protein synthetic pattern.

The behavior of particular sequences during the loss of oogenic messengers and replacement by embryonic messengers can now be analyzed by messenger hybridization to cloned probes. As we mentioned in the previous section, the presence of histone H3 and actin sequences in unfertilized eggs has been demonstrated by RNA blot hybridization. The changes in the amounts of these sequences during development have also been determined (Giebelhaus et al., 1983). These results are shown in Figures 12.22 (histone H3 sequences) and 12.23 (actin sequences). In both cases the sequences are present in substantial amounts in the egg, decrease at the two-cell stage, and show a progressive increase from the eight-cell to blastocyst stages. The degradation of oogenic messenger before accumulation of embryonic messenger is reminiscent of the situation for actin messengers in the sea urchin *Strongylocentrotus drobachiensis* (see previous discussion).

The data presented in this and previous sections of this chapter demonstrate that RNA synthesis occurs during cleavage of both mammalian and nonmammalian embryos. The major difference between these groups appears to be the *extent of involvement* of the newly synthesized RNA in development. In nonmammalian species, development will continue in the absence of transcription to a much more advanced stage than it will in the mammalian embryo. Thus, the embryonic genome intervenes in development of mammalian embryos during early

A B C D

FIGURE 12.22 RNA blot analysis of histone H3 RNA from progressive stages of early mouse development. *A*, RNA from 1000 unfertilized eggs. *B*, RNA from 1000 two-cell embryos. *C*, RNA from 1000 eight-cell embryos. *D*, RNA from 1000 blastocysts. (From D.H. Giebelhaus, J.J. Heikkila, and G.A. Schultz. 1983. Changes in the quantity of histone and actin messenger RNA during the development of preimplantation mouse embryos. Dev. Biol., *98:* 150.)

A B C   D

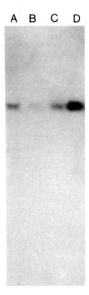

FIGURE 12.23 RNA blot analysis of actin RNA from progressive stages of early mouse development. *A*, RNA from 1000 unfertilized eggs. *B*, RNA from 1000 two-cell embryos. *C*, RNA from 1000 eight-cell embryos. *D*, RNA from 1000 blastocysts. (From D.H. Giebelhaus, J.J. Heikkila, and G.A. Schultz. 1983. Changes in the quantity of histone and actin messenger RNA during the development of preimplantation mouse embryos. Dev. Biol., *98:* 152.)

cleavage, whereas cleavage of nonmammalian embryos is directed primarily by the oocyte genome.

X Chromosome Gene Expression

The role of the genome in developmental regulation requires that the input of genomic information into the protein synthetic machinery be precisely regulated. In Chapter 4 we discussed a regulatory mechanism in female mammals that modulates the utilization of X chromosome information; this mechanism is dosage compensation. Female mammals have two X chromosomes, whereas males have but one. Dosage compensation ensures that only one X chromosome is active in cells of the female. According to the Lyon hypothesis (Lyon, 1961), one of the X chromosomes in the cells of a female becomes inactivated during development, and hence unavailable for transcription. The inactivated X becomes heterochromatic and is consequently highly condensed during interphase (the Barr body) and late replicating. The inactivation is heritable; every mitotic derivative has the same X inactivated. The initial inactivation is generally random; that is, either the paternal or maternal X may be inactivated in any cell. The Lyon hypothesis therefore predicts that a discrete regulatory event in embryonic cells inactivates one X and leaves the other transcriptionally active. Accordingly, both the maternal and the paternal X chromosomes should be euchromatic before this event. Since the embryonic genome is utilized early in mammalian development, it is also possible that both are transcriptionally active before inactivation.

Cytological evidence indicates that both X chromosomes are euchromatic during early cleavage, since neither X is condensed nor late replicating. A late-replicating X chromosome is first detected in the late blastocyst stage of mouse development (Mukherjee, 1976). Furthermore, one of the X chromosomes shows differential staining at approximately the same time—an indication of heterochromatization (Takagi, 1974). Therefore, it is possible that both X chromosomes are transcriptionally active during cleavage, although we should consider the possibility that functional inactivation precedes cytological inactivation. If both X chromosomes are active in the female embryo, she should produce twice as many proteins specified by X chromosome genes as her brothers.

In an elegant experiment Epstein et al. (1978) have examined the activity of hypoxanthine-guanine phosphoribosyl transferase (HGPRT), an X-linked enzyme, during the early blastocyst stage. To determine the sex of the blastocysts, Epstein and his associates used half embryos that had developed from intact embryos separated at the two-cell stage (Fig. 12.24). The twin half embryos were cultured to the blastocyst stage, and one half was used for HGPRT assay, whereas the other half was used for determination of sex by chromosome analyses. A comparison of HGPRT activity between male and female early blastocysts is shown in Figure 12.25. The ratio of female to male HGPRT activity from the combined data is 2.2:1.0. These results are consistent with the interpretation that both X chromosomes are active up to the early blastocyst stage. In additional experiments the investigators determined the enzyme activities of late blastocysts and found no difference between male and female embryos. This result suggests that X inactivation had occurred between the early and late blastocyst stages. A similar result has also been obtained by Kratzer and Gartler (1978).

Another means of determining whether both X chromosomes are active during early development is by cell transfer experiments (Gardner and Lyon, 1971). These experiments were conducted with embryos from a mouse strain heterozygous for X-linked pigmentation mutants. Single cells derived from blastocysts of this strain were injected into host blastocysts, which developed into chimeric mice with patches displaying the pigment of both X-linked genes of the donor cell. Since both kinds of pigment patches were derived from a *single donor cell*, it is obvious that X inactivation occurred in descendants of the donor cell *after* cell transfer.

Although we have described X chromosome inactivation in the embryo as being random, there is evidence that in certain extraembryonic tissues of the rodent embryo the maternal X chromosome is preferentially expressed (Takagi and Sasaki, 1975; Wake et al., 1976). This results from nonrandom inactivation of the paternal X chromosome in the cells of these extraembryonic tissues (Frels and Chapman, 1980). Preferential behavior of the X chromosomes in extraembryonic cells suggests that the maternal and paternal X chromosomes are somehow imprinted to indicate their origins. However, the imprinting of the X chromosomes is inconsequential for X inactivation in embryonic cells.

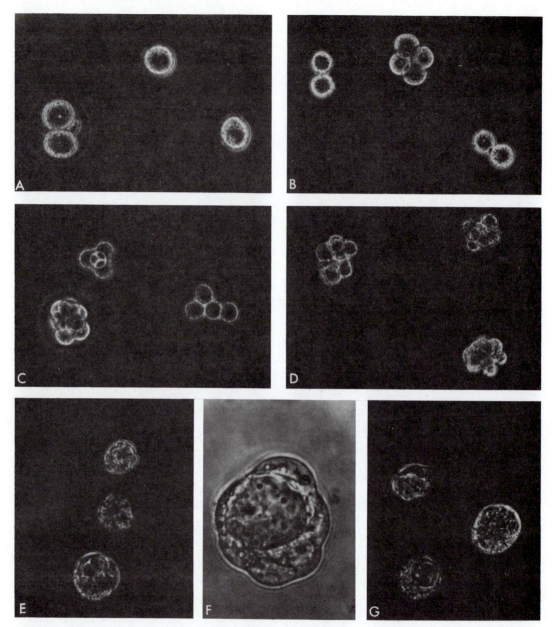

FIGURE 12.24 Development of mouse twin half-embryos in culture. Normal embryos cultured together with half-embryos are shown for comparison in each case. *A,* Single blastomeres shortly after separation; *B,* half-embryos after one cleavage; *C,,* half-embryos after two cleavages; *D,* half-embryos at morula stage; *E,* half-embryos at early blastocyst stage; *F,* magnified view of early half-blastocyst; *G,* early half-blastocysts. (From C.J. Epstein et al. 1978. Reprinted by permission from *Nature,* Vol. 274, No. 5670, pp. 500–503. Copyright © 1978 Macmillan Journals Limited.)

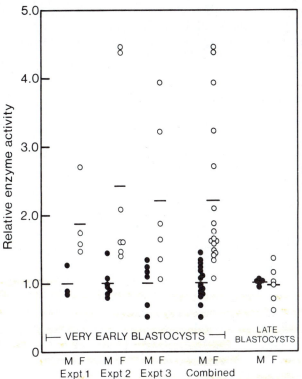

FIGURE 12.25 Relative activities of hypoxanthine-guanine phosphoribosyl transferase (HGPRT) in male and female blastocysts. After determination of sex, the individual half-blastocysts were pooled into groups of five of the same sex and were assayed for enzyme activity. To facilitate comparisons among experiments, the results of each activity were normalized to a mean male enzyme activity of 1.0. M: male; F: female. (After C.J. Epstein et al. 1978. Reprinted by permission from *Nature*, Vol. 274, No. 5670, pp. 500–503. Copyright © 1978 Macmillan Journals Limited.)

12–4. MACROMOLECULAR SYNTHESIS DURING SEED DEVELOPMENT AND GERMINATION

As we described in Chapter 10, the angiosperm zygote begins its development within the ovule. When the embryo is fully formed, it is enclosed by a seed coat, which is derived from the outer integument of the embryo sac. The embryo, nutrient storage tissues, and the seed coat constitute the **seed.**

At the completion of embryonic development in many plants, there is a water loss from seed tissues, which leads to desiccation of the embryo cells. The embryo then enters a **quiescent period,** during which development and metabolic activity cease until the seed germinates. For some seeds, the only requirement for germination processes to commence is imbibition of water. This is accompanied by a resumption of metabolic activity. In other seeds, imbibition of water is also necessary for metabolism to commence, but it is insufficient to trigger germination; additional environmental factors or internal changes are required. In the latter situation the ungerminated seeds are said to be **dormant.** Germination is completed when elongation of the embryonic axis leads to the rupture of the seed coat and protrusion of the elongating radicle from the seed; this is the commencement of seedling growth. In the discussion that follows our primary concern will be with the macromolecular events occurring during seed development and germination so as to foster an appreciation for the utilization of genomic information during these fascinating phases of plant development.

Gene Expression During Seed Development

Macromolecular synthesis during the interval between fertilization and the onset of quiescence has two roles: in the construction of the seed itself during development and in facilitating the rapid reinitiation of activity upon subsequent germination. As we shall see, the latter process is partially dependent upon transcripts that are produced during embryogenesis and stored in the seed for utilization after imbibition.

As with animal development, the study of gene expression during plant development is facilitated by monitoring the expression of genes encoding proteins that are developmentally regulated; that is, genes that are expressed either during a restricted phase of the life cycle or in particular tissues. Excellent candidates for such an analysis are genes encoding the seed proteins—proteins made in large amounts only during embryonic development. These proteins accumulate during seed maturation. This is the phase of seed development after the embryonic axis has been formed (see Chap. 10) when cell division has ceased and rapid cell enlargement occurs. The cotyledons show extensive growth during this phase. Among the seed proteins that are synthesized during maturation are the seed storage proteins, which accumulate primarily in the cotyledons and are the predominant proteins in the mature seed. In the soybean, for example, storage proteins constitute approximately 70% of total seed protein and account for 30 to 40% of seed weight (Meinke et al., 1981). Storage proteins are synthesized on the rough endoplasmic reticulum and deposited in membrane-enclosed protein bodies (for review, see Boulter [1981]).

After seed maturation there is a drastic reduction in metabolic activity as the embryo and its protective seed coat desiccate. After desiccation the seed is shed from the plant and remains quiescent until germination; during subsequent seedling growth the protein bodies containing the storage proteins are catabolized and used as a food source.

An advantage for studying gene expression during maturation is that this process can be controlled during *in vitro* culture. Normally, immature embryos removed from developing seeds do not mature when cultured; they germinate precociously. However, addition of the growth regulator abscisic acid (ABA) will prevent this precocious germination and allow maturation to proceed. ABA production during embryogenesis is thought to be the natural trigger of seed maturation and quiescence (Karssen et al., 1983). In some plants ABA production is stimulated by water stress. Therefore, as the seed begins to dry, ABA is produced, the seed matures and becomes quiescent (Bewley and Black, 1978).

As a consequence of their exposure to ABA, cultured immature embryos undergo a number of modifications in macromolecular synthesis that are characteristic of the maturation phase of development. For example, synthesis of seed proteins (including storage proteins) can be induced by addition of ABA to the culture medium (Crouch and Sussex, 1981; Triplett and Quatrano, 1982). The discovery of a regulator of the maturation phase of the life cycle with effects on the synthesis of embryo-specific proteins is an exciting one because of the potential for the analysis of the control over gene expression. Additional research should reveal whether ABA regulates gene expression directly and, if so, at what level(s) in information utilization.

The stage specificity of synthesis and the abundance of seed storage proteins make the analysis of expression of the genes encoding them an excellent system for developmental analysis. As we have learned, abundant proteins are products of translation of prevalent messengers, which present many technical advantages. Their patterns of synthesis and accumulation are relatively easy to monitor. Recently, cDNAs have been prepared to the soybean storage protein messengers; these cDNAs have been cloned and have become valuable probes of the transcripts. Furthermore, the cDNAs have been subjected to nucleotide sequence analysis and used as probes for complementary sequences in the genome (Beachy et al., 1981; Schuler et al., 1982a, b).

Recent RNA blot analyses with cloned soybean storage protein cDNAs (Fig. 12.26) demonstrate that polyadenylated RNA that hybridizes to these cDNAs is missing until the early cotyledon growth phase, increases dramatically in amount during cotyledon growth, and is reduced below the level of detection in mature, desiccated seeds. Thus, accumulation of these messengers is developmentally regulated. They are most abundant when the seeds are active in storage protein synthesis, suggesting that the primary control over storage protein synthesis is at the transcriptional level. Transcriptional-level regulation of the structural genes for abundant soybean seed proteins (including the storage proteins) has been confirmed by Goldberg et al. (1981b). Hybridization analysis with cloned cDNAs to the messengers for these proteins shows that the messengers accumulate during embryogenesis and decay before seed hydration. Furthermore, transcripts that hybridize to the DNA probes are missing from both the cytoplasmic and nuclear RNA of leaf cells. Thus, these genes are transcriptionally silent except during a short period in embryogenesis when they produce massive amounts of messenger primarily in cotyledon cells. They are analogous to genes in animals that encode abundant, developmentally regulated cell-specific proteins, such as globin and ovalbumin, whose genes are also regulated at the transcriptional level. Since the synthesis of the abundant seed proteins can be induced with ABA in many kinds of seeds, it will be interesting to learn whether transcription of their structural genes is regulated by this growth regulator.

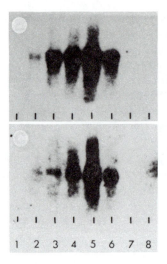

FIGURE 12.26 RNA blot analyses of soybean seed storage protein messengers during seed development. Polyadenylated RNA from seeds at consecutive stages of development was hybridized with probes for two seed protein sequences. Lane 1: early cotyledon stage, active mitosis occurring; lane 2: completion of cell division; lanes 3–5: seeds undergoing maturation; lane 6: mature, green seeds; lane 7: seeds undergoing desiccation; lane 8: dry seeds. (From D.W. Meinke, J. Chen, and R.N. Beachy. 1981. Expression of storage-protein genes during soybean seed development. *Planta, 153:* 134. Reprinted with permission of Springer-Verlag, Heidelberg.)

In addition to the highly abundant seed protein messengers, a variety of other messenger RNAs are synthesized and accumulate during embryogenesis. DNA-RNA hybridization experiments indicate that throughout their development, soybean embryos contain a nearly constant population of 14,000 to 18,000 diverse messenger RNAs. In the final stage of seed development, polysomes decay as messengers and ribosomes dissociate. Most of these mRNA sequences, however, persist in the dry seed (Goldberg et al., 1981a). As we shall discuss in the next section, these messengers stored in the quiescent seed are utilized for protein synthesis during germination. In contrast to the highly abundant seed protein messengers, these diverse sequences are present not only throughout embryogenesis and in the quiescent seed but are also detectable in leaf polysomes (Goldberg et al., 1981a). Thus, there appear to be two distinct categories of structural genes that are transcribed during embryogenesis: a small number encoding highly abundant seed protein messengers that are transcribed during a limited phase of embryonic development and a vast number of genes that are presumably encoding generalized proteins needed in small amounts throughout the life cycle and not subject to transcriptional modulation.

The Control of Protein Synthesis During Germination

Immediately upon imbibition ribosomes associate with messenger RNA and translation begins (Marcus and Freeley, 1964, 1965). The necessity for protein synthesis during germination is illustrated by treatment of imbibing wheat seeds with cycloheximide; protein synthesis is inhibited and germination is prevented (Marcus et al., 1975). Experiments with wheat seeds indicate that the initiation of protein synthesis does not require *de novo* RNA synthesis. When more than 90% of RNA synthesis is prevented with α-amanitin, polysome formation is not significantly reduced during the first 40 minutes of imbibition (Spiegel and Marcus, 1975). This result is interpreted to mean that early protein synthesis is directed by long-lived messengers that are synthesized during embryogenesis, stored in the seed, and recruited during imbibition. Translation of these messengers must utilize stored ribosomes and tRNA.

After a short period of reliance on preformed messengers, responsibility for protein synthesis shifts to newly synthesized messengers. This shift of responsibility results in changes in the pattern of protein synthesis (Cuming and Lane, 1979). The changes in the pattern of protein synthesis are eliminated when the imbibing seeds are exposed to α-amanitin (Thompson and Lane, 1980). This result suggests that the changes in the pattern of protein synthesis are dependent upon *de novo* messenger RNA synthesis. This interpretation is strengthened by the results of *in vitro* protein synthesis: Messengers isolated from quiescent embryos direct the synthesis of proteins resembling those formed during early imbibition, whereas those isolated from late postimbibition embryos direct the synthesis of proteins resembling those formed during late imbibition (Thompson and Lane, 1980; Lane and Tumaitis-Kennedy, 1981). Thus, as in postfertilization development of animals, a supply of preformed messengers directs protein synthesis until those messengers are degraded and are replaced by newly synthesized messengers.

REFERENCES

Anderson, D.M. et al. 1982. Sequence organization of the poly (A) RNA synthesized in lambrush chromosome stage *Xenopus laevis* oocytes. J. Mol. Biol., *155*: 281–309.

Bachvarova, R. et al. 1966. Activation of RNA synthesis associated with gastrulation. Proc. Natl. Acad. Sci. U.S.A., *55*: 358–365.

Bachvarova, R. and V. De Leon. 1977. Stored and polysomal ribosomes of mouse ova. Dev. Biol., *58*: 248–254.

Bachvarova, R. and V. De Leon. 1980. Polyadenylated RNA of mouse ova and loss of maternal RNA in early development. Dev. Biol., *74*: 1–8.

Bakken, A. H. 1973. A cytological and genetic study of oogenesis in *Drosophila melanogaster.* Dev. Biol., *33*: 100–122.

Ballantine, J.E.M., H.R. Woodland, and E.A. Sturgess. 1979. Changes in protein synthesis during the development of *Xenopus laevis.* J. Embryol. Exp. Morphol., *51*: 137–153.

Beachy, R.N., N.P. Jarvis, and K.A. Barton. 1981. In vivo and in vitro biosynthesis of subunits of the soybean 7S storage protein. J. Mol. Appl. Genet., *1*: 19–27.

Bewley, J.D. and M. Black. 1978. *Physiology and Biochemistry of Seeds in Relation to Germination. 1. Development, Germination, and Growth.* Springer-Verlag, Berlin.

Boulter, D. 1981. Biochemistry of storage protein synthesis and deposition in the developing legume seed. Adv. Bot. Res., *9*: 1–13.

Boycott, A.E. et al. 1930. The inheritance of sinistrality in *Limnaea peregra* (Mollusca, Pulmonata). Philos. Trans. R. Soc. Lond. (Biol.), *219*: 51–131.

Brachet, J., H. Denis, and F. deVitry. 1964. The effects of actinomycin D and puromycin on morphogenesis in amphibian eggs and *Acetabularia mediterranea.* Dev. Biol., *9*: 398–434.

Brachet, J., E. Hubert, and A. Lievens. 1972. The effects of α-amanitin and rifampicins on amphibian egg development. Rev. Suisse Zool., *79*: 47–63.

Brandhorst, B.P., and T. Humphreys. 1972. Stability of nuclear and messenger RNA molecules in sea urchin embryos. J. Cell Biol., *53*: 474–482.

Brandis, J.W., and R.A. Raff. 1978. Translation of oogenetic mRNA in sea urchin eggs and early embryos. Demonstration of a change in translational efficiency following fertilization. Dev. Biol., *67*: 99–113.

Brandis, J.W., and R.A. Raff. 1979. Elevation of protein synthesis is a complex response to fertilization. Nature (Lond.), *278*: 467–469.

Braude, P. et al. 1979. Post-transcriptional control in the early mouse embryo. Nature (Lond.), *282*: 102–105.

Briggs, R., and G. Cassens. 1966. Accumulation in the oocyte nucleus of a gene product essential for embryonic development beyond gastrulation. Proc. Natl. Acad. Sci. U.S.A., *55*: 1103–1109.

Briggs, R., E.U. Green, and T.J. King. 1951. An investigation of the capacity for cleavage and differentiation in *Rana pipiens* eggs lacking "functional" chromosomes. J. Exp. Zool., *116*: 455–499.

Briggs, R., and J.T. Justus. 1968. Partial characterization of the component from normal eggs which corrects the maternal effect of gene *o* in the Mexican axolotl (*Ambystoma mexicanum*). J. Exp. Zool., *167*: 105–116.

Brown, D.D., and J.B. Gurdon. 1964. Absence of ribosomal RNA synthesis in the anucleolate mutant of *Xenopus laevis.* Proc. Natl. Acad. Sci. U.S.A., *51*: 139–146.

Brown, D.D., and J.B. Gurdon. 1966. Size distribution and stability of DNA-like RNA synthesized during development of anucleolate embryos of *Xenopus laevis.* J. Mol. Biol., *19*: 399–422.

Brown, D.D., and E. Littna. 1964. RNA synthesis during the development of *Xenopus laevis*, the South African clawed frog. J. Mol. Biol., *8*: 669–687.

Brown, D.D., and E. Littna. 1966. Synthesis and accumulation of low molecular weight RNA during embryogenesis of *Xenopus laevis.* J. Mol. Biol., *20*: 95–112.

Busby, S., and A. Bakken. 1980. Transcription in developing sea urchins: Electron microscope analysis of cleavage, gastrula and prism stages. Chromosoma, *79*: 85–104.

Bushman, F.D., and W.R. Crain, Jr. 1983. Conserved pattern of embryonic actin gene expression in several sea urchins and a sand dollar. Dev. Biol., *98*: 429–436.

Childs, G., R. Maxson, and L.H. Kedes. 1979. Histone gene expression during sea urchin embryogenesis: Isolation and characterization of early and late messenger RNAs of *Strongylocentrotus purpuratus* by gene-specific hybridization and template activity. Dev. Biol., *73*: 153–173.

Collier, J.R. 1966. The transcription of genetic information in the spiralian embryo. *In* A.A. Moscona and A. Monroy (eds.), *Current Topics in Developmental Biology*, Vol. 1. Academic Press, New York, pp. 39–59.

Colot, H.V., and M. Rosbash. 1982. Behavior of individual maternal pA$^+$ RNAs during embryogenesis of *Xenopus laevis*. Dev. Biol., *94*: 79–86.

Costantini, F.D., R.J. Britten, and E.H. Davidson. 1980. Message sequences and short repetitive sequences are interspersed in sea urchin egg poly (A)$^+$ RNAs. Nature (Lond.), *287*: 111–117.

Crain, W.R., Jr., D.S. Durica, and K. Van Doren. 1981. Actin gene expression in developing sea urchin embryos. Mol. Cell. Biol., 1: 711–720.

Crouch, M.L., and I.M. Sussex. 1981. Development and storage-protein synthesis in *Brassica napus* L. embryos *in vivo* and *in vitro*. Planta (Berl.), *153*: 64–74.

Cuming, A.C., and B.G. Lane. 1979. Protein synthesis in imbibing wheat embryos. Eur. J. Biochem., *99*: 217–224.

Danilchik, M.V., and M.B. Hille. 1981. Sea urchin egg and embryo ribosomes: Differences in translational activity in a cell-free system. Dev. Biol., *84*: 291–298.

Darnbrough, C., and P.J. Ford. 1976. Cell-free translation of messenger RNA from oocytes of *Xenopus laevis*. Dev. Biol., *50*: 285–301.

Davidson, E.H. 1976. *Gene Activity in Early Development*, 2nd ed. Academic Press, New York.

Davidson, E.H., B.R. Hough-Evans, and R.J. Britten. 1982. Molecular biology of the sea urchin embryo. Science, *217*: 17–26.

DeLeon, D.V. et al. 1983. Most early variant histone mRNA is contained in pronuclei of sea urchin eggs. Dev. Biol., *100*: 197–206.

Denny, P.C., and A. Tyler. 1964. Activation of protein biosynthesis in non-nucleate fragments of sea urchin eggs. Biochem. Biophys. Res. Commun., *14*: 245–249.

Ecker, R.E., and L.D. Smith. 1971. The nature and fate of *Rana pipiens* proteins synthesized during maturation and early cleavage. Dev. Biol., *24*: 559–576.

Elsdale, T.R., M. Fischberg, and S. Smith. 1958. A mutation that reduces nucleolar number in *Xenopus laevis*. Exp. Cell Res., *14*: 642–643.

Epstein, C.J. et al. 1978. Both X chromosomes function before visible X-chromosome inactivation in female mouse embryos. Nature (Lond.), *274*: 500–503.

Feigenbaum, L., and E. Goldberg. 1965. Effect of actinomycin D on morphogenesis in *Ilyanassa*. Am. Zool., 5:198.

Flach, G. et al. 1982. The transition from maternal to embryonic control in the 2-cell mouse embryo. EMBO J., *1*: 681–686.

Ford, P.J. 1971. Non-coordinated accumulation and synthesis of 5S ribonucleic acid by ovaries of *Xenopus laevis*. Nature (Lond.), *233*: 561–564.

Freeman, G., and J.W. Lundelius. 1982. The developmental genetics of dextrality and sinistrality in the gastropod *Lymnaea peregra*. Wilh. Roux's Arch., *191*: 69–83.

Frels, W.I., and V.M. Chapman. 1980. Expression of the maternally derived X chromosome in the mural trophoblast of the mouse. J. Embryol. Exp. Morph., *56*: 179–190.

Fullilove, S.L., and R.C. Woodruff. 1974. Genetic, cytological, and ultrastructural characterization of a temperature-sensitive lethal in *Drosophila melanogaster*. Dev. Biol., *38*: 291–307.

Galau, G.A. et al. 1976. Structural gene sets active in embryos and adult tissues of the sea urchin. Cell, 7: 487–505.

Galau, G.A. et al. 1977. Synthesis and turnover of polysomal mRNAs in sea urchin embryos. Cell, *10*: 415–432.

Gardner, R.L., and M.F. Lyon. 1971. X chromosome inactivation studied by injection of a single cell into the mouse blastocyst. Nature (Lond.), *231*: 385–386.

Garen, A., and W. Gehring. 1972. Repair of the lethal developmental defect in *deep orange* embryos of *Drosophila* by injection of normal egg cytoplasm. Proc. Natl. Acad. Sci. U.S.A., *69*: 2982–2985.

Gerhart, J.G. 1980. Mechanisms regulating pattern formation in the amphibian egg and early embryo. *In* R.F. Goldberger (ed.), *Biological Regulation and Development.* Plenum Press, New York, pp. 133–315.

Giebelhaus, D.H., J.J. Heikkila, and G.A. Schultz. 1983. Changes in the quantity of histone and actin messenger RNA during the development of preimplantation mouse embryos. Dev. Biol., *98*: 148–154.

Golbus, M.S., P.G. Calarco, and C.J. Epstein. 1973. The effects of inhibitors of RNA synthesis (α-amanitin and actinomycin D) on preimplantation mouse embryogenesis. J. Exp. Zool., *186*: 207–216.

Goldberg, R.B. et al. 1981a. Abundance, diversity, and regulation of mRNA sequence sets in soybean embryogenesis. Dev. Biol., *83*: 201–217.

Goldberg, R.B., et al. 1981b. Developmental regulation of cloned superabundant embryo mRNAs in soybean. Dev. Biol., *83*: 218–231.

Goustin, A.S. 1981. Two temporal phases for the control of histone gene activity in cleaving sea urchin embryos (*S. purpuratus*). Dev. Biol., *87*: 163–175.

Grainger, J.L. et al. 1979. Intracellular pH controls protein synthesis rate in the sea urchin egg and early embryo. Dev. Biol., *68*: 396–406.

Griffith, J.K., B.B. Griffith, and T. Humphreys. 1981. Regulation of ribosomal RNA synthesis in sea urchin embryos and oocytes. Dev. Biol., *87*: 220–228.

Gross, P.R., and G.H. Cousineau. 1964. Macromolecule synthesis and the influence of actinomycin on early development. Exp. Cell Res., *33*: 368–395.

Gross, P.R., L.I. Malkin, and W.A. Moyer. 1964. Templates for the first proteins of embryonic development. Proc. Natl. Acad. Sci. U.S.A., *51*: 407–414.

Harvey, E.B. 1936. Parthenogenetic merogony or cleavage without nuclei in *Arbacia punctulata*. Biol. Bull., *71:* 101–121.

Harvey, E.B. 1940. A comparison of the development of nucleate and non-nucleate eggs of *Arbacia punctulata*. Biol. Bull., *79*: 166–187.

Hieter, P.A. et al. 1979. Histone gene switch in the sea urchin embryo. Identification of late embryonic histone messenger ribonucleic acids and the control of their synthesis. Biochemistry, *18*: 2707–2716.

Hille, M.B., and A.A. Albers. 1979. Efficiency of protein synthesis after fertilization of sea urchin eggs. Nature (Lond.), *278*: 469–471.

Hough-Evans, B.R. et al. 1977. Appearance and persistence of maternal RNA sequences in sea urchin development. Dev. Biol., *60*: 258–277.

Hultin, T. 1961. The effect of puromycin on protein metabolism and cell division in fertilized sea urchin eggs. Experientia, *17*: 410–411.

Humphreys, T. 1971. Measurements of messenger RNA entering polysomes upon fertilization of sea urchin eggs. Dev. Biol., *26*: 201–208.

Humphreys, T. 1973. RNA and protein synthesis during early animal embryogenesis. *In* S.J. Coward (ed.), *Developmental Regulation. Aspects of Cell Differentiation.* Academic Press, New York, pp. 1–22.

Infante, A.A., and L.J. Heilmann. 1981. Distribution of messenger ribonucleic acid in polysomes and nonpolysomal particles of sea urchin embryos: Translational control of actin synthesis. Biochemistry, *20*: 1–8.

Jeffery, W.R. 1983. Messenger RNA localization and cytoskeletal domains in ascidian embryos. *In* W.R. Jeffery and R.A. Raff (eds.), *Time, Space and Pattern in Embryonic Development.* Alan R. Liss, New York, pp. 241–259.

Jeffery, W.R., and D.G. Capco. 1978. Differential accumulation and localization of maternal poly (A)-containing RNA during early development of the ascidian, *Styela.* Dev. Biol., *67:* 152–166.

Jenkins, N.A., et al. 1978. A test for masked message: The template activity of messenger ribonucleoprotein particles isolated from sea urchin eggs. Dev. Biol., *63:* 279–298.

Karssen, C.M. et al. 1983. Induction of dormancy during seed development by endogenous abscisic acid: Studies on abscisic acid deficient genotypes of *Arabidopsis thaliana* (L.) Heynh. Planta (Berl.), *157:* 158–165.

Knowland, J., and C. Graham. 1972. RNA synthesis at the two-cell stage of mouse development. J. Embryol. Exp. Morphol., *27:* 167–176.

Kratzer, P.G., and S.M. Gartler. 1978. HGPRT activity changes in preimplantation mouse embryos. Nature (Lond.), *274:* 503–504.

Lane, B.G., and T.D. Tumaitis-Kennedy. 1981. Comparative study of levels of secondary processing in bulk mRNA from dry and germinating wheat embryos. Eur. J. Biochem., *114:* 457–463.

Laskey, R.A. et al. 1977. Protein synthesis in oocytes of *Xenopus laevis* is not regulated by the supply of messenger RNA. Cell, *11:* 345–351.

Levey, I.L., G.B. Stull, and R.L. Brinster. 1978. Poly(A) and synthesis of polyadenylated RNA in the preimplantation mouse embryo. Dev. Biol., *64:* 140–148.

Lockshin, R.A. 1966. Insect embryogenesis: Macromolecular syntheses during early development. Science, *154:* 775–776.

Lyon, M. 1961. Gene action in the X-chromosome of the mouse (*Mus musculus* L.). Nature (Lond.), *190:* 372–373.

Maisonhaute, C. 1977. Comparison des effets de deux inhibiteurs de la synthese d'ARN (Actinomycin D et α-amanitine) sur le développement embryonnaire d'un insecte déterminisme génique du début de l' embryogénèse de *Leptinotarsa decemlineata* (Coleoptera). Wilh. Roux's Arch., *183:* 61–77.

Manes, C. 1973. The participation of the embryonic genome during early cleavage in the rabbit. Dev. Biol., *32:* 453–459.

Manes, C. 1975. Genetic and biochemical activities in preimplantation embryos. *In* C.L. Markert and J. Papaconstantinou (eds.), *The Developmental Biology of Reproduction.* 33rd Annual Symposium of the Society for Developmental Biology. Academic Press, New York, pp. 133–163.

Marcus, A., and J. Feeley. 1964. Activation of protein synthesis in the imbibition phase of seed germination. Proc. Natl. Acad. Sci. U.S.A., *51:* 1075–1079.

Marcus, A., and J. Feeley. 1965. Protein synthesis in imbibed seeds. II. Polysome formation during imbibition. J. Biol. Chem., *240:* 1675–1680.

Marcus, A., S. Spiegel, and J.D. Brooker. 1975. Preformed mRNA and the programming of early embryo development. *In* R.H. Meints and E. Davies (eds.), *Control Mechanisms in Development.* Plenum Press, New York, pp. 1–19.

Mauron, A. et al. 1982. Accumulation of individual histone mRNAs during embryogenesis of the sea urchin *Strongylocentrotus purpuratus.* Dev. Biol., *94:* 425–434.

Maxson, R.E., Jr., and F.H. Wilt. 1982. Accumulation of the early histone messenger RNAs during the development of *Strongylocentrotus purpuratus*. Dev. Biol., *94*: 435–440.

Maxson, R. et al. 1983. Distinct organizations and patterns of expression of early and late histone gene sets in the sea urchin. Nature (Lond.), *301*: 120–125.

Meinke, D.W., J. Chen, and R.N. Beachy. 1981. Expression of storage-protein genes during soybean seed development. Planta (Berl.), *153*: 130–139.

Miller, L. 1974. Metabolism of 5S RNA in the absence of ribosome production. Cell, *3*: 275–281.

Mintz, B. 1964. Synthetic processes and early development in the mammalian egg. J. Exp. Zool., *157*: 85–100.

Moon, R.T., M.V. Danilchik, and M.B. Hille. 1982. An assessment of the masked message hypothesis: Sea urchin egg messenger ribonucleoprotein complexes are efficient templates for *in vitro* protein synthesis. Dev. Biol., *93*: 389–403.

Moon, R.T., K.D. Moe, and M.B. Hille. 1980. Polypeptides of nonpolyribosomal messenger ribonucleoprotein complexes of sea urchin eggs. Biochemistry, *19*: 2723–2730.

Moon, R.T. et al. 1983. The cytoskeletal framework of sea urchin eggs and embryos: Developmental changes in the association of messenger RNA. Dev. Biol., *95*: 447–458.

Mukherjee, A.B. 1976. Cell cycle analysis and X-chromosome inactivation in the developing mouse. Proc. Natl. Acad. Sci. U.S.A., *73*: 1608–1611.

Newport, J., and M. Kirschner. 1982a. A major developmental transition in early Xenopus embryos: I. Characterization and timing of cellular changes at the midblastula stage. Cell, *30*: 675–686.

Newport, J., and M. Kirschner. 1982b. A major developmental transition in early Xenopus embryos: II. Control of the onset of transcription. Cell, *30*: 687–696.

Newrock, K.M. et al. 1978. Histone changes during chromatin remodeling in embryogenesis. Cold Spring Harbor Symp. Quant. Biol., *42*: 421–431.

Perlman, S., and M. Rosbash. 1978. Analysis of *Xenopus laevis* ovary and somatic cell polyadenylated RNA by molecular hybridization. Dev. Biol., *63*: 197–212.

Petzoldt, U., P.C. Hoppe, and K. Illmensee. 1980. Protein synthesis in enucleated fertilized and unfertilized mouse eggs. Wilh. Roux's Arch., *189*: 215–219.

Pikó, L., and K.B. Clegg. 1982. Quantitative changes in total RNA, total poly (A), and ribosomes in early mouse embryos. Dev. Biol., *89*: 362–378.

Raff, R.A., et al. 1972. Oogenic origin of messenger RNA for embryonic synthesis of microtubule proteins. Nature (Lond.), *235*: 211–214.

Rice, T.B., and A. Garen. 1975. Localized defects of blastoderm formation in maternal effect mutants of *Drosophila*. Dev. Biol., *93*: 277–286.

Rosbash, M., and P.J. Ford. 1974. Polyadenylic acid-containing RNA in *Xenopus laevis* oocytes. J. Mol. Biol., *85*: 87–101.

Ruderman, J.V., and P.R. Gross. 1974. Histones and histone synthesis in sea urchin development. Dev. Biol., *36*: 286–298.

Ruderman, J.V., H.R. Woodland, and E.A. Sturgess. 1979. Modulations of histone messenger RNA during the early development of *Xenopus laevis*. Dev. Biol., *71*: 71–82.

Sawicki, J.A., T. Magnuson, and C.J. Epstein. 1981. Evidence for expression of the paternal genome in the two-cell mouse embryo. Nature (Lond.), *294*: 450–451.

Schafer, U. et al. 1982. Some somatic sequences are absent or exceedingly rare in *Xenopus* oocyte RNA. Dev. Biol., *94*: 87–92.

Scheller, R.H. et al. 1981. Organization and expression of multiple actin genes in the sea urchin. Mol. Cell. Biol., *1*: 609–628.

Schuler, M.A., E.S. Schmitt, and R.N. Beachy. 1982a. Closely related families of genes code for the α and α^1 subunits of the soybean 7S storage protein complex. Nucleic Acids Res., *10*: 8225–8244.

Schuler, M.A. et al. 1982b. Structural sequences are conserved in the genes coding for the α, α^1, and β-subunits of the soybean 7S seed storage protein. Nucleic Acids Res., *10*: 8245–8261.

Schultz, G.A. 1975. Polyadenylic acid–containing RNA in unfertilized and fertilized eggs of the rabbit. Dev. Biol., *44*: 270–277.

Shepherd, G.W. and M. Nemer. 1980. Developmental shifts in frequency distribution of polysomal mRNA and their posttranscriptional regulation in the sea urchin embryo. Proc. Natl. Acad. Sci. U.S.A., *77*: 4653–4656.

Shih, R.J. et al. 1978. Kinetic analysis of amino acid pools and protein synthesis in amphibian oocytes and embryos. Dev. Biol., *66*: 172–182.

Shiokawa, K., Y. Misumi, and K. Yamana. 1981a. Demonstration of rRNA synthesis in pre-gastrular embryos of *Xenopus laevis.* Dev. Growth Differ., *23*: 579–587.

Shiokawa, K. et al. 1981b. Non-coordinated synthesis of RNA's in pre-gastrular embryos of *Xenopus laevis.* Dev. Growth Differ., *23*: 589–597.

Showman, R.M. et al. 1982. Message-specific sequestration of maternal histone mRNA in the sea urchin egg. Proc. Natl. Acad. Sci. U.S.A., *79*: 5944–5947.

Showman, R.M. et al. 1983. Subcellular localization of maternal histone mRNAs and the control of histone synthesis in the sea urchin embryo. *In* G.M. Malacinski and W.H. Klein (eds.), *Molecular Aspects of Early Development.* Plenum Press, New York. In press.

Sinnott, E., L.C. Dunn, and T. Dobzhansky. 1958. *Principles of Genetics,* 5th ed. McGraw-Hill Book Co., New York.

Slater, I., D. Gillespie, and D.W. Slater. 1973. Cytoplasmic adenylation and processing of maternal RNA. Proc. Natl. Acad. Sci. U.S.A., *70*: 406–411.

Smith, L.D., and R.E. Ecker. 1965. Protein synthesis in enucleated eggs of *Rana pipiens.* Science, 150: 777–779.

Spiegel, S., and A. Marcus. 1975. Polyribosome formation in early wheat embryo germination independent of either transcription or polyadenylation. Nature (Lond.), *256*: 228–230.

Sturtevant, A.H., and G.W. Beadle. 1962. *An Introduction to Genetics.* Dover Publications, New York.

Summers, R.G. 1970. The effect of actinomycin D on demembranated *Lytechinus variegatus* embryos. Exp. Cell Res., *59*: 170–171.

Takagi, N. 1974. Differentiation of X chromosomes in early female mouse embryos. Exp. Cell Res., *86*: 127–135.

Takagi, N., and M. Sasaki. 1975. Preferential inactivation of the paternally derived X chromosome in the extraembryonic membranes of the mouse. Nature (Lond.), *256*: 640–642.

Tennent, D.H. 1914. The early influence of the spermatozoan upon the characters of echinoid larvae. Carnegie Inst. Wash. Publ., *182*: 129–138.

Thompson, E.W. and B.G. Lane. 1980. Relation of protein synthesis in imbibing wheat embryos to the cell-free translational capacities of bulk mRNA from dry and imbibing embryos. J. Biol. Chem., *255*: 5965–5970.

Triplett, B.A. and R.S. Quatrano. 1982. Timing, localization, and control of wheat germ agglutinin synthesis in developing wheat embryos. Dev. Biol., *91*: 491–496.

Venezky, D.L., L.M. Angerer, and R.C. Angerer. 1981. Accumulation of histone repeat transcripts in the sea urchin egg pronucleus. Cell, *24*: 385–391.

Wake, N., M. Takagi, and M. Sasaki. 1976. Non-random inactivation of X chromosomes in the rat yolk sac. Nature (Lond.), *262*: 580–581.

Wallace, H., and T.R. Elsdale. 1963. Effects of actinomycin D on amphibian development. Acta Embryol. Morphol. Exp., *6*: 275–282.

Wells, D.E. et al. 1981. Delayed recruitment of maternal histone H3 mRNA in sea urchin embryos. Nature (Lond.), *292*: 477–478.

Wilde, Ch. E., and R.B. Crawford. 1966. Cellular differentiation in the anamniota. III. Effects of actinomycin D and cyanide on the morphogenesis of *Fundulus*. Exp. Cell Res., *44*: 471–488.

Winkler, M.M. et al. 1980. Dual ionic controls for the activation of protein synthesis at fertilization. Nature (Lond.), *287*: 558–560.

Winkler, M.M., and R.A. Steinhardt. 1981. Activation of protein synthesis in a sea urchin cell-free system. Dev. Biol., *84*: 432–439.

Wold, B.J. et al. 1978. Sea urchin embryo mRNA sequences expressed in the nuclear RNA of adult tissues. Cell, *14*: 941–950.

Woodland, H.R. 1974. Changes in the polysome content of developing *Xenopus laevis* embryos. Dev. Biol., *40*: 90–101.

Woodland, H.R., and C.F. Graham. 1969. RNA synthesis during early development of the mouse. Nature (Lond.), *221*: 327–332.

Woods, D.E., and W. Fitschen. 1978. The mobilization of maternal histone messenger RNA after fertilization of the sea urchin egg. Cell Differ., *7*: 103–114.

PART FIVE
THE ORGANIZED GENERATION OF CELL DIVERSITY

13 Forming the Organizational Framework of the Multicellular Embryo

The phase of rapid cell division that follows fertilization produces the individual units from which the basic body plan of the adult organism is to be constructed. The strategies for organizing cells into a definitive body plan differ greatly between animals and plants. Morphogenesis in animals involves considerable rearrangement and displacement of cells, leading to the formation of groups of cells (**primordia**) that differentiate into specialized tissues and organs. In plants, on the other hand, cells are immobile because of the rigidity of cell walls. Elaboration of the plant body plan involves the formation of meristems, which produce new cells that differentiate *in situ*. In this chapter we shall discuss the contrasting mechanisms used by animals and plants in forming their basic body plans.

Like all other developmental processes, morphogenesis is directed by the genome. To discover how the genomic information is utilized to effect changes in the shape of the organism, we should examine the individual cells and their shape changes and ask how those changes occur and how they are coordinated. Ultimately, we would hope to discover the biochemical bases for shape changes and the genetic mechanisms that coordinate the biochemical activities.

The cellular approach to morphogenesis received its modern impetus from the pioneering work of Johannes Holtfreter and Paul Weiss. It has been followed with great success by numerous contemporary investigators. Our understanding of cell shape changes has improved markedly since time-lapse cinemicrography and electron microscopy have been applied to the investigation. But the most crucial era—understanding how cellular activities are coordinated—has barely begun. The cell surface has been recognized as the structure that is most intimately involved in cellular interactions. We are only beginning to understand the composition and functional organization of the cell surface, and we now have the technology to examine the distribution of individual molecules within it. These molecules—synthesized under the control of the genome—provide the link with the genome that may allow us to discover how morphogenesis is regulated. The promise for the future is enormous, and the groundwork laid by the pioneering investigators provides the direction for future research.

SECTION ONE
ESTABLISHMENT OF THE ANIMAL BODY PLAN

The marvelous diversity of adult body form is made possible by considerable plasticity in the developmental processes that elaborate the adult from the zygote. These processes involve a surprisingly small number of basic mechanisms. Furthermore, the diversity that distinguishes adults from one another is often indiscernible during early developmental stages. Von Baer (1828) and Haeckel (1868) observed that the general features of development appear first and the specializations appear later. The specializations are thus elaborated on structurally similar embryos. The early developmental program is highly conservative in evolution, and the processes utilized by the majority of animal species can be illustrated by using a few organisms as examples.

The zygote is subdivided into a large number of cells during cleavage and forms a blastula in which the cells are arranged in a sheet—the blastoderm. The body plan of the organism is established by reorganization and rearrangement of the blastoderm. The sequence of events that follows the blastula stage modifies the external appearance as well as the structural organization of the spherical, oval, or discoidal blastula so that it acquires the general form that is characteristic of an animal of its species. This process is called morphogenesis. The elegance of morphogenesis can best be appreciated by simply watching embryos develop. Perhaps the most impressive aspect of this phase of development is the rapidity of change. No matter how often one observes an embryo "going through its paces," it is impossible not to marvel at the events that are unfolding or be inquisitive about the mechanisms that produce these changes. The essence of development is *change of the whole organism*. Thus, although we frequently have to isolate single events occurring at only one time in only one part of the embryo, it is done to reconstruct the dynamic processes that occur in formation of the entire embryo. Hans Driesch in 1894 likened the study of the embryo to examining a masterpiece of art that when ". . . examined with a lens at a distance of 1 cm shows up quite differently than at 5 cm away. The first time we see only the blotches. Is then the study of blotches really the only task of the biologist?" (translated by Oppenheimer, 1967). By this rhetorical question, Driesch is exhorting us not to lose sight of the beauty and the dynamism of embryonic development by concentrating solely on minute details.

The initial modification of the blastoderm occurs at gastrulation, when portions of it are displaced inward. In the process three cell layers called germ layers are produced. The three are the outer ectoderm, the inner endoderm, and—between them—the mesoderm. The cells of these layers have approximately the same spatial relationships to one another that their descendants will have in the adult organism. The endoderm forms the lining of the gut. The process of displacement of the endoderm to the inside may simultaneously form the cavity of the gut as well. The ectoderm will form the epidermis and nervous system; the latter is delimited from the epidermis after gastrulation is complete. The mesoderm will form most of the intervening tissues. The mesoderm also gives rise to mesenchyme—a connective tissue that may associate with endodermal and ectodermal derivatives to form organs of compound germ layer origin.

Perhaps the most impressive aspect of gastrulation is the involvement of the entire blastoderm in the process. The pattern of gastrulation varies considerably from one phylogenetic group to another, but there are only a few primary mechanisms known. Gastrulation often occurs by a combination of these mechanisms. One of these is **epiboly**—the tendency of epithelial sheets of ectoderm to spread and surround inner sheets. In some instances, sheets may expand toward a single site; this is called **convergence.** Whereas ectoderm sheets tend to spread over the embryonic surface, other sheets tend to move to the inside of the embryo. There are three inward displacement mechanisms: **invagination, involution,** and **ingression.** Invagination is an infolding that resembles the pushing in of one side of a soft rubber ball. Involution is the turning in of an expanding outer layer so that it then spreads along the inner surface of the overlying layer. Ingression is the separation of small groups of individual cells from the blastoderm, followed by their migration into the embryonic interior.

The extent of movement during gastrulation depends, to a certain extent, on the number of cells in the blastoderm. The annelid gastrula, for example, is composed of approximately 30 cells and has very simple gastrulation. The frog gastrula, on the other hand, consists of 3×10^4 cells, necessitating very large-scale cell rearrangements. In general, there is less cell movement during gastrulation of invertebrates than during that of vertebrates. The mechanism of gastrulation is also complicated by the amount of yolk in the egg. Very yolky eggs have rather large cells in the vegetal hemisphere, which migrate only slightly. In the extremely yolky telolecithal egg the yolk is uncleaved, and modified mechanisms of gastrulation have evolved to accommodate this condition.

After the germ layers are formed, discontinuities appear in them as cells reorganize into discrete tissue and organ primordia. Reorganization of cells involves primarily the coordinated activity of sheets of cells (i.e., epithelia), although the migration of individual cells also occurs. The mechanisms utilized during reorganization are changes in cell adhesiveness, cell motility, and cell shape.

The formation of organ primordia occurs in a very definite pattern within the body. **Developmental integration** occurs along definite axes of symmetry that are directly influenced by egg polarity. Integration is achieved by the interplay of two mechanisms. One is the tendency of cells to acquire locations according to their predetermined fates. For example, the fates of cells may be determined by the cytoplasm parceled into them during cleavage, and they subsequently assume their correct position in the embryo. The second mechanism is the tendency of cells to differentiate according to their location within the embryo. In these cases, cell fate may be determined as a *consequence* of location. The imposition of cell fate as a consequence of location is called **induction.** Finally, both of these processes depend upon the abilities of cells to exchange information by **cell communication.**

13–1. *AMPHIOXUS:* GASTRULATION AND FORMATION OF THE BODY PLAN

The embryo that develops from the small isolecithal egg of the protochordate *Amphioxus* illustrates the most direct means of making a two-layered structure from a hollow ball—invagination on one side to form a cup-shaped structure. Gastrulation in *Amphioxus* was described by Conklin (1932). Figure 13.1 is a series of drawings based on that publication. As a prelude to invagination, the vegetal cells elongate and form a flattened **endodermal plate** consisting of the presumptive endoderm. The endodermal plate bends inward and ultimately disappears from the outer surface entirely. The invagination of the endoderm eventually obliterates the blastocoele and forms a cup-shaped structure with a double wall. The inner wall lines the newly created cavity—the archenteron—which opens to the exterior through the blastopore. The outer wall is ectoderm, consisting of presumptive epidermis and presumptive neural cells.

FIGURE 13.1 Stages of gastrulation of *Amphioxus*. The embryos in *A* to *G* are represented as cut in the median plane. *A*, Blastula. *B* and *C*, Beginning of invagination. *D*, Invagination advanced, the embryo attaining the structure of a double-walled cup with a broad opening to the exterior. *E* and *F*, Constriction of the blastopore. *G*, Completed gastrula. *H*, Middle gastrula, whole, viewed from side of blastopore. (After E.G. Conklin, 1932. From B.I. Balinsky. 1981. *An Introduction to Embryology*, 5th ed. Saunders College Publishing, Philadelphia, p. 203.)

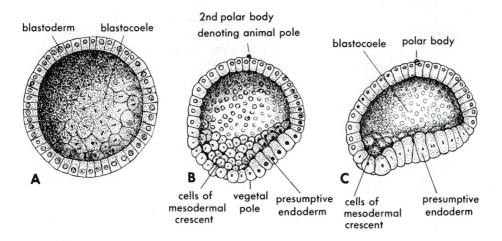

A — blastoderm, blastocoele

B — 2nd polar body denoting animal pole; cells of mesodermal crescent, vegetal pole, presumptive endoderm

C — blastocoele, polar body; cells of mesodermal crescent, presumptive endoderm

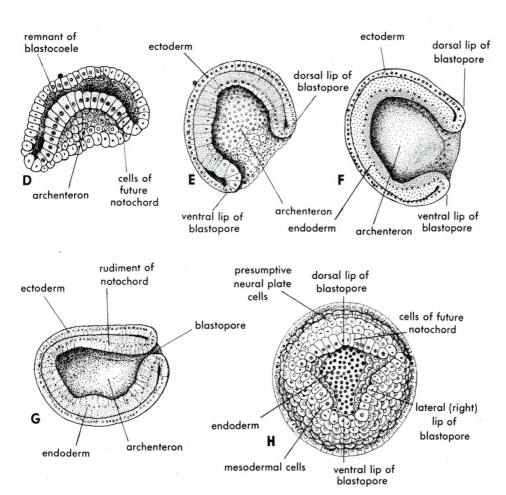

D — remnant of blastocoele; archenteron, cells of future notochord

E — ectoderm, dorsal lip of blastopore; ventral lip of blastopore, archenteron, endoderm

F — ectoderm, dorsal lip of blastopore; archenteron, ventral lip of blastopore

G — ectoderm, rudiment of notochord, blastopore; endoderm, archenteron

H — presumptive neural plate cells, dorsal lip of blastopore, cells of future notochord; lateral (right) lip of blastopore; endoderm, mesodermal cells, ventral lip of blastopore

During invagination the presumptive mesoderm and notochord occupy the rim of the cup, with the notochord forming the dorsal lip of the blastopore and the mesoderm forming the lateral and ventral lips. These cells are withdrawn to the interior as the blastopore contracts to become a small opening. Within the embryo the presumptive notochord and mesoderm shift their relative positions as the mesoderm converges toward the dorsal side of the embryo and straddles both sides of the notochord (Fig. 13.2). After contraction of the rim of the blastopore, the embryo lengthens in the anterior–posterior direction. As a result, the notochord is stretched into an elongate band flanked on either side by the mesoderm. On the outer surface the neural ectoderm is stretched over the notochord to form the **neural plate.** The notochord and the flanking mesoderm form the roof of the archenteron, whereas the endoderm forms the base and sides (Fig. 13.3A).

At the completion of gastrulation the cells that will form internal structures have been displaced to the inside, but there remains a great deal of rearrangement to be done, since the invaginated cells form one continuous sheet. The segregation of the notochord and mesoderm from the endoderm results in the endoderm becoming continuous and forming the lining of the prospective digestive tract. This process is illustrated by the drawings in Figure 13.3B to D, which are cross sections of progressively later stages. The segregation of the mesoderm occurs by the formation of a series of evaginations that separate from the archenteron to become mesodermal segments, or **somites,** each with its own cavity. These cavities later expand to form the body cavity, or **coelom.** Concurrently, on the outside of the embryo the neural plate separates from the presumptive epidermis. The latter advances over the neural plate from the sides and from the rear to become continuous around the embryo. At the same time, the neural plate rolls up on itself to form the **neural tube,** from which the central nervous system is derived.

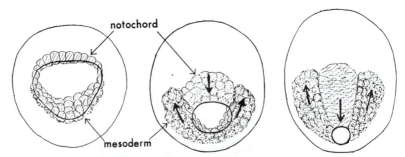

FIGURE 13.2 Change in the relative position of the presumptive mesoderm and presumptive notochord during closure of the blastopore in *Amphioxus* (diagrammatic). (After E.G. Conklin, 1932. From B.I. Balinsky. 1981. *An Introduction to Embryology*, 5th ed. Saunders College Publishing, Philadelphia, p. 204.)

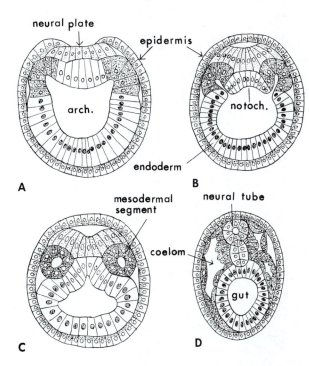

neural plate

epidermis

arch.

endoderm

A

notoch.

B

mesodermal segment

neural tube

coelom

gut

C

D

FIGURE 13.3 Elaboration of the body plan of *Amphioxus*. Cross sections of progressively later stages. (After E. Korschelt. 1936. *Vergleichende Entwicklungsgeschichte der Tiere*. Band I. Gustav Fischer Verlag, Jena, pp. 1142, 1143. From B.I. Balinsky. 1981. *An Introduction to Embryology*, 5th ed. Saunders College Publishing, Philadelphia, p. 205.)

13–2. THE SEA URCHIN: GASTRULATION AND FORMATION OF THE BODY PLAN

Gastrulation in sea urchin embryos (Fig. 13.4) also occurs by invagination of a single-layered blastoderm. However, in addition to invagination, which forms the archenteron, the descendants of the micromeres are involved in ingression, which begins before invagination. In the swimming blastula the vegetal pole begins to flatten to form the vegetal plate, and the embryo becomes pear-shaped. In the center of this plate a number of cells (derivatives of the micromeres; see Chap. 10) detach from the vegetal plate and enter the blastocoele to become the primary mesenchyme cells.

The presumptive primary mesenchyme cells in *Lytechinus* occupy a small depression at the vegetal pole where they surround a small central cluster of cells that do not ingress (Fig. 13.5). Scanning electron micrographs of fractured vegetal plates (Fig. 13.6A) reveal that the ingressing cells are bottle-shaped, with narrow apices and bulbous bases. At the completion of ingression the apical portions of the cells have completely withdrawn, and the internalized cells become rounded (Fig. 13.6B).

Electron micrographs of sectioned cells provide details about the ingression process (Katow and Solursh, 1980). The results of this study are summarized in the drawings of Figure 13.7. Prior to ingression, cells of the vegetal plate are indistinguishable from one another (Fig. 13.7A).

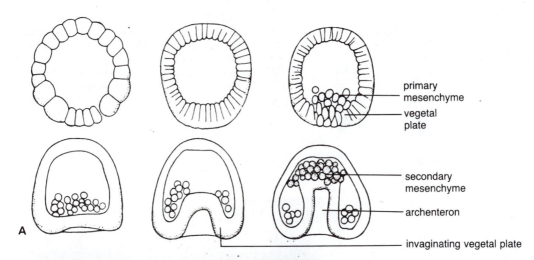

A

primary
mesenchyme

vegetal
plate

secondary
mesenchyme

archenteron

invaginating vegetal plate

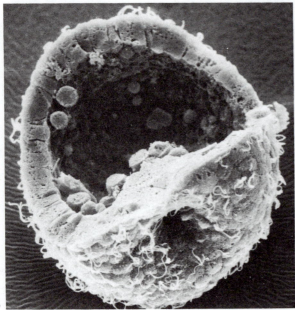

B

FIGURE 13.4 Sea urchin gastrulation. *A*, Drawings of the blastula and gastrula stages. (After T. Boveri, from H. Spemann. 1936. *Experimentelle Beiträge zu einer Theorie der Entwicklung.* Julius Springer, Berlin, p. 245. Reprinted with permission of Springer-Verlag, Heidelberg. Reproduced from B.I. Balinsky, 1975.) *B*, Scanning electron micrograph of sea urchin early gastrula. The depression at the bottom is due to the invagination of the vegetal plate to form the archenteron. The blastoderm has been fractured during preparation to reveal individual primary mesenchyme cells migrating in the blastocoele. A number of these cells are seen at the base of the archenteron. Note the cilia on the outer surface of blastoderm cells. × 1970. (Courtesy of Dr. W.J. Humphreys.)

They form an epithelium in which the cells are firmly associated by desmosomes near their apical ends and interdigitations on the basal side. All cells are ciliated and extend short processes into the hyaline layer. The basal surface is underlain by a basal lamina.

The basal lamina disappears from the basal surfaces of presumptive primary mesenchyme cells as their basal ends enlarge and bulge into the blastocoele. The apical ends are elongated, and the cells remain closely associated with their neighbors at this end (Fig. 13.7B and C). These cells are no longer ciliated, although the adjacent noningressing cells retain

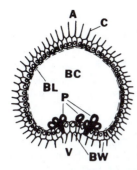

FIGURE 13.5 Diagram of a section through the animal (A)-vegetal (V) pole axis of the mesenchyme blastula of *Lytechinus*, showing primary mesenchyme cell (P, shaded) ingression. The ingressing cells surround a small central cluster of noningressing cells. The blastoderm (BW) is an epithelium composed of ciliated cells. A basal lamina (BL) underlies the blastoderm and marks the outer boundary of the blastocoele (BC). C: cilia. (From H. Katow and M. Solursh. 1980. Ultrastructure of primary mesenchyme cell ingression in the sea urchin *Lytechinus pictus*. J. Exp. Zool., *213:* 236.)

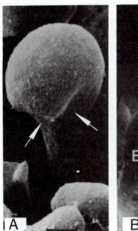
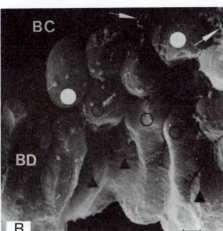

FIGURE 13.6 Scanning electron micrographs of fractured vegetal plate of *Lytechinus* mesenchyme blastula. *A,* Bottle-shaped ingressing presumptive primary mesenchyme cell. The fracture procedure has disrupted cellular contacts, thus exaggerating intercellular spaces. *B,* Bottle-shaped cells (open circles) are observed among adjacent non-ingressing cells (triangles) and post-ingression primary mesenchyme cells (solid circles), which are teardrop shaped. BC: blastocoel; BW: blastoderm. Scale bar equals 2 μm. (From H. Katow and M. Solursh. 1980. Ultrastructure of primary mesenchyme cell ingression in the sea urchin *Lytechinus pictus*. J. Exp. Zool., *213:* 234.)

their cilia. The elongate portions of the cells contain microtubules that are oriented in the apical-basal axes (Fig. 13.8). Interestingly, the apical cytoplasm of adjacent nonmesenchymal cells also contains microtubules on the sides facing the presumptive primary mesenchyme cells. These microtubules are also oriented parallel to the apical-basal axis.

Eventually, the apical junctions between presumptive primary mesenchyme cells and their neighbors disappear and spaces appear in the intercellular regions. The mesenchyme cells withdraw their apical processes, lose their microtubules, and become rounded as they are displaced into the blastocoele (Fig. 13.7D and E).

The appearance of the ingressing presumptive primary mesenchyme cells suggests that their basal ends relax to allow expansion, whereas the microtubules play a role in elongation of the neck region. Neighboring cells, which contain microtubules, might also be pressing against them, squeezing them out of the vegetal plate (Katow and Solursh, 1980). We have examined this process in detail, since we shall encounter additional examples of cellular ingression in this chapter, for which primary mesenchyme cell ingression may serve as a model.

Once inside, the primary mesenchyme cells become amoeboid and move about by means of long, filamentous pseudopodia, the **filopodia.** Time-lapse cinemicrography (Gustafson and Wolpert, 1961) shows that the filopodia move actively along the inner blastodermal surface, as if

FIGURE 13.7 Diagram showing primary mesenchyme cell ingression. Crosses: blastocoele material. A, The initial morphology of presumptive primary mesenchyme cells is identical to that of the neighboring blastomeres. They have cilia (C) and are covered by the hyaline layer (H) on the apical surface. The basal surface is adjoined by a thin discontinuous basal lamina (BL). The basal body (B) is surrounded by Golgi cisternae (G). Nuclei (N) are located near the basal surfaces. B, Elongation of presumptive primary mesenchyme cell (P) into the blastocoele. The presumptive primary mesenchyme cell and the neighboring blastomeres form apical processes (arrows). The basal lamina has disappeared from beneath the elongated presumptive primary mesenchyme cell and the neighboring blastomeres. C, Apical detachment of the primary mesenchyme cell. The presumptive primary mesenchyme cell elongates further (P), and the apical region of the cell forms convoluted processes (large arrow). Desmosomes between the presumptive primary mesenchyme cell and neighboring blastomeres have disappeared. Microtubules (small arrows) are found in the peripheral cytoplasm of the neck region in the presumptive primary mesenchyme cell and in the cytoplasm of neighboring blastomeres preferentially on the sides adjacent to the presumptive primary mesenchyme cells. D, Separation of the primary mesenchyme cell (P). Primary mesenchyme cells separate from the vegetal plate after the disappearance of desmosomes. The neighboring blastomeres extend apical cell processes toward each other, resulting in the occlusion of the intercellular spaces. Also, blastocoele material is observed in the intercellular spaces. E, Rounding of the primary mesenchyme cells. The neck region of the primary mesenchyme cells shortens toward the basal side, and the mesenchyme cells become round in shape (P). (From H. Katow and M. Solursh. 1980. Ultrastructure of primary mesenchyme cell ingression in the sea urchin *Lytechinus pictus.* J. Exp. Zool., *213*: 237.)

FIGURE 13.8 Electron micrograph of presumptive primary mesenchyme cell of *Lytechinus* embryo. Microtubules (M) are oriented longitudinally in the cell. Scale bar equals 0.1 μm. (From H. Katow and M. Solursh. 1980. Ultrastructure of primary mesenchyme cell ingression in the sea urchin *Lytechinus pictus*. J. Exp. Zool., *213*: 239.)

exploring its contours. Movement of the cells is caused by attachment, retraction, and reattachment of the filopodia to the blastoderm wall. The cells continue to move until they reach an area where adhesion is thought to be stronger and then become stationary. Thus, the final locations of these cells result from selective fixation of cellular bonds after random exploration. Fixation of primary mesenchyme cells occurs at the gastrula stage, at which time they form a ring at the base of the invaginating archenteron. The filopodia of the primary mesenchyme cells in the ring fuse with one another to form a syncytial cablelike structure (Fig. 13.9). Two branches then rise from the ring, one on each side of the archenteron. The cables formed by fusion of filopodia are the sites for the deposition of the skeletal matrix.

Two triradiate spicules form from skeletal rudiments deposited in the syncytial cables. The rays of the spicules then grow, bend, and branch

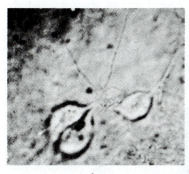
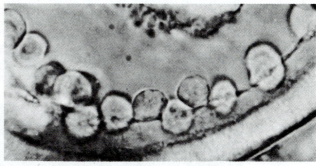

A   **B**

FIGURE 13.9 Fusion of migrating primary mesenchyme cells of the sea urchin gastrula. *A*, Migrating cells (note their long filopodia) meeting and fusing. × 860. *B*, A segment of the mesenchyme ring, with cells attached to a cable-like filopodial complex with attachments to the blastoderm wall. Skeletal matrix deposition is beginning on the left. × 1000. (From T. Gustafson and L. Wolpert. 1961. Studies on the cellular basis of morphogenesis in the sea urchin embryo. Directed movements of primary mesenchyme cells in normal and vegetalized larvae. Exp. Cell Res., *24*: 70, 71.)

in distinct ways to form definitive calcite spicules with species-specific morphology (see Fig. 13.16). The pattern of the spicules reflects the orientation of the cables formed by fusion of primary mesenchyme filopodia. Thus, cable orientation governs the skeletal pattern. Experimental evidence suggests that the pattern is determined by both innate properties of the primary mesenchyme cells and an influence from their association with the blastoderm wall.

Okazaki (1975) observed that isolated micromeres (the precursors of primary mesenchyme cells), cultured *in vitro,* fuse to form cables that elaborate spicules that partially resemble their *in situ* counterparts. Thus, some innate property of the primary mesenchyme cells generates a specific pattern of cable orientation that in turn governs the growth pattern of the calcite crystals. Harkey and Whiteley (1980) cultured primary mesenchyme cells that were isolated from the vegetal blastoderm. They observed that spicules formed by these cells have a characteristic morphology that differs from spicules formed in normal development (Fig. 13.10). Hence, the normal spicule morphology is not entirely dependent upon innate properties of the primary mesenchyme cells but is subject to exogenous influence. When dissociated blastoderm cells and isolated primary mesenchyme cells were combined, they often reconstituted gastrulalike structures (**gastruloids**) with an outer epithelium, within which the primary mesenchyme cells organized a syncytial ring much as in normal development. Gastruloids frequently formed spicules that approached the shape of normal spicules. This result suggests that the blastoderm epithelium of the gastruloids influences the spicule pattern, perhaps by allowing the primary mesenchyme cells to establish a more normal configuration of syncytial cables before spicule formation.

Since the geometry of the primary mesenchyme syncytium appears to play an important role in spicule pattern formation, we should examine the possible cellular basis for establishment and maintenance of that geometry as the primary mesenchyme cells migrate and form filopodia that fuse to form cables in a pattern that foreshadows spicule morphology. Ultrastructural studies of primary mesenchyme cells have suggested a role for microtubules in primary mesenchyme morphogenesis.

FIGURE 13.10 Comparison of spicules formed *in situ* (*A*) and by cultured primary mesenchyme cells (*B*). Scale bar equals 50 μm. (From M.A. Harkey and A.H. Whiteley. 1980. Isolation, culture, and differentiation of echinoid primary mesenchyme cells. Wilh. Roux' Archiv., *189:* 118. Reprinted with permission of Springer-Verlag, Heidelberg.)

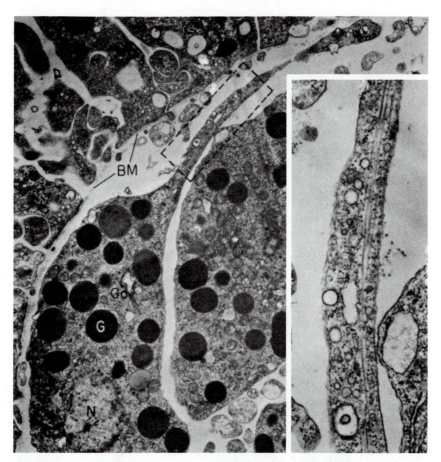

FIGURE 13.11 Electron micrograph of a primary mesenchyme cell fixed during its exploration of the sea urchin blastoderm wall (upper left). Extending from this cell is a long filopodium, which has been implicated in cell movement. Within the cell body one can easily distinguish the nucleus (N), yolk granules (G), the Golgi apparatus (Go), numerous mitochondria, and rough-surfaced endoplasmic reticulum. With the exception of a few mitochondria, none of these organelles is visible in the filopodium. BM: basement membrane. ×6500. Inset: Higher magnification of a portion (within rectangle) of the filopodium. In this process there are numerous ribosomes, small vesicles, and a large number of microtubules, which are oriented parallel to the long axis of the filopodium. ×27,500. (From J.R. Gibbins, L.G. Tilney, and K.R. Porter. 1969. Reproduced from *The Journal of Cell Biology*, 1969, Vol. 41, pp. 201–226 by copyright permission of the Rockefeller University Press.)

In migratory primary mesenchyme cells, large numbers of microtubules are oriented parallel to the long axes of the filopodia (Fig. 13.11). Following fusion of filopodia into cables, oriented microtubules are located in two distinct types of structures of the syncytium: the stalks that connect the cell bodies to the cable and the cable itself (Fig. 13.12). A summary of the shape changes undergone by primary mesenchyme cells before and

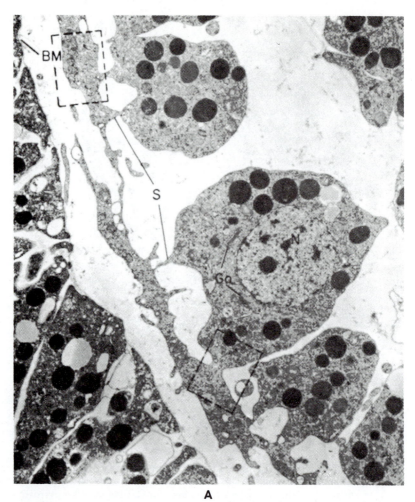

A

FIGURE 13.12 Primary mesenchyme cells of the sea urchin embryo after fusion of filopodia to form a cable syncytium. *A,* Low-power electron micrograph. Each cell body is connected to the cable by one or more stalks (S). Short processes extend from the cable cytoplasm toward the blastoderm wall. Within the cell body the nucleus (N) is in a central position. The Golgi zone (Go) lies on one side of the nucleus, and the other formed elements occupy the remainder of the cytoplasm of the cell body. Fine extracellular fibrils are present throughout the blastocoele. Squared-off areas are shown in *B* and *C* at higher magnification. × 8300.

after their ingression is shown in Figure 13.13. The lower drawings illustrate the orientation of microtubules within the cells at each stage. The orientation of microtubules parallel to the direction of cell asymmetry in every case strongly suggests that microtubules play a major role in development of the primary mesenchyme cells, particularly in their

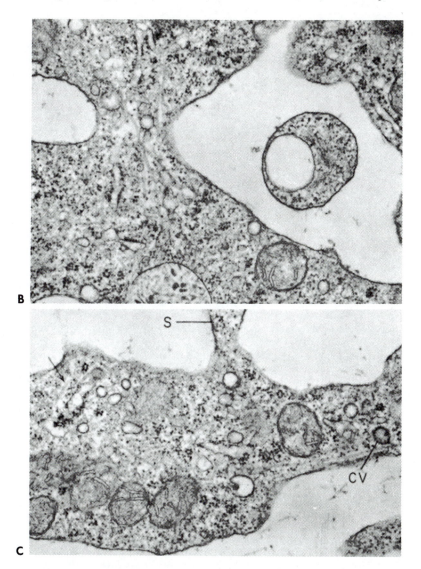

FIGURE 13.12 (*Continued*) *B,* This electron micrograph illustrates the distribution of microtubules in the narrow stalks that connect the cell bodies to the cable. Microtubules are common in this region and lie parallel to the long axis of the stalk. ×33,700. *C,* This illustrates the distribution of microtubules within the cable cytoplasm. Microtubules generally lie parallel to the long axis of the cable. The microtubules in the stalk (S) extend into the cytoplasm of the cable, so some appear at oblique angles to the long axis of the cable (note arrow). Present also within the cytoplasm of the cable are mitochondria, ribosomes, and vesicles. ×38,500. (From J.R. Gibbins, L.G. Tilney, and K.R. Porter. 1969. Reproduced from *The Journal of Cell Biology,* 1969, vol. 41, pp. 201–226 by copyright permission of the Rockefeller University Press.)

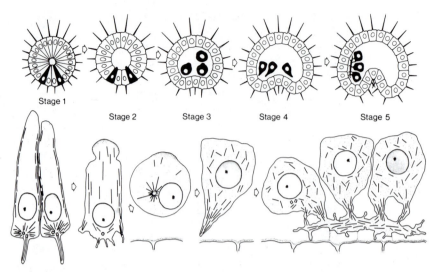

Stage 1

Stage 2   Stage 3   Stage 4   Stage 5

FIGURE 13.13 Role of microtubules in control of shape changes in primary mesenchyme cells during ingression. In the upper drawings are primary mesenchyme cells (black) exhibiting shape changes at different stages of ingression. Stage 1 = early blastula; stage 2 = late blastula; stage 3 = first primary mesenchyme; stage 4 = exploratory migrations; stage 5 = formation of cable syncytium before spicule formation. In the lower drawings the distribution of microtubules at each stage is shown. These organelles orient parallel to the asymmetrical axes of the cell except when the cell is spherical at stage 3, when they seem to radiate from a central nucleating site. (After J.R. Gibbins, L.G. Tilney, and K.R. Porter. 1969. Reproduced from *The Journal of Cell Biology*, 1969, vol. 41, pp. 201–226 by copyright permission of the Rockefeller University Press.)

migration and in cable orientation. The importance of microtubules in primary mesenchyme cell development is confirmed by the observation that treatment of sea urchin embryos with agents that disrupt microtubules prevents primary mesenchyme cell development (Tilney and Gibbins, 1969a). The implications of these observations for our understanding of skeletal formation are quite significant, since microtubular orientation is responsible for cable orientation, which in turn influences the skeletal pattern.

Those cells of the vegetal plate that are adjacent to the micromere derivatives do not escape into the blastocoele cavity but remain bound to their neighbors and to the hyaline layer. As the primary mesenchyme cells ingress, the remaining vegetal cells fill the gaps and, by doing so, cause the vegetal plate to flatten even more. The vegetal plate then bends inward, initially extending about a third of the distance across the blastocoele. At this point, invagination pauses. As is the case in *Amphioxus*, invagination causes the embryo to become cup-shaped, creating the archenteron cavity that communicates with the outside via a blastopore. Invagination is an autonomous property of the vegetal plate. This conclusion comes from the experiments of Moore and Burt (1939), who demonstrated that vegetal plates excised at the beginning of gastrulation will continue to invaginate although isolated from the remainder of the embryo. Gustafson and Wolpert (1967) have proposed that the property of the vegetal plate cells that causes invagination is a reduction of mutual contact between the columnar cells, which allows the inner borders of the cells to round up. The cells retain some lateral contact as well as their contact with the hyaline layer; as a result, they are held within the plate. Gustafson and Wolpert have pointed out that, if contact between columnar cells is reduced, the resultant rounding of the cells will cause the cell sheet to occupy a greater surface area. As a consequence, the sheet will curve in the direction of reduced contact (i.e., inward).

The forces that initiate invagination are sufficient to form the cup-shaped archenteron, but they are not capable of causing the archenteron to elongate. The force necessary to elongate the archenteron into a tube is provided by the pseudopodia that are formed by the **secondary mesenchyme cells** at the tip of the archenteron (Fig. 13.14). These pseudopodia extend to the inner surface of the gastrula wall, which they explore randomly. After specific cell adhesions are made, the pseudopodia attach to the wall at the junctions between the cells (Fig. 13.15) and shorten. The tension exerted by the pseudopodia is evident by the pulling in of the gastrula wall wherever the pseudopodia attach. The combined force exerted by the shortening pseudopodia is sufficient to cause invagination to resume and eventually to pull the archenteron tip to the gastrula wall near the animal pole (Fig. 13.16). As with the primary mesenchyme, the secondary mesenchyme pseudopodia have longitudinally oriented microtubules, which are probably involved in pseudopodial extension. The contractile force, however, is likely produced by microfilaments, which are also present within the pseudopodia (Fig. 13.17).

As the archenteron tip nears the gastrula wall, the secondary mesenchyme cells dissociate from it and migrate individually into the blastocoele. A sac forms at the tip of the archenteron and pinches off from

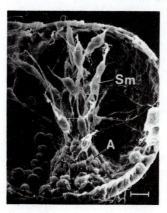

FIGURE 13.14 Scanning electron micrograph of fractured sea urchin gastrula, showing internal details. Secondary mesenchyme cells (Sm), which form at the archenteron (A) tip, extend long pseudopodia to the gastrula wall. Scale bar equals 10 µm. (From K. Akasaka, S. Amemiya, and H. Terayama. 1980. Scanning electron microscopical study of the inside of sea urchin embryos (*Pseudocentrotus depressus*). Effects of aryl β-xyloside, tunicamycin and deprivation of sulfate ions. Exp. Cell Res., *129:* 4.)

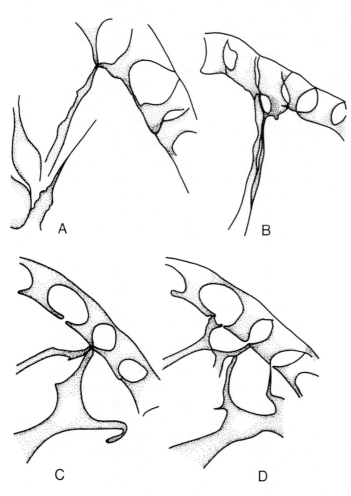

FIGURE 13.15 Attachments of pseudopodia, formed by secondary mesenchyme cells at the tip of the archenteron, to the gastrula wall. (Redrawn with permission from T. Gustafson and L. Wolpert, 1967, Cellular movement and contact in sea urchin morphogenesis, *Biological Reviews of the Cambridge Philosophical Society,* published by Cambridge University Press.)

the archenteron to become the **coelomic sac.** It is later constricted into a pair of coelomic pouches, one on each side of the archenteron. The pinching-off of the coelomic sac leaves the archenteron composed entirely of endoderm. At the point of contact of the archenteron tip with the gastrula wall, a mouth opening is formed, and a continuous tube—the gut—is produced. At the opposite end, the blastopore becomes the anus. The basic body plan of the **pluteus larva** has now been established.

13–3. VERTEBRATE GASTRULATION

Gastrulation of vertebrate embryos is complicated by the modified structure of the blastoderm. The presence of yolk in the vegetal hemisphere of the vertebrate egg alters cleavage to varying extents, depending upon the amounts. The resulting blastoderm is not a single-cell-layer–thick sheet surrounding a spherical blastocoele, but instead, shows varying degrees of deviation from this simple form.

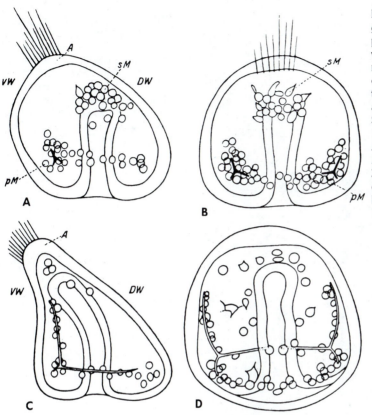

FIGURE 13.16 Sea urchin gastrulae, showing second phase of invagination and development of skeleton. pM: primary mesenchyme; sM: secondary mesenchyme; A: apical tuft of cilia; VW and DW: ventral and dorsal body walls. (After Schmidt. From H. Spemann. 1936. *Experimentelle Beiträge zu einer Theorie der Entwicklung.* Julius Springer, Berlin, p. 246. Reprinted with permission of Springer-Verlag, Heidelberg. Reproduced from B.I. Balinsky, 1975.)

Gastrulation of the Amphibian Embryo

The amphibian blastula has an eccentric blastocoele with a floor of large, yolk-laden cells running several cell layers deep and a thin roof of small, yolk-poor cells (see Fig. 1.9). The movements and developmental fates of these various cells have been mapped by the **vital dye marking method** developed by Vogt (Fig. 13.18A). Vital dyes (Nile blue sulfate and neutral red) stain cells without drastically affecting their viability. Pieces of agar impregnated with a dye are pressed against the outside of the embryo, and the displacements of the stained cells during gastrulation are charted. The resulting maps of the embryonic surface, which indicate the fates of different regions, are called **fate maps.** Vogt (1929) used this approach to derive fate maps of amphibian embryos. Figure 13.18B shows Vogt's fate map of the urodele embryo. It features three major regions that correlate with the areas of different pigmentation of the embryo. These are: (1) the prospective ectoderm (epidermis and neural plate) in the darkly pigmented animal hemisphere, (2) the prospective notochord and mesoderm in the marginal zone, and (3) the prospective endoderm in the pigment-free vegetal hemisphere. The surface location of cells that will form internal structures necessitates that these cells move to the interior of

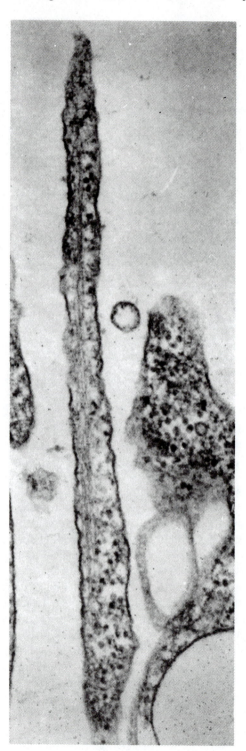

FIGURE 13.17 Longitudinal section of a pseudopodium extending from an archenteron apical cell. Within the center of the pseudopodium are microfilaments. ×70,000. (From L.G. Tilney and J.R. Gibbins. 1969b. Microtubules and filaments in the filopodia of the secondary mesenchyme cells of *Arbacia punctulata* and *Echinarachnius parma.* J. Cell Sci., *5:* 209.)

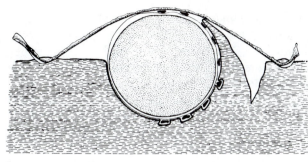

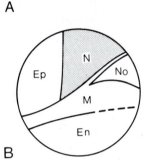

FIGURE 13.18 Fate mapping the amphibian embryo. A, Vogt's technique whereby the embryo is held against pieces of agar impregnated with vital dye. (After Vogt. From H. Spemann. 1936. *Experimentelle Beiträge zu einer Theorie der Entwicklung.* Julius Springer, Berlin, p. 6. Reprinted with permission of Springer-Verlag, Heidelberg.) B, Vogt's fate map of the urodele blastula. En: endoderm; Ep: epidermis; M: mesoderm; N: neural ectoderm; No: notochord. (After S. Løvtrup. 1975. Fate maps and gastrulation in amphibia—A critique of current views. Can. J. Zool., *53:* 475.)

A

B

the embryo during gastrulation, after which the exterior of the embryo will be covered by ectoderm.

Experiments on *Xenopus laevis* embryos by Keller (1975, 1976, 1978) have revealed that prospective mesoderm is not on the surface but in the subsurface layers in this species. The details of gastrulation must, therefore, be quite different in *Xenopus* than in urodeles. More recently, Keller (1980, 1981) has studied the cellular behavior that generates the cellular rearrangements during gastrulation. Because of the extensive information we now have on gastrulation in *Xenopus*, the details of germ layer formation presented in this section will pertain to this species. There are likely major deviations from this pattern in other amphibians, but there is general agreement that the *basic mechanisms* of gastrulation are the same in all amphibians. These are: epiboly, involution, and active cell migration.

The initial morphological indication of gastrulation is the appearance of a slitlike blastopore, which appears at the lower edge of the gray crescent. As we learned in Chapter 11, the gray crescent is on the future dorsal side of the embryo at the margin between the pigmented animal hemisphere and the unpigmented vegetal hemisphere. Its location has led to its designation as the **dorsal marginal zone.**

The blastopore is initially formed by the sinking below the surface of endodermal cells in a process that also produces a shallow groove that is the precursor of the archenteron cavity. As gastrulation continues, surface cells converge toward the blastopore and turn inward, causing the two ends of the blastopore groove to extend around the embryo in an increasingly greater arc (Fig. 13.19). The ends eventually meet ventrally, completing the blastopore circle. The dorsal rim of the blastopore, which is formed first, is called the **dorsal lip** of the blastopore. Lateral extension of the blastopore produces the **lateral lips,** whereas its completion produces the **ventral lip.** The endodermal cells encircled by the blastopore form the **yolk plug.** As gastrulation concludes, the blastopore closes over the yolk plug, and the endoderm withdraws into the interior.

The initial formation of the blastoporal groove is due to an indentation of the superficial cell sheet. This deformation appears to be caused by characteristic shape changes that are undergone by particular endodermal cells at this site. As with the ingressing presumptive primary mesenchyme cells of the sea urchin embryo, the inner borders of these cells round up, and the outer portions elongate perpendicular to the surface of the embryo. The net result of these contrasting behaviors at the opposite ends of the cells is a shape change: The cells develop long, narrow necks and bulbous bases and are called "bottle cells." In contrast to the presumptive primary mesenchyme cells of the sea urchin embryo, these bottle cells do not dissociate from their neighbors. On the contrary, they remain tightly bound to adjacent cells at their outer ends (Fig. 13.20). After the initial invagination, endodermal cells continue to be drawn inward (**involute**) as the archenteron enlarges. The small endodermal cells involuting over the blastoporal lip form the thin endodermal roof of the archenteron, whereas the large vegetal endodermal cells sink in to form the thicker archenteron floor (see Fig. 1.10).

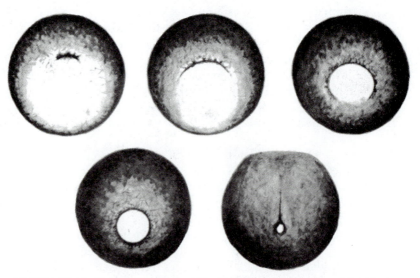

FIGURE 13.19 Changes in shape of the blastopore and closure of the blastopore during gastrulation in a frog. (From B.I. Balinsky. 1981. *An Introduction to Embryology*, 5th ed. Saunders College Publishing, Philadelphia, p. 210.)

The rearrangements of cells during *Xenopus* gastrulation have been described by Keller (1975, 1976, 1978) and Keller and Schoenwolf (1977). The animal half of the *Xenopus* blastula consists of presumptive ectoderm overlying the blastocoele. During gastrulation, this layer becomes thinner as it increases in surface area by about 50 percent (Keller, 1975). In doing so, it spreads vegetally to replace the endoderm (which is involuting) on the surface just above the blastopore. This spreading process is called epiboly. It is due to the expansion of the individual cells of the surface layer. This expansion, which occurs at the expense of the thickness of the surface layer, causes the ectoderm to increase in overall surface area (Keller, 1978).

The presumptive notochord and mesoderm cells in the *Xenopus* embryo are located entirely in the deep layers and undergo involution (to an even deeper position) while moving underneath the sheet of involuting endoderm (Fig. 13.21). The location of these cells at the beginning of gastrulation is indicated in Figure 13.21*B*. Note that they form a band in the marginal zone immediately below the surface endodermal cells lying just above the blastopore. The presumptive notochord involutes over the dorsal lip, the presumptive somite mesoderm over the lateral lips, and the presumptive lateral mesoderm over the ventral lip. Each of these regions begins its involution in succession, since the dorsal lip is formed first and the ventral lip last. As involution occurs, the endoderm and underlying (presumptive notochord and mesoderm) chordomesoderm move in unison, simultaneously producing the endodermal

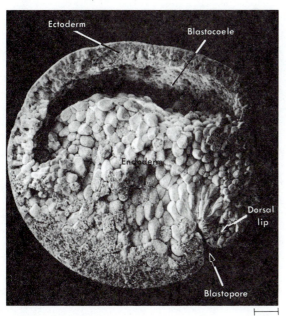

FIGURE 13.20 Scanning electron micrograph of a hemisected anuran gastrula, showing the cells of the interior. Note the bottle cells at the tip of the forming archenteron. Scale bar equals 0.1 mm. (Micrograph by J. Herkovits. From E.D.P. De Robertis and E.M.F. De Robertis, Jr. 1980. *Cell and Molecular Biology*, 7th ed. Saunders College Publishing, Philadelphia, p. 561.)

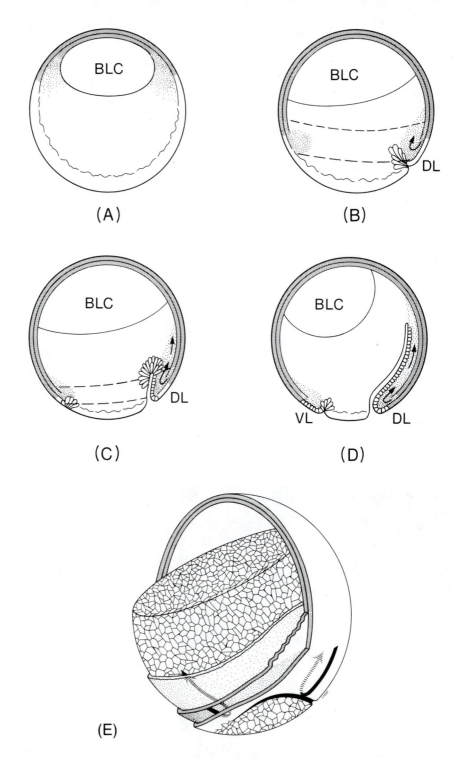

FIGURE 13.21 (*See legend on the opposite page.*)

606

◄**FIGURE 13.21** Schematic representations of *Xenopus* gastrulation. *A*, Blastula polar section. Presumptive ectoderm is shown shaded, while presumptive mesoderm is represented by sparse stippling and presumptive endoderm is clear. BLC: blastocoele. *B* to *D*, Progressively later stages of gastrulation. *B*, The blastopore is formed by invagination of presumptive endoderm. Animal hemisphere cells have spread toward the vegetal pole by epiboly. Mesoderm, which is presesnt in the lower layer, is involuting over the dorsal lip (DL) of the blastopore. Note that the mesoderm forms a band around the embryo (dashed lines). *C*, Involution of endoderm is continuing with the formation of an archenteron cavity. Bottle cells are beginning to form at the presumptive ventral lip of the blastopore. *D*, Yolk plug stage. Note the extension of the archenteron and the displacement of the blastocoele. Endoderm is involuting over the ventral lip (VL). *E*, Stereogram of a gastrula corresponding to the stage shown in *C*. The two outer layers have been cut away to show the involuting mesoderm (stippled as in the previous drawings), the inner endoderm, and the blastocoele. The arrow on the left represents the turning inward of the mesodermal band to a position overlying the endoderm. The arrow on the right represents the involution of endoderm over the dorsal lip. This endoderm will occupy the roof of the archenteron. (*A* to *D* after R.E. Keller and G.C. Schoenwolf. 1977. An SEM study of cellular morphology, contact, and arrangement, as related to gastrulation in *Xenopus laevis*. Wilh. Roux' Archiv., *82:* 168. Reprinted with permission of Springer-Verlag, Heidelberg.)

lining of the archenteron and a layer of chordomesoderm between it and the ectoderm.

According to Keller (1980, 1981), the behavior of the deep cells provides the key to understanding gastrulation (Fig. 13.22). The cells in the subsurface layers on the dorsal side of the embryo converge vegetally (i.e., toward the blastopore). The spreading of these cells is achieved by an interdigitation of cells of several layers into one layer that occupies an increased surface area. As the expanded cell layer converges toward the blastopore, cells of the overlying superficial layer flatten, spread, and divide. When the deep cells reach the blastoporal lip they reverse direction and involute to a deeper position. After involution, their behavior has changed; instead of interdigitating, which causes spreading, they begin to migrate actively toward the animal pole, using the noninvoluted deep marginal cells as their substrate (Nakatsuji, 1976; Keller and Schoenwolf, 1977). Attachment between these mesodermal cells and the accompanying endodermal cells causes the endoderm to be carried to the inside, thus enlarging the archenteron. The bottle cells, which formed the initial blastoporal groove, remain at the anterior tip of the archenteron and eventually flatten to become the anterior archenteron wall.

This interpretation of amphibian gastrulation differs in several respects from the classical one, particularly with respect to the roles of the bottle cells and the deep cells of the dorsal marginal zone. The classical view is that the bottle cells have the innate ability to sink below the surface and, afterward, draw adjacent cells into the interior (Holtfreter, 1944). Keller (1981) has shown, however, that bottle cell removal either before or during gastrulation does not prevent gastrulation. On the other hand, replacement of deep cells of the dorsal marginal zone with cells from the animal region causes a local blockage of movement of both the deep cells and the superficial layer. The deep cells of the dorsal marginal

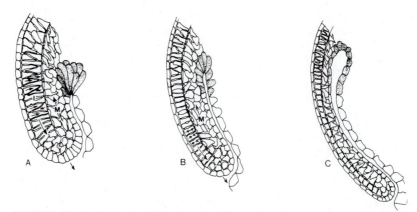

FIGURE 13.22 Behavior of deep cells of the marginal zone during *Xenopus* gastrulation. In the first half of gastrulation (*A*) the marginal zone extends toward the vegetal pole (large arrows) by interdigitation (I) of several layers of deep cells (see small arrows) to form one layer of greater area. Concurrently, superficial cells flatten, spread, and divide. On reaching the blastoporal lip, the deep cells undergo a change (Δ) in morphology and undoubtedly a related change in behavior that is presently undefined but essential for involution. After involution, the deep cells form the mesodermal cell stream and migrate (M) toward the animal pole, using the inner surfaces of the pre-involution deep marginal zone cells as a substratum. In the second half of gastrulation (*B*), the deep cells, which now form a columnar array that is one layer in thickness, flatten and spread as the marginal zone continues to extend vegetally (large arrows). Migration (M) of the mesodermal cell stream continues, and the endodermal cells, including the bottle cells, are pulled along toward the animal pole to form the roof of the archenteron. The bottle cells, which were in the form of a thick, compact mass in the early gastrula, begin to spread in the middle of gastrulation, and late in gastrulation they form a large area in the periphery of the archenteron (*C*). The mechanisms of motility of deep cells during interdigitation and migration are not known, but these processes and the transition from one to the other at the point of involution are the critical events in the behavior of the marginal zone and gastrulation as a whole. (From R.E. Keller. 1981. An experimental analysis of the role of bottle cells and the deep marginal zone in gastrulation of *Xenopus laevis*. J. Exp. Zool., *216:* 98.)

zone, by their ability to converge toward the blastopore, involute, and migrate, appear to play the major role in gastrulation, whereas the bottle cells are passively displaced by their attachment to the marginal zone cells.

Gastrulation of the Avian Embryo

Gastrulation of the avian blastoderm is complicated by its organization as a disc lying atop an uncleaved yolk mass (see Chap. 10). Consequently, invagination has been abandoned as a gastrulation mechanism. Unraveling the pattern of cell movement during avian gastrulation has proved to be a difficult problem, fraught with controversy. One of the difficulties is the inaccessibility of the lower layer of cells in the intact embryo. This layer has been made accessible to study by the development of procedures to culture the blastoderm *in vitro*. Classical studies of gastrulation movements involved the use of vital dyes and carbon particles to mark the cells of chick embryos. The use of dyes has drawbacks because the dyes stain both cell layers. Carbon particles mark the upper surfaces of the cells to which they are applied, but they may not be displaced uniformly as cells begin their movement. Thus, both procedures can complicate interpretation of data. More recently, transplantation of cells that can be identified in the host embryo has been used to chart cell movement. In one approach, cells whose nuclei have been labeled with ^{3}H-thymidine are transplanted from a portion of an embryo to the comparable region of an unlabeled embryo, and the ultimate locations of the labeled cells are determined by autoradiography (Fig. 13.23). In another approach, regions of quail and chick embryos are exchanged. Cells from these two species are distinguishable from one another, and the fates of transplanted cells are readily determined by light microscopy (Le Douarin, 1973).

The transplantation experiments have established that the epiblast is the source of the ectoderm, mesoderm, and endoderm. The mechanism by which the presumptive mesoderm and endoderm cells of the epiblast reach their internal locations has been variously called ingression, involution, or **immigration.** Internalization involves the formation of a structure called the **primitive streak.** The primitive streak makes its initial appearance as a triangular thickening at the posterior end of the area pellucida (Fig. 13.24). Formation of the primitive streak is apparently due to convergence of epiblast cells toward this region, causing the cells to pile up. The streak then elongates by a combination of forces. The anterior end pushes forward, and the posterior end of the area pellucida stretches backward. The stretching narrows and lengthens the primitive streak and also changes the shape of the area pellucida itself, which is modified from rounded to pear-shaped (Fig. 13.24*E*).

The function of the primitive streak during gastrulation is indicated by the following observations (summarized by Nicolet, 1971): (1) The mitotic index is no higher in the streak than elsewhere in the blastoderm; (2) cells in the epiblast lateral to the streak are displaced toward the streak, which they enter, and are finally found within the mesoderm and endoderm; (3) ^{3}H-thymidine-labeled cells from a primitive streak implant are replaced by unlabeled cells from the epiblast of the host. These observations suggest that the streak does not act as a proliferation center for the production of cells of the inner germ layers but merely serves as a passageway for the movement of cells to the interior.

The movement of cells via the primitive streak is shown in Figure 13.25. The center of the streak is a depression called the **primitive groove.** The cells at the bottom of the groove are elongated and are similar in shape to the ingressing presumptive primary mesenchyme cells of the sea urchin embryo (see Fig. 13.6). Their inner surfaces are highly active with reduced adhesion to adjoining cells, whereas their outer surfaces

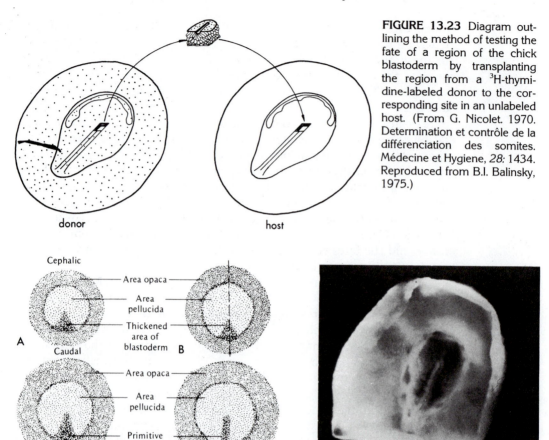

FIGURE 13.23 Diagram outlining the method of testing the fate of a region of the chick blastoderm by transplanting the region from a ³H-thymidine-labeled donor to the corresponding site in an unlabeled host. (From G. Nicolet. 1970. Determination et contrôle de la différenciation des somites. Médecine et Hygiene, *28:* 1434. Reproduced from B.I. Balinsky, 1975.)

donor   host

Cephalic

Area opaca

Area pellucida

Thickened area of blastoderm

Caudal

Area opaca

Area pellucida

Primitive streak taking shape

A   B   C   D   E

FIGURE 13.24 Chick gastrulation. *A–D,* Primitive streak formation. *A,* 3 to 4 hours incubation. *B,* 4 to 6 hours incubation. *C,* 7 to 8 hours incubation. *D,* 10 to 12 hours incubation. (Based on N.T. Spratt, Jr., 1946, from *Biology of Developing Systems* by Philip Grant. Copyright © 1978 by Holt, Rinehart and Winston. Reprinted by permission of Holt, Rinehart and Winston, CBS College Publishing.) *E,* Definitive primitive streak. Scale bar equals 1.0 mm. (Courtesy of N. Milos and S.E. Zalik.)

appear to adhere tightly to one another. Consequently, they are thought to pull neighboring cells into the groove as a migrating sheet. Within the groove the apical ends of the cells break contact with their neighbors and migrate into the blastocoele. Inward displacement from the surface epithelium is facilitated by the absence of a basement membrane in the streak region. Once inside, the cells round up and enter the migratory stream of mesenchyme cells (Solursh and Revel, 1978). The direction of movement is mainly downward from the epiblast toward the hypoblast, but the cell mass also spreads laterally and forward from the anterior end of the streak.

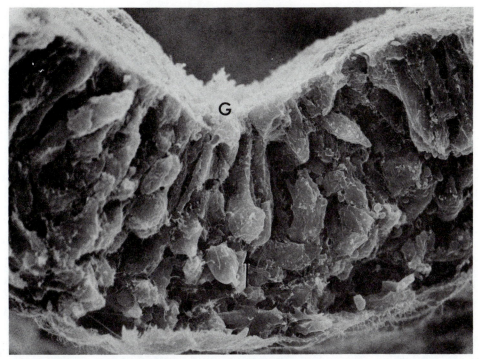

A

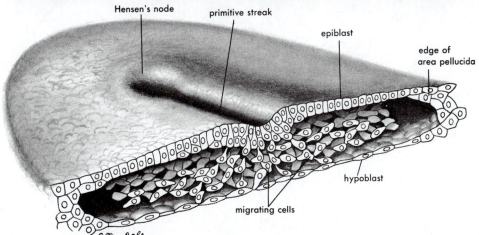

B

FIGURE 13.25 Immigration of cells via the primitive streak in chick gastrulation. *A,* Scanning electron micrograph of transversely fractured area pellucida. The columnar cells of the epiblast, still visible laterally, form bottle cells in the primitive groove (G). The expanded basal ends of these cells contact one another by cell processes (arrow). A round mesenchyme cell is to the left of the arrow. ×2100. (From M. Solursh and J.P. Revel. 1978. A scanning electron microscopy study of cell shape and cell appendages in the primitive streak region of the rat and chick embryos. Dif-ferentiation, *11:* 187.) *B,* Diagrammatic representation. The anterior half of the area pellucida is shown cut in cross section. (From B.I. Balinsky. 1981. *An Introduction to Embryology,* 5th ed. Saunders College Publishing, Philadelphia, p. 240.)

Once inside, the cells make contact with one another and again form cell sheets. Some of the ingressed cells form a sheet of mesoderm, while the rest enter the hypoblast and displace the original hypoblast cells to the periphery. The formation of the mesodermal layer is illustrated in the scanning electron micrographs of Figure 13.26, which show the lower surface of the area pellucida after removal of the hypoblast. The cells that colonize the hypoblast form most of the embryonic endoderm, whereas the original hypoblast cells form extraembryonic endoderm (Vakaet, 1962; Fontaine and Le Douarin, 1977). One of the most intriguing questions to emerge from this analysis of gastrulation is how the internalized cells become sorted out into presumptive mesoderm and presumptive endoderm.

At the anterior end of the primitive streak, a thickening develops, which has been termed **Hensen's node** (see Fig. 13.25). Presumptive notochord cells accumulate within the node. After the streak reaches its maximal length, it recedes toward the posterior end of the area pellucida, and as it does so, the presumptive notochord cells are "played out" in the middle sheet of cells in front of the retreating node. Spratt (1971) has

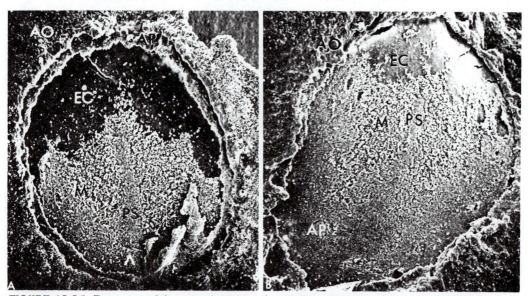

FIGURE 13.26 Formation of the mesoderm layer from cells ingressing via the primitive streak. Complete area pellucida and surrounding rim of area opaca viewed from the ventral surface after removal of the hypoblast. *A*, The mesoderm occupies the posterior half of the area pellucida and extends around the anterior end of the primitive streak. *B*, At a somewhat later stage, the mesoderm occupies the posterior three-quarters of the area pellucida. AO: area opaca; AP: area pellucida; EC: ectoderm; M: mesoderm; PS: primitive streak; arrowhead indicates the anterior-posterior axis of the embryo. ×40. (From M.A. England and J. Wakely. 1977. Scanning electron microscopy of the development of the mesoderm layer in chick embryos. Anat. Embryol., *150:* 292, 293. Reprinted with permission of Springer-Verlag, Heidelberg.)

likened this process to the drawing of a line with a piece of chalk. As the node retreats, the body axis is taking shape anterior to it. Thus, the formation of the anterior end of the body is proceeding well before gastrulation has been completed posteriorly.

The polarity of the avian embryo is evident before gastrulation, during the formation of the area pellucida (see Chap. 10). The thinning of the area pellucida begins at the presumptive posterior end, where the formation of the hypoblast epithelium also begins. This end of the area pellucida also retains its organizational capacity at gastrulation as demonstrated by experiments conducted by Spratt and Haas (1961a, 1961b). One of these experiments is outlined in Figure 13.27. If the entire area pellucida of an embryo is rotated 180°, the embryo develops with reversed polarity. On the other hand, if a central square—devoid of the posterior end—is cut out of the area pellucida and rotated, polarity is unaffected.

The significant difference between these two procedures is the rotation of the posterior margin of the area pellucida in the former but not in the latter. Rotation of the posterior margin displaces the organizational center of the embryo. Recent experiments by Eyal-Giladi and her associates indicate that this organizational center is localized in the posterior margin of the hypoblast (Azar and Eyal-Giladi, 1979, 1981; Mitrani and Eyal-Giladi, 1981). It performs two separate functions: induces the primitive streak to form in the overlying epiblast and directs the orientation of the streak. Spratt and Haas (1961a) suggested that the organi-

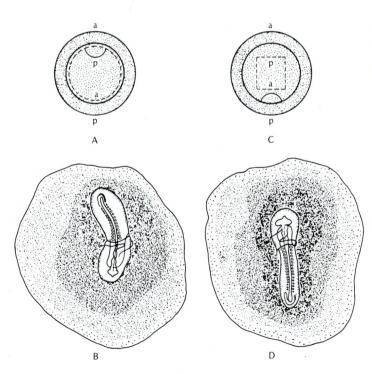

A

B

C

D

FIGURE 13.27 Evidence that the posterior margin of the area pellucida organizes the embryonic axis. This region is indicated by a semicircular line in *A* and *C*. *A*, The area pellucida is cut from the area opaca and rotated 180° so that posterior (p) and anterior (a) ends are interchanged. *B*, The resulting embryo has reversed polarity. *C*, Only the central portion of the area pellucida is rotated, leaving the posterior margin intact. *D*, The resulting embryo has normal polarity. (After N.T. Spratt, Jr., and H. Haas, 1961a, from *Biology of Developing Systems* by Philip Grant Copyright © 1978 by Holt, Rinehart and Winston. Reprinted by permission of Holt, Rinehart and Winston, CBS College Publishing.)

zational function is a consequence of the pattern of cell movement in the hypoblast. As shown in Figure 13.28, cells stream from the posterior end in a fountainlike pattern. The main stream is directed toward the anterior end, with less extensive and slower streaming directed laterally and posteriorly. As the primitive streak advances forward, the hypoblast cells in the center are being forced to the periphery by cells that immigrate from the primitive streak to form the embryonic endoderm (see previous discussion). According to Spratt and Haas, the hypoblast at the posterior end of the area pellucida functions as a "proliferation center," whose high mitotic rate provides the force that causes the cells to move away from this region. Their pattern of movement, which would be dictated by the geometry of the blastoderm, would determine the embryonic axis. Other investigators have failed to confirm that a proliferation center exists. Nevertheless, the patterns of cell movement described by Spratt and Haas do exist, and numerous experiments performed by them demonstrate that interference with the pattern of movement will modify the location or orientation of the axis (Spratt and Haas, 1960b, 1961a, 1961b). If these movements are not generated by mitosis within the posterior margin, other mechanisms must generate and coordinate the movement of the cell sheet, and thus organize the embryonic axis.

Gastrulation of the Mammalian Embryo

Before gastrulation, mammalian embryos bear a remarkable resemblance to the chick blastoderm. As we described in Chapter 10, a bilayered disc forms from the inner cell mass. The upper layer is the epiblast, and the lower layer is the hypoblast. The disc undergoes a shape change—much like that of the chick embryo—and forms a pear-shaped **embryonic shield,**

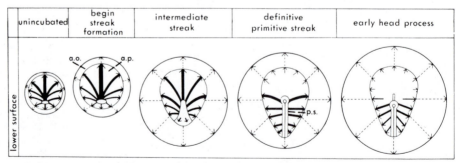

FIGURE 13.28 Diagrammatic representation of hypoblast cell movements in the chick blastoderm. General growth is indicated by dotted-line arrows, while directional cell movements are indicated by solid arrows. Heavier arrows indicate extensive and more rapid movements. a.o.: area opaca; a.p.: area pellucida. (After N.T. Spratt and H. Haas. 1960a. From B.I. Balinsky. 1981. *An Introduction to Embryology,* 5th ed. Saunders College Publishing, Philadelphia, p. 237.)

which possesses a primitive streak at the posterior end of the epiblast. The three-layered embryo is formed by immigration of epiblast cells via the primitive streak. Since no cell marking data are available, we must assume that mammalian gastrulation is similar to that in the chick. Thus, immigrating epiblast cells would form the mesoderm and may also enter the lower layer to form the embryonic endoderm.

Gastrulation of the Teleost Embryo

The teleost embryo faces a problem at gastrulation similar to that of the avian embryo: The blastoderm is a cap of cells sitting above a sphere of yolk (see Chap. 10). However, teleosts resolve this problem in quite a different way than birds do. Initially confined to the top of the egg, the blastoderm spreads over the surface of the yolk by epibolic expansion and eventually envelops it completely. The structure of the blastoderm is diagrammatically represented in Figure 13.29. It consists of a superficial layer of cells that tightly adhere to one another, which is called the **enveloping layer** and, below this, a loosely organized group of **deep cells.** The blastoderm overlies a blastocoele and, below that, the syncytial periblast (designated the **yolk syncytial layer** by Betchaku and Trinkaus, 1978). The cells of the enveloping layer do not contribute to tissues and organs of the embryo. The blastoderm deep cells, which give rise to the embryo proper, are highly active and move as individuals between the enveloping layer and the periblast. Movements of deep cells are of two types: epibolic spreading, which expands the circumference of the blastoderm, and convergence toward the presumptive posterior midline of the embryo (Ballard, 1973a). These movements cause a thinning in the center of the blastoderm and a buildup of cells at its peripheral margin to form the narrow germ ring and the embryonic shield (Fig. 13.30).

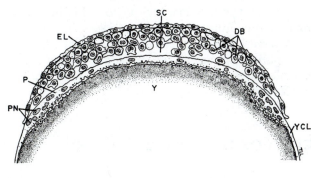

FIGURE 13.29 Diagram of an early gastrula-stage teleost embryo. The blastoderm is flattened and is extending over the yolk by epiboly. The highly flattened enveloping layer (EL) cells are closely united. The deep blastomeres (or deep cells, DB) move as individuals. Note the periblast (P) and yolk cytoplasmic layer (YCL), which are parts of a cytoplasmic continuum. The periblast is the thicker nucleated part (note periblast nuclei; PN) that lies between the blastoderm and the yolk (Y). SC: segmentation cavity or blastocoele. ×85. (From T.L. Lentz and J.P. Trinkaus. 1967. Reproduced from *The Journal of Cell Biology*, 1967, vol. 32, pp. 121–138 by copyright permission of the Rockefeller University Press.)

Within the shield the three germ layers are organizing. There is no evidence for invagination or involution of cells during this process (Ballard, 1973b). A fate map of the blastoderm in an early stage of embryonic shield formation is shown in Figure 13.31. Note that the mesodermal sheet is continuous throughout the blastoderm, whereas cells of the other germ layers are confined to the posterior half of the blastoderm (embryonic shield). In the midline of the embryonic shield the cells of the germ layers will organize the embryonic axis.

After the germ ring has been formed, it splits into an upper epiblast and a lower hypoblast. The epiblast spreads over the yolk and forms the ectodermal and mesodermal layers of the yolk sac. The yolk sac has quite a different origin from the yolk sacs of avian and mammalian embryos (see pp. 633–634 and Fig. 13.51). Hypoblast cells converge toward the embryonic shield and contribute to formation of the mesoderm of the elongating embryo (Ballard, 1973a).

The transparency of the teleost blastoderm has made it possible to observe directly the movements of the deep cells during their migration and has provided an unprecedented opportunity to study the *in vivo* migration of individual cells during the morphogenesis of a vertebrate embryo (Trinkaus, 1976; Trinkaus and Erickson, 1983). A simple procedure also yields preparations that can be used for *in vitro* analysis of deep cell movement. Removal of the blastoderm from the yolk syncytial layer leaves a number of deep cells remaining on the yolk syncytial layer. Embryos with the blastoderm removed in this way can be cultured and the movements of the adherent deep cells analyzed; the movement of such cells does not differ from that of their *in vivo* counterparts (Trinkaus and Erickson, 1983).

Movement of deep cells may be initiated by formation of a hemispherical "bleb" that extends in front of the cell body (Trinkaus, 1976). The cytoplasm then flows into the bleb, as evidenced by a corresponding reduction in the size of the cell body. Thus, the cell body is displaced in the direction of the protruding bleb. Another mode of deep cell movement is similar to the movement of fibroblasts *in vitro*. This involves formation of a spreading lamellipodium at the leading edge. This is preceded by formation of filopodia. Forward displacement of such cells occurs as the trailing edge is pulled forward by the spreading leading edge. The two modes of deep cell movement for the marine teleost *Fundulus* are illustrated in Figure 13.32.

13–4. CONSTRUCTION OF THE VERTEBRATE BODY AXIS

After gastrulation the basic body plan of the embryo is laid out along the anterior–posterior axis of body symmetry. The sheets of cells that constitute the germ layers undergo shape changes that appear to "mold" an elongated cylindrical embryo from the gastrula.

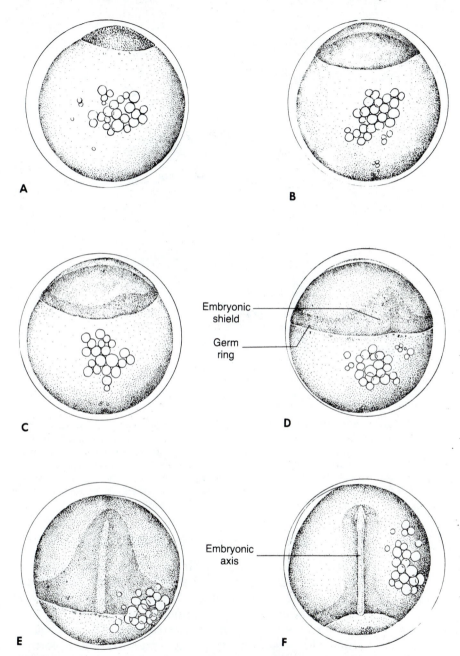

FIGURE 13.30 Gastrulation of *Fundulus heteroclitus* (a teleost). *A*, Midblastula. The blastoderm surmounts the large yolk sphere. *B*, Beginning of gastrulation. The blastoderm is starting to spread over the yolk. *C*, Early gastrula. The germ ring and early embryonic shield are visible. *D* and *E*, Mid- to advanced gastrulae. The shield is elongating. In *E* the presumptive embryonic axis can be seen. *F*, The embryonic axis is taking shape. The blastoderm has nearly covered the yolk. (From P.B. Armstrong and J.S. Child. 1965. Stages in the normal development of *Fundulus heteroclitus*. Biol. Bull., *128:* 157, 158, 159.)

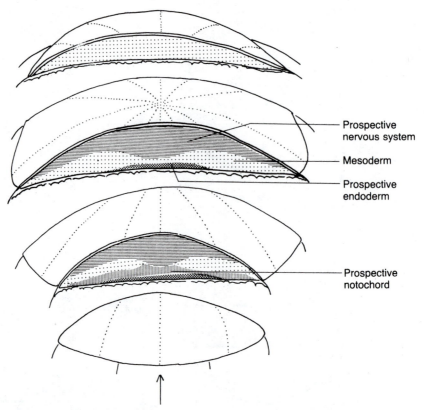

Prospective
nervous system

Mesoderm

Prospective
endoderm

Prospective
notochord

FIGURE 13.31 Stereogram of blastoderm of rainbow trout embryo. Fate map is indicated on transverse sections. The arrow marks axis of bilateral symmetry. The posterior end of the blastoderm, which contains the embryonic shield, is at the bottom. Mesoderm sheet is in light stipple, prospective nervous system is in horizontal lines, prospective notochord is in vertical lines, and prospective endoderm is in heavy stipple. (From W.W. Ballard. 1973b. A new fate map for *Salmo gairdneri*. J. Exp. Zool., *184:* 70.)

Formation of the Body Axis in the Amphibian

The formation of the vertebrate body plan is simplest in the amphibians, in which gastrulation results in a sphere that is surrounded with ectoderm and contains an inner archenteron lined with endoderm. Between the two layers a sheet of mesoderm is formed.

As gastrulation is completed with the disappearance of the yolk plug, the presumptive nervous system begins to form from the dorsal ectoderm. This process of **neurulation** so dominates the shape and appearance of the embryo that it is known as a neurula at this time (Fig. 13.33). The initial indication of neurulation is the flattening of the dorsal ectoderm to form the neural plate. The edges of the neural plate rise above the surface to form **neural folds,** which flank a central depression (the **neural groove**) that extends along the entire middorsal line of the

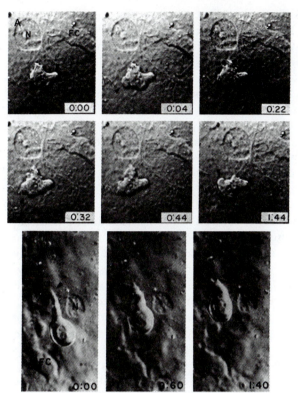

FIGURE 13.32 Movement of teleost (*Fundulus*) deep cells viewed with differential interference contrast optics. *A,* Frames from a time-lapse film showing a living deep cell engaged in blebbing movement on the surface of the yolk syncytial layer. Time is in minutes and seconds, with the first frame shown arbitrarily designated 0:00. N designates a large nucleus of the yolk syncytial layer (YSL), which serves as a stationary reference point, and FC designates another deep cell extensively flattened on the YSL substratum and not moving at this time. Certain irregularities of the surface of the YSL also serve as fixed reference points. The motile cell is moving toward the left. Note that the rounded trailing portion of the cell remains about the same size and morphology throughout the sequence (arrowhead in 0:00). At 0:04, a large bleb has just formed at the leading edge (arrowhead). The contour at its right is little changed, and the trailing edge has advanced slightly. At 0:22, cytoplasm has moved into and expanded the bleb, and at the same time the leading edge and trailing edge of the cell advance. The slightly irregular contour of the leading edge suggests that the leading edge of the now filled bleb is adhering to the yolk syncytial layer. At 0:32, a new large bleb has formed at the same region of the leading edge as the last one (arrowhead) and, with this, both the leading edge and the trailing edge have advanced. There has been little change in the contour of the right and left sides. At 0:44, cytoplasm has moved into the bleb, broadening it, but with little perceptible advance of the leading edge. The trailing edge, which has advanced, has the same form as at 0 time, but has thickened or broadened slightly. At 1:44, advance of both the leading and trailing edge has continued. A large bleb has just formed at the right leading edge (arrowhead) and a smaller one at the left. Note that this protrusive activity, which has broadened the leading front of the cell considerably, is accompanied by complete retraction of the prominent lamella at the right. *B,* This series shows a living deep cell engaged in a form of spreading movement. In this sequence, the forward movement of the cell body occurs mainly by shortening of the flattened leading protrusion. N designates a nucleus of the YSL; FC designates other deep cells extensively flattened on the YSL and not moving at this time. At 0:00, the cell is moving toward the top of the micrograph by means of a long protrusion whose leading portion is spread on the yolk syncytial layer. The leading edge is so thin that it can just barely be resolved (arrowhead). The cell body, including its trailing edge, is rounded. Note how the leading protrusion extends abruptly from the rounded cell body. At 0:60, the leading edge has moved forward only slightly and appears to have thickened; it has a more distinct contour. The long leading protrusion has shortened and thickened, and, with this, the cell body has moved forward. At 1:40, there has been only a slight forward movement of the leading edge, but it has broadened as the leading protrusion has shortened and thickened dramatically. With this the trailing edge has advanced. As the protrusion has shortened and thickened it has merged with the cell body. Scale bars equal 30 μm. (From J.P. Trinkaus and C.A. Erickson. 1983. Protrusive activity, mode and rate of locomotion and pattern of adhesion of *Fundulus* deep cells during gastrulation. J. Exp. Zool. In press.)

embryo. The neural folds eventually meet above the deepening neural groove, where they fuse to form the neural tube, which is the rudiment of the central nervous system. At its anterior end the neural tube expands and elaborates the brain, whereas in the trunk the neural tube becomes the spinal cord.

The most extensive analysis of neurulation has been conducted by Jacobson and his associates on the west coast newt, *Taricha torosa*. The neural plate of *Taricha* embryos initially constitutes the entire dorsal hemisphere of ectoderm. The hemisphere then flattens to form a disc. This is accomplished by a shrinkage in the surface area of the neural plate. The disc next undergoes a simultaneous elongation and further shrinkage as it is distorted into a keyhole shape. Finally, the neural plate elongates rapidly and rolls into the neural tube.

The behaviors of individual cells have been traced during neurulation by time-lapse cinemicrography and by cytological analysis. Cells of the neural plate tend to be displaced toward the midline and anteriorly along the midline as the neural plate changes from disc-shaped to keyhole-shaped (Burnside and Jacobson, 1968). Cell positions during neurulation are mapped by conducting a frame-by-frame analysis of time-lapse movies. A coordinate grid is superimposed on projected images of the early neural plate-stage embryo. The cells, traced as neurulation proceeds, are those at the intersections of grid lines. As these cells change position, the starting grid gradually becomes distorted (Fig. 13.34).

Cellular displacement can be partially accounted for by shape changes of individual cells. The cells undergo a concomitant apical shrinkage and increase in cell height. The apical shrinkage reduces the surface area occupied by cells and consequently causes shrinkage of the neural plate. However, changes in cell shape are not uniform for the entire neural plate; apical shrinkage of cells in some regions is more pronounced than in others. The degree of shrinkage is innate to cells of the neural plate, since cells transplanted to a heterologous site shrink an amount that is appropriate to their site of origin rather than to their new position. Ultrastructural studies suggest that elongation and apical constriction of these cells are due to activities of microtubules and microfilaments, respectively. Elongation coincides with the orientation of microtubules that are parallel to the axis of elongation (Fig. 13.35A), whereas during the apical constriction, bundles of microfilaments are seen to encircle the cell apex like a draw string (Fig. 13.35B). The drawing in Figure 13.36 illustrates the orientation of these two elements. The role of microtubules in neural plate cell elongation is supported by the observation that colchicine blocks the process (Burnside, 1973).

The proposal that drawstring constriction of cell apices is due to contraction of microfilaments is compatible with findings about the function of microfilaments in numerous other cells. The contractile function is also indicated by the results of cytochalasin B treatment, which simultaneously alters the microfilament network and blocks apical constriction (Burnside, 1973). Because of the side effects of cyto-chalasin, this experiment does not prove that the microfilaments cause constriction, but the results of the experiment certainly support this interpretation.

Another major factor in cellular displacement to produce the key-hole-shaped neural plate is the elongation of the neural plate along the midline. This portion of the neural plate has been termed the **notoplate,** since it overlies the notochord and is intimately attached to it. As the notochord is modified from a short, flat sheet to an elongate rod, the overlying notoplate changes from a broad sheet to a long column of cells stretched along the neural plate midline (Fig. 13.37). The behavior of

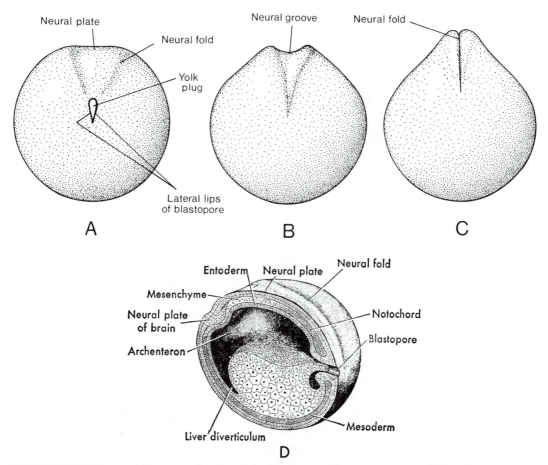

FIGURE 13.33 The amphibian neurula. *A* to *C,* Surface views of the posterior end. *A,* Neural plate stage. *B,* Neural fold stage. *C,* Fusion of neural folds. *D,* Stereogram of neural fold stage, with the left side of the embryo cut away. (Reprinted with permission of Macmillan Publishing Company, from *Fundamentals of Comparative Embryology of the Vertebrates,* Revised Edition, by A.F. Huettner. Copyright © 1949 by Macmillan Publishing Company, renewed 1977 by Mary R. Huettner, Richard A. Huettner, and Robert J. Huettner. *A* to *C* redrawn with permission.)

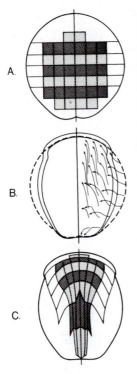

FIGURE 13.34 Neural plate formation in *Taricha torosa*, as monitored by time-lapse cinemicrography. *A*, Early neural plate-stage embryo (stage 13). A coordinate grid is superimposed over the embryo. A cell at each of the fifty points of intersection of grid lines is followed through successive stages of neurulation. *B*, This drawing of superimposed outlines of stage 13 (dashed line) and stage 15 (solid line) embryos shows the pathways of displaced cells. Cells were at the origins of arrows at stage 13 and at the heads of the arrows at stage 15. *C*, Stage 15 embryo. The transformation of the grid is shown. (After M.B. Burnside and A.G. Jacobson. 1968. Analysis of morphogenic movements in the neural plate of the newt *Taricha torosa*. Dev. Biol., *18:* 541, 543.)

these two elements is coordinate and interdependent. If either is isolated, neither will elongate. However, if the neural plate and underlying notochord are explanted together, the neural plate elongates normally and produces the keyhole shape *in vitro* (Jacobson and Gordon, 1976). The coordinate shrinkage and elongation of neural plate cells and extension of the notoplate region generate the rearrangement of cells within the neural plate that produces the keyhole shape.

Soon after the keyhole shape has been achieved the neural plate rapidly rolls into a tube. The same two mechanisms that contribute to neural plate shaping continue to operate. The apical surfaces of individual neural plate cells shrink even more than before. However, cellular elongation does not keep pace. Thus, instead of cells becoming long and thin, they become wedge-shaped with basal ends that are broader than the apical ends. The progressive shape changes of urodele neural ectoderm cells during neurulation are outlined in Figures 13.38 and 13.39. The latter figure contrasts the changes in neural ectoderm cells with presumptive epidermal cells that flatten to form a squamous epithelium at the same time that neural plate cells are undergoing elongation and apical shrinkage.

If formation of wedge-shaped cells were an active mechanism, it could cause neural tube formation by itself. However, as with earlier

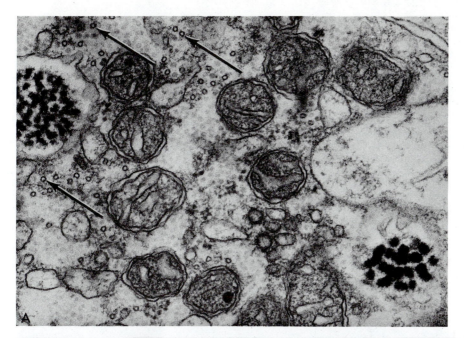

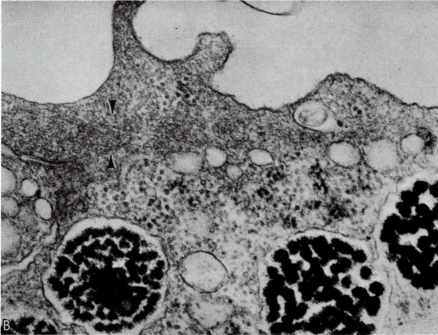

FIGURE 13.35 Electron micrographs of sections of urodele neural ectoderm cells. *A,* Section cut at right angle to the long axis. Numerous microtubules (arrows) are cut in cross section. ×61,000. *B,* Longitudinal section. Bundle of microfilaments is indicated by arrowheads. ×39,000. (From B. Burnside. 1971. Microtubules and microfilaments in newt neurulation. Dev. Biol., *26:* 422, 428.)

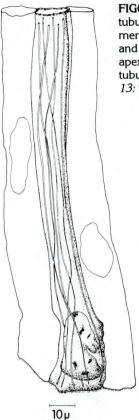

FIGURE 13.36 Schematic illustration of orientation of microtubules and microfilaments in elongating neural plate cells. Numerous microtubules are aligned parallel to the cells' long axis, and circumferential bundles of microfilaments encircle the cell apex in draw-string fashion. (From B. Burnside. 1973. Microtubules and microfilaments in amphibian neurulation. Am Zool., *13:* 994.)

10 μ

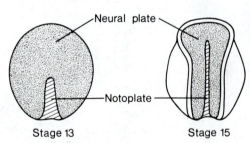

Neural plate

Notoplate

Stage 13   Stage 15

FIGURE 13.37 Drawings of stage 13 and stage 15 *Taricha torosa* embryos, showing the location of the notoplate at each stage. (After A.G. Jacobson. 1981. Morphogenesis of the neural plate and tube. *In* T.G. Connelly et al. (eds.), *Morphogenesis and Pattern Formation.* Raven Press, New York, p. 242.)

stages, notoplate elongation is also occurring—at an even faster rate than before—and must be considered as a possible driving force in neural tube formation. One could visualize the neural plate as an elastic sheet that when stretched would buckle the plate lateral to the midline out of the plane and roll it into a tube (Fig. 13.40). If stretching were the predominant mechanism causing neural tube formation, wedging of cells would be a passive response as the neural plate rolls into a tube. Further research should reveal which of these two mechanisms is predominant or whether both actively contribute to neural tube formation.

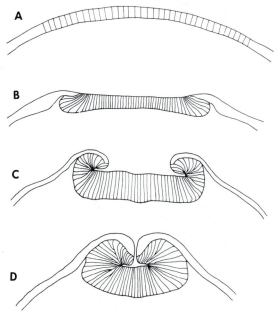

FIGURE 13.38 Diagrammatic representation of cellular shape changes of the urodele neural ectoderm during neurulation. *A,* Preplate stage. *B,* Neural plate stage. *C,* Neural fold stage. *D,* Prefusion stage. (From P. Karfunkel. 1974. The mechanisms of neural tube formation. Int. Rev. Cytol., *38:* 248.)

Fusion of the neural folds separates the neural ectoderm from the presumptive epidermis, which now completely encircles the embryo (Fig. 13.41). In this process a group of neural ectoderm cells is excluded from the neural tube and lies between it and the overlying epidermis along the middorsal line. Because of their location, these cells are called the **neural crest.** They do not remain in this location long, however; instead, they rapidly migrate laterally and ventrally to give rise to a variety of cell types in sites scattered throughout the body. Derivatives of the neural crest include pigment cells, sensory and autonomic ganglia, adrenal medulla, cartilage, skeletal and connective tissue in certain regions of the body, and the Schwann cells, which form nerve sheaths.

During the neurula stage the embryo undergoes a marked anterior–posterior elongation, and major internal organizational processes are occurring. The archenteron elongates to form the tubular gut, which later acquires anterior and posterior openings. The notochord separates from the mesoderm that flanks it on either side. It first forms a narrow band of cells underneath the neural plate and eventually becomes a cylindrical rod, which is round in cross section. The mesoderm, meanwhile, is undergoing a number of regional changes. The dorsal mesoderm on either side of the notochord becomes segmented and forms a paired series of somites (see Fig. 13.53). In the process each somite becomes separated from its adjoining somites but remains continuous with the lateral mesoderm. A split occurs in the lateral mesoderm, which forms the coelom, or primary body cavity.

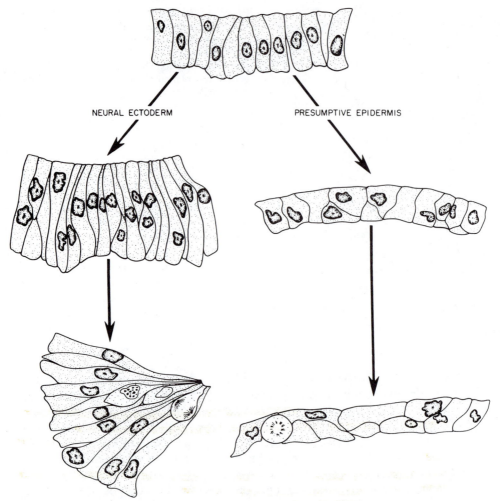

NEURAL ECTODERM

PRESUMPTIVE EPIDERMIS

FIGURE 13.39 Schematic illustration of cell shape changes exhibited by urodele ectoderm cells during neurulation. At the end of gastrulation (preplate stage, top), the entire ectoderm is a uniform simple columnar epithelium. During neurulation, neural ectoderm cells first elongate to form the neural plate (middle) and then become wedge-shaped as the neural folds roll into the neural tube (bottom). Concomitantly, epidermal cells flatten. (After B. Burnside. 1971. Microtubules and microfilaments in newt neurulation. Dev. Biol., *26:* 419.)

THE DORSAL LIP OF THE BLASTOPORE: THE PRIMARY ORGANIZER

The formation of the amphibian blastopore, which initiates gastrulation, also has important ramifications for the organization of the body axis. The blastopore marks the future posterior end of the embryo. The archenteron, which is formed by involution of the presumptive endoderm, extends forward from the blastopore to the future anterior end of

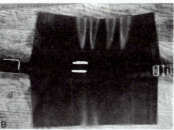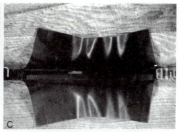

FIGURE 13.40 Model of neural tube formation by stretching an elastic sheet. A rubber sheet (*A*) is stretched along the midline, causing a fold to be raised on each side of the line (*B*). Additional stretching causes the folds to roll toward one another until they meet and form a tube (*C*). The white lines provide landmarks. (From A.G. Jacobson. 1978. Some forces that shape the nervous system. Zoon. *6*: 18.)

the embryo. The dorsal lip of the blastopore, which forms on the animal hemisphere margin of the blastopore, is the site of involution of the presumptive chordomesoderm. These cells spread underneath the overlying ectoderm and later form the notochord and somite mesoderm, whereas the dorsal ectoderm forms the neural plate. The remaining ectoderm, which spreads over the lateral and ventral surfaces of the embryo, becomes epidermis.

The dorsal lip of the blastopore has a key role to play in organization of the axis. The significance of the dorsal lip was demonstrated experimentally by Hans Spemann and Hilde Mangold in 1924. They found that the ability to involute and form an embryonic axis is retained by the dorsal lip of a urodele embryo when it is transplanted to the ventral or lateral region of another embryo. As a result, a secondary embryonic axis forms—complete with gut, neural tube, notochord, and somites. The formation of a secondary gut is not surprising, since the process of involution of the transplanted dorsal lip would produce a secondary archenteron. A notochord and somites would also be expected, since these structures are normally derived from dorsal lip mesoderm. However, the neural tube is not a derivative of the transplanted tissues, and its formation from ventro-lateral ectoderm that would normally form epidermis suggests that the dorsal lip material has *induced* the host ectoderm to develop into neural ectoderm. That this is, in fact, the case was demonstrated by the ingenious experimental procedure used by Spemann and Mangold. They used donor and host embryos that differed dramatically in the amount of pigment they contained, and in this way they were able to distinguish between donor and host cells in the secondary axis. Their results are shown in Figure 13.42. The donor dorsal lip was from an unpigmented embryo, whereas the host embryo was darkly pigmented. Most of the donor cells involuted, but a few remained on the surface, leaving a narrow white surface strip. In the interior the notochord developed from donor cells, and the somites were formed from both donor and host cells. The neural plate, on the other hand, was formed primarily of host cells, along with the few donor cells that did not involute. The dorsal lip had indeed induced the ectoderm to become neural ectoderm rather than epidermis.

The results of the Spemann and Mangold experiments indicate that during normal development the dorsal lip (or its derivatives) induces the

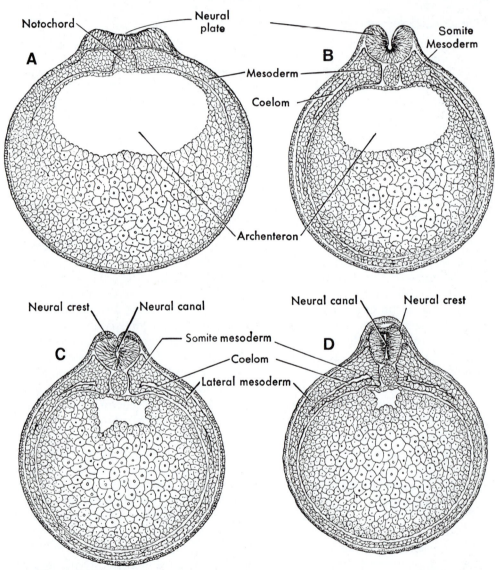

FIGURE 13.41 Cross sections of the amphibian neurula. *A,* Beginning of neural fold elevation. *B,* Advanced stage of neural fold elevation. *C,* Beginning of fusion. *D,* Neural tube is complete. (Reprinted with permission of Macmillan Publishing Company, from *Fundamentals of Comparative Embryology of the Vertebrates.* Revised Edition, by A.F. Huettner. Copyright © 1949 by Macmillan Publishing Company; renewed 1977 by Mary R. Huettner, Richard A. Huettner, and Robert J. Huettner.)

<mark>dorsal ectoderm to develop into neural ectoderm</mark>. This fixation of de-
<mark>velopmental fate</mark> is called **determination.** When does this neural induction
occur during normal development? An approximation has been obtained
from experiments that were also conducted by Spemann. He exchanged
presumptive epidermis and neural ectoderm in embryos that differed in
pigmentation, and he therefore was able to follow the fates of the transplanted tissue. If the exchange was made between early gastrulae, the
implants developed according to their new surroundings (Fig. 13.43).
Thus, presumptive epidermis developed into neural tissue, and vice
versa. Accordingly, the fates of these two regions are not fixed at the
early gastrula stage. Exchange of presumptive epidermis and neural ec-

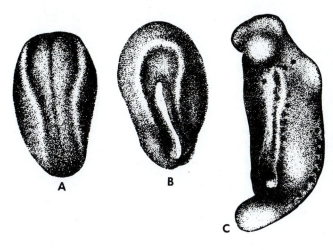

FIGURE 13.42 Induction of secondary
urodele embryo by means of transplanted
piece of dorsal lip. *A* and *B,* Recipient embryo in neural stage. *A,* Dorsal view with
host neural plate. *B,* Ventral view with induced neural plate. *C* and *D,* Same embryo in the tailbud stage with induced secondary embryo on the lower left side. *C,*
Surface view. *D,* Cross section. (From H.
Spemann. 1936. *Experimentelle Beiträge
zu einer Theorie der Entwicklung.* Julius
Springer, Berlin, pp. 93, 94. Reprinted with
permission of Springer-Verlag, Heidelberg.
Reproduced from B.I. Balinsky, 1975.)

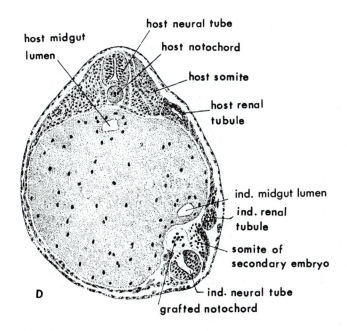

toderm at the end of gastrulation yielded an entirely different result. Transplanted neural ectoderm made an abortive attempt to form a neural tube and differentiate as neural tissue, whereas presumptive epidermis continued to develop as epidermis regardless of its new position. These results indicate that induction of the neural ectoderm occurs *during gastrulation.*

The role of the dorsal lip in establishing the anterior–posterior axis led Spemann to designate it the **primary organizer** of the amphibian embryo. The innate properties of the dorsal lip cause it to involute to form the archenteron. In addition, the presumptive chordomesoderm involutes over the dorsal lip, and the transplantation studies demonstrate that it has the ability to "self-differentiate" into notochord and somites. After involution the chordomesoderm cells spread along the inner surface of the dorsal ectoderm. Since neural induction occurs during this period, Spemann proposed that the chordomesoderm, which we now know is derived from the subsurface layers of the dorsal marginal zone, induces the overlying dorsal ectoderm to develop as neural ectoderm instead of as epidermis.

The role of the chordomesoderm in neural induction has been further investigated by implanting postinvolution chordomesoderm under ventral ectoderm. These implants function as inductors in the same way as the dorsal lip does (Spemann, 1938). Induction by the chordomeso-

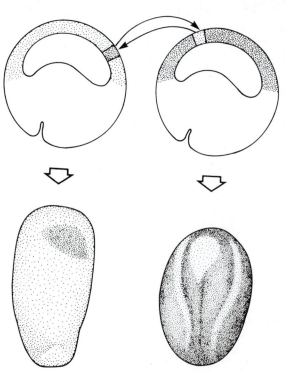

FIGURE 13.43 Exchange of presumptive epidermis and presumptive neural ectoderm between two *Triturus* early gastrulae differing in pigmentation. The lightly pigmented embryo (on the left) has had presumptive neural ectoderm implanted into its presumptive epidermal region, while the darkly pigmented embryo (on the right) has had presumptive epidermis implanted into its presumptive neural ectoderm region. In both cases, the grafts differentiate according to their new surroundings. (After H. Spemann. 1936. *Experimentelle Beiträge zu einer Theorie der Entwicklung.* Julius Springer, Berlin, p. 84. Reprinted with permission of Springer-Verlag, Heidelberg.)

derm is apparently not a simple all-or-none phenomenon that promotes generalized neural differentiation; rather, there are qualitative differences in the inductive influence exerted by different regions of the chordomesoderm. These differences appear to be responsible for regional specializations along the anterior–posterior axis (i.e., the formation of the head with brain and sense organs at the anterior end, followed by the trunk containing the spinal cord, and terminated at the posterior end by the tail). These qualitative aspects of induction have been demonstrated by transplantation of the dorsal lip at different times during gastrulation. We know that the first chordomesoderm to be involuted is displaced the greatest distance and reaches the future head region. Likewise, the last chordomesoderm to be involuted is displaced the least distance and remains at the posterior end. Therefore, the dorsal lip of the early gastrula contains the anterior chordomesoderm, and the dorsal lip of the late gastrula contains the posterior chordomesoderm. Transplantation of the former to the belly of a gastrula tends to induce a secondary head, whereas the latter tends to induce a secondary trunk and tail (Fig. 13.44).

Neural induction is an excellent example of extrinsic influence over cell differentiation, and therefore over differential gene expression. Consequently, induction would appear to be an ideal system for study of the control over cell differentiation. The discovery of induction by Spemann indeed triggered a flurry of research activity aimed at analyzing induction in the 1930s. However, induction proved to be a very difficult problem, and it remains today as one of the major unsolved problems of developmental biology. Even so, the considerable research activity has yielded a number of important results. One of the first was the discovery that the dorsal lip can still function as an inductor after it has been killed (Bautzmann et al., 1932). The simplest interpretation of this experiment is that a substance—the inductor—diffuses from the dead tissue. If true, it is likely that the inductor is a chemical. This hypothesis is supported by a number of observations. For example, placement of a membrane between dorsal lip and ectoderm allows passage (by diffusion) of the inductive influence (Toivonen et al., 1976).

The demonstration that induction occurs by transmission of a substance has led to numerous attempts to identify the substance. These efforts have been complicated by the discovery that a wide variety of adult tissues—dead and alive—can cause neural induction. This kind of result is difficult to evaluate. Does it mean that these various tissues liberate the same substance or that they produce different substances that cause the same effect? We still do not know the identity of the inductor or whether there are many inductors. The search continues.

Another approach to induction is to examine the responding tissue itself for clues as to how induction can influence the pathway of cell differentiation. This approach is particularly relevant because of the nonspecificity of the inductors. Perhaps the various inductors trigger an intrinsic response on the part of the responding tissue. According to this view, gastrula ectoderm has the capacity to differentiate as either epi-

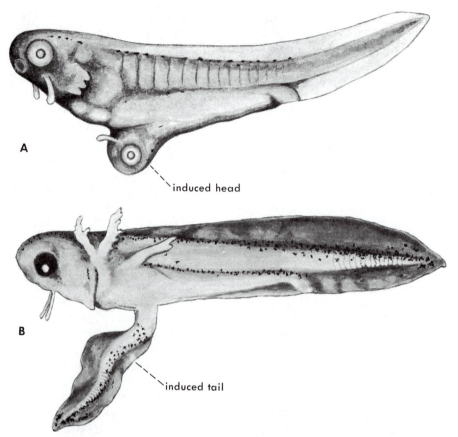

FIGURE 13.44 Regional induction. *A,* Head induction. *B,* Tail induction. (After O. Mangold, 1932, and H. Tiedemann, 1959. From B.I. Balinsky. 1981. *An Introduction to Embryology,* 5th ed. Saunders College Publishing, Philadelphia, p. 267.)

dermis or neural tissue, and induction from the chordomesoderm elicits the neurogenic capacity. Evidence in favor of this interpretation of induction was obtained by Holtfreter (1947), who demonstrated a technique that caused ectoderm to form neural tissue without exposure to inducing tissue. A short treatment with saline at either a slightly acidic or slightly alkaline pH is effective. According to Holtfreter, this treatment damages the ectoderm cells (by "sublethal cytolysis"), causing them to realize their neurogenic potential. Perhaps the reversible damage causes the release or activation of a specific neurogenic substance that already exists in these cells. Regardless of whether the inductor *causes* or *allows* neurulation, the effect on the responding cells is a permanent one, which results in the expression of the genes for the formation of the neural tube and differentiation of neurons. The ultimate solution to the problem of induction will be the discovery of the mechanisms that control the expression of these genes.

Formation of the Body Axis in the Chick

The cylindrical amphibian body plan is formed from a spherical gastrula by a relatively straightforward process. A much more complicated scheme occurs in the chick embryo, in which the body cylinder must be constructed from a flattened sheet. Gastrulation does not produce an archenteron cavity enclosed by endoderm. Instead, the forerunner of the gut is a space lying below the endoderm and above the yolk. The gut must be formed after gastrulation by a lateral infolding of the endoderm to enclose this space and surround it with endoderm (Figs. 13.45 and 13.46). As this process occurs, the endoderm is accompanied by an overlying layer of mesoderm that closely adheres to it. Ventrally, the gut initially remains in open communication with the yolk, and the endodermal wall is continuous with the extraembryonic endoderm surrounding the yolk. The extraembryonic endoderm and its overlying layer of mesoderm is called the yolk sac. The ventral connection between the yolk sac and the gut is gradually constricted to become the **yolk stalk,** which narrows as more and more of the gut is enclosed ventrally by endoderm. Furthermore, the body itself must be delimited from extraembryonic regions by the intervention of **body folds** that cause the embryo to rise above the surface of the yolk.

Another important aspect of development of the chick (indeed, of all terrestrial vertebrates) is that neurulation and accompanying processes of body plan organization begin at the anterior end, while the posterior regions are still undergoing gastrulation. These postgastrulation processes progressively spread to the posterior end as the primitive streak recedes. The anterior–posterior progression of development of the chick embryo is illustrated in Figure 13.47. The typical vertebrate axial structures are in the process of formation, including neural tube, notochord, and somites. The anterior sections show these structures in advanced stages, whereas the posterior sections show progressively earlier stages in axial organization. Note particularly the progression of neurulation. At the anterior end, fusion of the neural folds is imminent, whereas posteriorly the neural folds are just beginning to elevate in front of the receding primitive streak.

The organization of the avian embryonic axis is a function that has been ascribed to the anterior end of the primitive streak. This has been demonstrated by transplantation of portions of the anterior half of the streak of a donor duck embryo to a host chick embryo (and vice versa) below the epiblast, lateral to the existing primitive streak (Waddington and Schmidt, 1933). A secondary axis is formed, consisting of notochord and mesoderm, which are derived from the donor tissue, and a neural tube, which is induced to develop from the host ectoderm (Fig. 13.48). The anterior streak appears to be analogous to the dorsal lip of the amphibian embryo in that the tissue derived from it—the chordomesoderm—induces the ectoderm to undergo neurulation.

Formation of the Body Axis in the Mammal

In formation of the cylindrical body, the disc-shaped mammalian embryo undergoes folding processes that are comparable to those in the chick. The retention of this pattern of development, even though the mammalian egg is deficient in yolk, suggests that mammals have inherited the early developmental pattern of ancestors with yolky eggs. The formation of axial structures also occurs in the wake of the retreating Hensen's node. Compare the mammalian embryo in Figure 13.49 with the chick embryo in Figure 13.47.

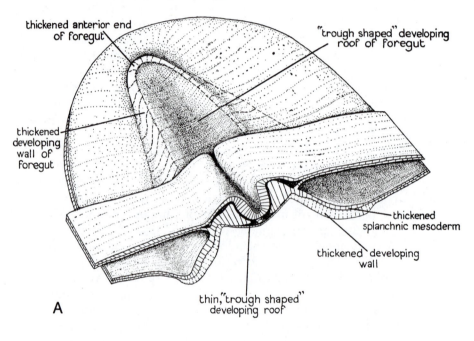

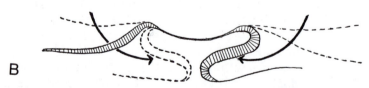

FIGURE 13.45 Foregut formation in the chick embryo. *A*, Stereogram of the anterior half of the area pellucida shown from above with most of the ectoderm and mesoderm removed. *B*, Diagram representing a cross section through the developing foregut region. On the left, the thickened wall is swinging ventromedially to form the foregut floor. On the right, the corresponding part of the endoderm has almost reached the midline. The dashed lines represent the stages in the displacement of the endoderm, and the arrows indicate the direction of movement. (From R. Bellairs. 1953. Studies on the development of the foregut in the chick blastoderm. I. The presumptive foregut area. J. Embryol. Exp. Morphol., *1:* 117. Reproduced from B.l. Balinsky, 1981.)

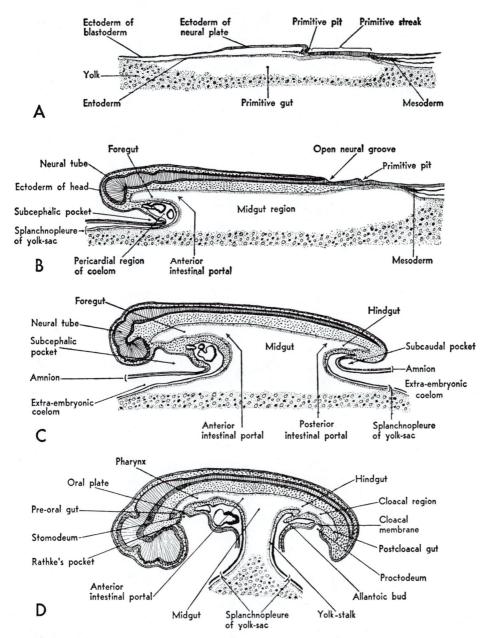

FIGURE 13.46 Schematic longitudinal-section diagrams of the chick, showing four stages in the formation of the gut tract. The embryos are represented as unaffected by torsion. *A,* Embryo toward the end of the first day of incubation; no regional differentiation of primitive gut is as yet apparent. *B,* Toward the end of the second day; foregut established. *C,* Embryo of about 2½ days; foregut, midgut, and hindgut established. *D,* Embryo of about 3½ days; foregut and hindgut increased in length at expense of midgut; yolk-stalk formed. (From *Early Embryology of the Chick,* 5th Edition, by B.M. Patten. Copyright © 1971 by McGraw-Hill Book Company. Used with permission of McGraw-Hill Book Company.)

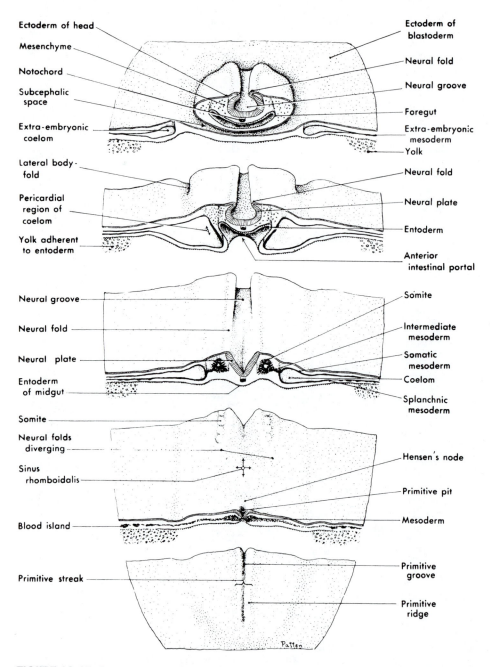

Ectoderm of head

Mesenchyme

Notochord

Subcephalic space

Extra-embryonic coelom

Ectoderm of blastoderm

Neural fold

Neural groove

Foregut

Extra-embryonic mesoderm

Yolk

Lateral body-fold

Pericardial region of coelom

Yolk adherent to entoderm

Neural fold

Neural plate

Entoderm

Anterior intestinal portal

Neural groove

Neural fold

Neural plate

Entoderm of midgut

Somite

Intermediate mesoderm

Somatic mesoderm

Coelom

Splanchnic mesoderm

Somite

Neural folds diverging

Sinus rhomboidalis

Blood island

Hensen's node

Primitive pit

Mesoderm

Primitive streak

Primitive groove

Primitive ridge

Patten

FIGURE 13.47 Stereogram of a 24-hour chick embryo. (After *Fundamentals of Comparative Embryology of Vertebrates,* Revised Edition, by A.F. Huettner. Copyright 1949 by Macmillan Publishing Company, renewed 1977 by M.R. Huettner, R.A. Huettner, and R.J. Huettner. This drawing is from *Early Embryology of the Chick,* 5th Edition, by B.M. Patten, 1971. Used with permission of McGraw-Hill Book Company.)

Organization of axial structures of the mammalian embryo is thought to be due to the anterior primitive streak. The inductive function of the streak is suggested by transplantation studies conducted by Waddington. In the first study (Waddington, 1934) he observed that a chick primitive streak implanted into the embryonic shield of a rabbit embryo will induce a secondary neural plate. Likewise—in the reciprocal experiment—a Hensen's node of a rabbit embryo induces a secondary neural plate in a chick embryo (Waddington, 1936).

13–5.   FORMATION OF ORGAN RUDIMENTS

After the organization of the body axis, the embryo undergoes a number of changes in overall body shape and gradually acquires the external appearance of a typical vertebrate embryo (Fig. 13.50). As the vertebrate embryo is taking shape, there is a clear chronological progression from formation of general vertebrate structures to the formation of more specific morphological characteristics. This is dramatically evident from

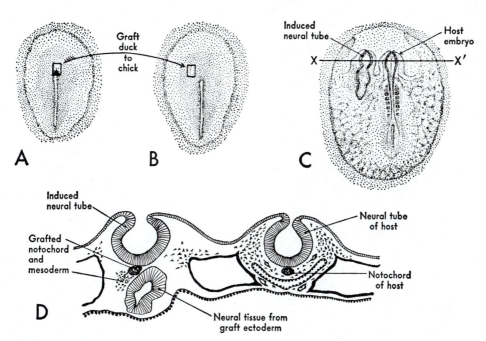

FIGURE 13.48 Induction of an accessory neural tube as a result of transplanting Hensen's node from a duck donor to a chick host. *A* and *B*, Locations of origin and site of placement of graft. *C* and *D*, Embryo cultured for 31½ hours after transplantation of node. *C*, Surface view. *D*, Cross section from level indicated by the line X–X' in *C*. (After C.H. Waddington and G.A. Schmidt. 1933. From *Early Embrology of the Chick*, 5th Edition, by B.M. Patten. Copyright © 1971 by McGraw-Hill Book Company. Used with permission of McGraw-Hill Book Company.)

the external appearance of the early embryos of higher vertebrates (i.e., reptiles, birds, and mammals). In spite of their highly divergent later appearances, these early embryos exhibit remarkable similarities, which are indicative of common developmental processes (Fig. 13.51). The major processes that occur during elaboration of the vertebrate body, as summarized by Balinsky (1981), are:

1. Elongation of the body
2. Formation of the tail
3. Subdivision of the body into head and trunk
4. Development of appendages
5. Separation of the embryo proper from the extraembryonic regions

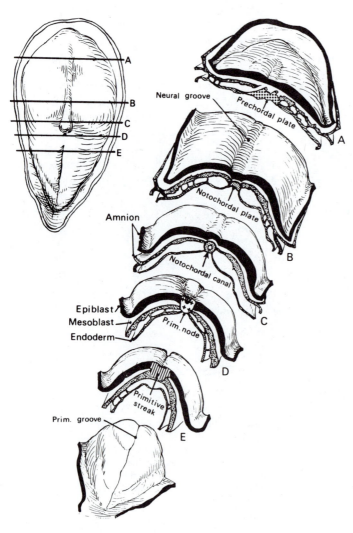

FIGURE 13.49 Stereogram of human embryo. The levels of the sections A to E are shown on a dorsal view of the intact embryo (inset). (From R. O'Rahilly. 1973. *Developmental Stages in Human Embryos. Part A: Embryos of the First Three Weeks (Stages 1 to 9)*. Carnegie Institution of Washington, Washington, D.C., p. 117.)

During the elaboration of the vertebrate body the continuity of the germ layers is lost as their cells reorganize to form clusters of cells that are the rudiments from which definitive tissues and organs will be formed. The sites of formation of some of the organ rudiments of the frog embryo are shown in Figure 13.52. Most of the rearrangements of the germ layers to form organ rudiments involve modifications to epithelial cell sheets. Some of the changes that can occur in epithelia are shown in Figure 13.53. These changes include: (1) **local thickenings,** which can occur (a) by a buildup of cells or (b) by the elongation of single cells; (2) **separation of epithelial layers** by the appearance of crevices that enlarge to form cavities; (3) **folding of epithelium,** which can take several forms, such as (a) linear folds that form grooves, (b) evaginations or outpocketings, and (c) invaginations or inpocketings; and (4) **dissociation of epithelium** into individually migrating cells.

After migrating, free cells may produce secondary aggregations to (1) surround an epithelium, (2) form a mass, (3) form a cord, or (4) form a sheet.

The condensation of free cells around epithelia is especially significant for the formation of organ primordia, since most vertebrate organs are formed from both epithelium and mesenchyme (free cells of meso-

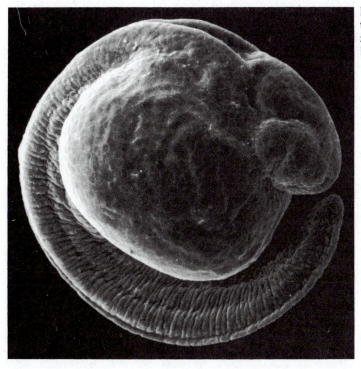

FIGURE 13.50 Scanning electron micrograph of tailbud-stage zebrafish embryo. (Courtesy of H.W. Beams and R.G. Kessel.)

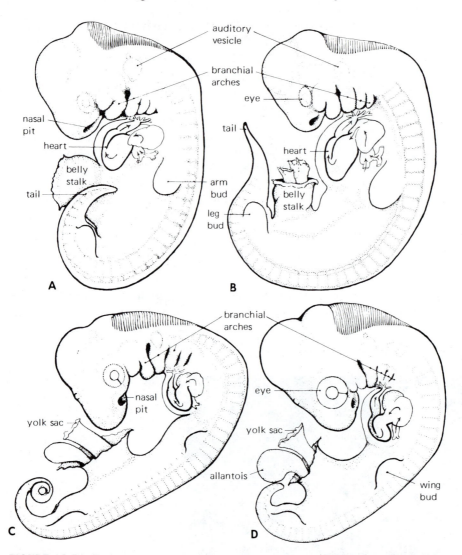

FIGURE 13.51 Embryos of (A) man, (B) pig, (C) reptile, and (D) bird at corresponding developmental stages. The striking similarities among the embryos indicate the fundamental similarities of their developmental processes. (From W. Patten. 1922. *Evolution.* Dartmouth College Press, Hanover, N.H. Reproduced from N.J. Berrill and G. Karp, 1976.)

FIGURE 13.52 Frog embryo, illustrating the sites of some organ primordia. *A*, External ▶ view. *B*, Same embryo with the skin on the left side removed. *C*, Same embryo cut in the median plane. (From B.I. Balinsky. 1981. *An Introduction to Embryology*, 5th ed. Saunders College Publishing, Philadelphia, p. 376.)

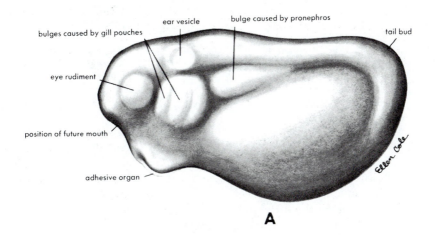

bulges caused by gill pouches
ear vesicle
bulge caused by pronephros
tail bud
eye rudiment
position of future mouth
adhesive organ

Ellen Cole

A

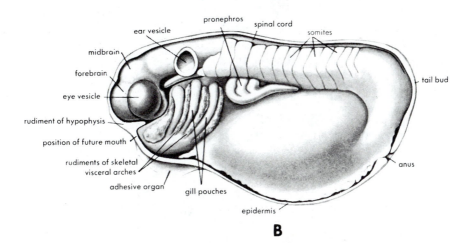

pronephros
spinal cord
somites
ear vesicle
midbrain
forebrain
eye vesicle
rudiment of hypophysis
position of future mouth
rudiments of skeletal
visceral arches
adhesive organ
gill pouches
epidermis
tail bud
anus

B

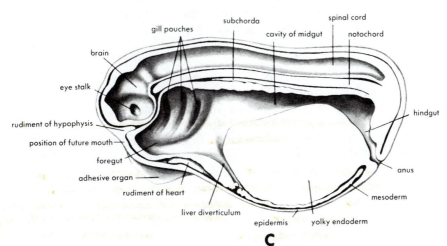

gill pouches
subchorda
cavity of midgut
spinal cord
notochord
brain
eye stalk
rudiment of hypophysis
position of future mouth
foregut
adhesive organ
rudiment of heart
liver diverticulum
epidermis
yolky endoderm
hindgut
anus
mesoderm

C

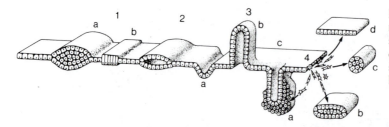

FIGURE 13.53 Examples of modifications of an epithelial cell sheet. Labels refer to processes discussed in the text. (Drawing suggested by a similar one in L.B. Arey, 1974.)

dermal origin). For example, epithelium of endodermal origin and mesodermal mesenchyme form gut-associated glands, and ectodermal epithelium and mesodermal mesenchyme give rise to peripheral structures. The condensation of mesenchyme around epithelium is also significant in promoting the differentiation of cells of the primordium into a definitive organ. An inductive stimulus provided by the mesenchyme is often essential for epithelial differentiation. This type of induction is called **secondary induction** to distinguish it from the primary induction that promotes formation of structures of the embryonic axis. The mesenchymal-epithelial interactions involved in organogenesis will be discussed in more detail in Chapter 14.

SECTION TWO
ESTABLISHMENT OF THE ANGIOSPERM BODY PLAN

The animal body plan is formed by the establishment of organ rudiments arranged along the axis of symmetry. The embryonic phase of development is completed with the differentiation of the rudiments into functional entities. During the ensuing postembryonic phase, the body is enlarged by growth of the individual components; once growth ceases, the functional components are maintained in a healthy state by cellular replacement within the tissues themselves. In plants, on the other hand, the complete body plan is not established during embryogenesis. Instead, two mitotically active formative regions—the apical meristems—are set aside. The shoot apical meristem produces the cells that will organize the expanding shoot, and the root apical meristem forms the root system. The apical meristems produce a continuously elongating body, which is referred to as the **primary body.** In many plants the whole plant is made up of the primary body. Other plants are also formed from the activity of two lateral meristems. These are the **vascular cambium,** which produces additional cells for the vascular conducting system, and the **cork cambium,** which produces the outer protective covering that replaces the embryonic epidermis. The lateral meristems and their derivatives constitute the **secondary body** of the plant. Meristems retain the capacity to elaborate an indefinite number of new structural components throughout the life of the organism. This developmental strategy has been called permanent embryogeny.

During animal development, the body plan is formed as the result of considerable movement and rearrangement of cells. Plant cells, on the other hand, are surrounded by a rigid cell wall, which restricts their mobility. Consequently, plant cells must differentiate *in situ* following their production by the meristematic tissues. The developmental fates of these cells are imposed upon them by virtue of their positions. Their positions in turn are a consequence of the pattern of cell production by the meristems. The meristems, therefore, assume critical roles in the establishment of the organizational framework of the plant body.

We shall now examine the two apical regions. The structure of the shoot tip is morphologically quite complex, since leaves and lateral buds are formed within a few hundred micrometers of the meristem. The root tip is morphologically simpler in that it does not give rise to lateral organs corresponding to those of the shoot tip. Consequently, we shall first analyze the root apex before dealing with the more complex shoot apex.

13–6. THE ROOT APEX

At germination of the seed the first visible developmental activity occurs at the root tip, where cell division and cell enlargement cause growth of the root, which splits the seed coat and allows the seedling to emerge. Root growth is a function of the root apex. The structure of this region is most easily understood by examining a longitudinal section of a root tip (Fig. 13.54). The mitotically active cells in the apex are in the apical meristem, which is the source of all cells that form the elongating root. The meristem is a bipolar structure, producing cells both proximal and distal to it. The proximal derivatives elongate and differentiate into the mature tissues of the root. As these proximal derivatives elongate, the root apical meristem is thrust forward into the soil. The distal derivatives function to protect the meristem from damage by the soil. These cells form a dome—the root cap—over the meristem. As the root grows into the soil, the root cap cells are subjected to the abrasive action of soil particles and are sloughed off. Continued cell production by the meristem counterbalances this attrition. The root cap cells perform additional functions. They produce a mucopolysaccharide that is thought to act as a lubricant to assist in soil penetration (Juniper and Roberts, 1965). Cells of the central core of the root cap are involved in gravity perception. These cells contain starch grains that act as statoliths, or gravity sensors. Displacement of the starch grains triggers chemical stimuli that cause asymmetrical elongation of cells that are proximal to the meristem, thus directing the root downward into the soil (Torrey and Feldman, 1977).

The distal and proximal derivatives of the meristem are produced by the distal meristem and the proximal meristem, respectively. These two functional regions are segregated from each other by an intervening group of cells called the **quiescent center (QC)**, in which cell division is infrequent. The quiescent center, which was discovered by Clowes (1956), can be demonstrated experimentally by feeding roots ³H-thymidine and detecting its incorporation into DNA by autoradiography. Only cells undergoing DNA synthesis, which is a prelude to cell division, will show silver grains over their nuclei. As shown in Figure 13.55, the cells at the quiescent center of the meristem remain unlabeled in such an experiment. Labeled cells, distal to the QC, are derived from the distal meristem, which is located at the margin of the QC. The proximal margin of the QC has the shape of an inverted cup. Cells along this margin divide to produce longitudinal rows (called **files**) that will form the main axis of the root.

In spite of its name, the quiescent center is not inactive in mitosis. Instead, the central cells are simply slower in completing the cell cycle (see Chap. 3) than the cells at the periphery. The difference in cycle times is due to variations in the length of the G1 phase of mitosis (Clowes, 1965). The central cells are not permanently repressed but have the potential to divide more rapidly, should conditions warrant. The

FIGURE 13.54 Median longitudinal section of the root tip of *Convolvulus arvensis* (field bindweed). The apical meristem (asterisk) is protected distally by the root cap. The cells of the root axis extend proximally from the meristem in longitudinal files. The files form the functional regions of the root. ×85. (From T.P. O'Brien and M.E. McCully. 1969. *Plant Structure and Development.* Macmillan Publishing Company, New York, p. 30.)

Root cap

potential of the QC cells to resume rapid divison is shown by experimental treatments that damage or remove the marginal meristem cells. The QC has the capacity to restore either one or both of the marginal meristematic regions (Torrey and Feldman, 1977). This potential to divide has also been demonstrated in roots that are recovering from growth inhibition induced by cold temperatures (Clowes and Stewart, 1967). During the recovery period, the quiescent center cells actively divide to reconstitute the other regions of the meristem. Once root growth is restored to normal levels, the central cells again become quiescent.

It has been proposed, therefore, that the quiescent center is the repository for the permanent meristem cells, or **apical initials,** of the root. According to this hypothesis, these slowly dividing central cells function as a reserve for the replacement of the peripheral meristematic cells under certain circumstances, or they may periodically replenish the meristematic cells at the margins, depending upon meristem requirements. The quiescent center is highly variable in size. It forms originally during early embryogenesis, recedes near the end of embryogenesis, and reinitiates shortly after germination. Thereafter, it is in flux, shrinking when the root ceases growing and expanding when the root is highly active. Thus, the quiescent center is not a morphological entity per se, but is a consequence of the growth pattern of the root. Clowes (1961) has proposed that the mitotic quiescence is not inherent to these cells but is imposed upon them as a result of their position within the apical meristem.

The proximal meristem is composed of several cell layers that are arranged in arcs over the proximal face of the quiescent center (Fig. 13.56). Longitudinal files of cells are "played out" behind the meristem. Cells of the files first elongate and then differentiate. Three concentric cylinders are formed from the files: the outer **epidermis,** the intermediate **cortex,** and the inner **central cylinder.** The central cylinder undergoes further differentiation to produce the vascular elements of xylem and phloem. The emergence of precise files of cells from the meristem indicates that the cellular pattern is determined by the meristem itself. The files extend into the quiescent center, suggesting that the pattern-determination function is actually within the quiescent center. This hypothesis is confirmed by the observation that a QC of corn can be removed and placed in culture, where it will directly elaborate a whole root (Feldman and Torrey, 1976). During the regeneration process, cell division within the QC reestablishes the proximal and distal meristems, which in turn form a new root cap and proximal root tissues, organized into the three concentric cylinders. Thus, the cells of the QC are not only able to regenerate the meristems but also to reestablish the normal pattern of tissue organization. This experiment indicates that the pattern for future root development resides within the initials of the QC. The regulation of the pattern must be exerted by the mechanisms that establish the orientation of the mitotic spindles within the meristematic tissue.

13–7. THE SHOOT APEX

The aerial portions of the plant are derived from the shoot apical meristem, which lays the foundation for the stem, leaves, axillary buds and their derivative lateral shoots, and the sexual reproductive organs. The structure of the shoot meristem must be considered in relation to the origin of lateral organs, particularly the leaves. The meristem and the emerging leaf primordia constitute the **shoot apex.** This is not a definite structure. It includes the whole of the terminal bud (with the meristem at the distalmost tip) and the youngest leaf primordia but has no specific lower limit. An overall appreciation of the organization of the shoot tip can be obtained by removing from it, under a dissecting microscope, successively younger leaf primordia, which are arrayed in order of their sequence of origin. Following the removal of the older (and therefore larger) leaf primordia that envelop the apex, one uncovers the terminal meristem surrounded by the youngest leaf primordia (Fig. 13.57A). The terminus varies considerably in size and shape, but in angiosperms it

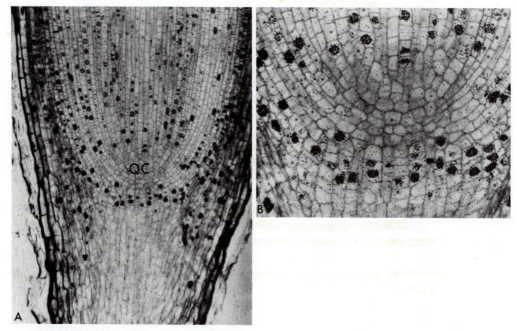

FIGURE 13.55 Autoradiographs of longitudinal sections of root apices of *Convolvulus arvensis*. A, Median longitudinal section of a root apex provided ³H-thymidine for 12 hours. Note the absence of label in a lens-shaped group of cells surrounded by labeled cells in the meristematic region. This is the quiescent center (QC) of the meristem. ×135. B, Enlarged view of A, showing the distribution of labeled nuclei in cells of the distal meristem below the QC and the proximal meristem above it. Stained nucleoli are evident in unlabeled nuclei of cells of the QC. ×320. (From H.L. Phillips, Jr., and J.G. Torrey. 1971. The quiescent center in cultured roots of *Convolvulus arvensis* L. Am. J. Bot., *58:* 667.)

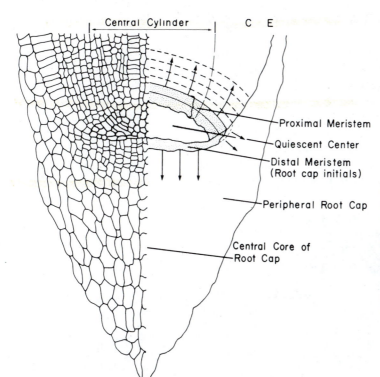

Central Cylinder

C E

- Proximal Meristem
- Quiescent Center
- Distal Meristem (Root cap initials)
- Peripheral Root Cap
- Central Core of Root Cap

FIGURE 13.56 This diagrammatic representation of a median longitudinal section of a root tip of corn shows the important functional regions of the root apex and their relationship to each other, based on evidence from experimental studies of the physiological activities of the cells. The cells are outlined on the left half of the diagram, and tissue layers or zones are indicated on the right half. Arrows indicate the direction of displacement of the cells derived from the meristems. C: cortex; E: epidermis. (After F.A.L. Clowes. 1959. Apical meristems of roots. Biol. Rev., *34:* 510. Reprinted by permission of Cambridge University Press. Courtesy of Dr. J.G. Torrey.)

usually has radial symmetry and is quite short—less than 50 micrometers in some species.

The internal structure of the shoot apex can be seen by examining longitudinal sections. Figure 13.57*B* illustrates the cellular organization of the apex of *Lupinus albus*, which is representative of the angiosperms. Note that the cells of the apical mound are stratified (Fig. 13.57*C*). The two superficial layers contain cells with a common orientation. This orientation is due to the division planes, which are perpendicular to the surface. These two layers constitute the **tunica.** Angiosperms commonly have one or more tunica layers. Below the tunica is the **corpus,** in which the cells are less orderly. Mitosis within the corpus can occur in all planes. At the base of the apical mound, the tunica-corpus stratification becomes obscured as leaf primordia are initiated. The apparent cellular homogeneity also gives way to increasing heterogeneity as cell differentiation begins.

The tunica and corpus are thought to be maintained by division of the initial cells contained within them. The layers of the tunica in turn provide the cells for construction of the organs and tissues of the shoot. Accordingly, the summit cells of the apex are the ultimate source of all cells of the shoot. To serve the apical initial function, the summit cells must be mitotically active. When cell division in the shoot apex is examined by autoradiography after incorporation of labeled thymidine, it can be seen that <mark>cell division within the summit is very infrequent</mark> (Fig. 13.58).

Although this was once interpreted to mean that these cells are mitotically inactive, and hence not involved in production of shoot cells, another interpretation is now favored. Quite convincing evidence is available that <mark>cell division, although infrequent, does occur in the summit</mark> (Steeves and Sussex, 1972). Wardlaw (1975b) has pointed out that only a small number of cells dividing infrequently would be necessary to provide cellular replenishment for the histogenic and organogenic regions below. This situation is reminiscent of that existing in the quiescent center of the root apex. As in the root, the <mark>initials may exist in a</mark>

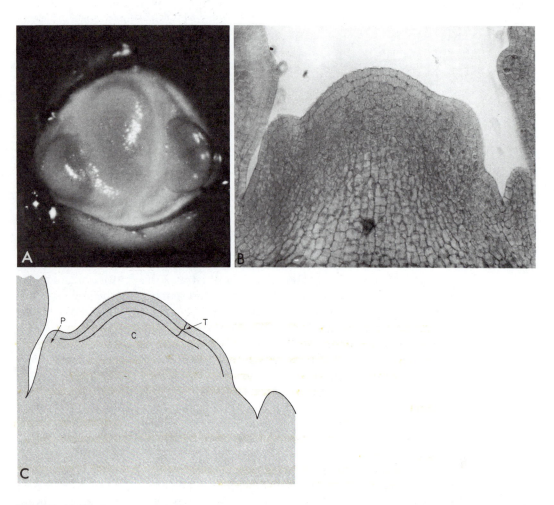

FIGURE 13.57 Structure of the shoot apex of *Lupinus albus*. *A*, Apex exposed by dissection. The youngest leaf primordium is at the top, the next youngest on the left. ×150. *B*, Median longitudinal section of an apex. ×215. *C*, Diagrammatic representation of the apex shown in *B*. C: corpus; T: tunica; P: leaf primordium. (From T.A. Steeves and I.M. Sussex. 1972. *Patterns in Plant Development.* Prentice-Hall, Inc., Englewood Cliffs, N.J., pp. 39, 41.)

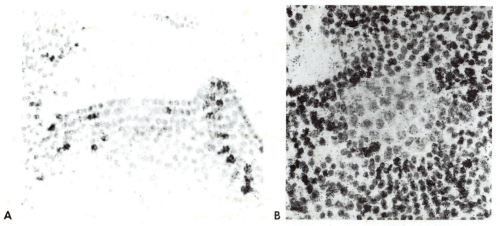

A   **B**

FIGURE 13.58 Autoradiographs of a ³H-thymidine-labeled shoot apex of *Helianthus annuus*. *A*, Median longitudinal section. *B*, Transverse section cut slightly below the surface of the apex. Note absence of labeled nuclei in central zone. ×260. (From T.A. Steeves et al. 1969. Analytical studies on the shoot apex of *Helianthus annuus*. Can. J. Bot., 47: 1370.)

region of slowly dividing cells. These cells in turn would be subtended by a peripheral region of more rapidly dividing cells, which intergrade into the differentiating regions. Thus, the shoot summit is analogous to a one-sided root quiescent center that is producing cells in only one direction.

The preceding concept of stem growth suggests that the structure of the shoot tip below the summit reflects the chronological history of the cells previously produced by the initials of the summit. Therefore, the tip is a dynamic structure in which the status of cells is changing with time. One model of the shoot apex is shown in Figure 13.59, in which the apex is represented as a system of tiers or zones emanating downward from the summit. These zones are an attempt to describe the approximate developmental status of cells of the shoot tip. It is important to keep in mind that these zones are not real entities. The margins of the zones represent the approximate levels at which cytological differences can be observed between cells above and below. The summit is called the **distal region** and contains the initials. Next comes the **subdistal region,** consisting of the more rapidly dividing meristematic cells. These two regions together constitute the apical meristem. Below them, differentiation is occurring at progressively more advanced stages at lower levels. The differentiating cells are in the following zones: the **organogenic region,** in which leaf primordia originate; the **subapical region,** which is characterized by continued cell division and elongation and by further differentiation of internal cells; and the **region of maturation,** in which the finishing touches are put on the lateral organs and the internal tissues.

The morphology of the shoot apex implies that its function is to produce the cells of the growing shoot and to organize them into a shoot in the proper spatial configuration. Since the derivatives of the apex are not capable of changing their spatial relationships, the fates of cells are more or less fixed by their location within the stem. Are fates imposed upon the cells by virtue of their location, or do they arise with their fates predetermined? This question is similar to that of regulation versus mosaicism in animal development, and it has been approached experimentally in much the same way. That is, if one perturbs the meristem, will the fates of cells be changed? The answer is yes. For example, if the meristem is bisected, each half will regenerate a whole meristem capable of producing a shoot (Pilkington, 1929; Ball, 1955). In fact, the meristem may be divided into a number of sectors, and each of them will produce an entire shoot.

Obviously, the developmental fates of the apical cells are not predetermined, and they have potentialities that are not expressed during normal development. This must mean that some mechanism is operating to integrate the development of the shoot derivatives, ensuring that developmental fates are realized in the correct spatial configuration. We have learned that in the root this function resides within the meristem itself; that is, an isolated quiescent center will regenerate an entire root. An equivalent experiment demonstrates that the shoot apex also is autonomous; whole plants can be regenerated *in vitro* from isolated shoot apical meristems (Fig. 13.60). Thus, the apical meristems at the opposite ends of the plant are autonomous structures that are capable of organizing functional tissues. Precise patterns of cell division produce highly organized structures in which the pattern of cell differentiation is regulated by the growth center itself.

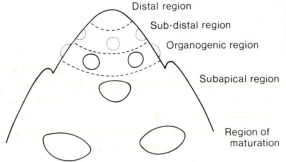

Distal region
Sub-distal region
Organogenic region
Subapical region
Region of maturation

FIGURE 13.59 Diagrammatic representation of a shoot apex. (After C.W. Wardlaw. 1957a. On the organization and reactivity of the shoot apex in vascular plants. Am. J. Bot., *44:* 179.)

FIGURE 13.60 Development of isolated shoot apical meristem of *Nicotiana tabacum.* ▶ *A,* Longitudinal section of intact shoot tip. ×367. *B,* Section of freshly isolated apical meristem. ×360. *C,* Section of meristem after 3 to 6 days in culture. *D,* Section of meristem after 6 to 9 days in culture. ×230. *E,* Plantlet from meristem culture. *F,* Mature flowering plant derived from meristem culture. (From R.H. Smith and T. Murashige. 1970. *In vitro* development of the isolated shoot apical meristem of angiosperms. Am. J. Bot., *57:* 563, 565.)

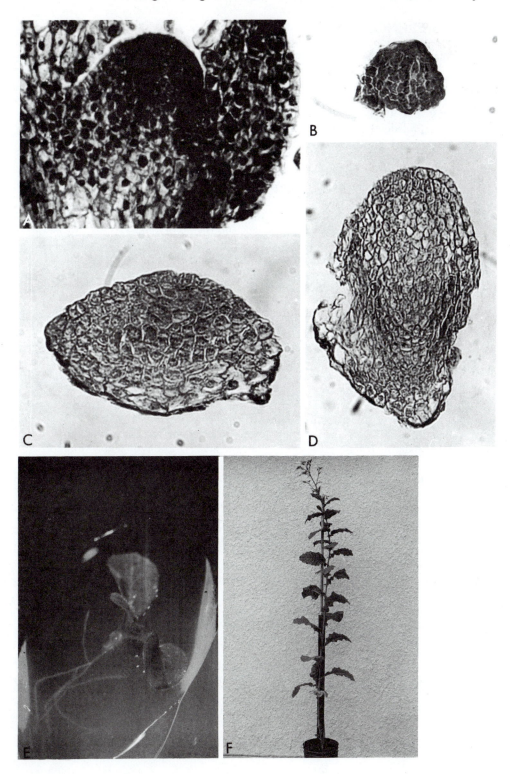

REFERENCES

Akasaka, K., S. Amemiya, and H. Terayama. 1980. Scanning electron microsopical study of the inside of sea urchin embryos (*Pseudocentrotus depressus*) Effects of aryl β-xyloside, tunicamycin and deprivation of sulfate ions. Exp. Cell Res., *129*: 1–13.

Arey, L.B. 1974. *Developmental Anatomy*, 7th ed. (Rev.). W.B. Saunders Co., Philadelphia.

Armsrong, P.B., and J.S. Child. 1965. Stages in the normal development of *Fundulus heteroclitus*. Biol. Bull., *128*: 143–168.

Azar, Y., and H. Eyal-Giladi. 1979. Marginal zone cells—The primitive streak-inducing component of the primary hypoblast in the chick. J. Embryol. Exp. Morphol, *52*: 79–88

Azar, Y., and H. Eyal-Giladi. 1981. Interaction of epiblast and hypoblast in the formation of the primitive streak and the embryonic axis, as revealed by hypoblast-rotation experiments. J. Embryol. Exp. Morphol, *61*: 133–144.

Balinsky, B.I. 1975. *An Introduction to Embryology*, 4th ed. W.B. Saunders Co., Philadelphia.

Balinsky, B.I. 1981. *An Introduction to Embryology*, 5th ed. Saunders College Publishing, Philadelphia.

Ball, E. 1955. On certain gradients in the shoot tip of *Lupinus albus*. Am. J. Bot., *42*: 509–521.

Ballard, W.W. 1973a. Morphogenetic movements in *Salmo gairdneri* Richardson. J. Exp. Zool., *184*: 27–48.

Ballard, W.W. 1973b. A new fate map for *Salmo gairdneri*. J. Exp. Zool., *184*: 49–74.

Bautzmann, H. et al. 1932. Versuche zur Analyse der Induktionsmittel in der Embryonalentwicklung. Naturwissenschaften, *20*: 971–974.

Bellairs, R. 1953. Studies on the development of the foregut in the chick blastoderm. I. The presumptive foregut area. J. Embryol. Exp. Morphol., *1*: 115–124.

Berrell, N.J., and G. Karp. 1976. *Development*. McGraw-Hill Book Co., New York.

Betchaku, T., and J.P. Trinkaus. 1978. Contact relations, surface activity, and cortical microfilaments of marginal cells of the enveloping layer and of the yolk syncytial and yolk cytoplasmic layers of *Fundulus* before and during epiboly. J. Exp. Zool., 206: 381–426.

Burnside, B. 1971. Microtubules and microfilaments in newt neurulation. Dev. Biol., *26*: 416–441.

Burnside, B. 1973. Microtubules and microfilaments in amphibian neurulation. Am. Zool., 13: 989–1006.

Burnside, M.B., and A.G. Jacobson. 1968. Analysis of morphogenic movements in the neural plate of the newt *Taricha torosa*. Dev. Biol., *18*: 537–552.

Clowes, F.A.L. 1956. Localization of nucleic acid synthesis in root meristems. . J. Exp. Bot., 7: 307–312.

Clowes, F.A.L. 1959. Apical meristems of roots. Biol. Rev., *34*: 501–529.

Clowes, F.A.L. 1961. *Apical Meristems*. Blackwell Scientific Publications, Oxford.

Clowes, F.A.L. 1965. The duration of the G_1 phase of the mitotic cycle and its relation to radiosensitivity. New Phytol., *64*: 355–359.

Clowes, F.A.L., and H.E. Stewart. 1967. Recovery from dormancy in roots. New Phytol., *66*: 115–123.

Conklin, E.G. 1932. The embryology of *Amphioxus*. J. Morphol., *54*: 69–118.

De Robertis, E.D.P., and E.M.F. De Robertis, Jr. 1980. *Cell and Molecular Biology*, 7th ed. Saunders College Publishing, Philadelphia.

England, M.A., and J. Wakely. 1977. Scanning electron microscopy of the development of the mesoderm layer in chick embryos. Anat. Embryol., *150*: 291–300.

Feldman, L.J., and J.G. Torrey. 1976. The isolation and culture *in vitro* of the quiescent center of *Zea mays*. Am. J. Bot., *63*: 345–355.

Fontaine, J., and N.M. Le Douarin. 1977. Analysis of endoderm formation in the avian blastoderm by the use of quail-chick chimaeras. The problem of the neuroectodermal origin of the cells of the APUD series. J. Embryol. Exp. Morphol., *41*: 209–222.

Gibbins, J.R., L.G. Tilney, and K.R. Porter. 1969. Microtubules in the formation and development of the primary mesenchyme in *Arbacia punctulata*. I. Distribution of microtubules. J. Cell Biol., *41*: 201–226.

Grant, P. 1978. *Biology of Developing Systems*. Holt, Rinehart and Winston, New York.

Gustafson, T., and L. Wolpert. 1961. Studies on the cellular basis of morphogenesis in the sea urchin embryo. Directed movements of primary mesenchyme cells in normal and vegetalized larvae. Exp. Cell Res., *24*: 64–79.

Gustafson, T., and L. Wolpert. 1967. Cellular movement and contact in sea urchin morphogenesis. Biol. Rev., *42*: 442–498.

Haeckel, E. 1868. Naturliche Schöpfungsgeschichte. Berlin.

Harkey, M.A., and A.H. Whiteley. 1980. Isolation, culture, and differentiation of echinoid primary mesenchyme cells. Wilh. Roux's Arch., *189*: 111–122.

Holtfreter, J. 1944. A study of the mechanics of gastrulation. Part II. J. Exp. Zool., *95*: 171–212.

Holtfreter, J. 1947. Neural induction in explants which have passed through a sublethal cytolysis. J. Exp. Zool., *106*: 197–222.

Huettner, A.F., 1949. *Fundamentals of Comparative Embryology of the Vertebrates*. Macmillan Inc., New York.

Jacobson, A.G. 1978. Some forces that shape the nervous system. Zoon, *6*: 13–21.

Jacobson, A.G. 1981. Morphogenesis of the neural plate and tube. In T.G. Connelly et al. (eds.), *Morphogenesis and Pattern Formation*. Raven Press, New York, pp. 233–263.

Jacobson, A.G., and R. Gordon. 1976. Changes in the shape of the developing vertebrate nervous system analyzed experimentally, mathematically and by computer simulation. J. Exp. Zool., *197*: 191–246.

Juniper, B.E., and R.M. Roberts. 1965. Polysaccharide synthesis and the fine structure of root cells. J. Roy. Micros. Soc. 85: 63–72.

Karfunkel, P. 1974. The mechanisms of neural tube formation. Int. Rev. Cytol., *38*: 245–271.

Katow, H., and M. Solursh. 1980. Ultrastructure of primary mesenchyme cell ingression in the sea urchin *Lytechinus pictus*. J. Exp. Zool., *213*: 231–246.

Keller, R.E. 1975. Vital dye mapping of the gastrula and neurula of *Xenopus laevis*. I. Prospective areas and morphogenetic movements of the superficial layer. Dev. Biol., *42*: 222–241.

Keller, R.E. 1976. Vital dye mapping of the gastrula and neurula of *Xenopus laevis*. II. Prospective areas and morphogenetic movements of the deep layer. Dev. Biol., *51*: 118–137.

Keller, R.E. 1978. Time-lapse cinemicrographic analysis of superficial cell behavior during and prior to gastrulation in *Xenopus laevis.* J. Morphol., *157:* 223–248.

Keller, R.E. 1980. The cellular basis of epiboly: An SEM study of deep-cell rearrangement during gastrulation in *Xenopus laevis.* J. Embryol. Exp. Morphol., *60:* 201–234.

Keller, R.E. 1981. An experimental analysis of the role of bottle cells and the deep marginal zone in gastrulation of *Xenopus laevis.* J. Exp. Zool., *216:* 81–101.

Keller, R.E., and G.C. Schoenwolf. 1977. An SEM study of cellular morphology, contact, and arrangement, as related to gastrulation in *Xenopus laevis.* Wilh. Roux's Arch., *82:* 165–186.

Korschelt, E. 1936. *Vergleichende Entwicklungsgeschichte der Tiere,* Band I. Gustav Fischer Verlag, Jena.

Le Douarin, N. 1973. A biological cell labelling technique and its use in experimental embryology. Dev. Biol., *30:* 217–222.

Lentz, T.L., and J.P. Trinkaus. 1967. A fine structural study of cytodifferentiation during cleavage, blastula, and gastrula stages of *Fundulus heteroclitus.* J. Cell Biol., *32:* 121–138.

Løvtrup, S. 1975. Fate maps and gastrulation in amphibia—A critique of current views. Can. J. Zool., *53:* 473–479.

Mangold, O. 1932. Autonome und komplementäre Induktionen bei Amphibien. Naturwissenschaften, *20:* 371–375.

Mitrani, E., and H. Eyal-Giladi. 1981. Hypoblastic cells can form a disk inducing an embryonic axis in chick epiblast. Nature (Lond.), *289:* 800–802.

Moore, A.R., and A.S. Burt. 1939. On the locus and nature of the forces causing gastrulation in the embryos of *Dendraster excentricus.* J. Exp. Zool., *82:* 159–171.

Nakatsuji, N. 1976. Studies on the gastrulation of amphibian embryos: Ultrastructure of the migrating cells of anurans. Wilh. Roux's Arch., *180:* 229–240.

Nicolet, G. 1970. Determination et contrôle de la différenciation des somites. Médecine et Hygiene, *28:* 1433–1437.

Nicolet, G. 1971. Avian gastrulation. Adv. Morphogen., *9:* 231–262.

Niu, M.C. 1956. New approaches to the problem of embryonic induction. *In* D. Rudnick (ed.), *Cellular Mechanisms in Differentiation and Growth.* Princeton University Press, Princeton, N.J., pp. 155–171.

Niu, M.C., and V.C. Twitty. 1953. The differentiation of gastrula ectoderm in medium conditioned by axial mesoderm. Proc. Natl. Acad. Sci. U.S.A., *39:* 985–989.

O'Brien, T.P. and M.E. McCully. 1969. *Plant Structure and Development.* Macmillan, Inc., New York.

Okazaki, K. 1975. Spicule formation by isolated micromeres of the sea urchin embryo. Am Zool., *15:* 567–581.

Oppenheimer, J. 1967. *Essays in the History of Embryology and Biology.* The M.I.T. Press, Cambridge, Mass.

O'Rahilly, R. 1973. *Developmental Stages in Human Embryos. Part A: Embryos of the First Three Weeks (Stages 1 to 9).* Publication 631. Carnegie Institution of Washington, Washington, D.C.

Patten, B.M. 1971. *Early Embryology of the Chick,* 5th ed. McGraw-Hill Book Co., New York.

Patten, B.M. and B.M. Carlson. 1974. *Foundations of Embryology*, 3rd ed. McGraw-Hill Book Co., New York.

Patten, W. 1922. *Evolution*. Dartmouth College Press, Hanover, N.H.

Phillips, H.L., Jr., and J.G. Torrey. 1971. The quiescent center in cultured roots of *Convolvulus arvensis* L. Am. J. Bot., *58*: 665–671.

Pilkington, M. 1929. The regeneration of the stem apex. New Phytol., *28*: 37–53.

Smith, R.H. and T. Murashige. 1970. *In vitro* development of the isolated shoot apical meristem of angiosperms. Am. J. Bot., *57*: 562–568.

Solursh, M. and J.P. Revel. 1978. A scanning electron microscopy study of cell shape and cell appendages in the primitive streak region of the rat and chick embryos. Differentiation, *11*: 185–190.

Spemann, H. 1936. *Experimentelle Beiträge zu einer Theorie der Entwicklung*. Julius Springer, Berlin.

Spemann, H. 1938. *Embryonic Development and Induction*. New Haven, Conn.: Yale University Press. Reprinted by Hafner Press (Macmillan, Inc.), New York. (1962).

Spemann, H., and H. Mangold. 1924. Induction of embryonic primordía by implantation of organizers from a different species. English translation of original German. *In* Willier, B.H. and J.M. Oppenheimer (eds.), *Foundations of Experimental Embryology*, 2nd ed. Hafner Press, New York, pp. 144–184 (1974).

Spratt, N.T., Jr. 1946. Formation of the primitive streak in the explanted chick blastoderm marked with carbon particles. J. Exp. Zool., *103*: 259–304.

Spratt, N.T., Jr. 1971. *Developmental Biology*. Wadsworth Publishing Co., Inc., Belmont, Calif.

Spratt, N.T., Jr. and H. Haas. 1960a. Morphogenetic movements in the lower surface of the unincubated and early chick blastoderm. J. Exp. Zool., *144*: 139–158.

Spratt, N.T., Jr. and H. Haas. 1960b. Integrative mechanisms in development of the early chick blastoderm. I. Regulative potentiality of separated parts. J. Exp. Zool., *145*: 97–137.

Spratt, N.T., Jr. and H. Haas. 1961a. Integrative mechanisms in development of the early chick blastoderm. II. Role of morphogenetic movements and regenerative growth in synthetic and topographically disarranged blastoderms. J. Exp. Zool., *147*: 57–93.

Spratt, N.T., Jr. and H. Haas. 1961b. Integrative mechanisms in development of the early chick blastoderm. III. Role of cell population size and growth potentiality in synthetic systems larger than normal. J. Exp. Zool., *147*: 271–293.

Steeves, T.A. et al. 1969. Analytical studies on the shoot apex of *Helianthus annuus*. Can. J. Bot., *47*: 1367–1375.

Steeves, T.A. and I.M. Sussex. 1972. *Patterns in Plant Development*. Prentice-Hall, Inc. Englewood Cliffs, N.J.

Tiedemann, H. 1959. Neue Ergebnisse zur Frage nach der chemischen Natur der Induktionsstoffe beim Organisatoreffekt Spemanns. Naturwissenschaften, *46*: 613–623.

Tilney, L.G. and J.R. Gibbins. 1969a. Microtubules in the formation and development of the primary mesenchyme in *Arbacia punctulata*. II. An experimental analysis of their role in development and maintenance of cell shape. J. Cell Biol., *41*: 227–250.

Tilney, L.G. and J.R. Gibbins. 1969b. Microtubules and filaments in the filopodia of the secondary mesenchyme cells of *Arbacia punctulata* and *Echinarachnius parma*. J. Cell Sci., 5: 195–210.

Toivonen, S., D. Tarin and L. Saxén. 1976. The transmission of morphogenetic signals from amphibian mesoderm to ectoderm in primary induction. Differentiation, 5: 49–55.

Torrey, J.G. and L.J. Feldman. 1977. The organization and function of the root apex. Am. Sci., 65: 334–344.

Trinkaus, J.P. 1973. Surface activity and locomotion of *Fundulus* deep cells during blastula and gastrula stages. Dev. Biol., 30: 68–103.

Trinkaus, J.P. 1976. On the mechanism of metazoan cell movements. *In* G. Poste and G.L. Nicolson (eds.), *The Cell Surface in Animal Embryogenesis and Development*. Elsevier/North Holland Bio-medical Press, Amsterdam, pp. 225–329.

Trinkaus, J. P. and C.A. Erickson. 1983. Protrusive activity, mode and rate of locomotion and pattern of adhesion of *Fundulus* deep cells during gastrulation. J. Exp. Zool. In press.

Vakaet, L. 1962 Some new data concerning the formation of the definitive endoblast in the chick embryo. J. Embryol. Exp. Morphol., 10: 38–57.

Vogt, W. 1929. Gestaltungsanalyse am Amphibienkeim mit örtlicher Vitalfärbung. II. Gastrulation und Mesodermbildung bei Urodelen und Anuren. Wilh. Roux's Arch., 120: 385–706.

von Baer, K. E. 1828. *Ueber Entwicklungsgeschichte der Tiere, Beobachtung, und Reflexion*. Königsberg.

Waddington, C.H. 1934. Experiments on embryonic induction. III. A note on inductions by chick primitive streak transplanted to the rabbit embryo. J. Exp. Biol., 11: 224–226.

Waddington, C.H. 1936. Organizers in mammalian development. Nature (Lond.), 138: 125.

Waddington, C.H. and G.A. Schmidt. 1933. Induction by heteroplastic grafts of the primtive streak in birds. Wilh. Roux's Arch. 128: 522–563.

Wardlaw, C.W. 1957a. On the organization and reactivity of the shoot apex in vascular plants. Am. J. Bot., 44: 176–185.

Wardlaw, C.W. 1957b. The reactivity of the apical meristem as ascertained by cytological and other techniques. New Phytol., 56: 221–229.

14 Organogenesis: Determination, Morphogenesis, and Differentiation

The elaboration of functional organs and tissues from their rudiments is the crowning achievement of the developmental process. The formation of these functional entities involves: (1) *morphogenesis*, which is dependent upon coordinated shape changes of individual cells, particularly epithelial cells, and (2) *differentiation* of the component cells, whereby they attain the ability to perform specialized functions. The elegance of this phase of development derives from its precision: Specialized cells form at an appropriate time and at the correct location to participate in the functional activity of the organism. An additional level of complexity is found in organs that are composed of more than one cell type, each of them differentiating along distinct pathways in the correct spatial relationships with their cohorts.

The chronological and spatial regulation of organ morphogenesis and cell differentiation is responsible for the orderly acquisition of adult form and function. The regulatory processes *determine* cell fate and *maintain* functional integrity of the cells. Functional specialization results from the utilization of a minute portion of the potential genetic repertoire, as protein synthetic capacity is severely restricted. As we discussed in Chapter 2, every cell in the body retains the genome that is established at fertilization, and cell specialization is made possible by differential utilization of portions of the genome. How does this selection process occur? It is clear that we must look outside the genome for these determinative influences. Once they are identified, we would hope to explain in molecular terms how they affect the expression of the genome.

In some instances we know that specialized regions of the egg cytoplasm determine the fates of cells: The pattern of cell division establishes the distribution of the cytoplasm, and hence the location of presumptive tissues and organs (see Chap. 11). Most functional entities, however, arise via a more complex route that requires interactions with neighboring regions of the embryo. The early embryo is a complex reaction system in which the possible fates of localized regions become progressively narrowed as a result of the interactions. The number of interactions that occur in the formation of even a simple organism is bewildering to contemplate, let alone comprehend. For this reason, investigators have chosen to study systems that offer certain experimental advantages, one of the most important of which is the ability to observe determination and differentiation *in vitro*. Other desirable characteristics are the ability to control determination *in vitro* and the availability of a simple assay system for cell-specific proteins. Once such developmental systems are identified, the most useful ones are often exploited by successive generations of investigators. This process leads to the gradual evolution of increasingly more sophisticated understanding of these systems. Although our understanding of the development of the vast majority of organs and tissues—even in popular experimental organisms—is negligible or even totally lacking, our comprehension of the determination of functional entities, their morphogenesis, and differentiation of their cells is expanding rapidly on the basis of this selective approach.

The discussion that follows is an analysis of four "developmental systems"; each has been chosen for its own unique properties, which illustrate different developmental principles.

14–1.   EPITHELIO-MESENCHYMAL INTERACTIONS

As we discussed in Chapter 13, mesenchyme and epithelia may combine to form organ rudiments of dual germ layer origin. The "condensation" of mesenchyme around a region of epithelium may be essential for the morphogenesis and differentiation of the epithelial derivative. Rudiments of this type illustrate very nicely how development depends upon interactions between cell populations. The effects of mesenchyme on an epithelium can be demonstrated by separating them and comparing the development of the isolated epithelium with its development when recombined with either homologous mesenchyme or mesenchyme from another source.

The dependence of epithelium on mesenchyme varies. At one extreme, a *specific* mesenchyme *determines* epithelial fate; at the other, *generalized* mesenchyme *permits* predetermined cells to realize their potential. The vertebrate epidermis is an example of epithelium whose fate is highly dependent upon the specific mesenchyme that underlies it. Unlike the epidermis, much of the internal glandular epithelium is restricted in its developmental potential and requires merely the presence of generalized mesenchyme to express its potential. However, as we shall see later, some glandular epithelia have a much more specific mesen-chymal requirement.

Mesenchymal Determination of Epidermal Fate

The vertebrate skin produces a variety of different elaborations (Fig. 14.1), including teeth, hair, feathers, scales, and glands. The skin is composed of the epidermis (derived from ectoderm) and the dermis (derived from mesoderm). The embryonic epidermis consists of two layers—the outer **periderm** and the inner **generative layer** (Fig. 14.2). The cells of the peri-derm become flattened and are transitory, eventually being sloughed from the surface. The generative layer consists of the mitotically active cells that replenish the periderm and produce the cells that form the skin derivatives. The dermis is a mesenchymal tissue that is separated from the epidermis by an acellular basal lamina.

Regional variation in the types of epidermal derivatives is respon-sible for much of the difference in the external appearance that is typical of regions of the body. The chick skin illustrates these differences and has been subjected to considerable experimental analysis. These studies have revealed that the variability in skin structure is due to local differ-ences in the mesoderm. The two types of skin derivatives that are pro-duced by the chick are scales and feathers. Scales form on the skin of the lower leg, whereas feathers cover much of the rest of the body.

Although they differ considerably in their final morphology, feath-ers and scales have quite similar embryonic origins (Fig. 14.3). The pri-mordia are distinguished by an elongation of cells of the generative layer. As the cells become columnar, they create a local epidermal thickening called the **placode.** Below the placode the mesenchyme condenses to form the **dermal papilla.** Growth of the papilla elevates the placode, form-

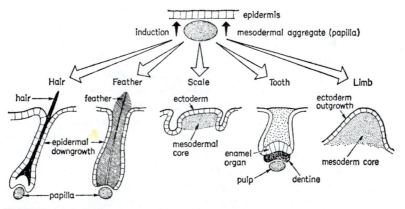

FIGURE 14.1 Elaborations of the vertebrate epidermis and their formation. (After E.M. Deuchar. 1975. *Cellular Interactions in Animal Development.* Chap-man & Hall Ltd., London, p. 149.)

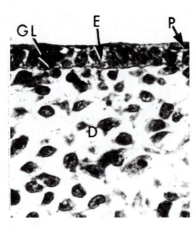

FIGURE 14.2 Transverse section of the skin of a 12-day mouse embryo. E: epidermis; P: periderm; GL: generative layer; D: dermis. (From P. Sengel, 1976. *Morphogenesis of Skin.* Cambridge University Press, Cambridge, p. 9.)

ing a bud. The bud then elongates and completes the morphological differentiation appropriate to either the feather or the scale.

Mesodermal determination of epidermal derivatives is demonstrated by the experiments outlined in Figure 14.4. In Figure 14.4*A*, thigh mesoderm is transplanted beneath ectoderm covering the embryonic wing. As a result, the wing feathers differentiate with the morphology and spatial arrangement of *leg feathers* (Cairns and Saunders, 1954). In Figure 14.4*B*, a piece of mesoderm from a feathered region of the body is combined with ectoderm from a nonfeathered region; *feathers* differentiate (Sengel et al., 1969). If a piece of ectoderm from a feathered region is combined with mesoderm from a scaled region (*C*), the ectoderm forms *scales* (Rawles, 1963).

The latter result is particularly noteworthy, since it indicates that the capacity to form scales exists in ectoderm that normally forms feathers. Although these two structures have quite distinct morphologies, they are fundamentally similar. Both of them are composed largely of keratin, the specialized protein produced by epidermal cells. Scales are the more primitive structures, being the predominant skin derivatives of reptiles, and are thought to be the evolutionary precursors of feathers. A feather may be thought of as an elongate, highly elaborate scale. Chicken epidermis has the inherent capacity to produce either structure, subject to the instructions of the mesoderm, which directs the epidermis to assemble either the scale or feather configuration from the keratin-producing cells. However, Rawles also observed that in the early stages of placode development, the scaled-region dermis is incapable of inducing scales when combined with feather epidermis; this combination results in formation of feathers only. Thus, the specificity of the inductive signal from the dermis is acquired after this stage. Recent evidence suggests that the dermis acquires its scale-specific inductive capacity by interacting with presumptive scale epidermis (Sawyer, 1979). This would mean that scale induction is a multistep process involving (initially) an influ-

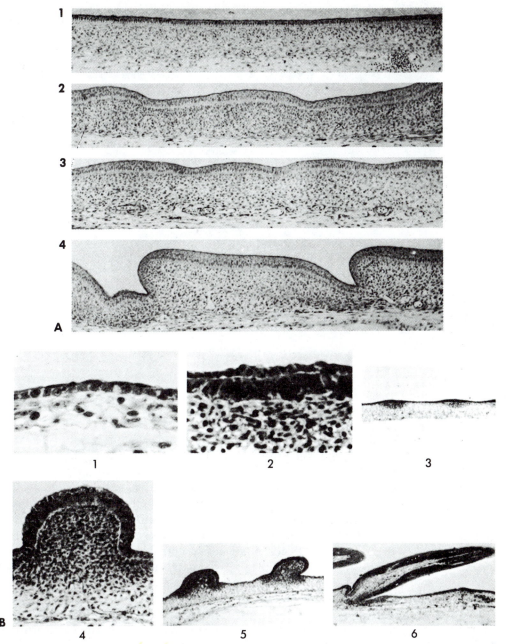

FIGURE 14.3 Stages in the formation of chick scales (*A*) and feathers (*B*). *A,* Scale placodes form above dermal papillae, and the scale buds elevate above the surface. × 100. (From M.E. Rawles. 1963. Tissue interactions in scale and feather development as studied in dermal-epidermal recombinations. J. Embryol. Exp. Morphol., *11:* 765–789.) *B,* Feather formation shows many similarities to the above sequence. *1,* Preplacodal stage. *2* and *3,* Placodal stage. Epidermis thickens above discrete dermal papillae. *4* and *5,* Feather buds form. *6,* Elongation stage. (Micrographs by Dr. B. Garber. From N.J. Berrill and G. Karp. 1976. *Development.* McGraw-Hill Book Co., New York, p. 445.)

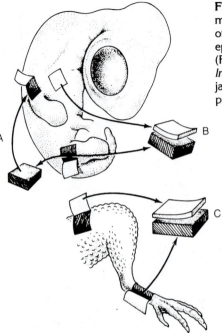

FIGURE 14.4 Outline of experiments demonstrating that the type of mesoderm controls the type of epidermal derivative that will form. (From N.K. Wessells. 1977. *Tissue Interactions and Development,* Benjamin/Cummings, Menlo Park, Calif., p. 56.)

A. Leg mesoderm in wing → leg feathers

B. Feather-area mesoderm plus
   nonfeather-ectoderm → feathers

C. Scale-area mesoderm plus
   feather-area ectoderm → scales

ence of the epidermis on the dermis and (subsequently) an influence of the dermis on the epidermis. The latter interaction fixes the fate of the epidermis.

Dermal-epidermal interactions are not restricted to the embryonic period; experimental evidence suggests that the maintenance of epidermal differentiation and the continued differentiation of the cells of the proliferating generative layer are dependent upon the continuing presence of the dermis. Recombinations of epidermis and dermis from different sites in the adult guinea pig indicate the continued dependence of the epidermis. For example, when epidermis from the ear is transplanted to the sole of the foot, the transplanted tissue develops the morphological characteristics of sole epidermis (Billingham and Silvers, 1967).

Tooth development in mammals provides an excellent experimental system for examining epithelio-mesenchymal interactions. Teeth are derived from the dental lamina, a region of epidermis in the oral cavity, and from mesenchyme that condenses below the dental lamina to form a series of dental papillae. The mesenchymal component of teeth is actually derived from the neural crest in amphibia and fish (and possibly in mammals). Therefore, it would be more precise to call it "ectomesenchyme." However, we shall use the more common term "mesenchyme" to refer to this tissue. The epidermal cells differentiate into **ameloblasts,** which secrete the tooth enamel, and the mesenchymal cells develop into **odontoblasts,** which secrete predentin, the organic matrix of dentin. The sequence of events in mouse tooth development is shown in Figure 14.5. As with other epithelio-mesenchymal interactions, such as in scale and feather development, there are reciprocal interactions between the epithelium and mesenchyme that lead to tooth formation. The earliest events are determinative in nature, whereas the later events are permissive.

The dental mesenchyme is the first of these two tissues to have its fate determined; the mechanism of its determination is poorly understood. Determination could occur during migration of mesenchyme cells to their final locations below the oral epithelium, or the oral epithelium itself may influence determination (Thesleff and Hurmerinta, 1981). Once in position, however, the mesenchymal dental papillae play decisive roles in determination of epithelial cell fate. As shown in Figure 14.5*B,* the epithelium envelops each dental papilla to form the **enamel organ.** Cells of the enamel organ eventually differentiate into ameloblasts. If the dental papilla is removed, the enamel organ will not be produced. Furthermore, if a papilla is placed below undetermined heterologous epidermis, it will induce enamel organ formation. An example of this is shown in Figure 14.6. In this experiment, a mouse dental papilla was combined with epidermis of the foot. The result is the formation of a well-developed tooth from foot epidermis. A dental papilla also relays instructions on the *kind of tooth* to be formed. Kollar and Baird (1970a) demonstrated that in reciprocal combinations of mesenchyme and epithelium of mouse incisor and molar tooth germs, the typical incisiform or molariform shape was determined entirely by the mesenchyme.

In addition to the instructive signal from mesenchyme to epithelium, the presence of the epithelium is essential to allow the mesenchymal cells to develop into odontoblasts, which precedes ameloblast differentiation. The final step in tooth formation is the differentiation of the enamel epithelial cells into ameloblasts. Underlying mesenchyme is necessary for ameloblast formation. However, this interaction is permissive; the early inductive event determines the fate of the epidermis, and the later event allows that fate to be realized (Thesleff and Hurmerinta, 1981).

Recent evidence implicates cell surface phenomena in mediating at least some of the interactions between epithelium and mesenchyme in tooth development. During differentiation of mesenchyme cells into odontoblasts, the epithelial and mesenchymal cells are separated by a basement membrane (Figs. 14.7 and 14.8*A*), consisting of a basal lamina under the epithelial cells and a fibrillar material associated with the mesenchymal cells. Molecules in the basement membrane include collagen, laminin, a heparan sulfate proteoglycan, and fibronectin (Thesleff

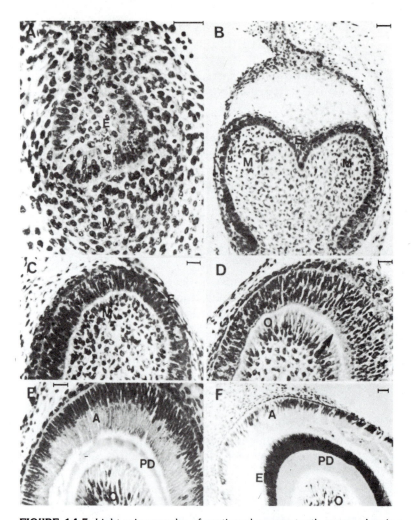

FIGURE 14.5 Light micrographs of sectioned mouse tooth germs showing progressive stages in tooth development. *A,* Bud stage. The epithelial tooth bud (E) is surrounded by a region of condensed mesenchyme. *B,* Early bell stage. The enamel epithelium (EE) has formed the enamel organ, which surrounds the dental papilla. The latter is composed of condensed mesenchyme. *C–F* are higher magnification micrographs of the epithelio-mesenchymal interface. *C,* Mesenchymal cells align under the enamel epithelium. *D,* The mesenchymal cells underlying the epithelium are now odontoblasts (O) that have initiated predentin secretion (arrow). *E,* Epithelial cells have differentiated into ameloblasts (A). The predentin (PD) is thicker. *F,* Enamel matrix (EM) has been produced by the ameloblasts. Scale bars equal 20 μm (*A, B,* and *F*) or 1.0 μm (*C, D,* and *E*). (From I. Thesleff and K. Hurmerinta. 1981. Tissue interactions in tooth development. *Differentiation, 18:* 77. Reprinted with permission of Springer-Verlag, Heidelberg.)

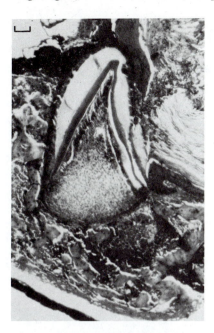

FIGURE 14.6 The result of combining a mouse dental papilla with foot epidermis. A well-developed tooth is organized from the epidermis. Scale bar equals 50 μm. (From E.J. Kollar and G.R. Baird. 1970b. Tissue interactions in embryonic mouse tooth germs. II. The inductive role of the dental papilla. *J. Embryol. Exp. Morphol., 24:* 177.)

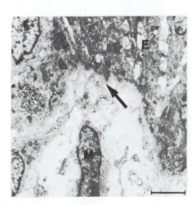

FIGURE 14.7 Electron micrograph of the epithelio-mesenchymal interface of a mouse tooth germ at the time of odontoblast differentiation (see Fig. 14.5*B*). Mesenchymal cell contact with the basal lamina under the enamel epithelium is indicated by the arrow. M: mesenchymal cell; E: epithelial cell. Scale bar equals 1.0 μm. (From I. Thesleff and K. Hurmerinta. 1981. Tissue interactions in tooth development. *Differentiation, 18:* 80. Reprinted with permission of Springer-Verlag, Heidelberg.)

and Hurmerinta, 1981; see also Chap. 5 for a discussion of these molecules). Processes from the mesenchymal cells make contact with the basal lamina. Grobstein (1953b) developed a widely used method to examine whether contacts of this type are essential to mediate induction. This technique involves culturing epithelium and mesenchyme on opposite sides of a thin filter. Induction across a filter that prevents contact is evidence for transmission of a diffusable inducing substance; the failure of induction suggests that physical contact between inducing and responding tissues is necessary. Nucleopore filters with a pore diameter of 0.2 to 0.6 μm allow penetration of fine mesenchymal cell processes that make contact with the basal lamina of the enamel epithelium. Under these conditions, differentiation of odontoblasts occurs. However,

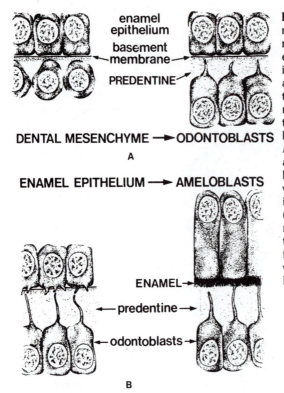

enamel epithelium
basement membrane
PREDENTINE

DENTAL MESENCHYME → ODONTOBLASTS

A

ENAMEL EPITHELIUM → AMELOBLASTS

ENAMEL →
← predentine →
← odontoblasts →

B

FIGURE 14.8 Schematic representations of the presumed mechanisms of epithelio-mesenchymal interactions involved in promotion of (*A*) odontoblast and (*B*) ameloblast differentiation in mouse tooth development. In *A*, an interaction between mesenchymal cells and basal lamina is essential, while in *B* dissolution of the basal lamina allows contact between odontoblasts and enamel epithelial cells, which mediates the odontoblast influence on the epithelial cells. (From I. Thesleff and K. Hurmerinta. 1981. Tissue interactions in tooth development. Differentiation, *18*: 84, 85. Reprinted with permission of Springer-Verlag, Heidelberg.)

filters with a pore size of 0.1 μm prevent the penetration of cell processes *and* differentiation. If induction were due to diffusion, pore size should have no effect upon induction, since even the smallest pore diameter would not impede the passage of diffusable substances. Thus, it seems likely that actual contact between mesenchymal cells and the basal lamina is a prerequisite for odontoblast differentiation (Thesleff et al., 1978).

This interpretation is supported by experiments in which the composition of the basement membrane is modified by various chemical means; such treatments also inhibit differentiation of mesenchymal cells into odontoblasts (Thesleff and Hurmerinta, 1981). The role of the epithelium in odontoblast differentiation may, therefore, be due to its participation in basement membrane formation. Contact between mesenchymal cells and the basement membrane appears to allow their differentiation into odontoblasts.

After the odontoblasts have initiated predentin secretion, cells of the enamel epithelium begin to differentiate into ameloblasts. This is preceded by the breakdown of the basal lamina, which is probably caused by the odontoblasts. The disappearance of the basal lamina allows for the formation of cell-cell contacts between the odontoblasts and enamel epithelial cells (Fig. 14.8B). Since the presence of odontoblasts is essential to allow terminal differentiation of the enamel epithelial cells, these contacts may mediate the odontoblast influence. Another possibility is that the predentin laid down by the odontoblasts mediates the odontoblast influence (Thesleff and Hurmerinta, 1981).

The necessity for epithelio-mesenchymal interactions at sequential steps in tooth development is now well documented and partially characterized. Additional research will undoubtedly reveal more about these interactions, particularly the role of cell surface phenomena in induction.

Figure 14.1 illustrates an additional vertebrate epidermal derivative that is dependent upon mesenchyme: the epidermis covering the embryonic limb bud. Early experiments by Harrison (1918) on amphibian embryos demonstrated that the limb bud mesoderm influences the initial outgrowth of the limb bud. If this mesoderm is transplanted below the ectoderm in other parts of the body, it will enlarge and cause the ectoderm to expand over it, thus producing a limb bud. Subsequent experiments with avian embryos have revealed that limb bud mesoderm also promotes formation of avian limb buds; however, the experiments also have shown that differentiation of the limb from the early limb bud involves continuing interactions between the limb bud ectoderm and mesoderm. The avian limb bud has served as the primary experimental material for the contemporary analysis of vertebrate limb development.

The cap of epidermis covering the mesenchyme of the avian limb bud thickens at the apex of the bud to form an elevation called the **apical ectodermal ridge (AER)**. As a result of the initial influence of the limb bud mesoderm, the ectoderm at the apex of the bud forms the AER, which in turn acquires the ability to promote differentiation of the mesodermal elements into the cartilage, muscle, and connective tissue of the limb. At the same time, the mesoderm exerts a reciprocal influence on the AER, maintaining it as a ridge. This mesodermal property is distinct from the one expressed earlier in development in initiating the ridge (Saunders, 1972).

The role of the mesoderm in limb morphogenesis is highly specific. For example, only limb bud mesoderm is capable of interacting with the ectoderm to form a limb; placement of mesoderm from other body regions below limb bud ectoderm leads to the degeneration of the ectodermal ridge and the absence of limb formation (Zwilling, 1972). Furthermore, the limb bud mesoderm determines the type of limb that forms; that is, it determines whether the limb bud develops into a front or a hind limb. Accordingly, presumptive thigh mesoderm implanted in the wing region will form foot parts (Saunders et al., 1957).

The formation of a limb is a highly ordered process and one that produces a structure in which the individual elements (e.g., cartilage, muscle) are arranged in a specific spatial pattern. The mechanisms involved in the patterning of limb elements will be discussed in section 14–4.

Interactions of Mesenchyme with Glandular Epithelia

The presence of mesenchyme is essential for the morphogenesis of internal glandular epithelia and for the biochemical specialization of their component cells. Examples of vertebrate glandular rudiments produced by condensation of mesenchyme around internal epithelia include the thyroid, salivary glands, and pancreas. The extent of dependence upon mesenchyme varies considerably among these glands. For example, the pancreas diverticulum can continue to develop *in vitro* if it is cocultured with virtually any mesenchyme. By way of contrast, salivary gland epithelium has rather strict mesenchymal requirements in order to develop its branched morphology and acquire its glandular function. The interaction between salivary gland epithelium and mesenchyme has been studied extensively, particularly with respect to the morphogenesis of the epithelial component.

The initial rudiments of the mammalian submandibular salivary glands are a pair of club-shaped epithelial buds that grow downward on either side of the tongue into the connective tissue layer. The buds soon become surrounded by mesenchyme, which condenses around them to form a connective tissue capsule. The epithelial portions of the glands then begin to branch and form knoblike thickenings (or **lobes**) at the tips of each branch. The lobes divide repeatedly to form the secretory acini, whereas the branches form the duct system. The branching of the lobes begins with the appearance of clefts in the rounded outer surface. As the clefts gradually widen, the adjacent portions of the lobe expand and become bifurcated by new clefts, producing a treelike structure (Fig. 14.9). It has been postulated that two morphogenic mechanisms assist in morphogenesis. Expansion of the tips of the lobes appears to result from an elevated rate of mitosis (Wessells, 1977), whereas the formation of clefts may be due to contraction of microfilaments (Spooner and Wessells, 1972).

The need for the mesenchyme in promoting epithelial morphogenesis is evident when the two components are separated and cultured *in vitro.* Both epithelium and mesenchyme are viable when cultured separately, but they do not undergo normal morphogenesis; the epithelium fails to branch, and the mesenchyme forms a sheet of proliferating cells. If, however, the components are positioned adjacent to one another in a culture dish, the mesenchyme surrounds the epithelial bud, which proceeds to branch and form secretory acini (Fig. 14.10). The capacity of salivary mesenchyme to promote the specialized branched morphology of the salivary glands is indicated by an experiment in which salivary mesenchyme was combined *in vitro* with mouse mammary epithelium. The mammary epithelium normally develops in a manner quite distinct from that of the salivary epithelium. However, in combination with salivary mesenchyme, it branches and forms knobs reminiscent of salivary tissue (Fig. 14.11). Although the mammary tissue assumes the *form* of salivary tissue, it is not known whether the cells undergo salivary *differentiation.* In other words, does salivary mesenchyme induce the mammary cells to produce salivary-specific proteins?

The ability of salivary mesenchyme to promote branching morphogenesis of heterologous epithelium suggests that this mesenchyme has

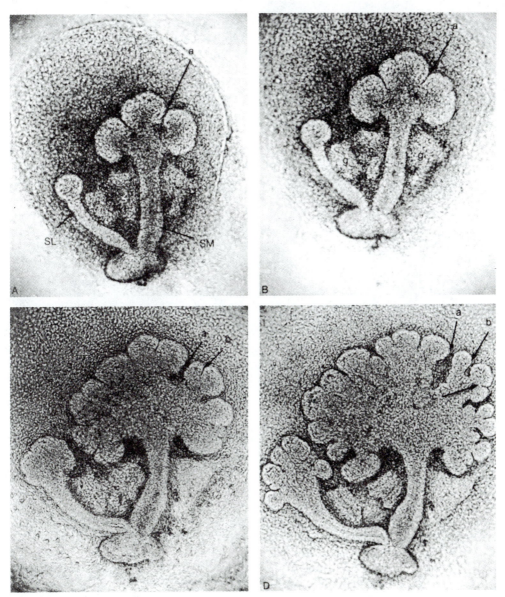

FIGURE 14.9 *In vitro* morphogenesis of a mouse salivary gland. Both the submandibular (SM) and sublingual (SL) glands are present. Follow the progressive scalloping of the peripheral margin of the submandibular gland. Note the progressive widening of cleft a, the appearance of cleft b, its widening, and the subsequent appearance of cleft c. (From N.K. Wessells. 1977. *Tissue Interactions and Development,* Benjamin/Cummings, Menlo Park, Calif., p. 95.)

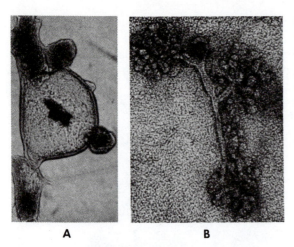

A   B

FIGURE 14.10 The effects of homologous mesenchyme on morphogenesis of the mouse submandibular gland *in vitro. A,* Epithelium alone. The gland fails to branch, and the epithelium begins to spread on the culture plate surface. *B,* Epithelium plus mesenchyme. Normal morphogenesis occurs. ×40. (From C. Grobstein. 1953a. Epithelio-mesenchymal specificity in the morphogenesis of mouse sub-mandibular rudiments *in vitro.* J. Exp. Zool., *124*: 407.)

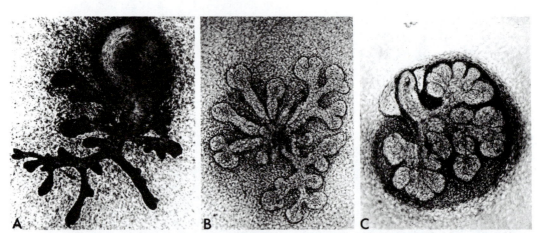

FIGURE 14.11 The effects of salivary mesenchyme on morphogenesis of mouse mammary epithelium. *A,* Typical mammary gland formation *in vitro.* Epidermal cyst and simple duct system have formed. *B,* Gland formed by combining mammary epithelium and submandibular gland mesenchyme *in vitro.* Morphogenesis resembles that of a typical submandibular gland with bifurcating branches of the epithelium. *C,* Typical submandibular gland formation *in vitro.* ×40. (From K. Kratochwil. 1969. Organ specificity in mesenchymal induction demonstrated in the embryonic development of the mammary gland of the mouse. Dev. Biol., *20:* 52, 60.)

a critical formative role in salivary development. Is this mesenchyme unique in its ability to promote salivary morphogenesis, or can other mesenchymes substitute for it? The specificity of the mesenchymal requirement has been examined by combining the submandibular epithelial bud with mesenchyme of various origins. Most mesenchyme fails to promote the growth and differentiation of the epithelium (Grobstein, 1953a). On this basis, the salivary mesenchyme has long been considered to be a specific promoter of salivary morphogenesis. Recent evidence, however, indicates that the requirement of salivary epithelium for its homologous mesenchyme is not so specific as was previously thought. Lawson (1974) has shown that mouse submandibular epithelium will branch and form acini when cocultured with mouse lung mesenchyme. However, lung mesenchyme must be present in larger amounts than submandibular mesenchyme to achieve the same amount of morphogenesis. This result suggests that the ability to branch in a characteristic way is inherent to salivary epithelium, and the role of the mesenchyme is to *permit* morphogenesis. The demonstrated role of salivary mesenchyme in promoting branching of heterologous epithelium and the refractoriness of salivary epithelium to most mesenchyme may indicate that (1) salivary mesenchyme is specialized to evoke branching of epithelia and (2) salivary epithelium will not express its capacity to branch unless it receives sufficient stimulus from branching-effective mesenchyme. Most mesenchyme will not satisfy this requirement; salivary mesenchyme will do so readily, but lung mesenchyme will do so only if present in greater amounts.

The nature of the interaction between salivary mesenchyme and epithelium has been studied extensively in efforts to clarify how epithelial differentiation is affected by the mesenchyme. One of the possibilities to be considered is that a diffusable inducing substance is transmitted to the epithelium from the mesenchyme. However, transfilter experiments (see p. 665) indicate that induction is not due to a diffusable substance (Saxén et al., 1976). Recent experiments have implicated the basal lamina in maintenance of the branched morphology and mesenchymal modification of the basal lamina in facilitating branching morphogenesis.

In the intact embryo, the basal lamina is located at the interface between the epithelium and mesenchyme (Fig. 14.12). It is not distributed uniformly over the epithelial surface. Instead, it is thinner at the distal tips of the lobes than in the clefts. Ultrastructurally, there are many discontinuities of the basal lamina at the tips (Coughlin, 1975). To evaluate the morphogenic roles of the basal lamina and mesenchyme, investigators have conducted experiments examining the *in vitro* morphogenesis of epithelia either in their absence or with one or both of these components present.

If mesenchyme alone is removed from the epithelium (leaving the basal lamina intact), the epithelium retains its branched morphology. Treatment of the mesenchyme-free epithelium with hyaluronidase degrades glycosaminoglycans (GAG), thus removing the basal lamina. In the presence of mesenchyme, the epithelium rounds up, and *after considerable delay* the basal lamina is regenerated, after which branching is reinitiated (Fig. 14.13).

The effects of basal lamina removal can be reversed by first culturing the epithelium in the absence of mesenchyme for a short time. Under

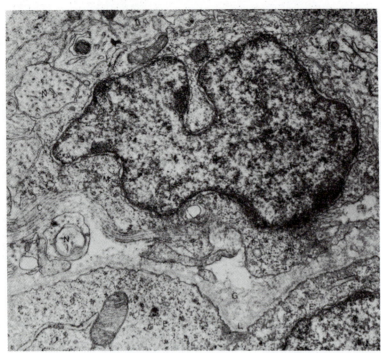

FIGURE 14.12 The epithelio-mesenchymal interface in the mouse embryonic salivary gland. Portions of a mesenchymal cell (M) and an epithelial cell (E) are shown, and between them are bundles of collagen fibrils (C), the basal lamina (L), and adjacent extracellular material that may be largely proteoglycans (G). (From N.K. Wessells. 1977. *Tissue Interactions and Development,* Benjamin/Cummings, Menlo Park, Calif., p. 222.)

FIGURE 14.13 The effects of the selective removal of mesenchyme and GAG on the morphogenesis of mouse salivary gland epithelium cultured with fresh salivary mesenchyme. The native mesenchymal capsule was removed enzymatically, followed by treatment with hyaluronidase to remove hyaluronic acid (a GAG). *A*, 0 hr. Epithelium is combined with fresh mesenchyme. *B*, 24 hr, Epithelium has rounded up. *C*, 48 hr, Branching morphogenesis has resumed. ×50. (From M.R. Bernfield, S.D. Banerjee, and R.H. Cohn. 1972. Reproduced from *The Journal of Cell Biology,* 1972, vol. *52*, pp. 674–689 by copyright permission of The Rockefeller University Press.)

these conditions the basal lamina is rapidly regenerated. When this epithelium is combined with mesenchyme, the branched morphology is maintained. These results indicate that (1) integrity of the basal lamina is dependent upon its glycosaminoglycans, (2) maintenance of branched morphology is dependent upon the basal lamina, and (3) mesenchyme delays the formation of basal lamina. As a consequence of (3), the epithelium rounds up and requires a long latent period before branching can resume (Bernfield, 1981). It is seemingly enigmatic that mesenchyme, which promotes branching morphogenesis, causes loss of branching and delays both basal lamina formation and branching morphogenesis.

What, then, is the role of mesenchyme in morphogenesis? Not surprisingly, if epithelium cleaned of both mesenchyme and basal lamina is cultured without mesenchyme, branching will not occur, even though basal lamina accumulates at the epithelial surface (Fig. 14.14). Thus, although basal lamina seems to be necessary for morphogenesis, it is not *sufficient*; the mesenchyme provides the proper conditions.

The presumed roles of the mesenchyme in salivary morphogenesis are summarized in the model in Figure 14.15. The mesenchyme appears to function in the generation of the branching morphology of the epithelium and in stabilizing the morphology once it appears. As we discussed

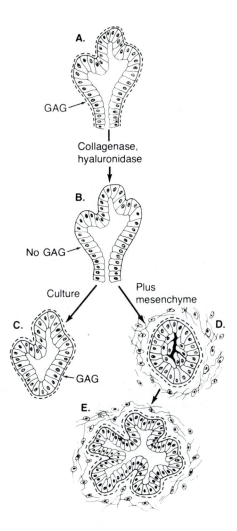

FIGURE 14.14 The combined effects of GAG and mesenchyme on salivary morphogenesis. *A*, Salivary epithelium with GAG layer intact. *B*, Enzymatic treatment has removed the GAG. One of two results will be obtained when this epithelium is cultured. In the absence of mesenchyme (*C*), GAG is replenished, but no morphogenesis occurs. In the presence of mesenchyme, the epithelium initially rounds up (*D*). GAG is replenished and branching morphogenesis resumes shortly thereafter (*E*). (From N.K. Wessells. 1977. *Tissue Interactions and Development*, Benjamin/Cummings, Menlo Park, Calif., p. 225.)

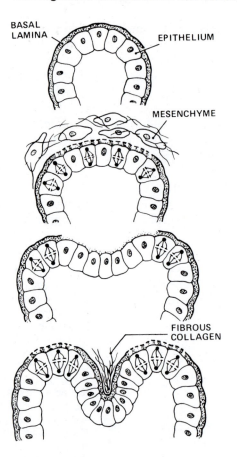

BASAL
LAMINA

EPITHELIUM

MESENCHYME

FIBROUS
COLLAGEN

FIGURE 14.15 Model depicting salivary morphogenesis. (From M.R. Bernfield. 1978. The cell periphery in morphogenesis. *In* J.W. Littlefield and J. deGrouchy (eds.), *Birth Defects*. Excerpta Medica International Congress Series No. 432. Excerpta Medica, Oxford, p. 123.)

previously, the epithelium expands by mitosis at the tips of the lobes. The GAG in the basal laminae covering the tips are undergoing rapid turnover; the rate of degradation at the tips actually exceeds the rate of synthesis (Bernfield and Banerjee, 1982). The mesenchyme's role in GAG turnover apparently involves producing enzymatic activity that degrades the existing GAG at the tips, which envelops the expanding epithelium (Bernfield, 1981). Restoration of GAG then occurs by the synthetic activity of the epithelial cells. However, before producing a new basal lamina, the epithelium, which has become free of this restrictive layer, invaginates by the contractile action of microfilaments within the cells. A cleft is thus produced.

The beginning cleft deepens by expansion of the flanking lobes (due to cell division). As it deepens, the mesenchyme helps to stabilize the cleft by depositing bundles of collagen. Collagen might stabilize the cleft by rendering the basal lamina more stable to mesenchyme-induced degradation. As a consequence, the basal lamina maintains the morphology of the epithelium in the cleft region (Bernfield, 1981). Meanwhile, at the tips of the expanding lobes the GAG are being degraded so that additional clefts can be formed.

Does mesenchyme also play a role in promoting cell differentiation as well as morphogenesis? The answer to this question might be found if investigators could ascertain whether the cells of mammary epithelium, in combination with salivary mesenchyme, synthesize salivary (rather than mammary) proteins.

14–2. AN ANALYSIS OF DEVELOPMENT: THE VERTEBRATE LENS

In this section we shall attempt to reconstruct the sequence of events in the formation of the vertebrate lens. The lens is an avascular tissue consisting of only one kind of differentiated cell, which makes it an excellent candidate for the study of cellular differentiation, particularly at the biochemical and molecular levels. Another advantage is the ready accessibility of the lens-forming cells on the surface of the embryo during the phase of cell determination. This accessibility has made the lens a target of numerous experimental studies on the mode of cell determination, dating back to the early 1900s. The lens, then, presents unprecedented opportunities for the study of both cell determination and cell differentiation.

Morphogenesis of the Lens

The relationship of the developing lens to the rest of the eye is shown in the micrographs of Figure 14.16 and the drawings of Figure 14.17. The primordia of the sensory portions of the eyes form as evaginations from the lateral walls of the presumptive brain. These evaginations are the **optic vesicles.** The distal portion of an optic vesicle expands and forms a cup-shaped structure, the **optic cup,** which gives rise to the pigment layer and the sensory layer of the retina. The constricting **optic stalk,** which connects the optic cup to the brain, is the forerunner of the optic nerve. The lens forms from the surface ectoderm that overlies the optic vesicle. The lens primordium first appears as a thickening of the surface ectoderm, which results from an elongation of these cells. The elongation of lens placode cells is apparently achieved by longitudinal alignment of cytoplasmic microtubules (Byers and Porter, 1964). In birds and mammals the primordium then forms a cup-shaped pocket (the **lens cup**), which eventually constricts from the surface ectoderm and forms a spherical **lens vesicle** (Fig. 14.18). The latter differentiates into the definitive lens. In amphibians (Fig. 14.19) and bony fishes, lens formation differs somewhat. In these groups the **inner layer** of epidermis thickens to form a solid mass of cells; the cells rearrange themselves into a vesicle, which then undergoes differentiation.

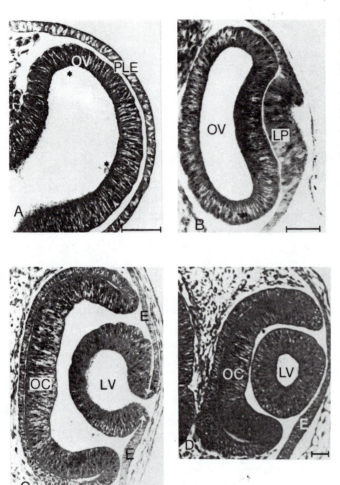

FIGURE 14.16 Development of the eye of the chick embryo. *A,* Section through the eye showing the optic vesicle (OV) underlying the presumptive lens epithelium (PLE). The future rim of the optic cup is marked with asterisks. Scale bar equals 50 μm. *B,* Section showing the early stages of both the optic vesicle (OV) and the lens placode (LP). Scale bar equals 50 μm. *C,* This section shows the optic cup (OC) and the open lens vesicle (LV). The latter is still connected to the ectoderm (E) by a short stalk (arrows). Scale bar equals 50 μm. *D,* Section showing the lens vesicle after closure and separation from the ectoderm, which is now the presumptive cornea. Scale bar equals 50 μm. (From C. Maloney and J. Wakely, 1982. Reproduced with permission from Experimental Eye Research. Copyright by Academic Press Inc. (London) Ltd.)

The differentiation of the vertebrate lens (Fig. 14.20) begins on the inner face of the lens vesicle with an elongation of the cells. The cells first become columnar and are later transformed into long "fiber cells," which occupy the center of the lens. The cells of the outer face of the lens vesicle remain epithelial and form a sheet covering the outer surfaces of the fiber cells. In the embryonic lens the entire epithelial layer remains mitotically active, producing new cells that differentiate into fiber cells. However, as the animal matures (Fig. 14.20*B*), the central region of the epithelium becomes mitotically quiescent, and mitosis is restricted to a zone that encircles the central region. The zone of cell division is called the **germinative region.** On the equatorial side of this region, differentiation of epithelial cells to form fiber cells is initiated in the **region of cellular elongation.** The elongate cells then pass toward the center of the lens, where they undergo further elongation and terminal

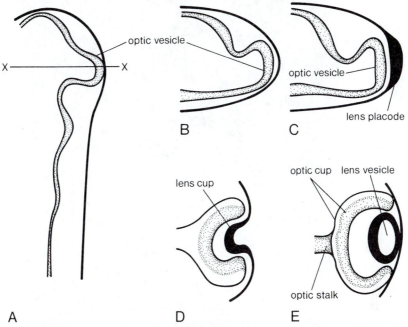

FIGURE 14.17 Development of the vertebrate eye. *A,* Dorsal view of the embryo at the stage of initial contact of optic vesicle with epidermis. X–X represents the plane of section shown at the same stage in *B* and in progressively later stages in *C* to *E.* Sections show the thickening of the lens placode (*C*), simultaneous invagination of optic cup and lens cup (*D*), and separation of the lens vesicle from the surface (*E*). (Adapted from "The Eye" by Alfred J. Coulombre, in *Organogenesis* edited by Robert L. DeHaan and Heinrich Ursprung. Copyright © 1965 by Holt, Rinehart and Winston, Inc. Reprinted by permission of Holt, Rinehart and Winston, CBS College Publishing.)

differentiation, losing their nuclei and acquiring a hard, transparent cytoplasm. This process of fiber cell formation continues throughout the life of the organism, with new fiber cells added around the central core (called the **nucleus**), layer upon layer. Thus, the first fiber cells formed during embryogenesis constitute the central core, whereas newly formed fiber cells are found at progressively more peripheral locations.

Determination of the Lens

The formation of the optic vesicle precedes the appearance of the lens rudiment; the latter begins to form after the optic vesicle contacts the overlying epidermis. Why does the lens begin to form after contact is made and at the position where the optic vesicle touches the epidermis? Is this merely the convergence of events in the self-differentiation of two independent structures, or is there a causal relationship between the development of the optic cup and the lens? The answer, at least for the

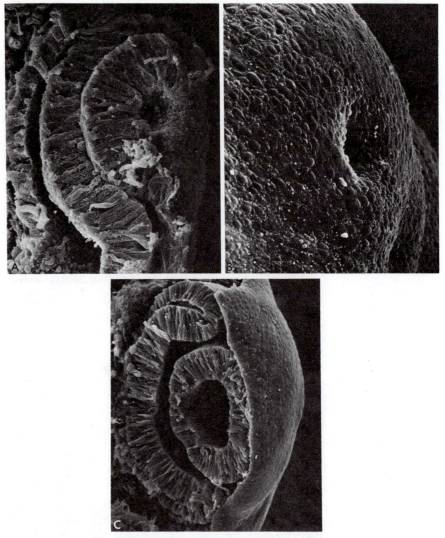

FIGURE 14.18 Scanning electron micrographs of lens formation in the chick embryo. *A*, Two-layered optic cup is on the left. On the right, the lens cup has invaginated from the surface epidermis. *B*, Hole in the surface formed by invagination of the lens cup. The cells remaining on the surface will cover the hole and form the cornea. *C*, Lens vesicle formation is complete, as the lens cup has constricted from the surface ectoderm. (From N.K. Wessells, 1977. *Tissue Interactions and Development.* Benjamin/ Cummings, Menlo Park, Calif., pp. 22, 44, 100. Courtesy of K.W. Tosney and Benjamin/ Cummings Publishing Co.)

chick and many amphibians, is clear: The optic vesicle induces the epidermis to form a lens. This conclusion is derived from the results of experiments in which the normal relationships between the optic vesicle and epidermis have been altered:

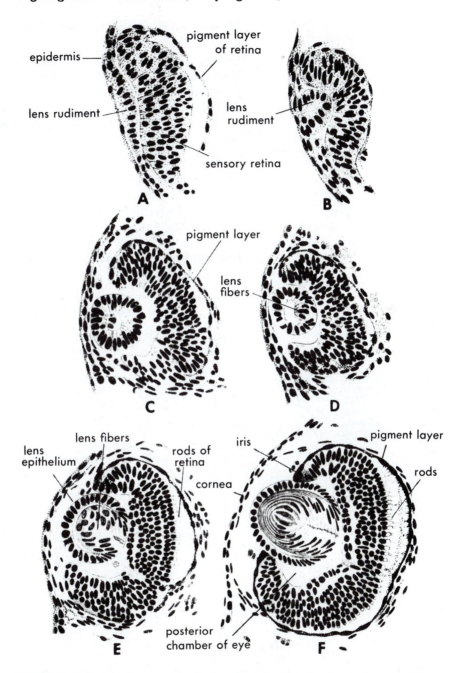

FIGURE 14.19 Drawings representing sections through the developing amphibian eye. Note in *A* that the lens placode forms in the inner layer of ectoderm. In *B*, the placode cells form the lens vesicle. *C* to *F*, Posterior cells of the lens vesicle elongate and form fiber cells. (After Rabl, 1898, from H. Spemann. 1936. *Experimentelle Beiträge zu einer Theorie der Entwicklung.* Julius Springer, Berlin, p. 27. Reprinted with permission of Springer-Verlag, Heidelberg. Reproduced from B.I. Balinsky, 1975.)

1. If the optic vesicle is removed before it reaches the head epidermis, the corresponding lens may not form (Spemann, 1912).
2. If the presumptive lens epidermis is removed and replaced with epidermis from another part of the body, the foreign ectoderm will form a lens when the optic vesicle contacts it (Lewis, 1904).
3. If the optic vesicle is transplanted below epidermis in another part of the body, the foreign epidermis may form a lens (Lewis, 1904).

The ability of epidermis to respond to the optic vesicle is called **competence.** This responsiveness is a developmentally acquired characteristic that, in the amphibians, depends upon the sequential inductive influence of the foregut endoderm and heart mesoderm, tissues that underlie the presumptive lens ectoderm before the optic vesicle makes contact with it (Jacobson, 1966). The sequence of inductive interactions involved in lens induction is shown in Figure 14.21. There is a great deal of species-specific variability in the extent to which the earlier inductive influences affect lens formation. In many species the lens can form only if the optic vesicle makes contact with competent epidermis. Thus, the

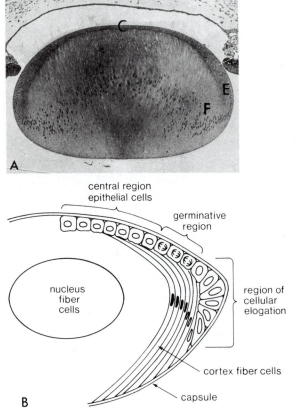

central region
epithelial cells

germinative
region

nucleus
fiber
cells

region of
cellular
elogation

cortex fiber cells

capsule

A

B

FIGURE 14.20 Structure of the vertebrate lens. *A,* Phase micrograph of an axial section from a 6-day chick embryo. C: central epithelium; E: region of cellular elongation; F: fiber region. The central core, or nucleus, of fiber cells is not well developed at this stage. (From J. Piatigorsky, H. deF. Webster, and S.P. Craig. 1972. Protein synthesis and ultrastructure during the formation of embryonic chick lens fibers *in vivo* and *in vitro.* Dev. Biol., *27:* 177.) *B,* Diagram of a vertebrate lens. Surrounding the lens is an acellular capsule. The epithelial cells are beneath the capsule on the outer face of the lens. The epithelial cells of the central region are mitotically quiescent. A ring of mitotically active cells surrounds the central region to form the germinative region. A region of cellular elongation is on the equatorial side of the germinative region. The fiber cells formed by this elongation constitute the cortex. The first fiber cells formed during lens ontogeny occupy the nucleus in the center of the lens. (After Papaconstantinou, J. 1967. Molecular aspects of lens cell differentiation. Science *156:* 338–346. Copyright 1967 by the American Association for the Advancement of Science.)

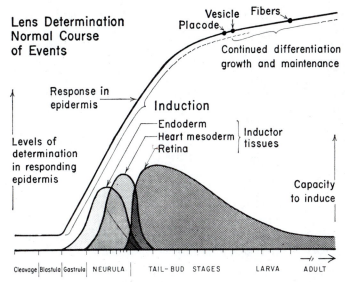

Lens Determination Normal Course of Events

Vesicle Fibers
Placode

Continued differentiation growth and maintenance

Response in epidermis

Induction

Endoderm
Heart mesoderm } Inductor tissues
Retina

Levels of determination in responding epidermis

Capacity to induce

Cleavage | Blastula | Gastrula | NEURULA | TAIL–BUD STAGES | LARVA | ADULT

FIGURE 14.21 Graphic model of lens induction in a urodele amphibian. The abscissa represents time, marked off in arbitrary stages. The response curve ordinate (on the left) is logarithmic; the level of response is a function of the sum of all past inductive influences. The inductive capacity ordinate (on the right) is linear and represents the capacities of the various inductors for lens determination (shaded regions). (From Jacobson, A. 1966. Inductive processes in embryonic development. *Science 152:* 25–34. Copyright 1966 by the American Association for the Advancement of Science.)

earlier inductors prepare the epidermis for the final induction, but they *are not sufficient* to cause determination. This is the classic pattern of lens induction. In other species, however, a lens can form in the absence of induction by the optic vesicle (Harrison, 1920; Spemann, 1938; Balinsky, 1951; Tahara, 1962). In these species the lens is said to "self-differentiate." Apparently the earlier inductive influences *are sufficient* to cause the epidermis to form a lens without an additional induction by the optic vesicle. The latter allows the overlying ectoderm to express this previously determined potential, thus localizing the lens-forming region. Even in amphibian species in which a lens self-differentiates, the retina (derivative of the optic cup) is essential for the continued well-being of the lens, since lenses will eventually regress in the absence of a retina. Consequently, lens induction is not a one-shot process that triggers an irreversible commitment for lens differentiation. The epidermis may require a prolonged exposure to various inductors before it is determined to form a lens, and once the lens is formed, the retina continues to play a role in its maintenance.

Sequential induction of the lens has also been demonstrated for the chick embryo. The initial inductive influence is exerted by the cephalic hypoblast upon the anterior epiblast during the primitive streak stage (Mizuno, 1972). After this early interaction the lens-forming property of the anterior ectoderm can be elicited by a variety of tissues (Karkinen-Jääskeläinen, 1978a). This is an example of two-step induction. The commitment of cells to a new pathway of development is **directive induction,** and subsequent nonspecific influences that allow them to express their potential is **permissive induction** (Saxén, 1977). The lens-forming potential that is triggered by the hypoblast is possessed by an extensive area of head ectoderm in early embryos (Barabanov and Fedt-sova, 1982). However, in normal development the lens forms only from the ectoderm overlying the optic vesicle. This localization of lens-forming potential is due to the optic vesicle, which is a strong (and the final) inducer of lens. In fact, optic vesicle can induce lens from heterotypic ectoderm, such as trunk ectoderm, which would otherwise never form lens. Thus, the optic vesicle has both directive and permissive inductive capacity (Karkinen-Jääskeläinen, 1978a).

The lens and optic vesicle are separated from one another by a basal lamina, which is formed from the fusion of the basal laminae produced by the two interacting epithelia (Johnston et al., 1979). To understand the mechanism of lens induction more fully, we wish to know whether contact of the lens with the basal lamina or with the optic vesicle by cellular processes penetrating the basal lamina is necessary for induction. Alternatively, induction might be mediated by a chemical emitted from the optic vesicle. Electron microscopic studies have failed to reveal any cellular contact across the basal lamina (Karkinen-Jääskeläinen, 1978b). Karkinen-Jääskeläinen has demonstrated that induction of lens formation in trunk ectoderm by optic vesicle can be transmitted across a Nucleopore filter with a pore size of 0.1 μm, and even across a dialysis membrane that would prevent passage of any molecule larger than 12,000 daltons. A 100-μm thick Millipore filter likewise fails to prevent induction. Cellophane, which blocks molecular transmission, prevents induction. Therefore, the inducer is likely to be a molecule with a molecular weight of less than 12,000 daltons that can be transmitted at least 100 μm.

As with the amphibian, chick lens requires the continued presence of neural retina for differentiation and maintenance of the differentiated state of lens cells (Philpott and Coulombre, 1965). The retina is separated from the lens by the gelatinous vitreous humor. This influence must therefore diffuse through the vitreous humor to influence the lens cells. A substance with these properties has recently been isolated from chick vitreous humor. It has been called "lentropin" (Beebe et al., 1980). Diffusion of the differentiation-promoting substance through the vitreous humor to contact the posterior aspect of the lens appears to account for the fact that all cell differentiation in the lens occurs in the posterior portion of the lens, which faces the retina (see Fig. 14.20).

Lens Cell Differentiation

Lens cell differentiation is readily studied in the chick embryo, which allows the investigator convenient access to lens cells at any stage of development. Another major advantage of using the chick lens is the ability to study lens fiber cell differentiation in explanted lens epithelia *in vitro*. The *in vitro* system takes advantage of the capability of the centralmost epithelial cells of the lens to elongate into differentiated lens fiber cells if the medium contains vitreous humor or fetal calf serum. The latter contains a substance that is similar (but not identical) to lentropin in its ability to promote lens fiber cell differentiation (Beebe et al., 1980). The experimental technique for preparing the epithelium for culture is shown in Figure 14.22. Elongation of the cuboidal epithelial cells in culture is illustrated in Figure 14.23.

The commitment to become lens confers upon the cells two major properties: the ability to elongate into fiber cells and the specialization for the synthesis of the lens-specific protein **δ-crystallin.** The elongation process in chick lens development is illustrated in Figure 14.24. Cell length increases in direct proportion to an increase in cell volume. As cell volume increases in the epithelium, cells bulge laterally against their neighbors. Since they are tightly bound to one another at their apices and to the lens capsule they can neither move laterally nor expand basally. Their only alternative is to expand apically (i.e., toward the anterior side of the lens), and therefore elongate into fiber cells (Beebe et al., 1982).

The lens-specific protein δ-crystallin accounts for 60 to 80% of the protein synthesized and accumulated in chick embryonic fiber cells and at least 80% of the soluble protein synthesized during development of the lens (Fig. 14.25; Genis-Galvez et al., 1968; Piatigorsky et al., 1972).

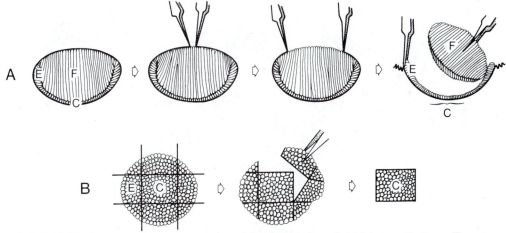

FIGURE 14.22 Diagrammatic representation of the explantation of chick lens epithelium. The upper row (A) represents axial sections through the lens, and the lower row (B) represents surface views of the epithelium. The lens from the eye of a 6-day chick embryo is removed and placed into sterile culture. A, The upper row shows the steps involved in isolation of the epithelium. The lens is oriented with the epithelium facing the surface of the dish. The lens capsule (represented by a dark line) is torn on the posterior (upper) side of the lens with microdissection forceps, and the fiber mass (F) is removed, leaving the central (C) and equatorial (E) epithelial cells still attached to the capsule. The approximate limits of the central region are represented by a bracket. B, In the lower row, the equatorial region is cut away, leaving the central epithelium with the cells adhering to the lens capsule, which is anchored to the dish. (After J. Piatigorsky. 1975. Lens cell elongation in vitro and microtubules. Ann. N.Y. Acad. Sci., 253: 334.)

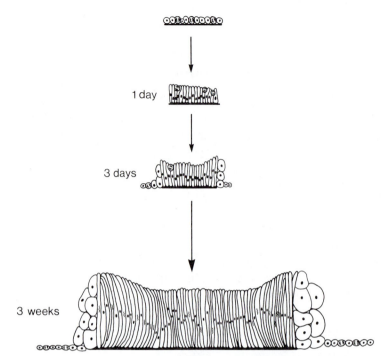

1 day

3 days

3 weeks

FIGURE 14.23 Diagrammatic representation of the behavior of cultured chick lens epithelium. The lens capsule is represented by a thick, black line. Note that the cells underlain by capsule elongate to form fiber cells; the peripheral cells elongate somewhat more than those in the center. Thin, wavy lines in some of these cells indicate pyknotic nuclei. Loss of the nucleus is part of the terminal differentiation of a fiber cell. Cells adjacent to the capsule produce an outgrowth of flattened cells. (After J. Piatigorsky. 1975. Lens cell elongation *in vitro* and microtubules. Ann. N.Y. Acad. Sci., *253*: 335.)

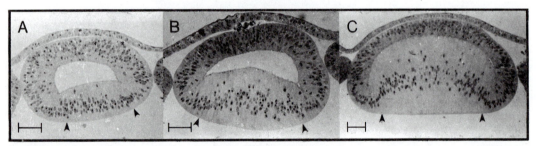

FIGURE 14.24 Light micrographs of sectioned chick embryo lenses at progressive stages of fiber cell elongation. Elongation occurs only in the posterior region of the lens, whereas the anterior cells remain epithelial. Scale bar equals 100 μm. (From D.C. Beebe et al. 1982. The mechanism of cell elongation during lens fiber cell differentiation. Dev. Biol., *92*: 55.)

δ-Crystallin is also known as the first important soluble crystallin (F.I.S.C.). Birds share with reptiles the property of accumulating δ-crystallin in their lens fiber cells. Two additional crystallins are also synthesized by bird and reptile lens cells. These crystallins (α and β) appear at somewhat later stages of development than does δ. The crystallin composition of the other vertebrate classes differs: Amphibians, mammals, and fishes synthesize α- and β-crystallins but also have an additional protein, γ-crystallin, which is not found in appreciable amounts in either birds or reptiles. The various crystallins account for approximately 80 to 90% of the soluble protein in lenses. Figure 14.26 compares the timetable of appearance of the various crystallins during lens development in the newt, chicken, and rat. It is evident that sequential appearance of crystallins is a common developmental strategy. However, these various vertebrates follow quite different sequences. The presence

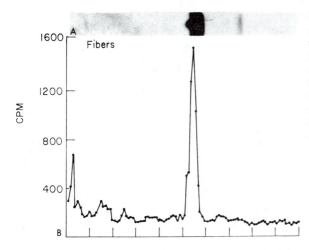

FIGURE 14.25 SDS-acrylamide-agarose gel electrophoresis of lens fiber proteins labeled with ³H-valine. A, Gel stained for protein. B, Radioactivity of gel regions. The major peak of radioactivity coincides with the major protein band in A. This represents δ-crystallin. (From P. Zelenka and J. Piatigorsky. 1974. Isolation and *in vitro* translation of δ-crystallin mRNA from embryonic chick lens fibers. Proc. Natl. Acad. Sci. U.S.A., *71:* 1898.)

DEVELOPMENTAL APPEARANCE OF THE CRYSTALLINS IN DIFFERENT ORGANISMS

Developmental Stage	Crystallin	Rat Ep.	Rat Fibers	Chicken Ep.	Chicken Fibers	Newt Ep.	Newt Fibers
Presumptive Fibers — Beginning lens invagination	α		−		−		−
	β		−		−		−
	γ		−		0		−
	δ		0		+ / −		0
Presumptive Fibers — Lens cup	α		+ / −		−		−
	β		−		−		−
	γ		−		0		−
	δ		0		+		0
Epithelia Fibers — Lens vesicle	α	+ / −	+ +	−	+ / −	−	−
	β	−	+ / −	+ / −	+	−	+ / −
	γ	−	+ / −	0	0	−	−
	δ	0	0	+	+ + +	0	0
Epithelia Fibers — Elongation of primary fibers	α	+	+ +	+	+	−	−
	β	−	+ +	+	+	−	+
	γ	−	+	0	0	−	+ +
	δ	0	0	+ +	+ + +	0	0
Epithelia Fibers — Embryonic lens	α	+ +	+ + +	+	+	−	+ +
	β	−	+ + +	+	+	+	+ +
	γ	−	+ + +	0	0	−	+ + +
	δ	0	0	+ +	+ + +	0	0

FIGURE 14.26 Relative amounts of crystallins observed by immunofluorescence in the epithelial and fiber cells of the developing lens of the rat, the chicken, and the newt *Notophthalmus viridescens.* The symbol (−) means no crystallin observed yet and (0) means that the crystallin is not present. (From J. Piatigorsky. 1981. Lens differentiation in vertebrates. A review of cellular and molecular features. Differentiation, *19:* 138. Reprinted with permission of Springer-Verlag, Heidelberg.)

of such large amounts of crystallins suggests that they are important for the optical properties of the lens. Each of the four classes of crystallin is composed of multiple polypeptides that are encoded by a family of related genes. With the exceptions of α-crystallin and one of the γ-crystallins, these polypeptides form multimeric proteins in the lens cell cytoplasm (Piatigorsky, 1981).

The specialization of cells for the synthesis of particular proteins is one of the principal aspects of cell differentiation. How do lens cells acquire the specialization for crystallin synthesis? We can at present describe some of the mechanics of this specialization, but we do not as yet understand why it occurs. In this analysis we concern ourselves only with the synthesis of δ-crystallin—the predominant protein of embryonic chick lens cells.

As we discussed in Chapter 4, the specialization for protein synthesis may result from the accumulation of the messengers for the cell-specific proteins in the cytoplasm. The following evidence will show that the lens is no exception. The first step in demonstrating the presence of a specific mRNA in differentiating cells involves extraction of the messenger RNA and its identification. Since many messengers are polyadenylated, a simple procedure to obtain the messenger population is to extract cytoplasmic RNA and apply it to a column of oligo (dT)-cellulose, which selectively binds the polyadenylated messengers. The messengers are then eluted from the column and analyzed. As shown in Figure 14.27, when cytoplasmic RNA of the chick lens fiber cell is subjected to this procedure, the bulk of the purified mRNA migrates as a single band during electrophoresis. When translated in an *in vitro* translation system (see Chap. 3 for discussion of this technique), 70 to 80% of the protein produced is δ-crystallin (Zelenka and Piatigorsky, 1974). Thus the specialization for δ-crystallin synthesis in these cells is reflected by a corresponding accumulation of δ-crystallin mRNA in the cytoplasm.

The purified δ-crystallin mRNA has been used to prepare cDNA, which is an excellent probe for quantification of mRNA, particularly in

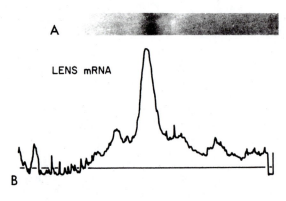

FIGURE 14.27 Acrylamide-agarose gel electrophoresis of lens mRNA. *A*, Stained gel. *B*, Spectrophotometric scan of gel. Peak and stained band represent δ-crystallin mRNA. (From P. Zelenka and J. Piatigorsky. 1974. Isolation and *in vitro* translation of δ-crystallin mRNA from embryonic chick lens fibers. Proc. Natl. Acad. Sci. U.S.A., *71:* 1898.)

measuring the changing levels of the messenger during development. Figure 14.28 summarizes the results of these experiments. Measurable hybridization is observed during lens placode formation (48 hours of development), and during later stages the number of molecules of δ-crystallin mRNA per cell increases. The initial detection of δ-crystallin synthesis coincides with the appearance of the messenger. The inset of Figure 14.28 shows an extrapolation back in time to estimate when δ-crystallin mRNA first begins to accumulate. The apparent initial accumulation of this messenger is at approximately 43 to 44 hours, which is 8 to 9 hours after the start of lens induction by the optic vesicle and 1 to 2 hours before the epidermal cells begin to elongate.

The results of hybridization studies indicate that increasing specialization for δ-crystallin synthesis, which occurs during lens cell differentiation, results from the accumulation of δ-crystallin mRNA. How

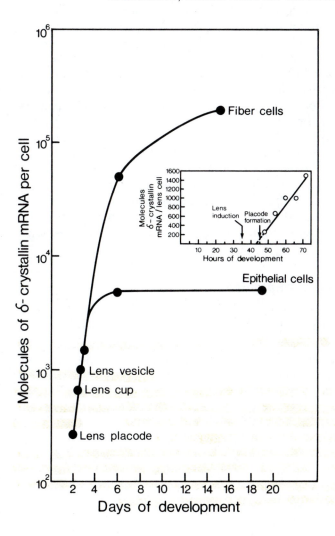

FIGURE 14.28 Increase in amount of δ-crystallin mRNA during chick lens development as determined by hybridization to δ-crystallin ^{3}H-cDNA. The broken line in the inset represents an extrapolation to 0 δ-crystallin mRNA molecules. (After J. Piatigorsky. 1981. Lens differentiation in vertebrates. A review of cellular and molecular features. Differentiation, *19:* 138. Reprinted with permission of Springer-Verlag, Heidelberg. The inset originally appeared in T. Shinohara and J. Piatigorsky. 1976. Quantitation of δ-crystallin messenger RNA during lens induction in chick embryos. Proc. Natl. Acad. Sci. U.S.A., *73:* 2811.)

do the cells accumulate the messenger? Is the control exerted at the transcriptional level by a specific activation of the gene for δ-crystallin? If so, hypomethylation of the δ-crystallin genes, which accompanies their expression in the embryonic lens (Jones et al., 1981), might play a role in transcriptional regulation (see Chap. 4 for a discussion of hypomethylation). Conversely, does the messenger accumulate because it is preferentially stabilized so that it is degraded at a lower rate? Does accumulation result from both increased synthesis and decreased degradation? Is messenger accumulation the only mechanism that accounts for the specialization for δ-crystallin synthesis? The lens system is well-suited for the examination of such questions, and with the rapid advances being made in molecular biology, the answers to these questions will not be long in coming. Once we know how the cell becomes specialized, we can begin to examine the role of the inducers in this process.

An interesting aspect of the data shown in Figure 14.28 is the early similarity between epithelial and fiber cells, which both accumulate δ-crystallin messenger, and the later leveling off of δ-crystallin mRNA accumulation in the epithelial cells. The rate of δ-crystallin synthesis in the epithelial cells declines about threefold between days 6 and 19 of development (Beebe and Piatigorsky, 1981). However, as shown in the graph, δ-crystallin mRNA does not decline in this interval. This suggests that posttranscriptional mechanisms limit the utilization of this mRNA in the epithelial cells.

After hatching, δ-crystallin synthesis in the lens declines and eventually becomes undetectable. This reflects a loss of δ-crystallin mRNA in both the epithelial and the fiber cells (Beebe and Piatigorsky, 1981; Tréton et al., 1982). Thus, after first becoming specialized for δ-crystallin synthesis these cells lose this property. The transitory nature of δ-crystallin gene expression adds another intriguing problem to be studied with this system.

Lens Regeneration in Amphibians

In addition to the normal embryonic differentiation of the lens, amphibians have the remarkable ability to replace a lost lens, using other elements of the eye. Urodeles can regenerate a lens from the dorsal iris epithelium, a source quite different from that of the normal embryonic lens (for reviews see Yamada [1967, 1977]). The iris epithelial cells are highly pigmented and nondividing. After the lens is removed, the iris cells lose their pigmentation, acquire the ability to divide, and then differentiate into lens cells, which synthesize crystallin. This conversion of one cell type into another is called metaplasia.

The ability of the iris to form a lens is acquired during the larval stage and is retained by the adult. Although in most types of regeneration the regenerate is assembled from damaged tissue, the urodele iris can replace a lens even though the iris itself is undisturbed. Furthermore, if the lens is manually replaced after its removal, regeneration does not occur. This observation suggests that the presence of a lens prevents formation of a new lens from the iris and that removal of the lens allows the iris to express its lens-forming potential.

Lens regeneration is but one aspect of the extensive regenerative capabilities of urodeles. In fact, the entire eye can be regenerated if a small part of the original eye is left. Other examples of regeneration include replacement of limbs, the tail, external gills, or the front part of the head, including the jaws. Limb regeneration has been studied extensively and will be discussed in more detail in section 14–4.

Regeneration by anuran amphibians is restricted to the larval stage of development. Frog and toad tadpoles are capable of regenerating their tails and their limbs. Regeneration of the lens has been demonstrated in *Xenopus laevis* tadpoles, but unlike in the urodele, the lens is produced by the corneal epithelium rather than the iris (Freeman, 1963).

14–3. AN ANALYSIS OF DEVELOPMENT: VERTEBRATE BLOOD CELLS

Differentiation of circulating blood cells (**hematopoiesis**) provides an opportunity to study a somewhat more complicated developmental scheme than occurs in lens cell differentiation. Hematopoiesis results in the production of different types of blood cells, which perform specialized functional roles. The three major types of blood cells are: red blood cells (**erythrocytes**), platelets (derived from **megakaryocytes**), and white blood cells (**leukocytes**). The latter category includes the **lymphocytes, granulocytes,** and **monocytes**. Thus, blood cell differentiation provides an excellent system for studying the generation of cell diversity during development.

The principal hematopoietic tissue in the adult vertebrate is the **bone marrow.** Marrow cells are transported via the circulation to a variety of hematopoietic organs in which they differentiate into functional blood cells. Hematopoietic organs are categorized as **myeloid** and **lymphoid,** based upon the types of cells they produce. The myeloid cells are erythrocytes, platelets, granulocytes and monocytes; they are produced in the bone marrow and spleen. The two major types of lymphoid cells are the B and T lymphocytes. As we shall discuss later, each of these cell types differentiates in a different lymphoid tissue.

The Ontogeny of Blood Cells

Although the bone marrow provides the blood cells that differentiate within the adult hematopoietic organs, the marrow itself is not the original source of these cells. The adult cells are derivatives of embryonic precursor cells that originate elsewhere and enter the marrow, which perpetuates the cell line. In birds and mammals, the initial blood cells originate during early embryonic development from the extraembryonic **blood islands.** Blood islands are established by mesenchyme cells that proliferate to form colonies, or foci, of hematopoietic cells. Cells at the periphery of colonies form epithelia, which organize a capillary network that encloses blood cells within vascular sinusoids. These sinusoids eventually establish continuities with the embryonic vascular network. In the chick the blood islands form in the extraembryonic mesoderm in the proximal portion of the area opaca (Fig. 14.29). In the mammalian embryo they originate from yolk sac mesoderm (Fig. 14.30). The blood islands are a temporary source of blood cells and produce primarily rather short-lived "primitive" erythrocytes; platelets, granulocytes, and lymphocytes are not formed.

In the avian embryo the blood islands may also participate for a short time in the production of the first definitive erythrocytes. However, all subsequent blood cells of both the myeloid and lymphoid lineages are derived from mesenchyme cells that originate *within the embryo* before the vascular system develops. These **hematopoietic stem cells** are thought to colonize the hematopoietic tissues and act as the source of blood cells throughout the remainder of embryonic and adult life (Lassila et al., 1982a, 1982b). Since the initial hematopoietic stem cells appear before vascularization, they could not have entered the embryo from the blood islands via the circulatory system. It is conceivable, however, that precursors migrate from the blood islands through tissues before the circulatory system develops. Thus, the original site of origin of these cells is uncertain.

In the mammalian embryo the blood islands are the initial source of the hematopoietic stem cells, which migrate into the developing embryo and colonize the subsequent hematopoietic tissues. This conclusion is supported by experiments conducted by Moore and Metcalf (1970). When they cultured mouse embryos at the 1 to 4 somite stage, they developed to the 10 to 20 somite stage after two days, at which time the circulatory system is well developed. However, if they removed the yolk sac before culturing the embryos, the embryos lacked blood cells; the fluid that circulated in the blood vessels was acellular. The isolated yolk sacs, on the other hand, supported differentiation of erythrocytes. In another experiment, mouse blood island cells were implanted into mouse hosts that had been irradiated to destroy the hosts' blood cell line. Both the myeloid and lymphoid cells were restored by the blood island implants. The mammalian fetal liver, which first supplements and later replaces the blood islands as a site of erythrocyte production, also possesses hematopoietic stem cells. These cells, which apparently originate in the yolk sac, reside temporarily in the fetal liver, where they proliferate, before taking up residence in the bone marrow. The marrow retains and amplifies the stem cells throughout adult life.

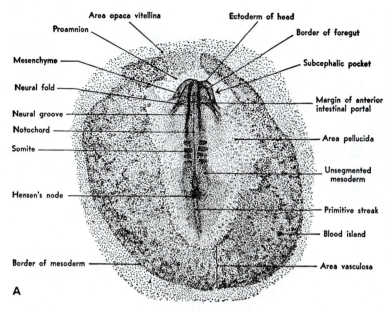

Area opaca vitellina
Proamnion
Mesenchyme
Neural fold
Neural groove
Notochord
Somite
Hensen's node
Border of mesoderm
A

Ectoderm of head
Border of foregut
Subcephalic pocket
Margin of anterior intestinal portal
Area pellucida
Unsegmented mesoderm
Primitive streak
Blood island
Area vasculosa

FIGURE 14.29 Blood islands of the chick embryo. *A,* Dorsal view of 24-hour blastoderm. Note the blood islands in the proximal portion of the area opaca. *B* to *D,* Drawings illustrating the cellular organization of the blood islands at 18 hours (*B*), 24 hours (*C*), and 33 hours (*D*). (From *Early Embryology of the Chick,* 5th Edition, by B.M. Patten. Copyright © 1971 by the McGraw-Hill Book Company. Used with permission of McGraw-Hill Book Company.)

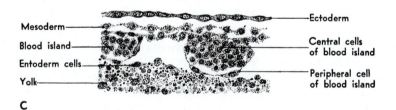

Ectoderm
Yolk-granule
B

Mesoderm
Blood island
Yolk

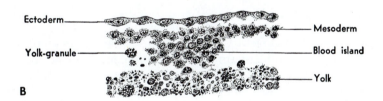

Mesoderm
Blood island
Entoderm cells
Yolk
C

Ectoderm
Central cells of blood island
Peripheral cell of blood island

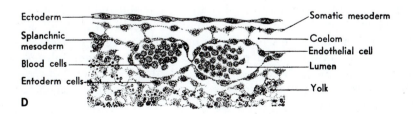

Ectoderm
Splanchnic mesoderm
Blood cells
Entoderm cells
D

Somatic mesoderm
Coelom
Endothelial cell
Lumen
Yolk

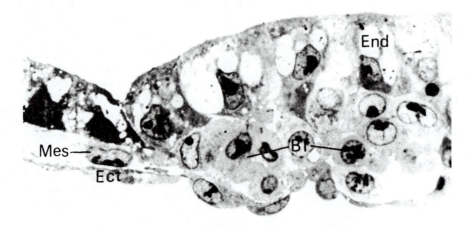

FIGURE 14.30 Light micrograph of eight-day mouse yolk sac. Early blood islands (Bl) are visible as small clusters of cells in the mesenchymal (Mes) layer, between the endoderm (End) and ectoderm (Ect). ×375. (From P.A. Marks, R.A. Rifkind, and A. Bank. 1974. Mammalian erythroid cell differentiation. *In Biochemistry of Cell Differentiation* [MTP International Review of Science, Biochemistry Section, Vol. 9]. University Park Press, Baltimore, p. 131.)

The Pluripotent Stem Cell

The production of several divergent blood cell types from the hematopoietic stem cell line poses an intriguing logistical problem: How do the stem cells embark on the various developmental pathways? Is there a heterogeneous population of precursors, each with a single fate, or a single kind of precursor that has different developmental potentials? The available evidence suggests that there is a **pluripotent stem cell** that is capable of differentiating along distinct developmental pathways, depending upon the influences to which it is subjected (Till and McCulloch, 1961).

The pluripotent stem cell persists throughout adult life and serves as a constant source of new blood cells. In the adult mammal the stem cells are located mainly in the bone marrow. The existence of such cells is demonstrated by so-called cell reconstitution experiments, which involve injection of bone marrow cells into a recipient mouse whose own hematopoietic system has been destroyed by x-rays. The injected cells populate the spleen of the recipient animal and form colonies, each of which is thought to have been established by a single cell and is therefore a clone. The evidence for the clonal relatedness of cells in a colony is provided by an experiment in which donor bone marrow cells are treated with low doses of x-rays to produce random chromosomal abnormalities prior to transplantation. All cells in an individual spleen colony in the host will have the same chromosomal composition, indicating that each colony is derived from a single precursor (Becker et al., 1963).

After colonies have been established for some time, they usually contain cells belonging to three different blood cell lines—erythrocyte, granulocyte-monocyte, or platelet. The stem cells proliferate to expand the stem cell population, which can subsequently become irreversibly committed to developing along one of the three developmental pathways. The committed cells in turn initially divide to expand the population of committed cells. These mitotically active cells are called **committed progenitors.** The population of committed cells eventually loses the ability to divide, and the cells undergo terminal differentiation to form functional blood cells.

Bone marrow cells transplanted into irradiated mice also repopulate lymphoid tissues, forming lymphocytes. In reconstitution experiments using donor bone marrow cells with radiation-induced chromosomal markers, myeloid spleen colonies and B and T lymphocytes were found to contain the chromosome marker of a common precursor (Abramson et al., 1977). Thus, the pluripotent stem cell gives rise to both the myeloid and lymphoid cell lines.

The factors controlling the commitment of pluripotent stem cells to differentiate along specific developmental pathways are uncertain. One theory is that differences in the cellular environments surrounding uncommitted stem cells determine the developmental pathway to be followed. These determinative influences are referred to as **hematopoietic inductive microenvironments,** or **HIM** (Trentin et al., 1974). According to this theory, a spleen colony established by stem cells transplanted to irradiated host mice initially encounters a single microenvironment, and as a result, a single developmental pathway is entered. However, as the colony increases in size, new microenvironments are encountered, and pluripotent stem cells, which have proliferated in the colony, may embark on different lines of differentiation. The result is a heterogeneous colony consisting of distinct differentiated cells derived from a single cell type. An alternative theory is that commitment of pluripotential stem cells is random, and the only effect of extrinsic influences is to shift the probability that cells would differentiate in a certain way (Till et al., 1964). Recent *in vitro* studies have demonstrated that a single stem cell may be placed in culture and divide to produce a clone containing several blood cell types. This result suggests that hematopoietic inductive microenvironments are not necessary for determination of stem cell fate (at least *in vitro*). It does not, however, rule out an *in vivo* role for them. Analysis of the clones also reveals that commitment of the stem cells to cell lineages may not be random, but may follow a predetermined pattern (Johnson and Metcalf, 1979). The ability to culture stem cells and obtain commitment *in vitro* provides a useful experimental system for the study of the process of stem cell commitment and analysis of the possible effects of exogenous substances on determination. Through exploitation of this *in vitro* system, the important problem of determination of the fates of pluripotent stem cells may be resolved.

Differentiation of Blood Cells from Committed Progenitors

The committed progenitors have developmental fates restricted to individual developmental pathways—erythropoiesis, granulopoiesis, thrombopoiesis (platelet-producing), or lymphopoiesis. Once their fates are determined, their proliferation and ultimate differentiation into functional blood cells are regulated by hormones called **hematopoietins.** Furthermore, blood cell differentiation appears to involve specific interactions among cells belonging to the various blood cell lines (Cline and Golde, 1979). We shall briefly discuss erythropoiesis and lymphopoiesis as examples of blood cell differentiation.

DIFFERENTIATION OF ERYTHROCYTES

Differentiation of erythrocytes from committed progenitors (Fig. 14.31) is a gradual process that proceeds through a number of recognizable intermediate stages (Rifkind, 1974). The most immature cell in this pathway is the **proerythroblast.** The nucleus of this cell contains diffuse chromatin with little heterochromatin and has a well-developed nucleolus that is active in the synthesis of ribosomal RNA. The transport of RNA synthesized at this stage to the cytoplasm apparently accounts for the intense basophilic staining properties that characterize the cytoplasm of the **basophilic erythroblast,** the next stage. During this stage the nucleus begins to undergo the changes that culminate in the loss of nuclear function and eventually in the loss of the nucleus itself. The basophilic erythroblast nucleus contains increasing amounts of heterochromatin, and the nucleolus begins to regress. Concomitant with these changes, RNA synthesis begins to decline. The initial cytochemical detection of the erythrocyte-specific protein, hemoglobin, is at the **polychromatophilic erythroblast** stage, although it is likely that globin synthesis is first initiated during the basophilic stage. The nucleus undergoes further regression during the polychromatophilic stage, as evidenced by additional heterochromatization and a reduction in the rate of RNA synthesis. This is the final stage at which cell division can occur. Ironically, hemoglobin synthesis is at its peak during this stage, even though the cytoplasmic concentration of RNA is declining. The highly pyknotic nucleus is expelled during the final erythroblast stage, the **orthochromatic** stage, which is also characterized by intense cytoplasmic staining for hemoglobin. After nuclear expulsion the cells, which now are called **reticulocytes,** enter the circulation. They are obviously incapable of cell division or RNA synthesis, although they continue to produce hemoglobin for awhile, sustained by synthetically active polysomes. As ribosomes disappear, hemoglobin synthesis ceases, and reticulocytes become mature **erythrocytes.**

The production of erythrocytes in the adult is regulated by the circulating glycoprotein hormone **erythropoietin,** whose levels in the circu-

Proerythroblast

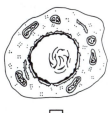

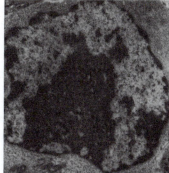

Basophilic
erythroblast

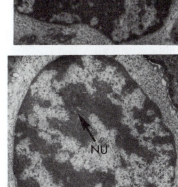

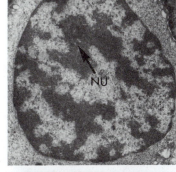

NU

Polychromatophilic
erythroblast

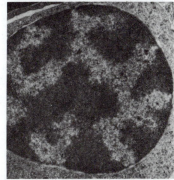

Orthochromatic
erythroblast

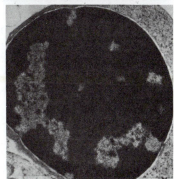

Reticulocyte

FIGURE 14.31 Mammalian erythropoiesis. On the left is a series of drawings depicting the erythropoietic stages. (After R.A. Rifkind. 1974. Erythroid cell differentiation. *In* J. Lash and J.R. Whitaker (eds.), *Concepts of Development.* Sinauer Associates, Inc., Sunderland, Mass., p. 156.) On the right are electron micrographs of nuclei at each nucleated stage. NU: nucleolus. Magnifications of micrographs (from top to bottom): ×4300, ×5000, ×8700, ×10,800. (Micrographs from D. Orlic. 1970. Ultrastructural analysis of erythropoiesis. *In* A.S. Gordon (ed.), *Regulation of Hematopoiesis,* vol. 1. Appleton-Century-Crofts, New York, p. 283. Copyright © Plenum Publishing Corp.)

lation are controlled by the kidney. This hormone reaches high levels in the blood during periods of reduced oxygen tension or anemia. The effects of erythropoietin on erythropoiesis are readily studied *in vitro*, usually by means of explants of bone marrow or fetal liver. Erythropoietin influences erythropoiesis in the proerythroblast stage. Under the influence of erythropoietin, proerythroblasts undergo mitosis, which increases the number of cells that have the potential to synthesize hemoglobin. Erythropoietin also promotes the differentiation of cells of the expanded proerythroblast population, which then enter the basophilic erythroblast stage. Recent *in vitro* studies have revealed that the response of erythroid cells to erythropoietin is mediated by an interaction between them and T lymphocytes (Cline and Golde, 1979).

Our ability to regulate erythropoiesis with erythropoietin provides an excellent way to study the hormonal control of the molecular events of cell differentiation. The earliest detected effect of erythropoietin on macromolecular synthesis is a stimulation of synthesis of 4S, 5S, and ribosomal RNA (Krantz and Goldwasser, 1965; Djaldetti et al., 1972; Maniatis et al., 1973). No globin mRNA is detectable at this time when probed with globin cDNA. Hence, the early effects of erythropoietin are not sufficient to promote the transition from proerythroblasts to hemoglobin-producing erythroblasts. With additional exposure to erythropoietin, the proerythroblasts undergo DNA synthesis and mitosis, and the daughter cells subsequently differentiate into erythroblasts, which accumulate globin mRNA and initiate globin synthesis (Ramirez et al., 1975).

The DNA synthesis that intervenes between the early period of RNA synthesis and the period of globin mRNA accumulation is apparently essential for the latter differentiation process. This conclusion comes from experiments in which DNA synthesis is inhibited by hydroxyurea; the presence of the inhibitor prevents the subsequent synthesis of globin by the cells (Rifkind et al., 1976). However, as with all inhibitor studies, this evidence is merely suggestive and does not *prove* that DNA synthesis is essential for the transition to globin mRNA production and globin synthesis.

The most fundamental event in differentiation of erythroid cells is the specialization for globin synthesis. Hemoglobin is actually a family of molecules, each of which is composed of heme and four polypeptide subunits. In the human, for example, these subunits, the globins, are encoded by two families of genes—the α-globin and β-globin gene families. Members of each gene family are closely linked and have related nucleotide sequences. Two α-like and two β-like polypeptides combine with heme to form a hemoglobin molecule. Regulation of globin gene expression is complicated by the fact that erythrocytes differentiating during different phases of the life cycle may exhibit selective expression of members of the globin gene families. The β-globin gene family, for example, consists of ε-, γ-, δ-, and β-globin genes as well as a β-like pseudogene (Fig. 14.32). In the embryo the ε gene is expressed in the yolk sac-derived primitive erythrocytes. In the early fetus the ε gene is no longer expressed, but the γ gene is expressed, whereas in the later stages of fetal life and in the adult the δ- and β-globin genes are expressed. Very little γ-globin gene expression occurs in the adult. These changes in globin gene expression during development are called **globin gene switching.**

Thus, not only must erythroid cells become specialized for globin synthesis, but determination must be made as to *which* globin genes are to be expressed. Recent evidence suggests that intrinsic changes in progenitor cells during development as well as factors in the hematopoietic environment are both involved in this selection (Cudennec et al., 1981; Papayannopoulou et al., 1982). The discovery that extrinsic factors influence globin gene switching opens up the possibility of studying the effects of these factors at the cellular and molecular levels.

The specialization of protein synthesis in erythroid cells involves not only an increase in synthesis of globins but also a reduction in the synthesis of nonglobin proteins. This specialization may be partially accounted for by transcriptional restriction. However, it has been demonstrated that posttranscriptional mechanisms are also involved, since the synthesis of nonglobin proteins becomes even more restricted during the aging of reticulocytes, which lack nuclei. Young reticulocytes predominantly synthesize globins, but another protein, designated **peptide I,** is also produced in significant amounts (Fig. 14.33). As reticulocytes age, globin synthesis continues, but synthesis of peptide I ceases. When RNA is extracted from young and old reticulocytes, it is found that RNA from young cells will direct the synthesis of both globin and peptide I *in vitro*, but that RNA from aged cells will only direct the synthesis of globin (Lodish and Small, 1976). Hence, the older cells lack the messenger for peptide I. This result must mean that the messengers for peptide I have a shorter life span than the globin messengers. The demonstrated long life span of globin messengers may be an important mechanism for protein synthetic specialization, particularly since nuclear function is lost early in the developmental pathway. Messengers with short life spans would be destroyed shortly after their production, and globin messengers would account for an increasingly larger percentage of the messenger available for translation as differentiation proceeds.

5′ ——— ε ——— Gγ Aγ ψβ1 ——— δ ——— β ——— 3′

FIGURE 14.32 The β-globin gene family. (After N.J. Proudfoot et al. 1980. Structure and *in vitro* transcription of human globin genes. Science *209*: 1329–1336. Copyright 1980 by The American Association for the Advancement of Science.)

FIGURE 14.33 Protein synthetic patterns of young (*A*) and old (*B*) reticulocytes. Cells were labeled with ^{35}S-methionine; cell supernatants were electrophoresed and autoradiographs made of gels to localize proteins that incorporated the labeled precursor. Peptide I and globin are designated. Old reticulocytes have lost the capacity to produce peptide I. (From H.F. Lodish and B. Small. 1976. Different lifetimes of reticulocyte messenger RNA. Cell, 7: 61. Copyright © Massachusetts Institute of Technology; published by the MIT Press.)

— I

— Globin

A B

DIFFERENTIATION OF LYMPHOCYTES

Lymphocytes are the blood cells that mediate the immune response, serving to protect the organism against (1) foreign macromolecules (antigens) and (2) foreign cells. Protection against foreign antigens is provided by the **humoral response system,** which is based on circulating antibodies that combine specifically with antigens. Protection against foreign cells, on the other hand, is provided by the **cell-mediated response system,** by which immunocompetent cells recognize and eliminate foreign cells without the direct involvement of a circulating antibody.

Two kinds of lymphocytes are known—**T** and **B lymphocytes.** Their designation refers to the histological site in which the stem cells originating in the bone marrow begin their differentiation. T lymphocytes are derived from the thymus gland, whereas B lymphocytes are so named because they emanate from the bursa of Fabricius in birds. Since the bursa does not exist in mammals, alternate sites of B lymphocyte derivation must be present in this vertebrate class. After acquiring the characteristics of either B cells or T cells, lymphocytes are released into the circulation and take up residence in the spleen, lymph nodes, and bone marrow.

T cells, upon release from the thymus, are characterized by the presence of a specific antigen, known as theta (θ), on their surface. T cells play a primary role in the cell-mediated response, recognizing and destroying foreign cells. T cells also interact in some unknown way with B cells to induce them to produce an antibody. The latter macromolecules are localized on the surfaces of B cells and serve as receptors for antigen. When antigen combines with the surface antibody, B cells are stimulated to proliferate and differentiate into a clone of antibody-producing cells. The antibody they produce is released into the circulation for neutralization of the foreign antigen.

Because an individual may encounter a wide variety of antigens in a lifetime, the potential to produce antibodies to counter these antigens must be immense. Recent estimates are that from 10^6 to 10^8 *different* antibodies can be produced by an individual (Tonegawa, 1983). How can the genome, which possesses the information to produce antibodies, cope with this incredible requirement? It is now evident that mechanisms exist to diversify the genetic information for antibody production and that this diversification is generated during B-cell development. As a consequence of these diversification processes, the genes encoding the antibody molecules differ among B lymphocytes and from those in the genome of the zygote. As we discussed in Chapter 4, it is an exception to the rule that the genome remains unaltered during development.

Each antibody, or immunoglobulin (Ig) molecule (Fig. 14.34), is a tetramer consisting of two identical light (L) chains and two identical heavy (H) chains. The great diversity among Ig molecules depends upon the production of diverse light and heavy chains, which combine to form tetramers. The L and H chains are composed of two segments each: The amino-terminal ends are quite variable among Ig molecules and are called

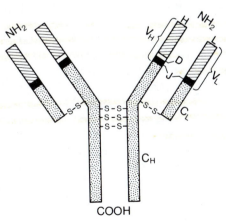

FIGURE 14.34 Model of an immunoglobulin molecule. Each immunoglobulin molecule is composed of two identical light chains and two identical heavy chains. Each chain consists of a V region (V_L or V_H) and a C region (C_L or C_H). The polypeptides are linked to one another by disulfide bonds.

the **variable (V) regions,** whereas the carboxy-terminal regions are the **constant (C) regions.** The light and heavy chain V regions form the combining site of an antibody molecule. Together they constitute the **V domain,** which recognizes the antigen and determines the specificity of an Ig molecule. The constant regions mediate the effector functions that produce the immune response. As we shall now outline, the polypeptides that comprise immunoglobulin molecules are encoded by three distinct genomic regions.

In the discussion that follows, in which immunoglobulin gene structure is correlated to the Ig molecules, we shall focus on the situation in mice, for which most data have been obtained. Important tools in these investigations have been **myeloma cells,** which are tumors of lymphoid cells. Since a lymphocyte produces a single species of immunoglobulin, a lymphocyte that becomes malignant can divide to produce a large number of progeny, each of which produces the same immunoglobulin. The myelomas can be maintained *in vitro,* providing an indefinite number of cells that are dedicated to the synthesis of a single molecular species and that can be subjected to molecular analysis.

There are two kinds of murine (i.e., mouse) light chains: κ and λ. All kappa (κ) chains have the same C region and one of several potential V regions. Lambda (λ) chains have one of two possible C regions but very little diversity among V regions. Distinct genomic regions encode the κ and λ chains; they are on separate chromosomes (i.e., unlinked). The heavy chains are polypeptides that may be composed of one of a large set of V regions and one of several C regions. The latter are classified into five types: C_μ, C_δ, C_γ, C_ϵ, and C_α. An immunoglobulin molecule is named for the type of heavy chain C region it contains; that is, IgM, IgD, IgG, IgE, or IgA, respectively. IgG actually consists of four different kinds of polypeptides, bringing the total number of immunoglobulin types to eight. The heavy chain genes are unlinked to either of the regions encoding the two kinds of light chains.

As we mentioned earlier, individual B cells make antibodies with only a single type of antigen-combining site. Thus, during development, the cell becomes committed to producing either kappa or lambda light chains and expressing either the maternal or paternal allele at the light and heavy chain loci. The expression of only one of the alleles at Ig gene loci is called **allelic exclusion.** Since an antibody is composed of two light chains and two heavy chains, the production of only one light chain type and allelic exclusion ensure that homologous chains in a single immunoglobulin molecule have the same variable regions, and consequently the same antigenic specificity.

The variability among immunoglobulin molecules depends upon the association of diverse light and heavy chains. These diverse chains are generated by the combination of a limited number of C regions with many potential V regions. Dryer and Bennett (1965) proposed that the V and C regions of an Ig polypeptide are encoded by separate genes. However, each Ig chain is a covalently continuous polypeptide that is translated on a single mRNA molecule. This suggests that DNA is rearranged by **somatic recombination** during B-cell development to bring a V gene into proximity with a C gene.

Evidence for somatic rearrangement of the V and C genes was first provided by a comparison of fragments of embryo and lymphocyte (myeloma cell) DNA produced by treatment with a restriction endonuclease (see Chap. 3 for a discussion of these enzymes). Hozumi and Tonegawa (1976) isolated full-length kappa mRNA, which they used as a probe to identify both the V and C genes. They then prepared a fragment of the 3' end of the mRNA, which is a specific probe for fragments of DNA containing the C gene. When embryo DNA was treated with the restriction enzyme *Bam* H1, two large fragments (9 and 6 kilobases [kb] long, respectively) were generated that hybridized with the full-length mRNA. However, only the 6 kb fragment hybridized with the C-gene probe. This must mean that the V gene is on the 9 kb fragment, whereas the C gene is only on the 6 kb fragment. Therefore, these genes are located at a distance from one another in embryo cell nuclei. By way of contrast, when myeloma DNA was restricted with *Bam* H1, a single 3.5 kb fragment was produced that hybridized with both probes. Thus, during lymphocyte differentiation, the V and C genes must be joined by recombination.

The V genes code for only a portion of the complete V region of an Ig chain. The remainder of the V region of lambda chains is encoded by the **joining, or J, gene segment,** which is adjacent to the C gene, from which it is separated by an intron. During B-cell development the V gene is translocated so that it abuts the J segment, resulting in a V-J-intron-C sequence. The situation is even more complicated for the kappa- and H-chain genomic regions. The kappa region contains five J segments (one of the five is nonfunctional). Any one of the V genes may be joined to any of the four functional J segments, thus increasing the diversity within kappa-chain V regions. The diversity is further increased by variation in the exact site of V-J joining (Sakano et al., 1979; Max et al., 1979). The steps in the generation of a kappa light chain from the germ-line DNA, through somatic recombination, transcription, messenger RNA processing, and, finally, translation are illustrated in Figure 14.35.

H-chain V regions consist of portions derived from the V gene, a J segment, and an additional segment called a **D** (for "diversified") **segment.** The heavy chain genomic region consists of 100 to 200 different V segments, approximately 12 D segments, four functional J segments, and eight C segments—one for each immunoglobulin class or subclass (Tonegawa, 1983).

During B-cell development, the heavy chain genomic region is rearranged such that one each of the V segments, D segments, and J segments is conjoined to form a complete V region. As with the kappa gene recombination, joining sites may be variable, thus increasing the diversity

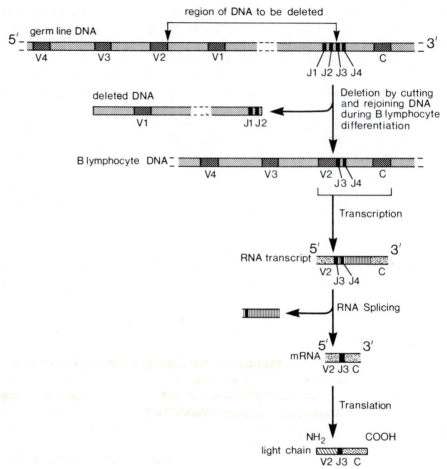

FIGURE 14.35 Summary of steps leading to synthesis of a kappa light chain. (After B. Alberts et al. 1983. *Molecular Biology of the Cell.* Garland Publishing, Inc., New York, p. 982.)

of potential V regions. The V region that is formed by recombination is initially expressed in all B cells along with C_μ to produce IgM. Later in B-cell development, the cells switch to production of one of the other heavy chain classes or subclasses. This is called **heavy chain class switching.** Interestingly, the same V region is expressed in both the original and subsequent heavy chains. Thus, the antibody-binding site is not modified, even though the effector region changes.

Why is C_μ expressed first, and how does the cell effect a switch from expression of C_μ to one of the other C genes? The initial expression of C_μ is apparently due to its position in the heavy chain genomic region. The order of heavy chain C genes is $5'\text{-}C_\mu\text{-}C_\delta\text{-}C_{\gamma^3}\text{-}C_{\gamma^1}\text{-}C_{\gamma^{2b}}\text{-}C_{\gamma^{2a}}\text{-}C_\epsilon\text{-}C_\alpha\text{-}3'$ (Alt et al., 1982). The V region that is formed by recombination would, therefore, initially be adjacent to C_μ. Transcription of V-D-J-C_μ and translation of the messenger would result in IgM.

Two mechanisms have been reported for the subsequent class switch. One involves posttranscriptional processing of multiclass transcripts to remove the RNA encoding the inappropriate heavy chain, whereas the other involves rearrangement of the genome to move the V-D-J region into proximity with the gene for a different heavy chain class and simultaneously to delete intervening genes (Tonegawa, 1983).

We have outlined several mechanisms that contribute to antibody diversity. Multiple V region combinations and several combinations of variable and constant regions of either a light or heavy chain can be generated by somatic recombination. The production of various light and heavy chains results in additional diversity when they are assembled into a functional antibody molecule. That is, a given light chain will form a different antibody when combined with heavy chain X than with heavy chain Y, and vice versa. There is one additional mechanism that we have not mentioned previously: **somatic mutation,** through which individual bases in the V region genes are modified (Weigert and Riblet, 1976). Clearly the genome encoding the immunoglobulins is malleable. It will be exciting to learn how this malleability is regulated and harnessed during B-cell development.

14–4. AN ANALYSIS OF DEVELOPMENT AND REGENERATION: THE VERTEBRATE LIMB

The vertebrate limb is formed from the limb bud, which is an outgrowth of a loose network of mesenchyme, surrounded by an epithelial sheet of ectoderm. The pattern of cartilage, muscle, and connective tissue of the limb is formed out of the mesenchyme. Although all vertebrate limbs are composed of these tissues, their forms vary, and the special characteristics of each limb are primarily due to the spatial pattern of cell differentiation. This principle applies to other organs as well. The differences between homologous organs in related species are primarily due to how the cell types within the organ are organized.

Development of the Chick Wing

Much of our understanding of vertebrate limb development is due to extensive analyses on the formation of the chick wing. The attractiveness of the chick wing for study by developmental biologists is partially due to the ready accessibility of the limb bud, which juts out from the body axis. It is also due to the ease with which the limb bud can be manipulated by the investigator: The bud can be disassembled and reconstructed readily, and it can be grafted. Another reason for studying the wing is that its components are tissues that differentiate in other parts of the body from a variety of embryonic sources. Hence, the implications of pattern formation in the wing go far beyond the formation of the wing itself.

The organization of the mesodermal derivatives in the chick wing provides an excellent model system for studying the mechanisms that regulate formation of spatial patterns of cell differentiation. The wing is a self-differentiating system containing all the information that is necessary to specify the differentiation of its cell types in the correct spatial pattern. The utilization of this information in forming the pattern of cell differentiation in the limb has been studied extensively, primarily by John Saunders and his associates.

The pattern of cell differentiation is evident in the wing by the positions of the muscle, cartilage, and connective tissue. The cartilage (most of which is later replaced by bone) is particularly easy to monitor, since it readily takes up dye (e.g., alcian green) in fixed preparations. The cartilaginous pattern is clearly evident when the remaining tissues are cleared in a methyl salicylate solution.

The proximo-distal sequence of cartilage in the chick wing (i.e., proceeding from the point of attachment of the wing toward the tip) is humerus, radius and ulna, and finally wrist and digits (Fig. 14.36). As the limb bud elongates, proximal structures begin to differentiate while the tip of the outgrowth remains undifferentiated. More distal structures gradually appear in sequence as elongation continues (Saunders, 1948). Thus, the humerus develops long before the digits appear.

The outgrowth of the limb is dependent upon the apical ectodermal ridge. As we discussed earlier in this chapter, the limb bud mesoderm promotes formation of the AER and is responsible for maintaining the ridge after it has formed. The role of the AER in limb outgrowth is illustrated by inserting a second AER adjacent to the first: An additional limb emerges under the grafted AER (Saunders and Gasseling, 1968). Conversely, removal of the AER from a limb bud leads to defective limb development, and the extent of the defects depends upon the time of its removal. If it is removed at early stages of development, the upper arm

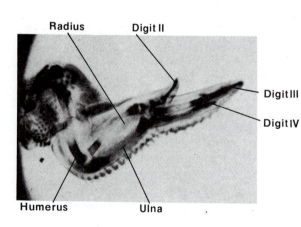

FIGURE 14.36 Photomicrograph of chick wing that has been fixed, stained with alcian green, and cleared in methyl salicylate. (From D. Summerbell and J.H. Lewis. 1975. Time, place, and positional value in the chick limb-bud. J. Embryol. Exp. Morphol., *33:* 632.)

may continue to form, but distal structures fail to appear. At later stages, the upper arm and forearm may be formed, but the digits fail to appear (Saunders, 1948; Summerbell et al., 1973). Thus, presumptive wing parts are determined in the same proximo-distal sequence as they differentiate, and the AER is essential for this process.

The proximo-distal sequence of cartilage differentiation is a consequence of the influence of the AER on the subjacent mesenchyme. As shown in Figure 14.37A, this loosely packed mesenchyme is undifferentiated. At more proximal locations the mesodermal cells aggregate to form cartilage and muscle. The region of undifferentiated mesenchyme has been called the **progress zone** (Fig. 14.37B) by Wolpert and his associates (Summerbell et al., 1973). Only after they have left the progress zone (by further outgrowth of the distal tip) can the cells differentiate. *In vitro* analyses have demonstrated that the AER has two measurable effects upon the subjacent mesenchyme: stimulation of cell division and the delay of its differentiation (Reiter and Solursh, 1982; Solursh et al., 1981). The stimulation of cell division is presumably the driving force that causes distal limb outgrowth.

After the distance between mesodermal cells and the AER reaches a critical value, the cells aggregate and begin differentiation. Presumably, the absence of aggregation within the progress zone is due to some influence from the AER. As we discussed in Chapter 5, extracellular matrices play important roles in cell aggregation. Recent studies on extracellular matrix components in the wing bud tip reveal tantalizing regional and stage-specific differences in the distribution of extracellular matrix components that could be involved in mediating cell aggregation (Dessau et al., 1980; Kosher et al., 1982; Tomasek et al., 1982). This avenue of investigation has much promise in providing an explanation for the recruitment of undifferentiated mesenchyme cells into differentiating tissue.

The emergence of mesenchyme cells from the progress zone to form cartilage in the correct proximo-distal sequence is an intriguing problem in determining the pattern of cell differentiation. Since limb development is dependent upon the AER, does the AER control the proximo-distal pattern of chondrogenesis? One possibility is that the AER emits qualitative information at different stages of limb development; that is, it emits a level-specific stimulus at each successive limb level. If this were true, differentiation of the mesodermal elements could be controlled by exposing mesoderm to AERs from different limb stages; thus, a young AER should promote differentiation of both proximal and distal elements, whereas older AERs should promote the differentiation of only distal elements. However, experiments in which mesoderm is capped with ectoderm of various ages show that the mesodermal derivatives develop in proper proximo-distal sequence *regardless of the age of the AER* (Rubin and Saunders, 1972). Thus, the information for the proper sequence of differentiation resides in the mesoderm itself. The AER is necessary for this information to be utilized.

Wolpert and his associates (Summerbell et al., 1973) have proposed that differentiation along the proximo-distal axis is determined within the progress zone. As cells leave the progress zone during elongation of the limb bud, those that emerge first will form proximal structures,

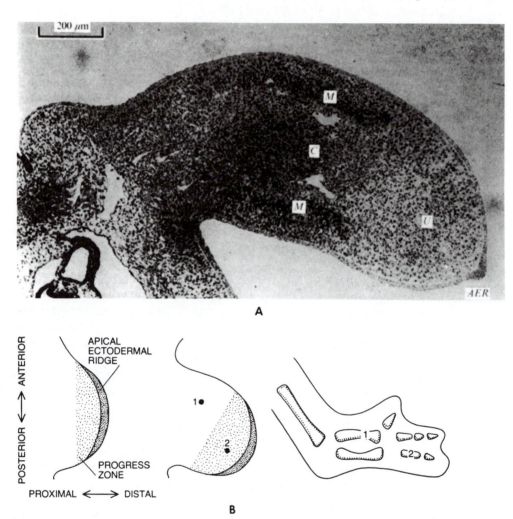

FIGURE 14.37 The chick wing bud. *A*, Photomicrograph of a sectioned wing bud. The AER is seen at the tip as thickened ectoderm. A region of undifferentiated mesenchyme (*U*) is present below the AER. In the proximal portion of the limb, muscle (*M*) and cartilage (*C*) are beginning to differentiate. (From D. Summerbell and J.H. Lewis. 1975. Time, place, and positional value in the chick limb-bud. *J. Embryol. Exp. Morphol., 33:* 627.) *B,* Diagram of the wing bud at progressive stages of development. According to Wolpert's progress zone model, cells proximal to the progress zone have had their positions specified while the position of cells within the progress zone is still undetermined. The former will differentiate into proximal structures whereas the latter will form more distal structures. (From Wolpert, L. 1978. Pattern formation in biological development. *Scientific American 239*(4): 154–164. Copyright © 1978 by Scientific American, Inc. All rights reserved.)

whereas those that emerge later will form more distal structures. Thus, the longer a cell remains in the zone, the more distal its ultimate location will be.

According to the progress zone model, no interaction between the progress zone and more proximal regions of the limb is necessary to establish the pattern of differentiated elements. This has been tested by experiments in which distal tips of limb buds (consisting of AER and the progress zone) are grafted onto the ends of limb buds that have had their own tips removed (Fig. 14.38A). In these experiments, the distal tips behave autonomously. Consider, for example, the results of grafting a young wing bud onto the stump of an older bud (Fig. 14.38B). Development proceeds, and the stump forms a humerus, radius, and ulna. However, the donor tip behaves independently and produces an additional set of cartilages that are arranged in the proper proximo-distal sequence. In the reverse experiment (transfer of an old tip onto a younger stump) the proximal elements differentiate from the stump mesoderm, distal elements are formed from the tip, but the intermediate elements are missing (Summerbell and Lewis, 1975).

Additional evidence for the progress zone model has been obtained by Wolpert et al. (1979). Chick wing buds were irradiated with high doses of x-rays. The irradiated wing buds were then transplanted to host embryos to determine their developmental potential. The irradiated buds developed into limbs having their proximal portions reduced or lost, whereas the digits were more or less normal. According to Wolpert et al., the irradiation destroyed a large percentage of cells in the progress zone and prevented the early exodus of cells from the zone. The surviving cells proliferated to repopulate the zone and remained there for a long time before exiting. Therefore, they specified only distal elements.

The progress zone model is controversial and highly dependent upon the interpretations given to the experiments. Saunders (1977) has pointed out some of the difficulties with the model. There is also some evidence that regulatory interactions between different regions of the limb do indeed occur during early stages of limb development (Searls and Janners, 1969; Stark and Searls, 1974; Kieny, 1977). Our understanding of proximo-distal regulation of limb bud organization will undoubtedly become more refined as additional investigations are conducted.

Considerable research has also been conducted on the control of the antero-posterior (A-P) axis of the wing. Consider the digits and their placement on the wing. The three digits and their order from front to back, as shown in Figure 14.36, are: II, III, and IV. What controls the pattern of digit differentiation, thus assuring that they will be positioned in the proper order?

As with the proximo-distal axis, the control of the A-P axis is also quite controversial. The original evidence on control of the A-P axis came from experiments conducted in John Saunders' laboratory. One of these experiments is outlined in Figure 14.39. If a small block of meso-

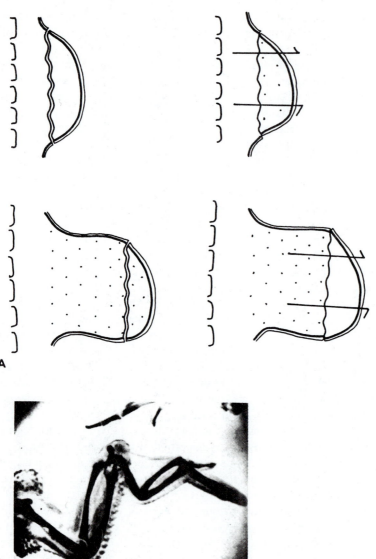

FIGURE 14.38 Exchange of distal tips between chick wing buds of different ages. *A,* Outline of experiment. *B,* Result obtained when young tip is placed on an older stump. (From D. Summerbell and J.H. Lewis. 1975. Time, place, and positional value in the chick limb-bud. J. Embryol. Exp. Morphol., *33:* 631, 632.)

derm near the posterior junction of an early limb bud and the body wall is grafted to a notch in the anterior edge of the wing bud, the wing tip shows a duplication. The significant characteristic of the duplication is that its orientation is reversed: Its posterior digits are closest to the mesodermal implant. From front to back, the digits of the two hands are: IV-III-II-III-IV. The extra hand is thus a mirror-image of the other.

The mesodermal region at the posterior edge of the limb bud has been designated the **zone of polarizing activity (ZPA)**. It is presumed to

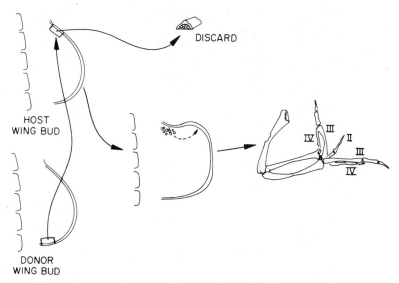

FIGURE 14.39 Transfer of a block of mesoderm from the posterior margin of a chick wing bud to the anterior margin of a host wing bud. The result (on the right) is a partial duplication of the wing tip. The two portions show mirror-image symmetry, and digit II is shared by them in the mirror plane. (From J.W. Saunders, Jr. 1972. Developmental control of three-dimensional polarity in the avian limb. Annals of the New York Academy of Sciences, *193:* 34.)

regulate the antero-posterior axis of the limb. Wolpert and his associates (Tickle et al., 1975; Summerbell, 1979) have proposed that the ZPA is the source of a substance (called a **morphogen**) that diffuses into other parts of the limb and influences the pattern of differentiation, which depends upon the parts' distance from the ZPA. The regions closest to the ZPA differentiate into posterior parts, but this tendency weakens with increasing distance from the ZPA, presumably due to decreasing concentration of morphogen.

The role of the ZPA has been questioned by Saunders (1977). His concerns are based on the following observations:

1. Removal of the ZPA at an early stage of limb development does not interfere with establishment of a normal A-P axis.
2. Various nonlimb tissues (e.g., somite or flank tissue) can produce the same effects on the A-P axis as the ZPA. Thus, the specificity of this effect is questionable.

These results cast some doubt on the role of the ZPA in normal development. Saunders therefore cautions against formulation of models of limb development based upon ZPA function.

Iten and her colleagues also discount the role of a gradient of morphogen emanating from the ZPA in specifying the pattern of limb components. They propose a model in which local cell-to-cell interactions determine the pattern of cell differentiation in the limb (for review, see Iten [1982]). See page 719 for more discussion of pattern formation by short range cellular interactions.

The experiments on chick wing development dramatize the difficulties encountered in understanding pattern formation. The pattern of cartilage in the wing is relatively simple. However, the definitive experiments that demonstrate conclusively how this pattern emerges have not been conducted. One difficulty in understanding pattern formation is assessing the effects of experimental manipulation on the pattern. Grafting and deficiency experiments are open to subjective interpretation, since it may be unclear to what extent the observed effects are due to mechanical interference or unnatural tissue combinations resulting from the manipulation. The investigation of ZPA is a case in point. Its transfer to the anterior limb bud margin clearly demonstrates an effect on digit placement. But does this effect reflect its *normal* function? That remains to be demonstrated definitively.

Regeneration of the Urodele Limb

Limb development in birds (as well as in reptiles and mammals) is a one-time occurrence that is restricted to the embryonic phase. Amphibians, however, have varying abilities to replace a limb lost in later stages of life. Although anurans can replace a limb only during the tadpole stage, urodeles can regenerate an amputated limb during either the larval or adult phase. The remarkable regenerative capacity of the urodele limb makes it an excellent system for the study of regeneration.

Regeneration has fascinated scientists for several generations. This fascination has produced a voluminous literature on the subject, dating back to the eighteenth century. Much of the philosophical framework for recent studies of regeneration was formulated by T. H. Morgan (1901). The major problems of regeneration identified by Morgan remain under active investigation today. These include the origins and developmental potentials of the cells in the regenerating part, the roles of the adjacent tissues in reconstruction, and the reasons for the great variation in regenerative capabilities among animals.

Morgan recognized two primary mechanisms for reconstitution of missing parts. One is morphallaxis, which involves reorganization of the remaining portions of the body to produce the missing structures. The other is epimorphosis, which occurs by formation of an outgrowth of new tissue from the wound surface. The well-known ability of the coelenterate *Hydra* to reconstitute itself out of a piece cut from a whole animal is an example of morphallaxis. Limb regeneration is the most exhaustively studied example of epimorphosis. However, we should not adhere to strict usage of these terms, since regeneration actually may involve both mechanisms. For example, in limb regeneration there is considerable reorganization of the tissues within the stump (morphallaxis) before the epimorphic outgrowth can occur.

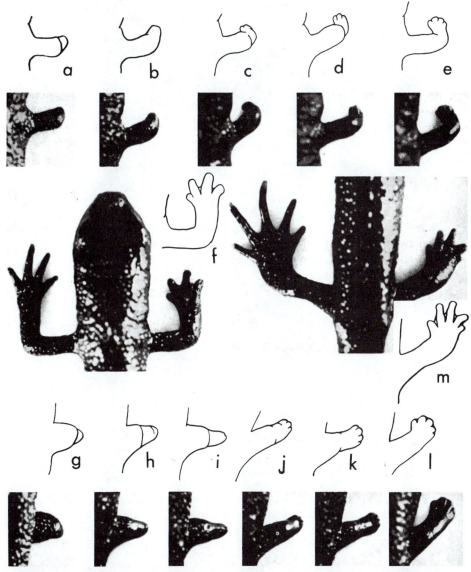

FIGURE 14.40 Regeneration in the limbs of *Triturus cristatus*. Stages a, b, c, d, e, f: consecutive stages of forelimb regeneration. Stages g, h, i, j, k, l, m: consecutive stages of hind limb regeneration. (From G. Schwidefsky. 1934. Entwicklung und determination der Extremitätenregenerate bei den Molchen. Wilh. Roux' Arch. für Entwicklungsmechanik, *132:* 64, 65. Reprinted with permission of Springer-Verlag, Heidelberg. Reproduced from B.I. Balinsky, 1975.)

The favorite subject of investigators studying limb regeneration is the adult newt. Regeneration of newt limbs is illustrated in Figures 14.40 and 14.41. The processes involved in limb regeneration include (Singer, 1973):

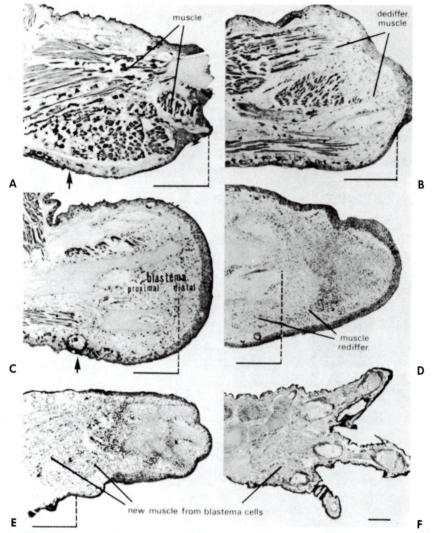

FIGURE 14.41 Series of light micrographs depicting muscle dedifferentiation and redifferentiation in the limb of the newt *Notophthalmus viridescens* following amputation. The dotted line indicates the approximate level of amputation. The black bar (0.5 mm long) calls attention to the same region of the limb stump in parts *A* to *E*. The plane of the section in A shows to good advantage the large amount of muscle present in the forearm at the time of amputation. The process of muscle dedifferentiation will extend proximally to the level indicated by the arrow. By 18 days (*B*), the distal 0.5 mm of the stump contains blastema cells derived from dedifferentiating (dediffer.) muscle. By 21 days (*C*), the level of dedifferentiation extends to the arrow (compare *C* and *A*). The blastema derives from various formed tissues of the old stump (proximal blastema). The proliferating regeneration cells form an outgrowth, the distal blastema. In the most proximal part of the blastema (arrow), dedifferentiation of muscle is incomplete, and these fibers redifferentiate by reorganization of their internal structure. In the blastema proper, cartilage differentiates first, then muscle (muscle rediffer., *D*). The sections at *E* and *F* show the large amount of new muscle that forms from blastema cells during the later periods of limb redifferentiation. (From E.D. Hay. 1974. Cellular basis of regeneration. *In* J. Lash and J.R. Whittaker (eds.), *Concepts of Development.* Sinauer Associates, Inc., Sunderland, Mass., p. 413.)

1. *Wound healing*. Repair of the wound begins with the spreading of epidermis from the edges of the wound to cover the open surface. This is a rapid process, usually accomplished in one or two days. The closure of the wound does not involve cell division. However, once closure is accomplished, epidermal cells proliferate to produce a multilayered mass of cells, which forms a conical bulge at the tip of the limb. This structure is the **apical epidermal cap.**

Soon after amputation of the limb, an inflammatory reaction occurs in the wound. This involves the appearance of white blood cells, cell and tissue debris, scattered erythrocytes, and fluid. As this reaction subsides, scar tissue forms under the epidermis.

2. *Tissue destruction (histolysis)*. Tissues within the stump undergo massive histolysis. In the process, differentiated tissues, such as muscle, cartilage, bone, and connective tissue, release individual mesenchymal cells that lack the characteristics of the cells of the differentiated tissue that produced them. The mesenchymal cells are identical to one another when examined with the electron microscope, *regardless of the tissue of their origin* (Hay, 1966). This process is called **dedifferentiation.** The scar tissue that formed during wound healing also releases its cells at this time.

3. *Formation of the blastema.* Mesenchyme cells released during dedifferentiation accumulate below the epidermis. These cells proliferate rapidly and cause the epidermis to bulge further. This mass of mesenchyme cells is called the **regeneration blastema.** As the blastema forms, histolysis of underlying tissues declines. It is intriguing that the regenerating limb forms an epithelio-mesenchymal reaction system much like that involved in differentiation of so many embryonic organs, including the embryonic limb.

4. *Morphogenesis and differentiation*. The first tissue to differentiate in the blastema is cartilage. It first appears at the ends of the persisting bone, which is completed by progressive addition to its distal end. Next, the more distal skeletal elements are added. When the cartilaginous reconstruction is completed, the regenerated skeleton is transformed into bone.

Muscle is formed by both *de novo* appearance of muscles around the cartilage and terminal addition to persisting muscles. Blood vessels are not obvious in early stages of reconstruction. They later extend into the regenerate to reproduce the original pattern of vascularization.

Many nerve fibers are cut during amputation. Very soon after amputation their axons grow into the wound and reconstruct the original nerve pattern. As we shall see later, the nerves play a significant role in controlling regeneration of the limb.

THE SOURCE AND ROLE OF BLASTEMAL CELLS

We have stated that blastemal cells arise from local dedifferentiation of stump tissue during histolysis. An alternative source of blastemal cells could be reserve cells that are mobilized from elsewhere as a consequence of amputation. The origin of blastemal cells is pinpointed by use of x-rays. Irradiation with x-rays prevents regeneration, possibly because mitosis is impaired in irradiated tissue (Wertz and Donaldson, 1980). If a portion of a limb is irradiated and amputated, that portion is incapable of regeneration. However, unirradiated portions of the same limb are capable of supporting regeneration when the limb is amputated there (Fig. 14.42). Clearly the cells necessary for regeneration must be supplied by the region of amputation.

Since the local cells support regeneration, which of these cells differentiate into the various limb tissues? The alternatives are listed in Table 14–1. Do differentiated cells (e.g., muscle) lose their specialized properties, proliferate, and subsequently redifferentiate solely according to their previous differentiated state? Or do differentiated cells truly dedifferentiate to pluripotent stem cells that are capable of forming a variety of different cells? Finally, are there local populations of reserve cells that retain the embryonic property of pluripotency and can form

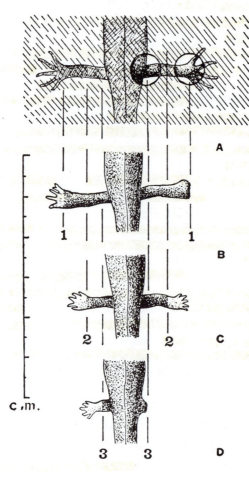

FIGURE 14.42 Local irradiation of parts of the limb of *Triturus cristatus* and the effects on amputation. *A,* Hatch marks indicate a lead shield that protects the entire left limb and parts of the right limb from radiation. X-rays could pass through only two holes above the base and the tip of the right limb. *B,* When amputated through an irradiated region, the right limb fails to regenerate. The control limb on the left regenerates normally. *C,* At a later date, the right limb was amputated through a nonirradiated region. Regeneration is normal. *D,* Still later, amputation through the irradiated base of the limb fails to elicit regeneration. (From V.V. Brunst. 1950. Influence of x-rays on limb regeneration in urodele amphibians. Q. Rev. Biol., *25:* 17. Reproduced from B.I. Balinsky. 1975.)

TABLE 14–1
Theoretical Possibilities for the Origin of Limb Regenerate Cells

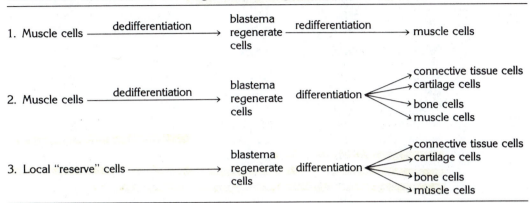

Source: Reprinted by permission from M Singer. 1973. Limb regeneration in the vertebrates. *Addison-Wesley Module in Biology,* no. 6. Addison-Wesley, Reading, MA, pp. 6–11.

the various differentiated cells of the limb? These alternatives are very difficult to test experimentally since the tissues of the limb are not pure populations of a single cell type. Most tissues contain connective tissue cells in addition to the specialized cells. Hence, the closest we can come to testing developmental potential during regeneration is to examine the potential of a *tissue* to participate in formation of alternate tissue types.

The standard experimental approach for examining differentiation potential of tissues is to irradiate a host diploid animal, which limits the ability of the host cells to participate in regeneration; implant tissue from a triploid donor; amputate; and conduct histological analyses on the regenerate. The presence of triploid nuclei in a tissue indicates that it is derived from the donor tissue. In some experiments the donor triploid cells are also labeled with ^{3}H-thymidine. Donor cells are detected in the regenerate by their triploid and radioactive nuclei.

When donor tissue is cartilage, donor cells are found in cartilage, perichondrium (the connective tissue layer that surrounds cartilage), connective tissue of joints, and fibroblasts. However, no donor cells are found in muscle or epidermis. On the other hand, when donor tissue is muscle, donor cells are found in all mesodermal derivatives (Steen, 1968; Namenwirth, 1974). Hence, the differentiation potential of cells in muscle tissue appears to be greater than that of cells in cartilage tissue. Epidermis is not capable of producing mesodermal tissues, although skin dermis has the ability to produce several mesodermal tissues (Hay and Fischman, 1961; Namenwirth, 1974; Dunis and Namenwirth, 1977). The dermis consists of connective tissue cells. Muscle—the other tissue capable of producing alternate cell types—also contains considerable connective tissue. We must, therefore, consider the likelihood that many of the mesodermal derivatives of the regenerate are produced by cells of connective tissue origin. No mesenchymal cell can produce epidermis, and epidermis can produce only epidermis.

REGULATION OF REGENERATION

We have examined the cellular contributions that adult tissues make to the regenerating limb and shall now consider some of the factors that may regulate formation of the regenerate: nerves of the stump, the skin, and hormones. One of the best characterized influences on regeneration is that produced by the nervous system. Neurons invade the blastema very soon after amputation and reconstitute the normal neural pattern. If the stump is denervated, regeneration is interrupted. If nerve fibers do not regrow to the amputation site, the tissues degenerate. However, if axons are permitted to regrow into the stump, regeneration may be reinitiated, particularly if the amputation wound is once again disturbed. These results suggest that nerves promote regeneration. This interpretation is supported by observations that experimental deviation of a limb nerve to a skin wound at the base of the limb will promote formation of a supernumerary limb. The neural influence is thought to be due to a chemical released by neural tissue, which is called the **neurotropic agent** (for review, see Singer, [1978]).

The neurotropic effect is produced by all nerves (motor, sensory, or central), regardless of whether they have functional association with the central nervous system. There is a threshold amount of nervous tissue that is necessary for regeneration. Below this threshold, regeneration does not occur. Above the threshold, regeneration occurs but is not affected by further increases in the amount of neural tissue. Nerves are thought to exert their effects on limb regeneration by promoting DNA synthesis and mitosis of mesodermal cells, which form the blastema (Mescher and Tassava, 1975; Loyd and Tassava, 1980).

The neurotropic effect is transitory, as demonstrated by sustained differentiation and morphogenesis of urodele limbs following denervation at stages after the onset of redifferentiation of the blastema. Growth of the denervated regenerates is retarded, however (Singer and Craven, 1948; Grim and Carlson, 1979). Apparently, nerves are necessary for the initiation of regeneration, but the redifferentiation and morphogenesis of the limb are independent of the nerves.

Vertebrate species that are incapable of limb regeneration have fewer nerve fibers per unit area of amputation wound than does *Triturus*. It is possible that the failure of these species to regenerate is at least partially due to an inadequate nerve supply, since regeneration has been induced in the limb of the lizard, the frog, and the opossum by supplementing the nervous supply of the amputation wound (Singer, 1974).

Soon after amputation skin epidermis rapidly covers the wound and proliferates to form the multilayered apical epidermal cap. Apical caps transplanted to the base of the blastema induce supernumerary limb regeneration. Thus, the apical epidermal cap may be necessary to promote regeneration. The specificity of the apical cap in this process is indicated by the fact that other epidermal grafts are without effect (Thornton, 1968).

One means for demonstrating the necessity for epidermis in promotion of regeneration is insertion of severed, amputated limbs into the coelomic cavity. In the coelom a wound epidermis is not formed and regeneration fails to occur (Goss, 1956a). If, however, epidermal wound healing is allowed to occur before insertion into the coelom, limbs form a blastema and regenerate inside the coelom (Goss, 1956b). Loyd and Tassava (1980) took advantage of this experimental procedure to compare the incorporation of ^{3}H-thymidine and the level of mitosis in regenerating (epidermis intact) and nonregenerating (epidermis absent) limbs. They demonstrated that DNA synthesis and the level of mitosis are substantially higher when epidermis is present. These results suggest that (like the nerves) the wound epidermis stimulates cell division in the underlying tissue, which results in formation of the blastema and subsequently in regeneration of the limb. A similar role for the epidermis has been reported by Globus et al. (1980). The analogy between the apical epidermal cap of the regenerate and the apical ectodermal ridge of the chick embryonic limb bud is obvious. They both may play similar roles in causing outgrowth of the limb by stimulating cellular proliferation, which provides the cells that differentiate proximally to produce definitive limb tissues.

Although epidermis is necessary for regeneration, the process can be prevented if a fresh wound is covered with a transplant of skin (which consists of epidermis *and dermis*). This result may be correlated with the electrical properties of the skin and provide clues to a possible significant factor in regeneration—electric current. Measurements with a sensitive vibrating probe have revealed that small currents enter the newt limb. Work with isolated amphibian skin has demonstrated that the skin maintains a large Na^+-dependent potential difference, which generates this current. After amputation, the direction and intensity of the current change dramatically at the tip of the stump. High levels of current exit the stump (Fig. 14.43). By the time of blastema appearance the magnitude of current flow has declined to less than 5 percent of the original values and transient shifts in its direction are observed (Borgens, 1982).

Borgens et al. (1977a) have demonstrated that amiloride, which blocks Na^+ transport, reduces the outward flow of current at the tip of the stump. This result suggests that the intact skin of the stump and perhaps the rest of the body is driving the current by moving sodium ions from the medium to the inside (the normal direction of Na^+ transport), then out through the tip of the stump. Amiloride has also been shown to inhibit regeneration, presumably through its effect on the stump current (Borgens et al., 1979b). These results are consistent with

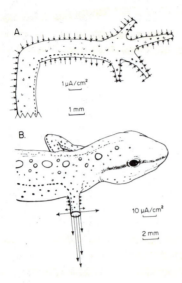

FIGURE 14.43 Current patterns around (*A*) the intact forelimb and (*B*) the stump of amputated forelimb of the newt. Arrows indicate the direction and magnitude of current density. Note the tenfold difference in current density scales. This enables representation of the small current densities of the intact limb. (From R.B. Borgens, J.W. Vanable, Jr., and L.F. Jaffe. 1979a. Bioelectricity and regeneration. Bioscience, *29:* 470. Copyright © 1979 by the American Institute of Biological Sciences.)

the hypothesis that the large outflow of current at the tip of the stump is necessary for regeneration of the limb.

If current flow is responsible for promoting regeneration, why is it that adult anuran amphibians (whose skins also transport Na$^+$ to the inside) cannot regenerate severed limbs? Borgens et al. (1979c) have measured the current at the frog's limb stump and have found the pattern of current density to be substantially different from that of the urodele. In the latter, current densities are always highest in the center of the stump, but in the frog the highest densities exit the limb around the periphery of the cut surface. Borgens et al. propose that this difference is due to the presence of lymph spaces below frog skin, which shunt the current to the periphery and prevent the high current density in the core tissues, which is needed to promote regeneration. Urodeles lack lymph spaces, and consequently their skin is tightly applied to underlying tissues. This would allow the current to exit in the core tissues of the urodele stump.

If high current density is a prerequisite for regeneration, will an enhancement of current cause a frog to regenerate a limb? Borgens et al. (1977b) have shown that regeneration of amputated frog limbs can be initiated (but not completed) by driving a small steady current through the stump with a battery. These results are suggestive that the failure of the frog limb to regenerate is at least partially due to the absence of sufficiently large core currents in the stump. Perhaps other factors (such as nerve density) prevent the current-enhanced frog stumps from completing their regeneration.

Regeneration is also affected by the endocrine system (Thornton, 1968). Removal of the anterior pituitary (hypophysectomy) prevents regeneration by adult urodeles; the effects are most pronounced if hypophysectomy is performed at the time of amputation. If hypophysectomy is delayed, the extent of regeneration is dependent upon the length of the delay. If delayed for at least 13 days, no effect on regeneration is observed. These results suggest that pituitary hormones act only during the initial stages of regeneration. The hormones involved are likely growth hormone and prolactin. The thyroid hormone thyroxine is also thought to be involved in regulating limb regeneration.

RESTORATION OF THE PATTERN OF CELL DIFFERENTIATION

One of the most intriguing aspects of regeneration is the fact that it mimics the pattern of cell differentiation that is established during embryonic limb development. Bryant and her associates have proposed a model to explain how the pattern of cell differentiation is reestablished during regeneration (French et al., 1976; Bryant et al., 1981). Cells of the limb are thought to possess positional information. During wound healing after amputation, cells with different positional values are brought into contact and interact with one another. This interaction leads to reinitiation of the pattern of cell differentiation.

The positional information is presumed to be specified along polar coordinates in a cell layer. Thus, each cell in the layer would have a positional value with an angular and a radial component (Fig. 14.44). It is assumed that positional values are properties of the connective tissue cells that dedifferentiate and regenerate the limb (see earlier discussion). The radial positional values represent the proximo-distal axis of the limb, with the proximal portion at the base of a cone and the distal point at the tip of the cone. The angular positional component corresponds to the circumferential location on cross sections at each proximo-distal level of the limb.

As we mentioned previously, when cells with disparate positional values confront one another during wound healing they interact to reinitiate the preexisting tissue pattern. This process is thought to result from intercalation. The discontinuity in positional values stimulates cell division, and cell growth generates the positional values missing between the confronting values. If angular positional values are missing, intercalation occurs by the shorter, rather than the longer, of the two routes; that is, if values 2 and 5 confront one another, 3 and 4 would be formed

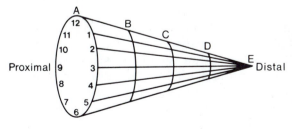

FIGURE 14.44 Polar coordinates of positional information on the surface of a limb. Each cell is thought to have information with respect to its position on a circumference (0–12) and on a radius (A–E). Note that circumferential values 0 and 12 are identical. (After S.V. Bryant, V. French, and P.J. Bryant. 1981. Distal regeneration and symmetry. Science *212*: 993–1002. Copyright 1981 by the American Association for the Advancement of Science.)

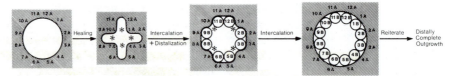

FIGURE 14.45 Model of distal outgrowth from a wound surface. Amputation removes B, C, D, and E levels of the pattern, leaving A level (shaded). The wound edge is outlined by the circle. During healing different circumferential values come into contact, resulting in circumferential intercalation (indicated by asterisks). The new cells have positional values identical to those of preexisting adjacent cells, however, forcing the new cells to the next more distal level (B). The B level is completed by subsequent intercalation, and reiteration of the process continues until the most distal values (E) are generated. (After S.V. Bryant, V. French, and P.J. Bryant. 1981. Distal regeneration and symmetry. Science *212:* 993–1002. Copyright 1981 by the American Association for the Advancement of Science.)

rather than 6, 7, 8, 9, 10, 11, 12(0), and 1. This is the **shortest intercalation rule.**

Regeneration of an amputated limb always produces parts that are distal to the cut, regardless of the level of the cut. This result implies that cell growth generates more distal positional values. This could occur as shown in Figure 14.45. After amputation the wound heals by contraction of tissue toward the wound center. This causes cells with different angular positional values to confront one another. Intercalation occurs, but since no angular positional values are missing at the level of amputation, the new cells must duplicate angular positional values belonging to preexisting cells. The same angular values cannot be duplicated at that proximo-distal level. Therefore, new cells are forced to adopt a more distal positional value. This is called the **distalization rule.** Thus, as shown in Figure 14.45 amputation at A level generates new cells with a B level value. The process of distalization continues until the most distal positional values are generated.

14–5. ORGANOGENESIS: PROSPECTUS

Recent advances in our understanding of cell-cell interactions and the development of technology to monitor cell differentiation and its control with more precision have placed us on the threshold of a dramatic improvement in our understanding of the coordinated cellular activities that constitute organogenesis. The literature in this area is now providing more precise and sophisticated answers to long-standing questions involving the regulation and the mechanisms of organogenesis. It is hoped that the reader will have been stimulated by this brief survey of organogenesis to follow the literature as the myriad stories of organogenesis begin to unfold and, hopefully, to participate actively in what will undoubtedly be exciting quests for new levels of understanding of this—the ultimate developmental process.

REFERENCES

Abramson, S., R.G. Miller, and R.A. Phillips. 1977. The identification in adult bone marrow of pluripotent and restricted stem cells of the myeloid and lymphoid systems. J. Exp. Med., *145*: 1567–1579.

Alberts, B. et al. 1983. *Molecular Biology of the Cell.* Garland Publishing, Inc., New York.

Alt, F.W. et al. 1982. Immunoglobulin heavy-chain expression and class switching in a murine leukaemia cell line. Nature (Lond.), *296*: 325–331.

Balinsky, B.I. 1951. On the eye cup–lens correlation in some South African amphibians. Experientia, 7: 180.

Balinsky, B.I. 1975. *An Introduction to Embryology,* 4th ed. W.B. Saunders Co., Philadelphia.

Barabanov, V.M., and N.G. Fedtsova. 1982. The distribution of lens differentiation capacity in the head ectoderm of chick embryos. Differentiation, *21*: 183–190.

Becker, A.J., E.A. McCulloch, and J.E. Till. 1963. Cytological demonstration of the clonal nature of spleen colonies derived from transplanted mouse marrow cells. Nature (Lond.), *197*: 452–454.

Beebe, D.C., D.E. Feagans, and H.A.H. Jebens. 1980. Lentropin: A factor in vitreous humor which promotes lens fiber cell differentiation. Proc. Natl. Acad. Sci. U.S.A., *77*: 490–493.

Beebe, D.C., and J. Piatigorsky. 1981. Translation regulation of δ-crystallin synthesis during lens development in the chicken embryo. Dev. Biol., *84*: 96–101.

Beebe, D.C. et al. 1982. The mechanism of cell elongation during lens fiber cell differentiation. Dev. Biol., *92*: 54–59.

Bernfield, M.R. 1978. The cell periphery in morphogenesis. *In* J.W. Littlefield and J. de Grouchy (eds.), *Birth Defects.* Excerpta Medica International Congress Series No. 432. Excerpta Medica, Oxford, pp. 111–125.

Bernfield, M.R. 1981. Organization and remodeling of the extracellular matrix in morphogenesis. *In* T.G. Connelly, L.L. Brinkley, and B.M. Carlson (eds.), *Morphogenesis and Pattern Formation.* Raven Press, New York, pp. 139–162.

Bernfield, M.R., and S.D. Banerjee. 1982. The turnover of basal lamina glycosaminoglycan correlates with epithelial morphogenesis. Dev. Biol., *90*: 291–305.

Bernfield, M.R., S.D. Banerjee, and R.H. Cohn. 1972. Dependence of salivary epithelial morphology and branching morphogenesis upon acid mucopolysaccharide-protein (proteoglycan) at the epithelial surface. J. Cell Biol., *52*: 674–689.

Berrill, N.J., and G. Karp. 1976. *Development.* McGraw-Hill Book Co., New York.

Billingham, R.E., and W.K. Silvers. 1967. Studies on the conservation of epidermal specificities of skin and certain mucosas in adult animals. J. Exp. Med., *125*: 429–446.

Borgens, R.B. 1982. What is the role of the naturally produced electric current in vertebrate regeneration and healing? Int. Rev. Cytol., *76*: 245–298.

Borgens, R.B., J.W. Vanable, Jr., and L.F. Jaffe. 1977a. Bioelectricity and regeneration: Large currents leave the stumps of regenerating newt limbs. Proc. Natl. Acad. Sci. U.S.A., *74*: 4528–4532.

Borgens, R.B., J.W. Vanable, Jr., and L.F. Jaffe. 1977b. Bioelectricity and regeneration. I. Initiation of frog limb regeneration by minute currents. J. Exp. Zool., *200*: 403–416.

Borgens, R.B., J.W. Vanable, Jr., and L.F. Jaffe. 1979a. Bioelectricity and regeneration. Bioscience, *29*: 468–474.

Borgens, R.B., J.W. Vanable, Jr., and L.F. Jaffe. 1979b. Reduction of sodium dependent stump currents disturbs urodele limb regeneration. J. Exp. Zool., *209*: 377–386.

Borgens, R.B., J.W. Vanable, Jr. and L.F. Jaffe. 1979c. Role of subdermal current shunts in the failure of frogs to regenerate. J. Exp. Zool., *209*: 49–55.

Brunst, V.V. 1950. Influence of x-rays on limb regeneration in urodele amphibians. Q. Rev. Biol., *25*: 1–29.

Bryant, S.V., V. French, and P.J. Bryant. 1981. Distal regeneration and symmetry. Science, *212*: 993–1002.

Byers, B., and K.R. Porter. 1964. Oriented microtubules in elongating cells of the developing lens rudiment after induction. Proc. Natl. Acad. Sci. U.S.A., *52*: 1091–1099.

Cairns, J.M., and J.W. Saunders, Jr. 1954. The influence of embryonic mesoderm on the regional specification of epidermal derivatives in the chick. J. Exp. Zool., *127*: 221–248.

Cline, M.J. and D.W. Golde. 1979. Cellular interactions in haematopoiesis. Nature (Lond.), *277*: 177–181.

Coughlin, M.D. 1975. Early development of parasympathetic nerves in the mouse submandibular gland. Dev. Biol., *43*: 123–139.

Coulombre, A.J. 1965. The eye. *In* R.L. DeHaan and H. Ursprung (eds.), *Organogenesis.* Holt, Rinehart and Winston, New York, pp. 219–251.

Coulombre, J.L., and A.J. Coulombre. 1971. Metaplastic induction of scales and feathers in the corneal anterior epithelium of chick embryo. Dev. Biol., *25*: 464–478.

Cudennec, C.A., J.-P. Thiery, and N.M. Le Douarin. 1981. *In vitro* induction of adult erythropoiesis in early mouse yolk sac. Proc. Natl. Acad. Sci. U.S.A., *78*: 2412–2416.

Dessau, W. et al. 1980. Changes in the patterns of collagens and fibronectin during limb-bud chondrogenesis. J. Embryol. Exp. Morphol., *57*: 51–60.

Deuchar, E.M. 1975. *Cellular Interactions in Animal Development.* Chapman & Hall Ltd., London.

Djaldetti, M. et al. 1972. Erythropoietin effects on fetal mouse erythroid cells. II. Nucleic acid synthesis and the erythropoietin-sensitive cell. J. Biol. Chem., *247*: 731–735.

Dreyer, W.J., and J.C. Bennett. 1965. The molecular basis of antibody formation. A paradox. Proc. Natl. Acad. Sci. U.S.A., *54*: 864–869.

Dunis, D.A., and M. Namenwirth. 1977. The role of grafted skin in the regeneration of X-irradiated axolotl limbs. Dev. Biol., *56*: 97–109.

Freeman, G. 1963. Lens regeneration from cornea in *Xenopus laevis.* J. Exp. Zool., *154*: 39–65.

French, V., P.J. Bryant, and S.V. Bryant. 1976. Pattern regulation in epimorphic fields. Science, *193*: 969–981.

Genis-Galvez, J.M., J.M. Castro, and E. Battaner. 1968. Lens soluble proteins: Correlation with the cytological differentiation in the young adult organ of the chick. Nature (Lond.), *217*: 652–654.

Globus, M., S. Vethamany-Globus, and Y.C.I. Lee. 1980. Effect of apical epidermal cap on mitotic cycle and cartilage differentiation in regeneration blastema in the newt, *Notophthalmus viridescens*. Dev. Biol., *75*: 358–372.

Goss, R.J. 1956a. Regenerative inhibition following limb amputation and immediate insertion into the body cavity. Anat. Rec., *126*: 15–28.

Goss, R.J. 1956b. The regenerative responses of amputated limbs to delayed insertion into the body cavity. Anat. Rec., *126*: 283–298.

Grim, M., and B.M. Carlson. 1979. The formation of muscles in regenerating limbs of the newt after denervation of the blastema. J. Embryol. Exp. Morphol., *54*: 99–111.

Grobstein, C. 1953a. Epithelio-mesenchymal specificity in the morphogenesis of mouse sub-mandibular rudiments *in vitro*. J. Exp. Zool, *124*: 383–413.

Grobstein, C. 1953b. Morphogenetic interaction between embryonic mouse tissues separated by a membrane filter. Nature (Lond.), *172*: 869–871.

Harrison, R.G. 1918. Experiments on the development of the forelimb of *Amblystoma*, a self-differentiating equipotential system. J. Exp. Zool., *25*: 413–461.

Harrison, R.G. 1920. Experiments on the lens in *Amblystoma*. Proc. Soc. Exp. Biol. Med., *17*: 199–200.

Hay, E.D. 1966. *Regeneration*. Holt, Rinehart and Winston, New York.

Hay, E.D. 1974. Cellular basis of regeneration. *In* J. Lash and J.R. Whittaker (eds.), *Concepts of Development*. Sinauer Associates, Inc. Sunderland, Mass., pp. 404–428.

Hay, E.D. and D.A. Fischman. 1961. Origin of the blastema in regenerating limbs of the newt, *Triturus viridescens*. Dev. Biol., *3*: 26–59.

Hozumi, N., and S. Tonegawa. 1976. Evidence for somatic rearrangement of immunoglobulin genes coding for variable and constant regions. Proc. Natl. Acad. Sci. U.S.A., *73*: 3628–3632.

Iten, L.E. 1982. Pattern specification and pattern regulation in the embryonic chick limb bud. Am. Zool., *22*: 117–129.

Jacobson, A. 1966. Inductive processes in embryonic development. Science, *152*: 25–34.

Johnson, G.R., and D. Metcalf. 1979. The commitment of multipotential hemopoietic stem cells: Studies *in vivo* and *in vitro*. *In* N. Le Douarin (ed.), *Cell Lineage, Stem Cells and Cell Determination*. INSERM Symposium No. 10. Elsevier/North Holland Biomedical Press, Amsterdam, pp. 199–213.

Johnston, M.C. et al. 1979. Origins of avian ocular and periocular tissues. Exp. Eye Res., *29*: 27–43.

Jones, R.E., D. DeFeo, and J. Piatigorsky. 1981. Transcription and site-specific hypomethylation of the δ-crystallin genes in the embryonic chicken lens. J. Biol. Chem., *256*: 8172–8176.

Karkinen-Jääskeläinen, M. 1978a. Permissive and directive interactions in lens induction. J. Embryol. Exp. Morphol., *44*: 167–179.

Karkinen-Jääskeläinen, M. 1978b. Transfilter lens induction in avian embryo. Differentiation, *12*: 31–37.

Kieny, M. 1977. Proximo-distal pattern formation in avian limb development. *In* D.A. Ede, J.R. Hinchliffe, and M. Balls (eds.), *Vertebrate Limb and Somite Morphogenesis*. Cambridge University Press, Cambridge, pp. 87–103.

Kollar, E.J., and G.R. Baird. 1970a. Tissue interactions in embryonic mouse tooth germs. I. Reorganization of the dental epithelium during tooth-germ reconstruction. J. Embryol. Exp. Morphol., *24*: 159–171.

Kollar, E.J., and G.R. Baird. 1970b. Tissue interactions in embryonic mouse tooth germs. II. The inductive role of the dental papilla. J. Embryol. Exp. Morphol., *24*: 173–186.

Kosher, R.A., K.H. Walker, and P.W. Ledger. 1982. Temporal and spatial distribution of fibronectin during development of the embryonic chick limb bud. Cell Differentiation, *11*: 217–228.

Krantz, S.B., and E. Goldwasser. 1965. On the mechanism of erythropoietin-induced differentiation. Biochim. Biophys. Acta, *103*: 325–332.

Kratochwil, K. 1969. Organ specificity in mesenchymal induction demonstrated in the embryonic development of the mammary gland of the mouse. Dev. Biol., *20*: 46–71.

Lassila, O. et al. 1982a. Erythropoiesis and lymphopoiesis in the chick yolk-sac–embryo chimeras: Contribution of yolk sac and intraembryonic stem cells. Blood, *59*: 377–381.

Lassila, O. et al. 1982b. Migration of prebursal stem cells from the early chicken embryo to the yolk sac. Scand. J. Immunol., *16*: 265–268.

Lawson, K.A. 1974. Mesenchyme specificity in rodent salivary gland development: The response of salivary epithelium to lung mesenchyme *in vitro*. J. Embryol. Exp. Morphol, *32*: 469–493.

Lewis, W.H. 1904. Experimental studies on the development of the eye in amphibia. I. On the origin of the lens in *Rana palustris*. Am. J. Anat., *3*: 505–536.

Lodish, H.F., and B. Small. 1976. Different lifetimes of reticulocyte messenger RNA. Cell, *7*: 59–65.

Loyd, R.M., and R.A. Tassava. 1980. DNA synthesis and mitosis in adult newt limbs following amputation and insertion into the body cavity. J. Exp. Zool, *214*: 61–69.

Maloney, C., and J. Wakeley. 1982. Microfilament patterns in the developing chick eye: Their role in invaginations. Exp. Eye Res., *34*: 877–886.

Maniatis, G.M. et al. 1973. Early stimulation of RNA synthesis by erythropoietin in cultures of erythroid precursor cells. Proc. Natl. Acad. Sci. U.S.A., *70*: 3189–3194.

Marks, P.A., R.A. Rifkind, and A. Bank. 1974. Mammalian erythroid cell differentiation. *In Biochemistry of Cell Differentiation* (MTP International Review of Science, Biochemistry Section, Vol. 9). University Park Press, Baltimore, pp. 129–160.

Max, E.E., J.G. Seidman, and P. Leder. 1979. Sequences of five potential recombination sites encoded close to an immunoglobin κ constant region gene. Proc. Natl. Acad. Sci. U.S.A., *76*: 3450–3454.

Mescher, A.L., and R.A. Tassava. 1975. Denervation effects on DNA replication and mitosis during the initiation of limb regeneration in adult newts. Dev. Biol., *44*: 187–197.

Mizuno, T. 1972. Lens differentiation *in vitro* in the absence of optic vesicle in the epiblast of chick blastoderm under the influence of skin dermis. J. Embryol. Exp. Morphol., *28*: 117–132.

Moore, M.A.S., and D. Metcalfe. 1970. Ontogeny of the haematopoietic system: Yolk sac origin of *in vivo* and *in vitro* colony forming cells in the developing mouse embryos. Br. J. Haematol., *18*: 279–296.

Morgan, T.H. 1901. *Regeneration*. Macmillan, Inc., New York.

Muthukkaruppan, V. 1965. Inductive tissue interaction in the development of the mouse lens *in vitro*. J. Exp. Zool., *159*: 269–288.

Namenwirth, M. 1974. The inheritance of cell differentiation during limb regeneration in the axolotl. Dev. Biol., *41*: 42–56.

Orlic, D. 1970. Ultrastructural analysis of erythropoiesis. *In* A.S. Gordon (ed.), *Regulation of Hematopoiesis*, Vol. 1. Appleton-Century-Crofts, New York, pp. 271–296.

Papaconstantinou, J. 1967. Molecular aspects of lens cell differentiation. Science, *156*: 338–346.

Papayannopoulou, T. et al. 1982. Hemoglobin switching in culture: Evidence for a humoral factor that induces switching in adult and neonatal but not fetal erythroid cells. Proc. Natl. Acad. Sci. U.S.A., *79*: 6579–6583.

Patten, B.M. 1971. *Early Embryology of the Chick*, 5th ed. McGraw-Hill Book Co., New York.

Philpott, G.W., and A.J. Coulombre. 1965. Lens development. II. The differentiation of embryonic chick lens epithelial cells *in vitro* and *in vivo*. Exp. Cell Res., *38*: 635–644.

Piatigorsky, J. 1975. Lens cell elongation *in vitro* and microtubules. Ann. N.Y. Acad. Sci., *253*: 333–347.

Piatigorsky, J. 1981. Lens differentiation in vertebrates. A review of cellular and molecular features. Differentiation, *19*: 134–153.

Piatigorsky, J., H. deF. Webster, and S.P. Craig. 1972. Protein synthesis and ultrastructure during the formation of embryonic chick lens fibers *in vivo* and *in vitro*. Dev. Biol., *27*: 176–189.

Proudfoot, N.J. et al. 1980. Structure and in vitro transcription of human globin genes. Science, *209*: 1329–1336.

Rabl, C. 1898. Über den Bau und die Entwicklung der Linse. I. Selachier und Amphibien. Zeitschr. für wissenschaftliche Zool., *63*: 496–572.

Ramirez, R. et al. 1975. Changes in globin messenger RNA content during erythroid cell differentiation. J. Biol. Chem., *250*: 6054–6058.

Rawles, M.E. 1963. Tissue interactions in scale and feather development as studied in dermal-epidermal recombinations. J. Embryol. Exp. Morphol., *11*: 765–789.

Reiter, R.S., and M. Solursh. 1982. Mitogenic property of the apical ectodermal ridge. Dev. Biol., *93*: 28–35.

Rifkind, R.A. 1974. Erythroid cell differentiation. *In* J. Lash and J.R. Whittaker (eds.), *Concepts of Development*. Sinauer Associates, Inc., Sunderland, Mass., pp. 149–162.

Rifkind, R.A. et al. 1976. Erythroid differentiation and the cell cycle: Some implications from murine fetal and erythroleukemic cells. Ann. Immunol. (Paris), *127*: 887–893.

Rubin, L., and J.W. Saunders, Jr. 1972. Ectodermal-mesodermal interactions in the growth of limb buds in the chick embryo: Constancy and temporal limits of the ectodermal induction. Dev. Biol., *28*: 94–112.

Sakano, H. et al. 1979. Sequences at the somatic recombination sites of immunoglobulin light chain genes. Nature (Lond.), *280*: 288–293.

Saunders, J.W., Jr. 1948. The proximo-distal sequence of origin of the parts of the chick wing and the role of the ectoderm. J. Exp. Zool., *108*: 363–404.

Saunders, J.W., Jr. 1972. Developmental control of three-dimensional polarity in the avian limb. Ann. N.Y. Acad. Sci., *193*: 29–42.

Saunders, J.W., Jr. 1977. The experimental analysis of chick limb bud development. *In* D.A. Ede, J.R. Hinchliffe, and M. Balls (eds.), *Vertebrate Limb and Somite Morphogenesis*. Cambridge University Press, Cambridge, pp. 1–24.

Saunders, J.W., Jr., J.M. Cairns, and M.T. Gasseling. 1957. The role of the apical ridge of ectoderm in the differentiation of the morphological structure and inductive specificity of limb parts in the chick. J. Morphol., *101*: 57–87.

Saunders, J.W., Jr., and M.T. Gasseling. 1968. Ectodermal-mesenchymal interactions in the origin of limb symmetry. *In* R. Fleischmajer and R.E. Billingham (eds.), *Epithelio-Mesenchymal Interactions.* 18th Hahnemann Symposium. Williams & Wilkins Co., Baltimore, pp. 78–97.

Sawyer, R.H. 1979. Avian scale development: Effect of the scaleless gene on morphogenesis and histogenesis. Dev. Biol., *68*: 1–15.

Saxén, L. 1977. Directive and permissive induction. *In* J. Lash and M. Burger (eds.), *Cell and Tissue Interactions.* Raven Press, New York, pp. 1–9.

Saxén, L. et al. 1976. Are morphogenetic tissue interactions mediated by transmissible signal substances or through cell contacts? Nature (Lond.), *259*: 662–663.

Schwidefsky, G. 1934. Entwicklung und determination der Extremitätenregenerate bei den Molchen. Wilh. Roux' Arch. für Entwicklungsmechanik, *132*: 57–114.

Searls, R.L., and M.Y. Janners. 1969. The stabilization of cartilage properties in the cartilage-forming mesenchyme of the embryonic chick limb. J. Exp. Zool., *170*: 365–376.

Sengel, P. 1976. *Morphogenesis of Skin.* Cambridge University Press, Cambridge.

Sengel, P., D. Dhouailly, and M. Kieny. 1969. Aptitude des constituants cutanés de l'aptérie médioventrale du Poulet à former des plumes. Dev. Biol., *19*: 436–446.

Shinohara, T., and J. Piatigorsky. 1976. Quantitation of δ-crystallin messenger RNA during lens induction in chick embryos. Proc. Natl. Acad. Sci. U.S.A., *73*: 2808–2812.

Singer, M. 1973. Limb regeneration in the vertebrates. *Addison-Wesley Module in Biology,* no. 6. Addison-Wesley, Reading, Mass.

Singer, M. 1974. Neurotropic control of limb regeneration in the newt. Ann. N.Y. Acad. Sci., *228*: 308–321.

Singer, M. 1978. On the nature of the neurotrophic phenomenon in urodele limb regeneration. Am. Zool., *18*: 829–841.

Singer, M., and L. Craven. 1948. The growth and morphogenesis of the regenerating forelimb of adult *Triturus* following denervation at various stages of development. J. Exp. Zool., *108*: 279–308.

Solursh, M., C.T. Singley, and R.S. Reiter. 1981. The influence of epithelia on cartilage and loose connective tissue formation by limb mesenchyme cultures. Dev. Biol., *86*: 471–482.

Spemann, H. 1912. Zur Entwicklung des Wirbeltierauges. Zool. Jahrb. Abt. f. allg. Zool. u. Phys. D. Tiere, *32*: 1–98.

Spemann, H. 1936. *Experimentelle Bieträge zu einer Theorie der Entwicklung.* Julius Springer, Berlin.

Spemann, H. 1938. *Embryonic Development and Induction.* Yale University Press, New Haven, Conn. Reprinted by Hafner Press (Macmillan, Inc.), New York. (1962).

Spooner, B.S., and N.K. Wessells. 1972. An analysis of salivary gland morphogenesis: Role of cytoplasmic microfilaments and microtubules. Dev. Biol., 27: 38–54.

Stark, R.J., and R.L. Searls. 1974. The establishment of the cartilage pattern in the embryonic chick wing, and evidence for a role of the dorsal and ventral ectoderm in normal wing development. Dev. Biol., *38*: 51–63.

Steen, T.P. 1968. Stability of chondrocyte differentiation and contribution of muscle to cartilage during limb regeneration in the axolotl (*Siredon mexicanum*). J. Exp. Zool., *167*: 49–71.

Summerbell, D. 1979. The zone of polarizing activity: Evidence for a role in normal chick limb morphogenesis. J. Embryol. Exp. Morphol., *50*: 217–233.

Summerbell, D., and J.H. Lewis. 1975. Time, place, and positional value in the chick limb-bud. J. Embryol. Exp. Morphol., *33*: 621–643.

Summerbell, D., J.H. Lewis, and L. Wolpert. 1973. Positional information in chick limb morphogenesis. Nature (Lond.), *244*: 492–496.

Tahara, Y. 1962. Formation of the independent lens in Japanese amphibians. Embryologia, *7*: 127–149.

Thesleff, I. and K. Hurmerinta. 1981. Tissue interactions in tooth development. Differentiation, *18*: 75–88.

Thesleff, I., E. Lehtonen, and L. Saxén. 1978. Basement membrane formation in transfilter tooth culture and its relation to odontoblast differentiation. Differentiation, *10*: 71–79.

Thornton, C.S. 1968. Amphibian limb regeneration. Adv. Morphogen., *7*: 205–249.

Tickle, C., D. Summerbell, and L. Wolpert. 1975. Position signalling and specification of digits in chick limb morphogenesis. Nature (Lond.), *254*: 199–202.

Till, J.E., and E.A. McCulloch. 1961. A direct measurement of the radiation sensitivity of normal mouse bone marrow cells. Radiat. Res., *14*: 213–222.

Till, J.E., E.A. McCulloch, and L. Siminovitch. 1964. A stochastic model of stem cell proliferation, based on the growth of spleen colony-forming cells. Proc. Natl. Acad. Sci. U.S.A., *51*: 29–36.

Tomasek, J.J., J.E. Mazurkiewicz, and S.A. Newman. 1982. Nonuniform distribution of fibronectin during avian limb development. Dev. Biol., *90*: 118–126.

Tonegawa, S. 1983. Somatic generation of antibody diversity. Nature (Lond.), *302*: 575–581.

Trentin, J.J., J.M. Rauchwerger, and M.T. Gallagher. 1974. Regulation of hemopoietic stem cell differentiation and proliferation by hemopoietic inductive microenvironments. *In* B. Clarkson and R. Baserga (eds.), *Control of Proliferation in Animal Cells.* Cold Spring Harbor Laboratories, Cold Spring Harbor, N.Y., pp. 927–932.

Tréton, J.A., T. Shinohara, and J. Piatigorsky, 1982. Degradation of δ-crystallin mRNA in the lens fiber cells of the chicken. Dev. Biol., *92*: 60–65.

Weigert, M., and R. Riblet. 1976. Genetic control of antibody variable regions. Cold Spring Harbor Symposium Quant. Biol., *41*: 837–846.

Wertz, R.L., and D.J. Donaldson. 1980. Early events following amputation of adult newt limbs given regeneration inhibitory doses of X irradiation. Dev. Biol., *74*: 434–445.

Wessells, N.K. 1977. *Tissue Interactions and Development.* Benjamin/Cummings Publishing Co., Inc., Menlo Park, Calif.

Wolpert, L. 1978. Pattern formation in biological development. Sci. Am., *239*(4): 154–164.

Wolpert, L., C. Tickle, and M. Sampford (with an appendix by J.H. Lewis). 1979. The effect of cell killing by X-irradiation on pattern formation in the chick limb. J. Embryol. Exp. Morphol., *50*: 175–198.

Yamada, T. 1967. Cellular and subcellular events in Wolffian lens regeneration. Curr. Top. Dev. Biol., *2*: 247–283.

Yamada, T. 1977. Control mechanisms in cell-type conversion in newt lens regeneration. *In* A. Wolsky (ed.), *Monographs in Developmental Biology*, vol. 13. Karger, Basel.

Zelenka, P., and J. Piatigorsky. 1974. Isolation and *in vitro* translation of δ-crystallin mRNA from embryonic chick lens fibers. Proc. Natl. Acad. Sci. U.S.A., *71*: 1896–1900.

Zwilling, E. 1972. Limb morphogenesis. Dev. Biol., *28*: 12–17.

Index/Glossary

Page number in *italics* refer to illustrations; page numbers in **bold face** refer to definitions of terms; page numbers followed by (t) refers to tables.

729